T0176017

CALCULUS
OF
ONE
VARIABLE

Joseph W. Kitchen, Jr.

Department of Mathematics
Duke University

Dover Publications, Inc.
Mineola, New York

To David V. Widder

Copyright

Copyright © 1968, 1996 by Joseph W. Kitchen, Jr.
All rights reserved.

Bibliographical Note

This Dover edition, first published in 2020, is an unabridged republication of the work originally published by Addison-Wesley Publishing Company, Reading, Massachusetts, in 1968.

Library of Congress Cataloging-in-Publication Data

Names: Kitchen, Joseph W., author.
Title: Calculus of one variable / Joseph W. Kitchen, Jr. (Department of Mathematics, Duke University).
Description: Dover edition. | Mineola: Dover Publications, Inc., 2020. | Originally published: Reading, Mass. : Addison-Wesley Pub. Co., 1968. | Summary: "Richly textured and versatile text characterizes real numbers as a complete, ordered field. Rigorous development of the calculus, plus thorough treatment of basic topics of limits and inequalities"—Provided by publisher.
Identifiers: LCCN 2019034739 | ISBN 9780486838069 (trade paperback)
Subjects: LCSH: Calculus. | Geometry, Analytic—Plane.
Classification: LCC QA303 .K63 2020 | DDC 515—dc23
LC record available at https://lccn.loc.gov/2019034739

Manufactured in the United States by LSC Communications
83806401
www.doverpublications.com

2 4 6 8 10 9 7 5 3 1

2019

PREFACE

The main concern of this book is one-variable calculus. Plane analytic geometry and infinite series are also treated. The book is intended for those who have the ability, leisure, and inclination to gain a thorough mastery of calculus. More particularly, it is designed for use in a first-year honors course at the university level. The typical reader is imagined to be a bright college freshman, who possibly has already been introduced to calculus in high school; it is assumed that he has a solid command of high school mathematics (algebra, geometry, and trigonometry), and that he likes to work on challenging problems. Hopefully the book will also provide some fresh insights for jaded professionals who regularly teach calculus and quite properly regard themselves as experts in the subject.

As the size of the book should suggest, there is much more material here than one should attempt to cover in a standard one-year course. My objective has been to provide a richly textured, flexible text with a great variety of attractive options, in addition to a solid core of standard material. I have also included many more exercises than one could expect to work; these range from the trivial to the downright unreasonable. An instructor using the book must therefore exercise considerable discretion: if he pursues too many peripheral topics, his students may lose a sense of continuity and they may fail to master essential techniques; if he gives too little thought to the selection of problems, he runs the risk of either stultifying or discouraging the students.

The author of any good calculus book must make some compromises. One could, in principle, start from axioms for sets and proceed to develop calculus with uncompromising rigor. Yet, Serge Lang is quite right in maintaining that no one in his right mind would do such a thing. Nevertheless, in this book we make fewer compromises than most. Our starting point is the characterization of the real numbers as a complete ordered field; the development of calculus which then follows is quite rigorous. It would be a mistake, however, to suppose that any sequence of theorems constitutes the whole of calculus. Viewed historically, calculus is a body of techniques which were devised to solve certain problems in the physical sciences. Calculus divorced from its applications is simply not calculus. Thus, while I have taken considerable pains to make the logical structure of the subject crystal clear, I have also tried to provide intuitive insights and to develop a feeling for the way calculus is applied. For this reason I have chosen not to sever ties with traditional, but often logically imprecise, notations and locutions. Similarly, the reader is expected to learn to extract solid mathematics from the careless derivations one often sees, where, for example, differentials are used abusively. In short, I have tried to maintain high mathematical standards without adopting a narrow and self-righteous dogmatism.

Several other features of the book deserve comment.

Stashed away in the exercises is a wide assortment of mathematical tidbits. One will find, for example, information on the Fibonacci numbers, continued fractions, the calculus of finite differences, the Chebyshev and Legendre polynomials, damped harmonic motion, Young's inequality, the Bernoulli numbers, a version of the inclusion-exclusion principle, and Bliss' formula. The exercises also contain important counterexamples. Users of the book are therefore urged to read the exercises as well as the text.

Throughout the book I have made an effort to show what can be done with the machinery at hand. Thus in the chapter on the transcendental functions, power series expansions for e^x, $\sin x$, $\cos x$, $\ln (1 + x)$, $\text{Tan}^{-1}x$, and $(1 + x)^a$ are all established without Taylor's theorem and without a single theorem on infinite series. In later chapters the same results are established by several other methods. In this way, I have tried to show both the power and the limitations of different techniques. I have also frequently treated a given topic at several different levels of sophistication. This is true, for example, of the trigonometric functions and the notion of area. In such cases an understanding of a topic at its simplest level suffices for later use in the book.

Because the book is intended as an introduction not only to calculus but also to analysis more generally, limits and inequalities receive unusually thorough coverage. Although plenty of ϵ's and δ's are scattered across the pages, I have tried to avoid direct appeals to the limit definitions, whenever possible, by invoking a handful of easily remembered limit theorems including a Pinching Theorem, Substitution Rule, and Linking Limit Lemma.

Perhaps the most unorthodox feature of the book is the treatment of integration. Of all the topics in elementary calculus, definite integrals—when to introduce it and how—is the most troublesome. In certain circles it is held that the Riemann integral has outlived its usefulness and that we should allow it to die a quick and natural death. Mathematicians of this disposition are inclined to feel that until such time as the Lebesgue integral is needed, it suffices to consider definite integrals of *regulated* functions (uniform limits of step functions). Nevertheless, among current calculus texts the Riemann integral reigns supreme. Furthermore, it has become somewhat fashionable to introduce the Riemann integral at an early stage, often before differentiation.

After much thought and experimentation I have come to the following conclusions:

1) Nothing is gained by the early introduction of integration.
2) The traditional approach to Riemann integrals as limits of sums is inelegant theoretically, awkward to apply, and impossible to motivate convincingly.

3) In a course in which foundational questions are not considered (to the extent, say, of discussing epsilonics, the completeness of the reals, and uniform continuity) the Riemann integral is a luxury item if not a waste of time.
4) The Riemann integral is nevertheless intrinsically interesting, and its study can be a very valuable part of one's mathematical training.

In this book integration first appears in two rather lengthy chapters, Chapters 7 and 8. In a sense these chapters have cross-purposes. Chapter 7 shows how inessential the Riemann integral is to the applications which are usually made of it (the calculation of areas, volumes, path lengths, moments, etc.); a simpler type of definite integral, the Newton integral, is used in its place. Chapter 8, on the other hand, places the Riemann integral in a new perspective, thereby giving it renewed vitality. This latter chapter begins with an abstract definition of a definite integral for which the Newton and Riemann integrals provide nontrivial examples. Certain questions concerning these definite integrals immediately arise; in the course of answering these questions step functions, upper and lower Riemann integrals, partitions, norms and refinements of partitions, Riemann sums—in short, all the usual paraphernalia of the Riemann theory—present themselves in an orderly and natural fashion. As a result, the Riemann integral emerges as an object of genuine mathematical interest. (Historically, of course, it is assured a place of honor in the world of ideas, since it is the first precise formulation of the vague but tantalizing concept of continuous summation.)

Chapters 7 and 8 suggest the need for a general reform in the teaching of elementary calculus. For "problem-oriented" or "cookbook style" courses, the Newton integral is a far more appropriate tool than the Riemann integral. If, on the other hand, the Riemann theory is taught, then it should be taught for the right reasons.

Throughout the book many sections and problems are starred. Apart from acting as a warning to the reader, these stars have no uniform meaning. Usually they indicate difficult and/or tangential material. Sometimes they indicate that previously starred material is prerequisite. For this reason the reader should not be surprised if some of the starred exercises turn out to be easier than some of unstarred ones (especially when copious hints are supplied).

The first ten chapters contain all the material usually considered necessary in a one-year course. For those who prefer not to stray too far from essentials, we shall next make suggestions regarding the use of these chapters.

The main objectives of Chapter 1 are twofold: to acquaint students with inductive proofs and definitions, and to pave the way for epsilonics by some drill in logic and inequalities. Accordingly, Sections 1–3 and 1–4 should be stressed, while the first two sections can be gone over quickly. The essential points to be gained from the material on logic are

a) an awareness of the need for quantification;
b) a clear understanding of the difference in meaning produced by switching the positions of existential and universal quantifiers;

c) an understanding of when a statement of the form $P \Rightarrow Q$ is true;

d) the mechanics of negating a statement which begins with a string of quantifiers and concludes with a statement of the form $P \Rightarrow Q$.

(A good deal of drill in the construction of truth tables and the proof of tautologies by Boolean algebra is largely wasted effort for the purposes of this book.) Section 1–8 and part of Section 1–9 (inductive definitions and the well-ordering property of the positive integers) should be stressed; the definition of the positive integers and the derivation of their basic properties from the axioms for an ordered field might be skipped.

Hopefully most users of the book will be able to skip over Chapter 2 entirely. It should be pointed out, however, that the treatment of straight lines in this chapter is by no means standard; in particular, the usual fudging and prestidigitation attending discussions of slopes is neatly bypassed.

Chapter 3 should be studied carefully. I have found it best to stress the use of the basic theorems on limits and to assign at first just enough exercises involving ϵ's and δ's to fix the basic definitions in mind. Further exercises in epsilonics from this chapter can be assigned at a later time.

The only material one might choose to omit from Chapter 4 is that pertaining to logarithmic differentiation and the calculation of nth-order derivatives.

Although Chapter 5 appears in its proper place logically, one may wish to postpone a careful study of its contents. If one knows a couple of facts about one-sided limits and the statements of the intermediate-value theorem and the theorem on extreme values, then one can safely study the next two chapters.

Chapter 6 should be studied with great care. Although Section 6–6 is optional, it is highly recommended. A possible condensation of this section, which is still very worthwhile, consists of the following: a proof that every function with a positive second derivative is convex, Jensen's inequality, and applications of Jensen's inequality.

One should plan to do most of Chapter 7. If Chapter 5 has not been studied by this time, then it will be necessary to omit or postpone the discussion of global properties of path length.

At this point I would recommend a careful study of Chapter 5. This should then be followed by a proof of the fact that continuous functions have antiderivatives. To this end one can use either the development given in the first four sections of Chapter 8 or the hints for Exercise 1, Section 8–4. Sections 8–5 and 8–6 are strictly optional in any case.

I regard the first six sections of Chapter 9 as mandatory. The rest of the chapter is optional though highly recommended.

I shall offer no opinions concerning Chapter 10, since there is no general agreement among present-day mathematicians as to the importance of formal integration.

I would like to thank all those who have helped in the production of this book. In particular, I would like to thank David V. Widder, who suggested the project; Lynn H. Loomis, who read all drafts of the manuscript and gave me much sound advice; John A. Kelingos and Frank Dorsey, who helped test the material in the class-room; Jim Lucke, who helped prepare the problem solutions; Fred Brauer, Thad Dankel, William Jenner, and David Lively, who read sizable portions of the manu-script and made helpful criticisms. I also wish to thank all my colleagues at Duke and most especially Joseph Shoenfield, who completely subverted my former ap-proach to integration. I am also grateful to the Duke Mathematics Department for bearing some of the cost of manuscript preparation and to the department sec-retaries, especially Edith Minton, for their help. I wish to thank the superb staff at Addison-Wesley, and most of all, my wife, Dorothy, who typed the manuscript and assisted in the proofreading.

Durham, North Carolina   J. W. K., Jr.
October 1967

CONTENTS

1 □ PRELIMINARIES

The chief aim of this book is a careful and thorough study of calculus. The material is to be so structured that each new result is proved from previous ones. The reader is expected to participate actively in this endeavor; sometimes he will be asked to supply proofs himself or to flesh out proofs which are given in skeletal form. Even when proofs are given in full, he will be expected to examine them critically. In the present chapter we set down the basic assumptions which we shall make regarding the real number system. We also introduce some of the language and notations of set theory and logic. All this is in preparation for an assault upon the more subtle aspects of calculus and, more generally, of analysis.

1-1 Sets and set operations

The notion of a set is an appropriate starting point for just about any text in mathematics. It is a primitive concept in the sense that all objects presently studied in mathematics can be defined in terms of sets. Consequently, we shall make no attempt to define a set in terms of anything simpler. Instead, the reader will have to be content with a few synonyms (collection, aggregate, class, family) and some examples. While the notion of a set is in a very literal sense abstract, it is not abstruse; it is present whenever we speak of a bunch of grapes, a group of people, or a herd of cows. A term used in some of the older geometry texts which is very nearly synonomous with the word "set" is the word "locus." Such books speak, for instance, of "the *locus* of points equidistant from two given points"; more modern texts speak instead of "the *set* of all points equidistant from two given points."

There are two methods commonly used to define sets. The simplest way to define a set is to list its members, in which case the listed elements are enclosed in curly brackets. Thus, if we write

$$S = \{1, 2, 3\},$$

we shall mean that S is the set consisting of the numbers 1, 2, and 3. The order of the listing is immaterial: $\{3, 2, 1\}$ and $\{2, 3, 1\}$ denote the same set. This method of defining sets can be used if a set contains only a finite number of elements. It can also be used to define some infinite sets. For example, we would interpret $\{2, 4, 6, 8, \ldots\}$ to be the set of all positive even integers. Most infinite sets, however, are defined by *property*. In this case, the following notation is used:

$$\{x \mid \ldots\},$$

which may be read, "the set of all x such that . . ." For example, $\{x \mid x > 0\}$ denotes the set of all x such that $x > 0$, in other words, the set of all positive numbers. This second method of defining sets may also be used for finite sets. Thus

$$\{x \mid x^2 = x\} = \{0, 1\} \quad \text{and} \quad \{x \mid 2x = 6\} = \{3\}.$$

We now introduce the standard notation for set membership.

Definition 1. If S is a set, then $p \in S$ means "p is a member of S" or "p belongs to S." By $p \notin S$ we shall mean that p **does not belong to** S.

Examples. $1 \in \{0, 1, 7\}$, whereas $2 \notin \{0, 1, 7\}$. Also, $30 \in \{x \mid x > 0\}$, whereas $-3 \notin \{x \mid x > 0\}$.

We consider next the notion of a subset.

Definition 2. Let A and B be sets. We say that A is a **subset** of B if and only if every member of A is also a member of B. If A is a subset of B, we write $A \subset B$ (which may also be read, "A **is contained in** B").

Figure 1–1 gives us an example: if A consists of all points inside the small oval and B consists of all points inside the large oval, then $A \subset B$. We cite a second example: $\{1, 2, 3\} \subset \{x \mid x > 0\}$.

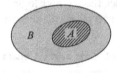

A set is completely determined by its members. Consequently, we define two sets to be *equal* if and only if they have

FIG. 1–1

the same members. A trivial, though frequently useful, consequence of these definitions is a criterion for the equality of two sets.

Proposition 3. *Let A and B be sets. Then $A = B$ if and only if $A \subset B$ and $B \subset A$.*

Proof. Suppose first that $A \subset B$ and $B \subset A$. By definition, this means that every member of A is a member of B and, conversely, that every member of B is a member of A. But this simply says that A and B have the same members, and this is precisely what is meant by the equation $A = B$. (In this text equality always means identity. Thus, $a = b$ means that a and b are two symbols for the same thing.)

It is clear from our definition of inclusion, on the other hand, that every set is a subset of itself. Hence, if $A = B$, then it is certainly true that $A \subset B$ and $B \subset A$.

Equality of sets may be regarded as a trivial or "improper" case of set inclusion. If A is a subset of a set B without actually being equal to B, we often say that A is a *proper subset* of B. For example, $\{0, 1, 5\}$ is a proper subset of $\{0, 1, 3, 5, 7\}$.

There are several important ways of combining given sets to produce new ones. Given two sets A and B, we define

$$A \cup B = \{x \mid \text{either } x \in A,\ x \in B,\ \text{or both}\},$$
$$A \cap B = \{x \mid x \in A \text{ and } x \in B\},$$
$$A \setminus B = \{x \mid x \in A \text{ but } x \notin B\}.$$

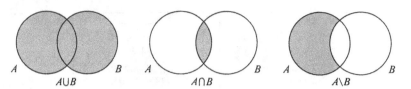

FIG. 1-2

In other words, the set $A \cup B$ consists of all elements which belong to at least one of the two sets A and B; it is called their *union*. The set $A \cap B$ consists of the elements common to both sets; it is called their *intersection*. The set $A \setminus B$ is called the *relative complement of B in A*. For example, if $A = \{1, 5, 6, 9\}$ and $B = \{5, 6, 10\}$, then $A \cup B = \{1, 5, 6, 9, 10\}$, $A \cap B = \{5, 6\}$, and $A \setminus B = \{1, 9\}$. It often happens that in a given discussion all the sets which are being considered are subsets of some large set U. The set U is then sometimes called the *universal set*, and if we have $A \subset U$, the set $U \setminus A$ is simply called the *complement of A* and is denoted by A' or A^c.

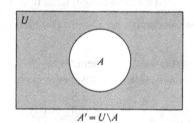

FIG. 1-3

$$A' = U \setminus A$$

The set operations which we have defined are easily pictured. If we portray the sets A and B as the sets of points inside two circles, then the sets $A \cup B$, $A \cap B$, and $A \setminus B$ are as indicated in Fig. 1-2. If U is a universal set, then it is customary to picture U as the set of points inside a rectangle (see Fig. 1-3). Subsets of U are then pictured as the insides of circles (provided that no more than three subsets of U are involved). Such *Venn diagrams* are useful since they often suggest correct set identities. The two diagrams in Fig. 1-4, for instance, suggest the identity

$$(A \cup B)' = A' \cap B'$$

(known as one of *De Morgan's rules*). Pictorial evidence for the identity

$$(A \cup B) \cap C = (A \cap C) \cup (B \cap C)$$

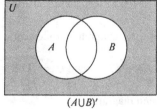

FIG. 1-4

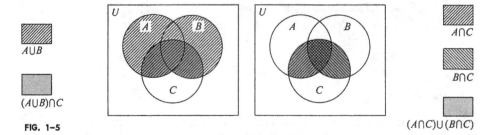

FIG. 1–5

can be obtained similarly (see Fig. 1–5). The systematic study of such identities comes under the heading of *Boolean algebra*. It is pursued to a limited extent in the exercises, but it need not concern us now. (Some work with set identities is prerequisite, however, for the section of Chapter 8 dealing with Jordan content.)

For the operations of intersection and relative complementation to be defined in all cases, we must allow the possibility that a set be empty. (Consider the intersection of $\{1, 3\}$ and $\{2, 4\}$ or the relative complement of $\{1, 3, 5\}$ in $\{1, 5\}$.)

Definition 4. The **null** (**void** or **empty**) **set** is the set which contains no elements. It is denoted by $\varnothing$.

Observe that if S is any set, then every element in $\varnothing$ is an element of S. (Since $\varnothing$ has no elements, this admittedly does not say very much.) Hence the null set is a subset of every set.

We mention another set operation of a different type. If S is a set, we denote the set consisting of all subsets of S by $\mathcal{S}(S)$; thus

$$\mathcal{S}(S) = \{A \mid A \subset S\}.$$

[Note, then, that $\mathcal{S}(S)$ is a set whose members are themselves sets. Although we could speak of $\mathcal{S}(S)$ as being a *set* of sets, it is customary, for reasons of logical clarity as well as euphony, to speak of it as a family of sets.] This $\mathcal{S}(S)$ is called the *power set* of S. If, for example, $S = \{3, 7\}$, then $\mathcal{S}(S) = \{\varnothing, \{3\}, \{7\}, S\}$.

Exercises

1. Let $A = \{1, 5, 7, 10\}$ and $B = \{2, 7, 11, 14\}$. Determine $A \cup B, A \cap B, A \setminus B, B \setminus A$, $(A \setminus B) \cup B$, and $(A \setminus B) \cup (A \cap B)$.

2. The preceding exercise should suggest two set identities. Write them, and illustrate them with diagrams.

3. Let $S = \{1, 2, 3\}$. List all the members of $\mathcal{S}(S)$. If S is a set containing n elements, make a guess as to the number of elements in $\mathcal{S}(S)$.

4. Let $S = \{1, 2, 3, 4\}$ be the universal set. Let $A = \{1, 2, 4\}, B = \{3, 4\}$, and $C = \{1, 2\}$. Determine each of the following sets: $A \cap B, B \cap C, A \cup B, A' \cap B$, and $(A' \cup C) \cap B$.

5. Illustrate the De Morgan law, $(A \cap B)' = A' \cup B'$, via Venn diagrams.

6. Illustrate the distributive law, $(A \cap B) \cup C = (A \cup C) \cap (B \cup C)$, via Venn diagrams.

7. Explain what differences (if any) there are in the following sets: $\emptyset$, $\{\emptyset\}$, $S(\emptyset)$, and $S(\{\emptyset\})$.

8. In this exercise we shall confine our attention to subsets of a given universal set U. One method for proving set identities is the construction of *in-out tables*. As an illustration, we construct an in-out table for $A \cup B$ and $A \cap B$. We consider an arbitrary element p of our universal set U. In general, there are four logical possibilities:

A	B	$A \cup B$	$A \cap B$
i	i	i	i
i	o	i	o
o	i	i	o
o	o	o	o

FIG. 1–6

a) p lies inside A and inside B;
b) p lies inside A but outside B;
c) p lies outside A but inside B;
d) p lies outside A and outside B.

These four possibilities give us the first two columns of the in-out table shown in Fig. 1–6; the abbreviations are self-explanatory. To see how the rest of the table is filled in, consider the third line of the table. If p lies outside A but inside B, then p lies inside $A \cup B$ but outside $A \cap B$. Thus the entry under $A \cup B$ is i, while the entry under $A \cap B$ is o.

A	B	$A \cup B$	$(A \cup B)'$	A'	B'	$A' \cap B'$
i	i	i	o	o	o	o
i	o	i	o	o	i	o
o	i	i	o	i	o	o
o	o	o	i	i	i	i

FIG. 1–7

We shall now construct an in-out table to prove the De Morgan rule, $(A \cup B)' = A' \cap B'$ (see Fig. 1–7). We see that the fourth and last columns are the same. This means that a point $p \in S$ belongs to $(A \cup B)'$ if and only if it belongs to $A' \cap B'$. Therefore, $(A \cup B)' = A' \cap B'$.

	A	B	C	$B \cup C$	$A \cap (B \cup C)$	$A \cap B$	$A \cap C$	$(A \cap B) \cup (A \cap C)$
1	i	i	i	i	i	i	i	i
2	i	i	o	i	i	i	o	i
3	i	o	i	i	i	o	i	i
4	i	o	o	o	o	o	o	o
5	o	i	i	i	o	o	o	o
6	o	i	o	i	o	o	o	o
7	o	o	i	i	o	o	o	o
8	o	o	o	o	o	o	o	o

FIG. 1–8

Similarly, we can construct an in-out table to prove the distributive law,

$$A \cap (B \cup C) = (A \cap B) \cup (A \cap C);$$

this table is given in Fig. 1–8. In constructing such an in-out table, we first list the eight in-out possibilities for the sets A, B, and C. The remaining columns are set up in such a way that the entries in any one column are determined by the entries in previous columns. In this case, the crucial columns are the fifth and the last. Since these columns have the same entries, it follows that $A \cap (B \cup C) = (A \cap B) \cup (A \cap C)$.

Prove the following identities by in-out tables.

a) $(A \cap B)' = A' \cup B'$ b) $A \cup (B \cap C) = (A \cup B) \cap (A \cup C)$

c) $(A')' = A$ d) $(A \cup B) \cup C = A \cup (B \cup C)$

1-2 The real numbers as a field

It is not possible to overestimate the importance of the real number system both to pure and applied mathematics. Anywhere measurements appear, we are certain to find the real numbers in action. Furthermore, we shall see that all the truly important theorems in calculus involve the real number system in an essential way: any attempt to get the same results with other number systems is doomed to failure.

The simplest real numbers are the "whole" or "counting" numbers 1, 2, 3, 4, ... These numbers are also known as the *positive integers* or the *natural numbers*. As is now customary, we shall use $\mathbf{Z}^+$ to denote the set of all positive integers. The set

$$\{\ldots, -3, -2, -1, 0, 1, 2, 3, \ldots\}$$

consisting of the natural numbers, their negatives, and zero will be denoted by $\mathbf{Z}$; its elements are called *integers*. The next most important class of real numbers is the *rationals*. By definition, a rational number is the ratio of two integers. Thus, $\frac{7}{8}$, $-\frac{11}{16}$, $3.51 = \frac{351}{100}$, and $0.363636\ldots = \frac{4}{11}$ are rational numbers. We shall use $\mathbf{Q}$ to denote the set of all rational numbers. (Note that $\mathbf{Z}^+ \subset \mathbf{Z} \subset \mathbf{Q}$.)

Rational numbers first arose in the measurement of line segments. In fact, the Greek geometers thought of numbers as being (in some rather fuzzy sense) "ratios" of line segments. At first they thought that all numbers were rational, or to use old-fashioned terminology, that any two line segments are "commensurate." They assumed, in other words, that given any two line segments AB and $A'B'$, one can always find a third line segment MN and natural numbers n and n' such that if the segment MN is laid end to end n times, one gets a line segment whose length is that of AB; whereas by laying MN end to end n' times, one gets a line segment whose length is that of $A'B'$ (see Fig. 1–9). (Thus the ratio of AB to $A'B'$ would be the

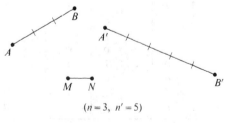

$(n = 3, \ n' = 5)$

FIG. 1–9

rational number n/n'.) The discovery of the Pythagorean theorem shattered this belief and caused the Greek mathematicians to revise their thinking about numbers. One particular consequence of the Pythagorean theorem is the fact that an isosceles right triangle whose legs are of unit length has a hypotenuse of length $\sqrt{2}$ (see Fig. 1-10). Mathematicians of the day were no doubt distressed when they failed to find a rational number whose square was two, and they found themselves facing a real dilemma when it was proved that no such number could exist. The proof of this fact is as beautiful as it is ancient.

FIG. 1-10

Theorem 1. *There is no rational number whose square is two.*

Proof. Let us assume, to the contrary, that there is a rational number r such that $r^2 = 2$. By definition, r is of the form m/n, where m and n are integers. We may assume that the fraction is in "lowest terms," by which we mean that m and n have no common factor. (For example, the rational number $\frac{20}{25}$ in lowest terms is $\frac{4}{5}$.) Thus we have the equation

$$m^2/n^2 = 2,$$

or equivalently, $m^2 = 2n^2$. Since the number 2 is a factor of the right-hand side of the latter equation, it must also be a factor of the left-hand side, namely m^2. This implies that 2 must be a factor of m. (More generally, if a prime number p is a factor of the product ab of two integers, then either p is a factor of a or it is a factor of b.) Hence m can be written

$$m = 2m',$$

where m' is an integer. Substituting this expression for m into the previous equation, we get

$$4(m')^2 = 2n^2 \quad \text{or} \quad n^2 = 2(m')^2.$$

However, it follows now from the above argument that 2 is a factor of n. But this means that 2 is a common factor of m and n, contrary to the assumption that m and n have no common factor. This contradiction shows that our original assumption was false. Hence the proof of the theorem is complete.

A real number which is not rational, such as $\sqrt{2}$, is called (with an unimaginativeness characteristic of many definitions in mathematics) an *irrational number*. However, while $\sqrt{2}$ is not rational, it does satisfy an algebraic equation with integer coefficients (namely, $x^2 - 2 = 0$). In more general terms, a number x which is the root of an equation of the form

$$a_0 x^n + a_1 x^{n-1} + \cdots + a_{n-1}x + a_n = 0,$$

where the coefficients $a_0, a_1, \ldots, a_n$ are integers, is called an *algebraic number*. In particular, every rational number is algebraic. (Why?) Even the algebraic numbers do not suffice, however, for the purposes of measurement. The area, for instance, of a circle of unit radius is π. Not only is π irrational, but it fails to be algebraic. Such a number is said to be *transcendental*. (The proof that π is irrational is by no means trivial; the proof that π is transcendental requires a small volume!)

The rationals, $\sqrt{2}$, and π are all examples of real numbers. The reader may feel that a definition of the real numbers is now in order. If so, he will be disappointed. It is possible to define the real numbers precisely in terms of the rational numbers (which in turn can be defined in terms of the integers). Such a construction of the real number system is lengthy, however, and while the overall procedure is certainly interesting and important, the details are, to be perfectly honest, a bit dull, and they contribute little to an understanding of calculus. In a sense the philosophical question of what a real number "is" is relatively unimportant for our purposes. Far more to the point is an understanding of those properties of the real numbers which are needed in the development of calculus. We shall therefore regard the real numbers as *undefined objects* satisfying certain *axioms*, in much the same way that "points" and "lines" are taken as undefined terms in the development of synthetic plane geometry. (The reader should not regard these axioms as "self-evident truths." When the reals are constructed from the rationals, these "axioms" actually become theorems which *must* be proved.) The present chapter contains only a partial listing of these axioms; the so-called axiom of completeness is deferred until Chapter 5. At that time we shall be able to show that the real numbers can be identified with formal decimal expansions, that is, with expressions of the form

$$\pm N.a_1 a_2 a_3 \ldots,$$

where N is a nonnegative integer written in decimal notation and the a_k's are the decimal digits $0, 1, 2, \ldots, 9$. As the reader may already know, the identification of real numbers with formal decimal expansions has one peculiar twist. To each formal decimal expansion there corresponds a real number, and, for the most part, this correspondence is one-to-one; however, there are some numbers with two different decimal expansions. For instance, the decimal expansions 1.769999... and 1.77000... represent the same real number. Formally, it is clear that this must be so. If we let $x = 1.769999\ldots$, then $10x = 17.6999\ldots$, so that (subtracting) $9x = 15.93000\ldots$, and thus $x = 1.77000\ldots$ In summary, we can eventually think of the real numbers as formal decimal expansions (and if the reader finds it helpful, he may secretly think of them in this way now). Our official position, however, is that the real numbers are objects which satisfy certain axioms presently to be enumerated and that the entire logical structure of calculus must be built upon them.

Throughout the book we shall use **R** to denote the set of all real numbers. We shall assume that **R** comes equipped with an algebraic structure. This structure includes two *operations*, one called *addition*, the other *multiplication*. The operation of addition (multiplication) enables us to associate with any pair of real numbers x and y an object denoted by $x + y$ (xy). The behavior of these operations is described by our first set of axioms. For devious reasons, we shall temporarily use F rather than **R** to denote the set of real numbers.

A1. *The set F is closed under addition. That is, if x and y are in F, then x + y is also in F.*

A2. *Addition in F is associative. That is, for all x, y, and z in F,*

$$x + (y + z) = (x + y) + z.$$

A3. *The set F contains an element denoted by* 0 *such that*

$$x + 0 = 0 + x = x$$

for all $x \in F$.

A4. *For each* $x \in F$, *there exists an element in F denoted by* $-x$ *such that*

$$x + (-x) = (-x) + x = 0.$$

A5. *Addition in F is commutative. That is, for all x and y in F,*

$$x + y = y + x.$$

The axioms above are concerned exclusively with the operation of addition. We list next a parallel set for multiplication.

M1. *The set F is closed under multiplication. That is, for all x and y in F, the product xy is in F.*

M2. *Multiplication is associative. That is, for all x, y, and z in F,*

$$x(yz) = (xy)z.$$

M3. *The set F contains an element denoted by* 1 *such that*

$$1 \cdot x = x \cdot 1 = x$$

for all $x \in F$.

M4. *For each x in F different from zero, there exists an element in F denoted by* x^{-1} *such that*

$$xx^{-1} = x^{-1}x = 1.$$

M5. *Multiplication is commutative. That is, for all x and y in F,*

$$xy = yx.$$

The next axiom provides an important link between the operation of addition and the operation of multiplication.

AM. The Distributive Law. *For all x, y, and z in F,*

$$x(y + z) = xy + xz.$$

Finally, we have what is sometimes known as the axiom of nontriviality (despite the fact that as a statement about the real numbers it is certainly trivial).

NT. $1 \neq 0$.

Using the language of abstract algebra, we can summarize this list of axioms for the real numbers by saying that the real numbers form a *field*.

Definition 2. Suppose that F is a set and that we have two operations $+$ and $\cdot$ defined for elements in F such that the above listed properties hold. Then F is called a **field**.

The set of axioms listed up to now fall far short of *characterizing* the real numbers. This is but to say that there exist fields which are in many ways quite unlike the field of real numbers. Two such fields will now be described.

x	y	z	$x + y$	$(x + y) + z$	$y + z$	$x + (y + z)$
a	a	a	a	a	a	a
a	a	b	a	b	b	b
a	b	a	b	b	b	b
a	b	b	b	a	a	a
b	a	a	b	b	a	b
b	a	b	b	a	b	a
b	b	a	a	a	b	a
b	b	b	a	b	a	b

FIG. 1-11

Example 1. Let $F = \{a, b\}$, where a and b are any two distinct elements. We define addition and multiplication by the following display:

$$a + a = a, \qquad aa = a,$$
$$a + b = b, \qquad ab = a,$$
$$b + a = b, \qquad ba = a,$$
$$b + b = a, \qquad bb = b.$$

One can quickly check that the field axioms are satisfied. One can prove that M2 holds, for example, by constructing a table showing all possibilities for the variables x, y, and z (see Fig. 1-11). It is also clear that

$$a + x = x + a = x$$

for all $x \in F$. In other words, a is the zero element of F. Similarly, it is easily verified that b has the property stated in M3, so we may write $b = 1$.

This particular field is called the *field of integers modulo* 2. In more conventional notation, $\mathbf{Z}_2 = \{0, 1\}$; addition and multiplication are given by the tables in Fig. 1-12.

$x + y$

x \ y	0	1
0	0	1
1	1	0

xy

x \ y	0	1
0	0	0
1	0	1

FIG. 1-12

Example 2. Our second example is the *field of integers modulo* 3. Let $\mathbf{Z}_3 = \{0, 1, 2\}$. The elements of $\mathbf{Z}_3$ are the remainders which are possible upon division of integers by 3. For example, the remainder obtained by dividing 17 by 3 is 2. Addition of elements in $\mathbf{Z}_3$ is defined by the following rule: to add two elements x and y in $\mathbf{Z}_3$, first form their sum in the usual sense and then find its remainder upon division by 3; this remainder is called the sum of x and y. For example, $2 + 1 = 0$, since the ordinary sum of 2 and 1 is 3 and its remainder upon division by 3 is 0. Similarly, $2 + 2 = 1$, since the usual sum is 4 and its remainder upon division by 3 is 1.

Multiplication is defined similarly: the product of x and y in $\mathbf{Z}_3$ is the remainder obtained by dividing the ordinary product of x and y by 3. Figure 1–13 shows the addition and multiplication tables for $\mathbf{Z}_3$.

$x + y$

x \ y	0	1	2
0	0	1	2
1	1	2	0
2	2	0	1

xy

x \ y	0	1	2
0	0	0	0
1	0	1	2
2	0	2	1

FIG. 1–13

As in the case of the integers modulo 2, the verification that $\mathbf{Z}_3$ is a field can be carried out by the construction of tables; the only essential difference is that the tables must now be larger. For example, to verify the associative laws or the distributive law, we need tables with twenty-seven rows. With a little more knowledge of algebra and elementary number theory, such unilluminating brute force methods can be replaced by briefer and more conceptually interesting ones.

The student will find further examples of fields in the exercises. We mention here that the rational numbers and the complex numbers are also important examples of fields.

Although we shall not attempt a systematic study of fields, we shall prove some elementary consequences of the field axioms. Throughout the remainder of this section, F will denote a field, and, to borrow P. Halmos' phrase, all apparently homeless elements will be assumed to belong to F. Thus, all the theorems which we shall prove in this section apply in particular not only to the real numbers but also to $\mathbf{Z}_2$, $\mathbf{Z}_3$, the rationals, and the complex numbers.

The number 0 postulated in A3 is unique. This is implicitly assumed in A4, which would otherwise be ambiguous. Also implicit in the statement of A4 is the assumption that $-x$ (called the *negative* of x or the *additive inverse* of x) is unique. We now justify these assumptions.

Proposition 3. *The elements of F whose existence is postulated in* A3 *and* A4 *are unique. Moreover, the element* 0 *is characterized by the fact that it is idempotent under addition; that is, it is the only element of F which satisfies the equation*

$$x + x = x.$$

Proof. Let us suppose that both 0 and $0'$ satisfy the condition in A3, namely,

$$x + 0 = 0 + x = x \qquad (*)$$

and

$$x + 0' = 0' + x = x \qquad (**)$$

for all $x \in F$. Then

$$0' = 0' + 0 \quad \text{(by *)}$$
$$= 0 \quad \text{(by **)},$$

which proves the uniqueness of 0.

Let us next suppose that for some $x \in F$, there exist elements x' and x'' in F such that

$$x + x' = x' + x = 0 \quad \text{and} \quad x + x'' = x'' + x = 0.$$

We must show that $x' = x''$. But

$$x' = x' + 0 = x' + (x + x'') = (x' + x) + x'' = 0 + x'' = x''.$$

(The reader should supply the justifications for the preceding string of equalities.)

Suppose now that $x \in F$ satisfies the equation $x + x = x$. Then

$$x = x + 0 = x + [x + (-x)] = (x + x) + (-x) = x + (-x) = 0.$$

Note that our field axioms say nothing concerning the operations of subtraction and division. There are two reasons for this: the operations of subtraction and division do not have very nice properties, and also these operations can be defined in terms of addition and multiplication. Because of Proposition 3, we may now define subtraction in the following way.

Definition 4. If x and y are two elements in F, we define their difference $x - y$ to be the element $x + (-y)$.

This is probably not the definition of subtraction which the reader first encountered. Most likely, $7 - 3$, for example, was defined to be that number which when added to 3 gives 7. The two definitions are easily reconciled.

Proposition 5. *The difference $a - b$ is the unique solution x of the equation*

$$b + x = a.$$

Proof. The expression $x = a - b$ clearly satisfies the above equation, since

$$b + x = b + [a + (-b)] = b + [(-b) + a]$$
$$= [b + (-b)] + a = 0 + a = a.$$

Suppose, on the other hand, that x satisfies the equation $b + x = a$. Then

$$x = 0 + x = [(-b) + b] + x$$
$$= (-b) + (b + x) = (-b) + a = a + (-b) = a - b.$$

We turn next to some facts about multiplication. In particular, it is extremely important to know how the element zero behaves with respect to multiplication.

Proposition 6. The elements of F whose existence is postulated in M3 and M4 are unique. For all $x \in F$, $0 \cdot x = 0$. If x and y are elements of F such that $xy = 0$, then either $x = 0$ or $y = 0$. The elements 0 and 1 are the only two elements of F which are idempotent with respect to multiplication.

Proof. The proof of the first assertion is analogous to the proof of Proposition 3, and is therefore left as an exercise for the reader.

To prove that $0 \cdot x = 0$, it suffices to show that $0 \cdot x$ is idempotent under addition. This, however, follows quickly from the distributive law:

$$0 \cdot x + 0 \cdot x = (0 + 0)x = 0 \cdot x.$$

The third assertion of the proposition may be equivalently stated as follows: if $xy = 0$ and $x \neq 0$, then $y = 0$. This is clear since

$$y = 1 \cdot y = (x^{-1}x)y = x^{-1}(xy)$$
$$= x^{-1} \cdot 0 = 0 \cdot x^{-1} = 0.$$

Suppose finally that $x \in F$ is idempotent under multiplication, that is,

$$x \cdot x = x^2 = x.$$

If $x \neq 0$, then

$$x = 1 \cdot x = (x^{-1}x)x = x^{-1} \cdot x^2 = x^{-1}x = 1.$$

If $x \neq 0$, the element x^{-1} is called the *reciprocal* or *multiplicative inverse* of x. We can now define division.

Definition 7. *If x and y are in F and $y \neq 0$, we define x/y to be the element xy^{-1}.*

We leave as an exercise for the reader the verification of the following analog of Proposition 5.

Proposition 8. If a and b are in F and $b \neq 0$, then a/b is the unique solution x of the equation $bx = a$.

Note that we have not defined a/b in the event that $b = 0$. In this case we cannot define a/b to be ab^{-1}, since b^{-1} has no meaning. (The set F contains no element x such that $bx = 0 \cdot x = 1$. For by Proposition 6, $0 \cdot x = 0$ for all $x \in F$, and the axiom of nontriviality states that $1 \neq 0$.) Similarly, if $b = 0$, we cannot define a/b to be the unique solution x of the equation $bx = a$. For if $a \neq 0$, the equation has no solution, while if $a = 0$, every element of F is a solution. Hence *in any field, division by zero is meaningless.* Furthermore, the inadvertent division by zero can lead to ridiculous conclusions. Consider, for example, this classical "proof" that $2 = 1$. Let $x = y$ be any nonzero real number. Then $y^2 = xy$, so that

$$x^2 - y^2 = x^2 - xy,$$

or factoring, we get

$$(x + y)(x - y) = x(x - y).$$

Dividing both sides by $x - y$, we get

$$x + y = x$$

or $2x = x$. Finally, by dividing both sides by x, we deduce that $2 = 1$. Each step of this "proof" is legitimate except for the step in which we divided both sides of an equation by $x - y = 0$.

In freshman high school algebra one is taught the rules for manipulating signed numbers. For example, the product of two negative numbers is, by definition, positive. The next proposition provides a rationale for this: it shows, among other things, that if the operations with signed numbers were not defined as they are, the set of real numbers would not form a field.

Proposition 9. *In any field the following identities hold:*

1) $-(-x) = x$,
2) $(-x) + (-y) = -(x + y)$,
3) $(-x)y = -(xy)$,
4) $(-x)(-y) = xy$.

Proof. By definition of $-x$, we have $x + (-x) = 0$. Also by definition, $-(-x)$ is the unique element in F which when added to $-x$ gives zero. Clearly, then $-(-x) = x$.

To prove (2), we shall temporarily use x' and y' to denote $-x$ and $-y$, respectively. This will prevent an excessive accumulation of parentheses. Thus

$$\begin{aligned}
(x + y) + (x' + y') &= [(x + y) + x'] + y' \\
&= [x + (y + x')] + y' \\
&= [x + (x' + y)] + y' \\
&= [(x + x') + y] + y' \\
&= (0 + y) + y' \\
&= y + y' = 0.
\end{aligned}$$

Hence $x' + y' = (-x) + (-y)$ is the additive inverse of $x + y$, or, in other words,

$$-(x + y) = (-x) + (-y).$$

To prove (3), we note that

$$xy + (-x)y = [x + (-x)]y = 0 \cdot y = 0.$$

Since xy has a unique additive inverse, it is clear from the above equation that $-(xy) = (-x)y$.

Identity (4) easily follows from (1) and (3):

$$(-x)(-y) = -[x(-y)] = -[(-y)x] = -[-(yx)] = yx = xy.$$

The time has now come for the student to gain for himself some practice in working with the axioms for a field.

Exercises

1. Prove the first statement in Proposition 6.

2. Prove Proposition 8.

3. Prove that in any field the following identities hold.

   a) $a(b - c) = ba - ca$
   b) $(-a) + (-a) = (-2)a$ (What interpretation should be given to -2?)
   c) $(a + b)^2 = a^2 + 2ab + b^2$ d) $(a + b)(a - b) = a^2 - b^2$
   e) $(-a)^{-1} = -(a^{-1})$ f) $(ab)^{-1} = (a^{-1})(b^{-1})$
   g) $(a^{-1})^{-1} = a$ h) $(a/b)/c = a/(bc)$

4. The *cancellation law for addition* states that if $a + b = a + c$, then $b = c$. Prove that this is valid in any field. Does the equation $ab = ac$ necessarily imply that $b = c$?

5. Complete the verification that $\mathbf{Z}_2$ is a field.

6. Verify A4 and M4 for the field $\mathbf{Z}_3$.

7. If F is a field, we say that a subset $K \subset F$ is a *subfield* of F if K is itself a field with respect to the operations of addition and multiplication which are defined for elements of F.

   a) Prove that a subset $K \subset F$ is a subfield of F if and only if K contains at least one nonzero element and is closed under the operations of addition, multiplication, and the formation of negatives and reciprocals.
   b) Using (a), show that the rational numbers form a subfield of the field of real numbers.
   c) Show that the set of all numbers of the form $a + b\sqrt{2}$, where a and b are rational, is a subfield of $\mathbf{R}$.
   d) Show that the set of all numbers of the form $a + b\sqrt[3]{2}$, where a and b are rational, is not a subfield of $\mathbf{R}$. (In this problem feel free to assume that the integers are closed under addition. Also assume the existence of $\sqrt{2}$ and $\sqrt[3]{2}$.)

8. Suppose that F satisfies all the conditions for a field except NT. Prove that F contains only one element.

9. Which of the axioms for a field are satisfied by the set of integers (with the usual operations of addition and multiplication)?

10. Some of the older books list the following "axiom" for addition: if $a = b$ and $c = d$, then $a + c = b + d$. ("Equals added to equals give equals.") Why does our understanding of equality make such an axiom superfluous?

11. Let m be any positive integer. By the *integers modulo m* we mean the set

$$\mathbf{Z}_m = \{0, 1, 2, \ldots, m - 1\},$$

with addition and multiplication being defined by analogy with Example 2 ($\mathbf{Z}_3$). Show that Axiom M4 does not hold for $\mathbf{Z}_4$. In general, it can be shown that all the field axioms, with the possible exception of M4, are satisfied by $\mathbf{Z}_m$. Moreover, $\mathbf{Z}_m$ is a field if and only if m is a prime number.

12. Show that in any field the expressions $[(a + b) + c] + d$, $[a + (b + c)] + d$, $(a + b) + (c + d)$, $a + [(b + c) + d]$, and $a + [b + (c + d)]$ are always equal. Their common value is denoted by $a + b + c + d$. A similar remark is true for products.

In the remaining exercises you may bring in intuitive notions about the real numbers. Do not, in other words, limit yourself to use of the field axioms, but answer the questions in the spirit of the text prior to the introduction of the field axioms.

13. Prove that $\sqrt[3]{2}$ is irrational, but algebraic.

14. Show that the repeating decimal 1.2176176176... represents a rational number. (More generally, any repeating decimal is a rational number, and conversely, every rational number has a repeating decimal expansion.)

15. Show that $\sqrt[3]{1 + \sqrt{2}}$ is algebraic.

16. If $2^x = 3$, show that x is irrational.

17. Determine which of the following are true. Prove the true statements and give examples to show that the false statements are false. (Such examples are known as "counterexamples.")

    a) If x is rational and y is irrational, then $x + y$ is irrational.
    b) If x is rational and y is irrational, then xy is irrational.
    c) The sum of two irrational numbers is irrational.

18. In proving that $\sqrt{2}$ is irrational, we needed the following fact: if n is an integer such that n^2 is even, then n itself must be even. By expanding $(2m + 1)^2$, show that the square of any odd number is odd. Deduce the assertion in the first sentence.

1–3 The order axioms

It has been said (by some admittedly biased mathematicians) that boys work with equalities and men work with inequalities. Inequalities are tricky—so much so that it seems a bit unfair that they should also be so important. Although they seldom get top billing in this book, they are nearly always present, performing thankless but indispensable tasks.

We shall list presently some of the basic rules for handling inequalities. These rules cannot be proved from the field axioms; in fact, before inequalities are meaningful at all, one needs a more specialized structure than that of a field. We shall now state additional axioms for the real numbers which will permit a study of inequalities.

Order Axioms. *There exists a subset* $\mathbf{R}^+$ *of* $\mathbf{R}$ *with the following properties.*

O1. If x is any member of $\mathbf{R}$, then exactly one of the following is true: either $x \in \mathbf{R}^+$, $x = 0$, or $(-x) \in \mathbf{R}^+$.

O2. The set $\mathbf{R}^+$ is closed under addition.

O3. The set $\mathbf{R}^+$ is closed under multiplication.

Later, after we have added one final axiom, we shall be able to show that the set $\mathbf{R}^+$ is unique. In any case we shall call $\mathbf{R}^+$ the set of *positive real numbers*. We now define the basic order relation in terms of $\mathbf{R}^+$.

Definition 1. If a and b are real numbers, we say that a **is less than** b (or that b **is greater than** a) and write $a < b$ (or $b > a$) if and only if $(b - a) \in \mathbf{R}^+$.

Observe that $a > 0$ if and only if $a - 0 = a \in \mathbf{R}^+$. Thus the set $\mathbf{R}^+$ of positive numbers may also be described as the set of all numbers greater than zero.

We list next some basic facts about inequalities.

1) *Given any two real numbers x and y, exactly one of the following is true: either $x < y$, $x = y$, or $y < x$.*

This is similar in form and logically equivalent to O1; both are called the *law of trichotomy*. To deduce the present version of the law from O1, we let $z = y - x$. According to O1, exactly one of the following is true: $z \in \mathbf{R}^+$, $z = 0$, or $(-z) \in \mathbf{R}^+$. In the first case we have $x < y$; in the second, $x = y$, and since $-z = x - y$, we have in the third case $y < x$.

2) *If $x < y$ and $y < z$, then $x < z$.*

Because of this property we say that the relation "$<$" is *transitive*. The proof follows easily from O2 and the identity

$$z - x = (y - x) + (z - y).$$

3) *If $x < y$, then $x + a < y + a$.*

Proof. The expression $(y + a) - (x + a) = y - x$ belongs to $\mathbf{R}^+$, since by hypothesis $x < y$. Thus $x + a < y + a$.

3') *If $x < y$ and $a < b$, then $x + a < y + b$.*

Proof. By (3), we may write $x + a < y + a$ and $y + a < y + b$. Hence by transitivity, $x + a < y + b$.

Roughly speaking, (3) and (3') state that inequalities may be added. Greater care is needed with multiplication.

4) *If $x < y$ and $a > 0$, then $ax < ay$.*

In other words, multiplication by a positive number does not alter the sense of an inequality. The proof is left to the reader.

4') *If $x < y$ and $a < 0$, then $ay < ax$.*

Numbers which are less than zero are called *negative*. According to the present rule, multiplication by a negative number reverses the sense of an inequality.

Proof. By the law of trichotomy, $(-a) \in \mathbf{R}^+$. Since $y - x$ is also in $\mathbf{R}^+$,

$$ax - ay = (-a)(y - x)$$

is in $\mathbf{R}^+$ by O3. Thus $ay < ax$.

4") *If $0 < a < b$ and $0 < x < y$, then $ax < by$.*

The proof of this is analogous to the proof of (3').

5) *If $x \neq 0$, then $x^2 > 0$.*

Proof. If $x \neq 0$, then either $x \in \mathbf{R}^+$ or $(-x) \in \mathbf{R}^+$. In the first case, $x^2 = x \cdot x$ is positive by O3. If $(-x) \in \mathbf{R}^+$, then $x^2 = (-x) \cdot (-x)$ is positive also by O3.

Since $1 = 1^2$, we see in particular that

6) $1 > 0$.

The last two rules concern reciprocals.

7) *If $a > 0$, then $a^{-1} > 0$.*

Proof. Suppose that a^{-1} is not positive. Then (since it cannot be zero) we must have $a^{-1} < 0$. Multiplication by the positive number a then yields

$$1 = a \cdot a^{-1} < a \cdot 0 = 0,$$

contrary to (6). This proves that a^{-1} must be positive.

8) *If $0 < a < b$, then $b^{-1} < a^{-1}$.*

Thus, taking reciprocals of positive numbers reverses inequalities. Upon examination of the identity

$$a^{-1} - b^{-1} = (ab)^{-1}(b - a),$$

this rule is easily seen to follow from (7) and O3.

All of the preceding rules hold true for any *ordered field*. By the latter we mean a field F possessing a subset F^+ closed under addition and multiplication and satisfying the trichotomy condition (that is, Axiom O1 with "F" replacing "$\mathbf{R}$" and "F^+" replacing "$\mathbf{R}^+$"). The rational numbers $\mathbf{Q}$ form an ordered field, where

$$\mathbf{Q}^+ = \{m/n \mid m, n \in \mathbf{Z}^+\}.$$

On the other hand, the fields $\mathbf{Z}_2$, $\mathbf{Z}_3$, and the complex numbers are not ordered fields. In the case of the complex numbers, this is clear because of the following result.

In an ordered field the equation $x^2 + 1 = 0$ has no solution.

Proof. If $x = 0$, then $x^2 + 1 = 1$. If $x \neq 0$, then $x^2 > 0$. This implies that $x^2 + 1 > 1$, and since $1 > 0$, we get by transitivity $x^2 + 1 > 0$. In no case then can $x^2 + 1$ equal 0. (Note that in the field $\mathbf{Z}_2$, $1^2 + 1 = 0$. Consequently, $\mathbf{Z}_2$ is not an ordered field.)

So far we have dealt only with *strong inequalities*. We define the weak inequality $a \leq b$ to mean that either $a < b$ or $a = b$. So, for example, the inequality $x^2 \geq 0$ holds for all real numbers. (The inequality "$a \leq b$" is read "a is less than or equal to b.") We define $x \geq y$ similarly. If $a < b$ and $b < c$, we also frequently write $a < b < c$, and say that b is between a and c.

We shall next illustrate how these basic rules may be applied.

Example 1. We shall show that for each positive number x,

$$x + (1/x) \geq 2.$$

In setting out to prove an inequality such as this one, it is often helpful to engage in some scratch work. In fact, it is frequently convenient to do precisely what is forbidden in a proof: to assume what is to be proved and work backwards to what is known. Under no circumstances will this constitute a proof, but it may be helpful in getting one's bearings. (If such shenanigans constituted a proof, there would be no end to the nonsense which could be irrefutably established. One could prove, for example, that $2 = 1$. Assume that $2 = 1$. Then $1 = 2$, and by adding these equations, we get $3 = 3$. Since the latter is obviously true, we have "proved" that $2 = 1$.)

Suppose now that $x + (1/x) \geq 2$. Since we are also assuming that $x > 0$, we may multiply both sides by x. This gives

$$x^2 + 1 \geq 2x.$$

Adding $-2x$ to both sides of this inequality, we get

$$x^2 - 2x + 1 \geq 0.$$

But $x^2 - 2x + 1 = (x - 1)^2$. Thus, if $x \neq 1$, then $(x - 1)^2 > 0$, whereas if $x = 1$, then $(x - 1)^2 = 0$. Hence the inequality

$$x^2 - 2x + 1 \geq 0$$

is valid.

Having done this scratch work, we attempt next to reverse the steps of this argument to get a proof.

Suppose that $x \neq 1$. Then

$$x^2 - 2x + 1 = (x - 1)^2 > 0$$

by rule (5). Adding $2x$ to both sides gives

$$x^2 + 1 > 2x.$$

Since $x > 0$, it follows that $x^{-1} > 0$, and we can multiply both sides of this inequality by x^{-1} without changing its sense. Thus

$$x + (1/x) > 2.$$

If $x = 1$, then $x + (1/x) = 2$. Thus we have proved that for all $x > 0$,

$$x + (1/x) \geq 2.$$

Moreover, we have also shown that equality holds if and only if $x = 1$.

Example 2. If a, b, and c are positive numbers which are not all equal, show that

$$(a + b + c)(bc + ca + ab) > 9abc.$$

This time we do not show the scratch work:

$$(a + b + c)(bc + ca + ab) - 9abc = a(b^2 + c^2) + b(c^2 + a^2) + c(a^2 + b^2) - 6abc$$
$$= a(b - c)^2 + b(c - a)^2 + c(a - b)^2.$$

The last three terms are visibly nonnegative. Since two of the numbers a, b, c are unequal, one of the terms is strictly positive and this causes the entire expression to be positive. The desired inequality follows.

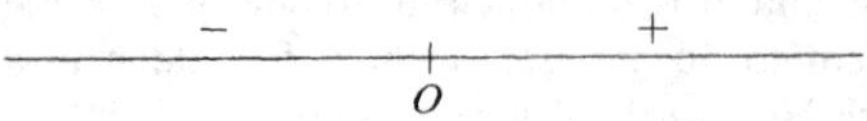

FIG. 1–14 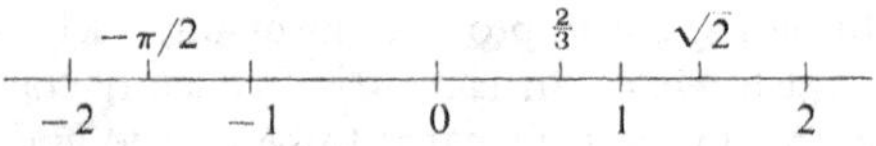   **FIG. 1–15**

In thinking about the real numbers, it is often helpful to regard them as points on a line. Although the reader may already be acquainted with the procedure for identifying points and numbers, we shall describe how this is done. Suppose that we are given a straight line in the usual Euclidean sense. First we select a point O on the line, henceforth to be called the *origin*. This point divides the line into two half-lines or rays, one of which we designate as *positive*, the other, *negative* (see Fig. 1–14). (It is conventional with a figure such as Fig. 1–14 to designate the right half-ray as positive.) Finally we choose a unit of length. Once these choices have been made, we can set up a correspondence between points on the line and real numbers. If $x > 0$, we assign to it that point P on the positive half-line whose distance from O is x. If $x < 0$, we assign to it that point P on the negative half-line whose distance from O is $-x$. The number 0 is matched with the origin. In this way we are able to match numbers with points, as shown in Fig. 1–15. If P is a point on the line, then the number x matched to it is called the *coordinate* of P. We shall assume that the correspondence between points on the line and real numbers is *one-to-one*. By this we mean that each point on the line has exactly one coordinate, and conversely, each number is the coordinate of just one point. We shall make no attempt to prove this even though it is of fundamental importance in linking algebra to geometry. In the first place, we cannot prove it now since our description of the real numbers is yet incomplete. In the second place, such a proof would also entail extensive probing into the foundations of synthetic Euclidean geometry. This means, of course, that any knowledge of Euclidean geometry which we may have can only be used descriptively; geometric arguments may be used to suggest, motivate, or elucidate analytic proofs, but never to supplant them.

Inequalities are easily interpreted geometrically. If coordinates are introduced in the conventional manner, then $a < b$ if and only if the point corresponding to a lies to the left of the point corresponding to b.

In the study of calculus we shall be especially interested in those subsets of the real line which, intuitively, have no gaps or breaks. The set $\{x \mid 2 \le x \le 3\}$, which is shown in Fig. 1–16, is one such example. These "connected" sets are called *intervals*. There are nine different kinds of intervals. If a and b are real numbers, with $a \le b$, then the set $\{x \mid a \le x \le b\}$ is called the *closed interval* from a to b and is denoted by $[a, b]$. The *open interval* from a to b is the set

$$(a, b) = \{x \mid a < x < b\}.$$

There are also half-open (or half-closed) intervals defined by

$$[a, b) = \{x \mid a \le x < b\},$$
$$(a, b] = \{x \mid a < x \le b\}.$$

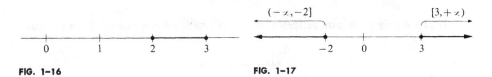

FIG. 1–16   FIG. 1–17

These *bounded* intervals, thought of as points on a line, are simply line segments, with or without endpoints. The *unbounded* intervals consist of the half-lines, with or without endpoints, and the entire line. The notations for these are the following:

$$[a, +\infty) = \{x \mid x \geq a\}, \qquad (a, +\infty) = \{x \mid x > a\},$$
$$(-\infty, a] = \{x \mid x \leq a\}, \qquad (-\infty, a) = \{x \mid x < a\}, \qquad (-\infty, +\infty) = \mathbf{R}.$$

Example 3. "Solve" the inequality $2x + 1 < 5x - 4$.

What is being asked for is a simple description of the *truth set* of the inequality, namely

$$S = \{x \mid 2x + 1 < 5x - 4\}.$$

Suppose that x satisfies the inequality. By adding $-2x + 4$ to both sides of the inequality we get $5 < 3x$. If we then multiply by the positive number $\frac{1}{3}$, we get $\frac{5}{3} < x$. Since the steps are reversible, we see that S is the open half-line $(\frac{5}{3}, +\infty)$.

Example 4. For what values of x is the expression $\sqrt{x^2 - x - 6}$ defined (only real numbers being considered)?

We shall assume now that *every nonnegative real number has a square root.* We shall be able to prove this later. Moreover, when we use $\sqrt{y}$ it *always* denotes the nonnegative square root. Thus, $\sqrt{(-2)^2} = +2$.

Clearly, $\sqrt{x^2 - x - 6}$ is defined if and only if

$$x^2 - x - 6 = (x + 2)(x - 3) \geq 0.$$

Equality holds if $x = -2$ or $x = +3$. If $x < -2$, both $x - 3$ and $x + 2$ are negative and therefore their product is positive. If $-2 < x < 3$, then $(x + 2)(x - 3) < 0$, since the first factor is positive and the second is negative. If $x > 3$, then $(x + 2)(x - 3)$ is positive since both factors are positive. Thus $\sqrt{x^2 - x - 6}$ is defined if and only if x belongs to the set $(-\infty, -2] \cup [3, +\infty)$, shown in Fig. 1–17.

Exercises

1. Let a, b, c, d be positive numbers such that $a/b < c/d$. Prove that

$$\frac{a}{b} < \frac{a + c}{b + d} < \frac{c}{d}.$$

Deduce that $\frac{1}{2} < \frac{7}{11} < \frac{4}{5}$.

2. Given two positive real numbers x and y, we define their arithmetic and geometric means as follows:

$$A = \tfrac{1}{2}(x + y), \qquad G = \sqrt{xy}.$$

The harmonic mean is the reciprocal of the arithmetic mean of the reciprocals. Thus,

$$\frac{1}{H} = \frac{1}{2}\left(\frac{1}{x} + \frac{1}{y}\right).$$

Prove that $A \geq G \geq H$. When can equality hold?

*3. Prove that

$$\frac{a+b+c}{3} \geq \sqrt{\frac{ab+bc+ca}{3}} \geq \sqrt[3]{abc},$$

where $a, b, c \geq 0$.

4. If $x, y, z > 0$, show that

$$(x + y + z)(1/x + 1/y + 1/z) \geq 9.$$

Define appropriately the arithmetic mean A and harmonic mean H of x, y, and z and deduce that $A \geq H$.

5. Prove that

$$x^3 + 1/x^3 \geq x^2 + 1/x^2$$

for all $x > 0$. When does equality hold?

6. If $a, b \geq 0$, show that

$$\sqrt{a+b} \leq \sqrt{a} + \sqrt{b}.$$

7. If $a, b \geq 0$, show that $a < b$ if and only if $a^2 < b^2$.

8. Let a, b, c be fixed real numbers with $a > 0$. By completing the square, show that the smallest value of the quadratic polynomial

$$Q(x) = ax^2 + 2bx + c$$

is $(ac - b^2)/a$. Suppose you know that $Q(x) \geq 0$ for *every* real number x. What can you then say about the coefficients a, b, c?

9. In working Example 4 we have assumed the familiar rules for multiplying signed numbers. (We have assumed, for instance, that the product of two negative numbers is positive.) Deduce these rules from the rules given in the text.

10. Prove rule (4) and also (4'').

11. Fill in the details for the proof of rule (8).

12. Show that $\mathbf{Z}_3$ is not an ordered field.

13. Let P^+ be the set of positive elements in an ordered field F, and let F^* be any subfield of F. Show that F^* may be regarded as an ordered field if $P^* = F^* \cap P^+$ is taken to be its set of positive elements. Thus, any subfield of the real numbers is an ordered field.

14. If $a = b$, what are the intervals $[a, b]$ and (a, b)?

15. Give simple descriptions of each of the following sets of real numbers. (In the case of finite sets, list their elements. In other cases, write them as intervals or finite unions of intervals.)

a) $\{x \mid 2x - 7 > 7x - 2\}$

b) $\{x \mid 2x - 7 \geq 7x - 2\}$

c) $\{x \mid x < x^2\}$

d) $\{x \mid \sqrt{x^2} = x\}$

e) $\{x \mid \sqrt{x^2 + 1} = x\}$

*f) $\{x \mid \sqrt{x + 6} > \sqrt{x + 1} + \sqrt{2x - 5}\}$

1-4 Absolute values

In this section we define and give the basic properties of absolute values.

Definition 1. The **absolute value** of a real number x is defined by the following rule:

$$|x| = \begin{cases} x & \text{if } x > 0, \\ 0 & \text{if } x = 0, \\ -x & \text{if } x < 0. \end{cases}$$

Examples. $|7| = 7$, since $7 > 0$. Also $|-3| = -(-3) = 3$, since $-3 < 0$.

If the real numbers are thought of as points on a line, then the absolute value of a number is simply its distance from the origin (see Fig. 1–18).

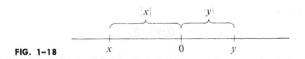

FIG. 1–18

Theorem 2. *Absolute values have the following properties.*

1) *For all $x \in \mathbf{R}$, $|x| = |-x| \geq 0$ and equality holds iff $x = 0$.*

2) *For all x and y in $\mathbf{R}$,*

$$|xy| = |x|\,|y|.$$

3) *For each $a > 0$, the inequality $|x| \leq a$ is equivalent to $-a \leq x \leq a$.*

4) Triangle Inequality. *For all real numbers x and y,*

$$|x + y| \leq |x| + |y|.$$

Proof. The proofs of (1) and (2) are left as exercises for the reader.

Suppose $|x| \leq a$. There are two cases to consider.

Case I. $x \geq 0$. Then $x = |x| \leq a$. On the other hand, $-a \leq 0 \leq x$. Thus, we have $-a \leq x \leq a$.

Case II. $x < 0$. Then $-x = |x| \leq a$, so that $x \geq -a$. On the other hand, $x < 0 \leq a$. Thus we have, once again, $-a \leq x \leq a$.

By considering the same two cases, one can show that, conversely, $-a \leq x \leq a$ implies that $|x| \leq a$.

We next prove (4). Since $|x| \leq |x|$ (trivially), it follows from (3) that

$$-|x| \leq x \leq |x|.$$

Similarly,

$$-|y| \leq y \leq |y|.$$

Adding these inequalities, we get

$$-(|x| + |y|) \leq x + y \leq |x| + |y|.$$

Hence by (3),

$$|x + y| \le |x| + |y|.$$

We shall frequently meet expressions of the form $|y - x|$, where x and y are real numbers. We present now two ways of interpreting such a quantity. If x and y are real numbers such that $x < y$, then $y - x = |y - x|$ is equal to the distance between x and y (see Fig. 1–19). If $x > y$, then the distance between x and y is

$$x - y = -(y - x) = |y - x|.$$

Thus, in both cases we see that $|y - x|$ *is the distance between x and y on the real line.* In numerical work, on the other hand, we frequently approximate a number x by a number y, in which case $|y - x|$ may be regarded as the error committed in making the approximation. For example, if we use 3.142 as an approximation to π, then the error which we commit is

$$|\pi - 3.142| = 0.000407...$$

We shall consider now some examples to illustrate the handling of absolute values.

Example 1. Describe the set

$$S = \{x \in \mathbf{R} \mid |x - 5| < 2\}.$$

By the previous theorem, the inequality $|x - 5| < 2$ is equivalent to

$$-2 < x - 5 < 2.$$

Adding 5 to this inequality, we get $3 < x < 7$. Since the steps are reversible, we get

$$S = (3, 7).$$

This is also clear from the fact that S is the set of all points on the line whose distance from 5 is less than 2, as shown in Fig. 1–20.

FIG. 1–19   FIG. 1–20

More generally, the inequality $|x - a| < \delta$ is equivalent to

$$a - \delta < x < a + \delta.$$

The set $\{x \in \mathbf{R} \mid |x - a| < \delta\}$ will be called the δ-*neighborhood of a.*

Example 2. In this problem we shall consider how many decimal places of $\sqrt{2}$ and π are needed to compute $\sqrt{2}\,\pi$ accurately to 75 decimal places.

If x is any real number, we shall temporarily use x_n to denote the value of x rounded to n decimal places. For example,

$$\pi_2 = 3.14, \qquad \pi_3 = 3.142, \qquad \pi_4 = 3.1416, \qquad \dots$$

In each case the error cannot exceed 5 in the $(n + 1)$th decimal place. For instance,

$$|\pi - \pi_2| = 0.0019... < 0.005,$$
$$|\pi - \pi_3| = 0.0004173... < 0.0005.$$

In general, $|x - x_n| \leq 5 \times 10^{-(n+1)}$.

Suppose now that we round π to m places and $\sqrt{2}$ to n places. Then the errors are respectively

$$|\pi - \pi_m| \leq 5 \times 10^{-(m+1)} \quad \text{and} \quad |\sqrt{2} - (\sqrt{2})_n| \leq 5 \times 10^{-(n+1)}.$$

We shall, of course, use the product $\pi_m(\sqrt{2})_n$ as an approximation to $\pi\sqrt{2}$. Our problem then is to find values of m and n so that $\pi_m(\sqrt{2})_n$ is the value of $\pi\sqrt{2}$ correct to 75 decimal places. In other words, we need to make the error $|\pi\sqrt{2} - \pi_m(\sqrt{2})_n|$ less than 5×10^{-76}.

First we must relate the error $|\pi\sqrt{2} - \pi_m(\sqrt{2})_n|$ to the round-off errors $|\pi - \pi_m|$ and $|\sqrt{2} - (\sqrt{2})_n|$. To do this we employ the frequently used device of adding and subtracting a term. We write

$$\pi\sqrt{2} - \pi_m(\sqrt{2})_n = \pi\sqrt{2} \times -\pi_m\sqrt{2} + \pi_m\sqrt{2} - \pi_m(\sqrt{2})_n$$
$$= (\pi - \pi_m)\sqrt{2} + \pi_m[\sqrt{2} - (\sqrt{2})_n].$$

Applying the triangle inequality, we get

$$e = |\pi\sqrt{2} - \pi_m(\sqrt{2})_n|$$
$$\leq |(\pi - \pi_m)\sqrt{2}| + |\pi_m[\sqrt{2} - (\sqrt{2})_n]|$$
$$= \sqrt{2}|\pi - \pi_m| + \pi_m|\sqrt{2} - (\sqrt{2})_n|.$$

Since $5\sqrt{2} < 8$ and $5\pi_m < 16$,

$$e < 0.8 \times 10^{-m} + 1.6 \times 10^{-n}.$$

Now the inequality

$$0.8 \times 10^{-m} + 1.6 \times 10^{-n} < 5 \times 10^{-76}$$

will hold if $m \geq 76$ and $n \geq 76$, and we cannot get by with smaller values for either m or n.

Thus, to compute $\pi\sqrt{2}$ accurately to 75 decimal places, we should round both π and $\sqrt{2}$ to 76 places.

Exercises

1. Show that for every real number x, $\sqrt{x^2} = |x|$. Use this fact to prove (1) and (2) of Theorem 2.

2. Try to find brief proofs for each of the following.

   a) For all real numbers x and $y \neq 0$,
   $$|x/y| = |x|/|y|.$$

   b) For all x and y,
   $$|x - y| \geq ||x| - |y||$$

   (reverse triangle inequality).

3. Describe the values of the fraction $x/|x|$ for different values of x.

4. Give simple descriptions of each of the following sets.

a) $\{x \mid |x| < 2\}$

b) $\{x \mid |x - 3| \leq \frac{1}{2}\}$

c) $\{x \mid |7x - 4| > 5\}$

d) $\{x \mid |7x + 3| + |3 - x| \geq 6|x + 1|\}$

e) $\{x \mid |x^2 - x - 6| \geq |x + 2| |x - 3|\}$

f) $\{x \mid |x^2 - x - 6| = |x + 2| |x - 3|\}$

g) $\{x \mid |x - 1| = |x - 4|\}$

h) $\{x \mid |x - 1| + |x - 5| = 4\}$

i) $\{x \mid x + |x| < 1\}$

j) $\{x \mid x - |x| > 2\}$

k) $\{x \mid |x^2 - x| + x > 1\}$

5. Suppose that x^* is an approximation of x and y^* is an approximation of y. If $2x^* + y^*$ is used as an approximation of $2x + y$, show that the error cannot exceed $2e_x + e_y$, where e_x and e_y are the errors of the approximation of x and y, respectively. How many decimal places of π and $\sqrt{2}$ are needed to calculate $2\pi + \sqrt{2}$ accurately to 10 decimal places?

1–5 Quantifiers

We turn next to logic. By a *proposition* we shall mean a sentence which is true or false as it stands. For example, the equation '$2 + 2 = 4$' is a proposition (which is true). The inequality '$3 > 5$' is also a proposition (which is false). 'The billioneth-trillioneth decimal digit of π is a 7' is a proposition which can be classified only hypothetically as true or false. That is to say, the statement as it stands is certainly either true or false, although from a practical point of view it is impossible to determine which is the case. Countless statements can be cited which are not propositions.

> 'Twas brillig, and the slithy toves
> Did gyre and gimbel in the wabe

is an obvious example. A less fanciful, but far more pertinent, example is the equation '$x + 1 = 3$.' As it stands it is neither true nor false, although it becomes true or false when values are substituted for the variable 'x.' For instance, if we replace 'x' by '2,' we get a true proposition, whereas if we replace 'x' by '3,' we get a false proposition. Statements such as this equation are called *open sentences, propositional functions*, or *statement frames*. The statements '$3x - 5y + 7 = 0$' and '$x < y$' are also examples of propositional functions.

Let $P(x)$ (read "P of x") be the propositional function '$x + 1 = 0$.' We can obtain propositions from $P(x)$ by substituting various values for the variable 'x.' For instance, by substituting the number '5' for 'x,' we get the false proposition $P(x)$, namely, '$5 + 1 = 0$.' However, we can also obtain a proposition from $P(x)$ by prefixing it with one of the following two phrases: 'for each x,' or 'there exists an x such that.' In the first case we get a false statement (namely, 'for each x, $x + 1 = 0$'), in the second case, a true statement [since $P(-1)$ is true]. These prefixing phrases are called *quantifiers*, and the mechanics of handling them is known as *quantification theory*.

We shall use $(\forall x)$ as an abbreviation for the *universal quantifier* 'for each x.' Thus $(\forall x)(x + 1 > x)$ may be translated: 'for each x, $x + 1 > x$.' Similarly, the symbol $(\exists y)$ represents the *existential quantifier* 'there exists a y such that.' Thus $(\exists y)(y^2 + 1 = 0)$ is the proposition 'there exists a y such that $y^2 + 1 = 0$.' The appropriate meaning of a quantified statement depends somewhat on its context. For example, the statement above, $(\forall x)(x + 1 > x)$, if occurring in a discussion involving the real numbers, probably should be interpreted to mean that the inequality $x + 1 > x$ is valid for *all real numbers* x. In this case the author most likely does not mean to assert that the inequality is true if x is a centurion, the Empire State Building, or a one-legged minotaur! Similarly, the statement $(\exists z)(z^2 + 1 = 0)$ is false if only real numbers are being considered, but true if complex values are allowed for the variable 'z.' In other words, whenever these quantifiers are used, there is present some universal set S (although it may not be mentioned explicitly) which contains all possible values of the variable which are being considered. If we wish to call attention to the universal set S, we can use such notations as $(\forall x \in S)$ or $(\exists x \in S)$. For instance, $(\exists x \in \mathbf{R})(x^2 + 1 = 0)$ falsely asserts that there exists a real number which satisfies the equation $x^2 + 1 = 0$.

The proposition $(\forall x)((x + 1)^2 = x^2 + 2x + 1)$ still involves the variable 'x,' but we are no longer *free* to give it values. We say that 'x' is a *bound* variable in this case. (In other contexts the term "dummy variable" is occasionally used.) We shall use the notation '$P(x)$' only when 'x' is a free variable.

Let us consider next the propositional function 'x is the father of y,' where the universal set S is the set of all human beings living or dead. We can first quantify the variable 'x' to form the statement

$$(\exists x)(x \text{ is the father of } y). \tag{$*$}$$

This statement is also a propositional function, but it has only one free variable, namely, 'y.' For example, if we substitute 'Johann Christian Bach' for 'y,' we get a (true) proposition, which asserts, namely, that Johann Christian Bach has a father. (In this case 'x' must be 'J. S. Bach.') We can also form a proposition from $(*)$ by quantifying y. For example, we can assert that

$$(\forall y)[(\exists x)(x \text{ is the father of } y)],$$

or, simplifying the notation in a conventional fashion,

$$(\forall y)(\exists x)(x \text{ is the father of } y). \tag{a}$$

This simply says that each man y has a father x, a true statement (if one makes possible allowance for Adam). Consider, on the other hand, the statement

$$(\exists x)(\forall y)(x \text{ is the father of } y), \tag{b}$$

which is shorthand for

$$(\exists x)[(\forall y)(x \text{ is the father of } y)].$$

This asserts something quite different. This says that there is one man x who is the father of all men. [The reader should under no circumstances read further until he is thoroughly convinced that this is the content of (b).] Note that (b) is a much stronger assertion than (a)—too strong, in fact. In (a), 'x' could vary with 'y'; in (b), it is fixed once and for all. More generally, if $P(x, y)$ is a propositional function involving the variables 'x' and 'y,' then the statement $(\exists x)(\forall y)P(x, y)$ is a stronger assertion than the statement $(\forall y)(\exists x)P(x, y)$. The second statement asserts that for each value of 'y' it is possible to find a corresponding value of 'x' which makes $P(x, y)$ true. In this case the values of 'x' can change to accommodate different values of 'y.' The first statement, however, asserts that one fixed value of 'x' may be found, say x_0, such that $P(x_0, y)$ is a true statement for all values of 'y.'

Let us consider a few more examples.

Example 1

$$(\forall x \in \mathbf{R})(\exists y \in \mathbf{R})(x + y = 0).$$

In translation this says: 'for each real number x, there exists a real number y such that $x + y = 0$.' In other words, the above statement is simply a symbolic rendering of Axiom A4 for the real numbers. Note that the stronger statement

$$(\exists y \in \mathbf{R})(\forall x \in \mathbf{R})(x + y = 0)$$

is false, since there is no single real number y which serves as an additive inverse for the entire set of real numbers.

Example 2

$$(\forall x \in \mathbf{R})(\forall y \in \mathbf{R})(x^2 - y^2 = (x + y)(x - y)).$$
$$(\forall y \in \mathbf{R})(\forall x \in \mathbf{R})(x^2 - y^2 = (x + y)(x - y)).$$

These two statements assert the same thing, namely, that the equation

$$x^2 - y^2 = (x + y)(x - y)$$

is valid for all real numbers x and y. Technically speaking, the equation

$$x^2 - y^2 = (x + y)(x - y),$$

without accompanying quantifiers, is as incomplete as the phrase "is running." On the other hand, it is troublesome always to include the appropriate quantifiers. Consequently, if the reader bumps up against an identity such as the one considered here, which stands in naked splendor with no quantifiers in sight, he should assume that it is being asserted for all appropriate values of the variables.

This example illustrates a general principle of quantification, namely, that in contrast to the case of quantifiers of mixed type, quantifiers of the same type can be interchanged at will. Thus, $(\forall x)(\forall y)P(x, y)$ and $(\forall y)(\forall x)P(x, y)$ state the same thing. For this reason quantifiers of the same type are occasionally lumped together.

The symbol $(\forall x, y)$, for instance, may be used as an abbreviation for $(\forall x)(\forall y)$ or $(\forall y)(\forall x)$, and may be read 'for all x and y.' Thus, the associative law of addition may be written

$$(\forall x, y, z)(x + (y + z) = (x + y) + z).$$

Similar remarks hold for existential quantifiers.

Example 3. Consider the statements

$$(\exists x \in S)(\forall y \in S)(x \geq y)$$

and

$$(\forall y \in S)(\exists x \in S)(x \geq y),$$

where S is a subset of the real numbers.

The first statement asserts that S has a largest member. Thus the statement is true if $S = [-3, 7]$, but false if $S = [-3, 7)$. The second statement is much weaker than the first and is true, in fact, for every subset S (since we can always let x be equal to y).

Mathematics is sometimes called a language. Without endorsing this viewpoint, one can safely say that mathematicians often use language in a way peculiarly their own. A sentence beginning 'for each x, there exists a y such that . . .' is something of a rarity in ordinary discourse, and yet mathematicians emit such utterances with a fluency and frequency that unnerves most laymen. One should realize, however, that many definitions and theorems cannot be couched in colloquial English without sacrificing clarity.

> Every number is less than some integer.
> Some integer is greater than every number.

Do these statements have the same meaning? If put to a vote, opinion would very likely be divided. Furthermore, those answering affirmatively would probably disagree as to whether the statements are true or false. The statements

$$\text{`}(\forall x \in \mathbf{R})(\exists y \in \mathbf{Z})(x < y)\text{'}$$

and

$$\text{`}(\exists y \in \mathbf{Z})(\forall x \in \mathbf{R})(x < y),\text{'}$$

though admittedly less idiomatic than the statements above, do at least have the marked advantage of being unambiguous. Even Aristotle, the founder of logic, was occasionally careless about quantification. Consider the opening sentence from his *Nicomachean Ethics:*

> Every art and every inquiry, and similarly every action and pursuit, is thought to aim at some good; and for this reason the good has rightly been declared to be that at which all things aim.

Do the two clauses say the same thing? If so, what is the point of the redundancy? If not, does the second clause follow logically from the first (as "for this reason" would seem to imply), or do you think Aristotle tried to pull a fast one?

Exercises

1. Listed below are a number of propositional functions in the variables x and y. Consider all possible propositions which one can form from them by using existential and universal quantifiers. List all the true propositions so obtained and indicate the relative strength of these statements. [For example, if '$x < y$' is the propositional function, then each of the following combinations of quantifiers, $(\forall x)(\exists y)$, $(\forall y)(\exists x)$, and $(\exists x)(\exists y)$, yields a true proposition, the first two being stronger than the third.]

a) $x^2 + y^2 = -1$ b) $x^2 + y^2 = 0$
c) $x^2 + y^2 = 1$ d) $x = y^2$
e) $x < y^2$ f) $|3x - y + 4| \le 3|x + 1| + |y - 1|$
g) $y^2 - x^2 \le 1$ h) $y^2 - x^2 \ge 1$
i) $\sqrt{|xy|} \le \dfrac{|x| + |y|}{2}$

2. Consider the following statements:

$$(\exists x \in \mathbf{R})(\forall y \in S)(y < x), \qquad (\exists x \in \mathbf{R})(\forall y \in S)(y > x), \qquad (\exists x \in \mathbf{R})(\forall y \in S)(|y| < x).$$

Determine which of these statements is true for each of the following choices of S:

a) $S = [-3, 10)$,   b) $S = (6, +\infty)$,   c) $S = (-\infty, 3]$,
d) $S = \mathbf{Z}^+$,   e) $S = \mathbf{Q}$.

3. a) Consider the following sets:

$$S_1 = \{1, 2, 5\}, \qquad S_2 = \{0, 2, 6\}, \qquad S_3 = \{2, 3, 7\}.$$

Determine each of the following sets:

$$A = \{x \mid (\exists i)(x \in S_i)\}, \qquad B = \{x \mid (\forall i)(x \in S_i)\}.$$

b) Let $\{S_a\}$ be a family of sets in which the index a ranges over a set A. Based upon your experience in (a), define appropriately the union and the intersection of this family of sets.

1-6 Logical connectives

We shall now consider various ways of combining sentences to form new ones.

We can negate statements.

Definition 1. The **negation** of a statement P is the statement '**not** P,' denoted by $\sim P$, which is understood to be true if P is false, and false if P is true.

For example, if P is the proposition '7 is not a prime number,' then $\sim P$ is the proposition '7 is a prime number.'

We can join two statements by the correlative conjunction 'both ... and.'

Definition 2. The **conjunction** of two statements P and Q is the statement '**both** P **and** Q,' denoted by $P \wedge Q$, which is understood to be true when both P and Q are true, and false otherwise.

For example, let P be the proposition '3 is a prime,' and let Q be the proposition '6 is a prime.' Then $P \wedge Q$ is the compound sentence '3 is a prime, and 6 is a prime.' According to our definition, $P \wedge Q$ is a false proposition.

We can join two statements by the correlative conjunction 'either . . . or.' In this case, however, there is possible ambiguity, since, in English usage, the word "or" can be used in the *exclusive* or the *inclusive* sense. For example, if a waitress tells you that you may have either soup or fruit juice as an appetizer, it is more than likely that she is using "or" in the exclusive sense; that is, she means to exclude the possibility of your having both soup and fruit juice before the entree. In ordinary usage, this is the most frequent meaning of "or." In mathematics, on the other hand, "or" is usually used in the inclusive sense. Earlier, for instance, we proved the following:

If a and b are two real numbers whose product is zero, then either $a = 0$ or $b = 0$.

Here "or" is used in the inclusive sense, since we wish to include the possibility that both a and b are equal to zero. (In legal documents "and/or" is often used when "or" is to be understood in the inclusive sense.) In any case, we make the following definition.

Definition 3. The **disjunction** of two statements P and Q is the statement '**either** P **or** Q,' denoted by $P \vee Q$, which is understood to be true when at least one of the two statements is true, and false when both are false. (In other words, "or" is used here in the inclusive sense.)

P	$\sim P$
T	F
F	T

P	Q	$P \wedge Q$	$P \vee Q$
T	T	T	T
T	F	F	T
F	T	F	T
F	F	F	F

FIG. 1–21

The definitions which we have given for the *logical connectives* $\vee$, $\wedge$, and $\sim$ can be summarized by the *truth tables* given in Fig. 1–21. These logical connectives are analogous to the set operations discussed earlier. This should not come as a surprise, since the definitions of $\cup$, $\cap$, and $'$ directly involve the connectives $\vee$, $\wedge$, and $\sim$ respectively:

$$A \cup B = \{p \in S \mid (p \in A) \vee (p \in B)\},$$
$$A \cap B = \{p \in S \mid (p \in A) \wedge (p \in B)\},$$
$$A' = \{p \in S \mid \sim(p \in A)\}.$$

We consider next the very important, though somewhat troublesome, logical connective 'if . . . , then.' If P and Q are statements, we shall denote by $P \Rightarrow Q$ the *conditional* statement 'if P, then Q'; in this case, P is called the *hypothesis* and Q, the *conclusion* of the conditional statement. Mathematics is strewn with statements of the form $P \Rightarrow Q$. Usually the hypothesis and conclusion are propositional functions, and universal quantification, if not explicitly stated, is understood. In constructing a truth table for the connective '$\Rightarrow$,' we shall therefore examine assertions of the form $(\forall x)(P(x) \Rightarrow Q(x))$. We shall regard '$x < 2 \Rightarrow 2x < 10$' as a true statement, at least so far as the real numbers are concerned. More precisely, then, we consider $(\forall x \in \mathbf{R})(x < 2 \Rightarrow 2x < 10)$ a true proposition. We are thereby committed to classifying the following statements as true:

$$\text{'}1 < 2 \Rightarrow 2 < 10,\text{'} \qquad \text{'}3 < 2 \Rightarrow 6 < 10,\text{'} \qquad \text{and} \qquad \text{'}6 < 2 \Rightarrow 12 < 10.\text{'}$$

This yields three entries in the truth table for '$\Rightarrow$' (see Fig. 1–22). On the other hand, we do not regard the statement '$2x < 10 \Rightarrow x < 1$' as true in all cases. In particular, we regard it as false if $x = 3$. From the fact that '$6 < 10 \Rightarrow 3 < 1$' is false, we see that the remaining entry in our truth table should be an F.

P	Q	$P \Rightarrow Q$
T	T	T
T	F	—
F	T	T
F	F	T

FIG. 1–22

P	Q	$P \Rightarrow Q$
T	T	T
T	F	F
F	T	T
F	F	T

FIG. 1–23

P	Q	$P \Leftrightarrow Q$
T	T	T
T	F	F
F	T	F
F	F	T

FIG. 1–24

Definition 4. If P and Q are statements, we shall denote by $P \Rightarrow Q$ the **conditional** statement '**if P, then Q**.' Its truth values are given in Fig. 1–23.

(The last two lines of the truth table for $P \Rightarrow Q$ might be misconstrued to imply that one can prove anything at all if one starts from a false assumption. In this connection, there is a story told about the great English mathematician G. H. Hardy. Once, at a party, Hardy asserted that from a false premise, one can prove anything one pleases. A member of Parliament who overheard this rash remark issued Hardy a challenge. "$4 = 7$. Prove that I am the Pope." Hardy thought a moment and then replied. "$4 = 7$. Subtract 1 from both sides of the equation. That gives $3 = 6$. Dividing this equation by 3, we get $1 = 2$, or, equivalently, $2 = 1$. Now it is well known that the Pope and you are two. But $2 = 1$. Therefore, the Pope and you are one.")

Definition 5. If P and Q are statements, we denote by $P \Leftrightarrow Q$ the **biconditional** statement 'P **if and only if Q**,' or 'P **is logically equivalent to Q**.' The truth values of the biconditional are given in Fig. 1–24.

We shall also use the standard abbreviation 'iff' for 'if and only if.'

Using the logical connectives just defined, one can build up complicated expressions such as $[(P \lor \sim Q) \Rightarrow R] \Leftrightarrow (\sim S \Rightarrow \sim P)$. Expressions of this sort are called *statement forms*. It is clear that the logical connectives can be used to join statement forms. For example, if φ is the statement form $(\sim P \lor Q) \Leftrightarrow R$, and if ω is the statement form $R \land \sim Q$, then

$$\varphi \lor \omega = [(\sim P \lor Q) \Rightarrow R] \lor (R \land \sim Q).$$

We shall now single out a class of statement forms which are of special interest.

Definition 6. A **tautology** is a statement form (that is, an expression consisting of letters joined by logical connectives) which is true no matter what propositions are substituted for the letters appearing in the form.

Perhaps the simplest example of a tautology is the statement form $P \lor \sim P$. No matter what proposition is substituted for P, the resulting proposition will be true. This is clear from an examination of the truth table for $P \lor \sim P$, given in Fig. 1–25.

A second example of a tautology is the statement form

P	$\sim P$	$P \lor \sim P$
T	F	T
F	T	T

FIG. 1–25

$$[(P \Rightarrow Q) \land (Q \Rightarrow R)] \Rightarrow (P \Rightarrow R),$$

which might be read, 'if P implies Q and Q implies R, then P implies R.' Once again we can verify that this is a tautology by constructing a truth table (see Fig. 1–26). Let $\varphi = P \Rightarrow Q$, $\omega = Q \Rightarrow R$, $\Delta = P \Rightarrow R$. Since the last column contains only

P	Q	R	φ	ω	$\varphi \land \omega$	Δ	$(\varphi \land \omega) \Rightarrow \Delta$
T	T	T	T	T	T	T	T
T	T	F	T	F	F	F	T
T	F	T	F	T	F	T	T
T	F	F	F	T	F	F	T
F	T	T	T	T	T	T	T
F	T	F	T	F	F	T	T
F	F	T	T	T	T	T	T
F	F	F	T	T	T	T	T

FIG. 1–26

[(P	⇒	Q)	∧	(Q	⇒	R)]	⇒	(P	⇒	R)
T	T	T	T	T	T	T	T	T	T	T
T	T	T	F	T	F	F	T	T	F	F
T	F	F	F	F	T	T	T	T	T	T
T	F	F	F	F	T	F	T	T	F	F
F	T	T	T	T	T	T	T	F	T	T
F	T	T	F	T	F	F	T	F	T	F
F	T	F	T	F	T	T	T	F	T	T
F	T	F	T	F	T	F	T	F	T	F

FIG. 1–27

T's, the statement form $[(P \Rightarrow Q) \wedge (Q \Rightarrow R)] \Rightarrow (P \Rightarrow R)$ is a tautology. A somewhat differently styled truth table which does the same job is shown in Fig. 1–27.

Definition 7. Two statement forms φ and ω are **logically equivalent** if and only if the statement form $\varphi \Leftrightarrow \omega$ is a tautology.

To put it somewhat differently, two statement forms are logically equivalent if they have the same "truth value" whenever propositions are substituted for the letters appearing in the forms. We shall consider several very important examples.

Example 1. The form $P \Rightarrow Q$ is logically equivalent to $\sim P \vee Q$.

This is clear from everyday usage. One might say, for example, "If that record doesn't sell, I'll eat my hat." The sense would be the same, however, if one said, "Either that record sells, or I'll eat my hat." In any case, we can prove the logical equivalence of these forms by the construction of a truth table (see Fig. 1–28).

Example 2. The form $\sim(P \Rightarrow Q)$ is equivalent to $P \wedge \sim Q$.

We shall make repeated use of this later in the text. From the truth table for '$\Rightarrow$,' one readily sees that $\sim(P \Rightarrow Q)$ is false except when P is true and Q is false. From the truth table for '$\wedge$,' one sees that $P \wedge \sim Q$ is false except when P and $\sim Q$ are both true. However,

(P	⇒	Q)	⇔	(~P	∨	Q)
T	T	T	T	F	T	T
T	F	F	T	F	F	F
F	T	T	T	T	T	T
F	T	F	T	T	T	F

FIG. 1–28

$\sim Q$ is true if and only if Q is false. Thus we see (without actually writing out the truth tables) that $\sim(P \Rightarrow Q)$ and $P \wedge \sim Q$ have the same truth values.

Example 3. The form $P \Rightarrow Q$ is logically equivalent to $\sim Q \Rightarrow \sim P$.

The form $\sim Q \Rightarrow \sim P$ is called the *contrapositive* of the form $P \Rightarrow Q$. The fact that these two forms are equivalent provides the basis for one type of indirect proof. Thus, if one is required to prove a theorem of the form $P \Rightarrow Q$, one can do this by assuming that the conclusion Q is false and deducing logically from this that the hypothesis P is false. The verification of this equivalence is left to the reader.

Example 4. The form $P \Leftrightarrow Q$ is logically equivalent to $(P \Rightarrow Q) \wedge (Q \Rightarrow P)$. In other words, a theorem of the form $P \Leftrightarrow Q$ really amounts to two theorems: one is simultaneously asserting $P \Rightarrow Q$ and also the *converse* statement $Q \Rightarrow P$.

Exercises

1. Determine which of the following statement forms are tautologies:

$$(P \wedge (P \Rightarrow Q)) \Rightarrow Q, \qquad P \Rightarrow (P \vee Q), \qquad P \Rightarrow (P \wedge Q),$$
$$(P \wedge Q) \Rightarrow P, \qquad (P \vee Q) \Rightarrow P.$$

2. Prove the equivalence in Example 3.

3. Prove the following equivalences:
   a) $P \Leftrightarrow \sim(\sim P)$,
   b) $\sim(P \vee Q) \Leftrightarrow (\sim P) \wedge (\sim Q)$,
   c) $\sim(P \wedge Q) \Leftrightarrow (\sim P) \vee (\sim Q)$,
   d) $[P \Rightarrow (Q \vee R)] \Leftrightarrow [(P \wedge \sim Q) \Rightarrow R]$.

4. Are the statement forms $(P \wedge Q) \Rightarrow R$ and $(P \Rightarrow R) \wedge (Q \Rightarrow R)$ logically equivalent?

5. A commuter waiting for a local train was pleasantly surprised when an express train unexpectedly stopped at the station. As he jumped aboard, the conductor rushed out and said, "This train doesn't stop here, so you can't get on." The commuter replied, "If the train doesn't stop here, then I'm not on it," and walked inside. Was the commuter's contention sound? (Taken from Stabler, *An Introduction to Mathematical Thought*, Addison-Wesley, Reading, Mass., 1953, p. 51.)

6. What is the set operation that corresponds to the exclusive 'or.' Express the exclusive 'or' in terms of the logical connectives $\vee$, $\wedge$, and $\sim$.

7. In this problem the universal set under consideration is the set of real numbers. In each case a propositional function involving one free variable 'x' is given. Find the sets of values for 'x' for which the following statements are true.
   a) $(x < 1) \Rightarrow (x < 3)$
   b) $(x < 1) \Leftrightarrow (x < 3)$
   c) $(x^2 < 0) \Rightarrow (x = 3)$
   d) $(x^2 < 0) \Rightarrow (x^4 = x^3)$
   e) $(\exists y)(x^2 + y^2 = 1)$
   f) $(\exists y)(x^2 + y^2 = -1)$
   g) $(\forall y)(|y| < x \Rightarrow 2y < 10)$

8. Now that we have discussed quantifiers and conditional statements, we can reformulate the definition of set inclusion as follows: $A \subset B$ iff $(\forall x)(x \in A \Rightarrow x \in B)$. Explain, according to this definition, why the null set is a subset of every set.

9. Explain why the tautology '$\sim(P \vee Q) \Leftrightarrow (\sim P) \wedge (\sim Q)$' implies the set identity $(A \cup B)' = A' \cap B'$. What are the set identities which correspond to the tautologies

$$\text{`}\sim(\sim P) \Leftrightarrow P,\text{'} \qquad \text{`}\sim(P \wedge Q) \Leftrightarrow (\sim P) \vee (\sim Q),\text{'}$$
$$\text{`}(P \vee Q) \wedge R \Leftrightarrow (P \wedge R) \vee (Q \wedge R)\text{'}?$$

1–7 Negation of quantified statements

Note the equivalence of the following two statements:

$$`\sim(\forall x)(x + 1 = 0)'$$

and

$$`(\exists x)(x + 1 \neq 0).'$$

More generally, '$\sim(\forall x)$' has the same meaning as '$(\exists x)\sim.$' In other words, to say that a statement is not always true is equivalent to saying that it is sometimes false. Similarly, '$\sim(\exists x)$' has the same meaning as '$(\forall x)\sim,$' as illustrated by the following equivalent statements:

$$`\sim(\exists x \in \mathbf{R})(x^2 + 1 = 0)'$$

and

$$`(\forall x \in \mathbf{R})(x^2 + 1 \neq 0).'$$

Let us now apply the observations of the preceding paragraph to negate the statement $(\exists x)(\forall y)(\exists z)P(x, y, z)$. We have

$$\sim(\exists x)(\forall y)(\exists z)P(x, y, z) \Leftrightarrow (\forall x)\sim(\forall y)(\exists z)P(x, y, z)$$
$$\Leftrightarrow (\forall x)(\exists y)\sim(\exists z)P(x, y, z)$$
$$\Leftrightarrow (\forall x)(\exists y)(\forall z)\sim P(x, y, z).$$

Note that we can obtain the last expression from the original one by interchanging the symbols '$\exists$' and '$\forall$' and placing the negation after the quantifiers. This illustrates a very useful rule for negating quantified statements. *To negate a statement beginning with a string of quantifiers, interchange the symbols '$\exists$' and '$\forall$' and place the negation symbol following the quantifiers.*

Example 1. Later we shall see that the statement '$\lim_{x \to a} f(x)$ exists' has the following meaning:

$$(\exists A)(\forall \epsilon > 0)(\exists \delta > 0)(\forall x)(0 < |x - a| < \delta \Rightarrow |f(x) - A| < \epsilon).$$

It will occasionally be necessary to unravel the meaning of the statement '$\lim_{x \to a} f(x)$ does not exist.' By our rule this must mean

$$(\forall A)(\exists \epsilon > 0)(\forall \delta > 0)(\exists x)(0 < |x - a| < \delta \text{ and } |f(x) - A| \geq \epsilon).$$

Note that we have used the fact that $\sim(P \Rightarrow Q)$ is equivalent to $P \wedge (\sim Q)$.

Example 2. Later we shall define a subset S of $\mathbf{R}$ to be *bounded* iff

$$(\exists x \in \mathbf{R})(\forall y \in S)(|y| \leq x).$$

(See Exercise 3, Section 1–5.) As an illustration of our rule for negating a quantified statement, we shall prove that every unbounded set contains an infinite number of elements.

Suppose that S is an unbounded set. To translate this into a quantified statement, we simply form the negation of the condition for a bounded set. We get

$$(\forall x \in \mathbf{R})(\exists y \in S)(|y| > x).$$

Setting $x = 1$, we are guaranteed that there exists a $y_1 \in S$ such that $|y_1| > 1$. Setting $x = |y_1|$, we see similarly that there exists a $y_2 \in S$ such that $|y_2| > |y_1|$. By the same reasoning (setting $x = |y_2|$), there exists a $y_3 \in S$ such that $|y_3| > |y_2|$. By continuing in this fashion, we see that S must contain an entire sequence of elements $y_1, y_2, \ldots$ such that $|y_1| < |y_2| < |y_3| < \cdots$ These inequalities imply that the numbers $y_1, y_2, \ldots$ are all distinct. Thus we see that S contains an infinite number of elements.

Example 3. Let $\{E_\lambda\}$ be a family of subsets of some universal set S, where the index λ ranges over some set Λ. We define the *union* and *intersection* of this family of sets to be

$$\bigcup_{\lambda \in \Lambda} E_\lambda = \{x \in S \mid (\exists \lambda \in \Lambda)(x \in E_\lambda)\}$$

and

$$\bigcap_{\lambda \in \Lambda} E_\lambda = \{x \in S \mid (\forall \lambda \in \Lambda)(x \in E_\lambda)\},$$

respectively. Thus for an element x to belong to the intersection, it must belong to each of the sets E_λ; on the other hand, x belongs to the union if it belongs to at least one of the sets E_λ. (Note that this is the answer to Exercise 4(b), Section 1–5.) Using the rule for negating quantified statements, one can prove *De Morgan's laws*

$$\left(\bigcup_\lambda E_\lambda\right)' = \bigcap_\lambda E_\lambda' \quad \text{and} \quad \left(\bigcap_\lambda E_\lambda\right)' = \bigcup_\lambda E_\lambda'.$$

Expressed somewhat more idiomatically, the first rule states that the complement of a union is the intersection of the complements.

Let $x \in S$. Then

$$x \in \left(\bigcup_\lambda E_\lambda\right)' \Leftrightarrow x \notin \bigcup_\lambda E_\lambda \Leftrightarrow \sim(\exists \lambda \in \Lambda)(x \in E_\lambda)$$

$$\Leftrightarrow (\forall \lambda \in \Lambda)(x \notin E_\lambda) \Leftrightarrow (\forall \lambda \in \Lambda)(x \in E_\lambda') \Leftrightarrow x \in \bigcap_\lambda E_\lambda'.$$

Thus

$$\left(\bigcup_\lambda E_\lambda\right)' = \bigcap_\lambda E_\lambda'.$$

The proof of the second law is left to the reader.

If we let $\Lambda = \{1, 2\}$, we get as a special case of De Morgan's laws:

$$(E_1 \cup E_2)' = (E_1') \cap (E_2') \quad \text{and} \quad (E_1 \cap E_2)' = (E_1') \cup (E_2').$$

Exercises

1. Write the following statement in symbolic form and then form its negation: 'For each $\epsilon > 0$, there exists an $N > 0$ such that

$$|f_n(x) - f(x)| < \epsilon$$

whenever $n > N$ and $x \in S$.'

2. Prove the second De Morgan law.

1-8 The principle of finite induction

The following identities are probably familiar to the student already:

$$a + (a + d) + (a + 2d) + \cdots + [a + (n - 1)d] = n[2a + (n - 1)d]/2,$$
$$a + ar + ar^2 + \cdots + ar^{n-1} = a(r^n - 1)/(r - 1), \quad \text{where} \quad r \neq 1,$$
$$(a + b)^n = a^n + C_{n,1}a^{n-1}b + C_{n,2}a^{n-2}b^2 + \cdots + b^n,$$

where

$$C_{n,k} = \frac{n(n - 1)(n - 2) \cdots (n - k + 1)}{1 \cdot 2 \cdot 3 \cdots k}.$$

In all cases n is a positive integer, and for small values of n (say $n = 1, 2, 3$, and 4), one can verify the identities directly. If we were ambitious enough, we could continue to verify the identities directly for the next hundred or so values of n, and doing so, we would probably cover most applications which we would make of them. Some-times this is the best one can do. A notable example of this is Goldbach's conjecture that every even number greater than 2 is the sum of two odd primes. Although no one has found an even number for which this is not true, no one has been able to prove the conjecture. On the other hand, many conjectures which were based upon such "observation" have later proved to be false (for example, Fermat's conjecture that $2^{2^n} + 1$ is always a prime).

Fortunately, there is a general method of proving the above identities and many others like them. This method is called *finite induction*, and is based upon a basic property of the set $\mathbf{Z}^+$ of all positive integers.

Definition 1. A subset S of $\mathbf{Z}^+$ will be called **inductive** if it has the following two properties:

1) 1 belongs to S;
2) S is closed under addition by 1; that is, $(\forall n)(n \in S \Rightarrow (n + 1) \in S)$.

Suppose, then, that $S \subset \mathbf{Z}^+$ is inductive. By definition, $1 \in S$. Since S is closed under addition by 1, S must contain the number $2 = 1 + 1$. Since $2 \in S$, it follows again from property (2) that $3 = 2 + 1$ is in S. Similarly, 4 must be in S, and so on. It would seem, then, that S must contain all the positive integers. This is our first formulation of the principle of induction.

Theorem 2. *If S is an inductive subset of $\mathbf{Z}^+$, then $S = \mathbf{Z}^+$.*

We offer no proof of this important theorem at present. (The phrase "and so on" prevents the preceding discussion from being a proof.) In the section following we shall give a proof based upon the assumption that the real numbers form an ordered field. From the viewpoint of foundations, this is a bit circular, since the principle of induction is really needed to prove the consistency of the axioms for an ordered field.

We shall now illustrate the use of Theorem 2 by proving the following identity:

$$1 + 2 + 3 + \cdots + n = n(n + 1)/2.$$

Let S be the set of all positive integers n for which the above equation is valid. We shall prove that $S = \mathbf{Z}^+$. By Theorem 2 it suffices to prove that the set S is inductive. Note that when $n = 1$, the equation is valid, since

$$1 = \frac{1 \cdot 2}{2}.$$

Thus $1 \in S$. We must next show that S is closed under addition by 1. Suppose that $k \in S$. We must show that $(k + 1) \in S$. To say that $k \in S$, means, of course, that the equation

$$1 + 2 + 3 + \cdots + k = \frac{k(k + 1)}{2}$$

is true. Add $k + 1$ to both sides of the equation. This gives

$$1 + 2 + 3 + \cdots + k + (k + 1) = \frac{k(k + 1)}{2} + (k + 1)$$
$$= \frac{(k + 1)[(k + 1) + 1]}{2}.$$

But to say that this equation is valid is equivalent to saying that $(k + 1) \in S$. This completes the proof.

We shall next state the principle of induction in a form which is perhaps better suited to proving identities such as the one above.

Theorem 3. *The Principle of Induction. Let $P(n)$ be a propositional function in which the variable n can assume all values in $\mathbf{Z}^+$. Suppose that*

a) *'$P(1)$' is true;*

b) *'$(\forall n \in \mathbf{Z}^+)(P(n) \Rightarrow P(n + 1))$' is true.*

Then '$P(n)$' is true for each $n \in \mathbf{Z}^+$.

Proof. Let

$$S = \{n \in \mathbf{Z}^+ \mid P(n)\}.$$

Then by (a), $1 \in S$. Furthermore, (b) amounts to saying that S is closed under addition by 1. Thus S is an inductive set, and hence $S = \mathbf{Z}^+$. In other words, $P(n)$ is true for each positive integer n.

Dominos provide a good analogy to the principle of finite induction. If a set of dominos are properly lined up in a row, and if the first domino is knocked over, then all the others will fall. For this to happen, however, the dominos must be lined up so that if one of them falls, this will cause the next one to fall. This is just the sort of thing which condition (b) guarantees.

Example 1. Let $P(n)$ be the statement

$$\text{'}x^{2n-1} + y^{2n-1} \text{ is divisible by } x + y.\text{'}$$

Since $x^{2n-1} + y^{2n-1} = x + y$ when $n = 1$, '$P(1)$' is clearly a true statement. Let us suppose now that $P(n)$ is true. We must prove on the basis of this assumption that $P(n + 1)$

is true. If $P(n)$ is true, then there exists a polynomial in x and y, call it $p(x, y)$, such that

$$x^{2n-1} + y^{2n-1} = (x + y)p(x, y).$$

Then

$$x^{2(n+1)-1} + y^{2(n+1)-1} = x^2(x^{2n-1} + y^{2n-1}) - y^{2n-1}(x^2 - y^2)$$
$$= [x^2p(x, y) - y^{2n-1}(x - y)](x + y).$$

Since $p(x, y)$ is a polynomial in x and y, so is the term in the square brackets. Thus, $x^{2(n+1)-1} + y^{2(n+1)-1}$ is divisible by $x + y$, or in other words, $P(n + 1)$ is true. By the principle of induction, '$P(n)$' must be true for all positive integers n.

Example 2. Prove the inequality

$$f(n) = 1 + \frac{1}{\sqrt{2}} + \frac{1}{\sqrt{3}} + \cdots + \frac{1}{\sqrt{n}} > 2(\sqrt{n + 1} - 1).$$

Since $2 < \frac{9}{4}$, we get, on taking square roots, $\sqrt{2} < \frac{3}{2}$. (How can this be justified in terms of our basic rules for handling inequalities?) After a slight amount of juggling, we get

$$1 > 2(\sqrt{2} - 1),$$

which is the desired inequality for the case in which $n = 1$.

Assume that the inequality holds for some value k of n. Thus

$$f(k) > 2(\sqrt{k + 1} - 1).$$

By adding $1/\sqrt{k + 1}$ to both sides, we get

$$f(k + 1) > 2(\sqrt{k + 1} - 1) + 1/\sqrt{k + 1}.$$

To complete the induction proof, it suffices to show that the right-hand side of the inequality is greater than $2(\sqrt{k + 2} - 1)$. Note, however, that

$$2(\sqrt{k + 1} - 1) + \frac{1}{\sqrt{k + 1}} > 2(\sqrt{k + 2} - 1) \Leftrightarrow 2\sqrt{k + 1} + \frac{1}{\sqrt{k + 1}} > 2\sqrt{k + 2}$$

$$\Leftrightarrow 2(k + 1) + 1 > 2\sqrt{(k + 1)(k + 2)}$$

$$\Leftrightarrow \sqrt{(k + 1)(k + 2)} < \frac{2k + 3}{2}.$$

We can easily verify the last inequality by using Exercise 2, Section 1–3:

$$\sqrt{(k + 1)(k + 2)} < \frac{(k + 1) + (k + 2)}{2} = \frac{2k + 3}{2}.$$

Exercises

1. Give induction proofs for each of the following:

a) $1^2 + 2^2 + 3^2 + \cdots + n^2 = \dfrac{n(n + 1)(2n + 1)}{6}.$

b) $1^3 + 2^3 + 3^3 + \cdots + n^3 = \left[\dfrac{n(n + 1)}{2}\right]^2.$

c) $\dfrac{1}{1 \cdot 2} + \dfrac{1}{2 \cdot 3} + \cdots + \dfrac{1}{n(n + 1)} = \dfrac{n}{n + 1}.$

     d) $x^n - y^n$ is divisible by $x - y$.

     e) If S is a set containing n elements, then $\mathcal{S}(S)$ contains 2^n elements.

     f) $(1 + h)^n \geq 1 + hn$ $(h > -1)$ *(Bernoulli's inequality)*.

     g) $(2n)! < 2^{2n}(n!)^2$.

*2. Three pegs are stuck in a board. On one of these pegs is a pile of disks graduated in size, as shown in Fig. 1–29, the smallest being on top. The object of this puzzle is to transfer the pile to one of the other two pegs by moving the disks one at a time from one peg to another in such a way that a disk is never placed on top of a smaller disk. Prove that this can be done for n disks in $2^n - 1$ moves.

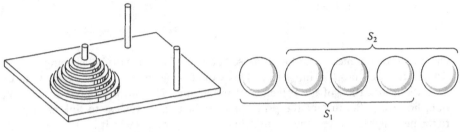

FIG. 1–29   **FIG. 1–30**

*3. **Theorem.** *All billiard balls have the same color.*

     Proof. Let $P(n)$ be the following statement: If S is a set containing n billiard balls, then all the billiard balls in S have the same color. We shall prove by induction that $P(n)$ is true for all $n \in \mathbf{Z}^+$. Certainly $P(1)$ is true. Assume now that for some n, $P(n)$ is true. Let S be any set containing $n + 1$ billiard balls. Then S is the union of two overlapping subsets S_1 and S_2 containing n billiard balls apiece. (The case when $n = 4$ is illustrated in Fig. 1–30.) According to the inductive hypothesis, all the balls in S_1 have the same color c_1, while all the balls in S_2 have the same color c_2. Since S_1 and S_2 overlap, however, we must have $c_1 = c_2$. Hence all balls in S have the same color. Q.E.D.

     Find the flaw in this proof.

   4. Prove the identity

$$1 + r + r^2 + \cdots + r^{n-1} = \frac{r^n - 1}{r - 1} \qquad (r \neq 1).$$

1-9 A deeper look at induction

In an effort to illustrate the principle of induction, a number of oversimplifications were made in the preceding section. It is time to set the record straight.

     Consider the identity

$$1 + 2 + 3 + \cdots + n = \frac{n(n + 1)}{2},$$

which we purportedly proved by induction. In point of fact, we did not prove it, nor could we have proved it since the left-hand side of the identity is meaningless at this stage of the game. So far the only arithmetic operations which have been discussed are the *binary* operations of addition and multiplication described by the field axioms, and the secondary (and also binary) operations of subtraction and division. The above identity cannot therefore be regarded as meaningful until the n-fold sum which appears there has been defined in terms of these basic binary operations. (Hopefully the student will have gathered from context and linguistic savvy that a binary operation is some sort of rule which enables one to associate with a *pair* of objects some third object.) To define these more general sums and products we need to use induction.

Let us consider first the matter of defining powers of a number. The first few powers are easily defined:

$$x^1 = x, \quad x^2 = x^1 \cdot x, \quad x^3 = x^2 \cdot x, \quad x^4 = x^3 \cdot x.$$

One might be tempted to add the phrase "et cetera" and regard x^n as being properly defined for each positive integer n. This would be no more nor less valid than our heuristic justification of the principle of induction. The problem has a simple solution, however. Because of the principle of induction, it suffices to link each power to the next highest power, that is, to define x^{n+1} in terms of x^n. It is obvious how this should be done: $x^{n+1} = x^n \cdot x$. This *recursion formula*, together with the definition of x^1 (namely, $x^1 = x$), completely determines x^n for positive integers n. (The number x^1 is well-defined, and, assuming that x^k is well-defined, x^{k+1} is then well-defined by the recursion formula. Thus, by the principle of induction, it follows that x^n is well-defined for each positive integer n.)

Before we give further examples of inductive definitions, it will be instructive to prove one of the exponent laws for positive integer powers. Specifically, let us prove that

$$(\forall m \in \mathbf{Z}^+)(\forall n \in \mathbf{Z}^+)(\forall x \in \mathbf{R})(x^m \cdot x^n = x^{m+n}).$$

Since two integer variables m and n are involved, one might suppose that a more general form of the principle of induction might be needed. This is not the case; it suffices to perform induction on just one of the variables, say "n." We let $P(n)$ be the propositional function

$$(\forall m \in \mathbf{Z}^+)(\forall x \in \mathbf{R})(x^m \cdot x^n = x^{m+n}).$$

$P(1)$ is true, since $x^m \cdot x^1 = x^m \cdot x = x^{m+1}$ by our inductive definition of powers of x. Assume now that $P(k)$ is true. Then for all $x \in \mathbf{R}$ and $m \in \mathbf{Z}^+$, we have

$$
\begin{aligned}
x^m \cdot x^{k+1} &= x^m \cdot (x^k \cdot x) && \text{(by definition of } x^n\text{)} \\
&= (x^m \cdot x^k) \cdot x && \text{(by the associative law)} \\
&= x^{m+k} \cdot x && \text{(by the inductive hypothesis)} \\
&= x^{(m+k)+1} && \text{(by definition of } x^n\text{)} \\
&= x^{m+(k+1)} && \text{(by the associative law).}
\end{aligned}
$$

Thus we have proved $P(k + 1)$ by assuming $P(k)$. By the principle of induction, $P(n)$ is therefore true for each positive integer n, and the proof is complete.

We shall denote the n-fold sum of the real numbers $a_1, a_2, \ldots, a_n$ by $a_1 + a_2 + \cdots + a_n$ or by the so-called sigma convention, namely, $\sum_{k=1}^{n} a_k$. The latter symbol is best thought of as denoting the sum of all numbers a_k obtained by allowing the variable "k" (usually called an index in this case) to take on all positive integer values from 1 to n. Thus "k" is a bound or dummy variable, and we may write, for instance,

$$\sum_{k=1}^{n} a_k = \sum_{j=1}^{n} a_j = \sum_{\Upsilon=1}^{n} a_\Upsilon.$$

We define n-fold sums inductively as follows:

$$\sum_{k=1}^{1} a_k = a_1, \qquad \sum_{k=1}^{n+1} a_k = \left(\sum_{k=1}^{n} a_k\right) + a_{n+1}.$$

Thus for $n = 2, 3$, and 4, we have

$$\sum_{k=1}^{2} a_k = \left(\sum_{k=1}^{1} a_k\right) + a_2 = a_1 + a_2,$$

$$\sum_{k=1}^{3} a_k = \left(\sum_{k=1}^{2} a_k\right) + a_3 = (a_1 + a_2) + a_3,$$

$$\sum_{k=1}^{4} a_k = \left(\sum_{k=1}^{3} a_k\right) + a_4 = ((a_1 + a_2) + a_3) + a_4.$$

Similarly, the n-fold product of $a_1, a_2, \ldots, a_n$, denoted either by $a_1 a_2 \cdots a_n$ or $\prod_{k=1}^{n} a_k$, is defined by the equations

$$\prod_{k=1}^{1} a_k = a_1, \qquad \prod_{k=1}^{n+1} a_k = \left(\prod_{k=1}^{n} a_k\right) a_{n+1}.$$

Let us reexamine the proof that

$$1 + 2 + 3 + \cdots + n = \frac{n(n + 1)}{2},$$

which may be more compactly written as

$$\sum_{k=1}^{n} k = \frac{n(n + 1)}{2}.$$

The equation is true when $n = 1$, since

$$\sum_{k=1}^{1} k = 1 = \frac{1 \cdot 2}{2}.$$

Assuming that it is true for n, we see that it is true for $(n + 1)$:

$$\sum_{k=1}^{n+1} k = \left(\sum_{k=1}^{n} k \right) + (n + 1) \qquad \text{(by definition of } \Sigma\text{)}$$

$$= \frac{n(n + 1)}{2} + (n + 1) \qquad \text{(by inductive hypothesis)}$$

$$= \frac{(n + 1)[(n + 1) + 1]}{2} \qquad \text{(by trivial algebra).}$$

This completes the inductive proof. The reader should note how the inductive definition of n-fold sums enters the proof; this is the only real difference between the present proof and the previous one.

We consider next a more fundamental deficiency in our treatment of induction. We have stated that everything must be proved from our axioms for the real numbers. This means that officially we must discard any intuitive notions about the integers until they have been established. Not only must we prove the principle of induction, but we must also *define* the positive integers from our axioms for the real numbers.

There should be no doubt as to the meaning of the number 1. It is the unique real number described by Axiom M3. We define the real number 2 to be $1 + 1$. It is worthwhile to note that since $1 > 0$, it follows that $2 > 1 > 0$. Thus the real number 2 is equal neither to 1 nor to 0. (Note that in the field $\mathbf{Z}_2$, we have $2 = 0$.) We define the real number 3 to be $2 + 1$, and we observe that 3 is distinct from 0, 1, or 2. (In $\mathbf{Z}_3$, we have $3 = 0$; in $\mathbf{Z}_2$, we have $3 = 1$.) Similarly, we define $4 = 3 + 1$. We can conceive of continuing this process indefinitely and letting $\mathbf{Z}^+$ be the set of all real numbers thus obtainable by repeated additions of the number 1. Unfortunately, the principle of induction, which ordinarily allows us to do such things, is itself in question. We shall therefore have to resort to another means of defining $\mathbf{Z}^+$.

Proposition 1. *There is a (unique) smallest set of real numbers which contains the number 1 and is closed under addition by 1.*

Proof. We let $\mathcal{F}$ be the family of *all* subsets of $\mathbf{R}$ which contain the number 1 and are closed under addition by 1. Then $\mathcal{F}$ is nonvoid, since the set $\mathbf{R}^+$ of positive real numbers has these properties and is therefore a member of $\mathcal{F}$. We now define S to be the set consisting of real numbers common to all sets in the family $\mathcal{F}$. Thus

$$S = \{x \in \mathbf{R} \mid (\forall A \in \mathcal{F})(x \in A)\}.$$

(The set S is called the *intersection* of the family of sets $\mathcal{F}$.) We shall now show that S satisfies the claim of the theorem.

We first show that S contains 1 and is closed under addition by 1. By the definition of $\mathcal{F}$, we have $(\forall A \in \mathcal{F})(1 \in A)$. Thus $1 \in S$. Suppose that $x \in S$. If A is *any* member of $\mathcal{F}$, then by definition of S, we have $x \in A$. Since (by definition of $\mathcal{F}$) A is closed under addition by 1, we also have $(x + 1) \in A$. Since $x + 1$ belongs to each $A \in \mathcal{F}$, it follows that $(x + 1) \in S$. This proves that S is closed under addition by 1.

If A is any member of $\mathfrak{F}$, then $S \subset A$. (For if $x \in S$, then x belongs to each member of $\mathfrak{F}$ and in particular $x \in A$.) Thus S is the smallest member of $\mathfrak{F}$, which is but to say that S is the smallest set containing 1 and closed under addition by 1.

Finally we observe that S is unique. If S^* were a second set with the required properties, then from the claims of smallness, we could conclude both that $S \subset S^*$ and that $S^* \subset S$, which implies, of course, that $S = S^*$.

Definition 2. We define the set of **positive integers** $\mathbf{Z}^+$ to be the smallest subset of $\mathbf{R}$ which contains 1 and is closed under addition by 1.

It is clear that $\mathbf{Z}^+$ must contain the numbers 1, 2, 3, and 4. Also, the principle of induction is an immediate consequence of our definition of $\mathbf{Z}^+$.

Theorem 3. Principle of Induction. *If S is an inductive subset of $\mathbf{Z}^+$, then $S = \mathbf{Z}^+$.*

Proof. If S is inductive, then by definition it contains the number 1 and is closed under addition by 1. Since by definition $\mathbf{Z}^+$ is the smallest set with this property, $\mathbf{Z}^+ \subset S$. However, we are given that $S \subset \mathbf{Z}^+$. Therefore, $S = \mathbf{Z}^+$.

Theorem 3 of Section 1–8, which may be thought of as a statement of the principle of induction in the language of propositional functions rather than sets, now follows as before. En route to a stronger version of the principle of induction, we shall prove several well-known properties of the positive integers.

1) 1 *is the smallest positive integer.*

Equivalently, $(\forall n \in \mathbf{Z}^+)(n \geq 1)$. This has a trivial proof by induction.

Let $S = \{n + 1 \mid n \in \mathbf{Z}^+\}$. Since $\mathbf{Z}^+$ is closed under addition by 1, S is a subset of $\mathbf{Z}^+$. It is clear, moreover, that $1 \notin S$. If we let $T = S \cup \{1\}$, then one can easily check that T is a subset of $\mathbf{Z}^+$ which contains 1 and is closed under addition by 1. Thus, $T = \mathbf{Z}^+$. This proves that every positive integer other than 1 is of the form $n + 1$, where $n \in \mathbf{Z}^+$. The following is an equivalent restatement of this result.

2) *If $n \in \mathbf{Z}^+$ and $n \neq 1$, then $(n - 1) \in \mathbf{Z}^+$.*

The next property includes the previous result.

3) *The set $\mathbf{Z}^+$ is closed under addition, multiplication, and the subtraction of a smaller number from a larger one.*

Proof. The proof that $\mathbf{Z}^+$ is closed under addition is virtually the same as our earlier proof of the exponent law for multiplication. The proof that $\mathbf{Z}^+$ is closed under multiplication is similar and is left to the reader. We shall prove the last assertion. Let $P(n)$ be the statement

$$(\forall m \in \mathbf{Z}^+)(m < n \Rightarrow (n - m) \in \mathbf{Z}^+).$$

Then $P(1)$ is "vacuously true," since the hypothesis $m < 1$ is always false by (1). Suppose that $P(k)$ is true when $k \in \mathbf{Z}^+$. If $m \in \mathbf{Z}^+$ and $m < k + 1$, then $m - 1 < k$.

If $m = 1$, then $(k + 1) - m = k \in \mathbf{Z}^+$. On the other hand, if $m \neq 1$, then $(m - 1) \in \mathbf{Z}^+$ by (2), and

$$(k + 1) - m = k - (m - 1) \in \mathbf{Z}^+$$

by the inductive hypothesis. Thus we have shown that $P(k + 1)$ is true [assuming, of course, that $P(k)$ is true]. By the principle of induction, '$P(n)$' is true for each $n \in \mathbf{Z}^+$, or, equivalently,

$$(\forall n, m \in \mathbf{Z}^+)(m < n \Rightarrow (n - m) \in \mathbf{Z}^+).$$

This completes the proof.

4) *If $n \in \mathbf{Z}^+$, then there are no positive integers strictly between n and $n + 1$.*

Proof. Suppose to the contrary that $n < k < n + 1$, where $k \in \mathbf{Z}^+$. Then $k - n < 1$. However, by (3) we know that $k - n$ is a positive integer. Since 1 is the smallest positive integer, we have a contradiction.

We are prepared now to prove two stronger (though logically equivalent) formulations of the principle of induction.

Theorem 4. Well-Ordering Property. *Every nonempty subset of $\mathbf{Z}^+$ has a smallest element.*

Proof. Suppose that S is a subset of $\mathbf{Z}^+$ which does not have a smallest element, and let $T = \{n \in \mathbf{Z}^+ \mid n \notin S\}$ be its complement in $\mathbf{Z}^+$. We shall prove that S is empty by showing that $T = \mathbf{Z}^+$.

For each $n \in \mathbf{Z}^+$, we set $I_n = \{k \in \mathbf{Z}^+ \mid k \leq n\}$, and we let $P(n)$ be the statement '$I_n \subset T$.' It obviously suffices to show that '$P(n)$' is true for each $n \in \mathbf{Z}^+$.

The number 1 must belong to T, since, if it belonged to S, it would be the smallest member of S. Thus, $I_1 = \{1\} \subset T$ so that the statement '$P(1)$' is true. Suppose that '$P(n)$' is true. Then S does not contain any integer less than or equal to n, and, consequently, if it did contain the integer $(n + 1)$, then $(n + 1)$ would be the smallest member of S. [Note that we are using the fact that there are no integers strictly between n and $(n + 1)$.] Thus $(n + 1) \in T$. This implies that

$$I_{n+1} = I_n \cup \{n + 1\} \subset T,$$

or equivalently, that '$P(n + 1)$' is true.

Theorem 5. Second Form of the Principle of Induction. *Let $P(n)$ be a propositional function such that*

1) '$P(1)$' *is true;*

2) *for each $n \in \mathbf{Z}^+$,* '$[P(1) \wedge P(2) \wedge \cdots \wedge P(n)] \Rightarrow P(n + 1)$' *is true.*

Then '$P(n)$' *is true for each $n \in \mathbf{Z}^+$.*

Notice how this relates to the first form of the principle of induction. In proving a statement by the first form of the principle of induction, one must prove that $P(n + 1)$ is true on the basis of the inductive hypothesis that $P(n)$ is true. In using the second form of the principle of induction, one can use a stronger inductive hypothesis: instead of assuming that $P(n)$ is true, one can, if need be, assume that all of the preceding statements $P(1), P(2), \ldots, P(n)$ are true. [We note a minor technical problem: in the statement of the theorem, the expression $P(1) \wedge P(2) \wedge \cdots \wedge P(n)$ requires an inductive definition. We trust the reader can supply it.]

Proof. Let $S = \{n \in \mathbf{Z}^+ \mid P(n)\}$. We wish to show that $S = \mathbf{Z}^+$. Suppose to the contrary that S is a proper subset of $\mathbf{Z}^+$. Then $T = \{n \in \mathbf{Z}^+ \mid n \notin S\}$ is nonempty. By the well-ordering property of $\mathbf{Z}^+$, T has a smallest element n_0. Since $1 \in S$, $n_0 \neq 1$. It follows that each of the integers $1, 2, \ldots, n_0 - 1$ belongs to S, or equivalently, that the statements $P(1), P(2), \ldots, P(n_0 - 1)$ are true. By (2), the statement $P(n_0)$ must also be true. But then $n_0 \in S$, which is a contradiction.

To illustrate the use of the well-ordering principle, we shall prove a theorem of elementary number theory.

Example 1. Prove that every positive integer $n \geq 2$ has a prime factor.

Recall that a positive integer p is *prime* iff its only factors are the trivial ones: 1 and p. Thus, 7, 11, and 23 are prime numbers. A positive integer which is not prime is called *composite*.

Proof. Suppose that the assertion is false. Then by the well-ordering property, there must exist a smallest positive integer $n \geq 2$ which does not have a prime factor. The number n cannot be prime. Hence we can write $n = ab$, where a and b are positive integers greater than one. But, then $2 \leq a < n$, which implies that a has a prime factor p. Since p must therefore be a factor of n, we have a contradiction.

By essentially the same argument, one can show that each positive integer $n \geq 2$ is a product of primes. It can also be shown that such a factorization is unique to within the order of the factors.

Example 2. *The Marriage Problem.* We shall now apply the second principle of induction to a combinatorial problem. Suppose that B is a finite set of boys and G is a finite set of girls. For each boy $b \in B$, we denote by G_b the set of his acquaintances in G. When is it possible for each boy to marry one of his acquaintances? (Only monogamous marriages are permitted.) We shall prove that such marriages can be arranged if for each $k \in \mathbf{Z}^+$, any k boys in B know collectively at least k girls in G. (To be a bit more precise, the marriage problem has a solution if for each nonempty subset F of B, the set $\bigcup_{b \in F} G_b$ has at least as many elements as the set F.)

We shall prove this by induction on the number n of elements of B. When $n = 1$ we have little to worry about. Suppose we know that it is true for all values of n less than some integer n_0, and consider the situation when $n = n_0$. It may happen that for each $k < n_0$, any k boys in B know collectively at least $k + 1$ girls in G. We can then marry one boy off and refer the rest of the boys to the inductive hypothesis. If, on the other hand, there is some set of k boys who know collectively exactly k girls, let them marry off (since the inductive hypothesis tells us this is possible) and consider the boys left over, all $n_0 - k$ of them. Now we shall see that for each $m \leq n_0 - k$, any m of the remaining boys know collectively at least m of the remaining girls, in which case we can apply the inductive hypothesis to complete the proof. Suppose to the contrary that for some $m \leq n_0 - k$, there are m of these boys who know fewer than m of the remaining girls. Then these m boys together with the original k know fewer than $m + k$ girls, which is contrary to assumption.

Example 3. We conclude this section with a proof of a famous inequality. If $a_1, \ldots, a_n$ are positive numbers, then

$$(a_1 + a_2 + \cdots + a_n)/n \geq \sqrt[n]{a_1 a_2 \cdots a_n}.$$

The left-hand side, which we denote by A_n, is called the *arithmetic mean* of $a_1, \ldots, a_n$; the right-hand side, which we denote by G_n, is called the *geometric mean* of $a_1, \ldots, a_n$. We shall also show that $A_n = G_n$ iff $a_1 = a_2 = \cdots = a_n$.

The inequality is trivial when $n = 1$. Also it was proved in one of the exercises for the case in which $n = 2$. Assume that it is true if $n = k$. Applying this inequality to $b_1 = a_{k+1}$ and $b_2 = b_3 = \cdots = b_k = A_{k+1}$, we get

$$\frac{a_{k+1} + (k-1)A_{k+1}}{k} = \frac{1}{k}\sum_{j=1}^{k} b_j \geq \sqrt[k]{b_1 b_2 \cdots b_k} \geq (a_{k+1}A_{k+1}^{k-1})^{1/k}.$$

Let

$$A = \frac{a_{k+1} + (k-1)A_{k+1}}{k}, \qquad G = (a_{k+1}A_{k+1}^{k-1})^{1/k}.$$

Applying the inequality for the case in which $n = 2$, we get

$$A_{k+1} = \tfrac{1}{2}(A_k + A) \geq (A_k A)^{1/2} \geq (G_k G)^{1/2} = (G_{k+1}^{k+1}A_{k+1}^{k-1})^{1/2k},$$

that is,

$$A_{k+1} \geq (G_{k+1}^{k+1}A_{k+1}^{k-1})^{1/2k}.$$

Thus $A_{k+1} \geq G_{k+1}$. This completes the proof of the inequality.

We turn next to the condition for equality. We let the reader verify the assertion if $n = 1$ or 2. Suppose the assertion concerning equality holds if $n = k$. If $A_{k+1} = G_{k+1}$, we see from the argument above that

$$A_k = A, \qquad A_k = G_k, \qquad A = G.$$

Since $A_k = G_k$, the induction hypothesis enables us to conclude that

$$a_1 = a_2 = \cdots = a_k.$$

Since $A = G$, we have

$$a_{k+1} = A_{k+1} = \frac{a_1 + \cdots + a_k + a_{k+1}}{k + 1},$$

whence

$$a_1 = a_2 = \cdots = a_k = a_{k+1}.$$

Exercises

1. Prove the following generalization of the triangle inequality:

$$\left|\sum_{k=1}^{n} a_k\right| \leq \sum_{k=1}^{n} |a_k|.$$

2. Prove that in any ordered field

$$\sum_{k=1}^{n} a_k^2 \geq 0$$

and that equality holds iff $a_1 = a_2 = \cdots = a_n = 0$.

3. Find numerical values of the following sums.

a) $\displaystyle\sum_{k=1}^{3} k^2$ b) $\displaystyle\sum_{j=1}^{4} (2j - 1)$ c) $\displaystyle\sum_{n=1}^{3} n^n$ d) $\displaystyle\sum_{k=1}^{2} 2^{2k-1}$

4. By examining values of the products

$$\prod_{k=1}^{n} \left(1 - \frac{1}{k+1}\right) \quad \text{and} \quad \prod_{k=1}^{n} \left(1 - \frac{1}{(k+1)^2}\right)$$

for small values of n, conjecture general formulas for these products and then prove your conjecture by induction.

5. If $a < 0$, show that a^n is positive or negative according as n is even or odd.

6. Show that

$$(1 - x) \prod_{k=1}^{n} (1 + x^{2^{k-1}}) = 1 - x^{2^n}.$$

*7. If $a_1, \ldots, a_n$ are all positive, show that

$$\prod_{k=1}^{n} (1 + a_k) \leq \sum_{k=1}^{n+1} \frac{S^{k-1}}{(k-1)!},$$

where $S = \sum_{k=1}^{n} a_k$.

8. Prove that $\mathbf{Z}^+$ is closed under addition and multiplication. Also prove the exponent laws

$$(a^m)^n = a^{mn} \quad \text{and} \quad (ab)^n = a^n b^n,$$

where m and n are positive integers.

9. Prove the following properties of sums.

a) $\displaystyle\sum_{k=1}^{n} (a_k + b_k) = \sum_{k=1}^{n} a_k + \sum_{k=1}^{n} b_k$ (Additive property)

b) $\displaystyle\sum_{k=1}^{n} c a_k = c \sum_{k=1}^{n} a_k$ (Homogeneous property)

c) $\displaystyle\sum_{k=1}^{n} (a_k - a_{k-1}) = a_n - a_0$ (Telescoping property)

10. State properties of products analogous to those for sums.

11. a) Apply $\sum_{k=1}^{n}$ to both sides of the identity $(k+1)^2 - k^2 = 2k + 1$, and use Exercise 9 to obtain the sum formula

$$\sum_{k=1}^{n} k = \frac{n(n+1)}{2}.$$

b) By applying the same technique to the identity

$$(k + 1)^3 - k^3 = 3k^2 + 3k + 1,$$

obtain a sum formula for $\sum_{k=1}^{n} k^2$.

c) Obtain a sum formula for $\sum_{k=1}^{n} k^3$.

d) Use the identity

$$\cos (k + \tfrac{1}{2})x - \cos (k - 1 + \tfrac{1}{2})x = -2 \sin kx \sin (x/2)$$

to obtain a formula for $\sum_{k=1}^{n} \sin kx$.

e) Using similar methods, show that

$$\tfrac{1}{2} + \cos x + \cos 2x + \cdots + \cos nx = \frac{\sin (n + \tfrac{1}{2})x}{2 \sin (x/2)}.$$

(This identity is very important in the study of Fourier series.)

f) Find a formula for $\sum_{k=1}^{n} 1/k(k + 1)$.

g) Find a formula for $\sum_{k=1}^{n} k2^k$.

12. We define max (x, y) to mean the larger of two numbers x and y. Thus

$$\max (x, y) = \begin{cases} y & \text{if } x \le y, \\ x & \text{if } x > y. \end{cases}$$

The notation $x \vee y = \max (x, y)$ is also used. Similarly, $x \wedge y = \min (x, y)$ is defined to be the smaller of the two numbers x and y. Verify the following identities:

$$|x| = x \vee (-x),$$
$$x \vee y = \tfrac{1}{2}(x + y + |x - y|),$$
$$x \wedge y = \tfrac{1}{2}(x + y - |x - y|),$$
$$x \vee y + x \wedge y = x + y,$$
$$x \vee y - x \wedge y = |x - y|.$$

What algebraic properties do the operations "$\vee$" and "$\wedge$" have? (Test for associativity, distributivity, etc.)

13. Give appropriate inductive definitions for

$$x_1 \vee x_2 \vee \cdots \vee x_n = \max (x_1, x_2, \ldots, x_n)$$

and

$$x_1 \wedge x_2 \wedge \cdots \wedge x_n = \min (x_1, x_2, \ldots, x_n).$$

Show that the following are always true:

$$\max (-x_1, \ldots, -x_n) = -\min (x_1, \ldots, x_n),$$
$$\min (x_1 + c, \ldots, x_n + c) = \min (x_1, \ldots, x_n) + c,$$
$$\max (x_1 + c, \ldots, x_n + c) = \max (x_1, \ldots, x_n) + c,$$
$$\min (cx_1, \ldots, cx_n) = c \min (x_1, \ldots, x_n) \quad (c > 0),$$
$$\max (cx_1, \ldots, cx_n) = c \max (x_1, \ldots, x_n) \quad (c > 0),$$
$$\min (x_1 + y_1, \ldots, x_n + y_n) \ge \min (x_1, \ldots, x_n) + \min (y_1, \ldots, y_n),$$
$$\max (x_1 + y_1, \ldots, x_n + y_n) \le \max (x_1, \ldots, x_n) + \max (y_1, \ldots, y_n).$$

14. a) Let S be a subset of $\mathbf{R}$. If $(\exists x \in S)(\forall y \in S)(x \geq y)$, then the number x (which must be unique) is called the *largest member* of S. If $(\exists x \in S)(\forall y \in S)(x \leq y)$, then x is called the *smallest member* of S. Determine the largest and smallest members of each of the following sets (when they exist): $\{2\}$, $\{-1, 2, 10\}$, $[3, 7)$, $(3, 7]$, $\mathbf{Z}^+$, $\mathbf{R}^+$.

   b) Prove by induction that $x_1 \vee \cdots \vee x_n$ and $x_1 \wedge \cdots \wedge x_n$ are respectively the largest and smallest members of the set $\{x_k \mid k \in \mathbf{Z}^+, 1 \leq k \leq n\}$. Thus every nonempty finite set of real numbers has a smallest and largest number.

15. We shall say that a subset S of $\mathbf{R}$ is *well-ordered* iff every nonempty subset of S has a smallest element. Determine which of the following sets are well-ordered. (Support your answers.)

   a) $\{1, 7, 16\}$ b) $\varnothing$
   c) $\{n \in \mathbf{Z}^+ \mid n \leq 3\}$ d) $\{-n \mid n \in \mathbf{Z}^+\}$
   e) $\mathbf{R}^+$ f) $\mathbf{Q}$
   g) $\{1 - 1/n \mid n \in \mathbf{Z}^+\}$ h) $\{1 - 1/n \mid n \in \mathbf{Z}^+\} \cup \{1\}$
   i) $\{1 - 1/n \mid n \in \mathbf{Z}^+\} \cup \{2 - 1/n \mid n \in \mathbf{Z}^+\}$
   j) $[0, 1)$
   k) Every subset of a well-ordered set
   l) The union of any two well-ordered sets

16. If $a_1, \ldots, a_n$ and $b_1, \ldots, b_n$ are positive numbers, show that

$$\min\left(\frac{a_1}{b_1}, \ldots, \frac{a_n}{b_n}\right) \leq \frac{a_1 + a_2 + \cdots + a_n}{b_1 + b_2 + \cdots + b_n} \leq \max\left(\frac{a_1}{b_1}, \ldots, \frac{a_n}{b_n}\right).$$

17. Let $a_1, \ldots, a_n$ and $b_1, \ldots, b_n$ be real numbers. Observe that for each real number x,

$$0 \leq \sum_{k=1}^{n} (a_k x + b_k)^2.$$

Expand the right-hand side and apply Exercise 8 of Section 1–3 to deduce the *Cauchy-Schwarz inequality*:

$$\left(\sum_{k=1}^{n} a_k b_k\right)^2 \leq \left(\sum_{k=1}^{n} a_k^2\right)\left(\sum_{k=1}^{n} b_k^2\right).$$

Show that equality holds iff the a_k's and b_k's are proportional.

18. a) Let $\{a_{ij}\}$ be a doubly indexed set of real numbers in which the index i ranges over the integers $1, 2, \ldots, m$, and j ranges over the integers $1, 2, \ldots, n$. Prove by induction that

$$\sum_{i=1}^{m}\left(\sum_{j=1}^{n} a_{ij}\right) = \sum_{j=1}^{n}\left(\sum_{i=1}^{m} a_{ij}\right).$$

Their common value is denoted by $\sum_{i,j=1}^{m,n} a_{ij}$.

   b) Prove the *Lagrange identity:*

$$\left(\sum_{k=1}^{n} a_k^2\right)\left(\sum_{k=1}^{n} b_k^2\right) = \left(\sum_{k=1}^{n} a_k b_k\right)^2 + \frac{1}{2}\sum_{i,j=1}^{n} (a_i b_j - a_j b_i)^2.$$

Using the identity, give another proof of the Cauchy-Schwarz inequality.

19. If $a_1, \ldots, a_n$ are positive numbers, use the Cauchy-Schwarz inequality to show that

$$\left(\sum_{k=1}^{n} a_k\right)\left(\sum_{k=1}^{n} \frac{1}{a_k}\right) \geq n^2.$$

20. a) Show that the statement forms $(P \wedge Q) \Rightarrow R$ and $(P \wedge Q) \Rightarrow (Q \wedge R)$ are logically equivalent.

   b) Using the first form of the principle of induction, prove the following result.

   Let $P(n)$ be a propositional function such that

   i) '$P(1)$' and '$P(2)$' are true;
   ii) for each $n \in \mathbf{Z}^+$, '$[P(n) \wedge P(n+1)] \Rightarrow P(n+2)$' is true.

   Then '$P(n)$' is true for every $n \in \mathbf{Z}^+$.

   [*Hint:* Let $Q(n) = P(n) \wedge P(n+1)$ and use (a).]

21. There is a great deal of mathematical (and some not-so-mathematical) literature concerning the *Fibonacci numbers*. This sequence begins 1, 2, 3, 5, 8, 13, 21, 34, $\ldots$, each term after the second being the sum of the preceding two terms. The formal inductive definition of these numbers is therefore given by the equations

$$a_1 = 1, \qquad a_2 = 2, \qquad a_{n+1} = a_{n-1} + a_n \quad (n > 1).$$

Use Exercise 20 to justify this definition and to prove that for each $n \in \mathbf{Z}^+$,

$$a_n < \left(\frac{1+\sqrt{5}}{2}\right)^n.$$

(For the student interested in reading about the Fibonacci numbers we recommend C. D. Olds, *Continued Fractions*, Random House, New York, 1962.)

22. Since our definition of $\mathbf{Z}$ was logically premature from a strictly logical point of view, we reiterate that $\mathbf{Z}$, the set of integers, consists of the positive integers, their additive inverses, and zero.

   a) Show that $x \in \mathbf{Z}$ iff x can be represented as the difference of two positive integers.

   b) Using (a) and the fact that $\mathbf{Z}^+$ is closed under addition and multiplication, prove that $\mathbf{Z}$ is closed under these operations plus the operation of subtraction.

23. The purpose of this exercise is to establish the exponent laws for rational powers. We shall regard the laws

$$a^m a^n = a^{m+n}, \qquad (a^m)^n = a^{mn}, \qquad (ab)^n = a^n b^n \qquad (*)$$

as having been proved for all positive integers m and n and for all real numbers a and b. In parts (a) and (b) we shall assume that a and b are nonzero real numbers. In parts (c) and (d) we shall further assume that they are positive.

   a) Let m, n, m', n' be positive integers such that $m - n = m' - n'$. Show that

$$a^m/a^n = a^{m'}/a^{n'}.$$

   If $x \in \mathbf{Z}$, we can therefore, without fear of ambiguity, define a^x to be a^m/a^n, where m and n are positive integers such that $x = m - n$. Observe that $a^0 = 1$ and that a^{-1} has its former meaning.

b) Prove that the exponent laws (∗) hold for all integers m and n. Also show that

$$a^{m-n} = a^m/a^n$$

for all integers m and n.

c) Suppose that m, n, m', n' are integers, where n and n' are positive, such that

$$m/n = m'/n'.$$

Show that

$$\sqrt[n]{a^m} = \sqrt[n']{a^{m'}}.$$

Use this fact to define appropriately a^x for rational numbers x. (In Chapter 5 we shall prove that every positive real number has an nth root. In the meantime we shall simply assume it.)

d) Prove that the exponent laws (∗) hold for all rational numbers m and n.

24. Let $a_1, a_2, a_3, \ldots$ be defined by

$$a_1 = 1, \qquad a_n = \frac{1}{n}\left(1 + \sum_{k=1}^{n-1} a_k\right)^{1/2} \quad \text{if} \quad n > 1.$$

Prove that for each $n \in \mathbf{Z}^+$, $a_n \leq 1/\sqrt{n}$.

The next five exercises are applications of the arithmetic-geometric inequality (Example 3).

25. Define the harmonic mean of n positive numbers $a_1, \ldots, a_n$ and prove that it is less than or equal to the geometric mean of these numbers.

26. Let $a_1, a_2, \ldots, a_n$ be positive numbers, and let $b_1, b_2, \ldots, b_n$ be any rearrangement of these numbers. Show that

$$\sum_{k=1}^{n} \frac{b_k}{a_k} \geq n.$$

[Observe that $\prod_{k=1}^{n} (b_k/a_k) = 1$.]

27. Show that for each positive integer n,

$$\left(1 + \frac{r}{n}\right)^n < \left(1 + \frac{r}{n+1}\right)^{n+1},$$

where r is any positive number. (Consider the arithmetic and geometric means of the numbers $a_1 = 1$, $a_2 = a_3 = \cdots = a_{n+1} = 1 + r/n$.) If P dollars are deposited in a bank account where interest is compounded n times a year at the rate of $100r\%$ per annum, then $(1 + r/n)^n P$ represents the amount of money in the account at the end of one year. Thus the preceding inequality implies that the more frequently interest is compounded, the more money one can make.

28. a) Show that if the value of $\sum_{k=1}^{n} a_k$ is specified (where the a_k's are nonnegative), then the maximum possible value of $\prod_{k=1}^{n} a_k$ occurs when $a_1 = a_2 = \cdots = a_n$. Similarly, if the value of the product is specified, then the sum is a minimum when $a_1 = \cdots = a_n$.

b) Show that among all rectangles having a given perimeter, the square has the largest area.

c) Show that among all triangles with a given area, the equilateral triangle has least perimeter. [Use Hero's formula: $A = \sqrt{s(s - a)(s - b)(s - c)}$, where A is the area of a triangle with side lengths a, b, and c, and $s = (a + b + c)/2$.]

29. For each positive integer n, show that

$$n^n \geq 1 \cdot 3 \cdot 5 \cdots (2n - 1)$$

and

$$n! < \left(\frac{n + 1}{2}\right)^n.$$

30. a) Define appropriately the symbol $\sum_{k=m}^{m+n} a_k$.
    b) Prove by induction that

$$\sum_{k=n+1}^{2n} \frac{1}{k} = \sum_{j=1}^{2n} \frac{(-1)^{j+1}}{j}.$$

31. Let x be a real number. Then *binomial coefficients*

$$\binom{x}{1}, \binom{x}{2}, \binom{x}{3}, \quad \cdots$$

are defined inductively as follows:

$$\binom{x}{1} = x, \qquad \binom{x}{n+1} = \frac{x - n}{n + 1}\binom{x}{n}.$$

We also define

$$\binom{x}{0} = 1.$$

a) Compute

$$\binom{5}{3}, \binom{2}{3}, \binom{4}{2}, \binom{\frac{1}{2}}{3}, \binom{-1}{2}.$$

b) Prove the following identities by induction:

$$\binom{x + 1}{n + 1} = \frac{x + 1}{n + 1}\binom{x}{n}, \qquad \binom{x}{n} + \binom{x}{n + 1} = \binom{x + 1}{n + 1}.$$

The second identity is called the *Pascal triangle property* for reasons that should soon be clear.

c) Using the Pascal triangle property, prove by induction that if n and k are nonnegative integers, then $\binom{n}{k}$ is a nonnegative integer. If $0 \leq k \leq n$, prove that

$$\binom{n}{k} = \frac{n!}{k!(n - k)!},$$

where $m!$ (read m *factorial*) has the following inductive definition: $0! = 1! = 1$, $(m + 1)! = (m!)(m + 1)$. Deduce that

$$\binom{n}{k} = \binom{n}{n - k}.$$

Show also that

$$\binom{n}{k} = 0 \quad \text{if} \quad k > n.$$

We can form an infinite triangular array of these coefficients as follows:

$$\binom{0}{0}$$

$$\binom{1}{0} \quad \binom{1}{1}$$

$$\binom{2}{0} \quad \binom{2}{1} \quad \binom{2}{2}$$

$$\binom{3}{0} \quad \binom{3}{1} \quad \binom{3}{2} \quad \binom{3}{3}.$$

After placing 1's along the sides, we can fill in the rest of the triangle by using the Pascal triangle property. We get

```
1
1   1
1   2   1
1   3   3   1
1   4   6   4   1
1   5  10  10   5   1.
```

d) Let S be a set having n elements. Using the Pascal triangle property, prove that the number of subsets of S having k elements is $\binom{n}{k}$.

e) Prove the binomial theorem

$$(x + y)^n = \sum_{k=0}^{n} \binom{n}{k} x^{n-k} y^k.$$

f) Deduce from the binomial theorem the following identities:

$$\sum_{k=0}^{n} \binom{n}{k} = 2^n \quad \text{and} \quad \sum_{k=0}^{n} (-1)^k \binom{n}{k} = 0.$$

g) Using (d) and (f), prove that a set having n elements has 2^n subsets.

32. Using the binomial theorem and the second form of the principle of induction, show that for each $r \in \mathbf{Z}^+$,

$$S_r(n) = \sum_{k=1}^{n} k^r$$

is a polynomial of degree $r + 1$ in n whose leading term is $[1/(r + 1)]n^{r+1}$.

The purpose of the remaining exercises is to establish some elementary facts of number theory.

33. *Euclidean Algorithm.* Let n and d be positive integers. Prove that there exist unique nonnegative integers q and r such that

$$n = dq + r$$

and $0 \le r < d$. (The numbers q and r, respectively, are called the *quotient* and *remainder* resulting from the division of n by d.) Prove this result by induction on n.

34. The number 12 is the greatest common divisor of 24 and 180, that is, 12 divides both 24 and 180, and if n is any such common divisor, then n divides 12. Note that the number 12 can be expressed in the form

$$12 = 24m + 180n,$$

where m and n are integers (namely, $m = 8$ and $n = -1$). The purpose of this exercise is to generalize this result. Let a and b be any two positive integers. Let S be the set of all *positive* integers of the form $na + mb$, where m and n are integers (possibly negative or zero). Observe that S contains both a and b and hence is nonempty. By the well-ordering property of $\mathbf{Z}^+$, the set S has a smallest element d. Prove that d divides both a and b. (Suppose d does not divide a. Show that the remainder resulting from division of a by d lies in S. Obtain a contradiction.) Prove that every common divisor of a and b also divides d. This shows that *a and b have a greatest common divisor d, and d can be written*

$$\gcd(a, b) = d = ma + nb,$$

where m and n are integers.

35. a) Let a, b and n be positive integers. Prove that if n divides ab and if $\gcd(n, a) = 1$, then n divides b. (Use the preceding exercise.)
    b) Deduce that if p is a prime number which divides the product ab of two positive integers, then either p divides a or p divides b.

36. a) Prove that every positive integer $n \geq 2$ can be written as a product of primes. (Essentially, the argument used in Example 1 applies.)
    b) Prove that the factorization of a positive integer into primes is unique to within the order of the factors. [Use the well-ordering property of $\mathbf{Z}^+$ and Exercise 35(b).]

2 □ ANALYTIC GEOMETRY
OF STRAIGHT LINES AND CIRCLES

In this chapter we deal with the rudiments of plane analytic geometry. The first section contains statements of the pertinent facts. A reader who finds he is already familiar with most of these may skip to Chapter 3. In subsequent sections of the present chapter these facts are established in a leisurely fashion.

2-1 A synopsis of basic formulas

Analytic geometry represents the marriage of pure synthetic geometry and algebra. In this hybrid branch of mathematics, one establishes correspondences between geometric objects (points, lines, curves) and algebraic objects (numbers, equations). As a result, algebra may be used to prove theorems in geometry, and, more importantly for us, geometric insights may be brought to the aid of analysis.

We have already seen how one can impose coordinates on a line, thereby achieving a correspondence between points on the line and real numbers. We now add a dimension and impose coordinates on a plane. Essential to this purpose is the notion of an *ordered pair*. An ordered pair may be defined set-theoretically as

$$(x, y) = \{\{x\}, \{x, y\}\}.$$

(This notation for ordered pairs conflicts, of course, with our notation for open intervals. We must therefore make sure that the proper interpretation for (x, y) is made clear by the context of its use.) One can then prove (Exercise 1, Section 2–2) the *basic property of ordered pairs*, namely, two ordered pairs (x, y) and (a, b) are equal iff $x = a$ and $y = b$. Thus, the ordered pairs $(2, -1)$ and $(-1, 2)$ are not the same.

Closely related to the notion of an ordered pair is the notion of a set operation. If A and B are sets, then the *Cartesian product* of A and B is defined to be the set

$$A \times B = \{(x, y) \mid x \in A \text{ and } y \in B\}.$$

For example, if $A = \{1, 2\}$ and $B = \{5, 7, 9\}$, then

$$A \times B = \{(1, 5), (1, 7), (1, 9), (2, 5), (2, 7), (2, 9)\}.$$

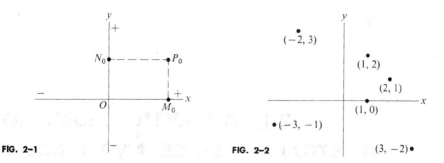

FIG. 2-1   **FIG. 2-2**

Let Π be a plane. We shall impose coordinates on Π so as to establish a one-to-one correspondence between points in Π and elements of $\mathbf{R}^2 = \mathbf{R} \times \mathbf{R}$. First we choose two perpendicular lines in Π, henceforth to be called the *coordinate axes*. We denote their point of intersection by O, and call it the *origin*. We also christen one axis the *x-axis* and the other axis the *y-axis*. Using a common unit of length and the point O as the common origin, we introduce coordinates onto the two axes. These choices determine the coordinate system. It is conventional (though not mandatory) to draw the x-axis in a horizontal position. It is also conventional, then, to take the positive x-axis to be the half-line to the right of O and the positive y-axis to be the half-line above O. We then assign coordinates to points in the plane as follows (see Fig. 2–1). If we are given a point P_0 in Π, we let M_0 and N_0 be the perpendicular projections of P_0 onto the x- and y-axes, respectively. If x_0 and y_0 denote the coordinates of M_0 and N_0, then we associate with the point P_0 the ordered pair (x_0, y_0); the numbers x_0 and y_0 are called the x- and y-coordinates of P_0. The resulting correspondence between points in Π and elements of $\mathbf{R}^2$ is one-to-one. Consequently, no harm will come from identifying a point P_0 with the corresponding ordered pair (x_0, y_0); thus we write $P_0 = (x_0, y_0)$ despite the fact that different types of objects appear on the two sides of the equality sign.

A number of points are shown in Fig. 2–2.

The sets

$$I = [0, +\infty) \times [0, +\infty) = \{(x, y) \mid 0 \le x, 0 \le y\},$$
$$II = (-\infty, 0] \times [0, +\infty),$$
$$III = (-\infty, 0] \times (-\infty, 0],$$
$$IV = [0, +\infty) \times (-\infty, 0],$$

which are called the first, second, third, and fourth *quadrants*, are shown in Fig. 2–3.

The distance between two points $P_1 = (x_1, y_1)$ and $P_2 = (x_2, y_2)$ is given by the formula

$$|P_1 P_2| = \sqrt{(x_2 - x_1)^2 + (y_2 - y_1)^2}.$$

From this formula one can readily show that if C is the circle of radius r with center (a, b), then

$$C = \{(x, y) \mid (x - a)^2 + (y - b)^2 = r^2\}.$$

We therefore call $(x - a)^2 + (y - b)^2 = r^2$ the *equation* of C.

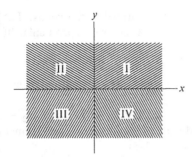

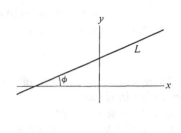

FIG. 2–3   **FIG. 2–4**

Straight lines have the simplest equations. If A, B, C are real numbers, with A and B not both zero, then

$$Ax + By + C = 0$$

is the equation of a line L. Conversely, each line has an equation of this form. The correspondence between these equations and straight lines is not quite one-to-one, since equations with proportional coefficients (such as $3x - y + 5 = 0$ and $-6x + 2y - 10 = 0$) determine the same line. If $B = 0$, then the equation of L can be written in the form $x = -C/A$, from which it is clear that L is parallel to the y-axis. If $B \neq 0$, we can solve for y to get

$$y = -(A/B)x - C/A.$$

The coefficient of x, namely $-A/B$, depends only on the line L and not on the coefficients of a particular equation for L. (This is because the coefficients of an equation for L are determined to within proportionality.) The number $\lambda = -A/B$ is called the *slope* of L. If $P_1 = (x_1, y_1)$ and $P_2 = (x_2, y_2)$ are distinct points on L, then

$$Ax_1 + By_1 + C = 0$$

and

$$Ax_2 + By_2 + C = 0.$$

Subtracting, we get

$$A(x_2 - x_1) + B(y_2 - y_1) = 0,$$

so that

$$\lambda = -\frac{A}{B} = \frac{y_2 - y_1}{x_2 - x_1}. \tag{*}$$

(Since P_1 and P_2 are distinct points, and since L is not parallel to the y-axis, $x_1 \neq x_2$.) Thus the slope of L is the difference quotient of the y- and x-coordinates of any two points on L. It follows, in particular, that $\lambda = \tan \phi$, where ϕ is the angle which L makes with the x-axis, as shown in Fig. 2–4. Another consequence of (*) is the *slope-point form* for the equation of a line. If L is a line with slope λ which passes through the point $P_1 = (x_1, y_1)$, then

$$y - y_1 = \lambda(x - x_1)$$

is the equation of L.

Let L and L' be two lines, neither of which is parallel to the y-axis. Let λ and λ' be their slopes. The lines are parallel iff $\lambda = \lambda'$. They are perpendicular iff

$$\lambda' = -1/\lambda.$$

There are two more formulas which we should consider, although we shall use them rather infrequently.

One of these is the *point of division formula*. Let $P_1 = (x_1, y_1)$, $P_2 = (x_2, y_2)$, and $0 \leq t \leq 1$ be given. Let $P = (x, y)$ denote the point on the line segment joining P_1 and P_2 such that $|P_1P|/|P_1P_2| = t$ (see Fig. 2–5). Then

$$x = (1 - t)x_1 + tx_2,$$
$$y = (1 - t)y_1 + ty_2.$$

Setting $t = \frac{1}{2}$, we see in particular that the midpoint of P_1P_2 is

$$\left(\frac{x_1 + x_2}{2}, \frac{y_1 + y_2}{2} \right).$$

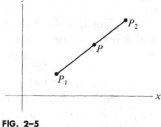

FIG. 2–5

We consider finally the formula for the distance from a point to a line. Suppose we are given the line

$$Ax + By + C = 0$$

and a point $P_1 = (x_1, y_1)$. Then the distance from P_1 to the line is

$$\frac{|Ax_1 + By_1 + C|}{\sqrt{A^2 + B^2}}.$$

2–2 Distance and point of division; circles

Proposition 1. *The distance between two points $P_1 = (x_1, y_1)$ and $P_2 = (x_2, y_2)$ is given by*

$$d = \sqrt{(x_2 - x_1)^2 + (y_2 - y_1)^2}.$$

Proof. This formula follows quickly from the Pythagorean theorem. We draw auxiliary lines as indicated in Fig. 2–6. By the Pythagorean theorem, we have

$$d^2 = |P_1Q|^2 + |QP_2|^2.$$

(We shall indicate the distance between two points A and B by $|AB|$.) But by Chapter 1,

$$|P_1Q| = |M_1M_2| = |x_2 - x_1|,$$
$$|P_2Q| = |N_1N_2| = |y_2 - y_1|.$$

Hence,

$$d = \sqrt{|x_2 - x_1|^2 + |y_2 - y_1|^2} = \sqrt{(x_2 - x_1)^2 + (y_2 - y_1)^2}.$$

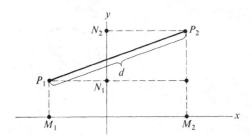

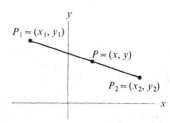

FIG. 2–6   **FIG. 2–7**

Example 1. The distance between $(-1, 2)$ and $(2, 6)$ is

$$d = \sqrt{(2 - (-1))^2 + (6 - 2)^2} = 5.$$

We consider next a set of formulas for the coordinates of a point on a line segment.

Proposition 2. *Let $P_1 = (x_1, y_1)$ and $P_2 = (x_2, y_2)$ be two points, and let λ be a real number in the interval $[0, 1]$. Set*

$$x = (1 - \lambda)x_1 + \lambda x_2, \qquad y = (1 - \lambda)y_1 + \lambda y_2.$$

Then the point $P = (x, y)$ lies on the line segment determined by P_1 and P_2, and

$$|P_1P|/|P_1P_2| = \lambda, \qquad |PP_2|/|P_1P_2| = 1 - \lambda.$$

Proof. By the distance formula,

$$|P_1P| = \sqrt{(x - x_1)^2 + (y - y_1)^2}$$
$$= \sqrt{\lambda^2(x_2 - x_1)^2 + \lambda^2(y_2 - y_1)^2}$$
$$= \lambda\sqrt{(x_2 - x_1)^2 + (y_2 - y_1)^2} = \lambda|P_1P_2|.$$

Similarly, $|PP_2| = (1 - \lambda)|P_1P_2|$. (Note that $0 \le 1 - \lambda \le 1$. For if $0 \le \lambda \le 1$, then $-1 \le -\lambda \le 0$, and adding 1 to both sides, we get, $0 \le 1 - \lambda \le 1$.) Since

$$|P_1P| + |PP_2| = \lambda|P_1P_2| + (1 - \lambda)|P_1P_2| = |P_1P_2|,$$

P must lie on the line segment joining P_1 and P_2, as indicated in Fig. 2–7. (Otherwise, $|P_1P| + |PP_2| > |P_1P_2|$.)

If $\lambda = \frac{1}{2}$, then

$$|P_1P| = |PP_2| = \frac{1}{2}|P_1P_2|.$$

Thus P is, in this case, the midpoint of P_1P_2.

Corollary 3. *If $P_1 = (x_1, y_1)$ and $P_2 = (x_2, y_2)$ are any two points, then the coordinates of their midpoint are*

$$x = \frac{1}{2}(x_1 + x_2), \qquad y = \frac{1}{2}(y_1 + y_2).$$

Example 2. Find the coordinates of the point which is two-thirds of the way from $P_1 = (-2, 5)$ to $P_2 = (1, -1)$ (see Fig. 2–8).

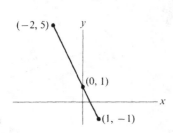

FIG. 2-8

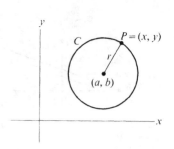

FIG. 2-9

Setting $\lambda = \frac{2}{3}$, we find from the formulas given in Proposition 2 that

$$x = \tfrac{1}{3}(-2) + \tfrac{2}{3}\cdot 1 = 0 \qquad \text{and} \qquad y = \tfrac{1}{3}\cdot 5 + \tfrac{2}{3}(-1) = 1$$

are the coordinates of the required point.

From the distance formula we can also easily get the equation of a circle. Let C be the circle whose center is the point (a, b) and whose radius is r, as shown in Fig. 2-9. By definition, C consists of all points whose distance from (a, b) is r. Thus a point $P = (x, y)$ lies on C iff the following equation is satisfied:

$$\sqrt{(x - a)^2 + (y - b)^2} = r.$$

This, however, is equivalent to the equation

$$(x - a)^2 + (y - b)^2 = r^2. \tag{$*$}$$

Thus,

$$C = \{(x, y) \mid (x - a)^2 + (y - b)^2 = r^2\}.$$

Equation ($*$) is called the *equation of C.*

By expanding the squares in ($*$) and collecting terms, we see that the equation of a circle can be written in the following form:

$$x^2 + y^2 + Ax + By + C = 0, \tag{$**$}$$

where A, B, and C are fixed real numbers. For example, the equation of the circle of radius 3 with center $(-1, 2)$ is

$$(x + 1)^2 + (y - 2)^2 = 3^2$$

or

$$x^2 + y^2 + 2x - 4y - 4 = 0.$$

Given an equation of the form ($**$), we can ask whether it is always the equation of a circle. The answer is no. One can show that if A, B, and C are fixed real numbers, then the set

$$\{(x, y) \mid x^2 + y^2 + Ax + By + C = 0\}$$

is either a circle, a point, or the null set. The proof of this is left as an exercise. We shall present here only a specific example. Consider the equation

$$x^2 + y^2 - 2x + 6y - 15 = 0.$$

Completing squares, we get

$$(x - 1)^2 + (y + 3)^2 - 1 - 9 - 15 = 0$$

or

$$(x - 1)^2 + (y + 3)^2 = 25 = 5^2.$$

Thus, in this case, we have the equation of the circle of radius 5 with center $(1, -3)$.

Exercises

1. Prove that $(x, y) = (x', y')$ iff $x = x'$ and $y = y'$.

2. If A contains m elements and B contains n elements, how many elements are in $A \times B$ and $B \times A$?

3. a) We define $(x, y, z) = ((x, y), z)$. Show that $(x, y, z) = (x', y', z')$ iff $x = x', y = y'$, and $z = z'$. This (x, y, z) is called an *ordered triple*.
   b) Define by induction $(x_1, x_2, \ldots, x_n)$ in such a way that two *n-tuples* $(x_1, x_2, \ldots, x_n)$ and $(y_1, y_2, \ldots, y_n)$ are equal iff $x_1 = y_1, \ldots, x_n = y_n$.

4. On separate graphs, plot the following sets of points.
   a) $\{(3, -2), (1, 4), (-1, -1), (-1, 2)\}$
   b) $A \times B$, where $A = [-1, 3]$ and B is the open interval $(1, 4)$
   c) $A \times B$, where $A = [-1, +\infty)$ and $B = (2, 3]$
   d) $\{(x, y) \mid (x - 1)^2 + y^2 = 16\}$
   e) $\{(x, y) \mid (x - 1)^2 + y^2 < 16\}$
   f) $\{(x, y) \mid (x - 1)^2 + y^2 > 16\}$
   g) $\{(x, y) \mid -1 \leq x \leq 1 \text{ and } y = \sqrt{1 - x^2}\}$
   h) $\{(x, y) \mid -1 \leq x < 0 \text{ and } y = -\sqrt{1 - x^2}\}$

5. Find the sets of points (x, y) satisfying the following equations:
   a) $x^2 + y^2 + 4x - 6y - 12 = 0$,　　b) $x^2 + y^2 + 4x - 6y + 13 = 0$,
   c) $x^2 + y^2 + 4x - 6y + 20 = 0$.

6. Let A, B, and C be real numbers. Show that the set

$$\{(x, y) \mid x^2 + y^2 + 2Ax + 2By + C = 0\}$$

is either a circle, a singleton set (a set containing just one point), or the null set.

7. a) Prove the *Minkowski inequality:*

$$\left(\sum_{k=1}^{n} (a_k + b_k)^2 \right)^{1/2} \leq \left(\sum_{k=1}^{n} a_k^2 \right)^{1/2} + \left(\sum_{k=1}^{n} b_k^2 \right)^{1/2}.$$

(Expand the square of the left-hand side of the inequality, and use the Cauchy-Schwarz inequality, Exercise 17, Section 1-9.)

b) Using the distance formula and part (a), prove analytically that if P_1, P_2, and P_3 are any three points in the plane, then

$$|P_1P_3| \leq |P_1P_2| + |P_2P_3|.$$

(Set $n = 2$; let the a's and b's be differences in the coordinates of the points.)

2-3 Equations of straight lines

In this section we shall obtain equations for the simplest and most important class of geometric figures, namely, straight lines. We begin with the following problem: given two points $P_1 = (x_1, y_1)$ and $P_2 = (x_2, y_2)$, find the equation satisfied by the coordinates of all points which lie on the perpendicular bisector L of P_1P_2 (see Fig. 2–10).

From plane geometry we know that L is the set of all points equidistant from P_1 and P_2. Thus a point $P = (x, y)$ lies on L iff $|P_1P| = |P_2P|$, or, using the distance formula, iff

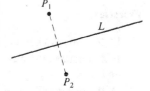

$$\sqrt{(x - x_1)^2 + (y - y_1)^2} = \sqrt{(x - x_2)^2 + (y - y_2)^2}.$$

Squaring both sides and expanding, we get

FIG. 2–10

$$x^2 - 2x_1x + x_1^2 + y^2 - 2y_1y + y_1^2 = x^2 - 2x_2x + x_2^2 + y^2 - 2y_2y + y_2^2$$

or

$$2(x_2 - x_1)x + 2(y_2 - y_1)y = x_2^2 + y_2^2 - x_1^2 - y_1^2. \tag{$*$}$$

Since we can reverse each of these steps, we may conclude that $P = (x, y)$ lies on L iff Eq. (∗) is satisfied.

Proposition 1. *Given two points $P_1 = (x_1, y_1)$ and $P_2 = (x_2, y_2)$, the equation of the perpendicular bisector of P_1P_2 is*

$$2(x_2 - x_1)x + 2(y_2 - y_1)y = x_2^2 + y_2^2 - x_1^2 - y_1^2.$$

Example 1. The equation of the perpendicular bisector of $(0, 1)$ and $(1, 0)$ is

$$2x - 2y = 0$$

or, equivalently,

$$x - y = 0.$$

In Eq. (∗) we set

$$a = 2(x_2 - x_1), \qquad b = 2(y_2 - y_1), \qquad \text{and} \qquad -c = x_2^2 + y_2^2 - x_1^2 - y_1^2.$$

The equation then becomes

$$ax + by + c = 0.$$

Since the points $P_1 = (x_1, y_1)$ and $P_2 = (x_2, y_2)$ are distinct, the numbers a and b are not both zero. Furthermore, since every straight line may be regarded as a perpendicular bisector, we have proved the first half of the following theorem.

Proposition 2. *Given any line L in the plane, there exist three real numbers a, b, and c, with $a^2 + b^2 \neq 0$, such that*

$$ax + by + c = 0$$

is the equation of L. Moreover, the triple (a, b, c) is unique to within a constant of proportionality. That is,

$$a'x + b'y + c' = 0$$

is also the equation of L iff there is a number $k \neq 0$ such that

$$a' = ka, \qquad b' = kb, \qquad c' = kc.$$

Proof. Let

$$S = \{(x, y) \mid ax + by + c = 0\},$$
$$S' = \{(x, y) \mid a'x + b'y + c' = 0\}.$$

By assumption, $S = S' = L$, $a^2 + b^2 \neq 0$, and $(a')^2 + (b')^2 \neq 0$.

Case I. $c = 0$. Since $(0, 0)$ belongs to S, it belongs to S'. This implies that $c' = 0$. Since $(-b, a)$ belongs to S, it belongs to S', so that

$$-a'b + b'a = 0.$$

Assuming that $b \neq 0$, then $k = b'/b$ has the desired properties. Assuming that $a \neq 0$, then $k = a'/a$ is seen to work.

Case II. $c \neq 0$, $a \neq 0$, $b \neq 0$. Since in this case $(0, -c/b)$ and $(-c/a, 0)$ belong to S, they belong also to S', so that

$$-b'c/b + c' = 0$$

and

$$-a'c/a + c' = 0.$$

If we set $k = b'/b$, then $b' = kb$, and by the first displayed equation, $c' = kc$. Hence, from the second displayed equation, we get $(a'/a)c = kc$. Since $c \neq 0$, it follows that $k = a'/a$ or $a' = ka$. The number k cannot be zero, since otherwise S' would be $\mathbf{R}^2$.

The remaining cases are left to the reader.

So far we have shown that every straight line has an equation of the form

$$ax + by + c = 0$$

(with $a^2 + b^2 \neq 0$) and that the coefficients a, b, and c are essentially unique. We shall next prove that every equation of this form is the equation of a straight line.

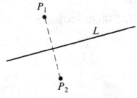

Definition 3. Suppose L is the perpendicular bisector of the line segment P_1P_2. Then we say that P_2 is the **reflection of P_1 in L**, and also that P_1 is the reflection of P_2 in L. In this situation we also say that P_1 and P_2 are **symmetric with respect to L** (see Fig. 2–11, which is identical to Fig. 2–10).

FIG. 2–11

The motivation for this definition is clear. If L were the surface of a mirror and P_1 an object in front of L, then P_2 would be the image of P_1 in the mirror.

Proposition 4. *Let a, b, and c be real numbers, with $a^2 + b^2 \neq 0$. Then*

$$ax + by + c = 0$$

is the equation of a straight line L. Moreover, if $P_1 = (x_1, y_1)$ is any point in the plane, then its reflection $P_2 = (x_2, y_2)$ in L can be determined by the equations

$$x_2 - x_1 = -ka, \qquad y_2 - y_1 = -kb,$$

where

$$k = 2\,\frac{ax_1 + by_1 + c}{a^2 + b^2}.$$

Also, the distance from P_1 to L is

$$d = \frac{|ax_1 + by_1 + c|}{\sqrt{a^2 + b^2}}.$$

Proof. We can show that $ax + by + c = 0$ is the equation of a line if we can find numbers x_1, y_1, x_2, and y_2 such that

$$a = 2(x_2 - x_1), \qquad b = 2(y_2 - y_1), \qquad \text{and} \qquad c = -x_2^2 - y_2^2 + x_1^2 + y_1^2.$$

For then, by Proposition 1, $ax + by + c = 0$ is the equation of the perpendicular bisector of $P_1 = (x_1, y_1)$ and $P_2 = (x_2, y_2)$. However, $ax + by + c = 0$ will also be the perpendicular bisector of P_1P_2 if there is some nonzero number λ such that

$$\begin{aligned} \lambda a &= 2(x_2 - x_1), \\ \lambda b &= 2(y_2 - y_1), \\ \lambda c &= -x_2^2 - y_2^2 + x_1^2 + y_1^2. \end{aligned} \qquad (**)$$

We now select numbers x_1 and y_1 such that

$$ax_1 + by_1 + c \neq 0.$$

(This can always be done unless $a = b = c = 0$. This possibility does not arise here since a and b are not both zero.) Define k, x_2, and y_2 by the equations in the statement of the theorem. It suffices to show that

$$2kc = x_2^2 + y_2^2 - x_1^2 - y_1^2$$

(for then, setting $\lambda = -2k$, we see that Eqs. $(**)$ are satisfied):

$$\begin{aligned} x_2^2 + y_2^2 - x_1^2 - y_1^2 &= (-ka + x_1)^2 + (-kb + y_1)^2 - x_1^2 - y_1^2 \\ &= k^2(a^2 + b^2) - 2kax_1 - 2kby_1 \\ &= k[k(a^2 + b^2) - 2ax_1 - 2by_1] \\ &= k[2(ax_1 + by_1 + c) - 2ax_1 - 2by_1] \\ &= 2kc. \end{aligned}$$

This proves that $ax + by + c = 0$ is the equation of the perpendicular bisector of P_1P_2.

The distance from P_1 to the line is

$$d = \tfrac{1}{2}|P_1P_2| = \tfrac{1}{2}\sqrt{(x_2 - x_1)^2 + (y_2 - y_1)^2}$$
$$= \tfrac{1}{2}\sqrt{k^2a^2 + k^2b^2} = \tfrac{1}{2}|k|\sqrt{a^2 + b^2}$$
$$= \frac{|ax_1 + by_1 + c|}{\sqrt{a^2 + b^2}}.$$

Example 2. Find the equation of a circle with center $(1, -2)$ which is tangent to the line $3x - 4y - 1 = 0$ (see Fig. 2-12).

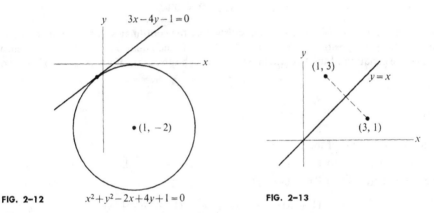

FIG. 2-12 $x^2+y^2-2x+4y+1=0$ FIG. 2-13

The radius of the circle will be the distance from $(1, -2)$ to the line. By the formula in Proposition 4,

$$r = \frac{3 \cdot 1 - 4(-2) - 1}{\sqrt{3^2 + (-4)^2}} = \frac{10}{5} = 2.$$

Hence the equation of the circle is

$$(x - 1)^2 + (y + 2)^2 = 2^2 \quad \text{or} \quad x^2 + y^2 - 2x + 4y + 1 = 0.$$

Example 3. Find the reflection of (x_1, y_1) in the line $y = x$ (see Fig. 2-13).

We rewrite the equation of the line as $x - y = 0$. Let (x_2, y_2) be the reflection of (x_1, y_1) in this line. Applying the formulas of Proposition 4, we get

$$k = 2\frac{x_1 - y_1}{1^2 + (-1)^2} = x_1 - y_1$$

and

$$x_2 = x_1 - k \cdot 1 = x_1 - (x_1 - y_1) = y_1,$$
$$y_2 = y_1 - k(-1) = y_1 + (x_1 - y_1) = x_1.$$

Hence to reflect a point in the line $y = x$, one just switches the positions of the coordinates.

Example 4. Find the equation of the line which passes through the points of intersection of the two circles

$$x^2 + y^2 - 2x + 6y = 0$$

and

$$x^2 + y^2 + 4x - 10y + 39 = 0.$$

Suppose that (x_1, y_1) is one of the points of intersection of the circles. Then

$$x_1^2 + y_1^2 - 2x_1 + 6y_1 = 0,$$
$$x_1^2 + y_1^2 + 4x_1 - 10y_1 + 39 = 0.$$

Subtracting, we get

$$6x_1 - 16y_1 + 39 = 0.$$

Thus the equation $6x - 16y + 39 = 0$ is satisfied by the coordinates of the points of intersection. But, by Proposition 4, $6x - 16y + 39 = 0$ is the equation of *some* line. Hence it must be the equation of the line which passes through the points of intersection of the two circles.

Exercises

1. Find the reflections of the point (x_1, y_1) in
   a) the x-axis,   b) the y-axis,   c) the line $x + y = 0$.
2. Let $P_1 = (x_1, y_1)$ and $P_2 = (x_2, y_2)$ be distinct points. Set

   $$x = (1 - \lambda)x_1 + \lambda x_2, \qquad y = (1 - \lambda)y_1 + \lambda y_2.$$

   We have seen that if $0 \le \lambda \le 1$, then (x, y) lies on the line segment determined by P_1 and P_2. By eliminating λ from these equations, show that x and y must satisfy an equation of the form $ax + by + c = 0$. Thus by Proposition 4, it follows that for *all* $\lambda \in \mathbf{R}$, the point $P = (x, y)$ with coordinates satisfying the above equations lies on the line determined by P_1 and P_2. Show, conversely, that for each point $P = (x, y)$ on this line, there exists exactly one value of λ such that the above equations are satisfied. Show further that

   a) P_1 lies between P and P_2 if $\lambda < 0$,   b) P_2 lies between P and P_1 if $\lambda > 1$,

   as shown in Fig. 2-14, and that in all cases

   $$|P_1P|/|P_1P_2| = |\lambda|,$$
   $$|PP_2|/|P_1P_2| = |1 - \lambda|.$$

   Thus we have seen that as t ranges over the set of all real numbers, the point (x, y) defined by the equations

   $$x = (1 - t)x_1 + tx_2,$$
   $$y = (1 - t)y_1 + ty_2$$

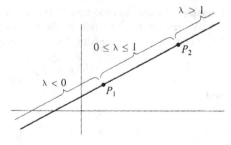

FIG. 2-14

traverses the line determined by P_1 and P_2. These are called *parametric equations* of the line. Those who value physical interpretations might like to think of the "parameter" t as time. The point $P = (x, y)$ can then be thought of as the position of a particle at time t moving with constant speed along the line.

3. Find the foot of the prependicular drawn from $(3, 0)$ to the line $3x + 4y + 1 = 0$.

4. Find the equation of the line which passes through $(0, 0)$ and the point of intersection of the lines $2x - y + 8 = 0$ and $x + 3y - 4 = 0$. [If k is any real number, infer that

$$(k - 2)x + (3k + 1)y - 4k - 8 = 0$$

is the equation of a line which passes through the point of intersection. Then find k so that the second condition for the line is satisfied.]

5. Let A and B be distinct points in a plane, and let k be a positive real number such that $k \neq 1$. Show that the set of all points P such that

$$|AP| = k|BP|$$

is a circle whose center lies on the line joining A and B. [To simplify the algebra, introduce coordinates so that A and B have coordinates $(0, 0)$ and $(1, 0)$.]

2–4 Slopes of lines

We discuss next the "slope" of a line, a concept which is important in the study of calculus.

Proposition 1. *Let L be a straight line in a plane where a coordinate system has been selected. If L is not parallel to the y-axis, then the ratio*

$$\lambda = \frac{y_2 - y_1}{x_2 - x_1}$$

is the same for all pairs of distinct points $P_1 = (x_1, y_1)$ and $P_2 = (x_2, y_2)$ on L. If $ax + by + c = 0$ is the equation of L, then $\lambda = -a/b$.

Proof. If P_1 and P_2 lie on L, then

$$ax_1 + by_1 + c = 0 \quad \text{and} \quad ax_2 + by_2 + c = 0.$$

Subtracting, we get

$$a(x_2 - x_1) + b(y_2 - y_1) = 0.$$

The coefficient b is nonzero, since otherwise L would be parallel to the y-axis. Similarly, $x_2 - x_1 \neq 0$. Hence, we get

$$a(x_2 - x_1) = -b(y_2 - y_1)$$

and finally,

$$\frac{y_2 - y_1}{x_2 - x_1} = -\frac{a}{b}.$$

Definition 2. The number λ above is called the **slope** of L.

It is easy to construct lines with a given slope. Let us, for instance, draw a line of slope $\frac{1}{2}$. Starting from any point P, we can move two units to the right and then one unit up to get a second point P' on the line (see Fig. 2–15). Alternatively, to get a second point on the line, we could move two units to the left and then one unit down. Or, again, we could move two units up and four units to the right.

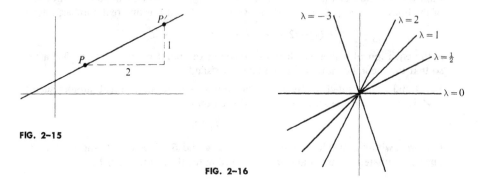

FIG. 2–15

FIG. 2–16

Figure 2–16 shows a number of straight lines through the origin and their slopes. Glancing at the figure, it is clear that the absolute value of the slope measures the steepness of the line. If we begin with a line which coincides with the x-axis and rotate it about the origin through 90° in a counterclockwise direction, then the slope of the line takes on all positive values in an increasing fashion. (Note that the slope of a vertical line is not defined.) Similarly, if we were to rotate the line through 90° in a clockwise direction, the slope would assume all negative values.

Geometrically, the slope of a line is equal to $\tan \phi$, where ϕ is the angle which the line makes with the x-axis (the so-called *angle of inclination*), as shown in Fig. 2–17. We shall make no use of this fact.

We consider next several special forms for the equation of a line.

1) *Slope-Point Form.* Let us find the equation of a line with slope λ that passes through the point $P_1 = (x_1, y_1)$.

Let P be any point on the line other than P_1. Then by Proposition 1,

$$\frac{y - y_1}{x - x_1} = \lambda.$$

Multiplying by $x - x_1$, we get

$$y - y_1 = \lambda(x - x_1).$$

We note that this equation is also satisfied when $P = P_1$. Hence this is the equation of the line.

2) *Slope-Intercept Form.* Let L be a line with slope λ, and let $(0, b)$ be the y-intercept of L (the point where the line intersects the y-axis), as shown in Fig. 2–18.

2-4

SLOPES OF LINES 71

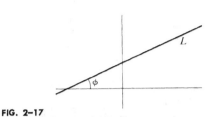

FIG. 2-17

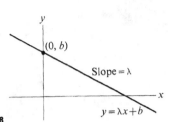

FIG. 2-18

The equation of this line, by the above formula, is

$$y - b = \lambda x$$

or

$$y = \lambda x + b.$$

3) *Point-Point Form.* Let L be the line determined by two distinct points

$$P_1 = (x_1, y_1) \quad \text{and} \quad P_2 = (x_2, y_2).$$

If L is not parallel to the y-axis, then the slope of L is

$$\lambda = \frac{y_2 - y_1}{x_2 - x_1}.$$

Hence, by the slope-intercept formula, the equation of L is

$$y - y_1 = \frac{y_2 - y_1}{x_2 - x_1}(x - x_1).$$

Multiplying both sides by $x_2 - x_1$, we get

$$(x_2 - x_1)(y - y_1) = (y_2 - y_1)(x - x_1).$$

In this particular form, this is also the equation of L when L is parallel to the y-axis. For memory purposes only, the best way to write the equation of L is the following:

$$\frac{y - y_1}{y_2 - y_1} = \frac{x - x_1}{x_2 - x_1}.$$

(Note that if L is parallel to either the x- or y-axis, the above equation is not valid.)

Example 1. Plot the line $3x - y - 2 = 0$.

Although there are many possible ways of going about such a problem, use of the slope-intercept formula is probably the best. We may write

$$y = 3x - 2.$$

Hence the line passes through $(0, -2)$ and has slope 3. From this information, we can easily draw the line (see Fig. 2-19).

Example 2. As an application of the point-point form, we shall derive a formula for the area of a triangle determined by three points

$$P_1 = (x_1, y_1), \quad P_2 = (x_2, y_2), \quad \text{and} \quad P_3 = (x_3, y_3).$$

(See Fig. 2-20.)

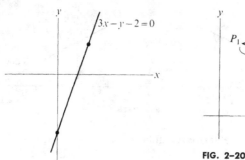

FIG. 2-19   **FIG. 2-20**

We use the familiar formula for the area of a triangle: $A = Bh/2$. Regarding P_1P_2 as the base, we get from our distance formula

$$B = \sqrt{(x_2 - x_1)^2 + (y_2 - y_1)^2}.$$

The height h is simply the distance from P_3 to the line determined by P_1 and P_2, which, according to Proposition 3, Section 2-3, may be written

$$h = \frac{|ax_3 + by_3 + c|}{\sqrt{a^2 + b^2}},$$

where $ax + by + c = 0$ is the equation of P_1P_2. The equation of P_1P_2 is

$$\frac{y - y_1}{y_2 - y_1} = \frac{x - x_1}{x_2 - x_1}$$

or

$$(y_2 - y_1)x - (x_2 - x_1)y + (y_1x_2 - y_2x_1) = 0.$$

Thus, we may set

$$a = y_2 - y_1, \qquad b = -(x_2 - x_1), \qquad c = y_1x_2 - y_2x_1.$$

Substituting into our previous expressions, we get

$$A = \pm\tfrac{1}{2}\{(y_2 - y_1)x_3 - (x_2 - x_1)y_3 + (y_1x_2 - y_2x_1)\}$$
$$= \pm\tfrac{1}{2}\{(y_2x_3 - y_3x_2) + (y_3x_1 - y_1x_3) + (y_1x_2 - y_2x_1)\}.$$

The sign is chosen so that A is nonnegative. Note that in the second expression for A, each term in parentheses can be obtained from the preceding one by cyclic interchange of subscripts. (Replace 1 by 2, 2 by 3, and 3 by 1.)

The formula may also be written in determinantal form:

$$A = \pm\tfrac{1}{2}\begin{vmatrix} x_1 & y_1 & 1 \\ x_2 & y_2 & 1 \\ x_3 & y_3 & 1 \end{vmatrix}.$$

We have seen (rather intuitively) that the slope determines the direction of a line. In support of this we shall now prove that two lines are parallel iff they have the same slope. We shall also prove that in the case of perpendicular lines, one slope is the negative reciprocal of the other.

Proposition 3. *Let $ax + by + c = 0$ be the equation of a line L, and let $P_1 = (x_1, y_1)$ be a point not on L. Then*

$$ax + by = ax_1 + by_1$$

is the equation of the line through P_1 parallel to L. Two distinct nonvertical lines are parallel iff they have the same slope.

Proof. The expression $ax + by = ax_1 + by_1$ is the equation of a line (since a and b are not both zero) which passes through P_1 (since the equation is visibly satisfied when $x = x_1$ and $y = y_1$). Suppose this line is not parallel to L. Then the two lines must intersect at some point (X, Y). Hence

$$aX + bY + c = 0 \quad \text{and} \quad aX + bY = ax_1 + by_1.$$

Subtracting these equations, we get

$$c = -ax_1 - by_1$$

or $ax_1 + by_1 + c = 0$. But this implies that $P_1 = (x_1, y_1)$ lies on L, contrary to hypothesis. If the lines are nonvertical, then their slope is $-a/b$. This proves that nonvertical parallel lines have the same slope.

Suppose now that two lines have the same slope λ. Using the slope-intercept form, we write the equations of these lines as

$$y = \lambda x + b, \quad y = \lambda x + b'.$$

If $b = b'$, the lines are the same. If $b \neq b'$, there can be no point of intersection, and the lines must be parallel.

Proposition 4. *Let $ax + by + c = 0$ be the equation of a line L, and let $P_1 = (x_1, y_1)$ be any point. Then*

$$bx - ay = bx_1 - ay_1$$

is the equation of the line through P_1 perpendicular to L. Two lines, neither of which is parallel to a coordinate axis, are perpendicular iff the slope of one is the negative reciprocal of the slope of the other.

Proof. Suppose that L is not parallel to either coordinate axis, and let $P_1 = (x_1, y_1)$ and $P_2 = (x_2, y_2)$ be two distinct points on L, as shown in Fig. 2–21. Then the slope of L is

$$\lambda = \frac{y_2 - y_1}{x_2 - x_1}.$$

We have seen that the perpendicular bisector L' of $P_1 P_2$ has an equation of the form

$$2(x_2 - x_1)x + 2(y_2 - y_1)y = C.$$

Hence the slope of L' is

$$\lambda' = -\frac{2(x_2 - x_1)}{2(y_2 - y_1)} = -\frac{1}{\lambda}.$$

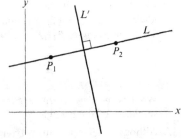

FIG. 2–21

A third line L^* is perpendicular to L iff L^* is parallel to L'. Hence, by the preceding theorem, L^* is perpendicular to L iff the slope of L^* is $-1/\lambda$. This proves the second part of the theorem.

Now $bx - ay = bx_1 - ay_1$ is the equation of a line L^* through P_1. If L is not parallel to either coordinate axis, the slope of L^* is b/a, which is the negative reciprocal of the slope of $L\,(-a/b)$. Hence, in this case, L^* is perpendicular to L. We shall let the reader worry about the case in which L is parallel to one of the coordinate axes.

Exercises

1. Plot the following sets of points.
   a) $\{(x, y) \mid 4x + 2y - 5 = 0\}$
   b) $\{(x, y) \mid 3x - y + 1 = 0\}$
   c) $\{(x, y) \mid (4x + 2y - 5)(3x - y + 1) = 0\}$
   d) $\{(x, y) \mid (4x + 2y - 5)^2 + (3x - y + 1)^2 = 0\}$
   e) $\{(x, y) \mid 4x + 2y - 5 > 0\}$
   f) $\{(x, y) \mid 3x - y + 1 < 0\}$
   g) $\{(x, y) \mid 4x + 2y - 5 > 0 \text{ and } 3x - y + 1 < 0\}$

2. Let L be the line with equation $2x - 3y + 7 = 0$, and let $P = (2, -1)$. Find the equation of the line through P which is
   a) parallel to L,   b) perpendicular to L.

3. Let $x^2 + y^2 + 2Ax + 2By + C = 0$ be the equation of a circle, and let $P_1 = (x_1, y_1)$ be a point on the circle. Prove that
$$x_1 x + y_1 y + A(x + x_1) + B(y + y_1) + C = 0$$
is the equation of the tangent to the circle at the point P_1.

4. In most books on analytic geometry the point-slope formula is the starting point for the treatment of straight lines. It is therefore crucial to know that the ratio
$$\frac{y_2 - y_1}{x_2 - x_1}$$
is the same for any two pairs of points $P_1 = (x_1, y_1)$ and $P_2 = (x_2, y_2)$ on a line L which is not parallel to the coordinate axes. The justification usually given is the following. Let $\overline{P}_1 = (\overline{x}_1, \overline{y}_1)$ and $\overline{P}_2 = (\overline{x}_2, \overline{y}_2)$ be a second pair of distinct points on L. Then the triangles $P_1 Q P_2$ and $\overline{P}_1 \overline{Q} \overline{P}_2$ are similar (see Fig. 2–22). Hence
$$\frac{P_2 Q}{P_1 Q} = \frac{\overline{P}_2 \overline{Q}}{\overline{P}_1 \overline{Q}} \quad \text{or} \quad \frac{y_2 - y_1}{x_2 - x_1} = \frac{\overline{y}_2 - \overline{y}_1}{\overline{x}_2 - \overline{x}_1}.$$

What is the flaw in this argument? What can and what cannot be deduced from the similarity of the triangles?

5. Although the argument in Exercise 4 is fallacious, one can develop the analytic geometry of straight lines in a rigorous fashion by starting with the point-slope formula.

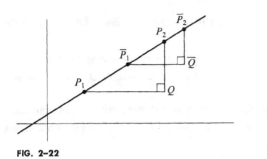

FIG. 2-22

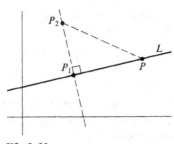

FIG. 2-23

We outline here how this can be done. (In doing this problem, you may not use any results of Section 2–3.)

a) Let L be a line not parallel to the y-axis. Let $P_1 = (x_1, y_1)$ be any point on L, let L' be the line through P_1 which is perpendicular to L, and let $P_2 = (x_2, y_2)$ be a point on L' distinct from P_1, as shown in Fig. 2–23. Using the Pythagorean theorem and its converse, prove that $P = (x, y)$ is a point on L distinct from P_1 iff

$$\frac{y - y_1}{x - x_1} = -\frac{1}{\lambda'},$$

where $\lambda' = (y_2 - y_1)/(x_2 - x_1)$.

b) Deduce that the ratio $(\bar{y}_2 - \bar{y}_1)/(\bar{x}_2 - \bar{x}_1)$ is the same (namely, $-1/\lambda'$) for any two distinct points $\bar{P}_1$ and $\bar{P}_2$ on L. This ratio is then defined to be the slope of L, and the slope-point and point-point formulas for a line follow immediately.

c) Deduce next that two lines are perpendicular iff one slope is the negative reciprocal of the other. Deduce that two lines are parallel iff they have the same slope.

One can then proceed to the proofs of the results in Section 2–3.

*6. Given a plane with a coordinate system, a *lattice point* is defined to be a point whose coordinates are integers. Prove that there is no equilateral triangle whose vertices are lattice points. (Assume that such a triangle exists, and prove that its area must be both rational and irrational.)

7. In much of what we have done in this chapter, the real line can be replaced by an arbitrary field. Given a field F, we can invent a plane geometry. We define the *affine plane over F* to be the set $F \times F = F^2$. The elements of F^2 (ordered pairs of elements of F) are called *points*. We define a *line* to be any set of points of the form

$$L = \{(x, y) \in F^2 \mid ax + by + c = 0\},$$

where a, b, and c are fixed elements of F such that a and b are not both zero. We say that two lines are *parallel* iff they have no point in common.

Let $F = J_3$. The plane over F consists of nine points indicated in Fig. 2–24. The line having the equation $2x + y = 1 = 0$ is the set $\{(1, 0), (2, 1), (0, 2)\}$. Find all lines,

FIG. 2-24

(0, 2)	(1, 2)	(2, 2)
(0, 1)	(1, 1)	(2, 1)
(0, 0)	(1, 0)	(2, 0)

and join by dashed lines those points which lie on the same line. Observe that the following are true:

A1. *If P and Q are distinct points, there is one and only one line joining them.*

A2. *If L is a line and P is a point not on L, then there exists one and only one line through P parallel to L.*

Observe also that every line contains three points and that every point lies on four lines.

*8. Prove that if F is any field, then A1 and A2 are satisfied. If F is a field containing n elements, show that

  a) the affine plane over F has n^2 points;
  b) every line has n points;
  c) every point lies on $n + 1$ lines;
  d) there are $n^2 + n$ lines in all.

*2–5 Applications to plane geometry

We shall illustrate in this section how analytic methods may be used to prove theorems in plane geometry.

Example 1. The diagonals of a parallelogram bisect each other.

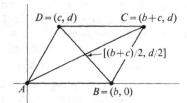

FIG. 2–25

  We are given a parallelogram $ABCD$, as shown in Fig. 2–25. To prove this theorem using analytic geometry, we must first introduce a coordinate system. We are free to do this in any way we see fit, and, in practice, we try to introduce coordinates so as to minimize the amount of algebra. In this case we can choose coordinate axes such that the x-axis coincides with the side AB extended and the origin is the point A. Thus the coordinates of B are of the form $(b, 0)$. (If we wished, we could have chosen the unit of length so that $b = 1$.) Let $D = (c, d)$. Since CD is parallel to AB, the y-coordinate of C is d. Let $C = (x, d)$. To find x, we use the fact that BC is parallel to AD. We consider two cases.

Case I. The parallelogram is a rectangle. Then $c = 0$ and $x = b = b + c$.

Case II. The parallelogram is not a rectangle. Then AD and BC are not parallel to the y-axis. Since they are parallel, their slopes must be equal. Hence

$$\frac{d}{c} = \frac{b}{x - b},$$

from which we get $x = b + c$.

In both cases then we get $C = (b + c, d)$. To show that the diagonals bisect each other, it suffices to show that the midpoint of AC is the same as the midpoint of BD. The midpoint of AC has the coordinates

$$x = \tfrac{1}{2} \cdot 0 + \tfrac{1}{2}(b + c) = \tfrac{1}{2}(b + c), \qquad y = \tfrac{1}{2} \cdot 0 + \tfrac{1}{2}d = \tfrac{1}{2}d.$$

The midpoint of BD has the coordinates

$$x = \tfrac{1}{2}b + \tfrac{1}{2}c = \tfrac{1}{2}(b + c), \qquad y = \tfrac{1}{2} \cdot 0 + \tfrac{1}{2}d = \tfrac{1}{2}d.$$

Hence the midpoints are the same.

Example 2. The perpendicular bisectors of the sides of a triangle are concurrent.

Given a triangle ABC, we can introduce coordinates so that the vertices are given by $A = (a, 0)$, $B = (b, 0)$, and $C = (0, c)$, as shown in Fig. 2-26. The equation of the perpendicular bisector of AB is

$$x = \tfrac{1}{2}(a + b).$$

The equations of the other two bisectors are

$$2ax - 2cy = a^2 - c^2$$

and

$$2bx - 2cy = b^2 - c^2.$$

To find the point of intersection of these second two bisectors, we solve their equations simultaneously. Subtracting, we get

$$2(a - b)x = a^2 - b^2 \qquad \text{or} \qquad x = \tfrac{1}{2}(a + b).$$

Hence this point of intersection lies on the third bisector, and this proves that the three bisectors are concurrent.

Example 3. The medians of a triangle are concurrent, and the distance from any vertex to the common point of intersection is two-thirds the length of the median drawn from that vertex.

We could introduce coordinates as in the previous problem. Instead, we shall let the coordinates of the vertices be $P_1 = (x_1, y_1)$, $P_2 = (x_2, y_2)$, and $P_3 = (x_3, y_3)$, as shown in Fig. 2-27. It will suffice to find for each vertex the point which is two-thirds the way down the median, and then to show that these points are the same for the three vertices.

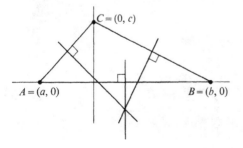

FIG. 2-26   FIG. 2-27

The midpoint of P_1P_2 has the coordinates

$$((x_1 + x_2)/2, (y_1 + y_2)/2).$$

Thus the point which is two-thirds the way down the median from P_3 has the coordinates

$$x = \tfrac{1}{3}x_3 + \tfrac{2}{3} \cdot \tfrac{1}{2}(x_1 + x_2) = (x_1 + x_2 + x_3)/3,$$
$$y = \tfrac{1}{3}y_3 + \tfrac{2}{3} \cdot \tfrac{1}{2}(y_1 + y_2) = (y_1 + y_2 + y_3)/3.$$

To find the corresponding point for P_1, we can simply interchange indices. (Replace 3 by 1, 1 by 2, and 2 by 3.) We do the same for P_2. But the above expressions for x and y are invariant under a permutation of the indices. This proves the theorem.

Note that in the course of the proof we have obtained an additional result: one can find the coordinates of the center of gravity of a triangle (the point of intersection of the medians) by taking the averages of the coordinates of the vertices.

Exercises

Use analytic methods to prove each of the following theorems.

1. A parallelogram is a rhombus iff the diagonals are perpendicular.
2. The altitudes of a triangle are concurrent.
3. Every angle inscribed in a semicircle is a right angle.
4. Suppose that two circles intersect in exactly two points. Prove that the line joining the points of intersection of these circles is perpendicular to the line joining their centers.

3 □ LIMITS

"When I use a word," Humpty Dumpty said in a rather scornful tone,
"it means just what I choose it to mean—neither more nor less."

Lewis Carroll, *Through the Looking Glass*

In scientific work, one is often faced with the problem of replacing a rather inexact "intuitive" concept (such as warmth, hardness, likelihood, or intelligence) by a more exact or precise one (such as temperature measured in degrees centigrade, the Brinell hardness number, numerical probability, or I.Q. scores on the Stanford-Binet tests). Rudolf Carnap has termed this process *explication*; he also termed the intuitive concept the *explicandum* and the corresponding exact concept the *explicatum*. The job immediately confronting us is that of explicating the notions of limit and continuity.

3-1 Functions

Basic to calculus and, more generally, to analysis is the notion of a limit. Basic to the notion of a limit, and indeed to all of mathematics, is the notion of a function.

Definition 1. Let A and B be sets, and let f be a rule which assigns to each element $x \in A$ exactly one element, which we denote by $f(x)$ (read f of x), in B. Then we say that f is a **function mapping A into B**, and we indicate this state of affairs by writing either

$$f: A \to B \qquad \text{or} \qquad A \xrightarrow{f} B.$$

Example 1. Let $A = B = \mathbf{R}$, and let f be the function which assigns to each real number its square. Thus

$$f(3) = 9, \qquad f(-2) = 4,$$

and, more generally,

$$f(x) = x^2.$$

Example 2. Let $A = B = \mathbf{Z}^+$, and let φ be the function which assigns to each $n \in \mathbf{Z}^+$ the least number of U.S. coins required to make n cents. Thus

$$\varphi(1) = \varphi(5) = \varphi(10) = \varphi(25) = \varphi(50) = 1,$$
$$\varphi(2) = \varphi(6) = \varphi(11) = \varphi(15) = \varphi(20) = \varphi(26) = \varphi(30) = \varphi(35)$$
$$= \varphi(51) = \varphi(55) = \varphi(60) = \varphi(75) = \varphi(100) = 2,$$

and so on.

If we have a function $f\colon A \to B$, then A is called the *domain* of f and is denoted by $\mathfrak{D}_f$. (The set B is somewhat less frequently called the *codomain* of f.) If $x \in A$, then $f(x)$ is sometimes called the *value of f at x*, or the *image of x under f*. The set of all values of f,

$$\mathfrak{R}_f = \{f(x) \mid x \in \mathfrak{D}_f\},$$

is called the *range* of f. Clearly, $\mathfrak{R}_f \subset B$. Should $\mathfrak{R}_f = B$, we say that f maps A *onto* B. In Example 1, we have, for instance,

$$\mathfrak{R}_f = \{x^2 \mid x \in \mathbf{R}\} = [0, +\infty),$$

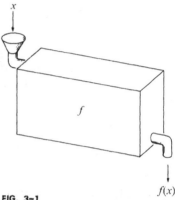

FIG. 3–1

while in Example 2, the function φ maps $\mathbf{Z}^+$ onto itself.

The reader may find it helpful to think of a function as a machine with an input and an output (see Fig. 3–1). If f is such a function, then we may think of $\mathfrak{D}_f$ as the set of all things which can be fed into the machine. If $x \in \mathfrak{D}_f$, then $f(x)$ is the output which results from feeding x into the machine. The only requirement we make on the machine is that it be deterministic: if the same thing is fed into the machine on two separate occasions, then the outputs on these two occasions must be the same. The range of f is the set of all possible outputs of the machine.

In applications of calculus, functions frequently arise in the following way. We are given two physical quantities x and y (such as temperature, length, mass, time, density, etc.) such that values of y are determined by values of x. We can then consider the function f which assigns to each value of x the corresponding value of y. Under these circumstances we speak of x as the *independent variable* and y as the *dependent variable*. Suppose, for example, that a baseball is thrust vertically upward with an initial velocity of 80 ft/sec. Then the height s (in feet) of the ball can be regarded as a function f of the time t (in seconds) which elapses from the moment the ball is released. Specifically, if we neglect air resistance, then

$$s = f(t) = 80t - 16t^2.$$

Note that in this case values of s do *not* determine values of t, so that we cannot regard s as the independent variable and t as the dependent variable. (Corresponding to the value 64 for s, we have, for instance, two values for t, namely, 1 and 4.)

If the domain and range of a function are subsets of **R**, as will usually be the case in this text, then the behavior of the function is best displayed through its graph.

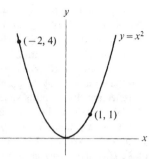

Definition 2. If f is a function, then by its **graph** we shall mean the set

$$G_f = \{(x, y) \mid x \in \mathfrak{D}_f \text{ and } y = f(x)\}.$$

For example, the graph of the function in Example 1 is the set of points $\{(x, x^2) \mid x \in \mathbf{R}\}$, which are plotted in Fig. 3–2.

FIG. 3–2

We consider now several more examples of functions and their graphs.

Example 3. Let f be the function with domain $[-1, +1]$ such that for each $x \in [-1, +1]$,

$$f(x) = \sqrt{1 - x^2}.$$

Sketch the graph of f.

If we let $y = f(x) = \sqrt{1 - x^2}$, then $y \geq 0$ (since $\sqrt{a}$ will *always* denote the nonnegative square root of a), and $x^2 + y^2 = 1$. Thus all points of the graph belong to both the upper half-plane and the circle of radius one with center at the origin. The graph is, in fact, the upper semicircle, as shown in Fig. 3–3.

The range of a function may be regarded as the projection of its graph onto the y-axis. Therefore, glancing at the graph of f, we see that its range is the interval $[0, 1]$.

In the last example, our description of the function is a bit long-winded. In the future, we might simply say, "Sketch the graph of the function $f(x) = \sqrt{1 - x^2}$." In general, we shall adopt the following convention. *If a function is defined by a formula, then unless otherwise specified, the domain and range are understood to be sets of real numbers, and the domain consists of all real numbers for which the formula is meaningful.* Thus, by "the function $1/x$" we shall mean the function which assigns to each nonzero real number its reciprocal.

In Chapter 6 we shall describe in some detail techniques for sketching the graphs of functions. In the meantime familiarity with the graphs of some of the simple functions will be helpful.

Example 4. Describe the behavior of the function $f(x) = 1/x$, and sketch its graph.

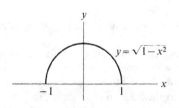

FIG. 3–3

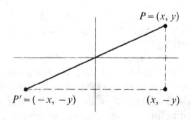

FIG. 3–4

Observe that $f(-x) = -f(x)$. Functions having this property are called *odd* functions. This property is reflected in the fact that the graph of f is *symmetric with respect to the origin*. In other words, if $P = (x, y)$ belongs to the graph of f, then so does the point $P' = (-x, -y)$; the origin is the midpoint of the line segment PP'. (See Fig. 3–4.) Consequently, it suffices to plot the points of G_f which lie in the right half-plane (quadrants I and IV), since we can obtain the remainder of the graph by reflecting these points through the origin (or equivalently, by reflecting these points successively in the x- and y-axes.)

We consider then the behavior of the function $1/x$ for positive values of x. If $0 < x_1 < x_2$, then $f(x_1) > f(x_2)$. Thus as we move to the right along the graph of f, the graph falls. We say that the function is *decreasing* on the interval $(0, +\infty)$. Also, we can make $f(x)$ arbitrarily close to zero by making x sufficiently large. Thus as we move to the right along the graph, we come arbitrarily close to the x-axis. For this reason the x-axis is called an *asymptote* of the graph. Similarly, we can make $f(x)$ arbitrarily large by making x sufficiently close to zero, so that as we move to the left, the graph becomes asymptotic to the y-axis. Thus, without plotting any individual points, we clearly see that the portion of the graph in the right half-plane must look somewhat like the graph given in Fig. 3–5. We get the rest of the graph by reflecting through the origin, as shown in Fig. 3–6.

We observe one further symmetry property:

$$(u, v) \in G_f \Leftrightarrow v = 1/u \Leftrightarrow u = 1/v \Leftrightarrow (v, u) \in G_f.$$

Hence G_f is symmetric about the line $y = x$.

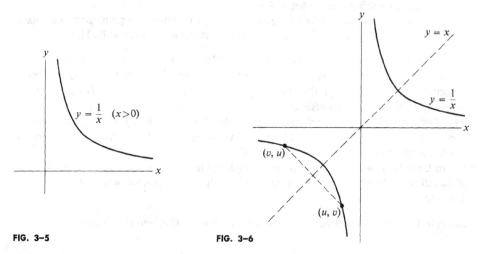

FIG. 3–5   FIG. 3–6

Example 5. Sketch the graph of the function $f(x) = \sin x$, and discuss its symmetry properties.

Once again we have an odd function. First we plot points on the graph for values of x in the interval $[0, \pi/2]$. As will customarily be done when working with the trigonometric functions, *all angles are measured in radians*. As x varies from 0 to $\pi/2$, $\sin x$ increases from $\sin 0 = 0$ to $\sin \pi/2 = \sin 90° = 1$. The graph of f in the interval is as shown in Fig. 3–7. Since the function is odd, we can obtain the graph of $\sin x$ for x in the interval $[-\pi/2, 0]$

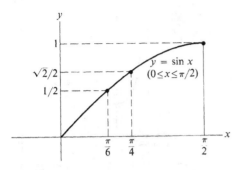

FIG. 3-7

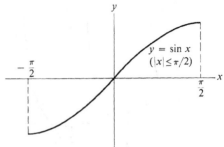

FIG. 3-8

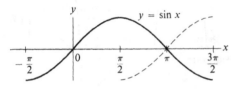

FIG. 3-9

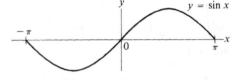

FIG. 3-10

by reflecting through the origin (see Fig. 3–8). Because of the identity

$$\sin (x + \pi) = -\sin x,$$

we can obtain the graph for x in the interval $[\pi/2, 3\pi/2]$ by shifting the graph in Fig. 3–8 π units to the right and reflecting it in the x-axis, as shown in Fig. 3–9. Finally, because $f(x + 2\pi) = f(x)$, the rest of the graph can be obtained by shifting the graph in Fig. 3–9 to the right or left by integral multiples of 2π (see Fig. 3–10). [In general, we say that a function g with domain $\mathbf{R}$ has *period* T iff $g(x + T) = g(x)$ for all $x \in \mathbf{R}$; if g has a period $T \neq 0$, we say g is a *periodic function*.]

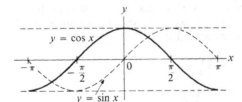

FIG. 3-11

Example 6. Sketch the graphs of the functions $\cos x$, $\tan x$, and $\sec x$.
   Since

$$\cos x = \sin (x + \pi/2),$$

we can obtain the graph of the cosine function by shifting the graph of the sine function $\pi/2$ units to the left, as indicated in Fig. 3–11. Since $\cos (-x) = \cos x$, the cosine function is said to be *even*, and like all even functions, its graph is symmetric about the y-axis. Like the sine function, its period is 2π.

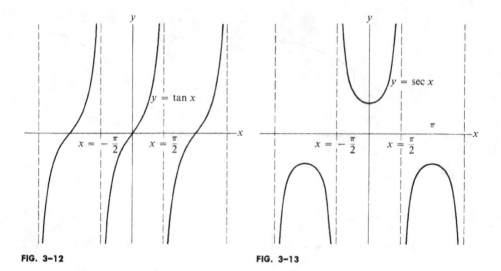

FIG. 3-12   FIG. 3-13

The tangent function has period π, since

$$\tan (x + \pi) = \frac{\sin (x + \pi)}{\cos (x + \pi)} = \frac{-\sin x}{-\cos x} = \tan x.$$

Since the tangent function is also odd, it suffices to graph $y = \tan x$ for $0 \leq x < \pi/2$. [Observe that the domain of $\tan x = (\sin x)/(\cos x)$ consists of all numbers where $\cos x$ is not zero, that is, all numbers except odd integral multiples of $\pi/2$.] For $0 < x < \pi/2$, both $\sin x$ and $\cos x$ are positive; as x increases in this interval, $\sin x$ increases and $\cos x$ decreases. Consequently, $\tan x = (\sin x)/(\cos x)$ is an increasing function on the interval $[0, \pi/2)$. Furthermore, $\tan x$ can be made arbitrarily large for values of x in $[0, \pi/2]$ which are sufficiently close to $\pi/2$. Thus the graph is asymptotic to the line $x = \pi/2$. From this information the graph is readily sketched (see Fig. 3–12).

The function $\sec x = 1/(\cos x)$ has the same domain as the tangent function. It is an even function with period 2π. Its graph, which we can easily sketch by referring to the graph of the cosine function, is shown in Fig. 3–13.

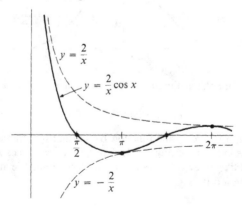

FIG. 3-14

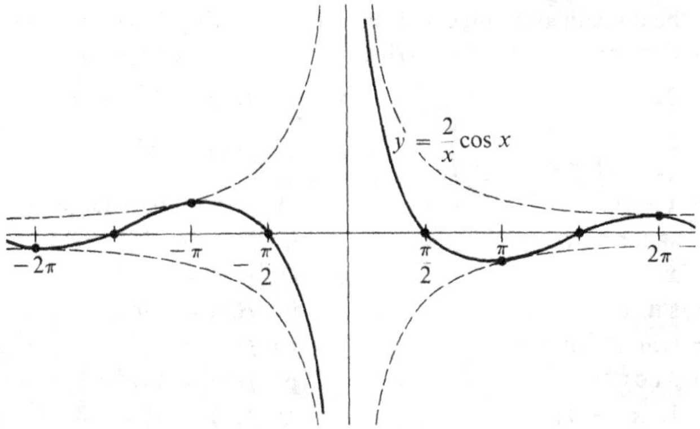

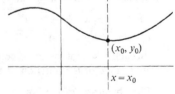

$$y = \frac{2}{x} \cos x$$

FIG. 3-15

Example 7. Sketch the graph of the function $f(x) = (2/x) \cos x$.

Consider first the behavior of $f(x)$ for positive values of x. The factor $2/x$ is always positive, while the factor $\cos x$ oscillates between -1 and $+1$. Consequently, the graph of f will oscillate between the graphs of $2/x$ and $-2/x$, as indicated in Fig. 3-14. Since f is an odd function, we can obtain the rest of the graph by reflecting through the origin, as shown in Fig. 3-15.

We close with a criterion for a set of points in the plane to be the graph of a function.

Proposition 3. *Let f be a function whose domain and range are subsets of the real numbers, and let S be the graph of f. Then S has the property that every line parallel to the y-axis intersects S in one point at most. Conversely, any set S with this property is the graph of a function.*

Proof. Let x_0 be a fixed real number. The line $x = x_0$ intersects S iff there exists a y such that $(x_0, y) \in S$. By the definition of a graph, $(x_0, y) \in S$ iff $x_0 \in \mathfrak{D}_f$ and $y = f(x_0)$. Thus the line $x = x_0$ intersects S iff $x_0 \in \mathfrak{D}_f$, in which case $(x_0, f(x_0))$ is the unique point of intersection (see Fig. 3-16).

FIG. 3-16

Suppose, conversely, that S is a set of points in the plane with the property stated above. Let A be the set of all $x_0 \in \mathbf{R}$ for which the line $x = x_0$ intersects S, and let f be the rule which assigns to each $x_0 \in A$ that unique number y_0 such that $(x_0, y_0) \in S$. Then S is the graph of f.

Exercises

1. Let f be the function with domain $\mathbf{Z}^+$ which assigns to each $n \in \mathbf{Z}^+$ the largest prime number dividing n. Thus $f(12) = 3$, since $12 = 2^2 \cdot 3$ is the factorization of 12 into primes. Find $f(6), f(7), f(16), f(75), f(13)$, and $f(52)$. What is the range of f?

2. Determine the domain and range of each of the following functions. Sketch their graphs. Which functions are even? odd? Which functions are periodic?

a) $f(x) = 2x - 3$

b) $f(x) = 4 - \sqrt{9 - x^2}$

c) $f(x) = \begin{cases} x & \text{if } x < 1, \\ 2x - 1 & \text{if } x \geq 1 \end{cases}$

d) $f(x) = |x|$

e) $f(x) = x - |x|$

f) $f(x) = \max(x, x^2)$

g) $f(x) = \operatorname{sgn} x = x/|x|$

h) $f(x) = x^2 - 2x - 1$

i) $f(x) = x^3$

j) $f(x) = \sqrt{x}$

k) $f(x) = \sin |x|$

l) $f(x) = |\sin x|$

m) $f(x) = (\sin x)/|\sin x|$

n) $f(x) = (\sqrt{x})^2$

o) $f(x) = 3 \cos 2x$

p) $f(x) = 1/x^2$

q) $f(x) = 1/(x^2 - 1)$

r) $f(x) = x|x|$

s) $f(x) = \sqrt{\cos x}$

t) $f(x) = \csc x$

u) $f(x) = x \cos x$

3. Determine the domains of the following functions.

a) $f(x) = \dfrac{x}{x^2 - 1}$

b) $f(x) = \sqrt{x(x - 1)(x - 2)}$

c) $f(x) = \sqrt{\dfrac{x}{x^2 - 1}}$

d) $f(x) = \sqrt{\sin x}$

e) $f(x) = \dfrac{1}{\sqrt{|x| - x}}$

f) $f(x) = \tan\left(\dfrac{\pi}{2} \cos x\right)$

g) $f(x) = \sqrt{x - 1} + \dfrac{1}{\sqrt{2 - x}}$

4. Which of the following subsets of $\mathbf{R}^2$ are graphs of functions?

a) $\{(x, y) \mid x^2 + y^2 = 4\}$

b) $\{(x, y) \mid x^2 + y^2 = 4 \text{ and } y \geq 2\}$

c) $\{(x, y) \mid y = \sqrt{4 - x^2}\}$

d) $\{(x, y) \mid |x| + |y| = 1\}$

In each case plot the set of points. If the set is the graph of a function, find the domain and range of the function.

5. Plot the graph of the function f such that

a) $f(x) = \begin{cases} x & \text{if } 0 \leq x \leq 1, \\ 2 - x & \text{if } 1 < x < 2; \end{cases}$

b) $f(x)$ is odd;   c) f has period 4.

3-2 Operations with functions

Our definition of a function is deficient in at least one respect. In certain cases we shall regard different rules as defining the same function. For instance, if f and g are defined by the formulas

$$f(x) = x^2 - \sin^2 x - \cos^2 x \quad \text{and} \quad g(x) = (x + 1)(x - 1),$$

then we shall regard f and g as the same function.

Definition 1. We say that two functions f and g are equal iff $\mathfrak{D}_f = \mathfrak{D}_g$ and for each $x \in \mathfrak{D}_f = \mathfrak{D}_g$, we have $f(x) = g(x)$.

Equivalently, $f = g$ iff the graphs of f and g are the same. A weaker condition is that the graph of f be a subset of the graph of g.

Definition 2. If f and g are functions, we say that f is a **restriction** of g or that g is an **extension** of f iff $\mathfrak{D}_f \subset \mathfrak{D}_g$ and for each $x \in \mathfrak{D}_f$, we have $f(x) = g(x)$.

If, for example, $f(x) = (\sqrt{x})^2$ and $g(x) = x$, then f is a restriction of g.

We now consider ways of forming new functions from old ones.

First we have the unary operation of restriction. If $f: A \to B$, and if $S \subset A$, then by the *restriction of f to S* we mean the function $g: S \to B$ such that $g(x) = f(x)$ for all $x \in S$. The function g (which is clearly a restriction of f) is denoted by $f \mid S$. This operation may seem to be rather unimportant. However, in dealing with a function such as $1/x$, it is often natural to treat separately its restrictions to the intervals $(-\infty, 0)$ and $(0, +\infty)$. [In more general terms, if f is any function and S is any set, then we can define $f \mid S$ to be the function with domain $\mathfrak{D}_f \cap S$ such that for all x in the latter set, $(f \mid S)(x) = f(x)$. This more general definition forces us to acknowledge the existence of the *null function* whose domain is $\varnothing$.]

We turn next to binary operations. The most important of these is *composition*. Suppose that we have three physical quantities x, y, and z which are related by the equations $y = g(x)$, $z = f(y)$, where f and g are functions. Then we may regard z as a function of x; specifically, $z = f(g(x))$. This function is called the composition product of f and g.

Definition 3. If we have two functions

$$g: A \to B \qquad \text{and} \qquad f: B \to C,$$

then the **composition product** of f and g is the function

$$f \circ g: A \to C$$

such that

$$(f \circ g)(x) = f(g(x))$$

for each $x \in A$. (See Fig. 3–17.)

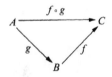

FIG. 3–17

[More generally, if f and g are any functions, then $f \circ g$ is the function with domain

$$\mathfrak{D}_{f \circ g} = \{x \mid x \in \mathfrak{D}_g \text{ and } g(x) \in \mathfrak{D}_f\}$$

such that $(f \circ g)(x) = f(g(x))$ for each x in the latter set.]

Example 1. Suppose $f(x) = x^2$ and $g(x) = \sin x$. Then

$$(f \circ g)(x) = f(g(x)) = f(\sin x) = (\sin x)^2,$$

whereas

$$(g \circ f)(x) = g(f(x)) = g(x^2) = \sin(x^2).$$

Note that $f \circ g \neq g \circ f$.

While composition is not a commutative operation, it is associative.

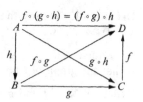

$$f \circ (g \circ h) = (f \circ g) \circ h$$

Proposition 4. *If*

$$h: A \to B, \qquad g: B \to C, \qquad f: C \to D,$$

then

$$f \circ (g \circ h) = (f \circ g) \circ h.$$

FIG. 3–18

Proof. The domains of $f \circ (g \circ h)$ and $(f \circ g) \circ h$ are both A. Furthermore, for each $x \in A$,

$$[f \circ (g \circ h)](x) = f((g \circ h)(x)) = f(g(h(x))) = (f \circ g)(h(x)) = [(f \circ g) \circ h](x).$$

Thus $f \circ (g \circ h) = (f \circ g) \circ h$ (see Fig. 3–18).

In terms of the machine analogy, we form $f \circ g$ by connecting the output of g to the input of f, as illustrated in Fig. 3–19.

Real-valued functions can also be combined by use of the arithmetic operations.

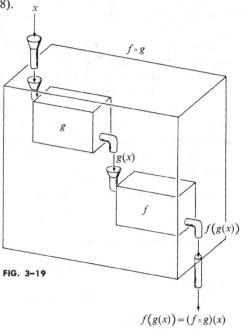

FIG. 3–19

Definition 5. *If*

$$f: A \to \mathbf{R} \qquad \text{and} \qquad g: A \to \mathbf{R},$$

then $f + g, f - g,$ *and* fg *are functions mapping* A *into* $\mathbf{R}$ *and are defined as follows: for each* $x \in A$,

$$(f + g)(x) = f(x) + g(x),$$
$$(f - g)(x) = f(x) - g(x),$$
$$(fg)(x) = f(x)g(x).$$

[In more general terms, if f and g are functions whose ranges are subsets of $\mathbf{R}$, then $f + g$ is the function with domain $\mathfrak{D}_f \cap \mathfrak{D}_g$ such that $(f + g)(x) = f(x) + g(x)$ for all x in $\mathfrak{D}_f \cap \mathfrak{D}_g$. We can define $f - g$ and fg in a similar way.]

Example 2. If $f(x) = x^2$ and $g(x) = \sin x$, then

$$(f + g)(x) = x^2 + \sin x \quad \text{and} \quad (fg)(x) = x^2 \sin x$$

for all $x \in \mathbf{R}$.

Division of functions is complicated somewhat by that ubiquitous spectre: possible division by zero.

Definition 6. Suppose that

$$f: A \to \mathbf{R} \quad \text{and} \quad g: A \to \mathbf{R}.$$

Let $S_g = \{x \in A \mid g(x) \neq 0\}$. Then f/g is the function with domain $A \cap S_g$ such that

$$\left(\frac{f}{g}\right)(x) = \frac{f(x)}{g(x)}$$

for all $x \in A \cap S_g$.

If $f, g, h: A \to \mathbf{R}$, then we have the identities

$$f + (g + h) = (f + g) + h, \quad f + g = g + h, \quad f(gh) = (fg)h,$$

etc. Each follows easily from the corresponding identity for real numbers.

Exercises

1. In our definition of a function the word "rule" might be thought ambiguous. We can avoid this difficulty by giving a set-theoretic definition of a function. Essentially this involves identifying a function (in the sense of the previous section) with its graph. We define a function to be a set f of ordered pairs such that if $(x, y) \in f$ and $(x, y') \in f$, then $y = y'$. The domain of f is then defined to be the set

$$\mathcal{D}_f = \{x \mid (\exists y)((x, y) \in f)\},$$

and the range of f is defined to be

$$\mathcal{R}_f = \{y \mid (\exists x)((x, y) \in f)\}.$$

If $(x, y) \in f$, we write $y = f(x)$.
a) Observe that if f is a function, then $f \subset \mathcal{D}_f \times \mathcal{R}_f$.
b) Let f and g be the sets

$$f = \{(0, 1), (2, 1), (3, -1)\} \quad \text{and} \quad g = \{(0, 1), (0, 2), (1, 3)\}.$$

Show that f is a function but that g is not. Find the domain and range of f. What is $f(3)$?

2. In each of the following cases determine whether f is an extension of g and vice versa.

a) $f(x) = \dfrac{x^2 - 1}{x^3 - 1}$, $g(x) = \dfrac{x + 1}{x^2 + x + 1}$ \qquad b) $f(x) = |x|$, $g(x) = \sqrt{x^2}$

c) $f(x) = \sqrt{\dfrac{x}{x + 1}}$, $g(x) = \dfrac{\sqrt{x}}{\sqrt{x + 1}}$

d) $f(x) = \sqrt{1 - x} + \sqrt{|x|}$, $g(x) = \sqrt{|1 - x|} + \sqrt{x}$

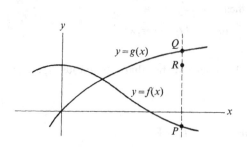

FIG. 3–20   FIG. 3–21

3. Let $f(x) = x^2$, $g(x) = 1/x$, $h(x) = \sin x$.

a) Compute $(f + g)(-2)$, $(fg)(\pi/3)$, $(h/g)(\pi/2)$, $(f \circ h)(\pi/6)$, $(g \circ h)(\pi/3)$.

b) Find the domains of the functions $f + g$, $g \circ h$, $h \circ g$, $g \circ g$, $g/(fh)$.

4. Suppose that we are given the graphs of two functions f and g and we wish to determine the graph of $f + g$. Suppose that $x_0 \in \mathfrak{D}_f \cap \mathfrak{D}_g$, and let P, Q, R be the points where the line $x = x_0$ meets the graphs of f, g, and $f + g$, as shown in Fig. 3–20. Show that the y-coordinate of R can be obtained by doubling the y-coordinate of the midpoint of P and Q.

a) Sketch fairly carefully the graphs of $y = 3 \sin 2x$ and $y = \cos 3x$. Then use the above observation to sketch the graph of $y = 3 \sin 2x + \cos 3x$.

b) By the same technique, sketch the graph of $y = x + (2/x)$.

5. Suppose that we are given the graphs of the functions f and g. Study Fig. 3–21 and then state explicitly a rule for constructing the graph of $f \circ g$. Then apply this procedure to sketch the graph of $y = \cos [(\pi/4) \sin x]$.

6. a) Express each of the following functions as the sum of an even function and an odd function:

$$3 - 2x + x^4 - 5x^7, \qquad (x + 2) \sin x - x^3 \sin 5x, \qquad \sin (x + \pi/3).$$

b) Suppose $f: \mathbf{R} \to \mathbf{R}$. Observe that $f = E + O$, where

$$E(x) = \tfrac{1}{2}[f(x) + f(-x)], \qquad O(x) = \tfrac{1}{2}[f(x) - f(-x)].$$

Show that E is even and O is odd. Prove that this way of expressing f as the sum of an even function and an odd function is unique.

c) Show that the sum of two even functions is even and the sum of two odd functions is odd.

d) What can you say regarding the product of two even functions? of two odd functions? of one even and one odd function?

e) Answer the same questions regarding composition products.

7. Suppose that $h: A \to B$ and $f, g: B \to \mathbf{R}$. Show that

$$(f + g) \circ h = f \circ h + g \circ h \qquad \text{and} \qquad (fg) \circ h = (f \circ h)(g \circ h).$$

Find an example, however, in which $f \circ (g + h) \neq f \circ g + f \circ h$.

8. Let S be a set, and let $\Re(S)$ be the set of all real-valued functions having domain S. Show that with the operations of addition and multiplication of functions defined in this section, $\Re(S)$ satisfies all the axioms for a field except M3. Such an algebraic structure is called a *commutative ring with identity*.

9. Let a, b, c, and d be fixed real numbers, and let f and g be polynomials defined by the equations

$$f(x) = ax + b, \qquad g(x) = cx + d.$$

Find necessary and sufficient conditions on the coefficients a, b, c, and d in order that $f \circ g = g \circ f$.

10. Let S be the real line with the numbers 0 and 1 deleted. Let f_0, f_1, f_2, f_3, f_4, and f_5 be functions with domain S defined by

$$f_0(x) = x, \qquad\qquad f_3(x) = 1/(1 - x),$$
$$f_1(x) = 1/x, \qquad\qquad f_4(x) = (x - 1)/x,$$
$$f_2(x) = 1 - x, \qquad\quad f_5(x) = x/(x - 1).$$

Let $G = \{f_0, f_1, f_2, f_3, f_4, f_5\}$. Make a table giving all possible composition products of elements in G. Using this table, show that G has the following properties:

G0. *G is closed under the formation of composition products.*

G1. *The operation is associative.*

G2. *G has an element e such that*

$$e \circ f = f \circ e = f$$

for all $f \in G$.

G3. *For each f in G, there exists an element in G denoted by f^{-1} such that*

$$f \circ f^{-1} = f^{-1} \circ f = e.$$

A set G with an operation $\circ$ satisfying these conditions is called a *group*, and the operation $\circ$ is called a *group operation*.

    Solve the equation $f_3 \circ f \circ f_4 = f_5$ for f.

11. The *identity function* $I \colon \mathbf{R} \to \mathbf{R}$ is defined by $I(x) = x$. Show that $g \circ g$ is a restriction of the identity function iff the graph of g is symmetric about the line $y = x$.

3–3 The limit concept for sequences

There is a natural relationship between sequences (as most people think of them intuitively) and functions whose domains are the set $\mathbf{Z}^+$. We can, for instance, associate with the sequence

$$1, \tfrac{1}{2}, \tfrac{1}{3}, \tfrac{1}{4}, \ldots$$

the function which assigns to each positive integer n the nth term of the sequence (in this case $1/n$). Conversely, if we are given a function f whose domain is $\mathbf{Z}^+$, then we can form the sequence $f(1), f(2), f(3), \ldots$ For example, if $f(n) = 10^n/n!$, we get the sequence

$$10, \ 10^2/2!, \ 10^3/3!, \ 10^4/4!, \ldots$$

In light of these observations the following definition should not seem unnatural.

Definition 1. A **sequence** is a function α whose domain is the set of positive integers. Given such a function, we use α_n rather than $\alpha(n)$ to denote the value which the function assigns to the integer n. Then α_n is called the **nth term** of the sequence.

We frequently indicate a sequence by simply writing down the first few terms, when the rule for finding the others is obvious. For example,

$$\tfrac{1}{2}, \tfrac{3}{4}, \tfrac{7}{8}, \tfrac{15}{16}, \ldots$$

might be used to indicate the sequence whose nth term is $1 - 2^{-n}$. This method is not, of course, foolproof. If we are given

$$1, 2, 3, \ldots,$$

for instance, it is reasonable to interpret this to be the sequence whose nth term is n. Nevertheless, there are other interesting sequences which begin this way. For example, we have the Fibonacci sequence

$$1, 2, 3, 5, 8, 13, 21, \ldots,$$

and also the sequence which begins

$$1, 2, 3, 4, 1, 2, 3, 4, 5, 1, 2, 3, 4, 5, 2, 3, 4, 5,$$
$$6, 2, 3, 4, 5, 6, 1, 2, 3, 4, 5, 2, 3, 4, 5, 6, 2, 3,$$
$$4, 5, 6, 3, 4, 5, 6, 7, 3, 4, 5, 6, 7, 1, 2, \ldots$$

In case the reader hasn't figured out the pattern of this last sequence, we suggest that he look back at Example 2, Section 3–1.

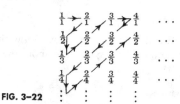

FIG. 3–22

For each of the sequences above, the range has been a rather sparse subset of the real line. Although it may seem unlikely at first, it is possible to construct a sequence whose terms include all the rational numbers. We begin by constructing a sequence whose range is the set of positive rationals. Since every positive rational number is of the form m/n, where m and n are positive integers, every such number is contained in the doubly infinite array of numbers shown in Fig. 3–22. We now pick a path which threads its way through each point of the array. Following this path, we get the following sequence of rational numbers:

$$1, 2, \tfrac{1}{2}, \tfrac{1}{3}, 1, 3, 4, \tfrac{3}{2}, \tfrac{2}{3}, \tfrac{1}{4}, \tfrac{1}{5}, \tfrac{1}{2}, \ldots$$

Each positive rational number appears in this sequence an infinite number of times. To get a sequence which contains each positive rational exactly once, we can simply delete from the above sequence each term which has previously appeared. This gives

$$1, 2, \tfrac{1}{2}, \tfrac{1}{3}, 3, 4, \tfrac{3}{2}, \tfrac{2}{3}, \tfrac{1}{4}, \tfrac{1}{5}, 5, 6, \ldots$$

It is now a simple matter to construct a sequence in which each rational number appears exactly once. The following does the trick:

$$0, 1, -1, 2, -2, \tfrac{1}{2}, -\tfrac{1}{2}, \tfrac{1}{3}, -\tfrac{1}{3}, 3, -3, \ldots$$

We shall not pretend to have equal interest in all sequences. Of special interest are those sequences which *converge*. The sequence

$$\tfrac{1}{2}, \tfrac{3}{4}, \tfrac{7}{8}, \tfrac{15}{16}, \ldots$$

is an example of a convergent sequence. We say in this case that the sequence converges to the number 1, or that it has the number 1 as a *limit*. Roughly speaking, we mean by this that as n gets larger and larger, the nth term of the sequence $\alpha_n = 1 - 2^{-n}$ gets closer and closer to 1; and by choosing n large enough, we can make α_n as close to 1 as we please. If $n \geq 20$, for instance, α_n differs from 1 only after the sixth decimal place, and if $n \geq 100$, then the difference between α_n and 1 occurs after the thirteenth decimal place. In general, we shall make the following provisional definition.

Provisional Definition I. *Explicandum.* A sequence α is said to have the **limit** A iff we can make the nth term of the sequence α_n as close to A as we please by making n sufficiently large.

This definition leaves a good deal to be desired. For instance, it leaves open the question of how close "as close as we please" is. As a first step in "hardening" this definition, let us observe that the expression $|\alpha_n - A|$ is a quantitative measure of how close α_n is to the number A. If we think of α_n as an approximation to A, then $|\alpha_n - A|$ is what we have previously called the error involved in making the approximation. As a second step toward explicating the notion of a limit, we offer the following provisional definition.

Provisional Definition II. Let the terms of a sequence α be regarded as approximations to a number A. We say that A is the limit of α if we can make the approximation error $|\alpha_n - A|$ less than any prescribed tolerance, $\epsilon > 0$, by making n sufficiently large.

If we strip this definition of the jargon of numerical analysis, it says this: given any number $\epsilon > 0$, the inequality $|\alpha_n - A| < \epsilon$ will hold if n is sufficiently large. Hence it now suffices to make precise what is meant by the phrase "if n is sufficiently large." Consider the inequality

$$n^2 > 10,000.$$

If n is sufficiently large, the inequality is valid. More accurately, it is valid whenever $n > 100$. In general, to say that a statement $P(n)$ is true for n sufficiently large means that there exists an integer N (100 in the case above) such that $P(n)$ is true whenever $n > N$.

Definition 2. *Explicatum.* Let α be a sequence of real numbers. We say that A is the **limit** of α iff for each $\epsilon > 0$, there exists a positive integer N such that the inequality

$$|\alpha_n - A| < \epsilon$$

holds for all integers $n > N$. If A is the limit of α, we indicate this by any of the following:

$$\lim \alpha = A, \qquad \lim_{n \to \infty} \alpha_n = A, \qquad \text{or} \qquad \alpha_n \to A.$$

We also say that the sequence **converges** to A.

Using quantifiers, we can write this definition more compactly as follows:

$$\alpha_n \to A \quad \text{iff} \quad (\forall \epsilon > 0)(\exists N \in \mathbf{Z}^+)(\forall n > N)(|\alpha_n - A| < \epsilon).$$

Example 1. Suppose that A is a fixed real number, and let α be that sequence each of whose terms is A. Thus $\alpha_n = A$ for all $n \in \mathbf{Z}^+$. We say that α is a *constant sequence* in this case. Since $|\alpha_n - A| = 0$ for each $n \in \mathbf{Z}^+$, it is clear from the definition of a limit that

$$\lim_{n \to \infty} \alpha_n = A.$$

To obtain less trivial examples of limits, we need to make an additional assumption about the field of real numbers, called the *Archimedean ordering property* of **R**. It may be stated as follows:

$$(\forall x > 0)(\exists n \in \mathbf{Z}^+)(n > x).$$

In other words, if we are given any positive real number, we can find an integer which is greater than the given number. In Chapter 5 we shall deduce this property of the real numbers from the completeness axiom. In the meantime we shall simply assume it.

Example 2. We shall prove that

$$\lim_{n \to \infty} (1 - 2^{-n}) = 1.$$

Given a number $\epsilon > 0$, we must, according to the definition, produce an integer N such that

$$|(1 - 2^{-n}) - 1| = 2^{-n} < \epsilon$$

whenever $n > N$. The inequality above is equivalent to $2^n > 1/\epsilon$. Note that $2^n > n$ for every positive integer n. (This can be readily proved by induction.) However, the inequality $n > 1/\epsilon$ certainly holds if n is sufficiently large, for by the Archimedean ordering property, there exists an $N \in \mathbf{Z}^+$ such that $N > 1/\epsilon$. Thus $n > N \Rightarrow n > 1/\epsilon$.

Having gone through the preceding analysis, we can now give a formal proof.

Let $\epsilon > 0$ be given. By the Archimedean ordering property, we can find an $N \in \mathbf{Z}^+$ such that $N > 1/\epsilon$. If $n > N$, then

$$2^n > n > 1/\epsilon$$

and, consequently,

$$|(1 - 2^{-n}) - 1| = 2^{-n} < \epsilon.$$

Hence, by the definition of a limit, we see that

$$(1 - 2^{-n}) \to 1.$$

Example 3. Consider the sequence

$$\tfrac{1}{3}, \tfrac{5}{6}, \tfrac{5}{9}, \tfrac{9}{12}, \tfrac{9}{15}, \tfrac{13}{21}, \tfrac{13}{24}, \cdots$$

in which the nth term is

$$\alpha_n = \frac{2n + (-1)^n}{3n}.$$

We shall prove that $\lim_{n \to \infty} \alpha_n = \tfrac{2}{3}$. In constructing a proof, the crucial inequality which we must consider is $|\alpha_n - \tfrac{2}{3}| < \epsilon$. But

$$\left|\alpha_n - \frac{2}{3}\right| = \left|\frac{2n + (-1)^n}{3n} - \frac{2}{3}\right| = \left|\frac{2}{3} + \frac{(-1)^n}{3n} - \frac{2}{3}\right| = \left|\frac{(-1)^n}{3n}\right| = \frac{1}{3n},$$

and the inequality $1/3n < \epsilon$ is equivalent to $n > 3/\epsilon$.

We are now prepared to give a formal proof. Let $\epsilon > 0$ be given. Let N be a positive number greater than $3/\epsilon$. (By the Archimedean ordering property, such an N exists.) If $n > N$, then

$$|\alpha_n - \tfrac{2}{3}| = 1/3n < \epsilon.$$

This completes the proof.

The sequences which we have considered so far have been extremely simple ones. In all cases it has been intuitively clear that the sequences converge, and finding these limits has been no trouble at all. To handle more complicated sequences with any sort of ease requires the development of some general theory concerning limits. (Constant recourse to the definition of a limit is an eventuality too horrible even to contemplate.) We shall state a couple of the more important theorems on limits and illustrate how they may be applied. The theorems will then be proved in the section which follows.

Our first theorem states in a precise way something which many people would regard as self-evident: if x^* is nearly equal to x, and if y^* is nearly equal to y, then $x^* + y^*$ and x^*y^* are nearly equal to $x + y$ and xy, respectively. Intuitively, the expressions $x + y$ and xy vary continuously with x and y.

Theorem 3. Continuity of the Arithmetic Operations. *Suppose that the sequences α and β have the limits A and B, respectively. Then the sequences $\alpha + \beta$, $\alpha - \beta$, and $\alpha\beta$ also have limits. Furthermore,*

$$\lim (\alpha + \beta) = A + B, \qquad \lim (\alpha - \beta) = A - B, \qquad \lim \alpha\beta = AB.$$

If $B \neq 0$, the sequence α/β also has a limit, namely, A/B.

Although this theorem gets used to death in analysis, it has no conventional name tag. Usually its content is alluded to by such relatively imprecise phrases as "the limit of a sum is equal to the sum of the limits," "the limit of a product is equal to the product of the limits," etc.

The Pinching Theorem. *Suppose that* α, β, *and* γ *are sequences of real numbers such that for all sufficiently large integers n,*

$$\alpha_n \leq \beta_n \leq \gamma_n.$$

Suppose also that

$$\lim_{n \to \infty} \alpha_n = \lim_{n \to \infty} \gamma_n = L.$$

Then the sequence β *converges and* $\lim_{n \to \infty} \beta_n = L$.

The hypotheses of the theorem are suggested by the following diagram.

$$\alpha_n \leq \beta_n \leq \gamma_n$$

$$\searrow_L \swarrow$$

The conclusion of the theorem then amounts to saying that a third arrow may be added to the diagram.

$$\alpha_n \leq \beta_n \leq \gamma_n$$

$$\searrow \downarrow \swarrow$$
$$L$$

We shall give numerous examples showing how these theorems may be applied.

Example 4. We will show that if r is any positive rational number, then

$$\lim_{n \to \infty} n^{-r} = \lim_{n \to \infty} 1/n^r = 0.$$

First we consider the case in which $r = 1/q$, where q is a positive integer. Given $\epsilon > 0$, there exists (by the Archimedean ordering property) an $N \in \mathbf{Z}^+$ such that $N > (1/\epsilon)^q$. Then

$$n > N \Rightarrow n > \left(\frac{1}{\epsilon}\right)^q \Rightarrow \sqrt[q]{n} > \sqrt[q]{\left(\frac{1}{\epsilon}\right)^q} = \frac{1}{\epsilon} \Rightarrow \left|\frac{1}{n^{1/q}}\right| = \frac{1}{\sqrt[q]{n}} < \epsilon.$$

Thus

$$\lim_{n \to \infty} \frac{1}{n^{1/q}} = 0.$$

We shall next consider the general case in which $r = p/q$, where p and q are positive integers. We proceed by induction on p. We have just discussed the case in which $p = 1$. Suppose, then, that

$$\lim_{n \to \infty} \frac{1}{n^{k/q}} = 0.$$

It follows that

$$\lim_{n \to \infty} \frac{1}{n^{(k+1/q)}} = \lim_{n \to \infty} \frac{1}{n^{k/q}} \frac{1}{n^{1/q}} = 0 \cdot 0 = 0,$$

since the limit of a product is the product of the limits. Hence we can terminate this demonstration with an appeal to the principle of induction.

Example 5. It is worthwhile to observe that constants can be moved in and out of limit signs with relative ease. If, for instance, c is a fixed real number and α is a convergent sequence of real numbers, then

$$\lim_{n\to\infty} c\alpha_n = c(\lim_{n\to\infty} \alpha_n) \quad \text{and} \quad \lim_{n\to\infty} (c + \alpha_n) = c + \lim_{n\to\infty} \alpha_n.$$

This quickly follows from the theorem on the continuity of the arithmetic operations. (Set $\beta_n = c$ for all positive integers n.) So, for example,

$$\lim_{n\to\infty} \frac{5}{n^7} = \lim_{n\to\infty} 5 \cdot \frac{1}{n^7} = 5 \cdot 0 = 0 \quad \text{and} \quad \lim_{n\to\infty} \left(\frac{6}{\sqrt[3]{n}} + 11\right) = 6 \cdot 0 + 11 = 11.$$

Example 6. The results in our theorem on the continuity of the arithmetic operations can easily be extended. Suppose, for example, that $\lim \alpha = A$, $\lim \beta = B$, and $\lim \gamma = C$. Then $\lim (\alpha + \beta) = A + B$ and

$$\lim (\alpha + \beta + \gamma) = \lim [(\alpha + \beta) + \gamma] = (A + B) + C = A + B + C.$$

Similarly, $\lim \alpha\beta\gamma = ABC$. Thus, using the previous examples, we may immediately write

$$\lim_{n\to\infty} \left(\frac{4}{n^3} + \frac{5}{\sqrt{n}} + 7\right) = 4 \cdot 0 + 5 \cdot 0 + 7 = 7.$$

Example 7. Find $\lim_{n\to\infty} (2n^2 + 1)/(3n^2 - 5n + 2)$.
     We get

$$\lim_{n\to\infty} \frac{2n^2 + 1}{3n^2 - 5n + 2} = \lim_{n\to\infty} \frac{2 + (1/n^2)}{3 - (5/n) + (2/n^2)} = \frac{2 + 0}{3 - 5 \cdot 0 + 2 \cdot 0} = \frac{2}{3}.$$

Example 8. Find $\lim_{n\to\infty} (\sqrt{n+1} - \sqrt{n})$.
     We use an algebraic device:

$$\sqrt{n+1} - \sqrt{n} = \frac{(\sqrt{n+1} - \sqrt{n})(\sqrt{n+1} + \sqrt{n})}{\sqrt{n+1} + \sqrt{n}} = \frac{1}{\sqrt{n+1} + \sqrt{n}} < \frac{1}{2\sqrt{n}}.$$

Thus for all $n \in \mathbf{Z}^+$,

$$0 < \sqrt{n+1} - \sqrt{n} < \frac{1}{2\sqrt{n}}.$$

Since $\lim_{n\to\infty} (1/2\sqrt{n}) = \frac{1}{2} \cdot 0 = 0$, it follows from the pinching theorem that

$$\lim_{n\to\infty} (\sqrt{n+1} - \sqrt{n}) = 0.$$

(In applying the pinching theorem, we can let $\alpha_n = 0$ for all $n \in \mathbf{Z}^+$.)

Example 9. Let $|r| < 1$. We will show that

$$\lim_{n\to\infty} r^n = 0.$$

Case I. $0 < r < 1$. Let $h = 1/r - 1$. Since $0 < r < 1$, it follows that $1/r > 1$ and $h > 0$. By the binomial theorem,

$$(1 + h)^n = 1 + nh + \cdots \geq 1 + nh.$$

Hence

$$0 < r^n = \frac{1}{(1 + h)^n} \leq \frac{1}{1 + nh}.$$

Now

$$\lim_{n \to \infty} \frac{1}{1 + nh} = \lim_{n \to \infty} \frac{1/n}{(1/n) + h} = \frac{0}{0 + h} = 0.$$

Hence by the pinching theorem,

$$\lim_{n \to \infty} r^n = 0.$$

Case II. $r = 0$. In this case our sequence is the constant sequence in which each term is zero. Hence its limit must be zero.

Case III. $-1 < r \leq 0$. We have for all $n \in \mathbf{Z}^+$

$$-|r|^n \leq r^n \leq |r|^n.$$

Since $0 \leq |r| = -r < 1$, we get from cases I and II

$$\lim_{n \to \infty} |r|^n = 0.$$

Thus

$$-|r|^n \leq r^n \leq |r|^n$$

Example 10. Find

$$\lim_{n \to \infty} \frac{5^n - 3^n + 1}{5^n + 3^n + (1/n)}.$$

By making generous use of the preceding examples, we are privileged to write

$$\lim_{n \to \infty} \frac{5^n - 3^n + 1}{5^n + 3^n + (1/n)} = \lim_{n \to \infty} \frac{1 - (3/5)^n + (1/5)^n}{1 + (3/5)^n + (1/n)(1/5)^n} = \frac{1 - 0 + 0}{1 + 0 + 0} = 1.$$

Example 11. Find $\lim_{n \to \infty} nr^n$, if $|r| < 1$. It is not immediately clear that this limit exists. The factor r^n tends to zero, while the factor n becomes unbounded. In such cases the product is capable of almost any sort of behavior. In this case the factor r^n wins out. In other words, $\lim_{n \to \infty} nr^n = 0$.

As in Example 9, it suffices to consider the case in which $0 < r < 1$. Let

$$h = (1/r) - 1.$$

If $n \geq 2$, it follows that

$$(1 + h)^n = 1 + nh + \frac{n(n-1)}{2} h^2 + \cdots \geq 1 + nh + \frac{n(n-1)}{2} h^2.$$

Then

$$0 < nr^n = \frac{n}{(1+h)^n} \leq \frac{n}{1 + nh + \dfrac{n(n-1)}{2}h^2} \cdot$$

By dividing numerator and denominator by n^2, one can show that the extreme right-hand term converges to zero. (Because of the n^2-term in the denominator, the numerator gets clobbered in the race to infinity.) By the pinching theorem, therefore,

$$\lim_{n \to \infty} nr^n = 0.$$

Example 12. Let x_0 be any positive number, and consider the sequence $x_1, x_2, x_3, \ldots$ defined by the recursion formula

$$x_n = \tfrac{1}{2}(x_{n-1} + 2/x_{n-1}).$$

Does the sequence have a limit?

Unlike the preceding examples, it is not at all clear that the sequence converges. It could conceivably converge for some values of x_0 but not for others. Let us suppose, however, that the sequence has a limit L. Assuming that $L \neq 0$, we can take the limit of each side of the recursion formula above to get the equation

$$L = \tfrac{1}{2}(L + 2/L).$$

Solving the equation, we get $L = \pm\sqrt{2}$. It is clear from the recursion formula that each term of the sequence is positive, and thus L cannot be $-\sqrt{2}$. This shows that *if* the sequence has a *nonzero* limit, then that limit must be $\sqrt{2}$. We shall now prove that the sequence does, in fact, converge to $\sqrt{2}$.

We must show that $x_n - \sqrt{2}$ tends to zero. In this particular instance, it is simpler to show first that

$$e_n = \frac{x_n - \sqrt{2}}{x_n + \sqrt{2}}$$

converges to zero. Observe that

$$\begin{aligned} e_n &= \frac{\tfrac{1}{2}(x_{n-1} + 2/x_{n-1}) - \sqrt{2}}{\tfrac{1}{2}(x_{n-1} + 2/x_{n-1}) + \sqrt{2}} \\ &= \frac{x_{n-1}^2 + 2 - 2\sqrt{2}\, x_n}{x_{n-1}^2 + 2 + 2\sqrt{2}\, x_n} \\ &= \left(\frac{x_{n-1} - \sqrt{2}}{x_{n-1} + \sqrt{2}}\right)^2 = (e_{n-1})^2. \end{aligned}$$

It follows that

$$e_n = (e_{n-1})^2 = (e_{n-2})^4 = (e_{n-3})^8 = \cdots = (e_0)^{2^n}.$$

[To be perfectly correct, one should prove the equation $e_n = (e_0)^{2^n}$ by induction.] It is easily checked that $|e_0| < 1$. Thus

$$0 < e_n = (e_0)^{2^n} \leq |e_0|^n.$$

It follows from Example 9 and the pinching theorem that $e_n \to 0$ as $n \to \infty$. From

$$e_n = \frac{x_n - \sqrt{2}}{x_n + \sqrt{2}},$$

we get

$$x_n = \frac{\sqrt{2} + \sqrt{2}\,e_n}{1 - e_n}.$$

Thus

$$\lim_{n \to \infty} x_n = \frac{\sqrt{2} + \sqrt{2} \cdot 0}{1 - 0} = \sqrt{2}.$$

The sequence $\{x_n\}$ converges rapidly enough to $\sqrt{2}$ to provide an efficient method for computing $\sqrt{2}$ to any desired number of decimal places. If, for instance, we set $x_0 = 2$, we get the sequence

$$x_1 = 1.5, \quad x_2 = 1.41666..., \quad x_3 = 1.41412..., \quad ...$$

More generally, if A and x_0 are any two positive numbers, then the sequence defined by the recursion formula

$$x_n = \tfrac{1}{2}(x_{n-1} + A/x_{n-1})$$

converges quite rapidly to $\sqrt{A}$. This recursion formula is widely used on digital computers for the computation of square roots.

Exercises

1. Calculate each of the following limits and amply justify your calculations in terms of the basic theorems on limits and the examples given in the text.

a) $\lim\limits_{n \to \infty} \dfrac{n^3 - 1}{3n^3 + n - 4}$

b) $\lim\limits_{n \to \infty} \dfrac{n \cos n}{n^2 + 24}$

c) $\lim\limits_{n \to \infty} \dfrac{2^n + 1}{2^n - n}$

d) $\lim\limits_{n \to \infty} (\sqrt[3]{n+1} - \sqrt[3]{n})$

e) $\lim\limits_{n \to \infty} (n!)/(n^n)$ (Exercise 29, Section 1–9, might be helpful.)

f) $\lim\limits_{n \to \infty} \sum\limits_{k=0}^{n} x^k$ $(|x| < 1)$

g) $\lim\limits_{n \to \infty} \sum\limits_{k=1}^{n} \dfrac{k^2}{n^3}$

h) $\lim\limits_{n \to \infty} \sum\limits_{k=1}^{n} \dfrac{1}{k(k+1)}$ (See Exercise 1(c), Section 1–8.)

i) $\lim\limits_{n \to \infty} \prod\limits_{k=1}^{n} (1 + x^{2^{k-1}})$ (See Exercise 6, Section 1–9.)

j) $\lim\limits_{n \to \infty} \dfrac{\sqrt[3]{n}\,\sin n!}{n+2}$

k) $\lim\limits_{n \to \infty} \left(\dfrac{n^3}{2n^2 - 1} - \dfrac{n^2}{2n + 1} \right)$

*l) $\lim\limits_{n \to \infty} \sqrt[n]{a^n + a^{-n}}$ $(a > 0)$

2. Using only the definition of a limit and the basic properties of the real numbers (including the Archimedean ordering property), prove the following.

a) $\lim\limits_{n\to\infty} \prod\limits_{k=1}^{n} (1000 - k) = 0$ b) $\lim\limits_{n\to\infty} \dfrac{n + \sin (n\pi/2)}{2n + 1} = \dfrac{1}{2}$

c) $\lim\limits_{n\to\infty} (\sqrt{n + 1} - \sqrt{n})\sqrt{n + 3} = \dfrac{1}{2}$

3. a) Show by two different methods that if $|r| < 1$, then

$$\lim_{n\to\infty} n^2 r^n = 0.$$

[First mimic the argument given in Example 11, but use an additional term in the expansion of $(1 + h)^n$. Next, observe that if $r > 0$, then $n^2 r^n = (nR^n)^2$, where $R = \sqrt{r}$. Then use Example 11, etc.]

b) Generalize Example 11 and part (a).

4. If $a > 0$, prove that $\lim_{n\to\infty} \sqrt[n]{a} = 1$. [Prove this first for the case in which $a > 1$. Let $h_n = \sqrt[n]{a} - 1$; show that $0 < h_n < (a - 1)/n$, and pinch. For the case in which $a < 1$, consider $1/a$ and apply the first case.]

5. Show that $\lim_{n\to\infty} \sqrt[n]{n} = 1$. (Set $b_n = \sqrt[2n]{n}$ and observe that $b_n > 1$ if $n > 1$. Set $b_n = 1 + h_n$ and deduce that $\sqrt{n} \geq 1 + nh_n$. Show by a pinching argument that $h_n \to 0$, etc.)

6. Prove that $\lim_{n\to\infty} \sqrt[n]{n^2} = 1$.

7. Find the limit of the sequence $\sqrt{2}, \sqrt{2\sqrt{2}}, \sqrt{2\sqrt{2\sqrt{2}}}, \ldots$ Carefully justify your answer.

8. Prove that the sequence $\sqrt{2}, \sqrt{2 + \sqrt{2}}, \sqrt{2 + \sqrt{2 + \sqrt{2}}}, \ldots$ converges to 2. [Obtain a recursion formula for the nth term a_n of the sequence. Set $h_n = 2 - a_n$ and prove by induction that

$$0 < h_n \leq \frac{2 - \sqrt{2}}{(2 + \sqrt{2})^{n-1}}.$$

Pinch.]

9. Show that $L = \sqrt{2} - 1$ satisfies the equation

$$L = \frac{1}{2 + L}$$

and hence that

$$L = \frac{1}{2 + L} = \frac{1}{2 + \dfrac{1}{2 + L}} = \frac{1}{2 + \dfrac{1}{2 + \dfrac{1}{2 + L}}} = \cdots$$

This suggests that the sequence

$$a_1 = \tfrac{1}{2}, \quad a_2 = \frac{1}{2 + a_1} = \frac{1}{2 + \tfrac{1}{2}}, \quad a_3 = \frac{1}{2 + a_2} = \frac{1}{2 + \dfrac{1}{2 + \dfrac{1}{2 + \tfrac{1}{2}}}}, \ldots$$

defined inductively by the equations

$$a_1 = \tfrac{1}{2}, \qquad a_{n+1} = \frac{1}{2 + a_n}$$

might conceivably converge to L. Prove that this is so. [Observe that

$$a_n - L = \frac{1}{2 + a_{n-1}} - \frac{1}{2 + L} = \frac{L - a_{n-1}}{(2 + a_{n-1})(2 + L)}.$$

Deduce from this that

$$|a_n - L| \le \frac{|a_1 - L|}{4^{n-1}}.$$

Pinch.]

In this case we say that the *continued fraction*

$$\cfrac{1}{2 + \cfrac{1}{2 + \cdots}}$$

converges to $\sqrt{2} - 1$.

10. What can you say regarding the convergence of the continued fraction

$$\cfrac{1}{3 + \cfrac{1}{3 + \cfrac{1}{3 + \cdots}}} \; ?$$

11. If $\lim_{n \to \infty} a_n = A \ne 0$ and $\lim_{n \to \infty} b_n = 0$, prove that $\lim_{n \to \infty} (a_n/b_n)$ does not exist. (Suppose to the contrary that the limit exists, and derive a contradiction using the theorems on limits.)

12. Consider the sequence defined by the recursion formulas

$$a_1 = 3, \qquad a_{n+1} = \tfrac{1}{2}(a_n - 1/a_n).$$

Show that the sequence does not converge. (The hint for the preceding problem applies. Also look over the first part of Example 12.)

13. a) Show that $\lim_{n \to \infty} (10^n/n!) = 0$.
    b) Generalize (a).

14. Let $\nu(n)$ denote the number of distinct prime factors of n. Prove that $\lim_{n \to \infty} [\nu(n)/n] = 0$. [*Hint:* Show that if $2^n < k \le 2^{n+1}$, then $\nu(k)/k < (n + 1)/2^n$. Observe that the latter converges to zero as $n \to \infty$. Then go back to the definition of a limit.]

*15. Show that

$$\lim_{n \to \infty} \sum_{k=1}^{n} \frac{1}{\sqrt{n^2 + k}} = 1.$$

16. a) Let l and w be the length and width of a rectangle. According to Greek geometers, the most pleasing rectangles are those for which

$$\frac{w}{l} = \frac{l}{l + w}.$$

(The term *golden mean* arose in this connection.) Prove that the above equation holds iff

$$L = \frac{l}{w} = \frac{1 + \sqrt{5}}{2}.$$

b) Show that the number L can be expressed as the simplest possible continued fraction, namely,

$$L = 1 + \cfrac{1}{1 + \cfrac{1}{1 + \cfrac{1}{1 + \cdots}}}.$$

In other words, show that the sequence defined by the recursion formula

$$r_1 = 1, \qquad r_{n+1} = 1 + 1/r_n$$

converges to L.

c) Show that if $n \geq 2$, then $r_n = a_n/a_{n-1}$, where a_n is the nth term of the Fibonacci sequence $1, 2, 3, 5, 8, 13, 21, \ldots$.

17. Find a sequence in which every terminating decimal appears at least once.

3-4 Proofs of the limit theorems

We shall now prove the theorems on limits which were stated and used in the previous section.

First of all, it will be convenient to generalize slightly the definition of a sequence. Since the limit of a sequence $\{\alpha_n\}$ depends only on the behavior of α_n for large values of n, no harm will come if there are a finite number of integers n for which α_n is undefined. Suppose, for instance, that we define

$$\alpha_n = \frac{2n^2 - n + 1}{(n - 1)(n - 3)}$$

for any positive integer n other than 1 or 3. Technically speaking, α is not a sequence, since its domain is not the *entire* set of positive integers. Yet there are two senses in which it is meaningful to say that $\lim_{n \to \infty} \alpha_n = 2$. First, if we define α_1 and α_3 in any way we like (for example, we could let $\alpha_1 = 10{,}000$ and $\alpha_3 = 1/100$), then the resulting sequence will have the limit 2. Second (and equivalently), the definition of a limit is meaningful for our nonsequence α, and the statement

$$\text{`}(\forall \epsilon > 0)(\exists N)(\forall n > N)(|\alpha_n - 2| < \epsilon)\text{'}$$

is true. We shall therefore generalize our notion of a sequence as follows.

Definition 1. A subset S of $\mathbf{Z}^+$ is **cofinite** if its complement in $\mathbf{Z}^+$ is a finite set. By a **sequence** we shall mean any function whose domain is a cofinite subset of $\mathbf{Z}^+$.

In other words, a subset S of $\mathbf{Z}^+$ is cofinite iff there exists an integer N such that $n \in S$ whenever n is a positive integer and $n \geq N$. It is clear that the intersection of two cofinite subsets of $\mathbf{Z}^+$ is again a cofinite subset. [This follows, for instance, from the De Morgan rule $(A \cap B)' = A' \cup B'$ and the fact that the union of two finite sets is finite.] Consequently, the sum or product of two sequences of real numbers is also a sequence.

We introduce two classes of sequences, $\mathfrak{N}$ and $\mathfrak{B}$. The set $\mathfrak{N}$ is the family of all *null sequences*, that is, sequences of real numbers which converge to zero. Note that $\alpha \in \mathfrak{N}$ iff $(\forall \epsilon > 0)(\exists N \in \mathbf{Z}^+)(\forall n \geq N)(|\alpha_n| < \epsilon)$. The set $\mathfrak{B}$ is the family of all *bounded sequences*.

Definition 2. A sequence α of real numbers is **bounded** iff there exists a real number M such that $|\alpha_n| \leq M$ for all $n \in \mathfrak{D}_\alpha$, in which case M is called a **bound** of the sequence.

Example 1. The sequence whose nth term is $(-1)^n + 1/n$ is bounded since for each $n \in \mathbf{Z}^+$,

$$|(-1)^n + 1/n| \leq 2.$$

On the other hand, the Archimedean ordering property tells us that the sequence whose nth term is

$$n + (-1)^n n = \begin{cases} 2n & \text{if } n \text{ is even,} \\ 0 & \text{if } n \text{ is odd} \end{cases}$$

is unbounded.

Proposition 3. $\mathfrak{N} \subset \mathfrak{B}$. *The sets $\mathfrak{N}$ and $\mathfrak{B}$ are closed under addition and multiplication. Moreover, if $\alpha \in \mathfrak{N}$ and $\beta \in \mathfrak{B}$, then $\alpha\beta \in \mathfrak{N}$.*

Proof. Suppose $\alpha \in \mathfrak{N}$. Then $(\forall \epsilon > 0)(\exists N)(\forall n > N)(|\alpha_n| < \epsilon)$. Setting $\epsilon = 1$, we see in particular that there exists a positive integer N such that

$$n > N \Rightarrow |\alpha_n| < 1.$$

Now the set $S = \{|\alpha_k| \mid 1 \leq k \leq N \text{ and } k \in \mathfrak{D}_\alpha\}$ is finite. If $S = \varnothing$, then the number one is a bound of the sequence α. If $S \neq \varnothing$, then S has a largest member M, and max $(M, 1)$ is a bound of the sequence. In either case, we see that α is bounded. Hence $\mathfrak{N} \subset \mathfrak{B}$.

Because of the relations

$$|\alpha_n + \beta_n| \leq |\alpha_n| + |\beta_n| \qquad \text{and} \qquad |\alpha_n\beta_n| = |\alpha_n| \, |\beta_n|,$$

it follows that if M_α and M_β are bounds of the sequences α and β, then $M_\alpha + M_\beta$ and $M_\alpha M_\beta$ are bounds of the sequences $\alpha + \beta$ and $\alpha\beta$, respectively. Thus $\mathfrak{B}$ is closed under addition and multiplication.

We prove next that $\mathfrak{N}$ is closed under addition. Let $\alpha, \beta \in \mathfrak{N}$. We must show that $(\alpha + \beta) \in \mathfrak{N}$ or, equivalently, that $(\forall \epsilon > 0)(\exists N)(\forall n > N)(|\alpha_n + \beta_n| < \epsilon)$. Given $\epsilon > 0$, there exist integers N_α and N_β such that

$$n > N_\alpha \Rightarrow |\alpha_n| < \epsilon/2, \qquad n > N_\beta \Rightarrow |\beta_n| < \epsilon/2.$$

Let $N = \max (N_\alpha, N_\beta)$. Then

$$n > N \Rightarrow n > N_\alpha \qquad \text{and} \qquad n > N_\beta \Rightarrow |\alpha_n + \beta_n| \leq |\alpha_n| + |\beta_n| < \epsilon/2 + \epsilon/2 = \epsilon.$$

This proves that $(\alpha + \beta) \in \mathfrak{N}$.

Finally we prove that if $\alpha \in \mathfrak{N}$ and $\beta \in \mathfrak{B}$, then $\alpha\beta \in \mathfrak{N}$. For variety, we couch the proof in terms of cofinite subsets. Observe that $\gamma \in \mathfrak{B}$ iff for *some* $\epsilon > 0$ the set

$$S_\epsilon = \{n \in \mathbf{Z}^+ \mid |\gamma_n| < \epsilon\}$$

is cofinite; on the other hand, $\gamma \in \mathfrak{N}$ iff for *each* $\epsilon > 0$ the set S_ϵ is cofinite.

Since $\beta \in \mathfrak{B}$, there exists an $M > 0$ such that

$$B = \{n \in \mathbf{Z}^+ \mid |\beta_n| < M\}$$

is cofinite. Let $\epsilon > 0$ be given. Since $\alpha \in \mathfrak{N}$, the set

$$A = \{n \in \mathbf{Z}^+ \mid |\alpha_n| < \epsilon/M\}$$

is cofinite. Now $A \cap B$ is cofinite, and

$$\{n \in \mathbf{Z}^+ \mid |\alpha_n\beta_n| < \epsilon\} \supset A \cap B.$$

Hence $\{n \in \mathbf{Z}^+ \mid |\alpha_n\beta_n| < \epsilon\}$ is cofinite. Since $\epsilon > 0$ was arbitrary, $\alpha\beta \in \mathfrak{N}$.

If A is a fixed real number, then the constant sequence in which each term is equal to A is a bounded sequence which we shall continue to denote by A. It is obvious that $\lim A = A$. Another immediate consequence of our definition of a limit is the following: $\lim_{n \to \infty} \alpha_n = A$ iff *the sequence $\alpha - A$ is a null sequence.*

Proposition 4. *Every convergent sequence is bounded.*

Proof. Suppose that α is a convergent sequence, and let $A = \lim \alpha$. Then the sequence $\alpha' = \alpha - A$ is a null sequence and therefore bounded, by the previous theorem. Thus $\alpha = \alpha' + A$, being the sum of two bounded sequences, is bounded.

Note that the contrapositive of this result is the following: *An unbounded sequence has no limit.* Thus, the sequence whose nth term is $n + (-1)^n n$ has no limit.

Theorem 5. *Sums and products of convergent sequences are convergent. If* $\lim \alpha = A$ *and* $\lim \beta = B$, *then*

$$\lim (\alpha + \beta) = A + B, \qquad \lim \alpha\beta = AB.$$

Proof. We can write

$$\alpha = A + \alpha' \qquad \text{and} \qquad \beta = B + \beta',$$

where α' and β' are null sequences. Then

$$\alpha + \beta = (A + B) + (\alpha' + \beta').$$

The sequence $\alpha' + \beta'$, being the sum of null sequences, is null. Thus,

$$\lim (\alpha + \beta) = A + B.$$

Also,

$$\alpha\beta = AB + A\beta' + \alpha'B + \alpha'\beta'.$$

Each of the sequences $A\beta'$, $\alpha'B$, and $\alpha'\beta'$ is null (why?), and hence their sum $A\beta' + B\alpha' + \alpha'\beta'$ is null. Hence, $\lim \alpha\beta = AB$. Q.E.D.

We have said nothing about differences of sequences because the difference $\alpha - \beta$ is easily treated as $\alpha + (-1)\beta$. Suppose, for instance, that $\lim \alpha = A$ and $\lim \beta = B$. Then

$$\lim (\alpha - \beta) = \lim [\alpha + (-1)\beta] = A + (-1)B = A - B,$$

by the previous theorem on limits. Thus, *the limit of a difference is equal to the difference of the limits.*

Quotients of sequences cannot be disposed of so easily. In the first place, the quotient of two sequences need not be a sequence even in our broadened sense of the term. For example, if α is the sequence beginning $1, 0, \frac{1}{2}, 0, \frac{1}{3}, 0, \frac{1}{4}, 0, \ldots$, then the domain of $1/\alpha$ is the set of odd positive integers, and therefore $1/\alpha$ is not a sequence. On the other hand, if a sequence α converges to a nonzero limit, then α can have only a finite number of zero terms, and thus $1/\alpha$ will be a sequence.

Lemma 6. *Suppose that the sequence α converges to a nonzero limit A. Then*

    1) for sufficiently large values of n, α_n has the same sign as A,

    2) $1/\alpha$ is a bounded sequence.

Proof. We consider two cases.

Case I. $A < 0$. Then $\epsilon = -A/2$ is a positive number, and by the definition of a limit, there must therefore exist an $N \in \mathbf{Z}^+$ such that

$$n > N \Rightarrow |\alpha_n - A| < -A/2.$$

This latter inequality is equivalent to

$$3A/2 < \alpha_n < A/2.$$

(See Fig. 3–23.) Thus if $n > N$, then α_n has the same sign as A and

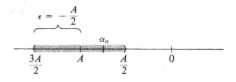

FIG. 3-23

$$|1/\alpha_n| = -1/\alpha_n < -2/A.$$

It follows that $1/\alpha$ is a bounded sequence. (We have considered all but at most N values of $1/\alpha$, and these neglected terms cannot affect the boundedness of the sequence.)

Case II. $A > 0$. One can either carry out a proof analogous to that for case I or reduce this second case to case I by considering the sequence $-\alpha$.

Theorem 7. *If $\lim \alpha = A$ and $\lim \beta = B \neq 0$, then*

$$\lim \alpha/\beta = A/B.$$

Proof

$$\frac{\alpha}{\beta} - \frac{A}{B} = \frac{B\alpha - A\beta}{\beta B} \quad \text{or} \quad \frac{\alpha}{\beta} = \frac{A}{B} + (B\alpha - A\beta) \cdot \frac{1}{B\beta}.$$

Now $\lim (B\alpha - A\beta) = BA - AB = 0$, or in other words, $(B\alpha - A\beta)$ is a null sequence. By the lemma, the sequence $1/(B\beta)$ is bounded. Hence the sequence $(B\alpha - A\beta)/(B\beta)$ is null. It follows that A/B is the limit of α/β.

Another consequence of Lemma 6 is the following.

Theorem 8. *Suppose that α and β are sequences such that*

$$\alpha_n \leq \beta_n$$

for sufficiently large values of n. If A and B are limits of α and β, respectively, then $A \leq B$.

Proof. $B - A = \lim (\beta - \alpha)$. If $B - A$ were negative, then $(\beta - \alpha)_n = \beta_n - \alpha_n$ would have to be negative for sufficiently large values of n. This, however, would violate the hypothesis. Hence $B - A$ must be nonnegative, or equivalently, $A \leq B$.

This theorem may be thought of as the weak companion of the pinching theorem, which we prove next.

Theorem 9. *Suppose that α, β, and γ are sequences such that*

1) for sufficiently large values of n, $\alpha_n \leq \beta_n \leq \gamma_n$,

2) $\lim \alpha = \lim \gamma = L$.

Then β converges, and $\lim \beta = L$.

Proof. We shall first establish a special case of the theorem. We shall then deduce the general result from the special one with the aid of previous limit theorems.

Case I. $\alpha \equiv 0$. Then $\lim \gamma = \lim \alpha = 0$, and for sufficiently large values of n, say $n > N_1$, it follows that $0 \leq \beta_n \leq \gamma_n$. Let $\epsilon > 0$ be given. Since γ is a null sequence, there exists an $N_2 \in \mathbf{Z}^+$ such that

$$n > N_2 \Longrightarrow |\gamma_n| < \epsilon.$$

Let $N = \max (N_1, N_2)$. Then

$$n > N \Longrightarrow |\beta_n| = \beta_n \leq \gamma_n = |\gamma_n| < \epsilon.$$

This proves that $\lim \beta = 0$.

Case II. General case. The inequality

$$0 \leq \beta_n - \alpha_n \leq \gamma_n - \alpha_n$$

holds for all sufficiently large values of n. Also, $\lim (\gamma - \alpha) = L - L = 0$. Hence by case I, $\lim (\beta - \alpha) = 0$, so that

$$\lim \beta = \lim [\alpha + (\beta - \alpha)] = L + 0 = L.$$

One might be tempted at first to regard this theorem as a corollary of the previous one. Indeed, under the added hypothesis that β is convergent, it is a corollary. Much of the usefulness of the pinching theorem, however, comes from the fact that one of its *conclusions* is the convergence of β.

If we are given a sequence $a_1, a_2, a_3, \ldots$, then sequences such as

$$a_2, a_4, a_6, a_8, a_{10}, \ldots \qquad \text{and} \qquad a_1, a_2, a_4, a_8, a_{16}, \ldots,$$

which we can obtain by deleting terms from the original sequence, are called *subsequences* of the original sequence.

Definition 10. Let $a_1, a_2, a_3, \ldots$ be a sequence. By a **subsequence** we mean any sequence of the form $a_{n_1}, a_{n_2}, a_{n_3}, \ldots$, where $n_1, n_2, n_3, \ldots$ is any sequence of positive integers which is strictly increasing ($n_1 < n_2 < n_3 < \cdots$).

Example 2. The sequence $1, \frac{1}{2}, \frac{1}{3}, \ldots$ is a subsequence of the sequence $1, 1/\sqrt{2}, 1/\sqrt{3}, 1/\sqrt{4}, \ldots$ (Take $n_k = k^2$ in the above definition.)

Our definition suffers the defect that it applies only to sequences whose domain is the entire set $\mathbf{Z}^+$. A better, though more sophisticated, definition is the following: By a subsequence of a sequence α we mean any sequence of the form $\alpha \circ \nu$, where ν is an increasing sequence of positive integers [that is, $\mathcal{R}_\nu \subset \mathbf{Z}^+$ and $(\forall j, k \in \mathcal{D}_\nu)$ $(j < k \Rightarrow \nu_j < \nu_k)$]. In keeping with our notation for sequences, we write, however, α_{ν_k} rather than $(\alpha \circ \nu)(k)$.

Proposition 11. *If a sequence $a_1, a_2, a_3, \ldots$ converges to a limit A, then any subsequence $a_{n_1}, a_{n_2}, a_{n_3}, \ldots$ also converges to A.*

Proof. Since $n_1, n_2, n_3, \ldots$ is a strictly increasing sequence of positive integers, one can easily prove by induction that $n_m \geq m$ for all $m \in \mathbf{Z}^+$.

Let $\epsilon > 0$ be given. By the definition of a limit,

$$(\exists N \in \mathbf{Z}^+)(\forall n > N)(|a_n - A| < \epsilon).$$

Consequently,

$$m > N \Rightarrow n_m \geq m > N \Rightarrow |a_{n_m} - A| < \epsilon.$$

This proves that $\lim_{m \to \infty} a_{n_m} = A$.

This theorem can be used effectively to prove that certain sequences do not have limits. One consequence of the theorem is that *if a sequence has two subsequences which converge to different limits, then the original sequence has no limit.* For example, the sequence $0, \frac{1}{2}, 0, \frac{3}{4}, 0, \frac{7}{8}, \ldots$ does not have a limit, since the subsequence consisting of the odd-numbered terms converges to zero while the subsequence consisting of the even-numbered terms converges to one.

Exercises

1. Prove that none of the following sequences have limits. Which of the sequences are unbounded?

a) $n!$ b) $\sin \dfrac{n\pi}{2}$ c) $\dfrac{2^n}{n^{1000}}$ d) $(-1)^n + \dfrac{1}{n}$ e) r^n ($|r| > 1$)

2. We have always spoken of *the* limit of a sequence as though it were impossible for a sequence to have more than one limit. Prove that this is so.

3. Let α be a sequence of real numbers. We say that a number A is a *limit point* of α iff for each $\epsilon > 0$ the inequality

$$|\alpha_n - A| < \epsilon$$

is true for infinitely many integers n. Show that the sequence whose nth term is sin $(n\pi/2)$ has three limit points: 0, -1, and $+1$. Show that A is a limit point of a sequence α iff α has a subsequence which converges to A.

4. Determine all the limit points of each of the following sequences.

    a) $1, 3, 5, 7, \ldots$ b) $2, 0, 4, 0, 6, 0, \ldots$ c) $+1, -1, +1, -1, \ldots$
    d) $1, 2, 1, 2, 3, 1, 2, 3, 4, 1, 2, 3, 4, 5, 1, \ldots$
    e) $1, \frac{1}{2}, \frac{2}{2}, \frac{1}{3}, \frac{2}{3}, \frac{3}{3}, \frac{1}{4}, \frac{2}{4}, \frac{3}{4}, \frac{4}{4}, \ldots$

5. a) Show that $\lim_{n \to \infty} \alpha_n \neq A$ iff there exists an $\epsilon > 0$ and a subsequence $\alpha_{n_1}, \alpha_{n_2}, \alpha_{n_3}, \ldots$ such that

$$|\alpha_{n_m} - A| \geq \epsilon$$

    for every positive integer m.

    b) Show that $\lim_{n \to \infty} \alpha_n = A$ iff every subsequence of α has a subsequence which converges to A.

6. Suppose that $\lim_{n \to \infty} a_n = A$ and $\lim_{n \to \infty} b_n = B$. Prove that the sequence obtained by interlacing these two sequences, namely, $a_1, b_1, a_2, b_2, \ldots$, is convergent iff $A = B$.

7. Consider a doubly indexed set of real numbers

$$a_{1,1}, \; a_{1,2}, \; a_{1,3}, \; \cdots$$
$$a_{2,1}, \; a_{2,2}, \; a_{2,3}, \; \cdots$$
$$a_{3,1}, \; a_{3,2}, \; a_{3,3}, \; \cdots$$
$$\vdots \qquad \vdots \qquad \vdots$$

Suppose that for each positive integer k, $\lim_{n \to \infty} a_{k,n} = A_k$. Suppose also that the sequence of limits $A_1, A_2, A_3, \ldots$ has itself a limit; call it A. Prove that there exists an increasing sequence of positive integers $n_1 < n_2 < n_3 < \cdots$ such that

$$\lim_{m \to \infty} a_{m,n_m} = A.$$

8. Suppose that α is a sequence and that $\lim_{n \to \infty} b_n = B$, where each of the terms of the sequence $b_1, b_2, \ldots$ is a limit point of the sequence α. Show that B is also a limit point of α.

*9. Show that the set of limit points of the sequence whose nth term is sin $n\pi x$ is finite iff x is rational. If x is irrational, show that the set of limit points is the interval $[-1, +1]$.

10. a) Let $n_1, n_2, n_3, \ldots$ be a sequence of positive integers such that no two terms of the sequence are the same. For each $N \in \mathbf{Z}^+$, infer that there can only be a finite number of integers m for which $n_m \leq N$, or equivalently, that $(\exists M \in \mathbf{Z}^+)(\forall m > M)(n_m > N)$.

    b) Let $n_1, n_2, n_3, \ldots$ be as in (a). If $\lim_{n \to \infty} \alpha_n = A$, prove that $\lim_{n \to \infty} \alpha_{n_m} = A$.

    c) Deduce from (b) that the convergence of a sequence is unaffected by a rearrangement of its terms.

3–5 Limits of functions of a continuous variable

In the last two sections we have discussed limits of functions whose domains were the set of positive integers. We now turn our attention to what are usually called functions of a continuous real variable. These are functions whose domains are typically either intervals or finite unions of intervals.

Consider the function f defined by the equation

$$f(x) = \frac{\sin x}{x}.$$

Although the function is not defined when $x = 0$, it turns out that the behavior of $f(x)$ for values of x close to zero is of considerable importance. A glance at the table below reveals that for values of x close to zero, $f(x)$ is close to the number one.

x	$(\sin x)/x$
$+0.10$	$0.998334...$
$+0.05$	$0.999583...$
$+0.01$	$0.999983...$

What is, in fact, true (although not provable from any function table, no matter how large) is that $f(x)$ can be made arbitrarily close to one by choosing x sufficiently close to zero. We indicate this state of affairs symbolically by the equation

$$\lim_{x \to 0} \frac{\sin x}{x} = 1,$$

which is usually read: "The limit of $(\sin x)/x$ as x approaches zero is one."

More generally, let us consider the behavior of a real-valued function $f(x)$ for values of x which are "close" but not equal to some number a. (The function may or may not be defined at a, and even if it is, the value of the function at a is, for the moment, of no interest.) We shall say that the limit of $f(x)$ is L as x approaches a if we can make $f(x)$ as close to the number L as we please by choosing x sufficiently close (but not equal) to a. We proceed immediately to the precise definition (explicatum).

Definition 1. Let f be a real-valued function, and let a be a real number. We say that **the limit of $f(x)$ as x approaches a is L,** and we write

$$\lim_{x \to a} f(x) = L,$$

if and only if for every $\epsilon > 0$ (no matter how small), there exists a number $\delta > 0$ (sufficiently small) such that

$$|f(x) - L| < \epsilon$$

whenever $0 < |x - a| < \delta$.

The phrases in this definition which appear in parentheses are dispensable from a strictly logical point of view; their sole function is a psychological one. Using the notations of Chapter 1, we can write the definition of a limit in a compact fashion which transparently exhibits its logical structure:

$$\lim_{x \to a} f(x) = L \Leftrightarrow (\forall \epsilon > 0)(\exists \delta > 0)(\forall x)(0 < |x - a| < \delta \Rightarrow |f(x) - L| < \epsilon)$$

(see Fig. 3–24). Note that the statements "$0 < |x - a|$" and "$x \neq a$" are equivalent.

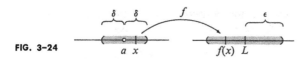

FIG. 3–24

While it would be altogether fitting and proper for us to prove that

$$\lim_{x \to 0} \frac{\sin x}{x} = 1,$$

we shall begin with simpler tasks and build toward this proof.

Example 1. Prove that

$$\lim_{x \to 2} (5x - 1) = 9.$$

According to the definition of a limit, we must therefore prove the following assertion:

$$(\forall \epsilon > 0)(\exists \delta > 0)(\forall x)(0 < |x - 2| < \delta \Rightarrow |(5x - 1) - 9| < \epsilon).$$

In other words, for any positive number ϵ which might be given to us, we must produce a number $\delta > 0$ such that

$$|(5x - 1) - 9| < \epsilon$$

whenever $0 < |x - 2| < \delta$. Note, however, that

$$|(5x - 1) - 9| = |5(x - 2)| = 5|x - 2|.$$

Consequently, the inequality

$$|(5x - 1) - 9| < \epsilon$$

will hold if $0 < |x - 2| < \epsilon/5$. Therefore, if we are given any $\epsilon > 0$, we can always set $\delta = \epsilon/5$ (or anything smaller), and the assertion

$$(\forall x)(0 < |x - 2| < \delta \Rightarrow |(5x - 1) - 9| < \epsilon)$$

will be true. The proof then is complete.

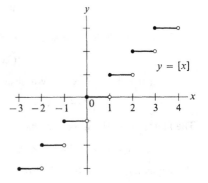

FIG. 3–25

It can be shown (Exercise 7, Section 3–5) that for each real number x there is a largest integer which does not exceed x. This integer, customarily denoted by $[x]$, is called the *greatest integer in x*. In particular, $[\frac{3}{2}] = [\sqrt{2}] = 1$, $[\pi] = 3$, and $[-\pi] = -4$. The graph of $[x]$ is shown in Fig. 3–25.

Example 2. Prove that

$$\lim_{x \to \pi/2} [\sin x] = 0.$$

To prove this we must show that

$$(\forall \epsilon > 0)(\exists \delta > 0)(\forall x)(0 < |x - \pi/2| < \delta \Rightarrow |[\sin x]| < \epsilon).$$

In this particular case we can actually prove something stronger, namely,

$$(\exists \delta > 0)(\forall \epsilon > 0)(\forall x)(0 < |x - \pi/2| < \delta \Rightarrow |[\sin x]| < \epsilon).$$

Set $\delta = \pi/2$, and let ϵ be any positive number. Suppose that $0 < |x - \pi/2| < \delta$. Then either $0 < x < \pi/2$ or $\pi/2 < x < \pi$. In either case,

$$|[\sin x]| = 0 < \epsilon,$$

and the proof is complete.

The graph of $y = [\sin x]$ is shown in Fig. 3–26.

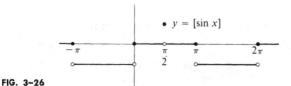

FIG. 3–26

In the definition of a limit, the dependence of δ upon ϵ can be quite complicated, as our next example (whose difficulty could be much greater) should suggest.

Example 3. Show that

$$\lim_{x \to 1} (3x^2 + 1) = 4.$$

First we shall do some scratch work. Given $\epsilon > 0$, we must produce a $\delta > 0$ such that

$$|(3x^2 + 1) - 4| < \epsilon$$

whenever $0 < |x - 1| < \delta$. We observe that

$$|(3x^2 + 1) - 4| = |3x^2 - 3| = 3|x + 1|\,|x - 1|.$$

The factor $|x + 1|$ will be less than 3 if $|x - 1| < 1$:

$$(|x - 1| < 1 \Rightarrow 0 < x < 2 \Rightarrow 1 < x + 1 < 3 \Rightarrow |x + 1| < 3).$$

Consequently, if $|x - 1| < 1$, it follows that

$$|(3x^2 + 1) - 4| < 9|x - 1|.$$

Suppose then that the inequality $|x - 1| < \epsilon/9$ also holds. We then have

$$|(3x^2 + 1) - 4| < \epsilon,$$

as desired. Hence we can let δ be the smaller of the two numbers 1 and $\epsilon/9$.

We now present the formal proof. Let $\epsilon > 0$ be given. Set $\delta = \min(1, \epsilon/9)$, and suppose that $0 < |x - 1| < \delta$. Since, in particular, $|x - 1| < 1$, we have by our previous calculations

$$|(3x^2 + 1) - 4| < 9|x - 1|.$$

However, since we also know that $|x - 1| < \epsilon/9$, it follows that

$$|(3x^2 - 1) - 4| < 9 \cdot \epsilon/9 = \epsilon.$$

We shall let the reader check for himself that $\delta = \min(2, \epsilon/12)$ also works.

Example 4. If a is any real number, prove that

$$\lim_{x \to a} \sin x = \sin a.$$

As a preliminary step, we observe that for every real number x, the inequality

$$|\sin x| \leq |x|$$

is valid.

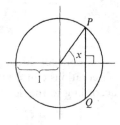

FIG. 3-27

Suppose first that $0 \leq x \leq \pi/2$. We construct an angle of x radians, as shown in Fig. 3-27. Then the length of the chord PQ is $2 \sin x$, whereas the length of the arc PQ is $2x$. Since "a straight line is the shortest distance between two points," it follows that $2 \sin x \leq 2x$, which implies that $|\sin x| \leq |x|$.

The inequality is trivial if $x \geq \pi/2$, since in this case

$$|\sin x| \leq 1 < \pi/2 \leq x = |x|.$$

Thus $|\sin x| \leq |x|$ whenever $x \geq 0$. If $x \leq 0$, then $-x \geq 0$ and

$$|\sin x| = |-\sin x| = |\sin(-x)| \leq |-x| = |x|.$$

Having established the inequality, we proceed to the proof of the original assertion. We use the following identity:

$$\sin x - \sin a = 2 \cos \tfrac{1}{2}(x + a) \sin \tfrac{1}{2}(x - a).$$

Since the values of the cosine function lie between -1 and $+1$,

$$|\sin x - \sin a| \leq 2|\sin \tfrac{1}{2}(x - a)| \leq |x - a|.$$

Let $\epsilon > 0$ be given. Set $\delta = \epsilon$. If $|x - a| < \delta$, then we have by the above inequality

$$|\sin x - \sin a| \leq |x - a| < \delta = \epsilon.$$

This proves the assertion.

Example 5. Show that for each number $x_0 > 0$,

$$\lim_{x \to x_0} \sqrt{x} = \sqrt{x_0}.$$

We use an algebraic trick which should now be familiar to the reader:

$$\sqrt{x} - \sqrt{x_0} = (\sqrt{x} - \sqrt{x_0})\frac{\sqrt{x} + \sqrt{x_0}}{\sqrt{x} + \sqrt{x_0}} = \frac{x - x_0}{\sqrt{x} + \sqrt{x_0}}.$$

If $x > 0$, then we have

$$|\sqrt{x} - \sqrt{x_0}| = \frac{|x - x_0|}{\sqrt{x} + \sqrt{x_0}} < \frac{|x - x_0|}{\sqrt{x_0}}.$$

The proof is now easy. Let $\epsilon > 0$ be given. Set $\delta = \min(x_0, \sqrt{x_0}\,\epsilon)$. If $0 < |x - x_0| < \delta$, then $x > 0$ (why?), and

$$|\sqrt{x} - \sqrt{x_0}| < \frac{|x - x_0|}{\sqrt{x_0}} < \frac{\sqrt{x_0}\,\epsilon}{\sqrt{x_0}} = \epsilon.$$

Before considering further examples, it is worthwhile to make a couple of general observations concerning limits. Suppose that $\lim_{x \to a} f(x) = L$. If we let $\epsilon = 1$, then according to the definition of a limit, there exists a $\delta > 0$ such that

$$|f(x) - L| < 1$$

whenever $0 < |x - a| < \delta$. In particular, $f(x)$ must be defined (that is, x must belong to the domain of f) whenever x belongs to the set

$$N'(a, \delta) = \{x \in \mathbf{R} \mid 0 < |x - a| < \delta\} = (a - \delta, a) \cup (a, a + \delta)$$

(see Fig. 3–28). Sets of the form $N'(a, \delta)$ are called *deleted neighborhoods* of a. Note also that the inequality

$$|f(x)| = |(f(x) - L) + L|$$
$$\le |f(x) - L| + |L| < |L| + 1$$

holds whenever $x \in N'(a, \delta)$. Thus we have proved the following proposition.

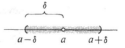

FIG. 3–28

Proposition 2. *If $f(x)$ has a limit as x approaches a, then there exists a deleted neighborhood of a throughout which f is both defined and bounded.*

Just for the record, a function f is said to be *bounded* on a set S iff

$$(\exists M)(\forall x \in S)(|f(x)| \le M).$$

We can also use the notion of a neighborhood to paraphrase the definition of a limit.

Proposition 3. $\lim_{x \to a} f(x) = L$ *iff for every neighborhood V of L there exists a deleted neighborhood U of a such that for all $x \in U$, $f(x) \in V$.*

Example 6. Show that $\lim_{x \to 1} \sqrt{1 - x^2}$ does not exist.

The domain of $f(x) = \sqrt{1 - x^2}$ is $[-1, +1]$, as shown in Fig. 3–29. Since the domain of f includes no points to the right of $+1$, there is no deleted neighborhood of 1 throughout which $f(x)$ is defined. Thus, $\lim_{x \to 1} f(x)$ cannot exist, by Proposition 2.

Example 7. Show that $\lim_{x\to 0} 1/x$ does not exist.

In this case the function is defined throughout a deleted neighborhood of 0, but in any such neighborhood the function is unbounded, as shown in Fig. 3–30. Consequently, the limit cannot exist.

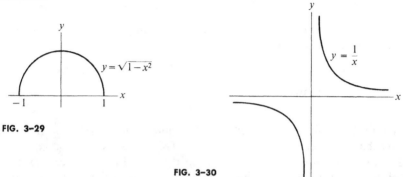

FIG. 3–29

FIG. 3–30

The efficient handling of limits requires the use of a small body of theorems. In particular, there are analogs to our theorem on the continuity of the arithmetic operations and to the pinching theorem for sequences. One can either prove these new theorems directly by modifying previous proofs, or one can try to deduce these new theorems from their analogs for sequences. We choose the latter course of action. The key to this line of development is the *linking limit lemma*, which is of interest in itself. Its statement requires a preliminary definition.

Definition 4. We say that a sequence $\alpha_1, \alpha_2, \alpha_3, \ldots$ **converges properly** to A iff

1) $\lim_{n\to\infty} \alpha_n = A$,

2) $\alpha_n \neq A$ for all n sufficiently large.

Thus the sequence $1, \frac{1}{2}, \frac{1}{3}, \ldots$ converges properly to zero, whereas the sequence $1, 0, \frac{1}{2}, 0, \frac{1}{3}, 0, \ldots$ does not. It should be clear that a sequence α converges properly to A iff $(\forall \epsilon > 0)(\exists N \in \mathbf{Z}^+)(\forall n > N)(0 < |\alpha_n - A| < \epsilon)$.

Theorem 5. Linking Limit Lemma. $\lim_{x\to a} f(x) = L$ *iff for each sequence* $x_1, x_2, \ldots$ *converging properly to* a, $\lim_{n\to\infty} f(x_n) = L$.

Proof. Suppose that $\lim_{x\to a} f(x) = L$, and let $x_1, x_2, \ldots$ be a sequence that converges properly to a. Given $\epsilon > 0$, we can choose a $\delta > 0$ such that $|f(x) - L| < \epsilon$ whenever $0 < |x - a| < \delta$. Since $x_1, x_2, \ldots$ converges properly to a, we may choose an integer N such that

$$0 < |x_n - a| < \delta$$

if $n > N$. Thus

$$n > N \Longrightarrow |f(x_n) - L| < \epsilon.$$

This proves that $f(x_n) \to L$. (See Fig. 3–31.)

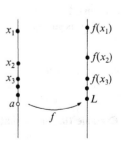

FIG. 3–31

To prove the converse, we shall suppose that $\lim_{x \to a} f(x) \neq L$, and we shall then deduce that there exists a sequence $x_1, x_2, \ldots$ converging properly to a for which $\lim_{n \to \infty} f(x_n) \neq L$. The statement $\lim_{x \to a} f(x) \neq L$ is equivalent to

$$(\exists \epsilon > 0)(\forall \delta > 0)(\exists x)(0 < |x - a| < \delta \text{ and } |f(x) - L| \geq \epsilon).$$

Let δ be of the form $1/n$, where $n \in \mathbf{Z}^+$. According to the above statement, there exists a number x_n such that

$$0 < |x_n - a| < 1/n \qquad \text{and} \qquad |f(x_n) - L| \geq \epsilon.$$

Observe, first of all, that

$$a - 1/n < x_n < a + 1/n$$

for all $n \in \mathbf{Z}^+$. Since $\lim_{n \to \infty} (a - 1/n) = \lim_{n \to \infty} (a + 1/n) = a$, it follows from the pinching theorem that $\lim_{n \to \infty} x_n = a$. Moreover, since $0 < |x_n - a|$ for all $n \in \mathbf{Z}^+$, the sequence $\{x_n\}$ converges properly to a. Finally, since $|f(x_n) - L| \geq \epsilon$ for all $n \in \mathbf{Z}^+$, it follows that $\lim_{n \to \infty} f(x_n) \neq L$. [For if $\lim_{n \to \infty} f(x_n) = L$, then the inequality $|f(x_n) - L| < \epsilon$ must hold for all n sufficiently large.]

This theorem can be used to prove the nonexistence of certain limits.

Example 8. Show that $\lim_{x \to 3} [x]$ does not exist.
Suppose to the contrary that the limit does exist and call the limit L. First we observe that the sequence

$$x_n = 3 - 1/2n$$

converges properly to 3. Thus by the linking limit lemma,

$$L = \lim_{n \to \infty} [x_n] = 2,$$

since $[x_n] = 2$ for all positive integers n.
Next we observe that the sequence defined by

$$y_n = 3 + 1/2n$$

also converges properly to 3. Thus

$$L = \lim_{n \to \infty} [y_n] = 3,$$

since $[y_n] = 3$ for all $n \in \mathbf{Z}^+$. This implies that $2 = 3$, a contradiction.

Example 9. Show that $\lim_{x \to 0} \sin (1/x)$ does not exist (see Fig. 3–32).
Let us suppose that the limit exists; call it L. The sequence

$$x_n = 2/n\pi$$

converges properly to zero, and thus by the linking limit lemma, L must be the limit of $\sin (1/x_n) = \sin (n\pi/2)$ as $n \to \infty$. However, the latter sequence is 1, 0, -1, 0, $+1$, 0, -1, 0, $\ldots$, and has no limit.

Example 10. Show that

$$\lim_{x \to 0} \frac{2 + x \sin (1/x)}{(x + 3)^2} = \frac{2}{9}.$$

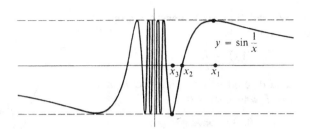

FIG. 3–32

Let $x_1, x_2, \ldots$ be any sequence which converges properly to zero. Then $x_n \sin(1/x_n)$, being the product of a null sequence and a bounded sequence, is null. From our theorems on limits of sequences, it therefore follows that

$$\lim_{n \to \infty} \frac{2 + x_n \sin(1/x_n)}{(x_n + 3)^2} = \frac{2 + 0}{(0 + 3)^2} = \frac{2}{9}.$$

An appeal to the linking limit lemma completes the proof.

We consider now some analogs of theorems on sequences.

Theorem 6. Continuity of the Arithmetic Operations. *Suppose that*

$$\lim_{x \to a} f(x) = A \qquad and \qquad \lim_{x \to a} g(x) = B.$$

Then

$$\lim_{x \to a} (f + g)(x) = A + B, \qquad \lim_{x \to a} (f - g)(x) = A - B,$$

and

$$\lim_{x \to a} (fg)(x) = AB.$$

If $B \neq 0$, then

$$\lim_{x \to a} f(x)/g(x) = A/B.$$

Theorem 7. Pinching Theorem. *Suppose that $f(x) \leq g(x) \leq h(x)$ for all x in some deleted neighborhood of a. Suppose also that*

$$\lim_{x \to a} f(x) = \lim_{x \to a} h(x) = L.$$

Then $\lim_{x \to a} g(x) = L$.

If we use the linking limit lemma, these theorems follow quickly from their analogs for sequences. Let us prove, for example, that the limit of a sum is equal to the sum of the limits. Let $x_1, x_2, \ldots$ be any sequence which converges properly to a. Then $f(x_n) \to A$ and $g(x_n) \to B$. Consequently,

$$\lim_{n \to \infty} (f + g)(x_n) = \lim_{n \to \infty} [f(x_n) + g(x_n)] = A + B.$$

Thus $\lim_{x \to a} (f + g)(x) = A + B$.

Another basic principle concerns limits of composed functions. It may be regarded as a continuous analog of the linking limit lemma.

Definition 8. We say that $f(x)$ converges properly to L as $x \rightarrow a$ iff

$$\lim_{x \to a} f(x) = L$$

and there exists a deleted neighborhood U of a such that $f(x) \neq L$ whenever $x \in U$.

Example 11. As $x \rightarrow 0$, $3x$ converges properly to zero. As $x \rightarrow 0$, $x \sin (1/x)$ also converges to zero (see Fig. 3–33). In this latter case, convergence is not proper since $x \sin (1/x)$ assumes the value zero infinitely often in each deleted neighborhood of zero.

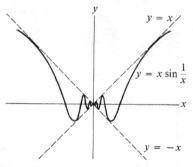

FIG. 3–33

Theorem 9. *Substitution Rule.* *Suppose that as $x \rightarrow a$, $g(x)$ converges properly to A and that*

$$\lim_{y \to A} f(y) = L.$$

Then

$$\lim_{x \to a} f(g(x)) = L.$$

Proof. By hypothesis, there exists a $\delta_g > 0$ such that

$$0 < |x - a| < \delta_g \Rightarrow g(x) \neq A.$$

Let $\epsilon > 0$ be given. Since $\lim_{y \to A} f(y) = L$, there exists an $\epsilon' > 0$ such that

$$0 < |y - A| < \epsilon' \Rightarrow |f(y) - L| < \epsilon.$$

Since $\lim_{x \to a} g(x) = A$, there exists a $\Delta > 0$ such that

$$0 < |x - a| < \Delta \Rightarrow |g(x) - A| < \epsilon'.$$

We set $\delta = \min (\delta_g, \Delta)$. Then

$$0 < |x - a| < \delta \Rightarrow 0 < |g(x) - A| < \epsilon' \Rightarrow |f(g(x)) - L| < \epsilon.$$

This proves that $\lim_{x \to a} f(g(x)) = L$.

Example 12. Show that

$$\lim_{x \to 0} \sin (\pi/2 + x^2) = 1.$$

Because of the continuity of the arithmetic operations,

$$\lim_{x \to 0} (\pi/2 + x^2) = \pi/2 + 0 \cdot 0 = \pi/2.$$

Furthermore, convergence in the latter case is clearly proper. By Example 4,

$$\lim_{y \to \pi/2} \sin y = \sin (\pi/2) = 1,$$

and the desired result follows by the substitution rule.

Example 13. Show that for each number a,

$$\lim_{x \to a} \cos x = \cos a.$$

While a proof *analogous* to that of Example 4 is possible, we shall deduce this result *from* Example 4. We use the identity

$$\cos x = \sin (\pi/2 - x).$$

As $x \to a$, $\pi/2 - x$ converges properly to $\pi/2 - a$, so that

$$
\begin{aligned}
\lim_{x \to a} \cos x &= \lim_{x \to a} \sin (\pi/2 - x) \\
&= \lim_{y \to \pi/2 - a} \sin y \\
&= \sin (\pi/2 - a) \\
&= \cos a.
\end{aligned}
$$

Example 14. Show that the hypothesis of proper convergence is an essential part of the substitution rule.

Let $g(x) \equiv 1$, and let

$$
f(x) = \begin{cases} 5 & \text{if } x \neq 1, \\ 2 & \text{if } x = 1. \end{cases}
$$

Then $\lim_{x \to 10} g(x) = 1$, but convergence is not proper. Since $f(g(x)) \equiv 2$, it follows that $\lim_{x \to 10} f(g(x)) = 2$, whereas $\lim_{y \to 1} f(y) = 5 \neq 2$.

Example 15. Show that

$$\lim_{x \to 0} \frac{\sin x}{x} = 1.$$

In Fig. 3–34 it is clear that the area of the triangle OPR is less than the area of the circular sector OPS, and that this in turn is less than the area of the triangle OQS:

$$
\begin{aligned}
\text{area } \triangle OPR &= \tfrac{1}{2} |OR| \cdot |RP| = \tfrac{1}{2} \cos x \sin x, \\
\text{area } \triangle ORS &= \tfrac{1}{2} |OS| \cdot |SQ| = \tfrac{1}{2} 1 \cdot \tan x.
\end{aligned}
$$

The area of the circular sector will be equal to the area of the circle, namely π, multiplied by the fraction $x/2\pi$. Thus we have

$$\tfrac{1}{2} \cos x \sin x \leq \tfrac{1}{2} x \leq \tfrac{1}{2} \tan x;$$

the inequalities are valid for $0 \leq x < \pi/2$.

From the inequality $\cos x \sin x \leq x$, we get

$$\frac{\sin x}{x} \leq \frac{1}{\cos x}.$$

Similarly, from the inequality $x \leq \tan x = \sin x/\cos x$, we get

$$\cos x \leq \frac{\sin x}{x}.$$

Hence

$$\cos x \leq \frac{\sin x}{x} \leq \frac{1}{\cos x} \qquad (*)$$

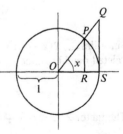

FIG. 3–34

if $0 < x < \pi/2$. Note that the terms appearing in this inequality are unchanged if x is replaced by $-x$. Hence (∗) holds for all x in the deleted neighborhood $N'(0, \pi/2)$.

By Example 13,

$$\lim_{x \to 0} \cos x = \lim_{x \to 0} \frac{1}{\cos x} = 1.$$

Hence, by the pinching theorem,

$$\lim_{x \to 0} \frac{\sin x}{x} = 1.$$

This result may be given a geometric interpretation. Referring to Fig. 3–35, note that

$$\frac{\text{length of the chord } PQ}{\text{length of the arc } PQ} = \frac{2 \sin x}{2x} = \frac{\sin x}{x}.$$

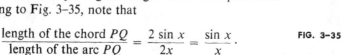

FIG. 3–35

As x tends to zero, the points P and Q move toward each other. Hence we see that as one point on a circle approaches another, the lengths of the chord and of the arc joining the points tend to zero at the same rate.

(A purist might justly charge that in our treatment of the trigonometric functions we have violated our general policy of deducing everything from the axioms for the real numbers. If so, he may find comfort in the promise that a purely analytic and neatly elegant treatment of these functions will be given later.)

Example 16. Find $\lim_{n \to \infty} n \sin (1/n)$.

The sequence $\{1/n\}$ converges properly to zero. Since

$$\lim_{x \to 0} \frac{\sin x}{x} = 1,$$

we have by the linking limit lemma

$$1 = \lim_{n \to \infty} \frac{\sin (1/n)}{1/n} = \lim_{n \to \infty} n \sin (1/n).$$

Example 17. Find

$$\lim_{x \to 1} \frac{\sin 2(x - 1)}{x^3 - 1}.$$

Observe that

$$\frac{\sin 2(x - 1)}{x^3 - 1} = \frac{\sin 2(x - 1)}{2(x - 1)} \cdot \frac{2}{x^2 + x + 1}.$$

Since $\lim_{x \to 1} 2(x - 1) = 0$, and since the expression $2(x - 1)$ is nonzero throughout each deleted neighborhood of the number 1, we see by the substitution rule that

$$\lim_{x \to 1} \frac{\sin 2(x - 1)}{2(x - 1)} = \lim_{y \to 0} \frac{\sin y}{y} = 1.$$

The factor $2/(x^2 + x + 1)$ tends to $\frac{2}{3}$ as $x \to 1$. (Why?) Hence

$$\lim_{x \to 1} \frac{\sin 2(x - 1)}{x^3 - 1} = \frac{2}{3}.$$

Example 18. Find

$$\lim_{x \to 0} \frac{1 - \cos \pi x}{x^2}.$$

Note that

$$\frac{1 - \cos \pi x}{x^2} = \frac{1 - \cos^2 \pi x}{x^2(1 + \cos \pi x)} = \left(\frac{\sin \pi x}{\pi x}\right)^2 \frac{\pi^2}{1 + \cos \pi x}.$$

By an argument similar to that given in Example 17,

$$\lim_{x \to 0} \frac{\sin \pi x}{\pi x} = 1.$$

Thus

$$\lim_{x \to 0} \frac{1 - \cos \pi x}{x^2} = \pi^2.$$

Example 19. Show that

$$\lim_{x \to 0} \frac{\sqrt{1 + \sin x} - \sqrt{1 - \sin x}}{\tan^2 x}$$

does not exist.

We multiply the numerator and denominator of the above expression by $\sqrt{1 + \sin x} + \sqrt{1 - \sin x}$. We get

$$f(x) = \frac{\sqrt{1 + \sin x} - \sqrt{1 - \sin x}}{\tan^2 x}$$

$$= \frac{2 \sin x}{\tan^2 x[\sqrt{1 + \sin x} + \sqrt{1 - \sin x}]}$$

$$= \frac{1}{\sin x} \cdot \frac{2 \cos^2 x}{\sqrt{1 + \sin x} + \sqrt{1 - \sin x}}.$$

As $x \to 0$, both $1 + \sin x$ and $1 - \sin x$ converge properly to 1. This implies that

$$\lim_{x \to 0} \sqrt{1 + \sin x} = \lim_{y \to 1} \sqrt{y} = 1 \quad \text{and} \quad \lim_{x \to 0} \sqrt{1 - \sin x} = \lim_{y \to 1} \sqrt{y} = 1$$

by the substitution rule and Example 5. It follows from Example 13 and the theorem concerning the continuity of the arithmetic operations that

$$\lim_{x \to 0} \frac{2 \cos^2 x}{\sqrt{1 + \sin x} + \sqrt{1 - \sin x}} = \frac{2 \cdot 1^2}{1 + 1} = 1.$$

Let us denote by $\varphi(x)$ the expression inside the preceding limit sign. Then

$$\varphi(x) = f(x) \sin x.$$

Suppose now that $f(x)$ has a limit as $x \to 0$; call the limit L. Since $\sin x \to 0$ as $x \to 0$, we have

$$1 = \lim_{x \to 0} \varphi(x) = \lim_{x \to 0} f(x) \sin x = L \cdot 0 = 0.$$

This is, of course, impossible. Hence $\lim_{x \to 0} f(x)$ does not exist.

Exercises

1. Give (ϵ, δ)-proofs of each of the following.

   a) $\lim_{x \to -2} (3x + 5) = -1$

   b) $\lim_{x \to 0} x \sin (1/x) = 0$

   c) $\lim_{x \to 3} (x^2 - 1) = 8$

   d) $\lim_{x \to 2} \dfrac{x}{x + 1} = \dfrac{2}{3}$

   e) $\lim_{x \to 3/2} [x] = 1$

   f) $\lim_{x \to x_0} [x] = [x_0]$ if x_0 is not an integer

   g) $\lim_{x \to 0} \sqrt{|x|} = 0$

   h) $\lim_{x \to a} \cos x = \cos a$

   i) $\lim_{x \to 3} f(x) = 6$, where $f(x) = \begin{cases} 2x & \text{if } x < 3, \\ 5000 & \text{if } x = 3, \\ 3x - 3 & \text{if } x > 3 \end{cases}$

   j) $\lim_{x \to 1} ([x] + \sqrt{x - [x]}) = 1$

   k) $\lim_{x \to x_0} \sqrt[3]{x} = \sqrt[3]{x_0}$

2. For which parts of Exercise 1 is convergence proper?

3. Show by any convenient method that none of the following limits exist. Describe qualitatively the behavior of the functions near the points in question.

   a) $\lim_{x \to 1} \sqrt{x(x - 1)}$

   b) $\lim_{x \to 2} \dfrac{x}{x - 2}$

   c) $\lim_{x \to 0} \sin^2 \dfrac{1}{x}$

   d) $\lim_{x \to 0} \dfrac{1}{x} \sin^2 \dfrac{1}{x}$

   e) $\lim_{x \to 1} f(x)$, where $f(x) = \begin{cases} x & \text{if } x < 1, \\ 3 & \text{if } x \geq 1 \end{cases}$

   f) $\lim_{x \to \pi} \dfrac{\sin x}{|\sin x|}$

   g) $\lim_{x \to 0} \tan \dfrac{1}{x}$

4. Prove that

   $$\lim_{x \to a} f(x) = 0 \qquad \text{iff} \qquad \lim_{x \to a} |f(x)| = 0.$$

5. Calculate each of the following limits, and amply justify your answer in terms of the theorems and examples in this section.

   a) $\lim_{x \to 2} \dfrac{x^2 + 5}{x^2 - 3}$

   b) $\lim_{x \to 1} \dfrac{x^2 - 2x + 1}{x^3 - x}$

   c) $\lim_{x \to 1} \dfrac{(x - 1)\sqrt{2 - x}}{x^2 - 1}$

   d) $\lim_{x \to 1} \left(\dfrac{1}{1 - x} - \dfrac{3}{1 - x^3} \right)$

   e) $\lim_{x \to 1} \dfrac{x^m - 1}{x^n - 1}$ (m, n are positive integers)

   f) $\lim_{x \to 0} \dfrac{\sqrt{1 + x^2} - 1}{x}$

   g) $\lim_{x \to 0} \dfrac{\sqrt{1 + x} - 1}{x}$

   h) $\lim_{x \to a} \dfrac{\sqrt{x - b} - \sqrt{a - b}}{x^2 - a^2}$ ($a > b$)

   i) $\lim_{x \to 0} \dfrac{\sin 5x}{x}$

   j) $\lim_{x \to 0} \dfrac{\tan 2x}{\sin 3x}$

   k) $\lim_{x \to 0} \dfrac{1 - \cos x}{x^2}$

l) $\lim\limits_{x \to 0} \left(\dfrac{1}{\sin x} - \dfrac{1}{\tan x} \right)$

m) $\lim\limits_{x \to \pi} \dfrac{\sin 3x}{\sin 2x}$

n) $\lim\limits_{x \to \pi/2} \left(\dfrac{\pi}{2} - x \right) \tan x$

o) $\lim\limits_{x \to \pi/4} \dfrac{\cos x - \sin x}{\cos 2x}$

p) $\lim\limits_{x \to 0} (\sin \sqrt{1 + (1/x)} - \sin \sqrt{1/x})$

q) $\lim\limits_{x \to 0} \dfrac{\sin (a + x) + \sin (a - x) - 2 \sin a}{x^2}$

r) $\lim\limits_{x \to 2} (x^2 - 4) \sin \dfrac{1}{x - 2}$

s) $\lim\limits_{x \to 0} \dfrac{\sqrt{2} - \sqrt{1 + \cos x}}{\sin^2 x}$

t) $\lim\limits_{x \to 0} \dfrac{\sqrt[3]{1 + x^2} - \sqrt[4]{1 - 2x}}{x + x^3}$

6. The purpose of this exercise is to present a treatment of the limit theorems for functions of a continuous variable analogous to our treatment of the limit theorems for sequences. This exercise will also be used in the next chapter as a basis for a proof of the chain rule.

Let x_0 be a fixed real number. We shall consider three classes of functions, $\mathfrak{F}_{x_0}$, $\mathfrak{B}_{x_0}$, and $\mathfrak{N}_{x_0}$. The class $\mathfrak{F}_{x_0}$ will consist of all real-valued functions defined near x_0. More precisely, $f \in \mathfrak{F}_{x_0}$ iff there exists a deleted neighborhood of x_0 throughout which f is defined, or equivalently, $(\exists \delta > 0)(\forall x)(0 < |x - x_0| < \delta \Rightarrow x \in \mathfrak{D}_f)$. The class $\mathfrak{N}_{x_0}$ is the set consisting of all functions f such that $\lim_{x \to x_0} f(x) = 0$. We define $f \in \mathfrak{B}_{x_0}$ iff there exists a deleted neighborhood of x_0 throughout which f is bounded, or equivalently, $(\exists \delta > 0)(\exists M)(\forall x)(0 < |x - x_0| < \delta \Rightarrow |f(x)| \leq M)$.

a) *Analog of Proposition 3, Section 3-4.* Show that $\mathfrak{N}_{x_0} \subset \mathfrak{B}_{x_0} \subset \mathfrak{F}_{x_0}$ and that the three classes of functions are closed under addition and multiplication. If $f \in \mathfrak{N}_{x_0}$ and $g \in \mathfrak{B}_{x_0}$, prove that $(fg) \in \mathfrak{N}_{x_0}$.

b) Observe that $\mathfrak{B}_{x_0}$ contains the constant functions. Also observe that

$$\lim\limits_{x \to x_0} f(x) = A \qquad \text{iff} \qquad (f - A) \in \mathfrak{N}_{x_0},$$

where in the latter expression A is to be interpreted as the constant function identically equal to A. Deduce that if $\lim_{x \to x_0} f(x)$ exists, then $f \in \mathfrak{B}_{x_0}$. (Compare this with the proof of Proposition 2.)

c) Suppose $\lim_{x \to x_0} f(x) = A$ and $\lim_{x \to x_0} g(x) = B$. Using (a) and (b), prove that

$$\lim\limits_{x \to x_0} (f + g)(x) = A + B \qquad \text{and} \qquad \lim\limits_{x \to x_0} (fg)(x) = AB.$$

d) If $\lim_{x \to x_0} g(x) = B \neq 0$, prove that $(1/g) \in \mathfrak{B}_{x_0}$. Then prove the theorem concerning limits of quotients.

e) State and prove the analogs of Theorems 8 and 9 of Section 3-4.

7. If x is any real number, prove that there exists a largest integer which does not exceed x. (If $x > 0$, show that there is a smallest positive integer N which exceeds x. Consider $N - 1$.)

8. Let P be the propositional function

$$0 < |x - 2| < \delta \Rightarrow |f(x) - 4| < \epsilon.$$

This may be quantified in the following ways.

A) $(\forall \epsilon > 0)(\forall \delta > 0)(\forall x)P.$ B) $(\forall \epsilon > 0)(\exists \delta > 0)(\forall x)P.$

C) $(\exists \epsilon > 0)(\forall \delta > 0)(\forall x)P.$ D) $(\exists \epsilon > 0)(\exists \delta > 0)(\forall x)P.$

E) $(\forall \delta > 0)(\exists \epsilon > 0)(\forall x)P.$ F) $(\exists \delta > 0)(\forall \epsilon > 0)(\forall x)P.$

a) Listed below are translations of the quantified statements above. Match the logically equivalent statements.

  I) f is defined and bounded on the set $(-\infty, 2) \cup (2, +\infty)$.

  II) $f(x) = 4$ for all real numbers x such that $x \neq 2$.

  III) There exists a deleted neighborhood of 2 throughout which f is bounded.

  IV) f is bounded throughout every deleted neighborhood of 2.

  V) $\lim_{x \to 2} f(x) = 4$.

  VI) There exists a deleted neighborhood of 2 throughout which f is identically equal to 4.

b) Note that (A) $\Rightarrow$ (E). Find all other possible implications.

c) For each of the following functions, determine which of the above quantified statements are valid.

  1) $f(x) = \dfrac{1}{x}$ 2) $f(x) = \begin{cases} 4 & \text{if } 0 \le x \le 5, \\ 0 & \text{otherwise} \end{cases}$ 3) $f(x) \equiv 4$

  4) $f(x) = \sin x$ 5) $f(x) = x$ 6) $f(x) = x^2$

9. We say that L is a *limit point* of a function f at the point a iff

$$(\forall \epsilon > 0)(\forall \delta > 0)(\exists x)(0 < |x - a| < \delta \text{ and } |f(x) - L| < \epsilon).$$

a) Find all limit points of the functions $1/x$, $\sin(1/x)$, and $\tan(1/x)$ at zero.

b) Prove that L is a limit point of f at a iff there exists a sequence $x_1, x_2, \ldots$ converging properly to a such that $\lim_{n \to \infty} f(x_n) = L$.

10. Let C be a circle of radius r, and denote by s_n and S_n the areas of regular n-sided polygons which are respectively inscribed in and circumscribed about C. Show that

$$s_n = nr^2 \sin \frac{\pi}{n} \cos \frac{\pi}{n}$$

and

$$S_n = nr^2 \tan \frac{\pi}{n}.$$

Show next that

$$\lim_{n \to \infty} s_n = \lim_{n \to \infty} S_n = \pi r^2.$$

Deduce finally that the area of C is πr^2.

11. Determine whether $\lim_{n \to \infty} [\cos(x/n)]^n$ exists.

12. Tangents are drawn to a circular arc at its midpoint and at its extremities, as indicated in Fig. 3-36. Let Δ and Δ' be the areas of the triangles ABC and $A'B'C$, respectively. Show that as the length of the arc tends to zero, the ratio Δ/Δ' approaches 4.

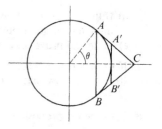

FIG. 3-36

3-6 Continuity

Not only have theologians debated over the number of angels that can dance on the head of a pin, but they have also debated whether an angel can move from point A to point B instantaneously, and whether in so doing the angel need pass through the space between. Without entering into this controversy, it is safe to say that the movements which we observe in our daily living do not display such discontinuities as those sometimes ascribed to angels. Consider a car traveling along the Pennsylvania Turnpike. For the purposes of discussion, we shall imagine the Turnpike to be a straight line on which a coordinate system has been introduced. Let $f(t)$ denote the coordinate of the car at time t. Suppose that at a certain time t_0 the car is midway through a tunnel of length 2ϵ. Then the car certainly must have been in the tunnel for some period of time (possibly quite short) prior to t_0, and it must remain in the tunnel for some period of time following t_0. In the preceding sentence the first clause may be formulated analytically as follows: There exists a $T < t_0$ such that

$$T < t \le t_0 \Rightarrow |f(t) - f(t_0)| < \epsilon.$$

The second clause has this translation: There exists a T' such that $t_0 < T'$ and

$$t_0 \le t < T' \Rightarrow |f(t) - f(t_0)| < \epsilon.$$

Set $\delta = \min(t_0 - T, T' - t_0)$. If $|t - t_0| < \delta$, then either

$$T < t \le t_0 \qquad \text{or} \qquad t_0 \le t < T.$$

In either case $|f(t) - f(t_0)| < \epsilon$. Since the tunnel was arbitrarily chosen, we have actually argued that

$$(\forall \epsilon > 0)(\exists \delta > 0)(\forall t)(|t - t_0| < \delta \Rightarrow |f(t) - f(t_0)| < \epsilon).$$

An equivalent assertion is that

$$\lim_{t \to t_0} f(t) = f(t_0).$$

Definition 1. A real-valued function f is said to be **continuous at a point** $a \in \mathbf{R}$ iff

$$\lim_{x \to a} f(x) = f(a),$$

or equivalently,

$$(\forall \epsilon > 0)(\exists \delta > 0)(\forall x)(|x - a| < \delta \Rightarrow |f(x) - f(a)| < \epsilon).$$

A function f is said to be **continuous on a set** $S \subset \mathbf{R}$ iff f is continuous at each point in the set.

Note that for a function f to be continuous at a point a, three conditions must be satisfied:

1) $f(a)$ must be defined (that is, $a \in \mathfrak{D}_f$);
2) $\lim_{x \to a} f(x)$ must exist;
3) the above limit must be $f(a)$.

Example 1. From Examples 4 and 13 of the preceding section, it follows that the sine and cosine functions are continuous on the entire real line. Also, by Example 5 of the same section, the square root function is continuous on the interval $(0, +\infty)$.

Example 2. The function $f(x) = [\sin x]$ is not continuous at $\pi/2$. In this case, conditions (1) and (2) are satisfied, but condition (3) is not, for we have shown that

$$\lim_{x \to \pi/2} f(x) = 0 \neq 1 = f(\pi/2).$$

Example 3. The function $\sin (1/x)$ is discontinuous at 0 since neither condition (1) nor condition (2) is satisfied.

From our limit theorems we can quickly deduce certain closure properties of the class of continuous functions. Suppose, for instance, that f and g are both continuous at a point x_0. Then

$$\lim_{x \to x_0} f(x) = f(x_0), \qquad \lim_{x \to x_0} g(x) = g(x_0),$$

and, consequently,

$$\lim_{x \to x_0} (f + g)(x) = f(x_0) + g(x_0) = (f + g)(x_0),$$

that is, the function $f + g$ is continuous at x_0. Products, differences, and quotients may be handled similarly.

Proposition 2. *If f and g are continuous on a set S, then so are the functions $f + g$, $f - g$, and fg. Moreover, the function f/g is continuous at all points of S where g is nonzero.*

The constant functions and the identity function $I(x) = x$ are continuous on the entire real line. Since polynomials may be built from these functions through a finite number of additions and multiplications, they too are everywhere continuous. A *rational function*, by which we mean a quotient of two polynomials, must therefore be continuous at all points where the denominator is nonzero.

The continuous functions are also closed under composition.

Proposition 3. Substitution Rule for Continuous Functions. *Suppose that f is continuous at x_0. Then*

1) $\lim_{n \to \infty} f(x_n) = f(x_0)$ *for each sequence $x_1, x_2, \ldots$ converging to x_0,*

2) $\lim_{t \to t_0} f(g(t)) = f(x_0)$ *whenever* $\lim_{t \to t_0} g(t) = x_0$. *In particular, if g is continuous at t_0 and $g(t_0) = x_0$, then $f \circ g$ is continuous at t_0.*

The proof, being very much like previous ones, is omitted. Note that hypotheses of *proper* convergence, present in the linking limit lemma and the substitution rule, do not appear in this theorem. So, for example,

$$\lim_{x \to 0} \cos [x \sin (1/x)] = \lim_{y \to 0} \cos y,$$

since the cosine function is continuous. The fact that $x \sin (1/x)$ does not converge *properly* to zero as $x \to 0$ is immaterial.

Exercises

1. a) Let n be a positive integer, and let x and x_0 be positive numbers. Show that

$$|\sqrt[n]{x} - \sqrt[n]{x_0}| = \frac{|x - x_0|}{y^{n-1} + y^{n-2}y_0 + \cdots + y_0^{n-1}} < \frac{|x - x_0|}{y_0^{n-1}},$$

where $y = \sqrt[n]{x}$ and $y_0 = \sqrt[n]{x_0}$. Prove that the function $f(x) = \sqrt[n]{x}$ is continuous on the interval $(0, +\infty)$.

b) Deduce from (a) that for each rational number r the function x^r is continuous on the interval $(0, +\infty)$.

c) For what rational numbers r is the function x^r defined and continuous on the entire line?

2. Show that $|x|$ is continuous on the entire real line. (Use the reverse triangle inequality $||x| - |a|| \le |x - a|$.)

3. Show that the function

$$f(x) = [x] + \sqrt{x - [x]}$$

is everywhere continuous.

4. Locate all points of discontinuity for each of the following functions. Sketch the graph of each function.

a) $f(x) = \dfrac{x}{x^2 - 4}$

b) $f(x) = \sec x$

c) $f(x) = \begin{cases} (\sin x)/x & \text{if } x \ne 0, \\ 1 & \text{if } x = 0 \end{cases}$

d) $f(x) = \begin{cases} 0 & \text{if } x < 0, \\ x & \text{if } 0 \le x < 1, \\ -x^2 + 4x - 2 & \text{if } 1 \le x < 3, \\ 4 - x & \text{if } 3 \le x \end{cases}$

e) $f(x) = x - [x]$

f) $f(x) = \dfrac{\sin x}{|\sin x|}$

g) $f(x) = \lim\limits_{n \to \infty} \dfrac{x^{2n} - 1}{x^{2n} + 1}$

h) $f(x) = (-1)^{[1/x]}$

i) $f(x) = (-1)^{[1/x]} \sin \dfrac{\pi}{x}$

5. For each of the functions below, the domain is the entire real line with a finite number of points deleted. Examine each of the points where the function is undefined, and determine whether the function can be defined at the point so as to be continuous there.

a) $\dfrac{x}{x - 2}$

b) $\dfrac{x^2 - 1}{x^3 - 1}$

c) $\sin \dfrac{1}{x}$

d) $x \sin \dfrac{1}{x}$

e) $\dfrac{x}{|x|}$

f) $\dfrac{\sin 2x}{x - x^2}$

g) $\dfrac{\sqrt[3]{1 + x} - 1}{x}$

6. Determine values of A and B so that the function

$$f(x) = \begin{cases} -2 \sin x & \text{if } x \le -\pi/2, \\ A \sin x + B & \text{if } -\pi/2 < x < \pi/2, \\ \cos x & \text{if } \pi/2 \le x \end{cases}$$

is continuous on the entire line.

7. If f and g are real-valued functions, we define the functions $f \lor g$ and $f \land g$ as follows:

$$\mathfrak{D}_{f \lor g} = \mathfrak{D}_{f \land g} = \mathfrak{D}_f \cap \mathfrak{D}_g$$

and

$$(f \lor g)(x) = f(x) \lor g(x) = \max \{f(x), g(x)\},$$
$$(f \land g)(x) = f(x) \land g(x) = \min \{f(x), g(x)\}$$

for all $x \in \mathfrak{D}_f \cap \mathfrak{D}_g$.

a) Graph the functions $f \lor g$ and $f \land g$ if $f(x) = x$ and $g(x) = x^2$.
b) Prove that if f and g are both continuous on a set S, then $f \lor g$ and $f \land g$ are continuous on S.

8. Sometimes the following interpretation of continuity is given: to say that f is continuous at a point x_0 means that the graph of the function immediately near the point $(x_0, f(x_0))$ can be traced by the single stroke of a pencil (i.e., without lifting the pencil). Consider the behavior of the following functions near zero, and then comment on the adequacy of this interpretation of continuity.

$$f(x) = \begin{cases} x \sin (1/x) & \text{if } x \neq 0, \\ 0 & \text{if } x = 0 \end{cases}$$

$$g(x) = \begin{cases} x & \text{if } x = 1/n \text{ for some } n \in \mathbf{Z}^+, \\ 0 & \text{otherwise} \end{cases}$$

9. Suppose that f and g are both continuous at x_0 and that $f(x_0) < g(x_0)$. Prove that there exists a $\delta > 0$ such that

$$|x - x_0| < \delta \Rightarrow f(x) < g(x).$$

Interpret this geometrically.

10. We say that f is *upper-semicontinuous at* x_0 iff

$$(\forall \epsilon > 0)(\exists \delta > 0)(\forall x)(|x - x_0| < \delta \Rightarrow f(x) < f(x_0) + \epsilon).$$

Determine the points at which each of the following functions is upper-semicontinuous.

a) $f(x) = [x]$

b) $f(x) = \begin{cases} 1/x & \text{if } x \neq 0, \\ 0 & \text{if } x = 0 \end{cases}$

c) $f(x) = \begin{cases} -1/x^2 & \text{if } x \neq 0, \\ 0 & \text{if } x = 0 \end{cases}$

d) $f(x) = \begin{cases} \sin (1/x) & \text{if } x \neq 0, \\ 0 & \text{if } x = 0 \end{cases}$

e) $f(x) = \begin{cases} \sin (1/x) & \text{if } x \neq 0, \\ 1 & \text{if } x = 0 \end{cases}$

f) $f(x) = \begin{cases} (1 + x^2) \sin (1/x) & \text{if } x \neq 0, \\ 1 & \text{if } x = 0 \end{cases}$

g) $f(x) = x - [x]$

11. Suppose that f and g are upper-semicontinuous at x_0.

a) Show that $f + g$ is upper-semicontinuous at x_0.
b) Show that $f \lor g$ is upper-semicontinuous at x_0.
c) Show that $f - g$, $f \land g$, and fg need not be upper-semicontinuous at x_0.

12. Prove that f is upper-semicontinuous at x_0 iff

1) f is bounded above near x_0;
2) $L \leq f(x_0)$, whenever L is a limit point of f at x_0.

13. We say that f is lower-semicontinuous at x_0 iff

$$(\forall \epsilon > 0)(\exists \delta > 0)(\forall x)(|x - x_0| < \delta \Rightarrow f(x_0) - \epsilon < f(x)).$$

    a) Prove that f is lower-semicontinuous at x_0 iff $-f$ is upper-semicontinuous at x_0.
    b) Prove that f is continuous at x_0 iff f is both upper-semicontinuous and lower-semi-continuous at x_0.
    c) Under what operations are lower-semicontinuous functions closed?

14. a) Give an (ϵ, δ)-proof that the function x^2 is continuous on the entire line.
    b) Deduce that if $x_n \to a$, then $x_n^2 \to a^2$.
    c) Suppose that $a_n \to A$ and $b_n \to B$. Use (a) and (b) and the identity

$$xy = \tfrac{1}{4}[(x + y)^2 - (x - y)^2]$$

to prove that $\lim_{n \to \infty} a_n b_n = AB$.

4 □ TECHNIQUES
OF DIFFERENTIATION

4-1 Definition of a derivative

Calculus centers upon two operations: differentiation and integration, both of which are defined in terms of limits. These operations may be studied separately to some extent. The relationship between the two forms the content of the so-called fundamental theorem of calculus.

Differential calculus was invented independently by Newton and Leibnitz. In Newton's work, differentiation arose in the study of motion. Suppose that a particle moves along a straight line on which a coordinate system has already been introduced. Let $f(t)$ be the coordinate of the particle at time t. [In physics, $f(t)$ is usually called the *displacement* of the particle.] The simplest type of motion occurs when $f(t)$ is a linear function of t. In this case f is of the form

$$f(t) = at + b,$$

where a and b are real constants. Observe that for any two values of t, say t_1 and t_2, we have

$$\frac{f(t_2) - f(t_1)}{t_2 - t_1} = a.$$

The constant a is called the *velocity* of the particle. According to the above equation, the velocity is equal to the change in displacement during any time interval $[t_1, t_2]$ divided by the time which has elapsed. In other words, the velocity is the change in displacement per unit time. The absolute value of the velocity is called the *speed*; it is a numerical measure of how fast the particle is moving. If the velocity is positive, the particle moves to the right, if negative, it moves to the left. (We are assuming, of course, that the usual conventions are observed in introducing coordinates.) If a particle moves with a constant velocity v, and if its displacement at time t_0 is x_0, we leave it to the reader to verify that

$$f(t) = x_0 + v(t - t_0).$$

Finally, we note that the graph of $f(t)$ is a straight line and that the slope of the line is the velocity.

We consider next the case in which $f(t)$ is not a linear function of t. In the case of a freely falling body, for example, we have $f(t) = 16t^2$, where time is measured in seconds and displacement is measured in feet. If $f(t)$ is not a linear function of t, then the ratio

$$v(t_1, t_2) = \frac{f(t_2) - f(t_1)}{t_2 - t_1}$$

is no longer constant, and the question of what we mean by the velocity of the particle then arises. The above ratio $v(t_1, t_2)$ is called the *average velocity* of the particle during the time interval (from t_1 to t_2); it clearly provides an over-all notion of the speed of the particle during a particular period of time. What we want, however, is a measure of the instantaneous velocity of the particle, or in other words, what a speedometer ought to register at some particular instant. The instantaneous velocity at time t_0 should clearly depend only on the behavior of the displacement $f(t)$ for values of t arbitrarily close to t_0. [More precisely, if $N(t_0, \delta)$ is any neighborhood of t_0, however small, then the instantaneous velocity at t_0 should be determined by the values of $f(t)$ for $t \in N(t_0, \delta)$.] Furthermore, the instantaneous velocity should in *some way* be related to the average velocity. This suggests that we define the instantaneous velocity to be a limit of the average velocity.

Definition 1. We define the **instantaneous velocity** at time t_0 to be

$$v(t_0) = \lim_{t \to t_0} v(t, t_0) = \lim_{t \to t_0} \frac{f(t) - f(t_0)}{t - t_0},$$

provided this limit exists.

Example 1. Let us calculate the velocity of a freely falling body at the end of three seconds. If we neglect air resistance, then we have $f(t) = 16t^2$, as previously observed. The desired velocity is

$$v(3) = \lim_{t \to 3} \frac{16t^2 - 16 \cdot 3^2}{t - 3} = \lim_{t \to 3} 16(t + 3) = 96 \text{ ft/sec.}$$

In more general terms,

$$v(t_0) = \lim_{t \to t_0} \frac{16t^2 - 16t_0^2}{t - t_0} = \lim_{t \to t_0} 16(t + t_0) = 32t_0 \text{ ft/sec.}$$

Limits of the form

$$\lim_{t \to t_0} \frac{f(t) - f(t_0)}{t - t_0}$$

arise in many contexts other than mechanics, and their study constitutes the subject of differential calculus.

Definition 2. Let f be a real-valued function, and let $x_0 \in \mathbf{R}$. We say that f is **differentiable** at x_0 iff the limit

$$\lim_{x \to x_0} \frac{f(x) - f(x_0)}{x - x_0}$$

exists. This limit, when it exists, is called the **derivative of f at** x_0 and is denoted by $f'(x_0)$.

Thus if $f(t)$ is the displacement of a particle, then $f'(t_0)$ is its velocity at time t_0.

There are a number of alternative notations for derivatives with which the reader should be familiar. According to our definition,

$$f'(x_0) = \lim_{x \to x_0} \frac{f(x) - f(x_0)}{x - x_0}.$$

We can make a "change of variables" by letting $h = x - x_0$. Then $x = x_0 + h$, and as x approaches x_0, h approaches zero. Thus we get

$$f'(x_0) = \lim_{h \to 0} \frac{f(x_0 + h) - f(x_0)}{h}.$$

Of course, in the above equation h is a "dummy variable," and we can replace it with anything we please. For instance, we may write

$$f'(x_0) = \lim_{\Delta x \to 0} \frac{f(x_0 + \Delta x) - f(x_0)}{\Delta x}.$$

In applications, we frequently have two physical quantities x and y related by an equation of the form $y = f(x)$. Suppose that x_0 is some fixed value of x. Then it is customary to let y_0 denote the corresponding value of y, that is,

$$y_0 = f(x_0).$$

If we set

$$\Delta x = x - x_0 \quad \text{and} \quad \Delta y = y - y_0,$$

then we may think of Δx as a change in the variable "x" and Δy as the corresponding change in "y." Thus

$$\Delta y = f(x) - f(x_0) = f(x_0 + \Delta x) - f(x_0),$$

so that the *difference quotient*

$$\frac{f(x_0 + \Delta x) - f(x_0)}{\Delta x} = \frac{\Delta y}{\Delta x}$$

may be interpreted as an average rate of change in y with respect to x, and the derivative

$$f'(x_0) = \lim_{\Delta x \to 0} \frac{\Delta y}{\Delta x}$$

represents the *instantaneous rate of change* in y with respect to x when $x = x_0$. The notation

$$\left. \frac{dy}{dx} \right|_{x=x_0} = \lim_{\Delta x \to 0} \frac{\Delta y}{\Delta x}$$

is also used.

So far we have spoken only of the "derivative at a point." This suggests that there ought to be something called a "derivative."

Definition 3. Let f be a real-valued function of a real variable. The **derivative** of f is the function f' whose domain $\mathfrak{D}_{f'}$ is the set of all points at which f is differentiable and which assigns to each $x \in \mathfrak{D}_{f'}$ the number

$$f'(x) = \lim_{\Delta x \to 0} \frac{f(x + \Delta x) - f(x)}{\Delta x}.$$

If y and x are related by the equation $y = f(x)$, then we may also write

$$f'(x) = \frac{dy}{dx} = \frac{df}{dx}.$$

Thus if s is the displacement at time t of a particle moving along a straight line, then ds/dt is the velocity of the particle. (Frequently ds/dt is called "the derivative of s with respect to t.")

The time has come for more examples.

Example 2. Let us find dy/dx if $y = x + 1/x$.

In calculating dy/dx, we shall use a recipe called the γ-process in most texts. First we write

$$y = x + \frac{1}{x}.$$

We next replace x by $x + \Delta x$ and y by $y + \Delta y$. This gives

$$y + \Delta y = x + \Delta x + \frac{1}{x + \Delta x}.$$

As a third step we subtract the first equation from the second and simplify. Thus we get

$$\Delta y = \Delta x + \frac{1}{x + \Delta x} - \frac{1}{x} = \Delta x - \frac{\Delta x}{(x + \Delta x)x}.$$

Step four involves dividing both sides by Δx:

$$\frac{\Delta y}{\Delta x} = 1 - \frac{1}{(x + \Delta x)x}.$$

Finally, we allow Δx to approach zero:

$$\frac{dy}{dx} = \lim_{\Delta x \to 0} \frac{\Delta y}{\Delta x} = 1 - \frac{1}{x^2}.$$

Example 3. Find f' if $f(x) = \sqrt{x}$.

By definition,

$$f'(x_0) = \lim_{x \to x_0} \frac{\sqrt{x} - \sqrt{x_0}}{x - x_0} = \lim_{x \to x_0} \frac{(\sqrt{x} - \sqrt{x_0})(\sqrt{x} + \sqrt{x_0})}{(x - x_0)(\sqrt{x} + \sqrt{x_0})}$$

$$= \lim_{x \to x_0} \frac{1}{\sqrt{x} + \sqrt{x_0}} = \frac{1}{2\sqrt{x_0}}.$$

Two things should be noted. The limit does not exist if $x_0 = 0$. Thus the domain of f' is the interval $(0, +\infty)$. In calculating the limit, we have tacitly assumed that the function $f(x) = \sqrt{x}$ is continuous for $x > 0$. (See Exercise 1, Section 3–6.)

Example 4. Find f' if $f(x) = |x|$.

By definition,

$$f'(x_0) = \lim_{\Delta x \to 0} \frac{|x_0 + \Delta x| - |x_0|}{\Delta x}.$$

We must consider three cases.

Case I. $x_0 > 0$. Since we are calculating the limit as Δx tends to zero, we need only consider values of Δx such that $|\Delta x| < x_0$. In the above expression for $f'(x_0)$ we may therefore replace $|x_0|$ by x_0 and $|x_0 + \Delta x|$ by $x_0 + \Delta x$. This immediately yields $f'(x_0) = 1$.

Case II. $x_0 < 0$. It is easily verified that in this case $f'(x_0) = -1$.

Case III. $x_0 = 0$. In this case

$$f'(0) = \lim_{\Delta x \to 0} \frac{|\Delta x|}{\Delta x}.$$

Now

$$\frac{|\Delta x|}{\Delta x} = \begin{cases} +1 & \text{if } \Delta x > 0, \\ -1 & \text{if } \Delta x < 0. \end{cases}$$

Hence the above limit does not exist.

In summary, the domain of f' is the real line with the point zero deleted, and

$$f'(x) = \begin{cases} +1 & \text{if } x > 0, \\ -1 & \text{if } x < 0. \end{cases}$$

Exercises

1. A ball is thrown straight up into the air with an initial velocity of 64 ft/sec. The displacement at time t will be given, to at least first approximation, by the equation

$$s = 64t - 16t^2.$$

a) Find the average velocity during the first half second, the first second, and the second second (from time $t = 1$ to $t = 2$).
b) Find the velocity of the ball at the end of $\frac{1}{2}$ sec, 1 sec, and 3 sec.
c) When will the ball reach its maximum height? When will it strike the ground?
d) Find the velocity of the ball as a function of t. What will the velocity of the ball be at the moment when it reaches its maximum height? What is the ball's velocity when it strikes the ground? How can you tell from the velocity whether the ball is going up or coming down?

2. Find dy/dx if
   a) $y = x^3 + 2x$,   b) $y = x^3 + 2x + 5$,   c) $y = (x + 1)/x$,
   d) $y = x - 1/x^2$,   e) $y = \sqrt{x + 5}$,   f) $y = 1/\sqrt{3x + 1}$.

3. Compute the following derivatives at the points indicated.
   a) $f'(\frac{1}{2})$ if $f(x) = [x]$ b) $f'(0)$ if $f(x) = \sin x$ c) $f'(0)$ if $f(x) = \cos x$
   d) $f'(0)$ if $f(x) = \begin{cases} x^2 \sin (1/x) & \text{if } x \neq 0, \\ 0 & \text{if } x = 0 \end{cases}$ e) $f'(\frac{3}{4})$ if $f(x) = x - [x]$

4. Calculate f' for each of the following. Indicate carefully the domain of f' in each case.

a) $f(x) = \sqrt[3]{x}$ b) $f(x) = [x]$ c) $f(x) = x - [x]$

d) $f(x) = \begin{cases} x & \text{if } x < 0, \\ 2x & \text{if } 0 \leq x \leq 3, \\ 9 - x & \text{if } x > 3 \end{cases}$ e) $f(x) = \begin{cases} x^2 & \text{if } x = 1/n \ (n \in \mathbf{Z}^+), \\ 0 & \text{otherwise} \end{cases}$

5. a) Let $f(x) = 2^x$. Assume that $\mathfrak{D}_f = \mathbf{R}$ and that f is differentiable at zero. Prove that f is differentiable everywhere and that

$$f'(x) = f'(0)2^x.$$

b) Generalize the preceding result.

6. Consider a thin metal rod AB. If P is any point along the rod, we denote by l its distance from A and by $\varphi(l)$ the mass of the piece of the rod from A to P (see Fig. 4–1). If there exists a constant l such that $\varphi(l) = \rho l$, we say that the rod is *homogeneous*, and we call ρ the *linear density* of the rod.

FIG. 4–1

a) In the case of a nonhomogeneous rod, define appropriately the average linear density of a segment of the rod. Also define appropriately the (instantaneous) linear density at a point.

b) Suppose that the rod is 20 cm long and that mass distribution in grams is given by

$$m = \varphi(l) = 2l + 7l^2.$$

Find the average linear density of the rod. Find the linear density at the point where $l = 5$ cm.

c) Suppose that the rod is made up of two 10 cm long homogeneous rods joined end to end. If the linear density of the first segment is 4 g/cm and that of the second is 7 g/cm, find φ. Is it meaningful to speak of the linear density of the rod at the juncture point where $l = 10$?

7. The subject of this exercise is the relationship between instantaneous velocity and constant velocity. It is recommended as an exercise in epsilonics.

Imagine two particles moving along a horizontal line on which a coordinate system has been introduced in the usual way. We denote by $f(t)$ and $v(t)$, respectively, the displacement and velocity of the first particle at time t. The second particle moves with a constant velocity v. Suppose further that at time t_0 the two particles are at the same point on the line.

a) If $v > v(t_0)$, prove that for some time immediately following t_0, the second particle will be to the east of the first. Show also that under these circumstances the second particle will be to the west of the first for some time immediately prior to t_0.

b) What is the situation if $v < v(t_0)$?

c) Is the number $v(t_0)$ completely determined by the conditions in (a) and (b)?

8. a) Show that the (instantaneous) rate of change of the area of a circle with respect to its radius is equal to the circumference.

b) Show that the rate of change of the volume of a sphere with respect to its radius is equal to the surface area.

c) What is the rate of change of the area of a square with respect to its side length?

4-2 Tangents to curves

In the preceding section the problem of determining the velocity of a particle led us to the notion of a derivative. Another problem which also leads directly to the notion of a derivative is the problem of finding tangent lines to curves.

First we must decide upon a suitable definition of a tangent line. In plane geometry a tangent line to a circle is customarily defined to be a line which intersects the circle in exactly one point. We would not want, however, to adopt this as the general definition of a tangent line to a curve. If we did, then every line parallel to the y-axis would be a tangent line to the parabola $y = x^2$. We observe, however, that the tangent line to a circle can be regarded as a limiting position of secants. Consider two points P and P' on a circle, and let $L(P, P')$ be the secant line through P and P' (see Fig. 4–2). If P' is then allowed to approach the point P, the line $L(P, P')$ approaches the tangent line to the circle at P. We shall take this property as our definition of a tangent line.

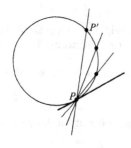

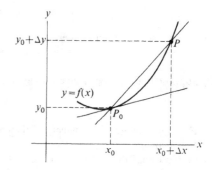

FIG. 4–2   FIG. 4–3

Let us consider the graph of a function f. Let us suppose that $y_0 = f(x_0)$ and that the graph of f has at the point $P_0 = (x_0, y_0)$ a nonvertical tangent line in the sense described above, as shown in Fig. 4–3. We shall determine the slope of this tangent line. Let P be a nearby point on the graph. If we denote the x-coordinate of P by $x_0 + \Delta x$, then the y-coordinate of P will be

$$y_0 + \Delta y = f(x_0 + \Delta x).$$

The slope $\lambda(P_0, P)$ of the secant line to the curve through P_0 and P will be

$$\lambda(P_0, P) = \frac{(y_0 + \Delta y) - y_0}{(x_0 + \Delta x) - x_0} = \frac{\Delta y}{\Delta x} = \frac{f(x_0 + \Delta x) - f(x_0)}{\Delta x}.$$

Suppose now that we allow Δx to approach zero. Then the point P on the graph will approach the point P_0, and the slope of the secant line $\lambda(P_0, P)$ will approach the slope of the tangent line at P_0. But

$$\lim_{\Delta x \to 0} \lambda(P_0, P) = f'(x_0).$$

Thus we see that $f'(x_0)$ is equal to the slope of the tangent line to the graph of the function f at the point $(x_0, f(x_0))$. It follows that the equation of the tangent line is

$$y - f(x_0) = f'(x_0)(x - x_0). \qquad (*)$$

The arguments in the preceding two paragraphs are not rigorous. In describing a tangent line to a curve as a limit of secants, we were not, for example, using the precise notion of a limit discussed in Chapter 3. Another difficulty of an even more fundamental nature is that of defining a curve. While these difficulties can be successfully resolved, this is certainly not the appropriate time to do so. Instead we shall take the easy way out. We shall simply *define* (*) to be the equation of the tangent line to the graph of f at the point $(x_0, f(x_0))$. The purpose of the preceding discussion is then purely heuristic: it serves only to make our definition appear to be a reasonable one.

Example 1. Find the equation of the tangent line to the parabola $y = x^2$ at the point where $x = 1$.

The slope of the tangent line will be

$$\left.\frac{dy}{dx}\right|_{x=1} = \lim_{\Delta x \to 0} \frac{(1 + \Delta x)^2 - 1}{\Delta x} = \lim_{\Delta x \to 0} (2 + \Delta x) = 2.$$

Hence, the equation of the tangent line is

$$y - 1 = 2(x - 1) \qquad \text{or} \qquad 2x - y - 1 = 0,$$

as shown in Fig. 4–4.

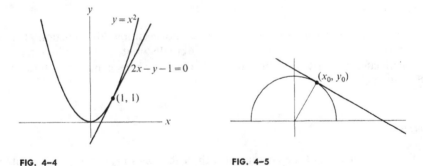

FIG. 4–4   FIG. 4–5

Example 2. We shall now show that the tangent line to a circle is perpendicular to the radius drawn to the point of contact.

Consider the circle $x^2 + y^2 = r^2$, and let (x_0, y_0) be a point on the upper semicircle (see Fig. 4–5). The upper semicircle is the graph of the function

$$f(x) = \sqrt{r^2 - x^2}.$$

Now the slope of the tangent line at (x_0, y_0) is

$$f'(x_0) = \lim_{x \to x_0} \frac{\sqrt{r^2 - x^2} - \sqrt{r^2 - x_0^2}}{x - x_0}.$$

If we multiply numerator and denominator by $\sqrt{r^2 - x^2} + \sqrt{r^2 - x_0^2}$ and simplify, we get

$$f'(x_0) = \lim_{x \to x_0} \frac{-(x + x_0)}{\sqrt{r^2 - x^2} + \sqrt{r^2 - x_0^2}} = -\frac{x_0}{\sqrt{r^2 - x_0^2}} = -\frac{x_0}{y_0}.$$

On the other hand, the slope of the radius drawn to the point (x_0, y_0) is y_0/x_0. Since the slopes are negative reciprocals of each other, the lines are perpendicular.

It should be intuitively clear that a curve cannot have a tangent line at a point if the curve has a break at that point. This suggests that differentiability implies continuity.

Proposition 1. *If f is differentiable at x_0, then f is continuous at x_0.*

Proof. By hypothesis,

$$f'(x_0) = \lim_{x \to x_0} \frac{f(x) - f(x_0)}{x - x_0}$$

exists. Let $Q(x) = [f(x) - f(x_0)]/(x - x_0)$. Then

$$f(x) = f(x_0) + Q(x)(x - x_0),$$

and by our theorems on limits, we have

$$\lim_{x \to x_0} f(x) = \lim_{x \to x_0} [f(x_0) + Q(x)(x - x_0)] = f(x_0) + f'(x_0) \cdot 0 = f(x_0).$$

Hence f is continuous at x_0.

The converse of the theorem is not true. For example, the function $f(x) = |x|$ is continuous but not differentiable at zero. Actually, matters can be infinitely worse. We shall later construct a function which is everywhere continuous but nowhere differentiable!

Exercises

1. Find the equations of the tangents to each of the following curves at the points indicated. In each case sketch the curve and the tangent line.
   a) $y = x^2$ at $(2, 4)$ and the origin   b) $y = \sqrt{x}$ at $(4, 2)$
   c) $y = 4x - x^2$ at $(2, 4)$

2. a) At what points on the curve $y = x^3$ is the tangent such as to cut the x-axis at $(1, 0)$?
   b) At what point on the curve is the tangent parallel to the line $12x - y = 16$?

3. Let $P_0 = (x_0, y_0)$ be a point on the hyperbola $xy = 1$, and let A_0 be the point where the tangent line at P_0 intersects the x-axis. Prove that the triangle OP_0A_0 is isosceles.

4. Show that the line $y = -x$ is tangent to the curve $y = x^3 - 6x^2 + 8x$. What is the point of tangency? Does the line intersect the curve at any other points?

5. Graph the function

$$f(x) = \begin{cases} x^2 \sin (1/x) & \text{if } x \neq 0, \\ 0 & \text{if } x = 0. \end{cases}$$

Show that the x-axis is the tangent line to the graph at the origin. Observe that in this case the tangent line intersects the curve an infinite number of times at points arbitrarily close to the point of tangency.

6. a) Consider the parabola $y = x^2$. Prove that a tangent line to the parabola lies entirely in the region beneath the parabola. In other words, every such tangent line is a subset of the set

$$S = \{(x, y) \,|\, y \leq x^2\}.$$

  b) Prove that if (x_0, y_0) is any point strictly beneath the parabola (that is, $y_0 < x_0^2$), then there are two distinct tangents to the parabola which pass through the point. In particular, find the tangent lines which pass through the point $(3, 5)$. Are these tangent lines perpendicular?

  c) Find the set of all points P beneath the parabola such that the two tangent lines passing through P are perpendicular.

7. For what values of a, b, and c do the graphs of

$$f(x) = x^2 + ax + b$$

and

$$g(x) = x^3 + cx$$

have a common tangent line at the point $(2, 2)$?

*8. Consider the parabola $y = 4px^2$, where p is a positive constant. Show that the area of a triangle determined by three distinct points in the parabola is twice the area of the triangle formed by the three tangent lines drawn from these three points.

9. In our discussion of tangent lines we did not consider the possibility of vertical tangent lines. Sketch carefully the curve $y = \sqrt[3]{x}$ near the origin. Should the y-axis be regarded as a tangent line to the curve at the origin? Explain. Is the function $\sqrt[3]{x}$ continuous at 0? differentiable at 0?

4-3 The differentiation of some basic functions

Throughout the rest of this chapter we shall be mainly concerned with the technique of finding derivatives. We shall get this program under way by finding the derivatives of certain basic functions. In the two sections following we shall show how to differentiate functions which can be formed from these basic functions by the operations of addition, multiplication, division, and composition.

Proposition 1. *The derivative of a constant function is zero.*

Proof. Suppose that $f(x) = c$ for all x in $\mathbf{R}$, where c is a real number. Then

$$f'(x) = \lim_{\Delta x \to 0} \frac{f(x + \Delta x) - f(x)}{\Delta x} = \lim_{\Delta x \to 0} \frac{c - c}{\Delta x} = 0.$$

Proposition 2. *If $y = x^n$, where n is a nonnegative integer, then*

$$dy/dx = nx^{n-1}.$$

Proof. Fix x_0. By the binomial theorem,

$$(x_0 + h)^n = x_0^n + \binom{n}{1} x_0^{n-1}h + \binom{n}{2} x_0^{n-2}h^2 + \cdots + h^n.$$

If we subtract x_0^n from both sides and divide by h, we get

$$\frac{(x_0 + h)^n - x_0^n}{h} = \binom{n}{1} x_0^{n-1} + \binom{n}{2} x_0^{n-2}h + \cdots + h^{n-1}.$$

The right-hand side of this equation may be regarded as a polynomial in h. Since polynomials are continuous, its limit as h tends to zero is the value of the polynomial when $h = 0$. Thus

$$\frac{d}{dx}(x^n)\Big|_{x=x_0} = \lim_{h\to 0} \frac{(x_0 + h)^n - x_0^n}{h} = \binom{n}{1} x_0^{n-1} = nx_0^{n-1}.$$

It follows from this theorem that

$$\frac{d}{dx}(x^5) = 5x^4$$

and

$$\frac{d}{dx}(x^{21}) = 21x^{20}.$$

Although we have only proved that the above differentiation formula is valid for nonnegative integer values of n, we shall later prove that it is valid for all real values of n. We note that the formula holds for $n = \frac{1}{2}$ and $n = \frac{1}{3}$, since, as we have previously seen,

$$\frac{d}{dx}(x^{1/2}) = \frac{d}{dx}(\sqrt{x}) = \frac{1}{2\sqrt{x}} = \tfrac{1}{2}x^{1/2-1}$$

and

$$\frac{d}{dx}(x^{1/3}) = \frac{d}{dx}(\sqrt[3]{x}) = \frac{1}{3\sqrt[3]{x^2}} = \tfrac{1}{3}x^{1/3-1}.$$

Proposition 3

$$\frac{d}{dx}(\sin x) = \cos x, \qquad \frac{d}{dx}(\cos x) = -\sin x.$$

Proof. By the addition formula,

$$\sin(x + h) = \sin x \cos h + \cos x \sin h.$$

Hence

$$\frac{\sin(x + h) - \sin x}{h} = \sin x \frac{\cos h - 1}{h} + \cos x \frac{\sin h}{h}.$$

In Chapter 3 we showed that

$$\lim_{h \to 0} \frac{\sin h}{h} = 1 \quad \text{and} \quad \lim_{h \to 0} \frac{\cos h - 1}{h} = 0.$$

Hence

$$\frac{d}{dx}(\sin x) = \lim_{h \to 0} \frac{\sin (x + h) - \sin x}{h} = (\sin x) \cdot 0 + (\cos x) \cdot 1 = \cos x.$$

The proof that $d(\cos x)/dx = -\sin x$ is similar and is therefore left to the reader.

Exercises

1. Prove that the differentiation formula

$$\frac{d}{dx}(x^n) = nx^{n-1} \qquad (x \neq 0)$$

   is valid if n is a negative integer.

2. a) Show that

$$\lim_{h \to 0} \frac{\tan h}{h} = 1.$$

   b) Use (a) and the addition formula for tangents to prove that

$$\frac{d}{dx}(\tan x) = \sec^2 x.$$

3. Prove that $d(\cos x)/dx = -\sin x$.

4. Let n be a positive integer. By definition,

$$\frac{d}{dx}(x^n)\Big|_{x=x_0} = \lim_{x \to x_0} \frac{x^n - x_0^n}{x - x_0}.$$

   Factor the numerator in this expression and calculate the limit to give a second proof of Proposition 2.

5. Note that

$$\frac{d}{dx}(\sin x)\Big|_{x=a} = \lim_{x \to a} \frac{\sin x - \sin a}{x - a}.$$

   Use the identity

$$\sin a - \sin b = 2 \cos \tfrac{1}{2}(a + b) \sin \tfrac{1}{2}(a - b)$$

   to calculate this limit. Prove the differentiation formula for the cosine function in a similar manner.

6. Calculate the derivatives of each of the following functions, using only the definition of a derivative. (This exercise is designed to inspire a true appreciation of the two following sections.)

   a) $\sec x$ b) $\sin^2 x$ c) $\sin x^2$

   d) $\sqrt{x}/(1 + x)$ e) $x \sin x$ f) $(\sin x)/x$

4-4 Differentiation of sums, products, and quotients

Theorem 1. *If f and g are real-valued functions which are differentiable at x_0, then $f + g$ is differentiable at x_0 and*

$$(f + g)'(x_0) = f'(x_0) + g'(x_0).$$

Proof. We find that

$$\lim_{h \to 0} \frac{(f + g)(x_0 + h) - (f + g)(x_0)}{h}$$

$$= \lim_{h \to 0} \left[\frac{f(x_0 + h) - f(x_0)}{h} + \frac{g(x_0 + h) - g(x_0)}{h} \right]$$

$$= f'(x_0) + g'(x_0),$$

since the limit of a sum is equal to the sum of the limits.

Loosely stated, we have shown that the derivative of a sum is equal to the sum of the derivatives. Thus, for example,

$$\frac{d}{dx}(x^4 + \sin x) = 4x^3 + \cos x.$$

Moreover, it follows easily by induction that the above result holds for any finite number of functions:

$$\frac{d}{dx}[u_1(x) + u_2(x) + \cdots + u_n(x)] = \frac{du_1}{dx} + \frac{du_2}{dx} + \cdots + \frac{du_n}{dx}.$$

For example,

$$\frac{d}{dx}(\sqrt{x} + 4 + x^5 + \tan x) = \frac{1}{2\sqrt{x}} + 0 + 5x^4 + \sec^2 x.$$

We have exhibited that in a good many cases limit processes commute with other operations. For example, we have shown that "the limit of a sum is equal to the sum of the limits," "the limit of a product is equal to the product of the limits," and "the derivative of a sum is equal to the sum of the derivatives." Following this rhetorical pattern, we might expect that the derivative of a product is equal to the product of the derivatives. This is false. Suppose for a moment that it were true. Then

$$\frac{d}{dx}(x^7) = \frac{d}{dx}(x^4 \cdot x^3)$$

$$= \frac{d}{dx}(x^4) \cdot \frac{d}{dx}(x^3)$$

$$= (4x^3)(3x^2) = 12x^5,$$

whereas we know that $d(x^7)/dx = 7x^6$. Let us then abandon rhetoric and substitute mathematics in our quest for a rule for differentiating products.

Let $y = u(x)v(x)$, where u and v are differentiable at x_0. We introduce the following notation:

$$u_0 = u(x_0), \qquad\qquad \Delta u = u(x_0 + \Delta x) - u_0,$$
$$v_0 = v(x_0), \qquad\qquad \Delta v = v(x_0 + \Delta x) - v_0,$$
$$y_0 = u(x_0)v(x_0), \qquad \Delta y = u(x_0 + \Delta x)v(x_0 + \Delta x) - y_0.$$

By the definition of a derivative,

$$\lim_{\Delta x \to 0} \frac{\Delta u}{\Delta x} = u'(x_0), \qquad \lim_{\Delta x \to 0} \frac{\Delta v}{\Delta x} = v'(x_0), \qquad \lim_{\Delta x \to 0} \frac{\Delta y}{\Delta x} = \frac{dy}{dx}\bigg|_{x=x_0},$$

assuming, for the moment, that the last limit exists. But

$$y_0 + \Delta y = (u_0 + \Delta u)(v_0 + \Delta v) = u_0 v_0 + u_0\,\Delta v + v_0\,\Delta u + \Delta u\,\Delta v.$$

Since $y_0 = u_0 v_0$,

$$\Delta y = u_0\,\Delta v + v_0\,\Delta u + \Delta u\,\Delta v \qquad \text{and} \qquad \frac{\Delta y}{\Delta x} = u_0 \frac{\Delta v}{\Delta x} + v_0 \frac{\Delta u}{\Delta x} + \Delta u \cdot \frac{\Delta v}{\Delta x}.$$

We now take the limit of each side as Δx approaches zero. Note first that

$$\lim_{\Delta x \to 0} \Delta u = \lim_{\Delta x \to 0} \frac{\Delta u}{\Delta x} \cdot \Delta x = u'(x_0) \cdot 0 = 0.$$

(The assertion that $\Delta u \to 0$ as $\Delta x \to 0$ is equivalent to saying that u is continuous at x_0.) Taking limits, we get

$$\frac{dy}{dx}\bigg|_{x=x_0} = u_0 v'(x_0) + v_0 u'(x_0) + 0 \cdot v'(x_0) = u(x_0)v'(x_0) + v(x_0)u'(x_0).$$

Thus we have proved the following theorem.

Theorem 2. *Let u and v be differentiable at x_0. Then uv is also differentiable at x_0 and*

$$(uv)'(x_0) = u(x_0)v'(x_0) + v(x_0)u'(x_0).$$

The rule for differentiating products may be written in less precise terms as

$$\frac{d}{dx}(uv) = u\frac{dv}{dx} + v\frac{du}{dx}.$$

Example 1. Let $y = x^2 \sin x$. Then

$$\frac{dy}{dx} = x^2 \cdot \frac{d}{dx}(\sin x) + \sin x \cdot \frac{d}{dx}(x^2)$$
$$= x^2 \cos x + (\sin x)(2x) = x^2 \cos x + 2x \sin x.$$

Let c be a constant. Since the derivative of a constant function is zero,

$$\frac{d}{dx}(cu) = c\frac{du}{dx} + u\frac{dc}{dx} = c\frac{du}{dx} + u \cdot 0 = c\frac{du}{dx}.$$

This proves the following corollary.

Corollary 3. The derivative of a constant times a function is equal to the constant times the derivative of the function.

It follows that the derivative of a difference is equal to the difference of the derivatives:

$$\frac{d}{dx}(u-v) = \frac{d}{dx}[u+(-1)v] = \frac{du}{dx} + \frac{d}{dx}[(-1)v] = \frac{du}{dx} + (-1)\frac{dv}{dx} = \frac{du}{dx} - \frac{dv}{dx}.$$

Example 2

$$\frac{d}{dx}(5x^3 - 7\tan x) = 5(3x^2) - 7\sec^2 x = 15x^2 - 7\sec^2 x.$$

The rule for differentiating products can easily be extended:

$$\frac{d}{dx}(uvw) = u\frac{d}{dx}(vw) + vw\frac{du}{dx}$$

$$= u\left[v\frac{dw}{dx} + w\frac{dv}{dx}\right] + vw\frac{du}{dx}$$

$$= uv\frac{dw}{dx} + u\frac{dv}{dx}w + \frac{du}{dx}vw.$$

More generally, it can be proved by induction that *one can obtain the derivative of a product of a finite number of factors by differentiating each factor one at a time, holding the other factors fixed, and then adding the resulting products.*

Example 3

$$\frac{d}{dx}(x^3\sin x\cos x) = 3x^2\sin x\cos x + x^3\cos x\cos x + x^3\sin x(-\sin x)$$

$$= 3x^2\sin x\cos x + x^3(\cos^2 x - \sin^2 x).$$

Next we shall obtain a rule for differentiating the reciprocal of a function. Let

$$y = 1/u(x).$$

Let us assume for the moment that y has a derivative and find out what that derivative must be. Now

$$u(x)y(x) = 1.$$

Differentiating both sides of the equation, we get

$$u\frac{dy}{dx} + y\frac{du}{dx} = 0.$$

Thus

$$\frac{dy}{dx} = -\frac{y(du/dx)}{u} = -\frac{(du/dx)}{u^2}.$$

Lemma 4. *Suppose that u is differentiable at x_0 and that $u(x_0) \neq 0$. Then $1/u$ is differentiable at x_0 and*

$$\frac{d}{dx}\left(\frac{1}{u(x)}\right)\bigg|_{x=x_0} = -\frac{u'(x_0)}{[u(x_0)]^2}.$$

Proof

$$\frac{d}{dx}\left(\frac{1}{u(x)}\right)\Bigg|_{x=x_0} = \lim_{x \to x_0} \frac{1/u(x) - 1/u(x_0)}{x - x_0} = \lim_{x \to x_0} -\frac{1}{u(x)u(x_0)} \frac{u(x) - u(x_0)}{x - x_0}$$

$$= -\frac{1}{[u(x_0)]^2} u'(x_0).$$

Note that we have used the fact that u is continuous at x_0. (Why is this true?)

Example 4. By the above formula,

$$\frac{d}{dx} (\sec x) = \frac{d}{dx}\left(\frac{1}{\cos x}\right) = -\frac{(-\sin x)}{\cos^2 x} = \tan x \sec x.$$

Theorem 5. *Let u and v be differentiable at x_0, and suppose that $v(x_0) \neq 0$. Then u/v is differentiable at x_0, and*

$$\left(\frac{u}{v}\right)'(x_0) = \frac{v(x_0)u'(x_0) - u(x_0)v'(x_0)}{[v(x_0)]^2}.$$

Less precisely,

$$\frac{d}{dx}\left(\frac{u}{v}\right) = \frac{v(du/dx) - u(dv/dx)}{v^2}.$$

This follows easily from the product rule and the lemma:

$$\frac{d}{dx}\left(\frac{u}{v}\right) = \frac{d}{dx}\left(u \cdot \frac{1}{v}\right) = u\frac{d}{dx}\left(\frac{1}{v}\right) + \frac{1}{v}\frac{du}{dx}$$

$$= u\frac{-dv/dx}{v^2} + \frac{1}{v}\frac{du}{dx} = \frac{v(du/dx) - u(dv/dx)}{v^2}.$$

Example 5

$$\frac{d}{dx} (\tan x) = \frac{d}{dx}\left(\frac{\sin x}{\cos x}\right)$$

$$= \frac{\cos x[d(\sin x)/dx] - \sin x[d(\cos x)/dx]}{\cos^2 x}$$

$$= \frac{\cos^2 x + \sin^2 x}{\cos^2 x} = \frac{1}{\cos^2 x} = \sec^2 x.$$

Example 6. We shall prove that the formula

$$\frac{d}{dx} (x^n) = nx^{n-1}$$

is valid if n is a negative integer and $x \neq 0$. Since $-n$ is a positive integer,

$$\frac{d}{dx} (x^{-n}) = (-n)x^{-n-1}$$

and

$$\frac{d}{dx} (x^n) = \frac{d}{dx}\frac{1}{x^{-n}} = -\frac{1}{(x^{-n})^2}\frac{d}{dx} (x^{-n}) = -\frac{(-n)x^{-n-1}}{x^{-2n}} = nx^{n-1}.$$

Thus, $d(1/x^6)/dx = d(x^{-6})/dx = -6x^{-7}$.

Let us summarize some of our results. First, we have proved the following general formulas for differentiation:

$$\frac{d}{dx}(u + v) = \frac{du}{dx} + \frac{dv}{dx},$$

$$\frac{d}{dx}(uv) = u\frac{dv}{dx} + v\frac{du}{dx},$$

$$\frac{d}{dx}(cu) = c\frac{du}{dx} \qquad (c \text{ a constant}),$$

$$\frac{d}{dx}\left(\frac{u}{v}\right) = \frac{v(du/dx) - u(dv/dx)}{v^2}.$$

All of these should be firmly committed to memory. In addition, the student should memorize the following special formulas:

$$\frac{d}{dx}(x^n) = nx^{n-1},$$

$$\frac{d}{dx}(\sin x) = \cos x, \qquad\qquad \frac{d}{dx}(\sec x) = \sec x \tan x,$$

$$\frac{d}{dx}(\cos x) = -\sin x, \qquad\qquad \frac{d}{dx}(\csc x) = -\csc x \operatorname{ctn} x,$$

$$\frac{d}{dx}(\tan x) = \sec^2 x, \qquad\qquad \frac{d}{dx}(\operatorname{ctn} x) = -\csc^2 x.$$

All but the last two formulas have been verified.

Exercises

1. Differentiate each of the following functions. [You may assume in this exercise that the formula

$$\frac{d}{dx}(x^n) = nx^{n-1}$$

holds for all rational numbers n.]

a) $x^4 - 3x^2 + 5x - 7$ b) $x + 1/x$

c) $\sqrt{x} + \sqrt{3}$ d) $3 \sin x + 1/\sqrt{x}$

e) $2\sqrt[3]{x} + \sqrt{2} \tan x - 11$ f) $x^4 \tan x$

g) $\sqrt{x} \sin x - 4\sqrt[3]{x^2}$ h) $(2x + 1) \sec x$

i) $\sin^2 x$ j) $\sin^3 x$

k) $\sin^4 x$ l) $\sqrt{x}(1 + 7x^2 \cos x)$

m) $x^5 \sec x \tan x$ n) $\sqrt{x} \cos^2 x \sin x$

o) $\sin 2x$ p) $\cos 2x$

q) $\cos\left(x + \dfrac{\pi}{3}\right)$

r) $\dfrac{\sin x}{x + 2}$

s) $\dfrac{1 + \sqrt{x}}{1 + x^2 + x^4}$

t) $\dfrac{x + \sin x}{x - \cos x}$

u) $\dfrac{x \sin x}{1 + x^2}$

v) $\dfrac{1 + \sqrt{x} + \sqrt[3]{x}}{1 - \sqrt[4]{x}}$

w) $\dfrac{x^7 \sin x \tan x}{1 + \sqrt{x} \cos^2 x}$

x) $\tan 2x$

y) $\tan^3 2x$

2. Verify the last two differentiation formulas listed in this section.

3. Using the product rule, prove by induction that

$$\frac{d}{dx}[f(x)]^n = n[f(x)]^{n-1}f'(x), \qquad (*)$$

where n is any positive integer.

a) Find the derivatives of $\sin^2 x$, $\sin^3 x$, and $\tan^{10} x$ from this formula.

b) Setting $f(x) = x$, we see once again that

$$\frac{d}{dx}(x^n) = nx^{n-1}, \qquad (**)$$

where n is a positive integer. Using the definition of a derivative, find the derivative of $f(x) = 1/x$, and by substitution into $(*)$, deduce that $(**)$ is also true if n is a negative integer and $x \neq 0$.

4. a) Let n be a positive integer. Prove that $\sqrt[n]{x}$ is differentiable on the entire half-line $(0, +\infty)$. Explain what use you make of the continuity of this function. (See Exercise 1, Section 3–6.)

b) Using (a) and $(*)$, prove that $(**)$ is valid whenever n is a rational number and $x > 0$.

c) For what rational numbers n is the function x^n differentiable on the entire line?

5. Show that, strictly speaking, the equation $(f + g)' = f' + g'$ is not always true. Show that the domain of $f' + g'$ can be a proper subset of the domain of $(f + g)'$.

6. Starting from the equation

$$1 + x + x^2 + \cdots + x^n = \frac{x^{n+1} - 1}{x - 1},$$

obtain formulas for

a) $1 + 2x + 3x^2 + \cdots + nx^{n-1}$,

b) $1^2x + 2^2x^2 + 3^2x^3 + \cdots + n^2x^n$.

(Simple differentiation will yield the first, and a little more, the second.)

*7. Show that the function

$$f(x) = \begin{cases} (\sin x)/x & \text{if } x \neq 0, \\ 1 & \text{if } x = 0 \end{cases}$$

has a derivative which is everywhere continuous.

4–5 The chain rule

Although the preceding section has increased enormously the repertoire of functions which we can now differentiate almost mechanically, note that we still lack the tools to differentiate such things as $\sin \sqrt{x}$ and $\sqrt{\sin x}$ without going back to the definition of a derivative. We now remedy this situation by presenting the chain rule, which is the rule for differentiating the composition product of two functions.

Suppose that $z = f(y)$ and that $y = g(x)$. Then $z = f(g(x)) = (f \circ g)(x)$. The chain rule asserts that under suitable conditions

$$\frac{dz}{dx} = \frac{dz}{dy} \frac{dy}{dx},$$

just as though the derivatives were fractions (when, in reality, they are limits of fractions). In somewhat more precise notation, the chain rule asserts that

$$\frac{dz}{dx} = f'(y)g'(x) = f'(g(x))g'(x).$$

Before entering into the proof of the chain rule (which involves some notorious technical difficulties), let us see how it can be applied.

Example 1. Differentiate $\sin \sqrt{x}$ and $\sqrt{\sin x}$.

First let $y = \sqrt{x}$, and let $z = \sin y$, so that $z = \sin \sqrt{x}$. Then

$$\frac{d}{dx}(\sin \sqrt{x}) = \frac{dz}{dy} \frac{dy}{dx} = (\cos y) \cdot \frac{1}{2\sqrt{x}} = \frac{\cos \sqrt{x}}{2\sqrt{x}}.$$

To differentiate $\sqrt{\sin x}$, let $y = \sin x$, and let $z = \sqrt{y}$. Then $z = \sqrt{\sin x}$. Differentiating, we get

$$\frac{d}{dx}(\sqrt{\sin x}) = \frac{dz}{dx} = \frac{dz}{dy} \frac{dy}{dx} = \frac{1}{2\sqrt{y}} \cdot \cos x = \frac{\cos x}{2\sqrt{\sin x}}.$$

We proceed now to the proof of the chain rule.

Theorem 1. Chain Rule. *Suppose that g is differentiable at x_0 and that f is differentiable at $g(x_0)$. Then $f \circ g$ is differentiable at x_0 and*

$$(f \circ g)'(x_0) = f'(g(x_0))g'(x_0).$$

Proof. The basic idea behind the proof of the chain rule is quite simple:

$$\frac{\Delta z}{\Delta x} = \frac{\Delta z}{\Delta y} \frac{\Delta y}{\Delta x}.$$

Taking limits, we get

$$\frac{dz}{dx} = \frac{dz}{dy} \frac{dy}{dx}.$$

However, the moment one starts to use more explicit notation, it becomes clear that more is involved.

Case I. We shall assume first that there is a deleted neighborhood V of x_0 throughout which $g(x) \neq g(x_0)$, or equivalently, that $g(x)$ converges properly to $g(x_0)$ as $x \to x_0$.

By definition,

$$(f \circ g)'(x_0) = \lim_{x \to x_0} \frac{(f \circ g)(x) - (f \circ g)(x_0)}{x - x_0} = \lim_{x \to x_0} \frac{f(g(x)) - f(g(x_0))}{x - x_0},$$

provided the limit exists. In computing this limit, we need to consider only numbers x which lie in the deleted neighborhood V. For such values of x, $g(x) \neq g(x_0)$, and we may therefore multiply and divide by $g(x) - g(x_0)$. Thus

$$(f \circ g)'(x_0) = \lim_{x \to x_0} \frac{f(g(x)) - f(g(x_0))}{g(x) - g(x_0)} \frac{g(x) - g(x_0)}{x - x_0}. \qquad (*)$$

By definition,

$$\lim_{x \to x_0} \frac{g(x) - g(x_0)}{x - x_0} = g'(x_0)$$

and

$$\lim_{y \to g(x_0)} \frac{f(y) - f(g(x_0))}{y - g(x_0)} = f'(g(x_0)).$$

The last equation, together with the substitution rule, enables us to write

$$\lim_{x \to x_0} \frac{f(g(x)) - f(g(x_0))}{g(x) - g(x_0)} = f'(g(x_0)).$$

Since the limit of a product is the product of the limits, we get from $(*)$

$$(f \circ g)'(x_0) = f'(g(x_0))g'(x_0).$$

Case II. We assume now that there is no deleted neighborhood of x_0 throughout which $g(x) \neq g(x_0)$. This implies that for each $n \in \mathbf{Z}^+$, the deleted neighborhood

$$N'(x_0, 1/n) = \{x \in \mathbf{R} \mid 0 < |x - x_0| < 1/n\}$$

must contain a point x_n such that $g(x_n) = g(x_0)$. Thus we get a sequence of points $\{x_n\}$, each different from x_0, such that

$$x_0 - 1/n < x_n < x_0 + 1/n.$$

From the pinching theorem, it is immediately clear that x_n converges properly to x_0. Hence, by the linking limit lemma, we have

$$g'(x_0) = \lim_{x \to x_0} \frac{g(x) - g(x_0)}{x - x_0} = \lim_{n \to \infty} \frac{g(x_n) - g(x_0)}{x_n - x_0} = 0.$$

We shall now prove that $(f \circ g)'(x_0) = 0$ also.

Since f is differentiable at $g(x_0) = y_0$, there must exist a deleted neighborhood of y_0 throughout which the difference quotient $[f(y) - f(y_0)]/(y - y_0)$ is bounded, say

$$\left|\frac{f(y) - f(y_0)}{y - y_0}\right| \leq M$$

whenever $0 < |y - y_0| < \epsilon$. Hence the inequality

$$|f(y) - f(y_0)| \leq M|y - y_0|$$

will hold whenever $|y - y_0| < \epsilon$. Since

$$\lim_{x \to x_0} g(x) = y_0 = g(x_0),$$

there exists a neighborhood V of x_0 such that

$$|g(x) - g(x_0)| < \epsilon$$

if $x \in V$. Consequently,

$$\left|\frac{f(g(x)) - f(g(x_0))}{x - x_0}\right| \leq M \left|\frac{g(x) - g(x_0)}{x - x_0}\right|$$

if $x_0 \neq x \in V$. Since

$$\lim_{x \to x} \frac{g(x) - g(x_0)}{x - x_0} = g'(x_0) = 0,$$

it follows from the pinching theorem that

$$(f \circ g)'(x_0) = \lim_{x \to x_0} \frac{f(g(x)) - f(g(x_0))}{x - x_0} = 0. \qquad \text{Q.E.D.}$$

Example 2. In Section 4-2 we managed with some difficulty to differentiate $\sqrt{r^2 - x^2}$. We shall now use the chain rule to do the same job.
Let $y = r^2 - x^2$, and let $z = \sqrt{y}$. Then $z = \sqrt{r^2 - x^2}$. Thus we get

$$\frac{dz}{dx} = \frac{dz}{dy}\frac{dy}{dx} = \frac{1}{2\sqrt{y}} \cdot (-2x) = -\frac{x}{\sqrt{r^2 - x^2}}.$$

Example 3. The chain rule can easily be extended. Suppose $w = f(z)$, $z = g(y)$, and $y = h(x)$. Then $w = (f \circ g \circ h)(x)$, and

$$\frac{dw}{dx} = \frac{dw}{dz}\frac{dz}{dx} = \frac{dw}{dz}\frac{dz}{dy}\frac{dy}{dx}$$

$$= f'(z)g'(y)h'(x)$$

$$= f'(g(h(x)))g'(h(x))h'(x).$$

Let us now differentiate $\sqrt{\sin(x^2)}$. Let $y = x^2$, $z = \sin y$, and $w = \sqrt{z}$. Then

$$\frac{d}{dx}(\sqrt{\sin(x^2)}) = \frac{dw}{dx} = \frac{dw}{dz}\frac{dz}{dy}\frac{dy}{dx} = \frac{1}{2\sqrt{z}}(\cos y)(2x) = \frac{x\cos(x^2)}{\sqrt{\sin(x^2)}}.$$

Example 4. After a little practice the reader should be able to use the chain rule without making substitutions. He should be able to write out without hesitation something like the following:

$$\frac{d}{dx}\left[\tan^3\left(\frac{1}{x}\right)\right] = 3\tan^2\left(\frac{1}{x}\right)\sec^2\left(\frac{1}{x}\right)\left(\frac{1}{-x^2}\right) = -\frac{3}{x^2}\tan^2\left(\frac{1}{x}\right)\sec^2\left(\frac{1}{x}\right).$$

Most of the older calculus books abound in problems on "related rates." Although they frequently seem contrived, these problems are rather fun and they do provide practice in using the chain rule.

Example 5. A spherical snowball is melting at the rate of 0.05 cc/sec. How fast is the surface area shrinking when the sphere is 10 cm in diameter?

We are given that

$$\frac{dV}{dt} = -0.05,$$

and we are asked to find dS/dt, where V and S are respectively the volume and surface area of the snowball. We use the well-known formulas

$$V = \tfrac{4}{3}\pi r^3 \qquad \text{and} \qquad S = 4\pi r^2.$$

By the chain rule, we get

$$-0.05 = \frac{dV}{dt} = \frac{dV}{dr}\frac{dr}{dt} = 4\pi r^2\frac{dr}{dt}.$$

Thus

$$\frac{dr}{dt} = -\frac{0.05}{4\pi r^2}$$

and

$$\frac{dS}{dt} = \frac{dS}{dr}\frac{dr}{dt} = 8\pi r\frac{dr}{dt} = -\frac{0.05}{r}.$$

When the diameter is 10 cm, then $r = 5$ cm, and thus

$$\frac{dS}{dt} = -0.01 \text{ cm}^2/\text{sec}$$

is the required rate.

Example 6. A beacon in a lighthouse a half mile off shore revolves at the rate of 2 revolutions per minute (rpm). Assuming that the shoreline is a straight line, how fast is the ray of light moving when it passes a point 1 mi from the lighthouse?

Let x and θ be as indicated in Fig. 4-6; x is measured in miles and θ in radians. Since the beacon revolves at the rate of 2 rpm, it follows that

$$\frac{d\theta}{dt} = 2(2\pi) = 4\pi \text{ rad/min}.$$

Now

$$x = \tfrac{1}{2}\tan\theta.$$

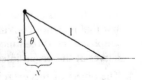

FIG. 4-6

Hence by the chain rule,

$$\frac{dx}{dt} = \frac{1}{2} \sec^2 \theta \frac{d\theta}{dt} = 2\pi \sec^2 \theta \text{ mi/min.}$$

In particular, for a point on the shore 1 mi from the lighthouse, $\theta = \pi/3$, and thus

$$\frac{dx}{dt} = \frac{2\pi}{\cos^2 (\pi/3)} = 8\pi \text{ mi/min.}$$

Exercises

1. Calculate the derivatives of the following functions.

a) $\cos 2x$

b) $\sqrt{1 + 3x}$

c) $(1 + 2x + x^3)^8$

d) $(1 + \sqrt[3]{x})^4$

e) $x\sqrt{1 - x^2}$

f) $\dfrac{1}{\sqrt{16 - x^2}}$

g) $5x \tan (x/3 + x^2)$

h) $\sqrt{x + \sqrt{x}}$

i) $\cos (\sin 3x)$

j) $\sqrt{1 + 3 \sin^2 x}$

k) $\sqrt[3]{1 + \tan (1 + \sqrt{x})}$

l) $(\sin x^2 + \cos 5x^3)^5$

m) $\sin^2 (7x - \tan (\cos x))$

n) $\sqrt[3]{x + x^3\sqrt{x - \cos x}}$

o) $\cos (x\sqrt{1 + \tan^5 (3x \sin x)})$

2. Let f and g be functions, and consider the graph of

$$y = f(x) \sin g(x).$$

Since the absolute value of the factor $\sin g(x)$ is less than or equal to one, the graph will oscillate between the graphs of $y = f(x)$ and $y = -f(x)$, as shown in Fig. 4-7. Prove that at the points where the curves touch, the curves have common tangents.

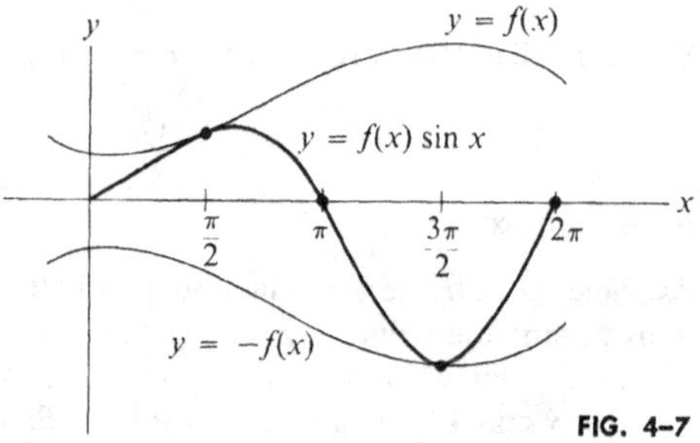

3. Let

$$f(x) = \begin{cases} x \sin (1/x) & \text{if } x \neq 0, \\ 0 & \text{if } x = 0, \end{cases}$$

$$g(x) = \begin{cases} x^2 \sin (1/x) & \text{if } x \neq 0, \\ 0 & \text{if } x = 0, \end{cases}$$

$$h(x) = \begin{cases} x^3 \sin (1/x) & \text{if } x \neq 0, \\ 0 & \text{if } x = 0. \end{cases}$$

FIG. 4-7

Prove that f is continuous but not differentiable at 0. Prove that g is differentiable at 0 but that g' is not continuous at 0. Prove that h has a continuous derivative at 0. Draw graphs of the three functions. Note that in the second two cases the tangent lines to the curves at the origin intersect the curves infinitely often.

4. The object of this exercise is to obtain a cleaner proof of the chain rule, one which circumvents the hazard of division by zero, and also which generalizes to higher dimensions.

We introduce three classes of functions. By an *infinitesimal* we shall mean a function f such that $f(0) = \lim_{x \to 0} f(x) = 0$. We denote by $\mathscr{I}$ the set of all infinitesimals. Observe that the identity function $I(x) = x$ is an infinitesimal. We shall be concerned with two subsets $\mathfrak{O}$ and $\mathfrak{o}$ (read "big oh" and "little oh"), consisting respectively of functions which approach zero *as* rapidly or *more* rapidly than the identity function. More precisely, $f \in \mathfrak{O}$ iff $f \in \mathscr{I}$ and there exists a deleted neighborhood throughout which the expression $f(x)/x$ is bounded; $f \in \mathfrak{o}$ iff $f \in \mathscr{I}$ and $\lim_{x \to 0} f(x)/x = 0$. Alternatively, to say that an infinitesimal f belongs to $\mathfrak{O}$ or $\mathfrak{o}$ means that f may be written as $f(x) = x\,g(x)$, where g belongs to $\mathfrak{B}_0$ or $\mathfrak{N}_0$, respectively. (See Exercise 6, Section 3–5.)

a) Decide which of the following functions are infinitesimals:

$$\sin x, \qquad \cos x, \qquad |x|, \qquad (\sin x)/(1 + x^2), \qquad \sqrt[3]{x}, \qquad \cos x - 1,$$

$$f(x) = \begin{cases} x^2 \sin(1/x) & \text{if } x \neq 0, \\ 0 & \text{if } x = 0, \end{cases}$$

$$g(x) = \begin{cases} x^{4/3}(-1)^{[1/x]} & \text{if } x = 0, \\ 0 & \text{if } x = 0. \end{cases}$$

Then decide which infinitesimals belong to $\mathfrak{O}$ and which belong to $\mathfrak{o}$.

b) Since $\mathfrak{N}_0 \subset \mathfrak{B}_0$, it follows that $\mathfrak{o} \subset \mathfrak{O}$. Since $\mathfrak{N}_0$ and $\mathfrak{B}_0$ are closed under addition and multiplication, it follows that $\mathfrak{o}$ and $\mathfrak{O}$ are closed under these operations. Prove that if $f \in \mathfrak{o}$ and $g \in \mathfrak{O}$, then $f \circ g$ and $g \circ f$ both belong to $\mathfrak{o}$.

c) Prove that f is differentiable at x_0 iff there exists a number $\lambda \in \mathbf{R}$ and a function $\varphi \in \mathfrak{o}$ such that

$$f(x_0 + h) - f(x_0) = \lambda h + \varphi(h),$$

in which case $\lambda = f'(x_0)$.

   Note the following two things. This provides a characterization of differentiation which does not involve division. Also, this says that if f is differentiable at x_0, then for small values of h, $f'(x_0)h$ is a good approximation to $f(x_0 + h) - f(x_0)$. In fact, among all approximations of the form λh, where λ is a constant, $f'(x_0)h$ is best in the sense that it is the only one for which the error $\varphi(h)$ tends to zero more rapidly than h.

d) If f is differentiable at x_0, apply (b) and (c) to deduce that the function Φ defined by

$$\Phi(h) = f(x_0 + h) - f(x_0)$$

belongs to $\mathfrak{O}$.

e) Fill in details and justifications for the following proof of the chain rule.

   Suppose that g is differentiable at x_0 and that f is differentiable at $y_0 = g(x_0)$. We wish to prove that

$$(f \circ g)(x_0 + h) - (f \circ g)(x_0) = f'(y_0)g'(x_0)h + \varphi(h),$$

where $\varphi \in \mathfrak{o}$. We can write

$$f(y_0 + k) - f(y_0) = f'(y_0)k + \varphi_f(k),$$
$$g(x_0 + h) - g(x_0) = g'(x_0)h + \varphi_g(h),$$

where φ_f and φ_g belong to $\mathfrak{o}$. Set

$$k = \Phi_g(h) = g(x_0 + h) - g(x_0).$$

Then

$$(f \circ g)(x_0 + h) - (f \circ g)(x_0) = f(y_0 + k) - f(y_0)$$
$$= f'(y_0)g'(x_0)h + f'(y_0)\varphi_g(h) + (\varphi_f \circ \Phi_g)(h).$$

If we set

$$\varphi(h) = f'(y_0)\varphi_g(h) + (\varphi_f \circ \Phi_g)(h),$$

then $\varphi \in \mathfrak{o}$, and the proof is complete.

5. Give an example showing how case II in the proof of the chain rule can arise.

6. A ladder 10 ft long stands vertically against a fence 4 ft high, as illustrated in Fig. 4–8. The bottom of the ladder is then pulled away from the fence at the rate of 1 ft/sec. Consider the projection of the tip of the ladder onto the ground. How fast is it moving when the bottom of the ladder is 3 ft from the fence?

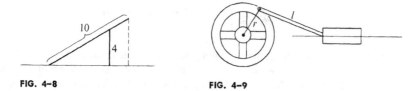

FIG. 4–8   FIG. 4–9

7. A piston which is free to slide along a straight line is attached by a shaft to a fixed point on the rim of a wheel, as shown in Fig. 4–9. The length of the shaft is l, and the radius of the wheel is r. Relate the speed of the piston to the rate of rotation of the wheel. If the wheel rotates with a constant angular velocity of a ω rad/sec, find the speed of the piston when the point on the rim where the shaft is attached is in its highest position.

8. A winch located at the end of a dock is 6 ft above the water and pulls in a rope at the rate of 2 ft/sec. Determine how fast a boat attached to the rope is being pulled toward the dock when 10 ft of the rope are out.

9. A street light is H ft high. A man h ft tall walks away from the light at the rate of F ft/sec. Find the rate at which the tip of his shadow moves when he is x ft from the base of the light.

10. Sand falls on the tip of a conical pile at the rate of 2 cu ft/sec. Assuming that the shape of the pile is always that of a right circular cone whose height is equal to the radius of its base, determine how fast the height is increasing when the pile is 6 ft high. How high must the pile be for the height to increase at a rate less than 10^{-3} ft/min?

11. A particle moves around the ellipse $16x^2 + 9y^2 = 400$ in a counterclockwise fashion. At what point(s) on the ellipse does the ordinate decrease at the same rate that the abscissa increases?

12. An airplane A flies in a straight line with a constant velocity v. Initially the airplane is directly over an antiaircraft gun positioned at 0. Determine the angular velocity of the gun if the gun is always kept pointed toward the plane.

13. Water is leaking from a hole in the vertex of a conical reservoir at the rate of 72 cu ft/min. If the reservoir is 20 ft deep and 30 ft across, find the rate at which the water level is being lowered when the water is 12 ft deep.

14. Let C be a circle of radius r whose center is at the point $(0, r)$ in the xy-plane. Imagine a motorcycle at night racing along C in the first quadrant toward the origin. Consider the point on the x-axis illumined by the headlight. Determine how fast this point moves toward the origin in terms of r, the distance s measured along C of the motorcycle from the origin, and the speed of the motorcycle.

4–6 Operators and higher-order derivatives

Most of our work has concerned functions whose domains and ranges are subsets of **R**. Let us denote by $\mathfrak{F}$ the family of all such functions. Although it might sound like a frightening or exotic idea at first, there is nothing to prevent us from stepping up a rung on the ladder of abstraction to contemplate a function whose domain and range are subsets of $\mathfrak{F}$. Such functions are usually called *operators*. Thus an operator is something which transforms functions (in the usual sense) into functions.

Example 1. The most pertinent possible example of an operator is the operator D which assigns to each function $f \in \mathfrak{F}$ its derivative f'. Thus, $D(f) = f'$. In particular, if f is the sine function, then $D(f) = f'$ is the cosine function. In symbols,

$$D(\sin) = \cos \quad \text{or} \quad [D(\sin)](x) = \cos x.$$

One more frequently sees the same state of affairs indicated by the equation

$$D(\sin x) = \cos x,$$

which, while less fussy in notation, is incorrect when interpreted literally. [Note that the equation $[D(\sin)](x) = \cos x$ says that $D(\sin)$ is that function which assigns to each real number x the number $\cos x$. The equation $D(\sin x) = \cos x$, on the other hand, says that D is a function which assigns to each number of the form $\sin x$ the number $\cos x$. Of course, no such function exists. For instance, $D(\frac{1}{2})$ would be ambiguous; since

$$\tfrac{1}{2} = \sin (\pi/6) = \sin (5\pi/6),$$

it follows that $D(\frac{1}{2})$ could be either $\cos (\pi/6) = \sqrt{3}/2$ or $\cos (5\pi/6) = -\sqrt{3}/2$.] We shall not break with tradition in this case, but simply trust that the reader will interpret properly such things as

$$D(x^3) = 3x^2 \quad \text{and} \quad D(|x|) = x/|x|.$$

The symbols d/dx, d/dt, etc., are often substituted for D. For example, we typically write

$$\frac{d}{dx}(x^4 + \cos x) = 4x^3 - \sin x \quad \text{and} \quad \frac{d}{dt}(\tan t) = \sec^2 t.$$

This notation is especially useful when unspecified constants are involved and it might otherwise be ambiguous as to how some expression is to be interpreted as a function. Thus

$$\frac{d}{dy}(ay^2 + by + c) = 2ay + b \quad \text{and} \quad \frac{d}{dx}(x^n) = nx^{n-1}.$$

Example 2. It is hardly an overstatement to describe calculus as the study of D and related operators. The so-called "calculus of finite differences" is centered upon the study of the operator Δ, which may be defined as follows:

$$[\Delta(f)](x) = f(x + 1) - f(x).$$

This Δ is called a *difference operator*. More generally, one can consider the operator Δ_h defined by

$$[\Delta_h(f)](x) = f(x + h) - f(x).$$

Let us consider a specific example. Let $f(x) = x(x - 1)$. Then

$$[\Delta(f)](4) = f(5) - f(4) = 20 - 12 = 8,$$

and, in general,

$$[\Delta(f)](x) = (x + 1)x - x(x - 1) = x[(x + 1) - (x - 1)] = 2x.$$

The latter equation is analogous to $D(x^2) = 2x$. In more general terms, we define for each positive integer n,

$$x^{(n)} = x(x - 1)(x - 2) \cdots (x - n + 1) = \prod_{k=1}^{n} (x - k + 1).$$

We can also express $x^{(n)}$ in terms of the binomial coefficients:

$$x^{(n)} = n! \binom{x}{n}.$$

Then, using the same sort of inexact notation as in the preceding example, we get from the Pascal triangle property (Exercise 31, Section 1–9)

$$\Delta(x^{(n)}) = n! \binom{x + 1}{n} - n! \binom{x}{n}$$

$$= n! \binom{x}{n - 1} = n \cdot (n - 1)! \binom{x}{n - 1}$$

$$= nx^{(n-1)},$$

which is, of course, analogous to $D(x^n) = nx^{n-1}$.

In the exercises, parallel properties of D and Δ will be developed in greater detail.

Example 3. We define an operator $\mathcal{L}$ as follows:

$$\mathcal{L}(f) = f'/f.$$

$\mathcal{L}(f)$ is called the *logarithmic derivative* of f. In particular,

$$\mathcal{L}(x^n) = \frac{nx^{n-1}}{x^n} = \frac{n}{x},$$

$$\mathcal{L}(\sin x) = \frac{\cos x}{\sin x} = \operatorname{ctn} x,$$

$$\mathcal{L}(\cos x) = \frac{-\sin x}{\cos x} = -\tan x.$$

The usefulness of logarithmic derivatives stems from the following properties:

1) $D(f) = f\mathcal{L}(f)$,
2) $\mathcal{L}(fg) = \mathcal{L}(f) + \mathcal{L}(g)$,
3) $\mathcal{L}(f/g) = \mathcal{L}(f) - \mathcal{L}(g)$,
4) $\mathcal{L}(f^r) = r\mathcal{L}(f)$ (r rational).

The latter three equations state that $\mathcal{L}$ acts on functions as logarithms act on real numbers. Without placing restrictions on g and f, none of the equations above is strictly true. In each instance the domain of the function appearing on the right-hand side can be a proper subset of the domain of the function appearing on the left-hand side. (Compare with Exercise 5, Section 4–4.) If, however, we assume that the functions f and g are differentiable at each point of their domains and that they never vanish, then the equations are valid as they stand. The proofs are easy. Property (1) follows trivially from the definition of $\mathcal{L}$. Property (2) follows from the product rule:

$$\mathcal{L}(fg) = \frac{(fg)'}{fg} = \frac{f'g + g'f}{fg} = \frac{f'}{f} + \frac{g'}{g} = \mathcal{L}(f) + \mathcal{L}(g).$$

Property (3) follows similarly from the rule for differentiating quotients. Property (4) follows from the chain rule:

$$\mathcal{L}(f^r) = \frac{rf^{r-1}f'}{f^r} = r\frac{f'}{f} = r\mathcal{L}(f).$$

Property (3) also follows from an application of (2) and (4):

$$\mathcal{L}(f/g) = \mathcal{L}(fg^{-1}) = \mathcal{L}(f) + \mathcal{L}(g^{-1}) = \mathcal{L}(f) + (-1)\mathcal{L}(g) = \mathcal{L}(f) - \mathcal{L}(g).$$

To illustrate the power of logarithmic differentiation, let us calculate the derivative of

$$f(x) = \frac{\sqrt[3]{x}\,\sin^2 x}{\sqrt{x - 2}\,\cos x} \qquad (0 < x < \pi/2).$$

First we calculate $\mathcal{L}(f)$:

$$[\mathcal{L}(f)](x) = \tfrac{1}{3}\mathcal{L}(x) + \tfrac{2}{3}\mathcal{L}(\sin x) - \tfrac{1}{2}\mathcal{L}(x - 2) - \mathcal{L}(\cos x)$$

$$= \frac{1}{3}\frac{1}{x} + \frac{2}{3}\operatorname{ctn} x - \frac{1}{2}\frac{1}{x - 2} + \tan x.$$

Thus

$$f' = f\mathcal{L}(f) = \left(\frac{1}{3x} - \frac{1}{2x - 4} + \frac{2}{3}\operatorname{ctn} x + \tan x\right)\frac{\sqrt[3]{x}\,\sin^2 x}{\sqrt{x - 2}\,\cos x}.$$

Exercise 16 is devoted to the algebra of operators. Perhaps the most important operation for operators is that of composition, which frequently is indicated by mere juxtaposition of symbols. Thus, if M and N are operators, MN denotes the operator $M \circ N$. As with ordinary functions, composition is usually noncommutative. For instance,

$$[(\mathcal{L}\Delta)(f)](x) = \frac{f'(x + 1) - f'(x)}{f(x + 1) - f(x)},$$

while

$$[(\Delta\mathcal{L})(f)](x) = \frac{f'(x+1)}{f(x+1)} - \frac{f'(x)}{f(x)}.$$

In keeping with the use of juxtaposition to indicate composition, we define the *n*th *iterate* L^n of an operator L to be the composition product of n copies of L. In other words, L^n is defined for positive integers n by the recursion formula

$$L^1 = L, \qquad L^{n+1} = L(L^n) = L \circ L^n.$$

For example,

$$D^2(x^4 + 2x) = D\big(D(x^4 + 2x)\big) = D(4x^3 + 2) = 12x^2$$

and

$$D^3(\sin x) = D\big(D^2(\sin x)\big) = D\big(D[D(\sin x)]\big)$$
$$= D\big(D(\cos x)\big) = D(-\sin x) = -\cos x.$$

In general, $D^n(f)$ is called the *n*th *derivative* of f and is denoted by $f^{(n)}$. For instance, if $f(x) = \sin x$, then

$$f^{(15)}(x) = -\cos x,$$

and in particular

$$f^{(15)}(\pi) = 1.$$

There are other frequently used notations for higher-order derivatives. If we write $y = f(x)$, then the *n*th derivative of f is often denoted by

$$\left(\frac{d}{dx}\right)^n y, \qquad \frac{d^n}{dx^n} y, \qquad \text{or} \qquad \frac{d^n y}{dx^n}.$$

When using the latter notation, we indicate the value of the *n*th derivative at a particular point x_0 by

$$\left.\frac{d^n y}{dx^n}\right|_{x_0} \qquad \text{or} \qquad \left.\frac{d^n y}{dx^n}\right|_{x=x_0}.$$

For example,

$$\left.\frac{d^2}{dx^2}(x^4 + 2x)\right|_{x=3} = 12x^2\Big|_{x=3} = 108.$$

The second and third derivatives of f are denoted by f'' and f''', respectively. The logic behind this notation should be clear:

$$D^2 f = D\big(D(f)\big) = (f')' = f'', \qquad D^3 f = D\big(D^2(f)\big) = (f'')' = f'''.$$

In this scheme of notation, however, the next few derivatives are denoted by $f^{\text{iv}}, f^{\text{v}}, f^{\text{vi}}$, etc.

Let us consider a particle moving along a straight line. We write $s = f(t)$, where s denotes displacement and t, time. We defined the derivative $v = ds/dt = f'(t)$ to be the (instantaneous) velocity of the particle. The second derivative of the displacement d^2s/dt^2, or equivalently dv/dt, is called the *acceleration* of the particle. In the case of a freely falling body, $s = 16t^2$ (roughly, the units being feet and seconds) and the acceleration is

$$a = \frac{d^2s}{dt^2} = \frac{d}{dt}(32t) = 32 \text{ ft/sec},$$

a constant. Much of the importance of the notion of acceleration stems from *Newton's second law of motion*, whose content is the equation

$$F = ma.$$

In this equation m represents the mass of the particle, and F represents the resultant force acting on the particle. (Newton used a "dot" notation for derivatives with respect to time. For example, $\dot{s}$ and $\ddot{s}$ were used to indicate the velocity and acceleration, respectively. This notation is still occasionally used in treatments of mechanics.)

Throughout the rest of the section we shall discuss techniques for calculating nth derivatives. First we state some basic rules. If $f^{(n)}(x_0)$ and $g^{(n)}(x_0)$ exist, then one can show by an easy induction argument that

$$(f + g)^{(n)}(x_0) = f^{(n)}(x_0) + g^{(n)}(x_0)$$

and

$$(cf)^{(n)}(x_0) = cf^{(n)}(x_0),$$

where c is a constant. The rule for calculating nth derivatives of products is more complicated. We begin with the product rule:

$$(fg)' = f'g + fg'.$$

Using the product rule again, we differentiate each side:

$$(fg)'' = [f''g + f'g'] + [f'g' + fg''] = f''g + 2f'g' + fg''.$$

Differentiating a third time, we get

$$(fg)''' = f'''g + 3f''g' + 3f'g'' + fg'''.$$

These equations are strongly reminiscent of the binomial theorem. We might expect by analogy that

$$(fg)^{(n)} = \sum_{k=0}^{n} \binom{n}{k} f^{(n-k)}g^{(k)},$$

where $f^{(0)} = f$. This equation is known as *Leibnitz's formula*.

Proposition 1. *If $f^{(n)}(x_0)$ and $g^{(n)}(x_0)$ exist, then $(fg)^{(n)}(x_0)$ exists and*

$$(fg)^{(n)}(x_0) = \sum_{k=0}^{n} \binom{n}{k} f^{(n-k)}(x_0)g^{(k)}(x_0).$$

Proof. We proceed by induction on n. When $n = 1$, we have the ordinary product rule. Assuming that it is true when $n = r$, we have (omitting arguments)

$$(fg)^{(r+1)} = D[(fg)^{(r)}]$$

$$= \sum_{k=0}^{r} \binom{r}{k} D[f^{(r-k)}g^{(k)}]$$

$$= \sum_{k=0}^{r} \binom{r}{k} [f^{(r+1-k)}g^{(k)} + f^{(r-k)}g^{(k+1)}]$$

$$= \sum_{k=0}^{r} \binom{r}{k} f^{(r+1-k)}g^{(k)} + \sum_{j=0}^{r} \binom{r}{k} f^{(r-k)}g^{(k+1)}$$

$$= \sum_{k=0}^{r} \binom{r}{k} f^{(r+1-k)}g^{(k)} + \sum_{k=1}^{r+1} \binom{r}{k-1} f^{(r+1-k)}g^{(k)}$$

$$= \binom{r}{0} f^{(r+1)}g^{(0)} + \sum_{k=1}^{r} \left[\binom{r}{k} + \binom{r}{k-1}\right] f^{(r+1-k)}g^{(k)} + \binom{r}{r} f^{(0)}g^{(r+1)}$$

$$= \binom{r+1}{0} f^{(r+1)}g^{(0)} + \sum_{k=1}^{r} \binom{r+1}{k} f^{(r+1-k)}g^{(k)} + \binom{r+1}{r+1} f^{(0)}g^{(r+1)}$$

$$= \sum_{k=0}^{r+1} \binom{r+1}{k} f^{(r+1-k)}g^{(k)},$$

which completes the induction argument. We note that the following pieces of information have been used in the calculation: product rule, Pascal triangle property (Exercise 31, Section 1–9), basic properties of n-fold sums (Exercise 9, Section 1–9), and the fact that

$$\binom{r}{0} = \binom{r+1}{0} = \binom{r}{r} = \binom{r+1}{r+1} = 1.$$

Example 4. If P is a polynomial of degree n, show that

$$P(x) = \sum_{k=0}^{n} \frac{P^{(k)}(0)}{k!} x^k,$$

and deduce from this the binomial theorem.

We begin by considering derivatives of x^n. We have

$$D(x^n) = nx^{n-1}, \qquad D^2(x^n) = n(n-1)x^{n-2}, \qquad D^3(x^n) = n(n-1)(n-2)x^{n-3}.$$

In general,

$$D^k(x^n) = n(n-1)\cdots(n-k+1)x^{n-k} = k!\binom{n}{k}x^{n-k}.$$

If n is a rational number, the above is valid for $x > 0$. If n is a positive integer, the above is valid for all x and we note in particular that

$$D^k(x^n) = \begin{cases} n! & \text{if } k = n, \\ 0 & \text{if } k > n. \end{cases}$$

These assertions are easily proved by induction.

Let

$$P(x) = a_0 + a_1 x + a_2 x^2 + \cdots + a_n x^n = \sum_{k=0}^{n} a_k x^k.$$

Taking the jth derivative of each side ($j \le n$), we get

$$P^{(j)}(x) = \sum_{k=0}^{n} a_k D^j(x^k) = \sum_{k=0}^{n} a_k j! \binom{k}{j} x^{k-j}$$

$$= \sum_{k=j}^{n} j! a_k \binom{k}{j} x^{k-j} = \sum_{i=0}^{n-j} j! a_{i+j} \binom{i+j}{j} x^i.$$

Setting $x = 0$ gives

$$P^{(j)}(0) = j! a_j.$$

Hence $a_j = P^{(j)}(0)/(j!)$, and thus

$$P(x) = \sum_{k=0}^{n} \frac{P^{(k)}(0)}{k!} x^k.$$

To deduce the binomial theorem, we set

$$P(x) = (a + x)^n.$$

Then P is clearly a polynomial of degree n and

$$P^{(k)}(x) = k! \binom{n}{k} (a + x)^{n-k}.$$

We set $x = 0$ and get

$$P^{(k)}(0) = k! \binom{n}{k} a^{n-k}.$$

Thus

$$(a + x)^n = P(x) = \sum_{k=0}^{n} \frac{P^{(k)}(0)}{k!} x^k = \sum_{k=0}^{n} \binom{n}{k} a^{n-k} x^k.$$

Example 5. If

$$f(x) = 1/(1 + x^2),$$

determine $f^{(n)}(0)$ for each positive integer n.

Using Leibnitz's rule, we shall take the nth derivative of each side of the equation

$$1 = (1 + x^2) f(x).$$

We get

$$0 = (1 + x^2) f^{(n)}(x) + n \frac{d}{dx} (1 + x^2) f^{(n-1)}(x) + \frac{n(n-1)}{2} \frac{d^2}{dx^2} (1 + x^2) f^{(n-2)}(x) + \cdots$$

$$= (1 + x^2) f^{(n)}(x) + 2nx f^{(n-1)}(x) + n(n-1) f^{(n-2)}(x).$$

In this way we obtain a recursion formula for $f^{(n)}(x)$. Setting $x = 0$, we get

$$f^{(n)}(0) + n(n-1) f^{(n-2)}(0) = 0 \qquad (n \ge 2).$$

To get things rolling, we observe that $f^{(0)}(0) = f(0) = 1$ and

$$f^{(1)}(0) = -\left.\frac{2x}{(1 + x^2)^2}\right|_{x=0} = 0.$$

We get succeeding derivatives from the recursion formula:

$$f^{(2)}(0) = -2 \cdot 1 f^{(0)}(0) = -2!,$$
$$f^{(3)}(0) = -3 \cdot 2 f^{(1)}(0) = 0,$$
$$f^{(4)}(0) = -4 \cdot 3 f^{(2)}(0) = 4!,$$
$$f^{(5)}(0) = 0,$$
$$f^{(6)}(0) = -6!.$$

In general,

$$f^{(2n-1)}(0) = 0, \qquad f^{(2n)}(0) = (-1)^n (2n)!.$$

Exercises

1. Use logarithmic differentiation to find the derivative of each of the following functions in the interval indicated.

a) $\dfrac{(10x^2 + 1)^2}{(3x - 4)^3 (x^2 - 1)^5}$ $(|x| < 1)$ b) $\sqrt[3]{\dfrac{x^2 + 1}{x(x + 2)^2}}$ $(x > -1)$

c) $\dfrac{\sqrt{x} \cos x}{\sqrt{1 - x^2}}$ $(|x| < 1)$ d) $\dfrac{(\sin x)^{3/4} (1 - \cos x)^{1/4}}{x^3 \sqrt{3x + 5}}$ $(0 < x < \pi)$

2. Calculate the higher derivatives indicated for each of the following functions.
   a) $f(x) = 3 \sin 2x$, $f'''(x) = ?$
   b) $f(x) = x^5 + 7x^3 - x^2$, $f^{(6)}(x) = ?$
   c) $f(x) = x \cos x$, $f''(x) = ?$
   d) $f(x) = \sin x + \sin x \cos x$, $f''(x) = ?$
   e) $f(x) = 1/(1 - x)$, $f^{(7)}(x) = ?$

3. A particle moves about the circle $x^2 + y^2 = r^2$ at the uniform rate of ω rad/sec. If at time $t = 0$ the particle is at the point $(r, 0)$, show that the projection of the particle at time t onto the x-axis is given by $x = r \cos \omega t$. Find the velocity and acceleration of this projection.

4. In the lighthouse problem (Example 6, Section 4-5), determine the acceleration of the light beam as it passes the point on the shore in question.

5. If $x = A \cos \omega t + B \sin \omega t + (E/2\omega) t \sin \omega t$, show that

$$\frac{d^2 x}{dt^2} + \omega^2 x = E \cos \omega t.$$

6. If $y = \sin (\sin x)$, show that

$$\frac{d^2 y}{dx^2} + \tan x \frac{dy}{dx} + y \cos^2 x = 0.$$

7. If $f(x) = \tan x$, show that

$$f^{(n)}(0) - \binom{n}{2} f^{(n-2)}(0) + \binom{n}{4} f^{(n-4)}(0) - \cdots = \sin \frac{n\pi}{2}.$$

[Write $\cos x f(x) = \sin x$ and apply Leibnitz's theorem.]

8. Prove that

$$\frac{d^n}{dx^n}\left(\frac{\sin x}{x}\right) = \frac{1}{x^{n+1}}\left[P \sin\left(x + \frac{n\pi}{2}\right) + Q \cos\left(x + \frac{n\pi}{2}\right)\right],$$

where P and Q are the polynomials

$$x^n - n(n-1)x^{n-2} + n(n-1)(n-2)(n-3)x^{n-4} + \cdots$$

and

$$nx^{n-1} - n(n-1)(n-2)x^{n-3} + \cdots,$$

respectively.

9. By forming in two different ways the nth derivative of x^{2n}, prove that

$$\sum_{k=0}^{n}\binom{n}{k}^2 = \frac{(2n)!}{(n!)^2}.$$

10. a) Let P be a polynomial of degree n, and let a be a real number. Prove by induction that there exists a unique polynomial Q of degree n such that

$$P(x) = Q(x - a)$$

for all $x \in \mathbf{R}$. Show that the coefficients of Q can be obtained through successive polynomial divisions by $x - a$.

b) Prove that

$$P(x) = \sum_{k=0}^{n} \frac{P^{(k)}(a)}{k!}(x - a)^k.$$

c) We say that a is a *zero of P of order r* if $P(x)$ is divisible by $(x - a)^r$ but not divisible by $(x - a)^{r+1}$. Prove that an equivalent condition is the following:

$$P(a) = P'(a) = \cdots = P^{(r-1)}(a) = 0 \quad \text{and} \quad P^{(r)}(a) \neq 0.$$

11. Let f be a function possessing an nth-order derivative throughout an open interval I. If $(1/x) \in I$, prove that

$$\frac{1}{x^{n+1}} f^{(n)}\left(\frac{1}{x}\right) = (-1)^n D^n\left[x^{n-1} f\left(\frac{1}{x}\right)\right].$$

12. Suppose that $y = f(x)$ satisfies a differential equation of the form

$$a_0(x)y'' + a_1(x)y' + a_2(x)y = 0,$$

where $a_0(x)$, $a_1(x)$, and $a_2(x)$ are polynomials, and $a_0(x)$ is of degree two or less, $a_1(x)$ of first degree or less, and $a_2(x)$ a constant. Prove that f possesses derivatives of all orders and that each of these derivatives satisfies a differential equation of the same form.

13. Suppose that the function f is such that $f^{(n)}(x) = 0$ for all $x \in \mathbf{R}$. Prove that f is a polynomial of degree $n - 1$ at most.

14. a) By differentiating each side of the equation

$$(1 + x)^n = \sum_{k=0}^{n} \binom{n}{k} x^k$$

and setting $x = 1$, obtain the equation

$$n2^{n-1} = \sum_{k=1}^{n} k \binom{n}{k}.$$

b) By similar techniques, find expressions for

$$\sum_{k=2}^{n} k(k-1)\binom{n}{k} \quad \text{and} \quad \sum_{k=1}^{n} k^2 \binom{n}{k}.$$

15. Let

$$P_n(x) = \frac{1}{2^n n!} \frac{d^n}{dx^n} (x^2 - 1)^n.$$

a) Show that $P_n(x)$ is a polynomial of degree n. (P_n is called the *Legendre polynomial* of order n.)

b) By applying Leibnitz's rule to

$$\frac{d^{n+2}}{dx^{n+2}} (x^2 - 1)^{n+1} = \frac{d^{n+2}}{dx^{n+2}} [(x^2 - 1) \cdot (x^2 - 1)^n]$$

and

$$\frac{d^{n+2}}{dx^{n+2}} (x^2 - 1)^{n+1} = \frac{d^{n+1}}{dx^{n+1}} [2(n+1)x(x^2 - 1)^n],$$

derive the relationships

$$P'_{n+1} = \frac{x^2 - 1}{2(n+1)} P''_n + \frac{(n+2)x}{n+1} P'_n + \frac{n+2}{2} P_n$$

and

$$P'_{n+1} = xP'_n + (n+1)P_n.$$

c) Deduce from (b) that P_n satisfies the differential equation

$$\frac{d}{dx} [(x^2 - 1)P'_n] - n(n+1)P_n = 0.$$

16. Let $\mathcal{P}$ be the set of all polynomials with real coefficients. By a *linear operator on* $\mathcal{P}$ we shall mean an operator L with domain $\mathcal{P}$ such that L maps $\mathcal{P}$ into itself and

$$L(f + g) = L(f) + L(g),$$
$$L(cf) = cL(f)$$

for all polynomials f and g and all real numbers c. In particular, the operators D, Δ, and Δ_h, when restricted to $\mathcal{P}$, are linear operators on $\mathcal{P}$. We denote by $\mathcal{L}(\mathcal{P})$ the set of all linear operators on $\mathcal{P}$.

a) Show that $\mathcal{L}(\mathcal{P})$ is closed under composition of operators. If L and M are in $\mathcal{L}(\mathcal{P})$, we define $L + M$ by

$$(L + M)(f) = L(f) + M(f).$$

Show that $L + M$ is also a linear operator on $\mathcal{P}$. One can show that with these operations $\mathcal{L}(\mathcal{P})$ satisfies all the field axioms with two exceptions: the commutative law of multiplication and the axiom concerning the existence of multiplicative inverses. Such a structure is called a *ring with identity*. The identity is the operator I defined by $I(f) = f$. Verify the distributive laws:

$$L(M + N) = LM + LN, \qquad (L + M)N = LN + MN.$$

For which of these laws is it essential that the operators be linear? If $L \in \mathcal{L}(\mathcal{P})$ and $c \in \mathbf{R}$, we define cL by $(cL)(f) = cL(f)$. Verify that cL is linear.
b) Prove that each member of $\mathcal{L}(\mathcal{P})$ is determined by the way in which it acts on the polynomials $1, x, x^2, x^3, \ldots$. Prove, in other words, that if L and M belong to $\mathcal{L}(\mathcal{P})$, and if $L(x^n) = M(x^n)$ for every nonnegative integer n, then $L = M$.
c) Given that P is a polynomial of degree n, show that P can be uniquely written as

$$P(x) = c_0 + c_1 x + c_2 x^{(2)} + \cdots + c_n x^{(n)}$$

and that the coefficients $c_0, \ldots, c_n$ can be obtained by successive divisions by x, $x - 1, x - 2$, etc. Prove that

$$P(x) = \sum_{k=0}^{n} \frac{\Delta^k P(0)}{k!} x^{(k)}$$

(Newton's formula). Use this formula to determine the polynomial P of degree three or less such that $P(0) = 2$, $P(1) = 2$, $P(2) = 8$, and $P(3) = 26$.
d) If L and M are linear operators on $\mathcal{P}$ such that $LM = ML$, then one can verify that the binomial formula

$$(L + M)^n = \sum_{k=0}^{n} \binom{n}{k} L^{n-k} M^k$$

continues to hold. Deduce that

$$f(x + n) = [(I + \Delta)^n f](x) = \sum_{k=0}^{n} \binom{n}{k} \Delta^k f(x)$$

for all $f \in \mathcal{P}$.
Show similarly that

$$\Delta^n f(x) = \sum_{k=0}^{n} \binom{n}{k} (-1)^k f(x + n - k).$$

e) An operator $L \in \mathcal{L}(\mathcal{P})$ is said to *have an inverse* iff there exists an operator $M \in \mathcal{L}(\mathcal{P})$ such that $LM = ML = I$. Show that such an M is unique. We call M the *inverse* of L and write $M = L^{-1}$. Show that the operator $E = I + \Delta$ has an inverse, and describe how it acts on a polynomial. Show further that E^{-1} has the expansion

$$E^{-1} = I - \Delta + \Delta^2 - \Delta^3 + \cdots$$

in the following sense: if we apply the right-hand side to any polynomial f, there will be only a finite number of nonzero terms and these will yield $E^{-1}(f)$. [If f

has degree n, then $\Delta^k(f) = 0$ for all $k \geq n + 1$, and so

$$
\begin{aligned}
I(f) &= (I + \Delta^{n+1})(f) \\
&= (I + \Delta)[I - \Delta + \Delta^2 - \cdots + (-\Delta)^n](f) \\
&= (I + \Delta)[f - \Delta(f) + \Delta^2(f) - \cdots] \\
&= (I + \Delta)[(I - \Delta + \Delta^2 + \Delta^3 + \cdots)(f)].
\end{aligned}
$$

Similarly, $(I - \Delta + \Delta^2 - \Delta^3 + \cdots)(I + \Delta) = I.$] Find in a similar fashion inverses for the operators $I - \Delta$, $I + D$, and $I - D$.

f) By rewriting the formula in Exercise 10(b), we see that if P is a polynomial, then

$$
P(x + h) = P(x) + \frac{P'(x)}{1!} h + \frac{P''(x)}{2!} h^2 + \frac{P'''(x)}{3!} h^3 + \cdots
$$

By setting $h = 1$, derive the expansion

$$
\Delta = D + \frac{D^2}{2!} + \frac{D^3}{3!} + \cdots
$$

Also show that

$$
D = \Delta - \frac{\Delta^2}{2} + \frac{\Delta^3}{3} - \frac{\Delta^4}{4} + \cdots
$$

4-7 Implicit differentiation

If $y = \sqrt{1 - x^2}$, then

$$
x^2 + y^2 = 1,
$$

and we say that the function $\sqrt{1 - x^2}$ is "defined implicitly" by the displayed equation. More generally, a function f is said to be defined implicitly by an equation of the form $F(x, y) = 0$ iff for each $x \in \mathfrak{D}_f$, $F(x, f(x)) = 0$. An equivalent assertion is that the graph of f is a subset of $\{(x, y) \in \mathbf{R}^2 \mid F(x, y) = 0\}$. An infinite number of functions can be defined implicitly by a single equation (which means that the term "defined" is misused). For instance, for each number $a \in (-1, +1)$, the function

$$
f_a(x) = \begin{cases} \sqrt{1 - x^2} & \text{if } -1 \leq x < a, \\ -\sqrt{1 - x^2} & \text{if } a \leq x \leq 1 \end{cases}
$$

is defined implicitly by the equation $x^2 + y^2 = 1$. The function

$$
f(x) = \begin{cases} \sqrt{1 - x^2} & \text{if } x \text{ is rational}, \\ -\sqrt{1 - x^2} & \text{if } x \text{ is irrational} \end{cases}
$$

is yet another example. It is true, of course, that among all the functions defined implicitly by $x^2 + y^2 = 1$, there are two functions which stand out, namely

$$
\sqrt{1 - x^2} \qquad \text{and} \qquad -\sqrt{1 - x^2},
$$

since these functions are continuous and their domains are "maximal." Furthermore, we can distinguish these two "natural" choices by specifying their values at zero. Thus, $f(x) = -\sqrt{1 - x^2}$ is "the" function defined implicitly by $x^2 + y^2 = 1$ which also satisfies the initial condition $f(0) = -1$.

We can formulate the following general problem: given an equation of the form $F(x, y) = 0$, what can be said concerning the existence, continuity, and differentiability of functions defined implicitly by this equation? The answer to this problem forms the content of the so-called "implicit function theorem." In Chapter 9 we shall settle this problem for the special case in which $F(x, y)$ is of the form $x - g(y)$. In the meantime we shall concern ourselves with solving problems of the following variety: assuming that a function $y = f(x)$ is defined implicitly by the equation $F(x, y) = 0$, and assuming that this function is differentiable, calculate the derivatives of the function dy/dx, d^2y/dx^2, ... in terms of x and y. If the hypothesis on f is not met, then of course the expression for dy/dx which we obtain in this way can be meaningless. Nonetheless, the implicit function theorem provides this reassurance: if the expression which we obtain formally for dy/dx is meaningful (e.g., does not involve zero in a denominator) at a point (x_0, y_0) where $F(x_0, y_0) = 0$, and if $F(x, y)$ is "nicely" behaved near (x_0, y_0), then there does indeed exist a function f such that

1) $\mathcal{D}_f$ is an open interval I containing x_0,

2) f is defined implicitly by the equation $F(x, y) = 0$,

3) f satisfies the initial condition $y_0 = f(x_0)$,

4) f is differentiable throughout I.

Let us consider some examples.

Example 1. Find the slope of the tangent line to the circle $x^2 + y^2 = 25$ at the point $(3, -4)$.

This problem can be worked without calculus. Since the tangent line is perpendicular to the radius drawn to the point $(3, -4)$, its slope must be the negative reciprocal of $-\frac{4}{3}$, or in other words, $\frac{3}{4}$.

We give two solutions using calculus, the first explicit and the second implicit.

Since the point $(3, -4)$ lies on the lower semicircle, we let

$$y = -\sqrt{25 - x^2}.$$

Then

$$\frac{dy}{dx} = \frac{x}{\sqrt{25 - x^2}},$$

so that the required slope is

$$\left.\frac{dy}{dx}\right|_{x=3} = \frac{3}{4}.$$

In using implicit differentiation we do not bother to solve the equation $x^2 + y^2 = 25$ to express y explicitly as a function of x. Instead we imagine that this has already been done, so we can treat the equation $x^2 + y^2 = 25$ as an *identity* in x. [More precisely, if $y = f(x)$

is defined implicitly by $x^2 + y^2 = 25$, then the identity which we have in mind is

$$x^2 + [f(x)]^2 = 25.]$$

We then differentiate both sides with respect to x. This yields

$$2x + 2y\frac{dy}{dx} = \frac{d}{dx}(25) = 0.$$

Thus

$$\frac{dy}{dx} = -\frac{x}{y},$$

so that the required slope is

$$\frac{dy}{dx}\bigg|_{\substack{x=3 \\ y=-4}} = \frac{3}{4}.$$

Our expression for dy/dx is meaningful for all points (x_0, y_0) on the circle $x^2 + y^2 = 25$ with two exceptions, namely, $(-5, 0)$ and $(+5, 0)$. If $y_0 > 0$, one can check that the function $f(x) = \sqrt{25 - x^2}$ ($|x| < 5$) satisfies properties (1), (2), (3), and (4) of the second paragraph in this section. If $y_0 < 0$, then

$$f(x) = -\sqrt{25 - x^2} \qquad (|x| < 5)$$

has these properties. If $y_0 = 0$, then there is no function f with these properties.

Example 2. Even when a function is defined explicitly, it may be convenient to calculate its derivatives by implicit differentiation. Consider, for instance, the job of calculating the first and second derivatives of

$$y = \sqrt[3]{x + \sqrt{x - 2}}$$

at the point $x = 6$.

We first get rid of a radical:

$$y^3 = x + \sqrt{x - 2}.$$

By the chain rule

$$3y^2\frac{dy}{dx} = 1 + \frac{1}{2\sqrt{x - 2}}. \qquad (*)$$

Setting $x = 6$, we get

$$\frac{dy}{dx}\bigg|_{x=6} = \frac{5}{48}.$$

Differentiating $(*)$ a second time, we get

$$6y\frac{dy}{dx} + 3y^2\frac{d^2y}{dx^2} = -\frac{1}{4(x - 2)^{3/2}}.$$

Setting $x = 6$ and using our previously calculated value for dy/dx, we get

$$\frac{d^2y}{dx^2}\bigg|_{x=6} = -\frac{41}{32}.$$

Example 3. Find dy/dx and d^2y/dx^2 if

$$x^{2/3} + y^{2/3} = a^{2/3} \qquad (a > 0).$$

Differentiating both sides with respect to x and multiplying by $\frac{3}{2}$, we get

$$x^{-1/3} + y^{-1/3}\frac{dy}{dx} = 0.$$

Thus

$$\frac{dy}{dx} = -\left(\frac{y}{x}\right)^{1/3}.$$

Since

$$x\left(\frac{dy}{dx}\right)^3 = -y,$$

we get, upon differentiation,

$$\left(\frac{dy}{dx}\right)^3 + 3x\left(\frac{dy}{dx}\right)^2\frac{d^2y}{dx^2} = -\frac{dy}{dx}.$$

Hence

$$\frac{d^2y}{dx^2} = -\frac{1 + (dy/dx)^2}{3x(dy/dx)} = \frac{1 + y^{2/3}/x^{2/3}}{3x^{2/3}y^{2/3}} = \frac{a^{2/3}}{3x^{4/3}y^{1/3}}.$$

Example 4. Find dy/dx if $x^3 + y^3 - 3xy = 0$.

In this case we cannot obtain an explicit expression for the functions defined implicitly by the equation. We shall simply have to accept on faith the fact that the problem is meaningful, and proceed formally. Thus we get

$$3x^2 + 3y^2\frac{dy}{dx} - 3y - 3x\frac{dy}{dx} = 0, \qquad \frac{dy}{dx} = \frac{y - x^2}{y^2 - x}.$$

Exercises

1. Find a single equation in x and y which defines implicitly each of the following functions:

$$f(x) = x + \sqrt{x^2 + 1}, \qquad g(x) = x - \sqrt{x^2 + 1}.$$

Determine $f'(2)$ and $g'(-1)$ directly and by implicit differentiation.

2. A curve C defined by an equation of the form

$$Ax^2 + By^2 + 2Cx + 2Dy + 2Exy + F = 0,$$

where A, B, C, D, E, and F are real constants, is called a *conic section*. Assuming that (x_1, y_1) is a point on C where the curve has a tangent line, show that the equation of the tangent line is

$$Ax_1x + By_1y + C(x_1 + x) + D(y_1 + y) + E(x_1y + y_1x) + F = 0.$$

[This is called the *scratch rule*. Note that if we remove the subscripts (i.e., the "scratches") from the equation for the tangent line, we get the equation for C.]

3. Show that the function

$$f(x) = \begin{cases} \cos x & \text{if } x \le \pi/4, \\ \sin x & \text{if } x > \pi/4 \end{cases}$$

is defined implicitly by the equation

$$2y^2 - 2(\sin x + \cos x)y + \sin 2x = 0.$$

Using implicit differentiation, find an expression for dy/dx in terms of x and y. Show that f is continuous but not differentiable at $x = \pi/4$. Is this fact somehow reflected in the general expression obtained for dy/dx?

4. Find dy/dx in each of the following cases.

a) $x^{1/2} + y^{1/2} = a^{1/2}$
b) $y^3 = 4(x^2 + y^2)$
c) $x \sin y - \cos y + \cos 2y = 0$
d) $x^p + y^p = a^p$
e) $x/(x - y) = 1/\sqrt{x + y}$
f) $x^m y^n = (x + y)^{m+n}$
g) $x \cos y = \sin (x + y)$

5. Find the equations of the tangent lines to the curves given below at the points indicated.

a) $x^2 - y^2 = 1$ at $(\sqrt{5}, 2)$
b) $2x^3 + 2y^3 = 9xy$ at $(3, 1)$
c) $x \cos y = \sin (x + y)$ at $(0, 0)$ and at $(\pi/2, \pi/2)$

6. Given $x^3 + y^3 - 3xy = 0$, find the values of dy/dx and d^2y/dx^2 when $x = \frac{2}{3}$ and $y = \frac{4}{3}$.

7. If $x = \tan \sqrt{y}$, show that

$$(x^2 + 1)^2 \frac{d^2y}{dx^2} + 2x(x^2 + 1)\frac{dy}{dx} = 2.$$

8. If $\sqrt[m]{y} + 1/\sqrt[m]{y} = 2x$, show that

$$(x^2 - 1)\frac{d^{n+2}y}{dx^{n+2}} + (2n + 1)x\frac{d^{n+1}y}{dx^{n+1}} + (n^2 - m^2)\frac{d^n y}{dx^n} = 0.$$

5 □ COMPLETENESS
OF THE REAL NUMBERS

One of the main objectives of this chapter is to explore some of the "global" properties of continuous functions. Specifically, we shall prove three important theorems: the intermediate-value theorem, what we shall call the theorem on extreme values, and a theorem on "uniform continuity." These theorems are deceptive. On the first encounter they are likely to seem self-evident and unimportant. In reality their proofs are reasonably sophisticated and require a more searching study of the real numbers than we have undertaken thus far. Furthermore, on these three results rest most of the really important theorems of calculus (such as the mean-value theorems and the fundamental theorem of calculus).

5-1 The least upper bound axiom and the Archimedean ordering property

It is occasionally suggested that mathematics is a kind of game. According to the rules of the game, one starts from a set of "axioms" and tries to work out their logical implications. Unlike the attitude in most elementary geometry courses, these axioms are not regarded as "self-evident truths," but rather they form part of the definition of some mathematical structure (such as a field or a group). So far we have played pretty much according to the rules. We have attempted to deduce everything from the axioms for an ordered field. We have cheated only once or twice. For instance, in proving that $\lim_{n \to \infty} 1/n = 0$, we made the additional assumption that the real numbers are Archimedean-ordered. Although this property seems obvious if we think of the real numbers as formal decimal expansions, it cannot be proved from the axioms for an ordered field. With only a few exceptions, however, the theorems which we have proved are valid if the real numbers are replaced by any ordered field. If, for instance, f is a function whose domain and range are subsets of an ordered field F, then the definitions of a limit and a derivative make perfectly good sense. Furthermore, our basic theorems concerning the differentiation of sums, products, etc., remain valid for such functions. By contrast, the real numbers will play an essential role in the proofs of the intermediate-value theorem and the theorem on extreme values: other ordered fields will not suffice.

What then is so special about the real numbers? The answer is that the real numbers have no "gaps." They are *complete* in a sense which we shall now attempt to make precise.

Definition 1. Let S be a subset of an ordered field F. We say that an element $M \in F$ is an **upper bound of** S iff $x \leq M$ whenever $x \in S$. We say that S is **bounded above** if it has an upper bound.

Example 1. Let $F = \mathbf{R}$. Then $S = [0, 1]$ is bounded above, since, for example, 3 is an upper bound of S. The numbers 10 and 1 are also upper bounds of S. More generally, the set of all upper bounds of S is the interval $[1, +\infty)$.

Example 2. Let $F = \mathbf{R}$, and let $S = \{(n - 1)/n \mid n \in \mathbf{Z}^+\}$. Then S is again bounded above, and the set of upper bounds is the interval $[1, +\infty)$.

Example 3. Let $F = \mathbf{R}$ and $S = \mathbf{Z}^+$. Then according to the Archimedean ordering property, S is not bounded above.

If a set S is bounded above, then among its upper bounds there may be a smallest element. Such an element is called the least upper bound of the set.

Definition 2. Let F be an ordered field, and let S be a subset of F which is bounded above. An element $L \in F$ is called **the least upper bound of** S (or **the supremum of** S) iff L is an upper bound of S and $L \leq M$ whenever M is an upper bound of S. If L is the least upper bound of S, we write

$$L = \text{lub } S = \text{sup } S.$$

In Examples 1 and 2 above the number 1 is the least upper bound of S. Note that the least upper bound of a set may or may not be a member of the set.

Proposition 3. *The least upper bound of a set is unique if it exists. If $L \in F$ is the least upper bound of S, then for each positive element ϵ in F, there exists an element $x \in S$ such that $x > L - \epsilon$.*

Proof. Suppose that L and L' are least upper bounds of S. Since L is a least upper bound and L' is an upper bound, $L \leq L'$. Similarly, $L' \leq L$. Hence $L = L'$, which proves that a set can have at most one least upper bound.

Suppose that L is the least upper bound of S. We shall prove that

$$(\forall \epsilon > 0)(\exists x \in S)(x > L - \epsilon).$$

Suppose that this is false. Then $(\exists \epsilon > 0)(\forall x \in S)(x \leq L - \epsilon)$, or equivalently, $L - \epsilon$ is an upper bound of S. However, since L is the least upper bound of S, it follows that $L \leq L - \epsilon$. But, of course, $L = (L - \epsilon) + \epsilon > L - \epsilon$, a contradiction.

Definition 4. An ordered field F is said to be **complete** iff the following condition is satisfied: every nonempty subset $S \subset F$ which is bounded above has a least upper bound.

This condition is called the *least upper bound axiom*. We can now state our one remaining assumption concerning the real numbers.

Axiom 5. *The real numbers form a complete ordered field.*

Not only do the real numbers form a complete ordered field, but in a certain sense they also constitute the *only* example of a complete ordered field.

The remainder of this chapter is devoted to consequences of the least upper bound axiom. The first of these is the Archimedean ordering property.

Theorem 6. Archimedean Ordering Property. *Given any real number x, there exists a positive integer n such that $n > x$.*

Proof. Suppose that this is false. Then the set $\mathbf{Z}^+$ would be bounded above, and according to the least upper bound axiom, $\mathbf{Z}^+$ would therefore have a least upper bound L. Since $1 > 0$, there exists, according to Proposition 3, an $n \in \mathbf{Z}^+$ such that

$$n > L - 1.$$

But this implies that

$$n + 1 > L,$$

and since $(n + 1) \in \mathbf{Z}^+$, this contradicts the assumption that L is an upper bound of $\mathbf{Z}^+$.

From this theorem it follows, as in Chapter 3, that

$$\lim_{n \to \infty} 1/n = 0 \quad \text{and} \quad \lim_{n \to \infty} r^n = 0,$$

where $|r| < 1$.

Theorem 7. *Between any two real numbers there exists a rational number.*

Proof. Let us consider first the case in which x and y are real numbers such that $0 < x < y$. Then $\epsilon = y - x > 0$. Since $1/n \to 0$, there exists an $n \in \mathbf{Z}^+$ such that

$$1/n < \epsilon = y - x.$$

By the Archimedean ordering property, there exists a positive integer m such that

$$m > nx,$$

and among such integers there is (by the well-ordering of $\mathbf{Z}^+$) a smallest one which we continue to denote by m. Hence

$$m - 1 \le nx.$$

Combining these inequalities, we get

$$x < \frac{m}{n} = \frac{m-1}{n} + \frac{1}{n} < x + (y - x) = y.$$

Thus $x < m/n < y$, which completes the proof for this case.

For the case in which $x < y < 0$, we note that $0 < -y < -x$. Hence by the previous case, there exists a rational number r such that $-y < r < -x$. Then $x < -r < y$, and since $-r$ is rational, this completes the proof for this second case. The remaining cases are left for the reader to worry about.

One often summarizes the content of this theorem by saying that the rational numbers are *dense* in **R**.

Definition 8. A set $S \subset \mathbf{R}$ is said to be **dense** in **R** iff between any two real numbers there exists an element of S.

We conclude with an additional consequence of the Archimedean ordering property.

Proposition 9. *If S is a nonempty subset of* **R** *which is bounded above, then a sequence* $x_1, x_2, \ldots$ *may be extracted from S such that*

$$\lim_{n \to \infty} x_n = \text{lub } S.$$

Proof. Let $n \in \mathbf{Z}^+$. Since $1/n > 0$, there exists a number $x_n \in S$ such that

$$L - 1/n < x_n,$$

where $L = \text{lub } S$. Since L is an upper bound of S, we also have $x_n \leq L$. Thus it is possible to get a sequence $x_1, x_2, \ldots$ of elements of S such that

$$L - 1/n < x_n \leq L$$

for each $n \in \mathbf{Z}^+$. Since $1/n \to 0$, it follows from the pinching theorem that

$$\lim_{n \to \infty} x_n = L.$$

Exercises

1. Is the null set bounded above? Does it have a least upper bound?
2. Define appropriately the terms lower bound and greatest lower bound. Formulate a greatest lower bound axiom, and prove that it is logically equivalent to the least upper bound axiom.
3. A *dyadic rational* is defined to be a number of the form $m/2^n$, where $m \in \mathbf{Z}$ and $n \in \mathbf{Z}^+$. Prove that the dyadic rationals are dense.
4. Let $S \subset \mathbf{R}$. Prove that S is dense in **R** iff for each real number L there exists a sequence $x_1, x_2, \ldots$ of elements of S which converge properly to L.
5. Let $S \subset \mathbf{R}$ be dense in **R**. Prove that between any two real numbers there are an infinite number of elements of S.
6. Let A and B be sets of real numbers, and let $c \in \mathbf{R}$. We define the sets $A + B$, cA, and $A \cdot B$ as follows:

$$A + B = \{x + y \mid x \in A \text{ and } y \in B\},$$
$$cA = \{cx \mid x \in A\},$$
$$A \cdot B = \{xy \mid x \in A \text{ and } y \in B\}.$$

   a) If A and B are bounded above, prove that $A + B$ is also bounded above and that $\text{lub } (A + B) = \text{lub } A + \text{lub } B$.
   b) If A is bounded above and $c \in \mathbf{R}$ is positive, show that cA is bounded above and that
$$\text{lub } cA = c \text{ lub } A.$$

c) If A and B are sets of nonnegative numbers which are bounded above, prove that $A \cdot B$ is bounded above and that

$$\text{lub } A \cdot B = (\text{lub } A)(\text{lub } B).$$

7. Let S be any subset of $\mathbf{R}$. Show that the set of upper bounds of S is either empty or else an interval of the form $[c, +\infty)$.

8. a) If K is an arbitrary field, then it can be shown (by the argument used in proving Proposition 1, Section 1-9) that K contains a unique smallest subset which contains the unit element 1 of K and which is closed under addition by 1. We denote this set by $(\mathbf{Z}^+)_K$. If $K = \mathbf{Z}_2$ or $\mathbf{Z}_3$, observe that $(\mathbf{Z}^+)_K = K$. If K is an ordered field, one can show that $(\mathbf{Z}^+)_K$ consists of positive elements of K, and that $(\mathbf{Z}^+)_K$ is closed under addition, multiplication, and subtraction of a smaller element from a larger one. One can also show that $(\mathbf{Z}^+)_K$ is well-ordered. All the proofs are the same as before. (Indeed they must be, since the only assumption we made about the real numbers was that they formed an ordered field.)

b) Let K be an arbitrary field, and let φ be any ordered field. We define a mapping φ of $(\mathbf{Z}^+)_F$ into K as follows:

$$\varphi(1) = 1', \qquad \varphi(n + 1) = \varphi(n) + 1',$$

where 1 and 1' denote the unit elements of F and K, respectively. Carefully justify this definition. (Is it important that F be ordered? Would the definition be valid if F were $\mathbf{Z}_2$ or $\mathbf{Z}_3$?) Prove that φ *preserves operations* in the sense that

$$\varphi(m + n) = \varphi(m) + \varphi(n), \qquad \varphi(mn) = \varphi(m)\varphi(n)$$

for all $m, n \in (\mathbf{Z}^+)_F$. Prove also that $\mathcal{R}_\varphi = (\mathbf{Z}^+)_K$. [Prove first by induction that $\mathcal{R}_\varphi \subset (\mathbf{Z}^+)_K$. Next show that $\mathcal{R}_\varphi$ contains 1' and is closed under addition by 1'.]

c) Prove that if K is also an ordered field, then φ is *order-preserving* in the sense that if $m, n \in (\mathbf{Z}^+)_F$ and $m < n$, then $\varphi(m) < \varphi(n)$. In particular, the mapping φ must be *one-to-one*, that is, distinct members of $(\mathbf{Z}^+)_F$ are mapped onto distinct members of $(\mathbf{Z}^+)_K$.

Let us summarize these results. If we are given two ordered fields F and K, then we can pair off members of $(\mathbf{Z}^+)_F$ and $(\mathbf{Z}^+)_K$: if $m \in (\mathbf{Z}^+)_F$, then we pair it with $m' = \varphi(m)$ in $(\mathbf{Z}^+)_K$. This pairing $m \leftrightarrow m'$ provides a one-to-one correspondence between the sets $(\mathbf{Z}^+)_F$ and $(\mathbf{Z}^+)_K$ [each element of $(\mathbf{Z}^+)_F$ is paired with exactly one element of $(\mathbf{Z}^+)_K$ and conversely]. Furthermore, paired elements behave in exactly the same way with respect to the field operations on F and K, since

$$(a + b)' = a' + b' \qquad \text{and} \qquad (ab)' = a'b'.$$

We say that $(\mathbf{Z}^+)_F$ and $(\mathbf{Z}^+)_K$ are *isomorphic* (abstractly the same) and that the mapping φ is an *isomorphism*.

*d) Observe that if $K = \mathbf{Z}_2$ or $\mathbf{Z}_3$, then the mapping φ is not one-to-one. If K is an arbitrary field, show that there are two possibilities:

   i) φ is one-to-one and $(\mathbf{Z}^+)_F$ and $(\mathbf{Z}^+)_K$ are isomorphic. (In this case K is said to be of *characteristic zero*.)

   ii) φ is not one-to-one and $(\mathbf{Z}^+)_K$ is a subfield of K consisting of a prime number p of elements. (In this case K is said to be of *characteristic p*.)

The following hints should get things rolling. Assume that $\varphi(m) = \varphi(n)$, where $m, n \in \mathbf{Z}^+$ and $m < n$. Show that $\varphi(n - m) = 0$. Thus

$$N_\varphi = \{n \in (\mathbf{Z}^+)_F \mid \varphi(n) = 0\}$$

is nonempty. Let p be the smallest member of N_φ. Show that p is a prime, i.e., the only elements of $(\mathbf{Z}^+)_F$ which divide p are p itself and 1. (Assume that this is false, use the fact that φ preserves operations, and get a contradiction via Proposition 6, Section 1–2.) Show that

$$N_\varphi = \{np \mid n \in (\mathbf{Z}^+)_F\}$$

and that $\varphi(m) = \varphi(n)$ iff m and n differ by an element of N_φ. Deduce that

$$(\mathbf{Z}^+)_K = \mathcal{R}_\varphi = \{\varphi(1), \varphi(2), \ldots, \varphi(p)\},$$

where the elements in the brackets are distinct. Now show that $(\mathbf{Z}^+)_K$ is a field. [In proving that $(\mathbf{Z}^+)_K$ satisfies M4, use Exercise 34, Section 1–9.]

9. An upshot of the previous exercise is that for any two ordered fields F and K, the sets of positive integers $(\mathbf{Z}^+)_F$ and $(\mathbf{Z}^+)_K$ are abstractly the same; thus no harm can come from dropping subscripts and regarding *the* set of positive integers $\mathbf{Z}^+$ as a subset of every ordered field F. As before, we let $\mathbf{Z}$ denote the subset of F consisting of elements of $\mathbf{Z}^+$, their additive inverses, and zero. We also let

$$\mathbf{Q} = \{m/n \mid m, n \in \mathbf{Z} \text{ and } n \neq 0\}.$$

$\mathbf{Q}$ is called the set of rational numbers in F. Prove that for any ordered field F, the following are logically equivalent:

a) F is Archimedean-ordered (that is, $\mathbf{Z}^+$ is not bounded above in F);
b) $\lim_{n \to \infty} 1/n = 0$;
c) $\lim_{n \to \infty} r^n = 0$ for each $r \in F$ such that $|r| < 1$;
d) $\mathbf{Q}$ is a dense subset of F.

10. We now give an example of a non-Archimedean-ordered field. We let F be the set of all rational functions with rational coefficients. Thus each $f \in F$ is of the form

$$f(x) = \frac{a_0 x^m + a_1 x^{n-1} + \cdots + a_m}{b_0 x^n + b_1 x^{n-1} + \cdots + a_n}, \qquad (*)$$

where $a_0, \ldots, a_m, b_0, \ldots, b_n$ are rational numbers. We regard two such rational functions f and g as the same iff $f(x) = g(x)$ for all $x \in \mathbf{R}$ with at most a finite number of exceptions. Thus if

$$f(x) = \frac{(x^2 + 1)(x^2 - 1)}{(x - 1)^2(x - 3)} \qquad \text{and} \qquad g(x) = \frac{(x^2 + 1)(x + 1)(x - 4)}{(x - 1)(x - 4)(x - 3)},$$

then f and g are the same.

a) Prove that F is a field with respect to the usual operations of addition and multiplication of functions. Also observe that F contains $\mathbf{Q}$ in the form of the constant functions.

b) If $f \in F$, we define f to be *positive* iff $f(x) > 0$ for x sufficiently large, or more precisely, iff $(\exists M \in \mathbf{R})(\forall x \in \mathbf{R})(x > M \Rightarrow f(x) > 0)$. Observe that $f \in F$ is positive iff f has a representation of the form $(*)$ above, in which the leading coefficients a_0 and b_0 have the same sign. Prove that with this definition of positive elements, F is an ordered field.

c) Let $f(x) = x$. Show that f is an upper bound of $\mathbf{Z}^+$ in F. Thus F is not Archimedean-ordered. Also show that if $g(x) = 1/x$, then g is "infinitely small" in that $g > 0$, and yet it is less than every positive rational number.

*11. Show that every non-Archimedean-ordered field K contains a subfield isomorphic to the field F of Exercise 8. (Let $x \in K$ be an upper bound of $\mathbf{Z}^+$ in K. Show that x is *transcendental* in the sense that it does not satisfy any equation of the form

$$a_0 x^n + a_1 x^{n-1} + \cdots + a_n = 0,$$

where $a_0, \ldots, a_n$ are integers. Let F' be the smallest subfield of K containing x, and show that there exists a mapping φ of F' onto F which is one-to-one and preserves both operations and order.)

5-2 The intermediate-value theorem

Three cities A, B, and C are located along a highway as indicated in Fig. 5–1. A driver leaves A at a certain time t_0, travels along the highway, and arrives at C at time t_1. Nothing else is known about the trip. The driver could have stopped along the way. He could have gone three-fourths of the way, realized that he had forgotten his brief case and returned to A before setting out again to C. About such matters we are totally ignorant. Yet we can be absolutely certain of one thing: the driver passed through B at least once during the trip.

FIG. 5-1

Let's see what this means mathematically. For purposes of the discussion, we shall regard A, B, and C as points. We introduce coordinates onto the line and use A, B, and C to designate the coordinates of the respective points. Thus we may assume that $A < B < C$. We also let $f(t)$ denote the coordinate of the car at time t. Then $f(t_0) = A$, and $f(t_1) = C$. Furthermore, the function f will be continuous. (Although it has been argued that angels can move from one point to another instantaneously, cars are not observed to behave in this manner.) To say that the car passed through B during the trip amounts to asserting that there exists a time T such that $t_0 < T < t_1$ and $f(T) = B$. Thus we are led to conjecture the following.

Theorem 1. Intermediate-Value Theorem. *Suppose that f is a real-valued function which is continuous on the interval $[a, b]$. Suppose also that $f(a) \neq f(b)$ and that Y is any number between $f(a)$ and $f(b)$. (See Fig. 5–2.) Then there exists a number X such that $a < X < b$ and*

$$Y = f(X).$$

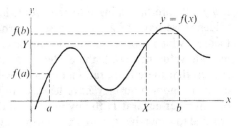

FIG. 5-2

Proof. Since $f(a) \neq f(b)$, there are two possibilities: either $f(a) < Y < f(b)$ or $f(a) > Y > f(b)$.

Case I. $f(a) < Y < f(b)$. Let us denote by I the interval $[a, b]$. Set

$$S = \{x \in I \mid f(x) < Y\}.$$

Then S is nonempty (since $a \in S$) and bounded above (by b). Hence S has a least upper bound X. We shall prove that $Y = f(X)$.

Since X is the least upper bound of S, we can find a sequence $x_1, x_2, \ldots$ of elements of S such that $\lim x_n = X$. Since f is continuous at X,

$$\lim_{n \to \infty} f(x_n) = f(X).$$

Since $f(x_n) < Y$ for each positive integer n, it follows that

$$f(X) = \lim_{n \to \infty} f(x_n) \leq Y. \tag{1}$$

Let $y_n = X + 1/n$. Then $\lim y_n = X$, and each y_n is an upper bound of S. Consequently, for n sufficiently large, y_n lies in I but not in S, and therefore $f(y_n) \geq Y$. Hence

$$f(X) = \lim_{n \to \infty} f(y_n) \geq Y. \tag{2}$$

From (1) and (2) we conclude that $f(X) = Y$.

Case II. $f(a) > Y > f(b)$. Let $g(x) = -f(x)$. Then g is continuous on I and $g(a) = -f(a) < -Y < -f(b) = g(b)$. Hence by case I, there exists an X such that $a < X < b$ and

$$-Y = g(X) = -f(X),$$

or $Y = f(X)$.

Corollary 2. Weierstrass Nullstellensatz. *Suppose that f is a real-valued function which is continuous on the interval $[a, b]$. Suppose also that $f(a)$ and $f(b)$ are of opposite sign. Then there exists a number X such that $a < X < b$ and $f(X) = 0$.*

We consider next some consequences of the intermediate-value theorem.

Proposition 3. *There exists a real number whose square is two. The rational numbers do not form a complete ordered field.*

Proof. Before launching into the proof, let us consider for a moment what sort of evidence we have of 2 having a square root. There is, of course, the Pythagorean theorem which supplies geometric evidence. Apart from this (and apart from the intermediate-value theorem), the evidence is mostly negative. We have already proved, in fact, that there is no rational number with the required property. In grade school or high school, you may have learned an algorithm for computing the square root of a number accurately to any number of decimal places. Yet it is unlikely that you are able to show that the algorithm works even for this special case of $\sqrt{2}$. This would require you to prove that the *infinite* decimal expansion obtained by the algorithm will, when squared, yield the number 2 *exactly*.

We turn now to the proof. Let $f(x) = x^2$, let $a = 1$, and let $b = 2$. Then since $f(x)$ is a polynomial, it is continuous everywhere, and in particular it is continuous on the interval $[a, b] = [1, 2]$. We observe that the number 2 lies between $f(a) = f(1) = 1$ and $f(b) = f(2) = 4$. Hence by the theorem there exists a number x between 1 and 2 such that $x^2 = f(x) = 2$.

It is a simple matter now to prove that the field of rational numbers $\mathbf{Q}$ is not complete. Let

$$S = \{x \in \mathbf{Q} \mid x < \sqrt{2}\}.$$

Then as a subset of $\mathbf{Q}$ the set S is both nonempty (since $1 \in S$) and bounded above (since 2 is an upper bound of S). Suppose that as a subset of $\mathbf{Q}$ the set S had a least upper bound L. Then L would also be the least upper bound of S as a subset of the field $\mathbf{R}$. (One must use the fact that the rationals are dense in $\mathbf{R}$.) But the least upper bound of S as a subset of $\mathbf{R}$ is $\sqrt{2}$. Thus $L = \sqrt{2}$, which is impossible since $\sqrt{2}$ is irrational.

Proposition 4. *The irrational numbers are dense.*

Proof. Let x and y be any two real numbers, say $x < y$. Then $x - \sqrt{2} < y - \sqrt{2}$, and there exists a rational number r between $x - \sqrt{2}$ and $y - \sqrt{2}$. It follows that $r + \sqrt{2}$ is between x and y, and it is easily checked that $r + \sqrt{2}$ is irrational.

So far we have only encountered functions whose discontinuities have been relatively few. From the fact that both the rationals and the irrationals are dense, it is easy to see, however, that the function

$$f(x) = \begin{cases} 1 & \text{if } x \text{ is rational,} \\ 0 & \text{if } x \text{ is irrational} \end{cases}$$

is everywhere discontinuous. [Suppose for a moment that f has a limit at some point, say $\lim_{x \to x_0} f(x) = L$. Then we can select a sequence of rational numbers $x_1, x_2, \ldots$ and a sequence of irrational numbers $y_1, y_2, \ldots$, both converging properly to x_0. It follows that

$$L = \lim_{n \to \infty} f(x_n) = 1 \quad \text{and} \quad L = \lim_{n \to \infty} f(y_n) = 0,$$

which is a contradiction.]

Proposition 5. *Let a be any positive real number, and let n be any positive integer. Then $\sqrt[n]{a}$ exists. That is, there exists a unique positive real number x such that $x^n = a$.*

Proof. We observe that for positive values of x, the function $f(x) = x^n$ is strictly increasing. That is, $x^n < y^n$ if $0 < x < y$. [Since

$$y^n - x^n = (y - x)(x^{n-1} + x^{n-2}y + \cdots + y^{n-1}),$$

$y^n - x^n$ is visibly a positive number.] Consequently, $\sqrt[n]{a}$ is unique if it exists.

Case I. $0 < a < 1$. The function $f(n) = x^n$ is continuous everywhere and hence, in particular, it is continuous on the interval $[0, 1]$. Moreover, the number a lies between $f(0) = 0$ and $f(1) = 1$. From the intermediate-value theorem, it follows that there exists a number x such that $0 < x < 1$ and $f(x) = x^n = a$.

Case II. $a \geq 1$. If $a = 1$, we can let $x = 1$. If $a > 1$, then $0 < 1/a < 1$. Hence by case I, there exists a number $y > 0$ such that $y^n = 1/a$. Let $x = 1/y$. Then $x > 0$ and

$$x^n = 1/y^n = a.$$

Proposition 6. *Every polynomial of odd degree with real coefficients has at least one real root.*

Proof. Let $P(x) = a_0 x^m + a_1 x^{m-1} + \cdots + a_m$, where $a_0 \neq 0$ and m is an odd positive integer. We may assume that $a_0 > 0$. [If not, we can consider the polynomial $-P(x)$.] We define the sequence $\{A_n\}$ by

$$A_n = \frac{P(n)}{n^m} = a_0 + a_1 \left(\frac{1}{n}\right) + a_2 \left(\frac{1}{n}\right)^2 + \cdots + a_m \left(\frac{1}{n}\right)^m.$$

Clearly, $\lim_{n \to \infty} A_n = a_0 > 0$. Consequently, for n sufficiently large,

$$A_n = \frac{P(n)}{n^m} > 0.$$

Since $n^m > 0$, $P(n)$ must be positive for n sufficiently large. Next we can consider the sequence $\{B_n\}$ defined by

$$B_n = \frac{P(-n)}{(-n)^m} = a_0 + a_1 \left(-\frac{1}{n}\right) + \cdots + a_m \left(-\frac{1}{n}\right)^m.$$

Once again $\lim_{n \to \infty} B_n = a_0 > 0$, and thus for n sufficiently large,

$$B_n = \frac{P(-n)}{(-n)^m} > 0.$$

Since m is odd, $(-n)^m$ is negative for all $n \in \mathbf{Z}^+$. Consequently, $P(-n) < 0$ for n sufficiently large.

The upshot is this: a positive number b can be found such that $P(b) > 0$, and a negative number a can be found such that $P(a) < 0$. Since polynomials are continuous everywhere, it follows from the intermediate-value theorem (or Corollary 2) that there exists a number X between a and b such that $P(X) = 0$.

Example 1. *The Lever of Mahomet.* Suppose that a train travels along a straight stretch of track from A to B. In the process, the train may speed up, slow down, back up, or halt. The only restrictions on the trip are that the motion of the train be continuous and that the trip be made in a finite length of time. Imagine that a rod is attached to a flat car on the train and is free to pivot back and forth under the influence of gravity and the motion of the train. (We shall also assume that once the rod hits the floor of the car, it remains there.) It might seem at first that if the trip takes a reasonable length of time, then no matter what position the rod had originally, by the end of the trip it will lie flat on the floor of the car. This is not true. We will prove, in fact, that if the schedule of the train is known in advance, then it is possible (in principle at least) to set the rod at the beginning of the trip in such a way that it will be in a vertical position at the moment the train arrives at B.

To prove this assertion, we consider the angle (measured in radians) which the rod makes with the floor of the car. If ϕ is the angle which the rod makes at the start of the trip, we shall denote by $f(\phi)$ the angle which it makes at the end of the trip (see Fig. 5-3). Intuition suggests that f is a continuous function, and theorems concerning differential equations can be brought to the defense of this assertion. (In other words, a small change in the initial position of the rod will result in a small change in the terminal position.) Also observe that $f(0) = 0$ and $f(\pi) = \pi$. By the intermediate-value theorem, it follows that there exists a number ϕ between 0 and π such that $f(\phi) = \pi/2$. In other words, there is an initial position for the rod which will result in its being vertical at the end of the trip. (This example appears in Courant and Robbins, *What is Mathematics?*, Oxford University Press, N.Y., 1941, p. 312.)

Example 2. The intermediate-value theorem is frequently used to check the accuracy of solutions to numerical problems. Suppose, for example, that we are required to find the roots of the equation

$$f(x) = x^3 - x^2 + 2x + 5 = 0$$

accurately to six decimal places. By Proposition 6 we know that this equation has at least one real root. A numerical analyst (using Newton's method perhaps) may come up with a number like $x = -1.13249432$ as an approximation to a root of the equation. The question then arises as to whether he may safely accept -1.132494 as the value of a root to six decimal places. To check the accuracy, he computes the following:

$$f(-1.1324945) = -0.000006, \qquad f(-1.1324935) = +0.000002.$$

From the intermediate-value theorem (or Corollary 2), there is a root of the equation which lies between -1.1324945 and -1.1324935. Hence to six places, -1.132494 is the value of a root of the equation.

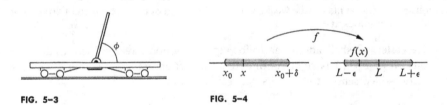

FIG. 5-3   **FIG. 5-4**

In our discussion of the intermediate-value theorem we have assumed that f is continuous throughout the interval $[a, b]$. In particular, this requires f to be continuous at each of the endpoints. Actually, this requirement is stronger than necessary. One-sided continuity will suffice.

Definition 7. We say that L is the limit of $f(x)$ as x approaches x_0 **from the right**, and we write

$$L = \lim_{x \to x_0+} f(x)$$

iff for every $\epsilon > 0$, there exists a $\delta > 0$ such that

$$|f(x) - L| < \epsilon$$

whenever $x_0 < x < x_0 + \delta$ (see Fig. 5-4). We say that f is **right-continuous** at x_0 iff

$$\lim_{x \to x_0+} f(x) = f(x_0).$$

The definition of the left-hand limit $\lim_{x \to x_0-} f(x)$ is the same except that the inequality $x_0 < x < x_0 + \delta$ is replaced by the inequality $x_0 - \delta < x < x_0$.

Example 3. Figure 5-5 gives an illustration of the above definitions.   Note that $\lim_{x \to 3+} [x] = 3 = [3]$, so that $[x]$ is continuous from the right at 3. On the other hand,

$$\lim_{x \to 3-} [x] = 2 \neq [3],$$

so that $[x]$ is not continuous from the left at 3.

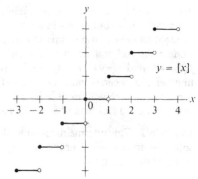

FIG. 5-5

We shall leave to the reader the proof of the following result:

Proposition 8. *The two-sided limit* $\lim_{x \to x_0} f(x)$ *exists iff the one-sided limits* $\lim_{x \to x_0-} f(x)$ *and* $\lim_{x \to x_0+} f(x)$ *exist and are equal, in which case*

$$\lim_{x \to x_0} f(x) = \lim_{x \to x_0-} f(x) = \lim_{x \to x_0+} f(x).$$

We can now state a strengthened form of the intermediate-value theorem.

Remark. In the statement of the intermediate-value theorem, the hypothesis that f is continuous at the endpoints can be weakened: it suffices to assume that f is right-continuous at a and left-continuous at b.

Hereafter we shall adopt the following convention: we shall say that f is continuous on $[a, b]$ iff f is right-continuous at a, left-continuous at b, and continuous in the two-sided sense at all points between a and b.

Proof. Define

$$g(x) = \begin{cases} f(x) & \text{if } a \le x \le b, \\ f(b) & \text{if } x > b, \\ f(a) & \text{if } x < a. \end{cases}$$

Then g is continuous on the open interval (a, b), and it is also continuous outside $[a, b]$. (g is in fact differentiable outside $[a, b]$.) Furthermore,

$$\lim_{x \to a+} g(x) = \lim_{x \to a+} f(x) = f(a) = g(a),$$

since f is right-continuous at a, and

$$\lim_{x \to a-} g(x) = \lim_{x \to a-} f(a) = f(a) = g(a).$$

Hence g is continuous at a. Similarly, g is continuous at b. Therefore, g satisfies

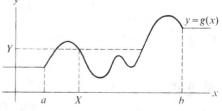

FIG. 5-6

the hypothesis of the intermediate-value theorem. Hence if Y is between $g(a) = f(a)$ and $g(b) = f(b)$, then there exists an X between a and b such that

$$Y = g(X) = f(X),$$

as shown in Fig. 5–6.

Exercises

1. If f is a function, and if $x \in \mathfrak{D}_f$ has the property that $f(x) = x$, then x is called a *fixed point* of f. Prove that every continuous mapping f of $[0, 1]$ into itself has a fixed point. [If $f(0) = 0$ or $f(1) = 1$, there is nothing to prove. In the remaining case, apply the intermediate-value theorem to the function $g(x) = f(x) - x$.]

*2. We say that a real-valued function f is nondecreasing on a set $S \subset \mathbf{R}$ iff

$$(\forall x, y \in S)(x \le y \Rightarrow f(x) \le f(y)).$$

Prove that every nondecreasing mapping of $[0, 1]$ into itself has a fixed point. [Let X be the least upper bound of the set $\{x \mid f(x) \ge x\}$. Show that X is a fixed point of f.]

3. Show that there exists a number x between zero and one such that $x^2 = \sin \pi x$. Write out a series of instructions for computing x to ten decimal places.

4. Let n be any positive integer. Show that the function $f(x) = \sqrt[n]{x}$ is right-continuous at zero.

5. A Buddhist monk lives in a hut at the foot of a mountain. Each night at sundown he leaves his hut, follows a well-worn path to the top of the mountain, meditates at various points along the way, and returns to his hut exactly at dawn the next morning. The villagers observed that there was considerable variation in this ritual—that it was virtually impossible to predict, for instance, where the monk might be at midnight. Nevertheless, one villager made the following conjecture: if one were given the schedules of any two of these nocturnal journeys, one would find at least one instance of agreement between the schedules, that is, one would find in at least one instance that the monk had been at the same point of the path at the same hour of night. The other villagers dismissed this as unlikely. Do you?

5–3 Some theorems on sequences

As preparation for the theorem on extreme values, we shall prove three theorems concerning limits of sequences.

Theorem 1. Monotone Sequence Property. *Let* $\{x_n\}$ *be a nondecreasing sequence of real numbers, that is,*

$$x_1 \le x_2 \le x_3 \le \cdots$$

Then the sequence has a limit iff it is bounded above.

Proof. We have already shown that a convergent sequence is bounded; in particular, it is bounded above.

Suppose now that $\{x_n\}$ is bounded above. This means, of course, that the range of the sequence S has an upper bound. Since S is also nonempty, S must have a least upper bound L. We will now prove that L is the limit of the sequence.

Let $\epsilon > 0$ be given. Then there exists an element of S, say x_N, such that

$$L - \epsilon < x_N.$$

Since L is an upper bound of S, and since the sequence is nondecreasing, it follows that

$$L - \epsilon < x_N \leq x_n \leq L$$

whenever $n \geq N$. Hence, if $n \geq N$, it follows that $|x_n - L| < \epsilon$. This proves that $\lim_{n \to \infty} x_n = L$ (see Fig. 5–7).

Example 1. Show that $\lim_{n \to \infty} (1 + 1/n)^n$ exists.

In Exercise 27, Section 1–9, a proof of the statement that the sequence $(1 + 1/n)^n$ is increasing is outlined. By the binomial theorem,

$$\left(1 + \frac{1}{n}\right)^n = 1 + n \cdot \frac{1}{n} + \frac{n(n-1)}{2!}\frac{1}{n^2} + \frac{n(n-1)(n-2)}{3!}\frac{1}{n^3} + \cdots + \frac{1}{n^n}$$

$$\leq 1 + 1 + \frac{1}{2!} + \frac{1}{3!} + \cdots + \frac{1}{n!}$$

$$\leq 1 + 1 + \frac{1}{2} + \frac{1}{4} + \cdots + \frac{1}{2^n}$$

$$= 1 + 2[1 - 2^{-(n+1)}] < 3.$$

Since $(1 + 1/n)^n$ is an increasing sequence bounded above (by 3), it follows that

$$\lim_{n \to \infty} (1 + 1/n)^n$$

exists. We shall later show that the limit is e, the base of the *natural logarithms*. It is one of the most important numbers in mathematics.

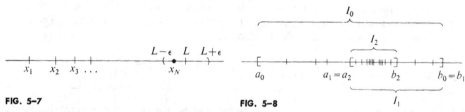

FIG. 5–7   FIG. 5–8

A given sequence may or may not have a convergent subsequence. For instance, the sequence 1, 2, 4, 8, ... does not have any convergent subsequence. (Every subsequence is unbounded and hence cannot have a limit.) On the other hand, the sequence 1, 0, 1, 0, ... has the convergent subsequences 1, 1, 1, ... and 0, 0, 0, ... If $\{x_n\}$ is a sequence whose range includes the rational numbers, then corresponding to each real number x_0 there will exist a subsequence which converges to x_0. Our next theorem guarantees the existence of convergent subsequences under certain conditions.

Theorem 2. Sequential Compactness Property.   *Every bounded sequence of real numbers has a convergent subsequence.*

Proof. Let $\{x_n\}$ be a bounded sequence of real numbers. Let a_0 be a lower bound and b_0 an upper bound of the sequence. Thus all elements of the sequence lie in the interval $I_0 = [a_0, b_0]$.

To extract a convergent subsequence, we shall use what is called the Weierstrass-Bolzano bisection method (see Fig. 5–8). First we bisect the interval I_0. One of the closed half-intervals so obtained must contain x_n for infinitely many integers n. Let $I_1 = [a_1, b_1]$ be such an interval. Then the length of I_1 is

$$b_1 - a_1 = \tfrac{1}{2}(b_0 - a_0).$$

We next bisect I_1, and we let $I_2 = [a_2, b_2]$ be a closed half-interval such that $x_n \in I_2$ for infinitely many integers n. Then

$$b_2 - a_2 = \tfrac{1}{2}(b_1 - a_1) = (\tfrac{1}{2})^2(b_0 - a_0).$$

We continue this process *ad infinitum*, and thus we obtain a nested sequence of intervals $I_0 \supset I_1 \supset I_2 \supset \cdots$ such that each interval $I_k = [a_k, b_k]$ contains x_n for infinitely many integers n and

$$b_k - a_k = (\tfrac{1}{2})^k(b_0 - a_0).$$

Observe that the left-hand endpoints of these intervals, namely the a_n's, form a non-decreasing sequence which is bounded above (by b_0). Hence by the monotone sequence property, there exists a real number L such that

$$\lim_{n \to \infty} a_n = L.$$

Since

$$b_n = a_n + (\tfrac{1}{2})^n(b_0 - a_0),$$

we see that

$$\lim_{n \to \infty} b_n = L + 0 \cdot (b_0 - a_0) = L.$$

We shall prove finally that the sequence $\{x_n\}$ has a subsequence which converges to L. Since I_1 contains x_n for infinitely many integers n, we may choose an integer n_1 such that $x_{n_1} \in I_1$. Since I_2 contains x_n for infinitely many n, there exists an integer $n_2 > n_1$ such that $x_{n_2} \in I_2$. Continuing in this fashion, we get a subsequence $\{x_{n_m}\}$ such that $x_{n_m} \in I_m$ for each m. Hence

$$a_m \leq x_{n_m} \leq b_m,$$

and it follows from the pinching theorem that $\lim_{m \to \infty} x_{n_m} = L$.

We now have all the machinery required for a proof of the theorem on extreme values. Nevertheless, before we proceed to a discussion of this theorem, we shall take this opportunity to present another standard tool which every aspiring young analyst should know how to handle, namely, the *Cauchy convergence criterion.*

Frequently, as in Example 1 of this section, one is given a sequence of real numbers $\{x_n\}$, and one is faced with the problem of proving that the sequence converges, before one can go about finding what the limit is. The monotone sequence property does give a sufficient condition for the sequence to converge, but, unfortunately, it applies only to nondecreasing sequences. The Cauchy convergence criterion is made to order for such situations, since it provides both a necessary and sufficient condition for a sequence to converge.

Definition 3. A sequence of real numbers $\{x_n\}$ is said to be **Cauchy** iff for every $\epsilon > 0$ there exists a positive integer N such that

$$|x_m - x_n| < \epsilon$$

whenever $m > N$ and $n > N$.

A shorthand which is sometimes used for the quantified statement in the definition above is

$$\lim_{m,n \to \infty} |x_m - x_n| = 0.$$

To say that a sequence is Cauchy means, roughly speaking, that eventually the terms of the sequence start to bunch closer and closer together.

Theorem 4. Cauchy Convergence Criterion. *A sequence of real numbers has a limit iff it is Cauchy.*

Proof. Let us assume first that $\{x_n\}$ is a convergent sequence; call its limit L. We shall show that $\{x_n\}$ is Cauchy. Let $\epsilon > 0$ be given. By the definition of a limit, there exists an N such that

$$|x_n - L| < \epsilon/2$$

for all $n > N$. Suppose now that $m > N$ and $n > N$. Then by the triangle inequality,

$$|x_m - x_n| = |(x_m - L) + (L - x_n)| \le |x_m - L| + |x_n - L| < \epsilon/2 + \epsilon/2 = \epsilon.$$

Hence $\{x_n\}$ is Cauchy.

Now suppose that $\{x_n\}$ is Cauchy. We shall prove first that the sequence is bounded. It follows from the definition of a Cauchy sequence (setting $\epsilon = 1$) that there exists an N such that

$$|x_m - x_n| < 1$$

if $m > N$ and $n > N$. In particular,

$$|x_n - x_{N+1}| < 1$$

for all $n > N$. Hence

$$|x_n| = |(x_n - x_{N+1}) + x_{N+1}| \le |x_n - x_{N+1}| + |x_{N+1}| < 1 + |x_{N+1}|$$

if $n > N$. Let M be the largest number in the set

$$\{|x_1|, |x_2|, \ldots, |x_N|, |x_{N+1}| + 1\}.$$

Then $|x_n| \le M$ for all $n \in \mathbf{Z}^+$.

Since the sequence $\{x_n\}$ is bounded, it has a convergent subsequence, say $x_{n_m} \to L$. We shall now prove that the entire sequence must, in fact, converge to L. Let $\epsilon > 0$ be given. From the definition of a Cauchy sequence we can find an N such that

$$|x_m - x_n| < \epsilon/2$$

whenever $m > N$ and $n > N$. We also know that the inequality

$$|x_{n_m} - L| < \epsilon/2$$

will hold for all but at most a finite number of positive integers m. Certainly, we can then choose an $m \in \mathbf{Z}^+$ such that $n_m \geq m > N$ and

$$|x_{n_m} - L| < \epsilon/2.$$

Hence, if $n > N$,

$$\begin{aligned} |x_n - L| &= |(x_n - x_{n_m}) + (x_{n_m} - L)| \\ &\leq |x_n - x_{n_m}| + |x_{n_m} - L| \\ &< \epsilon/2 + \epsilon/2 = \epsilon. \end{aligned}$$

Therefore, we have proved that $\lim_{n \to \infty} x_n = L$.

Exercises

1. Prove that the limit of the sequence $\sqrt{2}, \sqrt{2 + \sqrt{2}}, \sqrt{2 + \sqrt{2 + \sqrt{2}}}, \ldots$ exists and is equal to 2.

2. Let
$$a_n = 1 + \frac{1}{1!} + \frac{1}{2!} + \cdots + \frac{1}{n!}, \qquad b_n = 1 - \frac{1}{1!} + \frac{1}{2!} - \cdots + \frac{(-1)^n}{n!}.$$

   a) Show that
$$a_n \leq 1 + \sum_{k=0}^{n-1} \frac{1}{2^k} < 3$$

   for all positive integers n. Using the monotone sequence property, prove that the sequence $\{a_n\}$ has a limit.

   b) Show that for all positive integers m and n,
$$|b_m - b_n| \leq |a_m - a_n|.$$

   Deduce that $\{b_n\}$ is Cauchy and therefore has a limit.

   *c) Denote the limit of the sequence $\{a_n\}$ by e. Show that the limit of $\{b_n\}$ is $1/e$.

3. Let a_1 and b_1 be any two positive numbers such that $a_1 < b_1$. Subsequent terms of the sequences $\{a_n\}$ and $\{b_n\}$ are defined by the formulas
$$a_{n+1} = \sqrt{a_n b_n}, \qquad b_{n+1} = \tfrac{1}{2}(a_n + b_n).$$

   Prove that the sequences converge and have the same limit. (The limit is called the *arithmetic-geometric* mean of a_1 and b_1.)

4. Formulate and prove a necessary and sufficient condition for a nonincreasing sequence to converge.

5. A mapping φ of the positive integers into themselves is said to be a *permutation* of the positive integers if every positive integer occurs exactly once in the sequence $\varphi(1)$, $\varphi(2)$, $\varphi(3)$, ... For example, if

$$\varphi(n) = \begin{cases} n+1 & \text{if } n \text{ is odd,} \\ n-1 & \text{if } n \text{ is even,} \end{cases}$$

then φ is a permutation of the positive integers, since each positive integer appears exactly once in the sequence 2, 1, 4, 3, 6, 5, ...

Let $\{x_n\}$ be a sequence whose limit is L, and let φ be a permutation of the positive integers. Prove that $\lim_{n\to\infty} x_{\varphi(n)} = L$.

*6. Let I be a closed interval. [Thus I is an interval of one of the following types: $[a, b]$, $[a, +\infty)$, $(-\infty, b]$, $\mathbf{R}$.] Let f be a function mapping I into itself. (This means that $\mathcal{D}_f = I$ and $\mathcal{R}_f \subset I$.) We say that f is a *contraction mapping* iff there exists some number r such that $0 < r < 1$ and

$$|f(x) - f(y)| \le r|x - y|$$

for all $x, y \in I$. Prove that such a function has a unique fixed point $\xi \in I$ and that for each $x_0 \in I$ the sequence $x_1, x_2, \ldots$ defined by the recursion formula

$$x_n = f(x_{n-1})$$

converges to ξ. (This is a special case of a very useful theorem known as the *Picard fixed-point theorem*, or alternatively, *the contraction principle*. To prove it, show that the sequence $\{x_n\}$ is Cauchy, and let ξ be its limit. Prove that ξ is the required fixed point.)

7. *Decimal Expansions*

a) Let x be any positive real number (or, in more general terms, a positive element of any Archimedean-ordered field). Prove that x may be written as

$$x = N + \lim_{n\to\infty} \sum_{k=1}^{n} a_k 10^{-k},$$

where $N \in \mathbf{Z}^+$ and the a_k's are decimal digits (i.e., the numbers 0, 1, 2, ..., 9). [Let $N = [x]$. Let a_1 be the smallest nonnegative integer such that

$$(a_1 + 1)10^{-1} > x - N,$$

so that

$$x - 10^{-1} < N + a_1 10^{-1} \le x.$$

Show, more generally, by induction that decimal digits $a_1, \ldots, a_n$ can be chosen so that

$$x - 10^{-n} < N + \sum_{k=1}^{n} a_k 10^{-k} \le x.$$

Then pinch.]

b) In (a) we proved in effect that every real number has a decimal expansion. Using the monotone convergence property, prove, conversely, that to each formal decimal expansion there corresponds a real number.

*8. Let F be an ordered field. Prove that the following are logically equivalent:

   a) **Least Upper Bound Axiom.** *Every nonempty subset of F which is bounded above has a least upper bound.*
   b) **Monotone Convergence Property.** *Every nondecreasing sequence of elements of F which is bounded above has a limit.*
   c) **Sequential Compactness Property.** *Every bounded sequence in F has a convergent subsequence.*
   d) *F is Archimedean-ordered, and every Cauchy sequence converges.*

   [The proof that (a) $\Rightarrow$ (b) appears in the text. To prove (b) $\Rightarrow$ (c), one only needs to make one modification to the proof of Theorem 2. One must deduce from (b) that $\lim_{n \to \infty} 2^{-n} = 0$. The proof of Theorem 4 shows that (c) $\Rightarrow$ every Cauchy sequence converges. Show that (c) $\Rightarrow$ Archimedean ordering by observing that in a non-Archimedean-ordered field, the sequence of positive integers is bounded and yet contains no convergent subsequence. Finally, prove that (d) $\Rightarrow$ (a) by a bisection argument. Let S be a set which is bounded above, and choose an interval $[a_0, b_0]$ such that b_0 is an upper bound of S and a_0 is not. Bisect the interval, and let $[a_1, b_1]$ be a half-interval with the same properties, etc. Prove that the sequence $\{a_n\}$ so obtained is Cauchy, and let L be its limit. Prove that $L = \text{lub } S$.]

*9. Prove that every sequence of real numbers has a monotone subsequence. (A sequence is monotone if it is either nonincreasing or nondecreasing.) Use this fact to give an alternative proof of the sequential compactness property.

*10. Let F be an Archimedean-ordered field, and let K be a complete ordered field.

   a) Show first that our mapping φ of $(\mathbf{Z}^+)_F$ onto $(\mathbf{Z}^+)_K$ can be uniquely extended to an isomorphism of the rational numbers in F—call the set $\mathbf{Q}_F$—onto $\mathbf{Q}_K$, the rational numbers in K. [This means that you must show that there is exactly one function φ^* with the following properties: φ^* is a one-to-one mapping of $\mathbf{Q}_F$ onto $\mathbf{Q}_K$, φ^* preserves operations, and for each $n \in (\mathbf{Z}^+)_F$, $\varphi^*(n) = \varphi(n)$.] Show that this isomorphism also preserves order.
   b) For each $x \in F$, show that

$$x = \text{lub } \{y \in \mathbf{Q}_F \mid y < x\}$$

   and also that the set $\{\varphi(y) \mid y < x\}$ is bounded above in the field K, where φ denotes the unique isomorphism of $\mathbf{Q}_F$ onto $\mathbf{Q}_K$. Set

$$\varphi^*(x) = \text{lub } \{\varphi(y) \mid y < x\}.$$

   Prove that φ^* is an isomorphism of F onto a subfield of K.
   c) If F is itself complete, then $\mathfrak{R}_{\varphi^*} = K$. Thus F and K are isomorphic. Prove, moreover, that φ^* is the only isomorphism of F onto K. [Suppose that φ' is a second such isomorphism. Show first that $\varphi'(x) = \varphi^*(x)$ for all $x \in \mathbf{Q}_F$. Using the fact that squares are nonnegative and that in a complete field every nonnegative element is a square, prove next that φ' maps positive elements of F into positive elements of K. Deduce that φ' must be order-preserving. Show finally that $\varphi^*(x) = \varphi'(x)$ for all $x \in F$.] Thus we have shown that *there is at most one complete ordered field* (the field of real numbers) *to within isomorphism.* Also, it follows from (b) that *every Archimedean-ordered field is isomorphic to a subfield of the real numbers.*

5-4 The theorem on extreme values

We proceed directly to the main result.

Theorem 1. Theorem on Extreme Values. *Let f be a real-valued function which is continuous on a closed bounded interval $I = [a, b]$ (see Fig. 5–9). Then f assumes both a maximum and a minimum value on the interval. That is, there exist numbers X and X' in the interval I such that*

$$m = f(X) \leq f(x) \leq f(X') = M$$

for all $x \in [a, b]$.

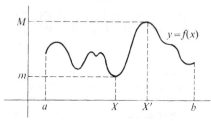

In such cases we write

$$M = \max \{f(x) \mid x \in I\} = \max_{x \in I} f(x)$$

and

$$m = \min \{f(x) \mid x \in I\} = \min_{x \in I} f(x).$$

FIG. 5–9

Proof. We shall prove only that $f(x)$ assumes a maximum value. [By considering the function $-f(x)$, one can show that $-\max \{-f(x) \mid x \in I\}$ is the minimum value of $f(x)$.]

First we will prove that the set

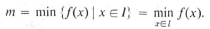
$$S = \{f(x) \mid x \in I\}$$

is bounded above. If not, then corresponding to each positive integer n we can find an $x_n \in I$ such that $f(x_n) > n$. Since $\{x_n\}$ is a bounded sequence of real numbers, it contains a convergent subsequence, say $\lim_{m \to \infty} x_{n_m} = x_0$. Then $x_0 \in I$, and since f is continuous at x_0,

$$\lim_{m \to \infty} f(x_{n_m}) = f(x_0).$$

However, it is quite clear that every subsequence of $\{f(x_n)\}$ is unbounded and hence cannot converge to anything. This contradiction shows that S is bounded above.

Let $M = \text{lub } S$. It suffices now to prove that there exists an $X \in I$ such that $f(X) = M$. Since $M = \text{lub } S$, we can choose $x_n \in I$ such that

$$M - 1/n < f(x_n) \leq M.$$

By the pinching theorem,

$$\lim_{n \to \infty} f(x_n) = M.$$

Now the sequence $\{x_n\}$ may not have a limit, but, being bounded, it at least has a convergent subsequence, say $x_{n_m} \to X$. Then $X \in I$, and since f is continuous at X, we have

$$f(X) = \lim_{m \to \infty} f(x_{n_m}) = M.$$

This completes the proof.

As in the case of the intermediate-value theorem, it is only necessary that $f(x)$ be right-continuous at a and left-continuous at b. It is important to note, however, the necessity of assuming that the interval $I = [a, b]$ is closed and bounded. For instance, the function $f(x) = 1/x$ is continuous on the interval $(0, 1]$, and yet it does not attain a maximum value.

Note one more thing. It follows from the intermediate-value theorem that

$$\{f(x) \mid x \in I\} = [m, M].$$

(Why?) Hence *a continuous real-valued function maps closed bounded intervals onto closed bounded intervals.*

Exercises

1. In the last paragraph of the proof of the theorem on extreme values it is stated that the sequence $\{x_n\}$ does not necessarily converge. Explain how this situation might arise.

2. The purpose of this exercise is to outline a proof of the theorem on extreme values which uses the least upper bound axiom in a more direct way.

   a) Show that if f is a real-valued function which is bounded above on each of two sub-sets A and B of $\mathfrak{D}_f$, then f is bounded above on $A \cup B$. [In saying that the function f is bounded above on A, we mean, of course, that the set of real numbers $\{f(x) \mid x \in A\}$ is bounded above, or equivalently, that there exists an $M \in \mathbf{R}$ such that $f(x) \leq M$ for all $x \in A$.]

   b) If f is right-continuous at x_0, show that there exists a $\delta > 0$ such that f is bounded above on the interval $[x_0, x_0 + \delta]$. What can be said if f is left-continuous at x_0? continuous (in the two-sided sense) at x_0?

   c) Prove that if f is continuous on $[a, b]$, then f is bounded above on $[a, b]$. [Let

   $$S = \{x \mid a \leq x \leq b \text{ and } f \text{ is bounded above on } [a, x]\}.$$

   Let $X = $ lub S. Using (a) and (b), argue that $X = b$. Then show that $X \in S$.]

   d) Prove finally that if f is continuous on $[a, b]$, then f assumes a maximum value on the interval. [Let

   $$M = \text{lub } \{f(x) \mid a \leq x \leq b\}.$$

   We need to show that there exists an $x \in [a, b]$ such that $f(x) = M$. Suppose this to be false Set.

   $$g(x) = 1/[M - f(x)],$$

   apply (c) to g, and obtain a contradiction.]

3. Prove that if f is upper-semicontinuous on $[a, b]$, then f assumes a maximum value on the interval. (See Exercise 10, Section 3-6. Modify either the proof of the theorem on extreme values given in the text or the proof outlined in the above exercise.) Show by example that such a function does not necessarily assume a minimum value on the interval $[a, b]$.

5-5 Uniform continuity

A function f is continuous on a set S iff

$$(\forall x_0 \in S)(\forall \epsilon > 0)(\exists \delta > 0)(\forall x)(|x - x_0| < \delta \Rightarrow |f(x) - f(x_0)| < \epsilon).$$

Because of the position of the quantifier $(\exists \delta > 0)$, it is clear that δ may depend not only on ϵ but also on x_0. As an illustration, let us consider the situation in which $S = (0, +\infty)$ and $f(x) = 1/x$. Since f is continuous on S, the above quantified statement is true. Thus for a fixed $x_0 > 0$ and $\epsilon > 0$, we are assured that at least one $\delta > 0$ exists such that

$$|x - x_0| < \delta \Rightarrow |1/x - 1/x_0| < \epsilon.$$

A little scratch work will reveal that among all such δ's, there is a largest, namely

$$\delta(x_0, \epsilon) = \frac{\epsilon x_0^2}{1 + \epsilon x_0}.$$

Thus we see that in this instance δ does depend in an essential way on both ϵ and x_0: for a fixed $\epsilon > 0$, arbitrarily small values of δ are required for values of x_0 sufficiently close to zero. More intuitively, for values of x close to zero, the expression $1/x$ is highly sensitive to changes in x.

For contrast, consider now the function $f(x) = \sin x$. Using the inequality

$$|\sin x - \sin x_0| \leq |x - x_0|,$$

we proved earlier that this function is continuous on the entire line. In this case the δ may be chosen independently of x_0; indeed, $\delta = \epsilon$ does the trick. In such cases we say that f is *uniformly continuous* on **R**.

Definition 1. A function f is **uniformly continuous on a set** S iff

$$(\forall \epsilon > 0)(\exists \delta > 0)(\forall x, y \in S)(|x - y| < \delta \Rightarrow |f(x) - f(y)| < \epsilon).$$

The essential difference between continuity on a set S and uniform continuity on S consists only of the relative order of the quantifiers. In the case of uniform continuity δ may depend only on ϵ.

Example 1. $1/x$ is continuous but not uniformly continuous on $(0, +\infty)$. On the other hand, $1/x$ is uniformly continuous on every interval of the form $[a, +\infty)$, where $a > 0$. [Set $\delta = \epsilon a^2/(1 + \epsilon a)$. Then if $x, y \in [a, +\infty)$ and $|x - y| < \delta$, one can verify that

$$\delta \leq \frac{\epsilon y^2}{1 + \epsilon y} = \delta(y, \epsilon),$$

and thus $|x - y| < \delta(y, \epsilon)$, which implies that $|1/x - 1/y| < \epsilon$.]

Example 2. $\sin x$ is uniformly continuous on the entire line.

We now state and prove the main result.

Theorem 2. Theorem on Uniform Continuity. *If f is continuous on a closed bounded interval $[a, b]$, then f is uniformly continuous on $[a, b]$.*

Proof. Suppose f is not uniformly continuous on $[a, b] = I$. Then

$$(\exists \epsilon > 0)(\forall \delta > 0)(\exists x, y \in I)(|x - y| < \delta \text{ and } |f(x) - f(y)| \geq \epsilon).$$

In particular, by taking δ to be of the form $1/n$, where $n \in \mathbf{Z}^+$, we see that there exist numbers x_n and y_n in I such that

$$|x_n - y_n| < 1/n \qquad \text{and} \qquad |f(x_n) - f(y_n)| \geq \epsilon.$$

Because the sequence $\{x_n\}$ is bounded, it has a convergent subsequence, say $\lim_{m \to \infty} x_{n_m} = x_0$. The limit x_0 must be in I, and it should be clear that $\lim_{m \to \infty} y_{n_m} = x_0$ also. [Since $|x_{n_m} - y_{n_m}| < 1/n_m \leq 1/m$, it follows that $\lim_{m \to \infty} (x_{n_m} - y_{n_m}) = 0$, and thus $\lim_{m \to \infty} y_{n_m} = \lim_{m \to \infty} [x_{n_m} - (x_{n_m} - y_{n_m})] = x_0 - 0 = x_0$.] Since f is continuous at x_0, we have

$$\lim_{m \to \infty} [f(x_{n_m}) - f(y_{n_m})] = f(x_0) - f(x_0) = 0.$$

This conflicts violently, however, with the assertion that

$$|f(x_n) - f(y_n)| \geq \epsilon \qquad \text{for all} \quad n \in \mathbf{Z}^+. \quad \text{Q.E.D.}$$

Exercises

1. a) Show that if f and g are uniformly continuous on a set S, then so are the functions $f + g$ and cf, where c is any constant.
   b) Show that the function x^2 is not uniformly continuous on **R**. Deduce that the product of two uniformly continuous functions need not be uniformly continuous.
   c) Suppose that g is uniformly continuous on a set S and that f is uniformly continuous on some set T containing $\mathfrak{R}_g$. Prove that $f \circ g$ is uniformly continuous on S.

2. Note that the function
$$f(x) = \begin{cases} 1 & \text{if } x \text{ is rational,} \\ 0 & \text{if } x \text{ is irrational} \end{cases}$$

is uniformly continuous on the set of rational numbers despite the fact that the function is nowhere continuous. Thus, according to our definitions, uniform continuity on a set does not necessarily imply continuity on the set. Show, however, that if S is an open interval, then uniform continuity on S does imply continuity on S. A weaker definition of continuity is occasionally used, according to which f is continuous on S iff

$$(\forall x \in S)(\forall \epsilon > 0)(\exists \delta > 0)(\forall y \in S)(|x - y| < \delta \Rightarrow |f(x) - f(y)| < \epsilon).$$

Note that uniform continuity does imply continuity in this weaker sense. Also observe that continuity in this weakened sense coincides with the convention which we established following the intermediate-value theorem, as to when a function is continuous on an interval of the form $[a, b]$.

3. a) Suppose that f is uniformly continuous on a set S. Show that if $\{x_n\}$ is a Cauchy sequence of elements of S, then $\{f(x_n)\}$ is a Cauchy sequence.

   b) Use (a) to show that the function $\sin(1/x)$ is not uniformly continuous on the interval $(0, 1)$.

4. Let I be an interval other than $\varnothing$ or a singleton set. We say that $S \subset I$ is *dense in I* iff between any two members of I there is a member of S. For example, the set of positive rational numbers is dense in each of the intervals $(0, +\infty)$ and $[0, +\infty)$. Prove the following extension theorem.

Suppose that S is a dense subset of an interval I and that f is uniformly continuous on S. Then f can be uniquely extended to a function f* which is continuous (in the weakened sense of Exercise 2) on I. The extended function is in fact uniformly continuous on I.

To prove uniqueness, suppose that g is a second such function. Consider an arbitrary $x_0 \in I$. Choose a sequence $\{x_n\}$ in S which converges to x_0. Deduce that

$$g(x_0) = \lim_{n \to \infty} g(x_n) = \lim_{n \to \infty} f(x_n) = \lim_{n \to \infty} f^*(x_n) = f^*(x_0).$$

To show existence, let $x_0 \in I$, and select, as before, a sequence $\{x_n\} \subset S$ converging to x_0. Apply Exercise 3 to conclude that $\lim_{n \to \infty} f(x_n)$ exists. Define

$$f^*(x_0) = \lim_{n \to \infty} f(x_n).$$

Show that f^* is well-defined, i.e., show that if $\{y_n\}$ is a second such sequence, then

$$\lim_{n \to \infty} f(y_n) = \lim_{n \to \infty} f(x_n).$$

(Consider the interlaced sequence $x_1, y_1, x_2, y_2, \ldots$; apply Exercise 3 again and Exercise 6, Section 3-4.) Observe that f^* is an extension of f [i.e., $f^*(x) = f(x)$ for all $x \in S$]. Show finally that f^* is uniformly continuous on I.

5. We shall now use the extension theorem above to develop the theory of exponential functions. In Chapter 9 a completely different approach will be taken. First let us review a bit. Let $a > 1$. In Exercise 23, Section 1-9, we defined a^x for all rational numbers (albeit our definition was somewhat premature since we had not shown the existence of nth roots.) The exponential laws, $a^x a^y = a^{x+y}$, $(a^x)^y = a^{xy}$, and $(ab)^z = a^z b^z$ were also proved. The problem which we now consider is that of defining a^x in some sensible manner for irrational values of x.

   a) Show that if $x > 0$, then $a^x > 1$. From this deduce that if x and y are rational numbers and $x < y$, then $a^x < a^y$.

   b) In Chapter 3 we showed that $\lim_{n \to \infty} a^{1/n} = \lim_{n \to \infty} \sqrt[n]{a} = 1$. Use this fact to prove that a^x is uniformly continuous on each set of the form $(-\infty, b) \cap \mathbf{Q}$. [If x and y are rational numbers with $x \le y \le b$, show first that $a^y - a^x \le a^b(a^{y-x} - 1)$.]

   c) By the extension theorem, a^x can be uniquely defined for irrational x in $(-\infty, b)$ in such a way that the function $f(x) = a^x$ is continuous on the interval $(-\infty, b)$. Argue that a^x may be regarded as being defined and continuous on $\mathbf{R}$. Show that the exponent law $a^x a^y = a^{x+y}$ holds for all real numbers x and y.

   d) Define a^x appropriately if $0 < a \le 1$, and prove the exponent laws $(a^x)^y = a^{xy}$ and $(ab)^z = a^z b^z$.

6 □ MEAN-VALUE THEOREMS
AND THEIR APPLICATIONS

6–1 A necessary condition for relative maxima and minima

One type of problem which we shall consider in this chapter is that of finding maximum or minimum values of a function. Problems of this sort arise frequently in applications. A manufacturer may wish to minimize the production cost of an item, or a chemist may wish to maximize the purity of yield of a chemical process. Calculus provides a technique for solving such problems.

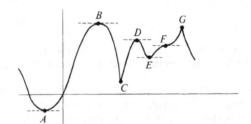

FIG. 6–1

First we need to introduce some definitions. Figure 6–1 shows the graph of a function. Note that the function takes on its greatest value at B. We say that the function has an "absolute maximum" at this point. The function has an "absolute minimum" at A. In addition to the absolute maximum occuring at B, the function also has peak values at D and G. We say that the function has relative maxima at these points. The function has relative minima at C and E. We now give the formal definitions.

Definition 1. We say that a function f has an **absolute maximum** at x_0 iff

$$f(x_0) \geq f(x)$$

for all $x \in \mathfrak{D}_f$, in which case $f(x_0)$ is called the **maximum value** of the function. We say that f has a **relative** (or **local**) **maximum** at x_0 iff there exists a neighborhood V of x_0 such that

$$f(x_0) \geq f(x)$$

for all $x \in V$. We say that f has a **strict relative maximum** at x_0 iff there exists a deleted neighborhood V of x_0 such that

$$f(x_0) > f(x)$$

for all $x \in V$.

Definitions of absolute minimum, relative minimum, and strict relative minimum can be obtained simply by reversing the inequalities in the above definition. We shall also use the term *extremum* to designate either a maximum or a minimum.

Example 1. The function $f(x) = \sqrt{1 - x^2}$ has an absolute maximum at zero. It has absolute minima at $+1$ and -1. The function also has a strict relative maximum at zero, but it does not have any relative minimum. The obvious candidates, $+1$ and -1, do not qualify, since the function fails to be defined throughout any neighborhood of these points. At $+1$ and -1 we have what are sometimes called "endpoint extrema" (see Fig. 6–2).

Example 2. The integer-value function $f(x) = [x]$ has neither an absolute maximum nor an absolute minimum. At $\frac{1}{2}$ the function has both a relative maximum and a relative minimum, since the inequalities

$$f(\tfrac{1}{2}) \geq f(x) \qquad \text{and} \qquad f(\tfrac{1}{2}) \leq f(x)$$

both hold for all x in the neighborhood

$$V = \{x \in \mathbf{R} \mid |x - \tfrac{1}{2}| < \tfrac{1}{2}\} = (0, 1)$$

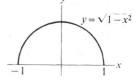

FIG. 6–2

of $\frac{1}{2}$. In fact, the function has relative maxima at all points and relative minima at all points except the integers. The function does not, however, have any strict extrema.

Now suppose that we are given a function and are asked to locate all its relative extrema. Let us observe that for the graph shown in Fig. 6–1 there are tangent lines at A, B, D, and E. Furthermore, it appears that the tangents at these points are parallel to the x-axis. This means that the derivative of the function must be zero at these points. We shall now prove that this is, in fact, the case.

Theorem 2. *Suppose that f has a relative extremum at x_0. If f is differentiable at x_0, then $f'(x_0) = 0$.*

Proof. Suppose that f has a relative maximum at x_0. By definition there exists an $\epsilon > 0$ such that

$$f(x_0) \geq f(x)$$

whenever $x_0 - \epsilon < x < x_0 + \epsilon$. Consider the difference quotient

$$Q(h) = \frac{f(x_0 + h) - f(x_0)}{h}.$$

It is clear that
$$Q(h) \leq 0 \quad \text{if} \quad 0 < h < \epsilon$$
and
$$Q(h) \geq 0 \quad \text{if} \quad -\epsilon < h < 0.$$
Hence
$$\lim_{h \to 0+} Q(h) \leq 0$$
and
$$\lim_{h \to 0-} Q(h) \geq 0.$$
Thus
$$f'(x_0) = \lim_{h \to 0} Q(h) = \lim_{h \to 0+} Q(h) = \lim_{h \to 0-} Q(h) = 0.$$

The moral of this theorem is: *To locate the relative extrema of a function, one should initially examine the points at which the derivative of the function is zero.* The student should be cautioned about two things. First, the function may have a relative extremum at a point where the function is not differentiable (for example, points C and G on the graph given in Fig. 6–1). Hence it is a good idea to examine also points where the function is not differentiable. Second, there may be points (such as F on the same graph) where the derivative of the function is zero, but where the function has neither a relative maximum nor a minimum. Also note that we do not yet have any algorithms for distinguishing between relative maxima and relative minima. We shall discuss these matters further after proving the mean-value theorem.

Exercises

1. Show that the function $f(x) = x^3$ has a zero derivative at zero, although it has neither a relative maximum nor a relative minimum there.
2. Is the derivative of $f(x) = x^{2/3}$ ever zero? Does the function have any relative extrema?
3. Prove or disprove the following. If a function is not differentiable at a point, then it has a relative extremum at the point.
4. Let
$$f(x) = \begin{cases} \sin 1/x & \text{if} \quad x \neq 0, \\ 1 & \text{if} \quad x = 0. \end{cases}$$

Is f differentiable at zero? Does f have a relative extremum at zero? If so, is it a strict relative extremum?

6–2 The mean-value theorem

Suppose that P and Q are two points on a curve C, and let L be the secant joining P and Q, as shown in Fig. 6–3. The mean-value theorem states that in certain cases one can find a point R on the curve between P and Q at which the tangent line is parallel to L.

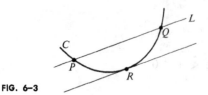

FIG. 6-3

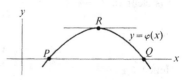

FIG. 6-4

First we consider the special case in which the curve is the graph of a function φ and the line L is the x-axis (see Fig. 6–4). We denote the x-coordinates of P and Q by a and b, respectively. By relabeling P and Q if necessary, we may assume that $a < b$. Clearly, $\varphi(a) = \varphi(b) = 0$. It is also obvious that the tangent line to the curve at the point $(x, \varphi(x))$ will be parallel to L iff $\varphi'(x) = 0$. Our first theorem, Rolle's theorem, states conditions on the function φ which guarantee the existence of such a tangent line.

Theorem 1. Rolle's Theorem. *Suppose that φ is a real-valued function with the following properties:*

1) *φ is differentiable on the open interval (a, b);*

2) *φ is right-continuous at a and left-continuous at b;*

3) *$\varphi(a) = \varphi(b) = 0$.*

Then there exists a number X such that $a < X < b$ and

$$\varphi'(X) = 0.$$

Proof. We consider two cases.

Case I. Let us assume first that φ is identically zero on $[a, b]$. Then φ' is zero at all points of (a, b).

Case II. For the case in which φ is not identically zero, the proof follows quickly from the theorem on extreme values and the necessary condition given in the previous section for relative extrema. Since φ is differentiable throughout the interval (a, b), it is, in particular, continuous there (see Fig. 6–5). Since we also have one-sided

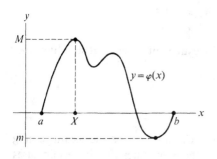

FIG. 6-5

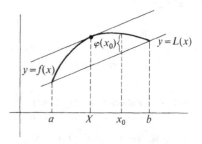

FIG. 6-6

continuity at the endpoints of the interval, we can apply the theorem on extreme values. Thus we can conclude that φ assumes both a maximum value M and a minimum value m on the interval $[a, b]$. Clearly, $M \geq 0$ and $m \leq 0$, with strict inequality in at least one instance. (For if both m and M were zero, then φ would have to be identically zero.) Assume that $M > 0$, and let X be a point in the interval $[a, b]$ where φ takes on the value M. Thus $\varphi(X) = M$, and $a < X < b$. (The point X cannot be either a or b, since the function is zero at these points.) Clearly, φ has a relative maximum at X, and since φ is differentiable at X, it follows that $\varphi'(X) = 0$. A similar argument can be made if $m \neq 0$.

Using Rolle's theorem, we can now prove the mean-value theorem (also called the law of the mean).

Theorem 2. Mean-Value Theorem. *Suppose that f is a real-valued function which satisfies the following conditions:*

1) *f is differentiable throughout the open interval (a, b);*

2) *f is right-continuous at a and left-continuous at b.*

Then there exists a number X such that $a < X < b$ and

$$f(b) - f(a) = (b - a)f'(X).$$

Proof. Before giving a proof of the theorem, we shall interpret it geometrically. The last equation in the statement of the theorem can be rewritten as

$$\frac{f(b) - f(a)}{b - a} = f'(X).$$

The left-hand side is the slope of the secant joining the points $(a, f(a))$ and $(b, f(b))$ on the graph of f. The right-hand side is the slope of the tangent line to the graph of f at the point $(X, f(X))$. Hence the geometric content of the equation is the assertion that these two lines are parallel, as shown in Fig. 6–6.

The proof of the theorem is typical of many proofs involving Rolle's theorem. The crucial step is the construction of a function to which Rolle's theorem can be applied. In this case, we can use the auxiliary function φ defined as

$$\varphi(x) = f(x) - f(a) - \frac{f(b) - f(a)}{b - a}(x - a).$$

One can easily verify that $\varphi(a) = \varphi(b) = 0$. We next observe that φ is the sum of the function f and a polynomial of degree one at most. Since f is right-continuous at a and left-continuous at b, and since a polynomial is everywhere continuous, it follows that φ is right-continuous at a and left-continuous at b. Moreover, φ is differentiable throughout the open interval (a, b); the derivative is given by the equation

$$\varphi'(x) = f'(x) - \frac{f(b) - f(a)}{b - a}.$$

Hence the function φ satisfies the hypotheses of Rolle's theorem. According to the theorem, then, there exists a number X such that $a < X < b$ and

$$0 = \varphi'(X) = f'(X) - \frac{f(b) - f(a)}{b - a},$$

or equivalently,

$$f(b) - f(a) = (b - a)f'(X).$$

Although the preceding proof is both brief and correct, it suffers from a lack of motivation in the choice of the function φ. While it is possible to acquire some measure of skill in selecting such functions, ingenuity (to say nothing of trial and error) is always needed. In this particular instance the function φ does have a simple geometric interpretation. The equation of the secant line through the points $(a, f(a))$ and $(b, f(b))$ is easily seen to be

$$y = L(x) = f(a) + \frac{f(b) - f(a)}{b - a}(x - a),$$

and thus we may think of the function

$$\varphi(x) = f(x) - L(x)$$

as representing the deviation of the graph of f from the secant line (see Fig. 6–6).

The remainder of this chapter is devoted to consequences of the mean-value theorem. These applications can give only a hint, however, of the power and effectiveness which this tool can have when handled by a skilled analyst. At present we shall only indicate by an example how it may be used to determine the number of real roots of an equation.

Example 1. Determine the number of roots of the equation $f(x) = 4x^5 - 5x^4 + 2 = 0$. The derivative of f,

$$f'(x) = 20x^3(x - 1),$$

is zero only when $x = 0$ or $x = 1$. Consequently, the equation $f(x) = 0$ can have at most one real root in each of the intervals $(-\infty, 0)$, $(0, 1)$, and $(1, \infty)$. [Suppose, to the contrary, that there were two roots in the interval $(0, 1)$. By Rolle's theorem, f' would have to be zero at some point in the interval, which is impossible.] We note that

$$f(0) = 2 > 0 \quad \text{and} \quad f(1) = 1 > 0.$$

We also know that for sufficiently large positive integers n,

$$f(-n) < 0 \quad \text{and} \quad f(n) > 0.$$

(See, for example, the proof of Proposition 6, Section 5–2.) Hence it follows from the intermediate-value theorem that there is one root in the interval $(-\infty, 0)$. On the other hand, there can be no roots in the interval $(0, 1)$. [The minimum value of f on the interval $[0, 1]$ is one. Otherwise, f would have a relative minimum at some point in the interval $(0, 1)$ and f' would have to be zero at that point.] One can argue similarly that there is no root in the interval $(1, +\infty)$.

We may conclude, therefore, that the equation $f(x) = 4x^5 - 5x^4 + 2 = 0$ has exactly one real root and that this root is negative.

Exercises

1. Let $\varphi(x) = (x - a)^m(x - b)^n$, where m and n are positive integers. Show that the X of Rolle's theorem divides the interval $[a, b]$ in the ratio m/n.

2. If f is a quadratic polynomial, show that the X of the mean-value theorem is the midpoint of the interval $[a, b]$.

3. Show by the construction of counterexamples that the hypotheses of the mean-value theorem cannot be significantly weakened.

4. Determine the number of real roots of the equation

$$3x^4 - 8x^3 + 6x^2 - 5 = 0.$$

*5. Suppose that P is the polynomial

$$P(x) = a_0x^n + a_1x^{n-1} + \cdots + a_n,$$

where $a_0 \neq 0$. We define

$$P(+\infty) = a_0 \quad \text{and} \quad P(-\infty) = a_0(-1)^n.$$

Suppose also that the equations $P(x) = 0$ and $P'(x) = 0$ have no real roots in common and that $x_1 < x_2 < \cdots < x_m$ are the real roots of the equation $P'(x) = 0$. Prove that the number of real roots of the equation $P(x) = 0$ is exactly equal to the number of changes of sign in the sequence

$$P(-\infty), P(x_1), P(x_2), \ldots, P(x_m), P(+\infty).$$

Apply this result to Example 1 in the text and to the preceding problem. Prove also that the equation

$$1 - x + \frac{x^2}{2} - \frac{x^3}{3} + \cdots + (-1)^n\frac{x^n}{n} = 0$$

has one real root if n is odd and no real root if n is even.

6. Suppose that f is continuous at x_0 and that $\lim_{x \to x_0} f'(x)$ exists. Use the law of the mean to prove that f' exists and is continuous at x_0.

7. We say that x_0 is a *zero* of a function f iff $f(x_0) = 0$. Thus the zeros of $x^2 - 1$ are $+1$ and -1. Suppose that f and g are mappings of the real line into itself with the property that

$$f(x)g'(x) - f'(x)g(x) \neq 0$$

for all $x \in \mathbf{R}$. Prove that the zeros of f and g separate each other in the following sense: between any two consecutive zeros of one function is to be found exactly one zero of the other function (see Fig. 6–7). (Suppose this to be false, and apply Rolle's theorem to

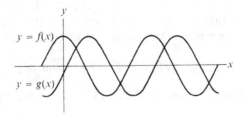

FIG. 6–7

either f/g or g/f, as appropriate.) This result is important in the study of certain differential equations. Note that it applies if $f(x) = \sin x$ and $g(x) = \cos x$.

8. Although the derivative of a function need not be continuous, it shares with continuous functions the intermediate-value property. This is the content of *Darboux's theorem*.

   a) If $f'(x_0) < 0$, prove that there exists a $\delta > 0$ such that $f(x_0) > f(x)$ whenever $x_0 < x < x_0 + \delta$. Also, if $f'(x_0) > 0$, prove that there exists a $\delta > 0$ such that $f(x_0) > f(x)$ whenever $x_0 - \delta < x < x_0$.

   b) Suppose that f' exists throughout the interval $[a, b]$, that $f'(a) < 0$, and that $f'(b) > 0$. Prove that there exists a number X such that $a < X < b$ and $f'(X) = 0$. (Let m be the minimum value of f on the interval $[a, b]$, and let X be any point in the interval where f takes on the value m.) Interpret this result geometrically.

   c) Prove the following theorem.

   Darboux's Theorem. *Suppose that f' exists on the interval $[a, b]$ and that K is any number strictly between $f'(a)$ and $f'(b)$. Then there exists a number X such that $a < X < b$ and $f'(X) = K$.*

   [Assume first that $f'(a) < K < f'(b)$. Apply (b) to the function $\varphi(x) = f(x) - Kx$.]

9. In view of Darboux's theorem, one might hope that a derivative also shares with continuous functions the extreme-value property. This is false. Show, for example, that the derivative of the function

$$f(x) = \begin{cases} x^{4/3} \sin (1/x) & \text{if } x \neq 0, \\ 0 & \text{if } x = 0 \end{cases}$$

exists at all points of the interval $[0, 1]$ but fails to be bounded there.

6–3 Significance of the first derivative

In this section we shall use the law of the mean to investigate the significance of the first derivative. In particular, we shall discuss conditions for a function to be constant, to be increasing, and to be decreasing.

We have already observed that the derivative of a constant function is zero. We shall now prove the converse.

Theorem 1. *Suppose that the function f has zero derivative throughout an open interval I. Then the function is constant on I.*

Proof. We select any number x_0 from I. We will prove that if x is any other element of I, then $f(x) = f(x_0)$. Since I is an interval, the closed interval having x_0 and x as endpoints will be a subset of I. We can then apply the mean-value theorem to deduce the existence of a number X between x_0 and x with the property that

$$f(x) - f(x_0) = f'(X)(x - x_0).$$

Since X also belongs to I, it follows that $f'(X) = 0$. Thus, $f(x) - f(x_0) = 0$ or $f(x) = f(x_0)$.

Corollary 2. *Suppose that f and g are functions such that $f'(x) = g'(x)$ for all x in some open interval I. Then throughout I the functions f and g differ by a constant.*

Proof. Set $h = f - g$. Then for all x in I,

$$h'(x) = f'(x) - g'(x) = 0.$$

By the preceding theorem, there is a real number C such that $h(x) = C$ for all x in I. Thus

$$f(x) = g(x) + C$$

for all x in I.

The preceding theorem and its corollary are frequently applied in the study of integration and differential equations. The following application is typical.

Example 1. If we let

$$y = y(x) = \sqrt{1 - x^2},$$

then by differentiating we can quickly verify that the *differential equation*

$$y\frac{dy}{dx} + x = 0$$

is satisfied. To be more precise, the equation

$$y(x)y'(x) + x = 0$$

holds for all x in the open interval $(-1, +1)$.

We can now ask whether $y(x) = \sqrt{1 - x^2}$ is the only solution of the differential equation above, and, if not, we can then seek the most general solution of the equation. First of all, one can easily check that the function

$$y(x) = -\sqrt{12 - x^2}$$

also satisfies the differential equation on the open interval $(-\sqrt{12}, +\sqrt{12})$. Hence the differential equation does not have a unique solution.

Suppose now that $y(x)$ is a function such that the equation

$$y(x)y'(x) + x = 0$$

is satisfied by all x in some open interval I. Then

$$\frac{d}{dx}[y(x)]^2 = 2y(x)y'(x) = -2x = \frac{d}{dx}(-x^2)$$

for all x in I. Hence by the preceding corollary, there exists a real number C such that

$$[y(x)]^2 = C - x^2$$

for all x in I. By considering signs, one can easily argue that C must be positive, I must be a subinterval of $(-\sqrt{C}, \sqrt{C})$, and either

$$y(x) = \sqrt{C - x^2}$$

for all $x \in I$ or else

$$y(x) = -\sqrt{C - x^2}$$

for all $x \in I$.

Note that the solution to the differential equation is uniquely determined if we specify the value of the solution at zero and require I to be as large as possible. For instance, if we add the "initial condition," $y(0) = -2$, to the differential equation, then the problem has just one solution, namely, $y = -\sqrt{4 - x^2}$ ($|x| < 2$).

Whether the graph of a function rises or falls depends on the sign of the derivative. Consider the function $f(x) = x^2$. For negative values of x, the graph falls; for positive values of x, it rises, as shown in Fig. 6–8. For negative values of x, the derivative $f'(x) = 2x$ is negative; for positive values of x, it is positive. The fact that the graph falls when the derivative is negative and rises when the derivative is positive is by no means coincidental. First we introduce some definitions.

Definition 3. A function f is said to be **nondecreasing** or, less precisely, **increasing** on a set S iff

$$f(x_1) \le f(x_2)$$

whenever x_1 and x_2 are elements of S such that $x_1 < x_2$. We say that f is **strictly increasing** on S iff

$$f(x_1) < f(x_2)$$

whenever x_1 and x_2 are elements of S such that $x_1 < x_2$.

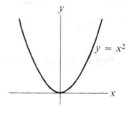

FIG. 6–8

The proper definitions of the terms *nonincreasing* and *strictly decreasing* are obvious. A function f is said to be *monotone* on a set S iff f is either nondecreasing or nonincreasing on S.

Example 2. The function $f(x) = x^3$ is strictly increasing on the entire real line.

Example 3. The function $f(x) = [x]$ is nondecreasing on the entire real line.

Example 4. The function $f(x) = x^2$ is strictly decreasing on the interval $(-\infty, 0]$ and strictly increasing on the interval $[0, +\infty)$.

Example 5. The dizzy-dancer function

$$f(x) = \begin{cases} 1 & \text{if } x \text{ is rational,} \\ 0 & \text{if } x \text{ is irrational} \end{cases}$$

fails to be monotone on any open interval.

Theorem 4. *Suppose that f is nondecreasing throughout an open interval I. If f is differentiable at a point $x_0 \in I$, then $f'(x_0) \ge 0$. Conversely, if $f'(x) \ge 0$ for all $x \in I$, then f is nondecreasing on I; moreover, unless f' vanishes throughout some open interval, the function f will be strictly increasing on I.*

Proof. Suppose that f is nondecreasing on I. Suppose also that $x_0 \in I$ and that $f'(x_0)$ exists. If $x \in I$ and $x \ne x_0$, then

$$Q(x) = \frac{f(x) - f(x_0)}{x - x_0} \ge 0.$$

[For if $x > x_0$, then the numerator of $Q(x)$ is nonnegative while the denominator is positive. If $x < x_0$, then the numerator of $Q(x)$ is nonpositive while the denominator is negative.] Since the inequality $Q(x) \geq 0$ holds for all x in a deleted neighborhood of x_0,

$$f'(x_0) = \lim_{x \to x_0} Q(x) \geq 0.$$

Let us suppose now that $f'(x) \geq 0$ for all $x \in I$. Let x_1 and x_2 be two points of I, with $x_1 < x_2$. By the mean-value theorem, there exists a number X between x_1 and x_2 (and hence in I) such that

$$f(x_2) - f(x_1) = f'(X)(x_2 - x_1).$$

The right-hand side, being the product of a nonnegative number and a positive number, is nonnegative. Hence $f(x_1) \leq f(x_2)$. This proves that f is nondecreasing on I. Suppose that f is not, however, strictly increasing on I. Then there exist two points in I, call them x_1 and x_2, such that $x_1 < x_2$ and $f(x_1) = f(x_2)$. Since f is nondecreasing on the interval $[x_1, x_2]$, it must follow that f is constant on the interval. Hence f' must be zero throughout the open interval (x_1, x_2). This completes the proof.

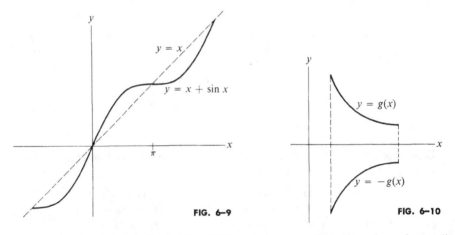

FIG. 6–9   FIG. 6–10

Example 6. Let $f(x) = x + \sin x$. Then $f'(x) = 1 + \cos x$. Since the values of $\cos x$ lie between -1 and $+1$, it is clear that $f'(x) \geq 0$ for all x. Furthermore, $f'(x)$ is strictly positive except where $\cos x$ is equal to -1, namely, at odd integral multiples of π. Hence, by the theorem, the function is strictly increasing on the entire line. Its graph is shown in Fig. 6–9.

Multiplication by -1 reverses inequalities. Consequently, a function g is decreasing on a set S iff the function $-g$ is increasing on S. Also, it is clear that g has a negative derivative at some point iff $-g$ has a positive derivative at that point (see Fig. 6–10). By this line of reasoning one can prove the "dual" of the preceding theorem.

Theorem 4'. *Suppose that f is nonincreasing throughout an open interval I. If f is differentiable at a point x_0 in I, then $f'(x_0) \leq 0$. Conversely, if $f'(x) \leq 0$ for all $x \in I$, then f is nonincreasing on I; moreover, unless f' vanishes throughout some open subinterval of I, the function f will be strictly decreasing.*

Example 7. Determine where the function

$$f(x) = x^3 - 3x + 1$$

is increasing and where it is decreasing. From the factored form of $f'(x)$,

$$f'(x) = 3x^2 - 3 = 3(x + 1)(x - 1),$$

it is apparent that $f'(x) > 0$ if $x > 1$ (since both factors $x + 1$ and $x - 1$ are then positive). If $-1 < x < 1$, then $f'(x) < 0$ (since one factor is positive and the other is negative). Finally, if $x < -1$, then $f'(x) > 0$. We summarize this information in the diagram shown in Fig. 6–11.

Hence f is strictly increasing on the intervals $(-\infty, -1)$ and $(+1, +\infty)$ and strictly decreasing on the interval $(-1, +1)$. The graph of f can be easily sketched from this information (see Fig. 6–12).

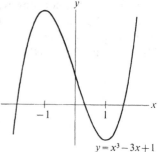

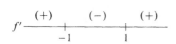

FIG. 6–11   **FIG. 6–12** $y = x^3 - 3x + 1$

Exercises

1. Determine where each of the following functions is increasing and where each is decreasing. Also sketch the graphs of the functions.

   a) $x^3 - 3x^2 + 2$ b) $x^3 - 3x^2 + 2x + 2$ c) $(1 - x^2)^2$

   d) $x + |x|$ e) $(\sin x)(1 + \cos x)$ f) $x\sqrt{4x - x^2}$

   g) $x + (1/x)$ h) $x + |\sin x|$ i) $\sqrt{x}/(1 + x)$

   j) $x - [x]$ k) $\sin x - 3\cos(x/3)$ l) $x^3/(1 + x^4)$

   m) $\sin x \cos 2x$ n) $x - 2\sin x$ o) $(1 - x + x^2)/(1 + x + x^2)$

2. Suppose that the displacement of a particle at time t is given by an equation of the form

$$s = f(t) = \tfrac{1}{2}at^2 + v_0 t + s_0,$$

   where a, v_0, and s_0 are constants.

   a) Show that the particle has constant acceleration a and that its displacement and velocity at time $t = 0$ are s_0 and v_0, respectively.

   b) Prove, conversely, that the function f is completely determined by the information given in (a).

3. Find the general solution of the differential equation

$$x \frac{dy}{dx} + y = 0.$$

4. a) Prove the following result.

   Suppose that $f(0) = g(0)$ and that $f'(x) \leq g'(x)$ for all $x \geq 0$. Then $f(x) \leq g(x)$ for all $x \geq 0$.

   [*Hint:* Consider $g - f$.] Interpret this result both geometrically and in terms of particle motion along a straight line.

   b) Starting from the inequality $\cos x \leq 1$, use (a) to deduce in succession the inequalities

   $$\sin x \leq x, \qquad \cos x \geq 1 - x^2/2, \qquad \sin x \geq x - x^3/6$$

   for $x \geq 0$.

   c) Set

   $$S_n(x) = x - \frac{x^3}{3!} + \frac{x^5}{5!} - \cdots + (-1)^{n+1} \frac{x^{2n-1}}{(2n-1)!},$$

   $$C_n(x) = 1 - \frac{x^2}{2!} + \frac{x^4}{4!} - \cdots + (-1)^{n+1} \frac{x^{2n-2}}{(2n-2)!}.$$

   Prove by induction that

   $$S_{2n}(x) \leq \sin x \leq S_{2n-1}(x) \qquad \text{and} \qquad C_{2n}(x) \leq \cos x \leq C_{2n-1}(x)$$

   for all $n \in \mathbf{Z}^+$ and $x \geq 0$.

   d) Using a common set of coordinate axes, graph the functions sin, S_2, and S_3.

   e) Using (c), prove that if $0 \leq x \leq \pi/4$, then the polynomials

   $$S_3(x) = x - \frac{x^3}{3!} + \frac{x^5}{5!} \qquad \text{and} \qquad C_3(x) = 1 - \frac{x^2}{2!} + \frac{x^4}{4!}$$

   can be used to compute $\sin x$ and $\cos x$ accurately to three decimal places.

   f) Find a polynomial which can be used to compute $\sin x$ accurately to seven decimal places for values of x in the interval $[0, \pi/4]$.

   g) Use the inequalities in (c) and pinching arguments to calculate the following limits:

   $$\lim_{x \to 0} (\sin x - x)/x^3, \qquad \lim_{x \to 0} (\cos x - 1)/x^2, \qquad \lim_{x \to 0} (x \cos x - \sin x)/(x \sin^2 x).$$

   h) Prove (once again) that the function

   $$f(x) = \begin{cases} (\sin x)/x & \text{if } x \neq 0, \\ 1 & \text{if } x = 0 \end{cases}$$

   has a continuous derivative at zero.

5. Prove that

$$\frac{\tan x_2}{\tan x_1} > \frac{x_2}{x_1}$$

if $0 < x_1 < x_2 < \pi/2$.

*6. Show that

$$\frac{\tan x}{x} > \frac{x}{\sin x}$$

if $0 < x < \pi/2$.

7. a) Let

$$f(x) = \begin{cases} x + x^{4/3} \sin{(1/x)} & \text{if } x \neq 0, \\ 0 & \text{if } x = 0. \end{cases}$$

Show that $f'(0) = 1$ and yet there is no neighborhood throughout which f is non-decreasing.

b) If $f'(x_0) > 0$, prove that there exists a $\delta > 0$ such that

$$x_0 - \delta < x < x_0 \Rightarrow f(x) < f(x_0), \qquad x_0 < x < x_0 + \delta \Rightarrow f(x_0) < f(x).$$

We say in such a situation that f is *strictly increasing at* x_0.

c) Explain what difference (if any) there is between saying that f is strictly increasing at x_0 and saying that f is strictly increasing throughout some neighborhood of x_0.

8. a) Let

$$f(x) = \begin{cases} 3x + x^2 \sin{(1/x)} & \text{if } x \neq 0, \\ 0 & \text{if } x = 0. \end{cases}$$

Show that f is strictly increasing on the interval $[0, 1]$. Observe, however, that the graph of f crosses the line $y = 3x$ infinitely often.

b) A tribe of cannibals captured two missionaries. They decided to eat only one of them. To determine which man should be eaten, the chief decided to conduct a race. The rules of the race were the following. Both missionaries must leave the starting line at the same time. They must both complete the course within one hour. (Otherwise the cannibals would alter their policy and eat them both.) Both missionaries must move steadily forward, and they must not arrive at the finish line at the same time. The first missionary to pass the other would be eaten.

After much thought the missionaries evolved a strategy which kept both of them from being eaten. Can you discover such a strategy?

6-4 Sufficient conditions for relative extrema

We have previously shown that if a function f has either a relative maximum or a relative minimum at some point, then the derivative of the function, if it exists, will be zero at that point. Thus for functions which are everywhere differentiable, the vanishing of the first derivative is a necessary condition for a relative extremum. Nevertheless, we have already cited several functions (for instance, x^3 and $x + \sin x$) whose first derivatives vanish at points where the functions have neither a relative maximum nor a relative minimum. In this section we shall prove two theorems which give sufficient conditions for relative extrema.

Theorem 1. First-Derivative Test for a Relative Maximum. *Suppose that f is a function such that*

1) *f is continuous at x_0;*

2) *$f'(x) < 0$ for x immediately to the right of x_0;*

3) *$f'(x) > 0$ for x immediately to the left of x_0.*

Then f has a strict relative maximum at x_0.

Proof. First we must state explicitly what we mean by conditions (2) and (3). We mean that there exists a $\delta > 0$ such that $f'(x) < 0$ if $x_0 < x < x_0 + \delta$ and $f'(x) > 0$ if $x_0 - \delta < x < x_0$.

Suppose that $x_0 < x < x_0 + \delta$. Then the mean-value theorem is applicable, and we can conclude from it that there exists a number X such that $x_0 < X < x$ and

$$f(x) - f(x_0) = f'(X)(x - x_0) < 0.$$

Thus $f(x_0) > f(x)$.

Suppose next that $x_0 - \delta < x < x_0$. Then by the mean-value theorem there exists a number X such that $x < X < x_0$ and

$$f(x_0) - f(x) = f'(X)(x_0 - x) > 0.$$

Thus $f(x_0) > f(x)$.

We have shown, therefore, that $f(x_0) > f(x)$ for all x in a deleted neighborhood of x_0 (namely, $\{x \mid 0 < |x - x_0| < \delta\}$). Thus f has a strict relative maximum at x_0.

Example 1. Let

$$f(x) = 1 - x^{2/3}.$$

The function is continuous but not differentiable at zero:

$$f'(x) = -\tfrac{2}{3}x^{-1/3} = -2/(3\sqrt[3]{x})$$

if $x \neq 0$. Clearly, $f'(x) < 0$ if $x > 0$ and $f'(x) > 0$ if $x < 0$. Thus the function has a strict relative maximum at zero (see Fig. 6–13).

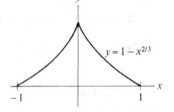

FIG. 6–13

If it has not already occurred to the reader, it probably will not come as a surprise to find that one can obtain sufficient conditions for a relative minimum by simply reversing the inequalities in conditions (2) and (3) of the preceding theorem.

Example 2. Let us consider once again the function

$$f(x) = x^3 - 3x + 1.$$

By previous calculations we know that the sign of f' can be shown by the diagram given in Fig. 6–14. Since the function is continuous, we can conclude that f has a relative maximum at -1 and a relative minimum at $+1$.

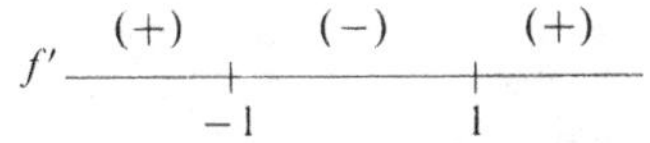

$$f' \quad \underset{-1}{\underline{\quad (+) \quad | \quad (-) \quad | \quad (+) \quad}} \underset{1}{}$$

FIG. 6–14

The next theorem, though weaker than the first, is frequently more convenient to apply.

Theorem 2. Second-Derivative Test for a Relative Maximum. *Suppose that*

$$f'(x_0) = 0 \quad and \quad f''(x_0) < 0.$$

Then f has a strict relative maximum at x_0.

Proof. Since

$$0 > f''(x_0) = \lim_{x \to x_0} \frac{f'(x) - f'(x_0)}{x - x_0} = \lim_{x \to x_0} \frac{f'(x)}{x - x_0},$$

there exists a $\delta > 0$ such that

$$0 < |x - x_0| < \delta \Rightarrow \frac{f'(x)}{x - x_0} < 0.$$

If $x_0 < x < x_0 + \delta$, it follows that $f'(x) < 0$. Similarly, if $x_0 - \delta < x < x_0$, then $f'(x) > 0$. Since f is differentiable at x_0, it is continuous there. The conditions of the first-derivative test are therefore satisfied, and we can conclude that f has a strict relative maximum at x_0.

Theorem 2′. Second-Derivative Test for a Relative Minimum. *Suppose that*

$$f'(x_0) = 0 \quad and \quad f''(x_0) > 0.$$

Then f has a strict relative minimum at x_0.

We can prove this by applying Theorem 2 to the function $-f$.

Example 3. We return to our old friend

$$f(x) = x^3 - 3x + 1.$$

We have seen that the derivative is zero at $+1$ and -1. Since $f''(1) = 6 > 0$, f has a strict relative minimum at $+1$. Since $f''(-1) = -6 < 0$, f has a strict relative maximum at -1.

Example 4. *Osgood's Barnyard Problem.* This problem is a classic and can be formulated in a variety of ways. One of the more whimsical statements of the problem is to be found in Osgood's book on analytic geometry.

A barnyard was bounded on one side by a river. Cattle entered the barnyard each evening through a gate A, went to the river, drank for a spell, and then entered the barn through a door B (see Fig. 6–15). Most of the cattle performed this ritual unthinkingly. There was an intellectual among them, however, who was somewhat lazy, more than a little compulsive, and therefore, not surprisingly, efficiency-minded. She decided that the only sensible way to perform this ritual was to do it in such a way as to minimize her steps. How did she manage this?

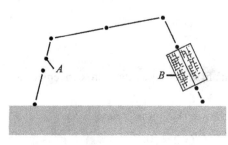

FIG. 6–15   **FIG. 6–16**

In order to solve the problem, we shall idealize it somewhat. We shall assume that A and B are points and that the river is a straight line R. Our job is to determine the point P on R for which $|AP| + |PB|$ is a minimum.

First we translate this geometric problem into an analytic one by introducing co-ordinates. We do this in such a way that the x-axis is R and the y-axis is the line through A perpendicular to R. We may then set $A = (0, a)$ and $B = (c, b)$. We now consider any point $P = (x, 0)$ on R (see Fig. 6–16). By the distance formula,

$$s = |AP| + |PB|$$
$$= \sqrt{x^2 + a^2} + \sqrt{(c - x)^2 + b^2}.$$

To find where s is a minimum, we first calculate the derivative of s:

$$\frac{ds}{dx} = \frac{x}{\sqrt{x^2 + a^2}} - \frac{c - x}{\sqrt{(c - x)^2 + b^2}}.$$

Since ds/dx exists for all $x \in \mathbf{R}$, it follows that ds/dx must be zero at any point where s has a relative minimum. Setting ds/dx equal to zero, we get

$$\frac{x}{\sqrt{x^2 + a^2}} = \frac{c - x}{\sqrt{(c - x)^2 + b^2}}. \qquad (*)$$

Before solving this equation for x, we observe that a solution of the equation must lie in the open interval $(0, c)$. [If x is a number outside this interval, then the two sides of $(*)$ have opposite signs.] To solve $(*)$, we first square both sides and then cross multiply. We get

$$x^2[(c - x)^2 + b^2] = (c - x)^2(x^2 + a^2),$$

which, upon cancellation of the term $x^2(c - x)^2$, yields

$$b^2x^2 = a^2(c - x)^2.$$

Thus

$$bx = \pm a(c - x).$$

Since x must lie in the interval $(0, c)$, only the plus sign is appropriate. If we then solve the resulting equation, we get finally

$$x = \frac{ac}{a + b}.$$

This suggests that the minimum value of x occurs when $x = x_{\min} = (ac)/(a + b)$. To prove that this is so, we calculate the second derivative:

$$\frac{d^2s}{dx^2} = \frac{a^2}{(x^2 + a^2)^{3/2}} + \frac{b^2}{[(c - x)^2 + b^2]^{3/2}}.$$

Clearly, d^2s/dx^2 is positive for all values of x. Hence s has a strict relative minimum when $x = x_{\min}$. Actually, s has an absolute minimum when $x = x_{\min}$. (Since d^2s/dx^2 is always positive, ds/dx is strictly increasing. Hence $ds/dx > 0$ if $x > x_{\min}$, and $ds/dx < 0$ if $x < x_{\min}$. Thus s is strictly increasing to the right of $x_{\min}$ and strictly decreasing to the left of $x_{\min}$. It follows that s has a strict absolute minimum when $x = x_{\min}$.)

We have succeeded in solving our problem analytically. Since our original problem was posed in purely geometric terms, it is appropriate now to seek a geometric interpretation of the solution. Let α and β be the angles indicated in Fig. 6–16. Then

$$\frac{x}{\sqrt{x^2 + a^2}} = \cos \alpha \quad \text{and} \quad \frac{c - x}{\sqrt{(c - x)^2 + b^2}} = \cos \beta,$$

so that Eq. (∗) becomes

$$\cos \alpha = \cos \beta.$$

Since α and β are acute angles, we must therefore have $\alpha = \beta$. In other words, our efficiency-minded cow should drink at that point P on the river where the river makes equal angles with the lines drawn from P to the gate A and to the barn door B. We can also inform the cow how to find this point geometrically. All that she needs to do is to find the reflection B' of B in R, and then find the point P where AB' intersects R. (In Fig. 6–17, observe that $\beta = \beta'$, since R is the perpendicular bisector of BB'. Observe also that $\alpha = \beta'$, since α and β' are vertical angles. Thus $\alpha = \beta$.)

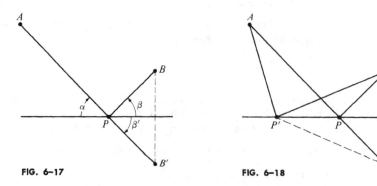

FIG. 6–17   **FIG. 6–18**

At the risk of destroying the reader's faith in the efficacy of calculus, we now present a purely geometric solution of the problem. Let P be constructed as we have described above, and let P' be any other point on R, as shown in Fig. 6–18. We must prove that

$$|AP'| + |P'B| > |AP| + |PB|.$$

Since P' lies on the perpendicular bisector of BB', it follows that $|BP'| = |B'P'|$. By construction, the points A, P, and B' are collinear and thus

$$|AP| + |PB'| = |AB'|.$$

In any nondegenerate triangle, the length of any one side is less than the sum of the lengths of the other two; therefore,

$$|AB'| < |AP'| + |P'B'|.$$

Combining these results, we get

$$\begin{aligned}
|AP'| + |P'B| &= |AP'| + |P'B'| \\
&> |AB'| = |AP| + |PB'| \\
&= |AP| + |PB|.
\end{aligned}$$

This completes the proof.

Although we posed this problem in a fanciful manner, its solution is important in optics, for, according to *Fermat's principle*, light behaves very much like our efficiency-minded cow. Being quite proud of its reputation of being the fastest thing around, it selects its travel routes so as to minimize the travel time. Imagine, then, that R is the surface of a mirror located in a vacuum (or some other medium with constant index of refraction). Suppose that a beam of light decides to move from A to B via the reflecting mirror. If travel time is to be minimized, the light must travel along the same route taken each evening by our cow. In this context angle α is called the *angle of incidence* and angle β is called the *angle of reflection*. Thus, assuming Fermat's principle, we have proved the *law of reflection: when a beam of light traveling through a vacuum is reflected off a mirror, the angle of incidence is equal to the angle of reflection.*

Example 5. Find the relative extrema of the function $f(x) = x\sqrt{ax - x^2}$ $(a > 0)$.

The domain of the function is the closed interval $[0, a]$. The function is zero at the endpoints and positive at all other points on the interval. By the theorem on extreme values, the function must assume an absolute maximum at least once in the open interval $(0, a)$. We will prove that this point is unique and that the function has no other relative extrema. Now

$$f'(x) = \frac{x(3a - 4x)}{2\sqrt{ax - x^2}} \qquad (0 < x < a).$$

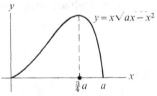

The only point where f' is zero is where $x = \frac{3}{4}a$. We can conclude, therefore, that the function f has a strict absolute (and relative) maximum at $x = \frac{3}{4}a$ and that it has no other relative extrema. The maximum value of the function is $f(\frac{3}{4}a) = 3\sqrt{3}\,a^2/16$. The graph of the function is shown in Fig. 6–19.

FIG. 6–19

The fact that the function has a strict relative maximum at $x = \frac{3}{4}a$ follows from the first-derivative test, since f' is easily seen to change sign from $+$ to $-$ as x passes through $\frac{3}{4}a$. In this case it would be tedious to apply the second-derivative test directly. However, there is a variant of the second-derivative test which is useful in this and similar situations. Suppose, in general, that

$$f'(x) = g(x)/h(x),$$

where $g(x_0) = 0$ and $h(x_0) \neq 0$. [Thus in our particular problem we could set

$$g(x) = x(3a - 4x), \qquad h(x) = 2\sqrt{ax - x^2},$$

and $x_0 = \frac{3}{4}a$.] Then

$$f''(x) = \frac{h(x)g'(x) - g(x)h'(x)}{[h(x)]^2}$$

and, in particular,

$$f''(x_0) = g'(x_0)/h(x_0).$$

It follows that we can limit our attention to the sign of $g'(x_0)/h(x_0)$, and if h is always positive (as it is in this particular problem), to that of $g'(x_0)$. Since in this particular case

$$g'(\tfrac{3}{4}a) = 3a - 8x\Big|_{x=3a/4} = -6a < 0,$$

we can conclude that f has a strict relative maximum at $\frac{3}{4}a$.

Example 6. Determine the most economical shape for a floorless conical tent.

The problem as stated is slightly ambiguous. If the volume of the tent were specified in advance, we would try to minimize the lateral area of the tent. If, on the other hand, the lateral area of the tent were specified, we would then try to maximize the tent's volume. It turns out that these two "dual" problems have the same solution.

For the sake of definiteness we shall assume that the volume V of the tent is specified. Thus

$$V = \tfrac{1}{3}\pi r^2 h, \tag{1}$$

where r and h are the radius and height of the tent, respectively, as indicated in Fig. 6–20. We wish to minimize the lateral area

$$S = \pi r \sqrt{r^2 + h^2}.$$

Clearly, we may just as well minimize

$$W = (S/\pi)^2 = r^4 + r^2 h^2. \tag{2}$$

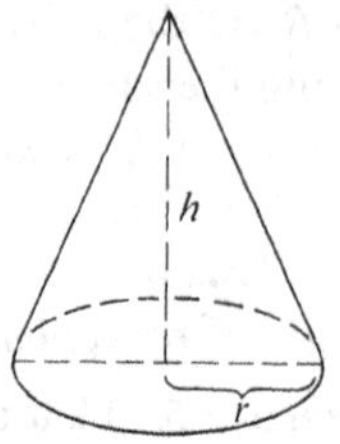

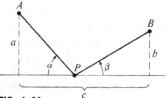

FIG. 6–20

(There is nothing shameful about avoiding radicals.) To minimize W, we could first solve Eq. (1) for h in terms of r, substitute this into our expression for W, calculate dW/dr, etc. Instead, we shall use implicit differentiation. Differentiating Eqs. (1) and (2) with respect to r, we get

$$0 = \frac{dV}{dr} = \frac{1}{3}\pi\left(2rh + r^2\frac{dh}{dr}\right) \quad \text{or} \quad \frac{dh}{dr} = -\frac{2h}{r}$$

and

$$\frac{dW}{dr} = 4r^3 + 2rh^2 + 2r^2 h\frac{dh}{dr} = 2r(2r^2 - h^2).$$

To find when W is a minimum, we set dW/dr equal to zero. This yields $h = r\sqrt{2}$. We can then check that

$$\frac{d^2W}{dx^2} = 12r^2 + 6h^2 > 0.$$

This proves that S has a unique absolute minimum when $h = r\sqrt{2}$.

Note that if we were to solve the dual problem, the solution would be the same, since we would be led once again (but for different reasons) to the simultaneous equations

$$\frac{dV}{dr} = \frac{dW}{dr} = 0.$$

Example 7. In the preceding problem we made use of the so-called "auxiliary variable" technique. As a second example of this technique, let us return to the barnyard problem. This time we use the angles α and β as our variables (see Fig. 6–21). They are related by the equation

$$a\,\mathrm{ctn}\,\alpha + b\,\mathrm{ctn}\,\beta = |A_0 P| + |PB_0| = c.$$

We now seek the minimum value of

$$s = a\csc\alpha + b\csc\beta$$

FIG. 6–21

subject to the above condition. We can treat either α or β as the "independent variable," and since α and β appear symmetrically in the two equations, it obviously is immaterial which we choose. We arbitrarily select α to play the role of independent variable. We now differentiate both equations with respect to α and set $ds/d\alpha$ equal to zero. We get

$$-a\csc^2\alpha - b\csc^2\beta\,\frac{d\beta}{d\alpha} = 0,$$

$$\frac{ds}{d\alpha} = -a\csc\alpha\,\text{ctn}\,\alpha - b\csc\beta\,\text{ctn}\,\beta\,\frac{d\beta}{d\alpha} = 0.$$

We next rearrange these equations:

$$a\csc^2\alpha = -b\csc^2\beta\,\frac{d\beta}{d\alpha},$$

$$a\csc\alpha\,\text{ctn}\,\alpha = -b\csc\beta\,\text{ctn}\,\beta\,\frac{d\beta}{d\alpha}.$$

Finally, we divide the second equation by the first, and obtain

$$\cos\alpha = \frac{\text{ctn}\,\alpha}{\csc\alpha} = \frac{\text{ctn}\,\beta}{\csc\beta} = \cos\beta,$$

as before.

Exercises

1. a) Locate the maxima and minima of the function $f(x) = x^3 - 12x$. Use both the first- and second-derivative tests.

   b) Use the first-derivative test to show that $1 - |x|$ has a relative maximum at zero. Note that the second-derivative test is not applicable.

   c) Let

$$f(x) = \begin{cases} x^2[2 + \sin(1/x)] & \text{if } x \neq 0, \\ 0 & \text{if } x = 0. \end{cases}$$

   Show that the function has a strict relative minimum at zero, and yet that neither the first- nor the second-derivative test is applicable.

2. Locate all relative extrema for each of the following functions. Apply either the first- or the second-derivative test, as convenient.

   a) $8x^3 - 9x^2 + 1$ b) $\sqrt{x(a-x)}$ $(a > 0)$ c) $x^2 + 16/x$

   d) $(\sin x)(1 + \cos x)$ e) $\tan x - 8\sin x$ f) $x^{3/2}(x-18)^{-1/2}$

3. Determine the maximum and minimum values of $x^3 - 3x + 2$ on the interval $[-3, 1.5]$.

4. Prove that among all rectangles with a given perimeter, the square has greatest area.

5. A farmer decides to fence off a rectangular pasture along a straight river. Since he has just 1000 ft of fencing to do the job with, and since he figures that the cows will not escape via the river, he decides not to erect a fence along the river itself. Under these circumstances, how can the farmer enclose the greatest area for pasture?

6. A wire of length L is to be cut into two pieces. One of these will then be bent to form a circle; the other will be bent into the shape of a square. How should the wire be cut so that the sum of the areas will be a maximum? a minimum?

*7. The dimensions of a rectangular swimming pool are a and b. A gentleman standing at one corner of the pool spots a girl wearing a bikini at the corner diagonally opposite. Assuming that his walking speed is v_w and that his swimming speed is v_s, what is the shortest length of time required for the gentleman to get over to the girl (assuming, of course, that the girl stays put)? Consider all possible cases.

8. Determine the maxima and minima of $x^3 + 3px + q$. Discuss the nature of the roots of the equation $x^3 + 3px + q = 0$.

9. Given n numbers $a_1, \ldots, a_n$, determine x so that

$$(a_1 - x)^2 + (a_2 - x)^2 + \cdots + (a_n - x)^2$$

is a minimum.

10. We shall consider the path of a beam of light moving from a point A_1 in medium M_1 (say air) to a point A_2 in medium M_2 (say water). We shall suppose that the surface separating these two media is a plane, and we shall denote by c_k the speed of light in medium M_k ($k = 1, 2$). Let P be the point where the light beam strikes the separating surface. Use Fermat's principle to prove the law of refraction:

$$(\sin \phi_1)/(\sin \phi_2) = c_1/c_2,$$

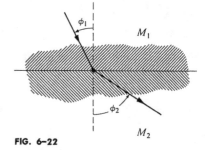

FIG. 6-22

where ϕ_1 and ϕ_2, respectively, are the angle between the normal line at P and the incidental ray, and that between the normal line and the refracted ray (see Fig. 6-22).

11. a) Let f be a function continuous on the entire real line, and let P_0 be an arbitrary point in the plane. Using the theorem on extreme values, prove that there is a point P^* on the graph of f which is closest to P_0. Show by example that P^* need not be unique.

b) Assume now that f is everywhere differentiable and that P_0 does not lie on the graph of f. Prove that the straight line joining P_0 to a point P^* with the above property is normal to the graph of f at the point P^*.

12. a) Find the point(s) on the parabola $y = x^2$ closest to the point $(2, 1)$.

b) Do the same for the points $(0, -4)$ and $(0, 5)$.

13. Among all right circular cylinders with a given volume, find the one with the least lateral area.

14. Among all right circular cylinders inscribed in a given sphere, show that the one with maximum volume is such that its altitude is $\sqrt{2}$ times the radius of the base.

15. A right circular cylinder is to be inscribed in a given right circular cone. Determine the dimensions which yield maximum volume and the dimensions which yield maximum surface area. (Include the areas of the ends of the cylinder.)

16. Prove that among all triangles with a given base and a given perimeter, the isosceles triangle has greatest area.

17. A 27-ft ladder is placed against a fence 8 ft high. The lower end of the ladder is pulled directly away from the fence. Determine the maximum horizontal overhang (that is, the greatest horizontal distance which the ladder projects over the fence). Choose as independent variable the angle which the ladder makes with the ground.

18. Find a point on the altitude of an isosceles triangle (and also inside the triangle) for which the sum of the distances from the point to the vertices is a minimum. In doing this problem, use the angle α as variable (see Fig. 6–23). Show that the minimum is attained when $\alpha = \pi/6$. What is the situation if the base angle is itself less than $\pi/6$?

*19. Given positive numbers $a_1, a_2, \ldots, a_n$, determine the minimum value of

$$\frac{a_1 + a_2 + \cdots + a_{n-1} + x}{n\sqrt[n]{a_1 a_2 \cdots a_{n-1}x}}$$

for $x > 0$. Use this result to prove by induction that

$$\sqrt[n]{a_1 a_2 \cdots a_n} \leq \frac{a_1 + a_2 + \cdots + a_n}{n}.$$

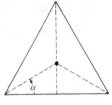

FIG. 6–23

6–5 The sign of the second derivative

The rise or fall of the graph of a function is an important feature which, as we have seen, is linked directly to the sign of the first derivative. Another important qualitative feature of a graph is its direction of bending. As we shall see, this latter property is linked to the sign of the second derivative.

Suppose that a function f has a second derivative which is everywhere positive. Then the first derivative is strictly increasing, and, consequently, as we move along the graph of the function (from left to right), the tangent line to the graph turns in a counterclockwise direction. Intuitively, this means that the graph of the function *bends upward* (as in Fig. 6–24). If, on the other hand, f has a negative second derivative, then the tangent line will rotate in a clockwise fashion, and we say that the graph *bends downward*.

Example 1. Determine the directions of bending of the graph of $f(x) = 2/(1 + x^2)$.
  We have

$$f'(x) = -4x/(1 + x^2)^2 \quad \text{and} \quad f''(x) = -5(1 - 3x^2)/(1 + x^2)^3.$$

The second derivative is negative on the interval $(-\sqrt{3}/3, \sqrt{3}/3)$, and hence the graph bends downward there; the second derivative is positive on the intervals $(-\infty, \sqrt{3}/3)$ and $(\sqrt{3}/3, +\infty)$, and hence the graph bends upward on these intervals, as shown in Fig. 6–25.

Points at which the function switches direction of bending, such as $-\sqrt{3}/3$ and $\sqrt{3}/3$, are called *points of inflection*.

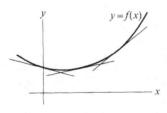

FIG. 6–24

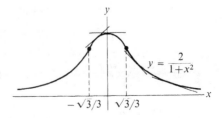

FIG. 6–25

Example 2. Discuss the bending of the sine curve $y = \sin x$.

Since $d^2y/dx^2 = -\sin x = -y$, the sine curve bends downward at all points in the upper half-plane and upward at all points in the lower half-plane. The points of inflection are the points where the curve intersects the x-axis.

Proposition 1. *Suppose that $f''(x_0) = 0$ and that $f'''(x_0) \neq 0$. Then the graph $y = f(x)$ has a point of inflection when $x = x_0$.*

Proof. Suppose $f'''(x_0) < 0$. Since

$$f'''(x_0) = \lim_{x \to x_0} \frac{f''(x)}{x - x_0},$$

it follows that there exists a δ such that

$$f''(x) < 0 \quad \text{if} \quad x_0 < x < x_0 + \delta,$$
$$f''(x) > 0 \quad \text{if} \quad x_0 - \delta < x < x_0.$$

Consequently, the graph bends upward on the interval $(x_0 - \delta, x_0)$ and downward on the interval $(x_0, x_0 + \delta)$. Thus there is a point of inflection when $x = x_0$.

A similar argument works when $f'''(x_0) > 0$.

Exercises

1. Determine the directions of bending and the points of inflection for each of the following functions. Sketch the graphs.

a) $f(x) = x^3$ b) $f(x) = \sqrt{x}$ c) $f(x) = x^3 - 3x + 1$

d) $f(x) = 2x/(1 + x^2)$ e) $f(x) = \tan x$ f) $f(x) = \sqrt{1 + x^2}$

g) $f(x) = x\sqrt{2 - x}$ h) $f(x) = (x^2 - 4)/(x^2 - 9)$

*6-6 Convexity

We shall now study in more detail the significance of the sign of the second derivative. We will show that if the second derivative of a function is everywhere positive, then the graph of f lies above each of its tangent lines and a chord joining any two points of the graph lies above the graph, as shown in Fig. 6-26. The second of

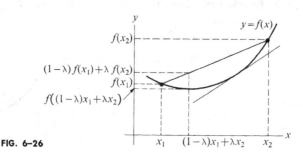

FIG. 6-26

these two properties is perhaps the more interesting, since it leads to rather remarkable inequalities; we take it as our definition of a *convex function*. From the point of division formula in analytic geometry, we know that the coordinates of any point (x, y) on the chord joining x_1 and x_2 may be written

$$x = (1 - \lambda)x_1 + \lambda x_2, \qquad y = (1 - \lambda)f(x_1) + \lambda f(x_2),$$

where $0 \leq \lambda \leq 1$. For this point to lie above the graph of f, we must have

$$f((1 - \lambda)x_1 + \lambda x_2) \leq (1 - \lambda)f(x_1) + \lambda f(x_2).$$

Definition 1. A function f is **convex** on an interval I iff

$$f((1 - \lambda)x_1 + \lambda x_2) \leq (1 - \lambda)f(x_1) + \lambda f(x_2)$$

whenever x_1 and x_2 belong to I and $0 \leq \lambda \leq 1$. If, in addition, we have strict inequality,

$$f((1 - \lambda)x_1 + \lambda x_2) < (1 - \lambda)f(x_1) + \lambda f(x_2)$$

whenever $x_1 \neq x_2$ and $0 < \lambda < 1$, then we say that f is **strictly convex** on I.

Example 1. Show that $f(x) = x^2$ is strictly convex on **R**.
   Suppose $x_1 \neq x_2$ and $0 < \lambda < 1$. Then $0 < 1 - \lambda$ and

$$0 < \lambda(1 - \lambda)(x_1 - x_2)^2 = (1 - \lambda)x_1^2 + \lambda x_2^2 - [(1 - \lambda)x_1 + \lambda x_2]^2.$$

Thus

$$f((1 - \lambda)x_1 + \lambda x_2) = [(1 - \lambda)x_1 + \lambda x_2]^2 < (1 - \lambda)x_1^2 + \lambda x_2^2$$
$$= (1 - \lambda)f(x_1) + \lambda f(x_2),$$

which proves the assertion.

Example 2. Let f be a polynomial of degree one, say $f(x) = ax + b$.
   It is easily verified that

$$f((1 - \lambda)x_1 + \lambda x_2) = (1 - \lambda)f(x_1) + \lambda f(x_2)$$

for *all* real numbers x_1, x_2, and λ. Thus f is convex, but not strictly convex, on **R**.

   Suppose that f is convex on I and that x_1, x_2, and x_3 are in I. Then

$$f(\tfrac{1}{3}(x_1 + x_2 + x_3)) = f(\tfrac{1}{3}x_1 + \tfrac{2}{3}(\tfrac{1}{2}x_2 + \tfrac{1}{2}x_3))$$
$$\leq \tfrac{1}{3}f(x_1) + \tfrac{2}{3}f(\tfrac{1}{2}x_2 + \tfrac{1}{2}x_3)$$
$$\leq \tfrac{1}{3}f(x_1) + \tfrac{2}{3}[\tfrac{1}{2}f(x_2) + \tfrac{1}{2}f(x_3)]$$
$$= \tfrac{1}{3}f(x_1) + \tfrac{1}{3}f(x_2) + \tfrac{1}{3}f(x_3).$$

More generally, f of any arithmetic mean is less than or equal to the arithmetic mean of the f's.

Proposition 2. Jensen's Inequality. *Suppose that f is convex on an interval I and that $x_1, \ldots, x_n$ are points in I. Let $\lambda_1, \ldots, \lambda_n$ be nonnegative numbers such that*

$$\lambda_1 + \lambda_2 + \cdots + \lambda_n = 1.$$

Then $(\lambda_1 x_1 + \cdots + \lambda_n x_n) \in I$ *and*

$$f(\lambda_1 x_1 + \cdots + \lambda_n x_n) \le \lambda_1 f(x_1) + \cdots + \lambda_n f(x_n).$$

Moreover, if the λ_k's *are all strictly positive, and if f is strictly convex on I, then equality holds iff* $x_1 = \cdots = x_n$.

We leave the proof to the reader. (The reader might find it helpful to prove the theorem first for the case in which $n = 3$. This should suggest how he can proceed with the inductive proof.) An expression of the form $\lambda_1 x_1 + \lambda_2 x_2 + \cdots + \lambda_n x_n$, where the λ_k's satisfy the above conditions, is called a *weighted arithmetic mean* (also a *convex combination*) of $x_1, \ldots, x_n$. In the exercises we shall give a physical interpretation of such means as centers of gravity. In particular, if

$$\lambda_1 = \lambda_2 = \cdots = \lambda_n = 1/n,$$

we have the ordinary arithmetic mean.

Corollary 3. *If f is convex on I and* $x_1, \ldots, x_n$ *are points in I, then*

$$f\left(\frac{1}{n}(x_1 + \cdots + x_n)\right) \le \frac{1}{n}[f(x_1) + \cdots + f(x_n)].$$

Moreover, if f is strictly convex on I, then equality occurs only if $x_1 = \cdots = x_n$.

By applying the corollary to the function x^2, we deduce that

$$(x_1 + x_2 + \cdots + x_n)^2 \le n(x_1^2 + x_2^2 + \cdots + x_n^2)$$

for all real numbers $x_1, \ldots, x_n$, and that equality holds only if $x_1 = \cdots = x_n$.
Convex functions are related to convex sets.

Definition 4. A subset K of $\mathbf{R}^2$ is **convex** iff it contains the line segment joining any two of its points (see Fig. 6–27).

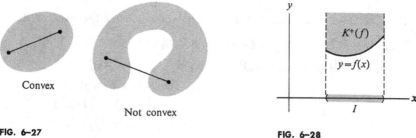

Convex

Not convex

FIG. 6–27   **FIG. 6–28**

Proposition 5. *A function f is convex on an interval I iff the strip of points above the graph of the function, namely,*

$$K^+(f) = \{(x, y) \mid x \in I \text{ and } y \ge f(x)\},$$

is convex (see Fig. 6–28).

The proof is easy and is left to the reader.

It follows from Example 1 that the set of points above the parabola $y = x^2$ is convex, as shown in Fig. 6–29.

We proceed now to a study of the properties of convex functions. We shall temporarily use $\lambda(AB)$ to denote the slope of the line determined by the points A and B.

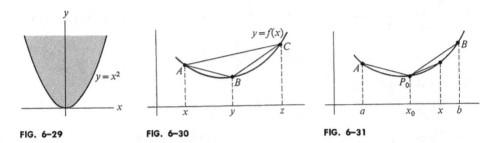

FIG. 6–29   FIG. 6–30   FIG. 6–31

Lemma 6. The Three Chords Lemma. *Suppose that A, B, and C are three points on the graph of a convex function f, where B is to the right of A and C is to the right of B. Then*

$$\lambda(AB) \leq \lambda(AC) \leq \lambda(BC),$$

with strict inequalities if f is strictly convex (see Fig. 6–30).

Proof. Let x, y, and z be the first coordinates of the points A, B, and C, respectively. Then $x < z$, and there exists a (unique) number λ such that $0 < \lambda < 1$ and

$$y = \lambda x + (1 - \lambda)z.$$

Since f is convex,

$$f(y) \leq \lambda f(x) + (1 - \lambda)f(z).$$

Thus

$$\lambda(AB) = \frac{f(y) - f(x)}{y - x}$$

$$\leq \frac{\lambda f(x) + (1 - \lambda)f(z) - f(x)}{\lambda x + (1 - \lambda)z - x}$$

$$= \frac{f(z) - f(x)}{z - x} = \lambda(AC).$$

One can show similarly that $\lambda(AC) \leq \lambda(BC)$.

Proposition 7. Continuity of Convex Functions. *If f is convex on an open interval I, then f is continuous on I.*

Proof. Let x_0 be an arbitrary point in I. Since I is an open interval, we can select points a and b in I such that $a < x_0 < b$ (Fig. 6–31). Suppose that $x_0 < x < b$.

By the three chords lemma,

$$m \leq \frac{f(x) - f(x_0)}{x - x_0} \leq M,$$

where m and M are the slopes of AP_0 and P_0B, respectively. Thus

$$f(x_0) + m(x - x_0) \leq f(x) \leq f(x_0) + M(x - x_0).$$

As $x \to x_0+$, the outer two members of the inequality converge to $f(x_0)$. By the pinching theorem, it follows that

$$\lim_{x \to x_0+} f(x) = f(x_0).$$

One can show similarly that

$$\lim_{x \to x_0-} f(x) = f(x_0).$$

Hence $\lim_{x \to x_0} f(x) = f(x_0)$, which means that f is continuous at x_0.

Proposition 8. *Suppose that f is convex on an open interval I. Then throughout the vertical strip*

$$V = I \times \mathbf{R} = \{(x, y) \mid x \in I\},$$

the graph of f [and hence also $K^+(f)$] lies above each of its tangent lines (see Fig. 6-32).

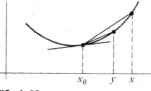

FIG. 6-32

Proof. Suppose that f is differentiable at $x_0 \in I$. Since the equation of the tangent line to the graph at $(x_0, f(x_0))$ is

$$y = f(x_0) + f'(x_0)(x - x_0),$$

it clearly suffices to prove that

$$f(x) \geq f(x_0) + f'(x_0)(x - x_0)$$

for all x in I. Let us consider first the case in which $x > x_0$. If $x_0 < y < x$, then by the three chords lemma

$$\frac{f(y) - f(x_0)}{y - x_0} \leq \frac{f(x) - f(x_0)}{x - x_0}.$$

Thus

$$f'(x_0) = \lim_{y \to x_0+} \frac{f(y) - f(x_0)}{y - x_0} \leq \frac{f(x) - f(x_0)}{x - x_0}$$

or

$$f(x) \geq f(x_0) + f'(x_0)(x - x_0).$$

A similar argument applies when $x < x_0$.

Proposition 9. *Suppose that f is convex on an open interval I. If f is differentiable at two points $x_1 < x_2$ in I, then $f'(x_1) \leq f'(x_2)$.*

Proof. Since the curve lies above the tangent lines at $(x_1, f(x_1))$ and $(x_2, f(x_2))$, we get

$$f(x_2) \geq f(x_1) + f'(x_1)(x_2 - x_1) \qquad \text{and} \qquad f(x_1) \geq f(x_2) + f'(x_2)(x_1 - x_2).$$

Hence

$$f'(x_1) \leq \frac{f(x_2) - f(x_1)}{x_2 - x_1} \leq f'(x_2).$$

Corollary 10. *If f is convex on an open interval I, then f'' is nonnegative at all points of I where the second derivative exists.*

Proof. Suppose that $f''(x_0)$ exists, where x_0 is in I. Then f' must be continuous at x_0 and hence, in particular, it must exist throughout some neighborhood V of x_0. Since f is convex, f' must be nondecreasing throughout V, and $f''(x_0)$ must therefore be nonnegative.

We have seen that for an everywhere differentiable function to be convex, it is necessary for the derivative to be nondecreasing. For a twice differentiable function to be convex, a nonnegative second derivative is necessary. We shall next prove that these conditions are also sufficient.

Proposition 11. *If f' exists and is nondecreasing (strictly increasing) on an open interval I, then f is convex (strictly convex) on I.*

Proof. Let x_1 and x_2 be points in I with $x_1 < x_2$, and let λ be any number between zero and one. Set

$$y = \lambda x_1 + (1 - \lambda)x_2.$$

Now

$$\lambda f(x_1) + (1 - \lambda)f(x_2) - f(y) = \lambda[f(x_1) - f(y)] + (1 - \lambda)[f(x_2) - f(y)].$$

By the mean-value theorem, there exist numbers X_1 and X_2 such that

$$x_1 < X_1 < y < X_2 < x_2$$

and

$$f(x_1) - f(y) = f'(X_1)(x_1 - y) = -f'(X_1)(1 - \lambda)(x_2 - x_1),$$
$$f(x_2) - f(y) = f'(X_2)(x_2 - y) = \lambda f'(X_2)(x_2 - x_1).$$

Thus

$$\lambda f(x_1) + (1 - \lambda)f(x_2) - f(y) = \lambda(1 - \lambda)(x_2 - x_1)[f'(X_2) - f'(X_1)].$$

Since f' is nondecreasing, the right-hand side of the equation is nonnegative. Therefore,

$$f(\lambda x_1 + (1 - \lambda)x_2) = f(y) \leq \lambda f(x_1) + (1 - \lambda)f(x_2).$$

This proves that f is convex on I.

Corollary 12. *If f'' exists and is nonnegative on an open interval I, then f is convex on I. In addition, unless f'' is identically zero throughout some open subinterval of I, f is strictly convex on I.*

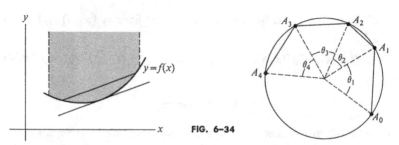

FIG. 6-33   FIG. 6-34

Corollary 13. *Suppose that f is differentiable throughout an open interval I, and also suppose that throughout the strip $V = I \times \mathbf{R}$, the graph of f lies above each of its tangent lines. Then f is convex on I.*

The argument used in proving Proposition 9 can be used in proving the second corollary. From the fact that the graph lies above each of its tangent lines, it follows that f' is nondecreasing.

We can now state the geometric significance of a nonnegative second derivative. Suppose that $f''(x) \geq 0$ for all x in some open interval I. Then, if we ignore what happens outside I, the following are true (see Fig. 6-33):

1) The set of points above the graph of f is convex.

2) A chord connecting two points of the graph lies above the graph.

3) The graph lies above each of its tangent lines.

Example 3. Let $f(x) = x^2$. Then $f''(x) = 2 > 0$. Hence f is strictly convex on the entire real line. A glance at the original proof of this fact (see Example 1) will convince the reader that we have come a long way.

Example 4. Given that $x_1, \ldots, x_n$ are positive numbers, prove that their harmonic mean is less than or equal to their arithmetic mean.

Let $f(x) = 1/x$. Then $f''(x) = 2/x^3 > 0$ if $x > 0$. Thus f is strictly convex on $(0, +\infty)$. By Jensen's inequality, we have

$$\frac{1}{(x_1 + \cdots + x_n)/n} = f\left(\frac{1}{n}(x_1 + \cdots + x_n)\right)$$

$$\leq \frac{1}{n}(f(x_1) + \cdots + f(x_n)) = \frac{1}{n}\left(\frac{1}{x_1} + \cdots + \frac{1}{x_n}\right),$$

with equality iff $x_1 = \cdots = x_n$. By taking reciprocals, we get the desired result. [In more general terms, the weighted harmonic mean

$$\frac{1}{\lambda_1(1/x_1) + \cdots + \lambda_n(1/x_n)}$$

is less than or equal to the weighted arithmetic mean $\lambda_1 x_1 + \cdots + \lambda_n x_n$.]

Example 5. Let $A_0, A_1, \ldots, A_{n-1}$ be the vertices of a convex polygon inscribed in a circle, and let the vertices be ordered in a counterclockwise direction (see Fig. 6-34). Prove that among such n-gons, a regular n-gon has the greatest area.

Let θ_k denote the central angle $\angle A_{k-1}OA_k$, where O is the center of the circle. Then the area of the n-gon is

$$\sum_{k=1}^{n} \frac{1}{2} \left(r \cos \frac{\theta_k}{2} \right) \left(2r \sin \frac{\theta_k}{2} \right) = \frac{1}{2} r^2 \sum_{k=1}^{n} \sin \theta_k,$$

where r is the radius of the circle. Since the function $-\sin x$ is convex in the interval $[0, \pi]$, it follows that

$$\frac{1}{n} \sum_{k=1}^{n} \sin \theta_k \leq \sin \frac{1}{n} \left(\sum_{k=1}^{n} \theta_k \right) = \sin \frac{2\pi}{n}$$

and that equality occurs iff $\theta_1 = \theta_2 = \cdots = \theta_n$. This gives the desired result.

We define a function f to be *concave* on an interval I iff

$$f((1 - \lambda)x + \lambda y) \geq (1 - \lambda)f(x) + \lambda f(y)$$

for all $x, y \in I$ and $\lambda \in [0, 1]$. It is obvious that f is concave on I iff $-f$ is convex on I. Moreover, if f is concave, then the graph of f lies above each of its chords and below each of its tangent lines, etc.

Exercises

1. Prove Jensen's inequality (Proposition 2).
2. a) Prove that a polynomial of odd degree (≥ 3) cannot be convex on the entire real line.
   b) Which polynomials of degree four are convex on the entire real line? Are such polynomials strictly convex?
3. Imagine n point-masses, with masses $m_1, \ldots, m_n$ distributed along a horizontal weightless rod. Let $x_1, \ldots, x_n$ be coordinates of these points. If the rod rests upon a fulcrum with coordinate $\bar{x}$, show that the rod will be in equilibrium iff

$$\bar{x} = \lambda_1 x_1 + \lambda_2 x_2 + \cdots + \lambda_n x_n,$$

where $\lambda_k = m_k/M$ ($k = 1, \ldots, n$) and $M = m_1 + m_2 + \cdots + m_n$. Note that $\lambda_k \geq 0$ ($k = 1, \ldots, n$) and $\lambda_1 + \lambda_2 + \cdots + \lambda_n = 1$. [The tendency of the rod to rotate is measured by the *total torque about the fulcrum*, namely,

$$T = \sum_{k=1}^{n} m_k(x_k - \bar{x}).$$

If $T > 0$, the rod will rotate clockwise; if $T < 0$, it will rotate counterclockwise.]
4. Prove Proposition 5.
5. Let

$$f(x) = \begin{cases} x & \text{if } 0 < x \leq 1, \\ 1 & \text{if } x = 0. \end{cases}$$

Show that f is convex on the interval $[0, 1]$ although it is not continuous at zero. Does this contradict Proposition 7?

6. Let
$$f(x) = \begin{cases} x^2 \sin(1/x) & \text{if } x \neq 0, \\ 0 & \text{if } x = 0. \end{cases}$$

Prove that there is no open interval containing zero throughout which f is either convex or concave.

7. In this exercise we will show that the class of convex functions is closed under certain operations.

   a) Suppose that f is convex on an interval I and that c is a positive real number. Prove that cf is convex on I.

   b) Prove that if f and g are convex on I, then $f + g$ is convex on I.

   c) Given that f and g are real-valued functions, we have defined $f \vee g$ to be the function with domain $\mathfrak{D}_f \cap \mathfrak{D}_g$ such that for all $x \in \mathfrak{D}_f \cap \mathfrak{D}_g$,

   $$(f \vee g)(x) = \max \{f(x), g(x)\}.$$

   Prove that if f and g are convex on I, then $f \vee g$ is convex on I.

   d) Suppose that f is convex and increasing on the entire line. Prove that if g is convex on an interval I, then $f \circ g$ is convex on I.

   e) Is the product of two convex functions convex?

8. a) Prove that the intersection of two convex sets is convex. (You may, if you wish, prove that the intersection of any family of convex sets is again convex.) Show by example that the union of two convex sets need not be convex.

   b) Let f and g be convex on an interval I. Show that

   $$K^+(f \vee g) = K^+(f) \cap K^+(g),$$

   and deduce that $f \vee g$ is convex on I.

9. Let p be a rational number greater than one. Deduce from Jensen's inequality that

$$(x_1 + x_2 + \cdots + x_n)^p \leq n^{p-1}(x_1^p + \cdots + x_n^p)$$

whenever $x_1, x_2, \ldots, x_n$ are positive real numbers. When does equality hold?

10. Let p be a rational number greater than one. Prove that the set

$$K = \{(x, y) \mid |x|^p + |y|^p \leq 1\}$$

is convex. [Show that the function $f(x) = (1 - |x|^p)^{1/p}$ is concave on the interval $[-1, +1]$. Note that $K = K^-(f) \cap K^+(-f)$ and apply Exercise 8.]

*11. A function f is said to be *weakly convex* on an interval I iff

$$f(\tfrac{1}{2}(x + y)) \leq \tfrac{1}{2}[f(x) + f(y)]$$

for all $x, y \in I$. Interpret this geometrically. Prove that if f is both weakly convex and continuous on I, then f is continuous on I. (Functions which are weakly convex but discontinuous exist, but they are very pathological in their behavior and also rather difficult to construct.)

12. Show that f is convex on an interval I iff

$$f(x_1)(x_3 - x_2) + f(x_2)(x_1 - x_2) + f(x_3)(x_2 - x_3) \geq 0$$

whenever $x_1, x_2,$ and x_3 are in I and $x_1 \leq x_2 \leq x_3$.

13. Let φ be twice differentiable on $(0, +\infty)$. Prove that $f(x) = x\varphi(x)$ is convex on $(0, +\infty)$ iff $g(x) = \varphi(1/x)$ is convex on $(0, +\infty)$.

14. Suppose that f is strictly convex and continuous on the closed bounded interval $[a, b]$. Prove that there is just one point in the interval where f has its minimum value. What can be said about the points of the interval where f assumes its maximum value?

15. Prove that the perimeter of an n-gon inscribed in a circle is maximum when the n-gon is regular.

6–7 Approaches to infinity

We shall now generalize our notion of a limit so that utterances such as "$f(x)$ tends to $-\infty$ as $x \to 3+$" or

$$\text{"}\lim_{x \to -\infty} f(x) = +\infty\text{"}$$

are meaningful. Rather than beginning with precise definitions, we shall attempt to convey the intuitive content of such statements through examples.

Consider the behavior of the function

$$f(x) = 2 + \frac{1}{x - 3}.$$

We can make $f(x)$ arbitrarily close to the number 2 by making x sufficiently large. We indicate this by writing

$$\lim_{x \to +\infty} \left(2 + \frac{1}{x - 3}\right) = 2.$$

Furthermore, as x is allowed to grow arbitrarily large, $f(x)$ approaches 2 through values which are greater than 2. This additional bit of information is customarily indicated as follows:

$$\lim_{x \to +\infty} \left(2 + \frac{1}{x - 3}\right) = 2+$$

or "as $x \to +\infty$, $f(x) \to 2+$." We can make the expression $f(x)$ take on arbitrarily large negative values by choosing values of $x < 3$ which are sufficiently close to the number 3 (see Fig. 6–35). In symbols,

$$\lim_{x \to 3-} f(x) = -\infty.$$

We also have

$$\lim_{x \to -\infty} \left(2 + \frac{1}{x - 3}\right) = 2-$$

and

$$\lim_{x \to 3+} \left(2 + \frac{1}{x - 3}\right) = +\infty.$$

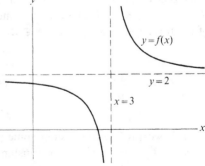

FIG. 6–35

We say that the line $y = 2$ is a *horizontal asymptote* and that the line $x = 3$ is a *vertical asymptote*.

We need to assign a precise meaning to the statement

$$\lim_{x \to a} f(x) = L,$$

where a and L can be any of the following symbols: $+\infty$, $-\infty$, b, $b+$, and $b-$ ($b \in \mathbf{R}$). Since there are twenty-five possibilities for a and L, one might think that twenty-five separate definitions are required. Actually, one suffices. Recall our translation of the statement "$\lim_{x \to a} f(x) = L$" in terms of neighborhoods: for each neighborhood V of L, there exists a deleted neighborhood U of a such that for all x,

$$x \in U \Rightarrow f(x) \in V.$$

We shall adopt this as our general definition. All that is needed is a suitable interpretation of the term "neighborhood" for each of the symbols $+\infty$, $-\infty$, etc.

It is convenient at this point to introduce the *extended real line*. By definition, it is the set $\overline{\mathbf{R}}$ formed by adjoining the symbols $+\infty$ and $-\infty$ to the set of real numbers. (Sometimes $+\infty$ and $-\infty$ are called the *ideal points* on $\overline{\mathbf{R}}$.) We extend the ordering on $\mathbf{R}$ to $\overline{\mathbf{R}}$ by arbitrarily decreeing that $-\infty < +\infty$, $-\infty < x$, and $x < +\infty$ for all $x \in \mathbf{R}$. We define intervals as before:

$$[a, b] = \{x \in \overline{\mathbf{R}} \mid a \leq x \leq b\}, \quad [a, b) = \{x \in \overline{\mathbf{R}} \mid a \leq x < b\},$$
$$(a, b] = \{x \in \overline{\mathbf{R}} \mid a < x \leq b\}, \quad (a, b) = \{x \in \overline{\mathbf{R}} \mid a < x < b\}.$$

Note then that $\overline{\mathbf{R}} = [-\infty, +\infty]$, while for $x \in \mathbf{R}$ such intervals as $(x, +\infty)$, $(-\infty, x]$, etc., have their former meaning. We then define neighborhoods as intervals of $\overline{\mathbf{R}}$. By a neighborhood of $+\infty$ we shall mean any interval of the form $(x, +\infty]$, where $x \in \mathbf{R}$; by a deleted neighborhood of $+\infty$ we shall mean any interval of the form $(x, +\infty)$. The neighborhoods of $-\infty$ are intervals of the form $[-\infty, x)$; the deleted neighborhoods are of the form $(-\infty, x)$. If x is a real number, then by a neighborhood of the symbol $x+$ we mean any interval of the form $[x, y)$, where $y > x$; the deleted neighborhoods of $x+$ are of the form (x, y), where $y > x$. We trust that the reader can now define appropriately the neighborhoods of $x-$.

Definition 1. Suppose that f is a function whose domain and range are subsets of $\overline{\mathbf{R}} = [-\infty, +\infty]$. Suppose that a and L are either members of $\overline{\mathbf{R}}$ or else symbols of the form $x+$ or $x-$, where $x \in \mathbf{R}$. Then

$$\lim_{x \to a} f(x) = L$$

iff for each neighborhood V or L, there exists a deleted neighborhood U of a such that for all $x \in \overline{\mathbf{R}}$,

$$x \in U \Rightarrow f(x) \in V.$$

It is worthwhile to consider some special cases of this definition. The statement "$\lim_{x \to a-} f(x) = +\infty$" can, for instance, be translated as follows:

$$(\forall M \in \mathbf{R})(\exists \delta > 0)(\forall x)(a - \delta < x < a \Rightarrow f(x) > M).$$

We also have

$$\lim_{x \to -\infty} f(x) = a+ \quad \text{iff} \quad (\forall \epsilon > 0)(\exists M \in \mathbf{R})(\forall x)(x < M \Rightarrow a \le f(x) < a + \epsilon),$$

$$\lim_{x \to +\infty} f(x) = -\infty \quad \text{iff} \quad (\forall M \in \mathbf{R})(\exists N \in \mathbf{R})(\forall x)(x > N \Rightarrow f(x) < M).$$

The reader is invited to enlarge this list.

Proofs are omitted in the examples which follow.

Example 1. Describe the behavior of $1/x$ at zero, $+\infty$, and $-\infty$.

We have

$$\lim_{x \to 0+} 1/x = +\infty, \quad \lim_{x \to 0-} 1/x = -\infty.$$

Note that $\lim_{x \to 0} 1/x$ does not exist even in our extended sense. We also have

$$\lim_{x \to +\infty} 1/x = 0+, \quad \lim_{x \to -\infty} 1/x = 0-.$$

The weaker assertions $\lim_{x \to +\infty} 1/x = 0$ and $\lim_{x \to -\infty} 1/x = 0$ are also true.

Example 2. Discuss the behavior of the functions $(\sin x)/x$, $\sin x$, and $x^2 \sin^2 x$ at $+\infty$.

We find that

$$\lim_{x \to +\infty} (\sin x)/x = 0,$$

although each of the stronger statements

$$\lim_{x \to +\infty} (\sin x)/x = 0+ \quad \text{and} \quad \lim_{x \to +\infty} (\sin x)/x = 0-$$

is false. The functions $\sin x$ and $x^2 \sin^2 x$ have no limits as $x \to +\infty$. The function $\sin x$ may be said to oscillate finitely as $x \to +\infty$. The function $x^2 \sin^2 x$ oscillates infinitely as $x \to +\infty$.

We can also generalize our notion of the limit of a sequence.

Definition 2. Let α be a sequence of extended real numbers. Let L be any of the symbols $+\infty$, $-\infty$, $a+$, or $a-$ ($a \in \mathbf{R}$). We define

$$\lim_{n \to \infty} \alpha_n = L$$

to mean that for each neighborhood V of L there exists an $N \in \mathbf{Z}^+$ such that

$$n > N \Rightarrow \alpha_n \in V.$$

If the above quantified statement is true with the term "neighborhood" replaced by "deleted neighborhood," we say that α **converges properly** to L.

We have, for example,

$$\lim_{n \to \infty} n! = +\infty, \quad \lim_{n \to \infty} \frac{2^n - 3^n}{2^n + 3^n} = -1-,$$

$$\lim_{n \to \infty} \frac{2^n}{n!} = 0+, \quad \lim_{n \to \infty} \frac{\sin^2 (\pi/2)n}{n} = 0+.$$

Convergence is proper in all cases except the last.

The notion of proper convergence can also be formulated for functions of a "continuous variable."

Definition 3. We say that $f(x)$ **tends properly** to L as $x \to a$ iff for each deleted neighborhood V of L there exists a deleted neighborhood U of a such that for all x,

$$x \in U \Rightarrow f(x) \in V.$$

Many of our earlier theorems can be generalized. Three of these follow.

Theorem 4. Pinching Theorem. *Suppose that the inequalities*

$$f(x) \le g(x) \le h(x)$$

hold for all x in some deleted neighborhood U of a and that

$$\lim_{x \to a} f(x) = \lim_{x \to a} h(x) = L.$$

Then $\lim_{x \to a} g(x) = L.$

Theorem 5. Linking Theorem. *It follows that*

$$\lim_{x \to a} f(x) = L$$

iff for each sequence $\{x_n\}$ converging properly to a,

$$\lim_{n \to \infty} f(x_n) = L.$$

Theorem 6. Substitution Rule. *If*

$$\lim_{x \to a} g(x) = A \quad \text{properly} \quad \text{and} \quad \lim_{y \to A} f(y) = B,$$

then

$$\lim_{x \to a} f\big(g(x)\big) = B.$$

Because of the substitution rule, it follows, for example, that

$$\lim_{x \to +\infty} f(x) = \lim_{y \to 0+} f\left(\frac{1}{y}\right)$$

whenever one of these limits exists.

We can give proofs of these theorems without considering endless cases if we make some preliminary observations about neighborhoods. Let x be either a point on the extended real line or a symbol of the form $a+$ or $a-$, where $a \in \mathbf{R}$. We observe the following.

1) *If U and V are (deleted) neighborhoods of x, then $U \cap V$ is a (deleted) neighborhood of x.* If, for instance, $x = +\infty$, $U = (a, +\infty)$, and $V = (b, +\infty)$, then $U \cap V = (c, +\infty)$, where $c = \max(a, b)$.

2) *Let U be a neighborhood of x. If $a \in U$, $c \in U$, and $a \le b \le c$, then $b \in U$.* This follows from the fact that neighborhoods (as we have defined them) are intervals.

3) *x has a countable fundamental system of (deleted) neighborhoods, that is, there exists a sequence of (deleted) neighborhoods of x*

$$V_1 \supset V_2 \supset V_3 \cdots$$

such that for each (deleted) neighborhood V of x there exists a $k \in \mathbf{Z}^+$ *such that* $V_k \subset V$. If, for instance, $x = a+$, then

$$[a, a + 1), \qquad [a, a + \tfrac{1}{2}), \qquad [a, a + \tfrac{1}{3}), \qquad \cdots$$

is a fundamental system of neighborhoods of $a+$, while

$$(a, a + 1), \qquad (a, a + \tfrac{1}{2}), \qquad (a, a + \tfrac{1}{3}), \qquad \cdots$$

is a fundamental system of deleted neighborhoods of a. If $x = +\infty$, then

$$(1, +\infty), \qquad (2, +\infty), \qquad (3, +\infty), \qquad \cdots$$

is a fundamental system of deleted neighborhoods of x.

Proof of Theorem 4. Let V be any neighborhood of L. Since $\lim_{x \to a} f(x) = L$ and $\lim_{x \to a} h(x) = L$, we can choose deleted neighborhoods U_f and U_h of a such that

$$x \in U_f \Rightarrow f(x) \in V \qquad \text{and} \qquad x \in U_h \Rightarrow h(x) \in V.$$

Then $U^* = U_f \cap U_h \cap U$ is a deleted neighborhood of a, and for all $x \in U^*$,

$$f(x) \in V, \qquad h(x) \in V, \qquad \text{and} \qquad f(x) \le g(x) \le h(x).$$

By (2) above, $g(x) \in V$ and the proof is complete.

Proof of Theorem 6. Let W be any neighborhood of B. Since $\lim_{y \to A} f(y) = B$, there exists a deleted neighborhood V of A such that

$$y \in V \Rightarrow f(y) \in W.$$

Since $g(x) \to A$ properly as $x \to a$, there exists a deleted neighborhood U of a such that

$$x \in U \Rightarrow g(x) \in V.$$

Thus

$$x \in U \Rightarrow g(x) \in V \Rightarrow f(g(x)) \in W,$$

thereby proving that $\lim_{x \to a} f(g(x)) = B$.

Proof of Theorem 5. The proof of "$\Rightarrow$" is virtually the same as the proof above. Consider "$\Leftarrow$." Suppose $\lim_{x \to a} f(x) \ne L$. Then there exists a neighborhood V of L such that for each deleted neighborhood U of a there exists an $x \in U$ such that $f(x) \notin V$. Let $U_1 \supset U_2 \supset U_3 \supset \cdots$ be a fundamental system of deleted neighborhoods of a. Then for each $n \in \mathbf{Z}^+$ we can select an $x_n \in U_n$ such that $f(x_n) \notin V$. It is easily shown that $x_n \to a$ properly. (If U is any deleted neighborhood of a, choose $N \in \mathbf{Z}^+$ such that $U_N \subset U$. Then $n > N \Rightarrow x_n \in U_n \subset U_N \subset U$.) It is equally clear, however, that $\lim_{n \to \infty} f(x_n) \ne L$.

It would be nice, of course, to extend our theorem on the continuity of the arithmetic operations. This can be done to a certain extent.

Proposition 7. *If*

$$\lim_{x \to a} f(x) = \lim_{x \to a} g(x) = +\infty,$$

then

$$\lim_{x \to a} [f(x) + g(x)] = +\infty.$$

Proof. Let $M \in \mathbf{R}$. We can find deleted neighborhoods U_f and U_g of a such that

$$x \in U_f \Rightarrow f(x) > M/2, \qquad x \in U_g \Rightarrow g(x) > M/2.$$

Then $U = U_f \cap U_g$ is a deleted neighborhood of a and

$$x \in U \Rightarrow f(x) + g(x) > M/2 + M/2 = M.$$

This proves that $\lim_{x \to a} [f(x) + g(x)] = +\infty$.

This result suggests that we define

$$(+\infty) + (+\infty) = +\infty.$$

For similar reasons we define

$$(-\infty) + (-\infty) = -\infty,$$

$$\pm\infty + x = x + (\pm\infty) = \pm\infty \quad \text{if} \quad -\infty < x < +\infty,$$

$$\frac{x}{\pm\infty} = 0 \quad \text{if} \quad -\infty < x < +\infty,$$

$$x(\pm\infty) = (\pm\infty)x = \begin{cases} \pm\infty & \text{if} \quad 0 < x \le +\infty, \\ \mp\infty & \text{if} \quad -\infty \le x < 0. \end{cases}$$

In other words, each of these equations corresponds to a true theorem on limits. For instance, the equation $x/(\pm\infty) = 0$ is a shorthand for the following: if

$$\lim_{t \to a} f(t) = x \qquad (-\infty < x < +\infty)$$

and

$$\lim_{t \to a} g(t) = \pm\infty,$$

then $\lim_{t \to a} [f(t)/g(t)] = 0$. The student should be warned, however, that while certain of the rules of arithmetic remain valid for these extended operations on $\overline{\mathbf{R}}$ (such as commutativity and associativity), others (such as the cancellation law) do not. Moreover, certain "indeterminate forms," such as $(+\infty) + (-\infty)$, $(\pm\infty)/(\pm\infty)$, $0/0$, and $0 \cdot (\pm\infty)$, must remain undefined. [One could, of course, define $(+\infty)/(+\infty)$ arbitrarily to be, say, 1492; the corresponding statement about limits, however, would be false.]

Example 3. Discuss the behavior of the function $f(x) = x^2/(x^2 - 1)$ and sketch its graph.
We find that

$$\lim_{x\to+\infty} \frac{x^2}{x^2-1} = \lim_{x\to+\infty} \frac{1}{1-1/x^2} = \frac{1}{1-0} = 1.$$

Thus the graph has the horizontal asymptote $y = 1$. The line $x = 1$ is a vertical asymptote. Since

$$\lim_{x\to1+} \frac{1}{x-1} = +\infty,$$

it follows that

$$\lim_{x\to1+} \frac{x^2}{x^2-1} = \lim_{x\to1+} \frac{1}{x-1} \frac{x^2}{x+1}$$

$$= (+\infty)\cdot \tfrac{1}{2} = +\infty.$$

Similarly, $\lim_{x\to1-} f(x) = -\infty$. Since

$$f'(x) = -\frac{2x}{(x^2-1)^2},$$

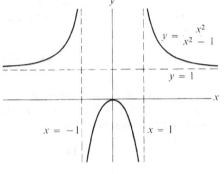

FIG. 6-36

it is clear that the function is strictly decreasing in the intervals $(0, 1)$ and $(1, +\infty)$ and has a relative maximum at 0. Combining this information with the fact that the function is even, we can easily sketch its graph (see Fig. 6-36).

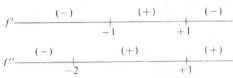

FIG. 6-37

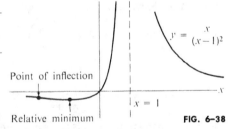

Point of inflection

Relative minimum

FIG. 6-38

Example 4. Graph the function $f(x) = x/(x-1)^2$.
We have

$$\lim_{x\to+\infty} f(x) = 0+, \quad \lim_{x\to-\infty} f(x) = 0-, \quad \lim_{x\to1} f(x) = +\infty.$$

Thus $y = 0$ is a horizontal asymptote, and $x = 1$, a vertical asymptote. Also,

$$f'(x) = -\frac{x+1}{(x-1)^3}, \quad f''(x) = \frac{2x+4}{(x-1)^4}.$$

The signs of f' and f'' are therefore as indicated in Fig. 6-37. Thus f has a relative minimum at $x = -1$ and a point of inflection when $x = -2$, as shown in Fig. 6-38.

Example 5. Graph the function $f(x) = x + 1/x$.

The y-axis is a vertical asymptote. In addition, the graph has the line $y = x$ as asymptote (see Fig. 6–39). The function has a relative minimum when $x = 1$ and a relative maximum when $x = -1$.

We have not yet given a precise definition of an asymptote to a curve. We do so now for the case in which the curve is the graph of a function f. We say that the line $y = mx + b$ is an asymptote to the graph of f if the distance from the point $(x, f(x))$ to the line tends to zero either as $x \to +\infty$ or as $x \to -\infty$. The distance in question is

$$D(x) = \frac{|f(x) - mx - b|}{\sqrt{1 + m^2}}.$$

Since the denominator is constant, $D(x)$ tends to zero iff $f(x) - mx - b$ tends to zero. If the line $y = mx + b$ is an asymptote, the numbers m and b can be determined in this order as follows:

$$m = \lim_{x \to \pm\infty} \frac{f(x)}{x}$$

and

$$b = \lim_{x \to \pm\infty} [f(x) - mx].$$

Conversely, if these limits exist, then $y = mx + b$ is an asymptote. In the case of Example 5, we have

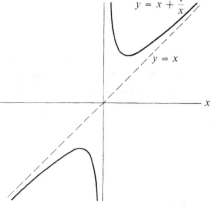

$$m = \lim_{x \to \pm\infty} \frac{f(x)}{x} = \lim_{x \to \pm\infty} \left(1 + \frac{1}{x^2}\right) = 1$$

and

$$b = \lim_{x \to \pm\infty} [f(x) - x] = \lim_{x \to \pm\infty} \frac{1}{x} = 0,$$

FIG. 6–39

thereby showing that $y = x$ is an asymptote.

An asymptote is frequently a limiting position of tangent lines. Consider first the case of a horizontal asymptote. Suppose that as $x \to +\infty$, the graph of f is asymptotic to the line $y = b$ [i.e., $\lim_{x \to +\infty} f(x) = b$]. To say that this asymptote is a limiting position of tangent lines amounts to asserting that $\lim_{x \to +\infty} f'(x) = 0$. The next theorem states a sufficient condition for this to occur.

Proposition 8. *Suppose that as $x \to +\infty$, $f(x)$ and $f'(x)$ both approach finite limits. Then $\lim_{x \to +\infty} f'(x) = 0$.*

Proof. Consider the sequence $a_1, a_2, \ldots$, where $a_n = f(n + 1) - f(n)$. If we let $b = \lim_{x \to +\infty} f(x)$, then clearly

$$\lim_{n \to \infty} a_n = b - b = 0.$$

By the mean-value theorem, we can write (for n sufficiently large) $a_n = f'(\xi_n)$, where $n < \xi_n < n + 1$. As $n \to \infty$, $\xi_n \to +\infty$, and therefore

$$\lim_{x \to +\infty} f'(x) = \lim_{n \to \infty} f'(\xi_n) = \lim_{n \to +\infty} a_n = 0.$$

[Note that the hypothesis on f' can be weakened; it suffices to assume that $\lim_{x \to +\infty} f'(x) = L$ for some $L \in \overline{\mathbf{R}}$.]

Corollary 9. *Suppose that as $x \to +\infty$, the graph of f is asymptotic to the line $y = mx + b$, and suppose that*

$$\lim_{x \to +\infty} f'(x) = L$$

for some $L \in \overline{\mathbf{R}}$. Then $L = m$.

Proof. Let $g(x) = f(x) - mx$. Then $g'(x) = f'(x) - m$. Also $\lim_{x \to +\infty} g(x) = b$ and $\lim_{x \to +\infty} g'(x) = L - m$. By the theorem, $L - m = 0$.

The hypothesis that $\lim_{x \to +\infty} f'(x) = L$ for some $L \in \overline{\mathbf{R}}$ is fulfilled if f'' exists and is of constant sign throughout some deleted neighborhood of $+\infty$. This follows from an analog of the monotone convergence principle for sequences.

Proposition 10. *Suppose that f is monotone (either nondecreasing or nonincreasing) throughout a deleted neighborhood V of $+\infty$. Then there exists an $L \in \overline{\mathbf{R}}$ such that*

$$\lim_{x \to +\infty} f(x) = L.$$

If f is bounded on V, then $L \in \mathbf{R}$. The same is true if $+\infty$ is replaced by $-\infty$, $a+$, or $a-$ ($a \in \mathbf{R}$).

The proof is left to the reader.

Example 6. Functions are frequently presented to us as solutions of equations of various sorts, rather than in closed form. One can often obtain quite a bit of qualitative information about the function without first expressing it in closed form.

Suppose, for instance, that we are told that a function f has the following two properties:

1) $f(x) > 0$ for all $x \in \mathbf{R}$;
2) $f'(x) = -x f(x)$ for all $x \in \mathbf{R}$.

Let us see what further information we can obtain about such a function.

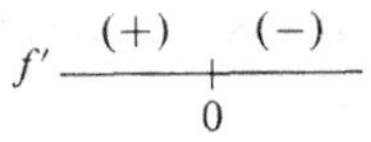

FIG. 6-40 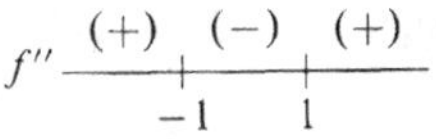   FIG. 6-41

The sign of f' must be as indicated in Fig. 6–40. Thus f is strictly decreasing to the right of zero and strictly increasing to the left of zero. The function f has a relative maximum at zero. The second derivative is

$$f''(x) = -f(x) - x f'(x) = (x^2 - 1) f(x),$$

so that its sign is that of $x^2 - 1$ (see Fig. 6–41). Thus, the graph of f has points of inflection when $x = \pm 1$. For $x > 0$, $f(x)$ is decreasing and bounded below by zero, so that $f(x)$ tends to a nonnegative limit $A \in \mathbf{R}$ as $x \to +\infty$. For $x > 1$, $f'(x)$ is increasing and bounded above by zero, so that $f'(x)$ tends to a nonpositive limit as $x \to +\infty$. Because of the last theorem, we have

$$\lim_{x \to +\infty} f'(x) = 0.$$

Hence

$$\lim_{x \to +\infty} f(x) = \lim_{x \to +\infty} -(1/x)f'(x) = 0 \cdot 0 = 0.$$

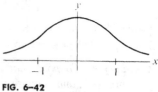

FIG. 6–42

Similarly, $\lim_{x \to -\infty} f(x) = 0$. Putting this information together, we see that the graph of f must be a bell-shaped curve having the x-axis as an asymptote, as shown in Fig. 6–42.

Example 7. Sketch the curve $x^3 + y^3 = 3x^2$.
    We solve for y:

$$y = [x^2(3 - x)]^{1/3}.$$

The sign of y is as shown in Fig. 6–43. Let us first search for asymptotes. We have

$$\lim_{x \to \pm\infty} \frac{y}{x} = \lim_{x \to \pm\infty} \left[\frac{3}{x} - 1\right]^{1/3} = -1,$$

$$\lim_{x \to \pm\infty} (y + x) = \lim_{x \to \pm\infty} \frac{y^3 + x^3}{y^2 - xy + x^2}$$

$$= \lim_{x \to \pm\infty} \frac{3x^2}{y^2 - xy + x^2}$$

$$= \lim_{x \to \pm\infty} \frac{3}{(y/x)^2 - (y/x) + 1} = 1.$$

Hence $y = -x + 1$ is an asymptote.

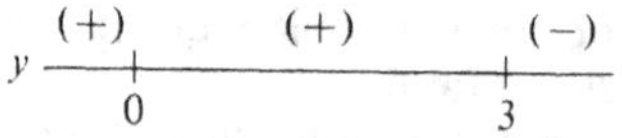

FIG. 6–43

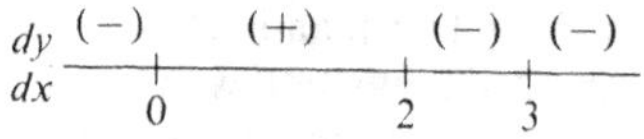

FIG. 6–44

We calculate dy/dx by implicit differentiation:

$$3x^2 + 3y^2 \frac{dy}{dx} = 6x, \qquad y^2 \frac{dy}{dx} = (2 - x)x.$$

Thus the sign of dy/dx is as indicated in Fig. 6–44. If x is not equal to 0 or 3, then $y \neq 0$, and so

$$\frac{dy}{dx} = \frac{(2 - x)x}{y^2} = \frac{2 - x}{\sqrt[3]{x(3 - x)^2}}.$$

Hence as $x \to 0$ or $x \to 3$, $|dy/dx| \to +\infty$. It follows that the graph has vertical tangents at $x = 0$ and $x = 3$ (see Fig. 6–45).

    In the remaining examples we sketch some famous classical curves by applying techniques developed in this chapter.

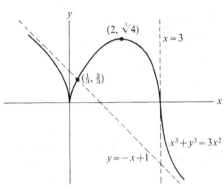

FIG. 6-45

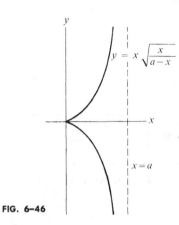

FIG. 6-46

Example 8. Sketch the *cissoid of Diocles:*

$$(x^2 + y^2)x - ay^2 = 0 \qquad (a > 0).$$

Clearly, the curve is the union of the graph of the function

$$f(x) = x\sqrt{\frac{x}{a-x}}$$

and the reflection of the graph in the x-axis. The expression under the radical sign exists and is nonnegative only for values of x in the interval [0, a), and

$$\lim_{x \to a-} f(x) = +\infty.$$

Thus x = a is a vertical asymptote. We have

$$f'(x) = \frac{(\frac{3}{2}a - x)}{a - x} f(x).$$

The above expression is visibly positive on the interval (0, a), and

$$\lim_{x \to 0+} f'(x) = \frac{\frac{3}{2}a}{a} \cdot 0 = 0.$$

Therefore, f is a strictly increasing function, and its graph has the x-axis as a one-sided tangent line at the origin. From this information the curve is readily sketched (see Fig. 6–46). The origin is called a *cusp* of the curve.

Example 9. The *folium of Descartes* is defined by the equation

$$x^3 + y^3 - 3xy = 0$$

(or, somewhat more generally, by $x^3 + y^3 - 3axy = 0$).

We begin by observing that the equation is symmetric with respect to x and y, and, consequently, that the curve is symmetric with respect to the line y = x. We also observe that the curve is unbounded, since for each fixed x ∈ **R**, we get a cubic equation in y which has at least one solution (but no more than three solutions). To study the curve in greater detail, it is convenient to introduce a third variable t defined by t = y/x. This is legitimate

	$-1 < t < 0$	$0 < t < 2^{-1/3}$	$2^{-1/3} < t < 1$
x	$-$	$+$	$+$
y	$+$	$+$	$+$
dx/dt	$+$	$+$	$-$
dy/dt	$-$	$+$	$+$

FIG. 6–47

for all points on the curve except the origin. If (x, y) is a point on the curve, we can now express x and y in terms of t. This is done as follows:

$$y = tx, \qquad x^3 + t^3 x^3 - 3tx^2 = 0.$$

$$x = \frac{3t}{1 + t^3}, \qquad y = \frac{3t^2}{1 + t^3}. \qquad (*)$$

Conversely, one can check that for each $t \neq -1$, the point (x, y) defined by $(*)$ lies on the curve. In particular, if $t = 0$, we get the origin. Equations $(*)$ are called *parametric equations* of the curve.

One can check that the replacement of t by $1/t$ results in the interchange of x and y. Thus it suffices to study the behavior of the curve for values of t in the interval $(-1, +1]$, since the rest of the curve can then be obtained by reflection in the line $y = x$. Now

$$\frac{dx}{dt} = \frac{3(1 - 2t^3)}{(1 + t^3)^2} \quad \text{and} \quad \frac{dy}{dt} = \frac{3t(2 - t^3)}{(1 + t^3)^2}.$$

As a result, we see that the signs shown in Fig. 6–47 hold.

Thus as t increases from -1 to 0, the point on the curve moves to the right and downward through the second quadrant. As t increases from 0 to 1, the point moves upward through the first quadrant. As t ranges from 0 to $2^{-1/3}$, the point moves to the right, and for t between $2^{-1/3}$ and 1, it doubles back to the left. When $t = 0$, we have a horizontal tangent; when $t = 2^{-1/3}$, we have a vertical tangent. This is indicated in Fig. 6–48.

We study next the behavior of the curve as $t \to -1+$. Clearly, $x \to -\infty$ and $y \to +\infty$. It turns out, moreover, that as $t \to -1+$, the point (x, y) approaches a fixed line. Our job is now to find this asymptote of the curve. Proposition 9 tells us that in certain cases an asymptote is the limiting position of tangent lines. This suggests that we calculate the equation of the tangent line to the point on the curve corresponding to t, and then observe what happens when $t \to -1+$. By implicit differentiation, we get

$$\frac{dy}{dx} = -\frac{x^2 - y}{y^2 - x}.$$

Thus at a point (x_0, y_0) on the curve the equation of the tangent line is

$$\frac{y - y_0}{x - x_0} = -\frac{x_0^2 - y_0}{y_0^2 - x_0}$$

or

$$(x_0^2 - y_0)x + (y_0^2 - x_0)y - x_0 y_0 = 0.$$

This equation, in contrast to the one immediately preceding, is valid for all points on the curve except the origin. It follows that at the point corresponding to $t \in (-1, 0)$

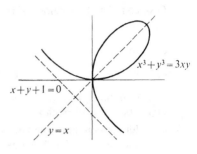

FIG. 6–48   **FIG. 6–49**

the tangent line is

$$[3t^2 - (1 + t^3)t^2]x + [3t^4 - 3(1 + t^3)t]y - 3t^3 = 0.$$

Taking the limit as $t \to -1+$, we get

$$x + y + 1 = 0$$

as a possible asymptote. To check that this line is in fact an asymptote, we show that the distance from the line to the point on the curve corresponding to t tends to zero as $t \to -1+$. The distance in question is

$$\frac{|x + y + 1|}{\sqrt{2}} = \frac{\left| \dfrac{3t}{1 + t^3} + \dfrac{3t^2}{1 + t^3} + 1 \right|}{\sqrt{2}}$$

$$= \frac{|t^3 + 3t^2 + 3t + 1|}{\sqrt{2}(1 + t^3)} = \frac{(1 + t)^3}{\sqrt{2}(1 + t^3)}$$

$$= \frac{(1 + t)^2}{\sqrt{2}(1 - t + t^2)}.$$

As $t \to -1+$, this distance clearly tends to zero, as shown in Fig. 6–49. The origin, incidentally, is called a *double point* of the curve.

Exercises

1. Determine by inspection each of the following limits. In which cases is convergence proper?

a) $\displaystyle\lim_{x \to +\infty} \frac{3x^5 - 2}{4x^5 + x^2 - 20}$

b) $\displaystyle\lim_{x \to -\infty} \frac{3x^5 - 2}{4x^5 + x^2 - 20}$

c) $\displaystyle\lim_{x \to +\infty} \frac{x \cos x}{x^2 - 4}$

d) $\displaystyle\lim_{x \to +\infty} \frac{x^4 - 2x}{3x^2 + \sin x}$

e) $\displaystyle\lim_{x \to 2+} \frac{x}{x^2 - 4}$

f) $\displaystyle\lim_{x \to -2-} \frac{x}{x^2 - 4}$

g) $\displaystyle\lim_{x \to -1+} \frac{\tan(\pi x/4)}{x^2 + 2x + 1}$

h) $\displaystyle\lim_{x \to -1-} \frac{\tan(\pi x/4)}{x^2 + 2x + 1}$

i) $\displaystyle\lim_{x \to 0} \frac{1}{[\cos x] - 1}$

j) $\displaystyle\lim_{x \to (\pi/2)-} \tan x$

k) $\displaystyle\lim_{x \to (\pi/2)+} \tan x$

2. Give complete translations of the following statements:

a) $\lim_{x\to a+} f(x) = -\infty$ $(a \in \mathbf{R})$.

b) $\lim_{x\to +\infty} f(x) = L$ $(L \in \mathbf{R})$.

c) $\lim_{x\to -\infty} f(x) = +\infty$.

d) $\lim_{x\to a-} f(x) = L+$ $(L \in \mathbf{R})$.

e) $\lim_{x\to -\infty} f(x) = L-$ $(L \in \mathbf{R})$.

f) f is bounded throughout some neighborhood of $a+$ $(a \in \mathbf{R})$.

g) f is bounded throughout some neighborhood of $-\infty$.

3. a) Exhibit a countable fundamental system of deleted neighborhoods of $10-$.

b) Exhibit a countable fundamental system of neighborhoods of $-\infty$.

4. Give a complete proof of the theorem on limits corresponding to the equation

$$x/(\pm\infty) = 0 \qquad (-\infty < x < +\infty).$$

5. Find functions f and g such that

$$\lim_{x\to -\infty} f(x) = \lim_{x\to -\infty} g(x) = +\infty$$

and

a) $\lim_{x\to -\infty} [f(x) - g(x)] = +\infty$;

b) $\lim_{x\to -\infty} [f(x) - g(x)] = 0$;

c) $\lim_{x\to -\infty} [f(x) - g(x)] = 5$;

d) $f(x) - g(x)$ is bounded throughout some deleted neighborhood of $-\infty$ but has no limit as $x \to -\infty$;

e) $f(x) - g(x)$ is unbounded throughout every deleted neighborhood of $-\infty$ but approaches neither $+\infty$ nor $-\infty$ as $x \to -\infty$.

6. Do Exercise 5 with $f(x) - g(x)$ replaced by $f(x)/g(x)$.

7. Find a function f such that $f(x)$ tends to a finite limit as $x \to +\infty$ and

a) f' is bounded throughout some deleted neighborhood of $+\infty$ but has no limit as $x \to +\infty$;

b) f' is unbounded throughout each deleted neighborhood of $+\infty$.

Why do these examples not contradict Theorem 8?

8. Suppose that $\lim_{x\to a+} f(x) = +\infty$ and that f' exists throughout some deleted neighborhood of $a+$. Show that f' is unbounded. What more can be said if f'' is of constant sign throughout some deleted neighborhood of $a+$?

9. Show that the hyperbola

$$\frac{x^2}{a^2} - \frac{y^2}{b^2} = 1$$

has the asymptotes

$$\frac{x}{a} + \frac{y}{b} = 0 \qquad \text{and} \qquad \frac{x}{a} - \frac{y}{b} = 0.$$

Draw a sketch for the case in which $a = 3$ and $b = 2$. Show that the hyperbola

$$\frac{y^2}{b^2} - \frac{x^2}{a^2} = 1$$

has the same asymptotes.

10. Consider the *astroid*

$$x^{2/3} + y^{2/3} = a^{2/3} \qquad (a > 0).$$

Show that the curve has two horizontal and two vertical tangents. Observe that the points at which these occur are cusps. (The figure is also known as a *four-cusp hypocycloid*.) Sketch the curve for the case in which $a = 1$. Let T be any tangent line to the curve which is neither horizontal nor vertical. Prove that the distance between the points of intersection of T and the coordinate axes is equal in all cases to the number a.

11. Sketch the graphs of the following rational functions. Label asymptotes, maxima, minima, and points of inflection.

a) $\dfrac{2x}{x+1}$ b) $\dfrac{x-1}{x(x-2)}$ c) $\dfrac{x}{(x-1)(x-2)}$ d) $\dfrac{1}{(x-1)^2}$

e) $\dfrac{x^3}{(x-1)^2}$ f) $\dfrac{x^2}{1+x^2}$ g) $\dfrac{x}{(x+1)(x-2)^2}$

12. a) Show that the *strophoid*

$$(x^2 + y^2)x - a(x^2 - y^2) = 0 \qquad (a > 0)$$

has one vertical asymptote, one vertical tangent line, two horizontal tangents, and a double point. Sketch the curve for the case in which $a = 1$.

b) Set $t = y/x$ and find parametric equations for the curve. On the sketch, indicate points on the curve corresponding to various values of t. Using the parametric equations of the curve, describe its behavior near the origin.

13. Describe the curve

$$y^2 = (x - a)(x - b)(x - c).$$

a) If $a < b < c$, show that the curve falls in two pieces, one bounded, the other unbounded. Show that the curve has three vertical tangents, two horizontal tangents, two inflection points, and no asymptotes. Sketch the curve for $a = 1, b = 2$, and $c = 3$.

b) If $a = b < c$, show that the curve has one vertical tangent and two inflection points. Sketch the curve for $a = b = 1$ and $c = 2$.

c) If $a < b = c$, show that the curve lies in one piece, has one vertical tangent, two horizontal ones, and a double point. Sketch the curve for $a = 0$ and $b = c = 1$.

d) If $a = b = c$, show that the curve has a cusp and a horizontal cuspidal tangent. Sketch the curve for $a = b = c = 0$.

Try to visualize the second two cases as limiting cases of the first one, and the last case as a limiting case of the second two.

14. Sketch the curve $y^2 x^2 = x^2 - 1$.

15. Trace the curve $y^2 = x^2(1 - x^2)$. Locate horizontal and vertical tangents. What is the situation at the origin with regard to tangents?

16. Sketch the curve $y^2(x - 1) = x^2(x + 1)$. Show that the curve falls in two pieces which have a pair of common asymptotes. Are there any inflection points?

*17. Sketch the curve $y^4 - x^4 + xy = 0$.

18. Sketch the curve $(y - 4x^2)^2 = x^7$.

19. As in Example 6, we shall assume that a function f satisfies the differential equation

$$f'(x) = -x f(x)$$

for all $x \in \mathbf{R}$. Instead of assuming, however, that $f(x) > 0$ for all $x \in \mathbf{R}$, we shall assume only that $f(0) > 0$. Similarly, we shall assume that there exists a function g such that

$$g'(x) = x g(x)$$

for all $x \in \mathbf{R}$ and $g(0) > 0$.

a) Prove that the product $f(x) g(x)$ is constant, and deduce from this that f and g are everywhere positive.
b) Sketch the graph of g.
c) Prove that if

$$h'(x) = -x h(x)$$

    for all $x \in \mathbf{R}$ and $h(0) = 0$, then h is identically zero.
d) Prove that f is an even function. [Let $h(x) = f(x) - f(-x)$ and apply (c).]

20. Assume that f satisfies the differential equation

$$f'(x) = f(x)$$

for all $x \in \mathbf{R}$ and that $f(0) = 1$.

a) Prove that $f(x) f(-x)$ is identically equal to one. Deduce that f is everywhere positive. What can you deduce about the qualitative behavior of f (increase, decrease, maxima, minima, etc.)?
b) Prove that $\lim_{x \to -\infty} f(x) = 0$ and hence that $\lim_{x \to +\infty} f(x) = +\infty$.
c) Sketch the graph of f.

21. Sketch the curve $y = x^{2/3}(x + 1)^2$. (To save you some algebraic computations, the following information is given: $y'' = 0$ iff $x \simeq -0.4$ or $x \simeq 0.057$, where the corresponding values of y are ~ 0.11 and ~ 0.14.)

22. Sketch the graph of a function f which has *all* of the following properties:

a) f is defined and continuous at all points except $+1$:

$$f(-1) = -3, \qquad f(0) = 0, \qquad f(2) = 1, \qquad f(3) = 2,$$

$$\lim_{x \to -\infty} f(x) = 0 \quad \text{and} \quad \lim_{x \to +\infty} f(x) = 3.$$

b) f' exists and is continuous at all points except ± 1:

$$\lim_{x \to -1+} f'(x) = +\infty, \qquad \lim_{x \to -1-} f'(x) = -\infty,$$

    and the sign of f' is given by the diagram in Fig. 6-50.
c) f'' exists and is continuous at all points except ± 1, and the sign of f'' is given by the diagram in Fig. 6-51.

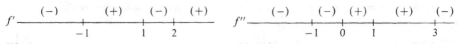

f' —— $(-)$ —— $(+)$ —— $(-)$ —— $(+)$ f'' —— $(-)$ —— $(-)$ $(+)$ —— $(+)$ —— $(-)$

        -1 1 2 -1 0 1 3

FIG. 6-50   **FIG. 6-51**

23. Sketch the graph of a function f which has all of the following properties:

a) f is defined and continuous at all points except -2:

$$f(-3) = 2, \qquad f(-1) = -2, \qquad f(0) = 0,$$

$$f(1) = 2, \qquad f(2) = 1, \qquad f(3) = 5/2,$$

$$\lim_{x \to -2-} f(x) = 0+, \qquad \lim_{x \to -2+} f(x) = +\infty,$$

and

$$\lim_{x \to -\infty} f(x) = 3.$$

b) As $x \to +\infty$, the graph of f is asymptotic to the line $y = x$.

c) f' exists and is continuous at all points except -2 and 0:

$$\lim_{x \to 0} f(x) = +\infty,$$

and the sign of f' is given by the diagram in Fig. 6–52.

d) f'' exists and is continuous at all points except -2 and 0, and the sign of f'' is given by the diagram in Fig. 6–53.

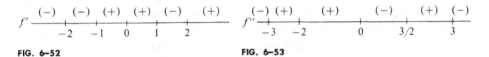

FIG. 6–52   FIG. 6–53

24. Prove the monotone convergence principle.

25. There is an obvious way of defining least upper bounds and greatest lower bounds for subsets of the extended real line. Prove that every subset S of $\overline{\mathbf{R}}$ has both a least upper bound and a greatest lower bound (where, of course, these may be $\pm\infty$). Is it always true that glb $S \leq$ lub S? (Be careful.)

26. The *one-point compactification* $\mathbf{R}^*$ *of* $\mathbf{R}$ is obtained by adjoining to $\mathbf{R}$ the symbol ∞. We define a deleted neighborhood of ∞ to be any set of the form

$$V = (-\infty, -a) \cup (a, +\infty),$$

where $a \in \mathbf{R}$. By a neighborhood of ∞ we mean any set of the form $V^* = V \cup \{\infty\}$, where V is a deleted neighborhood of ∞.

a) Verify that the neighborhoods of ∞ have properties (1) and (3), but not (2), given in this section.

b) Show that $\lim_{x \to \infty} f(x) = L$ iff $\lim_{x \to +\infty} f(x) = L$ and $\lim_{x \to -\infty} f(x) = L$.

c) Show that $\lim_{x \to a} f(x) = \infty$ iff $\lim_{x \to a} |f(x)| = +\infty$.

d) Prove that every sequence of elements of $\mathbf{R}^*$ has a convergent subsequence.

e) Note that (2) is not involved in the proofs of the linking limit lemma and the substitution rule. It follows that these rules are valid if the symbol ∞ is permitted. However, (2) is involved in the proof of the pinching theorem. Find a counterexample for the following assertion: if $f(x) \leq g(x) \leq h(x)$ for all x in some deleted neighborhood of a, and if $\lim_{x \to a} f(x) = \lim_{x \to a} h(x) = \infty$, then $\lim_{x \to a} g(x) = \infty$.

f) Formulate a correct version of the pinching theorem involving ∞.

7 □ ANTIDIFFERENTIATION
AND ITS APPLICATIONS

7-1 Antiderivatives

A typical problem in the past has been that of finding the derivative of a given function. As we shall presently see, there are many instances in which we need to solve the opposite problem: given a function f, we must find a function F whose derivative is f. Such problems bring us into the realm of *integral calculus*.

A couple of examples will illustrate how "antidifferentiation" can arise.

Problem 1. Find the area of the region in $\mathbf{R}^2$ bounded by the parabola $y = x^2$, the x-axis, and the vertical lines $x = 1$ and $x = 3$ (the shaded region in Fig. 7–1).

To solve this problem by calculus, we conceive of the region as being generated by a vertical line moving continuously to the right, and we consider the rate at which the area of the region increases. More specifically, we introduce an area function A defined on the interval $[0, +\infty)$. If $x_0 \geq 0$, then $A(x_0)$ will denote the area of the set $\{(x, y) \mid 0 \leq x \leq x_0 \text{ and } 0 \leq y \leq x^2\}$ (see Fig. 7–2). The desired area will clearly be equal to $A(3) - A(1)$. To determine the function A, we first consider its derivative. Let $x_0 \geq 0$, and let $h > 0$. Then $A(x_0 + h) - A(x_0)$ is equal to the area of the shaded region in Fig. 7–3. This area is less than the area of the circum-

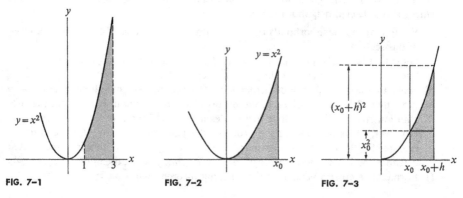

FIG. 7–1   FIG. 7–2   FIG. 7–3

244

scribed rectangle with width h and height $(x_0 + h)^2$. Thus,

$$A(x_0 + h) - A(x_0) < h(x_0 + h)^2.$$

Similarly, $hx_0^2 < A(x_0 + h) - A(x_0)$. Multiplication of these inequalities by h^{-1} yields

$$x_0^2 < \frac{A(x_0^2 + h) - A(x_0)}{h} < (x_0 + h)^2.$$

Taking limits as $h \to 0+$, we have, by the usual pinching argument,

$$\lim_{h \to 0+} \frac{A(x_0 + h) - A(x_0)}{h} = x_0^2.$$

In particular, it follows that A is right-continuous at zero. If $x_0 > 0$, $h < 0$, and $x_0 + h > 0$, then the inequality

$$x_0^2 > \frac{A(x_0 + h) - A(x_0)}{h} > (x_0 + h)^2$$

holds, so that

$$\lim_{h \to 0-} \frac{A(x_0 + h) - A(x_0)}{h} = x_0^2.$$

It follows that A has the following properties:

1) A is continuous on $[0, +\infty)$ and differentiable on $(0, +\infty)$;
2) $A'(x) = x^2$ for all $x > 0$;
3) $A(0) = 0$.

Thus the problem reduces essentially to finding an *antiderivative* of x^2. Any function of the form $\frac{1}{3}x^3 + C$, where C is a constant, satisfies conditions (1) and (2); moreover, it may be shown that these are the only such functions. The third condition is fulfilled, however, only when $C = 0$. Thus $A(x) = \frac{1}{3}x^3$. In conclusion, the desired area is equal to

$$A(3) - A(1) = \tfrac{26}{3}.$$

More generally, if $0 \le a < b$, then the area of the region bounded by the parabola $y = x^2$, the x-axis, and the vertical lines $x = a$ and $x = b$ is equal to

$$A(b) - A(a) = \tfrac{1}{3}(b^3 - a^3).$$

Problem 2. Consider two particles with masses m_1 and m_2. Calculate the work required to move the second particle along a straight line from a distance r_1 to a distance $r_2 > r_1$ from the first particle, as illustrated in Fig. 7–4.

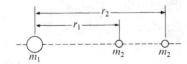

FIG. 7–4

If $r \geq r_1$, let $W(r)$ denote the work required to move the second particle from a distance r_1 to a distance r from the first. As we move the second particle from a distance r to a distance $r + h$ ($h > 0$), the gravitational force drops from $F(r) = Gm_1m_2/r^2$ to $F(r + h) = Gm_1m_2/(r + h)^2$. (Here G is a constant.) Consequently, the work done during that period, $W(r + h) - W(r)$, lies between $hF(r)$ and $hF(r + h)$. Thus

$$\frac{Gm_1m_2}{(r + h)^2} < \frac{W(r + h) - W(r)}{h} < \frac{Gm_1m_2}{r^2}.$$

If h is negative, the inequality signs are reversed. It follows that

$$\frac{dW}{dr} = \lim_{h \to 0} \frac{W(r + h) - W(r)}{h} = \frac{Gm_1m_2}{r^2}.$$

Once again our problem reduces to that of finding an "antiderivative." The function

$$W(r) = -\frac{Gm_1m_2}{r} + C,$$

where C is a constant, is easily seen to work. The appropriate value of C can be determined from the initial condition

$$0 = W(r_1) = -\frac{Gm_1m_2}{r_1} + C.$$

Hence, $C = Gm_1m_2/r_1$, and the required amount of work is

$$W(r_2) = Gm_1m_2\left(\frac{1}{r_1} - \frac{1}{r_2}\right).$$

Definition 1. We shall say that a function F is an **antiderivative** of f on an interval J iff F is continuous on J and $F'(x) = f(x)$ for all $x \in J$, with possibly a finite number of exceptions.

Example 1. If α is any rational number other than -1, then $x^{\alpha+1}/(\alpha + 1)$ is an antiderivative of x^α on the interval $(0, +\infty)$.

Example 2. Let J be a bounded interval with endpoints a and b ($a < b$). [Thus J is either $[a, b]$, $[a, b)$, $(a, b]$, or (a, b).] We define φ_J as follows:

$$\varphi_J(x) = \begin{cases} 1 & \text{if } x \in J, \\ 0 & \text{if } x \notin J. \end{cases}$$

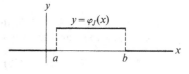

In this case there is no function whose derivative is everywhere equal to φ_J. (This would violate Darboux's theorem, Exercise 8, Section 6-2.) The function does, however, have an antiderivative. If we set

$$F(x) = \begin{cases} 0 & \text{if } x < a, \\ x - a & \text{if } a \leq x \leq b, \\ b - a & \text{if } x \geq b, \end{cases}$$

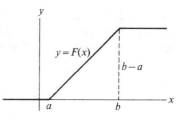

FIG. 7-5

then F is continuous on **R**, and except when $x = a$ and $x = b$, we have $F'(x) = \varphi_J(x)$. Thus F is an antiderivative of φ_J on the entire line. (See Fig. 7–5 for graphs of the functions.)

Example 3. It is possible for an unbounded function to have an antiderivative. For instance, if

$$f(x) = \begin{cases} x^{-2/3} & \text{if } x \neq 0, \\ 0 & \text{if } x = 0, \end{cases}$$

then $3x^{1/3}$ is an antiderivative of f on **R**, despite the fact that f is unbounded near zero.

It is obvious that if F is an antiderivative of f, then for any constant C, the function $F + C$ is also an antiderivative of f. Thus, $3x^2$, $3x^2 + 4$, and $3x^2 - \pi$ are all antiderivatives of $6x$. Apart from constant terms, however, a function can have only one antiderivative on a given interval.

Proposition 2. *If F and G are antiderivatives of f on an interval J, then there exists a constant C such that $G(x) = F(x) + C$ for all $x \in J$.*

Proof. It suffices to show that the function $H = G - F$ is constant on J. Suppose a and b are any two arbitrary points in J, say $a < b$. According to the hypothesis, H is continuous on J and

$$H'(x) = G'(x) - F'(x) = f(x) - f(x) = 0$$

for all $x \in J$, with possibly a finite number of exceptions. Suppose that the numbers

$$a = x_0 < x_1 < \cdots < x_n = b$$

include all elements of the set $\{x \mid a \leq x \leq b \text{ and } H'(x) \neq 0\}$. Since H is continuous on $[x_0, x_1]$ and differentiable on (x_0, x_1), the mean-value theorem tells us that there exists a $\xi \in (x_0, x_1)$ such that $H(x_1) - H(x_0) = H'(\xi)(x_1 - x_0)$. Since $H'(\xi) = 0$, it follows that $H(x_0) = H(x_1)$. Thus by applying the mean-value theorem to H on each of the intervals $[x_0, x_1], [x_1, x_2], \ldots, [x_{n-1}, x_n]$, we see that

$$H(a) = H(x_0) = H(x_1) = \cdots = H(x_n) = H(b).$$

Since a and b were chosen arbitrarily from J, it follows that H is constant on J.

Although the preceding theorem probably came as no surprise, it should be pointed out that there is good reason for us to limit our attention to intervals when discussing antiderivatives. Otherwise, it would be possible for a function to have antiderivatives which did not differ by a constant. For instance, the functions

$$f(x) = \begin{cases} 1/x + 5 & \text{if } x > 0, \\ 1/x - \pi & \text{if } x < 0, \end{cases}$$

and $1/x$ do not differ by a constant on the set $S = (-\infty, 0) \cup (0, +\infty)$ even though their derivatives are equal at all points of S.

It is convenient to have a notation for antiderivatives.

Definition 3. If F is an antiderivative of f, we write

$$\int f(x)\, dx = F(x) + C.$$

For example, we have

$$\int \cos x \, dx = \sin x + C \quad \text{and} \quad \int \sin t \, dt = -\cos t + C.$$

In these examples the symbols "dx" and "dt" appear to contribute little. They become useful, however, when the expression for a function involves unspecified constants. Thus

$$\int x^3 y^4 \, dx = \tfrac{1}{4} x^4 y^4 + C,$$

since dx singles out x as the variable, but

$$\int x^3 y^4 \, dy = \tfrac{1}{5} x^3 y^5 + C,$$

since y is the variable.

Exercises

1. The purpose of this problem is to present an alternative solution to Problem 1. The basic ideas can be traced to Archimedes, and although no calculus is directly involved, these ideas will recur in the chapter on the Riemann integral.

   Let P be the region bounded by the parabola $y = x^2$, the x-axis, and the vertical line $x = b$ $(b > 0)$. For each positive integer n, we split P into n strips $P_1, \ldots, P_n$ by constructing the vertical lines $x = h$, $x = 2h$, $x = 3h$, $\ldots$, $x = (n-1)h$, where $h = b/n$ (see Fig. 7–6). [Explicitly, P_k is the region bounded by the parabola, the x-axis, and the lines $x = (k-1)h$, $x = kh$.] If we use $A(S)$ to denote the area of a set S, then clearly

$$A(P) = \sum_{k=1}^{n} A(P_k).$$

   We next get upper and lower estimates on each term $A(P_k)$. As in the first solution, we get estimates by considering rectangles circumscribed about and inscribed in the strip P_k. Show that

$$h[(k-1)h]^2 \leq A(P_k) \leq h[kh]^2, \quad k = 1, 2, \ldots, n.$$

   By summing over these inequalities, we get

$$\sum_{k=1}^{n} h[(k-1)h]^2 \leq A(P) \leq \sum_{k=1}^{n} h[kh]^2.$$

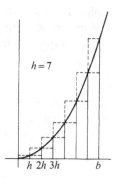

$h = 7$

   Now simplify the sums by using the formula

$$\sum_{k=1}^{n} k^2 = \frac{n(n+1)(2n+1)}{6}.$$

   Then take limits as $n \to \infty$, and show that the two sums converge to $b^3/3$. Deduce that $A(P) = b^3/3$.

$h \ \ 2h \ 3h \qquad\qquad b$

FIG. 7–6

2. Find the area of the trapezoid bounded by the lines $y = mx$, $y = 0$, $x = a$, and $x = b$, where $n > 0$ and $0 < a < b$. Do this by using

   a) the method of Problem 1 in the text, b) the method of Problem 1 in the exercises.

3. Determine whether the following functions have antiderivatives on the intervals indicated. If so, find the required antiderivatives.

   a) $[x]$ on the interval $[-1, 3]$ b) $1/\sqrt{x}$ on the interval $[0, +\infty)$
   c) $1/\sqrt[3]{x}$ on the interval $\mathbf{R} = (-\infty, +\infty)$ d) $1/x^2$ on the interval $[-1, +1]$
   e) $\varphi_{[-2,1]}$ and $\varphi_{(-2,1]}$ on the interval $\mathbf{R}$ f) $|x|$ on the interval $\mathbf{R}$

   g) $\varphi_Q(x) = \begin{cases} 1 & \text{if } x \text{ is rational} \\ 0 & \text{otherwise} \end{cases}$ on the interval $[0, 1]$

4. Calculate each of the indicated antiderivatives.

   a) $\int (4x^3 - 5x + 9)\, dx$ b) $\int (x^3 + 1)^2\, dx$ c) $\int (ax + b)^n\, dx$ $(n \in \mathbf{Z}^+)$

   d) $\int [(x - 1)/\sqrt{x}]\, dx$ e) $\int \sqrt{a + x}\, dx$ f) $\int \cos ax\, dx$

   g) $\int x \cos x^2\, dx$ h) $\int \sec^2 t\, dt$ i) $\int \tan^2 t\, dt$

   j) $\int (x/\sqrt{1 - x^2})\, dx$ k) $\int \cos x \sin^3 x\, dx$

*5. A piece of string of length 2π in. is wound around a circle of radius 1 in. One end of the string is secured to the circumference. The other end is unwound at the rate of 1 in./sec; the string is kept taut during the process (see Fig. 7-7). Find the rate at which the string sweeps out area. Find the area swept out at the end of 1 sec, π sec, 2π sec. (The curve that is traced out by the endpoint of the string is called an *involute* of the circle.)

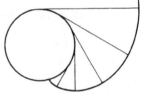

FIG. 7-7

6. A car moves with acceleration $d^2x/dt^2 = 0.016(t - 6)$ ft/sec². When $t = 1$, $v = 11$, and $x = 2$. Find the velocity $v = dx/dt$ and the displacement x in terms of t. Find the minimum speed of the car. How far does the car travel in the first 10 sec after reaching the minimum speed?

7-2 Finding antiderivatives

Generally speaking, antidifferentiation is considerably more difficult than differentiation. Although we shall postpone a systematic treatment of techniques for finding antiderivatives until a later chapter, we shall disclose a few tricks now, so that we can proceed to some applications.

There are a couple of general rules for calculating antiderivatives. If f and g both have antiderivatives on an interval I, then the same will be true of $f + g$, and

$$\int (f + g)(x)\, dx = \int f(x)\, dx + \int g(x)\, dx.$$

Also, if k is a constant, then

$$\int kf(x)\, dx = k \int f(x)\, dx.$$

[The reader who feels uneasy about the exact meaning of these two displayed equations is to be congratulated on well-placed skepticism. If F and G are antiderivatives of f and g, respectively, then $F + G$ is an antiderivative of $f + g$. This is a possible interpretation of the first equation. At any rate, it will do for the purposes we have in mind. Alternatively, the equation can be regarded as a set identity. We can, for instance, interpret $\int f(x)\,dx$ as the set of *all* antiderivatives of f. Thus

$$\int \cos x\,dx = \{\sin x + C \mid C \in \mathbf{R}\},$$

or, better yet,

$$\int \cos = \{\sin + C \mid C \in \mathbf{R}\}.$$

We also define the sum of two sets of functions $\mathfrak{F}$ and $\mathfrak{G}$ to be

$$\mathfrak{F} + \mathfrak{G} = \{F + G \mid F \in \mathfrak{F} \text{ and } G \in \mathfrak{G}\}.$$

The equation $\int (f + g)(x)\,dx = \int f(x)\,dx + \int g(x)\,dx$ may then be regarded as an assertion of the equality of two sets of functions. As this matter of interpretation will never be crucial, we prefer not to commit ourselves to either of the interpretations proposed.] There is, unfortunately, no general rule for finding the antiderivative of the product of two functions. The difficulty caused by the lack of such a rule is mitigated to some extent by a technique called "integration by parts," which will be discussed later.

Let us consider a few examples.

Example 1. Find $\int \sin^2 \theta\,d\theta$.

Here we can use a trigonometric identity:

$$\sin^2 \theta = \frac{1 - \cos 2\theta}{2}.$$

Thus

$$\int \sin^2 \theta\,d\theta = \tfrac{1}{2}\int d\theta - \tfrac{1}{2}\int \cos 2\theta\,d\theta$$

$$= \tfrac{1}{2}\theta - \tfrac{1}{4}\sin 2\theta + C.$$

Example 2. Find $\int x^2\sqrt{1 + x^3}\,dx$.

To work this problem, one must use the chain rule in reverse. Note that the factor x^2 is essentially the derivative of $1 + x^3$. This suggests that we make the substitution $u = 1 + x^3$. Then $du/dx = 3x^2$, so that

$$x^2\sqrt{1 + x^3} = \tfrac{1}{3}\sqrt{u}\,\frac{du}{dx}.$$

Now $\tfrac{1}{3}\int \sqrt{u}\,du = \tfrac{2}{9}u^{3/2} + C$, so that

$$\frac{d}{dx}\left(\tfrac{2}{9}u^{3/2} + C\right) = \tfrac{1}{3}\sqrt{u}\,\frac{du}{dx},$$

by the chain rule. Thus

$$\int x^2\sqrt{1 + x^3}\,dx = \tfrac{2}{9}u^{3/2} + C = \tfrac{2}{9}(1 + x^3)^{3/2} + C.$$

In working problems such as the preceding one, it is convenient to use the formalism of *differentials*. In Leibnitz's notation for a derivative, dy/dx, we have made no attempt to assign a meaning to the symbols "dy" and "dx" separately. Nevertheless, the chain rule suggests that we ought to be able to treat these "differentials" much like numbers. Accordingly, while we could dismiss the equation

$$dy = \frac{dy}{dx} dx$$

as a meaningless vulgarity, we shall countenance it at least provisionally in the hope that it may prove helpful.

If a variable y is a function of a variable x, say $y = f(x)$, we shall agree to write

$$dy = f'(x)\, dx.$$

For example, if $y = x^2$, then $dy = 2x\, dx$. Similarly, $d(\sin x) = \cos x\, dx$, and $d(\tan \theta) = \sec^2 \theta\, d\theta$.

An expression of the form $f(x)\, dx$ is called a *first-order differential*. We define the sum of two such differentials as follows:

$$f(x)\, dx + g(x)\, dx = [f(x) + g(x)]\, dx.$$

Also, we define the product of a function and a differential by

$$f(x)[g(x)\, dx] = [f(x)g(x)]\, dx.$$

Thus if u and v are functions of x, it follows that

$$d(uv) = \frac{d}{dx}(uv)\, dx = \left(u\frac{dv}{dx} + v\frac{du}{dx} \right) dx$$

$$= \left(u\frac{dv}{dx} \right) dx + \left(v\frac{du}{dx} \right) dx$$

$$= u\left(\frac{dv}{dx} dx \right) + v\left(\frac{du}{dx} dx \right) = u\, dv + v\, du.$$

Similarly,

$$d(u + v) = du + dv$$

and

$$d\left(\frac{u}{v} \right) = \frac{v\, du - u\, dv}{v^2}.$$

Note that in the equation $d(uv) = u\, dv + v\, du$, there is nothing to indicate that we were thinking of u and v as functions of an independent variable x. The equation would be the same if we thought of u and v as functions of, say, t or θ. This disappearance of variables might conceivably lead to difficulties. Suppose, for instance, that $y = x^2$ and $x = \sin \theta$. Then dy has two meanings according as we think of y as a function of x or of θ:

$$dy = \frac{dy}{dx} dx = 2x\, dx, \qquad dy = \frac{dy}{d\theta} d\theta = 2 \sin \theta \cos \theta\, d\theta.$$

Thanks to the chain rule, these two interpretations are consistent, since $dx = \cos\theta\,d\theta$, and thus

$$dy = 2\sin\theta\cos\theta\,d\theta = 2x\,dx.$$

It is this inner consistency, derived from the chain rule, which enables us to work fearlessly with differentials, without being overly concerned about their meaning.

We shall now illustrate how this formalism can be put to use.

Example 3. Find $d(\sin x^2)/dx$.

We have

$$d(\sin x^2) = (\cos x^2)\,d(x^2) = (\cos x^2)(2x\,dx) = 2x\cos x^2\,dx,$$

and therefore

$$\frac{d(\sin x^2)}{dx} = 2x\cos x^2.$$

Example 4. Find dy/dx if $x^3 + y^3 - 3xy = 0$.

Taking differentials of both sides, we get

$$0 = d(x^3) + d(y^3) + d(-3xy)$$
$$= 3x^2\,dx + 3y^2\,dy - 3(x\,dy + y\,dx).$$

Thus

$$\frac{dy}{dx} = -\frac{x^2 - y}{y^2 - x}.$$

Example 5. We return to a previous example. Find $\int x^2\sqrt{1 + x^3}\,dx$.

Let $u = 1 + x^3$. Then $du = 3x^2\,dx$. Thus

$$\int x^2\sqrt{1 + x^3}\,dx = \tfrac{1}{3}\int\sqrt{u}\,du = \tfrac{2}{9}u^{3/2} + C = \tfrac{2}{9}(1 + x^3)^{3/2} + C.$$

Example 6. Find $\int\sin^3\theta\,d\theta$.

Note that

$$\int\sin^3\theta\,d\theta = \int(1 - \cos^2\theta)\sin\theta\,d\theta.$$

Let $u = \cos\theta$. Then $du = -\sin\theta\,d\theta$. Therefore

$$\int(1 - \cos^2\theta)\sin\theta\,d\theta = -\int(1 - u^2)\,du = -u + \tfrac{1}{3}u^3 + C$$
$$= -\cos\theta + \tfrac{1}{3}\cos^3\theta + C.$$

Example 7. Find $\int\sqrt{\sin\theta\tan\theta}\,d\theta$, where $0 < \theta < \pi/2$.

Setting $u = \cos\theta$, we find that

$$\int\sqrt{\sin\theta\tan\theta}\,d\theta = -\int u^{-1/2}\,du = -2u^{1/2} + C$$
$$= -2\sqrt{\cos\theta} + C.$$

Let us pause to reflect on what has taken place in this series of examples. We have established certain conventions for handling symbols which are as yet meaningless. We have been able, nonetheless, to apply this empty formalism to establish results both meaningful and correct. It should be clear that in all cases we could have circumvented the use of differentials and that we can carefully establish each of the results by simply retracing the path of calculation, replacing differentials by derivatives, and applying our general theorems on differentiation (especially the chain rule). At the same time, it should be clear that differentials are handy; they allow us to make routine calculations with minimum mental effort.

We can now do one of two things. We can continue to use differentials naively as in the preceding examples. We would limit their use, then, to that of a computational aid, and take care that any computation done with the hocus-pocus of differentials could readily be transformed into respectable mathematics. Alternatively, we can give a precise meaning to differentials and develop their theory rigorously. Our course of action will be a compromise, weighted rather heavily toward the first alternative. That is, we shall continue to use differentials formally, without being overly concerned about their meaning. On the other hand, we shall outline how a careful treatment of differentials could be carried out.

We begin with a word about functions of two variables. If we have three sets A, B, C and a function $f: A \times B \to C$, then we shall abbreviate $f((x, y))$ to $f(x, y)$, and we shall think of f as a function of two variables, the first taking on values in A, the second taking on values in B.

Given a function f such that $\mathcal{D}_f$ and $\mathcal{R}_f$ are subsets of $\mathbf{R}$, we define the differential of f to be the mapping $df: \mathcal{D}_{f'} \times \mathbf{R} \to \mathbf{R}$ given by

$$df(x, h) = f'(x)h.$$

Geometrically, $df(x, h)$ is the approximation to $f(x + h) - f(x)$ which results from the replacement of the graph of f by the tangent line to the graph at the point $(x, f(x))$. More generally, we define a first-order differential form $f\,dg$ as follows:

$$[f\,dg](x, h) = f(x)\,g'(x)h.$$

One can then verify that these differentials behave as they should.

Consider first the equation $d(fg) = f\,dg + g\,df$. If f and g are differentiable at x_0, then

$$
\begin{aligned}
[d(fg)](x_0, h) &= (fg)'(x_0)h \\
&= [f(x_0)\,g'(x_0) + g(x_0)f'(x_0)]h \\
&= [f\,dg](x_0, h) + [g\,df](x_0, h) \\
&= [f\,dg + g\,df](x_0, h).
\end{aligned}
$$

Thus, if f and g have a common domain and are differentiable throughout it, then $d(fg)$ and $f\,dg + g\,df$ are equal as functions of two variables. Similar remarks apply to the linearity properties:

$$d(f + g) = df + dg \quad \text{and} \quad d(cf) = c\,df \quad (c \text{ a constant}).$$

Our initial understanding of differentials was this: if x and y are "variables" related by the equation $y = f(x)$, then $dy = f'(x)\,dx$. Let us see in what sense this is true for differentials as we have just defined them. First of all, we have defined differentials of functions, not variables. If x and y are functions, then the only reasonable interpretation of the equation $y = f(x)$ is that $y = f \circ x$. By the chain rule,

$$dy(u, h) = y'(u)h = f'(x(u))x'(u)h = [(f' \circ x)\, dx](u, h)$$

whenever $f'(x(u))$ and $x'(u)$ exist. Thus

$$dy = (f' \circ x)\, dx = f'(x)\, dx,$$

provided suitable differentiability assumptions are made concerning x and f. [In general, $f'(x)\, dx$ is merely a restriction of dy.]

Having pursued the matter this far, it ought to be clear that we could now justify all our previous juggling of differentials.

Differentials have an interesting history. They were used by the very first practitioners of calculus (though certainly not in the sense just defined). At that time, calculus was not based on a rigorous theory of limits, the latter being developed more than a century later. A derivative dy/dx, or "fluxion," was thought of as a rate of change of the "flowing quantity" y with respect to the flowing quantity x. What's more, dy/dx was actually regarded as the quotient of two infinitely small increments in the flowing quantities. It was assumed that the infinitesimals dx and dy were so small that no integral multiple of them could ever have measurable size! (Compare this with the notion of Archimedean ordering.) Although mathematicians of the period felt somewhat self-conscious about discussing infinitely small quantities, their achievements far outweighed their embarrassment at talking nonsense. Nevertheless, the need to put calculus on a sound basis was sharply pointed out by Bishop Berkeley in a polemic called *The Analyst* (1734). We quote several lines from Berkeley's essay.

> And what are these fluxions? The velocities of evanescent increments. And what are these same evanescent increments? They are neither finite quantities, nor quantities infinitely small, nor yet nothing. May we not call them the ghosts of departed quantities?

We are no longer able to think of differentials in the same way Newton did, and yet, however contemptuous we might be of the metaphysical fog from which they emerged, it is utter foolishness to deny their usefulness. We shall use differentials mainly to calculate antiderivatives. The reader is urged to gain some fluency in using differentials, although how he chooses to interpret them is largely his own affair. We close with this admonition: seek ye first a thorough understanding of the chain rule, and a proper understanding of differentials will be added unto you.

Exercises

1. Find dy when
   a) $y = 5x^4$,   b) $y = \sec \phi$,   c) $y = t \sin t$,   d) $y = u - v$,
   e) $y = u^2 v + v \sin u$,   f) $y = x \cos z^2$,   g) $y = uvw$,   h) $y = u^3 v - w \cos uv$.
2. Find dy/dx by the method of Example 4.
   a) $x^2 + y^2 = 1$ b) $x^3 + y^5 = 3xy^2$ c) $x \sin y + y^2 \sin x = 1$

3. Calculate the following antiderivatives.

a) $\int x\sqrt{1-x^2}\,dx$

b) $\int x\cos 3x^2\,dx$

c) $\int \dfrac{x\,dx}{\sqrt{1-x^2}}$

d) $\int x^3\sqrt{1+3x^4}\,dx$

e) $\int \cos\theta \sin^4\theta\,d\theta$

f) $\int \dfrac{x\,dx}{(7+5x^2)^{3/2}}$

g) $\int \cos^3\theta\,d\theta$

h) $\int \sin^2 x\cos^5 x\,dx$

i) $\int \dfrac{x}{\sqrt{1-x^2}}\sin\sqrt{1-x^2}\,dx$

j) $\int \sin 3x\cos 3x\,dx$

k) $\int \dfrac{\sin^3 x}{\cos^5 x}\,dx$

l) $\int \dfrac{\sin x\cos x}{\sqrt{1+2\cos^2 x}}\,dx$

4. Use half-angle formulas to evaluate each of the following antiderivatives.

a) $\int \cos^2\theta\,d\theta$

b) $\int \sin^2\theta\cos^2\theta\,d\theta$

c) $\int \sin^4\theta\,d\theta$

7-3 The Newton integral

The solution to a problem often reduces to the evaluation of a difference of the form $F(b) - F(a)$, where F is an antiderivative of some given function f. (Note that this is true of the two problems given at the beginning of this chapter.) Although antiderivatives are ambiguous (for they are determined only to within an additive constant), such differences are not.

Proposition 1. *If F and G are antiderivatives of f on $[a, b]$, then*

$$F(b) - F(a) = G(b) - G(a).$$

Proof. There exists a constant C such that $F(x) = G(x) + C$ for all $x \in [a, b]$. In particular,

$$F(b) = G(b) + C \quad \text{and} \quad F(a) = G(a) + C,$$

so that upon subtracting these equations, we get the desired result.

Definition 2. If f has an antiderivative F on the interval $[a, b]$, we say that f is **Newton-integrable** on $[a, b]$ and the **Newton integral** is defined to be the number

$$\int_a^b f = \int_a^b f(x)\,dx = F(b) - F(a).$$

[Because of the preceding result, this definition is legitimate, since the number $F(b) - F(a)$ depends only on the function f and the numbers a and b; importantly, it is independent of a particular choice of F.]

Example 1. Find $\int_{-2}^{3} |x|\, dx$.

The function $f(x) = \frac{1}{2}x|x|$ is an antiderivative of $|x|$. Hence

$$\int_{-2}^{3} |x|\, dx = f(3) - f(-2) = \frac{9}{2} - (-2) = 6\frac{1}{2}.$$

Example 2. Find $\int_{\pi/3}^{\pi/2} \sin\theta\, d\theta$.

In working such problems, the following notation is useful:

$$F(x)\Big|_a^b = F(x)\Big|_{x=a}^{x=b} = F(b) - F(a).$$

In this case, for example, we can write

$$\int_{\pi/3}^{\pi/2} \sin\theta\, d\theta = -\cos\theta\Big|_{\pi/3}^{\pi/2} = -\cos\frac{\pi}{2} - \left(-\cos\frac{\pi}{3}\right) = \frac{1}{2}.$$

Example 3. If J is a subinterval of $[a, b]$, show that $\int_a^b \varphi_J(x)\, dx = l(J)$, the length of J.

If c and d are the endpoints of J, then

$$F(x) = \begin{cases} 0 & \text{if } a \le x < c, \\ x - c & \text{if } c \le x \le d, \\ d - c & \text{if } d < x \le b \end{cases}$$

is an antiderivative of φ_J on $[a, b]$. Hence φ_J is Newton-integrable on $[a, b]$ and

$$\int_a^b \varphi_J(x)\, dx = F(b) - F(a) = (d - c) - 0 = l(J).$$

Definite integrals have a number of important properties which we proceed to derive.

Suppose that f has an antiderivative F on $[a, b]$ and an antiderivative G on $[b, c]$. If we define

$$H(x) = \begin{cases} F(x) & \text{if } a \le x \le b, \\ G(x) + F(b) - G(b) & \text{if } b \le x \le c, \end{cases}$$

then it is easily verified that H is an antiderivative of f on $[a, c]$. Moreover,

$$\int_a^b f + \int_b^c f = H(b) - H(a) + H(c) - H(b) = H(c) - H(a) = \int_a^c f.$$

We have therefore proved the following result.

Proposition 3. Additivity of Newton Integrals. *If f is Newton-integrable on $[a, b]$ and $[b, c]$, then f is Newton-integrable on $[a, c]$ and $\int_a^c f = \int_a^b f + \int_b^c f$.*

We leave the proof of the next property to the reader.

Proposition 4. Linearity of Newton Integrals. *If f and g are Newton-integrable on $[a, b]$, the same is true of the functions $f + g$ and cf, where c is a constant, and*

$$\int_a^b (f + g) = \int_a^b f + \int_a^b g \quad \text{and} \quad \int_a^b cf = c\int_a^b f.$$

By applying the linearity property to $f - g = f + (-1)g$, we see that the integral of a difference is equal to the difference of the integrals.

Corollary 5. *Let f and g be as above. Then $f - g$ is Newton-integrable on $[a, b]$, and*

$$\int_a^b (f - g) = \int_a^b f - \int_a^b g.$$

Proposition 6. Positivity of Newton Integrals. *Suppose that f is Newton-integrable on $[a, b]$ and that f is nonnegative on $[a, b]$. Then*

$$\int_a^b f(x)\, dx \geq 0.$$

Moreover, under these circumstances, equality can hold only when $f(x) = 0$ for all $x \in [a, b]$, with possibly a finite number of exceptions.

Proof. Let F be an antiderivative of f on $[a, b]$. Since $F'(x) = f(x) \geq 0$ for all $x \in [a, b]$, with possibly a finite number of exceptions, it follows, as in the proofs of Theorem 4, Section 6–3, and Proposition 2, Section 7–1, that F is a nondecreasing function on $[a, b]$. Thus $\int_a^b f = F(b) - F(a) \geq 0$. If $\int_a^b f = 0$, then F is a constant function and $F'(x) = 0$. Thus $f(x) = F'(x) = 0$ for all $x \in [a, b]$, with possibly a finite number of exceptions.

Corollary 7. Monotonicity of Newton Integrals. *Suppose that f and g are Newton-integrable on $[a, b]$ and that $f(x) \leq g(x)$ for all $x \in [a, b]$. Then*

$$\int_a^b f(x)\, dx \leq \int_a^b g(x)\, dx.$$

Proof. The difference $g - f$ is nonnegative on $[a, b]$, and thus

$$\int_a^b g - \int_a^b f = \int_a^b (g - f) \geq 0.$$

Therefore, $\int_a^b f \leq \int_a^b g$.

Corollary 8. Triangle Inequality for Definite Integrals. *If f and $|f|$ are both Newton-integrable on $[a, b]$, then*

$$\left| \int_a^b f(x)\, dx \right| \leq \int_a^b |f(x)|\, dx.$$

Proof. For all $x \in [a, b]$,

$$-|f(x)| \leq f(x) \leq |f(x)|.$$

Hence by the monotonicity property,

$$-\int_a^b |f| \leq \int_a^b f \leq \int_a^b |f|$$

or, equivalently, $|\int_a^b f| \leq \int_a^b |f|$.

Let us consider a few more examples.

Example 4. Find $\int_0^2 (x^2 - [x]) \, dx$.

Rather than trying to find an antiderivative of $x^2 - [x]$ on $[0, 2]$, we use the various properties of definite integrals:

$$\int_0^2 (x^2 - [x]) \, dx = \int_0^2 x^2 \, dx - \int_0^2 [x] \, dx,$$

$$\int_0^2 x^2 \, dx = \tfrac{1}{3}x^3 \Big|_0^2 = \tfrac{8}{3},$$

$$\int_0^2 [x] \, dx = \int_0^1 [x] \, dx + \int_1^2 [x] \, dx = \int_0^1 0 \cdot dx + \int_1^2 1 \cdot dx = 1.$$

Thus $\int_0^2 (x^2 - [x]) \, dx = \tfrac{8}{3} - 1 = \tfrac{5}{3}$.

Example 5. Estimate the sum of the square roots of the first one-hundred positive integers.

We consider the function $\sqrt{x}$. If k is a positive integer, then

$$\sqrt{k} \leq \sqrt{x} \quad \text{if} \quad k \leq x \leq k+1$$

and

$$\sqrt{k} \geq \sqrt{x} \quad \text{if} \quad k-1 \leq x \leq k.$$

Thus, by the monotonicity property,

$$\int_{k-1}^k \sqrt{x} \, dx \leq \int_{k-1}^k \sqrt{k} \, dx = \sqrt{k} = \int_k^{k+1} \sqrt{k} \, dx \leq \int_k^{k+1} \sqrt{x} \, dx.$$

Summing over these inequalities, we get from the additive property

$$\int_0^n \sqrt{x} \, dx = \sum_{k=1}^n \int_{k-1}^k \sqrt{x} \, dx \leq \sum_{k=1}^n \sqrt{k}$$

$$\leq \sum_{k=1}^n \int_k^{k+1} \sqrt{x} \, dx = \int_1^{n+1} \sqrt{x} \, dx.$$

Since

$$\int_0^n \sqrt{x} \, dx = \tfrac{2}{3}x^{3/2} \Big|_0^n = \tfrac{2}{3}n^{3/2}$$

and

$$\int_1^{n+1} \sqrt{x} \, dx = \tfrac{2}{3}[(n+1)^{2/3} - 1],$$

we get

$$\tfrac{2}{3}n^{3/2} \leq \sum_{k=1}^n \sqrt{k} \leq \tfrac{2}{3}[(n+1)^{3/2} - 1].$$

Setting $n = 100$ and $n = 99$, we get

$$\sum_{k=1}^{100} \sqrt{k} \geq 666\tfrac{2}{3} \quad \text{and} \quad \sum_{k=1}^{99} \sqrt{k} < 666.$$

Thus $\sum_{k=1}^{100} \sqrt{k}$ lies between $666\tfrac{2}{3}$ and $666 + 10 = 676$.

A different approach to the same problem will be found in the exercises.

To close this section, we consider the use of substitutions in evaluating definite integrals. Suppose that we are asked to evaluate the definite integral

$$I = \int_{-\pi/2}^{\pi/6} \cos^3 x \sin^4 x \, dx.$$

We can do this by first finding $\int \cos^3 x \sin^4 x \, dx$. This calls for the substitution $u = \sin x$. Then $du = \cos x \, dx$, $\cos^2 x = 1 - u^2$, and

$$\int \cos^3 x \sin^4 x \, dx = \int (1 - u^2) u^4 \, du$$

$$= \tfrac{1}{5} u^5 - \tfrac{1}{7} u^7 + C = \tfrac{1}{5} \sin^5 x - \tfrac{1}{7} \sin^7 x + C.$$

Thus

$$I = [\tfrac{1}{5} \sin^5 x - \tfrac{1}{7} \sin^7 x]\Big|_{x=-\pi/2}^{x=\pi/6} = \tfrac{229}{4480}.$$

Observe that

$$[\tfrac{1}{5} \sin^5 x - \tfrac{1}{7} \sin^7 x]\Big|_{x=-\pi/2}^{x=\pi/6} = [\tfrac{1}{5} u^5 - \tfrac{1}{7} u^7]\Big|_{u=\sin -\pi/2}^{u=\sin \pi/6}$$

$$= \int_{-1}^{1/2} (1 - u^2) u^2 \, du.$$

In other words, we could have obtained the same result by changing limits of integration when we made the substitution $u = \sin x$.

Consider a second example. Let us find

$$I = \int_{-1}^{2} x \sin \frac{\pi x^2}{4} \, dx.$$

We let $u = \pi x^2/4$. When $x = -1$, then $u = \pi/4$, and when $x = 2$, then $u = \pi$. Thus

$$\int_{-1}^{2} x \sin \frac{\pi x^2}{4} \, dx = \frac{2}{\pi} \int_{\pi/4}^{\pi} \sin u \, du = -\frac{2}{\pi} \cos u \Big|_{\pi/4}^{\pi}$$

$$= \frac{2}{\pi} \left(1 + \frac{\sqrt{2}}{2}\right).$$

For all Newton integrals $\int_a^b f$ considered thus far, we have required that $a < b$. We now relax this requirement before stating the change of variable rule. If $a \geq b$, we define as before

$$\int_a^b f = F(b) - F(a),$$

where, in this instance, F is an antiderivative of f on the interval $[b, a]$. Additivity and linearity continue to hold; that is,

$$\int_a^b f + \int_b^c f = \int_a^c f, \qquad \int_a^b (f + g) = \int_a^b f + \int_a^b g,$$

provided in each of these equations at least two of the integrals exist. In addition, we have the identities

$$\int_a^b f = -\int_b^a f, \qquad \int_a^a f = 0.$$

Proposition 9. Change of Variable Rule for Newton Integrals. *The formula*

$$\int_a^b f(u(x))u'(x)\,dx = \int_{u(a)}^{u(b)} f(u)\,du$$

is valid under the following circumstances:

1) *u is a continuous function mapping [a, b] into an interval I;*
2) *u'(x) exists and is nonzero for all x ∈ [a, b], with possibly a finite number of exceptions;*
3) *f is Newton-integrable on I.*

Proof. Let F be an antiderivative of f on I. Then $F \circ u$ is continuous on $[a, b]$, and by the chain rule,

$$\frac{d}{dx}[F(u(x))] = F'(u(x))u'(x) = f(u(x))u'(x)$$

for all $x \in [a, b]$, with possibly a finite number of exceptions. [Note that if $F'(x) = f(x)$ failed to hold at a point $x_0 \in I$, and if the function u assumed the value x_0 infinitely often, then the above assertion would be false. It follows from the mean-value theorem and hypotheses (1) and (2) that u can assume a given value only finitely often.] Thus $F \circ u$ is an antiderivative of $(f \circ u)u'$ on $[a, b]$, and we therefore have

$$\int_a^b f(u(x))u'(x)\,dx = (F \circ u)(b) - (F \circ u)(a) = F(u(b)) - F(u(a))$$
$$= \int_{u(a)}^{u(b)} f(u)\,du.$$

The remarkable thing about the substitution rule is that it enables us to calculate many Newton integrals without producing antiderivatives explicitly.

Example 6. Show that

$$\int_{\tan a}^{\tan b} \frac{du}{1 + u^2} = b - a$$

if $-\pi/2 < a \le b < \pi/2$.

Let $u(t) = \tan t$. Then u is continuous and has a positive derivative throughout the interval $(-\pi/2, \pi/2)$. It therefore maps the interval $[a, b]$ onto the interval $[\tan a, \tan b]$. The function

$$f(u) = \frac{1}{1 + u^2}$$

is continuous on the entire line and, in particular, on the interval $[\tan a, \tan b]$. We shall prove in the next chapter that continuous functions have antiderivatives. In the meantime, we simply take it for granted. This means that the hypotheses of the substitution rule are satisfied. Hence

$$\int_{\tan a}^{\tan b} \frac{du}{1 + u^2} = \int_a^b f(u(t))u'(t)\,dt.$$

But

$$f(u(t))u'(t) = \frac{1}{1 + \tan^2 t} \cdot \sec^2 t = 1.$$

Hence

$$\int_{\tan a}^{\tan b} \frac{du}{1 + u^2} = \int_a^b dt = b - a.$$

In particular, we have

$$\int_0^1 \frac{du}{1 + u^2} = \frac{\pi}{4} \quad \text{and} \quad \int_{-1}^{\sqrt{3}} \frac{du}{1 + u^2} = \frac{\pi}{3} - \left(-\frac{\pi}{4}\right) = \frac{7\pi}{12}.$$

Example 7. Calculate

$$\int_{-1/2}^{\sqrt{3}/2} \sqrt{1 - x^2}\, dx.$$

We make the substitution $x = \sin \theta$. The sine function has a positive derivative on the interval $[-\pi/6, \pi/3]$ and maps this interval onto the interval $[-1/2, \sqrt{3}/2]$. Also we have

$$\sqrt{1 - x^2} = \sqrt{1 - \sin^2 \theta} = +\cos \theta \quad \text{and} \quad dx = \cos \theta\, d\theta$$

for all $\theta \in [-\pi/6, \pi/3]$. Hence by the substitution rule,

$$\int_{-1/2}^{\sqrt{3}/2} \sqrt{1 - x^2}\, dx = \int_{-\pi/6}^{\pi/3} \cos^2 \theta\, d\theta$$

$$= \left[\frac{\theta}{2} + \frac{\sin 2\theta}{4}\right]\Bigg|_{-\pi/6}^{\pi/3}$$

$$= \left(\frac{\pi}{6} + \frac{\sqrt{3}}{8}\right) - \left(\frac{\pi}{12} - \frac{\sqrt{3}}{8}\right)$$

$$= \tfrac{1}{4}(\pi + \sqrt{3}).$$

Example 8. Calculate

$$\int_{-4}^{-2} \frac{dx}{x\sqrt{x^2 - 4}}.$$

We let $x = 2 \sec \theta$, and note that $-4 = 2 \sec (2\pi/3)$ and $-2 = 2 \sec \pi$. Moreover, the secant function has a positive derivative on the interval $[2\pi/3, \pi)$. (See Fig. 7-8.) Also

$$dx = 2 \sec \theta \tan \theta\, d\theta$$

and

$$\sqrt{x^2 - 4} = |2 \tan \theta| = -2 \tan \theta$$

for $\theta \in [2\pi/3, \pi]$. Thus

$$\int_{-4}^{-2} \frac{dx}{x\sqrt{x^2 - 4}} = \int_{2\pi/3}^{\pi} \frac{2 \sec \theta \tan \theta\, d\theta}{(2 \sec \theta)(-2 \tan \theta)}$$

$$= -\tfrac{1}{2} \int_{2\pi/3}^{\pi} d\theta = -\pi/6.$$

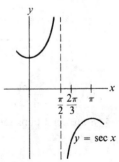

FIG. 7-8

The last three examples illustrate the use of the so-called trigonometric substitutions. For integrals involving a radical of the form $\sqrt{a^2 - x^2}$, the substitution $x = a \sin \theta$ often works. If radicals $\sqrt{a^2 + x^2}$ and $\sqrt{x^2 - a^2}$ are involved, then the substitutions $x = a \tan \theta$ and $x = a \sec \theta$, respectively, are recommended.

Exercises

1. Evaluate each of the following integrals by using directly the definition of the Newton integral.

a) $\int_{-1}^{2} (5x^2 - 2)\, dx$

b) $\int_{0}^{1} x^{-1/3}\, dx$

c) $\int_{-1}^{3/2} [x]\, dx$

d) $\int_{0}^{\pi} |\cos \theta|\, d\theta$

e) $\int_{-2}^{1} x\sqrt{1 + x^2}\, dx$

f) $\int_{0}^{1} (1 - x)^n\, dx$

2. Prove the linearity property of Newton integrals (Proposition 4).

3. Evaluate the following integrals by any convenient method.

a) $\int_{-1}^{3} (x - [x^2])\, dx$

b) $\int_{0}^{\sqrt{\pi}} |x \cos (x^2)|\, dx$

c) $\int_{0}^{5} (5\varphi_{[1,3]}(x) - x\varphi_{[2,4]}(x))\, dx$

d) $\int_{-1}^{3} (x^2 \sin \pi[x] - 3\varphi_{[\sqrt{2}, \sqrt{3}]}(x))\, dx$

e) $\int_{-1}^{2} \sqrt{x^2 + x^4}\, dx$

f) $\int_{-\pi/2}^{\pi/6} \sin^3 \theta\, d\theta$

g) $\int_{-\pi/4}^{\pi/6} \sqrt{\sin \theta \tan \theta}\, d\theta$

h) $\int_{-1}^{1/2} x^2\sqrt{1 - x^2}\, dx$

i) $\int_{0}^{2 \tan z} \frac{du}{4 + u^2}$

j) $\int_{-2}^{2\sqrt{3}} \frac{dt}{(4 + t^2)^{3/2}}$

k) $\int_{3}^{3\sqrt{2}} \frac{\sqrt{x^2 - 9}}{x}\, dx$

l) $\int_{0}^{1} \frac{dx}{\sqrt{4 - x^2}}$

m) $\int_{-1}^{1} \frac{du}{u^2 + 2u + 5}$ (First complete the square in the denominator.)

4. State conditions for the validity of the translation-invariance formula

$$\int_{a}^{b} f(t)\, dt = \int_{a+c}^{b+c} f(t - c)\, dt.$$

Prove the formula.

5. Do the same for the formula

$$\int_a^b f(t)\, dt = \int_a^b f(b + a - t)\, dt.$$

6. Suppose that f and g are respectively even and odd functions which are Newton-integrable on the interval $[0, a]$. Prove that f and g are Newton-integrable on the interval $[-a, a]$ and that

$$\int_{-a}^a f(t)\, dt = 2 \int_0^a f(t)\, dt \qquad \text{and} \qquad \int_{-a}^a g(t)\, dt = 0.$$

7. Estimate the sum of the cube roots of the first one-hundred integers.

8. Let r be any positive rational number. Show that

$$\lim_{n \to \infty} \frac{1}{n^{r+1}} \sum_{k=1}^{n} k^r = r + 1.$$

(Estimate $\sum_{k=1}^{n} k^r$ by the technique of Example 5.)

9. Show that $\int_1^{10} [x^2]\, dx = 990 - \sum_{k=1}^{99} \sqrt{k}$. By using $\int_1^{10} (x^2 - 1)\, dx$ and $\int_1^{10} x^2\, dx$ as lower and upper estimates of $\int_1^{10} [x^2]\, dx$, estimate $\sum_{k=1}^{100} \sqrt{k}$. Compare this result with that of Example 5.

7-4 Areas in rectangular coordinates

We shall consider the problem of finding areas of regions in a plane. Our approach will be somewhat more intuitive than usual; we shall postpone foundational questions until the next chapter.

We begin with a generalization of the first problem in this chapter. Suppose that f is a function which is nonnegative and continuous on the interval $[a, b]$. We are interested in finding a formula for the area of the region B bounded by the graph of f, the x-axis, and the vertical lines $x = a$, $x = b$ (see Fig. 7–9). Such regions are often called *ordinate sets*. We think of this region as being generated by the continuous motion of a vertical line moving to the right, and we study the rate at which the area increases. If, for $a \le x_0 \le b$, we let $A(x_0)$ denote the area of the ordinate set $\{(x, y) \mid a \le x \le x_0, 0 \le y \le f(x)\}$, then one can show that the area function A is continuous on $[a, b]$ and that $A'(x) = f(x)$ for all $x \in (a, b)$. (Later in this section we shall derive a more general result.) Thus the required area is

$$A(B) = A(b) = A(b) - A(a) = \int_a^b f(x)\, dx.$$

Example 1. Find the area of the region inside the ellipse

$$\frac{x^2}{a^2} + \frac{y^2}{b^2} = 1.$$

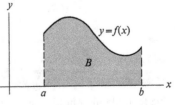

FIG. 7–9

By symmetry, the area A of the region is equal to four times the area of that portion of the region which belongs to the first quadrant (see Fig. 7–10). The latter region is an ordinate set, so our formula is applicable:

$$A = 4\int_0^a (b/a)\sqrt{a^2 - x^2}\, dx.$$

We make the substitution $x = a \sin \theta$, and apply the substitution rule:

$$A = 4ab\int_0^{\pi/2} \cos^2 \theta\, d\theta = 4ab\left[\frac{\theta}{2} + \frac{\sin \theta \cos \theta}{4}\right]_0^{\pi/2}$$

$$= 4ab(\pi/4) = \pi ab.$$

Setting $b = a$, we get as a special case the familiar formula for the area of a circle.

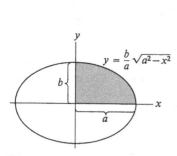

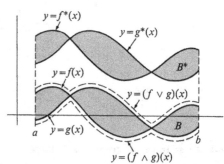

FIG. 7–10   **FIG. 7–11**

We consider next the problem of calculating the area of a region which is bounded by the graphs of two functions f and g, both continuous on an interval $[a, b]$, and the vertical lines $x = a$, $x = b$ (see Fig. 7–11). Specifically, the region in question is the set

$$B = \{(x, y) \mid a \le x \le b,\ (f \wedge g)(x) \le y \le (f \vee g)(x)\}.$$

The reader will recall the definitions of $f \wedge g$ and $f \vee g$:

$$(f \vee g)(x) = \max\,\{f(x), g(x)\}, \qquad (f \wedge g)(x) = \min\,\{f(x), g(x)\}.$$

If f and g are nonnegative, then the area of B will be equal to $\int_a^b (f \vee g)(x)\, dx$, the area of the ordinate set determined by $f \vee g$, minus $\int_a^b (f \wedge g)(x)\, dx$. Thus

$$A(B) = \int_a^b (f \vee g)(x)\, dx - \int_a^b (f \wedge g)(x)\, dx$$

$$= \int_a^b [(f \vee g)(x) - (f \wedge g)(x)]\, dx$$

$$= \int_a^b |f(x) - g(x)|\, dx. \qquad\qquad (*)$$

This formula also holds when f and g take on negative values. This can be shown as follows. Since f and g are bounded on the interval $[a, b]$, we can select a positive constant c so that the functions $f^* = f + c$ and $g^* = g + c$ are both nonnegative on $[a, b]$. Then the region

$$B^* = \{(x, y) \mid a \leq x \leq b, (f^* \wedge g^*)(x) \leq y \leq (f^* \vee g^*)(x)\}$$

is congruent to B, and thus

$$A(B) = A(B^*) = \int_a^b |f^* - g^*| = \int_a^b |(f + c) - (g + c)| = \int_a^b |f - g|.$$

Example 2. Find the area of the region bounded by the curves $y = \sin x$, $y = \cos x$, the y-axis, and the line $x = \pi$—the shaded region in Fig. 7–12.

From (∗), we get

$$A = \int_0^\pi |\sin x - \cos x|\, dx$$

$$= \int_0^{\pi/4} |\sin x - \cos x|\, dx + \int_{\pi/4}^\pi |\sin x - \cos x|\, dx$$

$$= \int_0^{\pi/4} (\cos x - \sin x)\, dx + \int_{\pi/4}^\pi (\sin x - \cos x)\, dx$$

$$= (\sqrt{2} - 1) + (\sqrt{2} + 1) = 2\sqrt{2}.$$

Suppose that a circle of radius a rolls along the x-axis without slipping. We consider the path which is traced out by a fixed point on the circumference. The path is called a *cycloid*. We shall be interested in finding the area of a region which is bounded by one arch of the cycloid and the x-axis (see Fig. 7–13). Although the cycloid is the graph of a continuous function, it is not possible to find a useful closed-form expression for the function. It is more natural to use parametric equations for the curve. We shall assume that when the rolling circle is tangent at the origin, the point P coincides with the origin. We select as our parameter the ratio θ of the x-coordinate of the point T at which the generating circle is tangent to the x-axis, divided by the radius a. Then (since the circle rolls without slipping) θ measures in radians the central angle determined by P and T. A little trigonometry then shows that the coordinates of P are given by

$$x = a(\theta - \sin \theta), \qquad y = a(1 - \cos \theta).$$

We observe that $dx/d\theta = y \geq 0$, with strict inequality except at the isolated points where $\theta = 2n\pi$ ($n \in \mathbf{Z}$). This shows that P moves steadily to the right as θ increases.

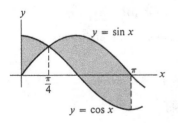

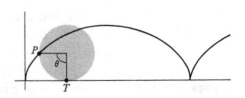

FIG. 7–12   **FIG. 7–13**

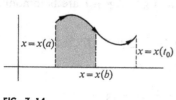

FIG. 7-14

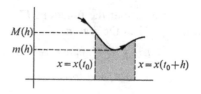

FIG. 7-15

We shall calculate the area mentioned by first developing a general formula. Consider an arc γ which has the parametric equations

$$x = x(t), \qquad y = y(t) \qquad (a \le t \le b),$$

where $y(t)$ and $x'(t)$ are continuous and nonnegative for $t \in [a, b]$. [We may think of $x(t)$ and $y(t)$ as the coordinates of a point on γ at time t.] We shall obtain a formula for the area of the region bounded by γ, the x-axis, and the lines $x = x(a)$, $x = x(b)$. If $a \le t \le b$, we define $A(t_0)$ to be the area of the region bounded by γ, the x-axis, and the lines $x = x(a)$, $x = x(t_0)$ (see Fig. 7-14). Thus $A(a) = 0$, and $A(b)$ is the desired area. Suppose that $a < t_0 < t_0 + h < b$. Since $dx/dt \ge 0$, it follows that $x(t_0) \le x(t_0 + h)$ and $A(t_0 + h) - A(t_0)$ is the area of the strip in Fig. 7-15. We estimate $A(x_0 + h) - A(x_0)$ by the areas of two rectangles. If $M(h)$ and $m(h)$ denote respectively the maximum and minimum values of $y(t)$ for $t_0 \le t \le t_0 + h$, then

$$m(h)[x(t_0 + h) - x(t_0)] \le A(t_0 + h) - A(t_0) \le M(h)[x(t_0 + h) - x(t_0)]. \quad (**)$$

We divide by h and take limits as $h \to 0+$. Since y is continuous at t_0, it follows that $M(h) \to y(t_0)$ and $m(h) \to y(t_0)$. Also

$$\lim_{h \to 0+} \frac{x(t_0 + h) - x(t_0)}{h} = x'(t_0).$$

By the pinching theorem, we get

$$\lim_{h \to 0+} \frac{A(t_0 + h) - A(t_0)}{h} = y(t_0)x'(t_0).$$

One can show similarly that the difference quotient $[A(t_0 + h) - A(t_0)]/h$ also approaches $y(t_0)x'(t_0)$ as $h \to 0-$. Hence

$$A'(t_0) = y(t_0)x'(t_0) \qquad (a < t_0 < b).$$

One can easily argue that the function A is continuous at the endpoints a and b. This yields the formula

$$A = A(b) - A(a) = \int_a^b A'(t)\, dt = \int_a^b y(t)x'(t)\, dt.$$

Example 3. Find the area beneath one arch of the cycloid

$$x = a(\theta - \sin \theta), \qquad y = a(1 - \cos \theta).$$

By the formula just derived, we get

$$A = \int_0^{2\pi} y \frac{dx}{d\theta} \, d\theta = a^2 \int_0^{2\pi} (1 - \cos\theta)^2 \, d\theta$$

$$= a^2 \int_0^{2\pi} (1 - 2\cos\theta + \cos^2\theta) \, d\theta \quad = a^2 \int_0^{2\pi} \left(\frac{3}{2} + \frac{\cos 2\theta}{2} - 2\cos\theta \right) d\theta$$

$$= a^2 \left[\frac{3}{2}\theta + \frac{\sin 2\theta}{4} - 2\sin\theta \right]\Big|_0^{2\pi}$$

$$= 3\pi a^2.$$

We shall now examine the expression $\int_a^b y(t)x'(t) \, dt$ in greater detail. If $y(t) \geq 0$ and $x'(t) \leq 0$ for $t \in [a, b]$, then the arc γ is traced from right to left as the parameter t increases, and the area of the region bounded by γ, the x-axis, and the lines $x = x(a)$, $x = x(b)$ is equal to $-\int_a^b y(t)x'(t) \, dt$ (see Fig. 7–16).

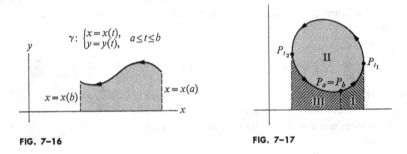

FIG. 7–16   FIG. 7–17

Suppose next that as t varies over the interval $[a, b]$, the point $P_t = (x(t), y(t))$ moves once around a simple loop in the upper half-plane in a counterclockwise direction, as indicated in Fig. 7–17. (Specifically, we assume that a line parallel to one of the axes intersects the loop in two points at most.) For definiteness, let us suppose that the point $P_a = P_b$ lies on the lower side of the loop. We let P_{t_1} and P_{t_2}, respectively, be the eastern- and westernmost points on the loop. Thus $a \leq t_1 < t_2 \leq b$ and $x'(t)$ is nonnegative on the intervals $[a_1, t_1]$ and $[t_2, b]$ and nonpositive on the interval $[t_1, t_2]$. We have then

$$\int_a^b y(t)x'(t) \, dt = \int_a^{t_1} y(t)x'(t) \, dt + \int_{t_1}^{t_2} y(t)x'(t) \, dt + \int_{t_2}^b y(t)x'(t) \, dt$$

$$= A(\mathrm{I}) - A(\mathrm{II}) + A(\mathrm{III})$$

$$= -(\text{area of the region inside the loop}).$$

The same holds if $P_a - P_b$ lies on the upper side of the loop. The formula

$$A = -\int_a^b y(t)x'(t) \, dt,$$

which we obtain in this way for the area of the region inside the loop, actually holds in more general circumstances. In the first place, our assumption that the loop lies in the upper half-plane is nonessential. For by adding a sufficiently large positive constant C to $y(t)$, we can always obtain a congruent loop which lies in the upper half-plane, and thus

$$A = -\int_a^b (y(t) + C)x'(t)\,dt$$

$$= -\int_a^b y(t)x'(t)\,dt - C\int_a^b x'(t)\,dt$$

$$= -\int_a^b y(t)x'(t)\,dt - C[x(b) - x(a)]$$

$$= -\int_a^b y(t)x'(t)\,dt.$$

Also, we can weaken the assumption that dx/dt exists and is continuous on $[a, b]$. No harm will come if dx/dt fails to exist at a finite number of points. [For by (∗∗) above, the continuity of $x(t)$ and $y(t)$ implies continuity of the area function A.] As a result, the loop can have a finite number of sharp corners. We now summarize our discussion.

Suppose that

$$x = x(t), \qquad y = y(t) \qquad (a \le t \le b)$$

are parametric equations of a simple loop γ *whose points are traced out once in a counterclockwise direction as t increases from a to b. If x(t) is an antiderivative on* $[a, b]$, *then*

$$A = -\int_a^b y(t)x'(t)\,dt$$

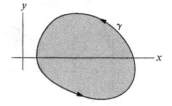

FIG. 7–18

is the area of the region inside the loop γ, the shaded region in Fig. 7–18.

If one assumes that $y(t)$ is an antiderivative on $[a, b]$, then one can show that the area inside γ is given by

$$A = \int_a^b x(t)y'(t)\,dt.$$

Hence, if both $x(t)$ and $y(t)$ are antiderivatives on $[a, b]$, then we also have the formula

$$A = \tfrac{1}{2}\int_a^b [x(t)y'(t) - y(t)x'(t)]\,dt.$$

Example 4. Compute once again the area of the region inside the ellipse

$$\frac{x^2}{a^2} + \frac{y^2}{b^2} = 1.$$

We use the parametrization

$$x = a\cos t, \qquad y = b\sin t \qquad (0 \le t \le 2\pi)$$

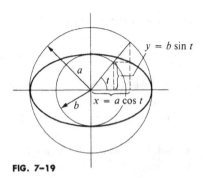

FIG. 7-19

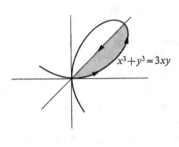

FIG. 7-20

(see Fig. 7-19). By the last formula,

$$A = \tfrac{1}{2}\int_0^{2\pi} [xy' - yx']\, dt$$

$$= \tfrac{1}{2}\int_0^{2\pi} [ab\cos^2 t + ab\sin^2 t]\, dt$$

$$= \tfrac{1}{2}ab\int_0^{2\pi} dt = \pi ab.$$

Example 5. Compute the area of the loop in the folium of Descartes: $x^3 + y^3 = 3xy$.

Because of symmetry about the line $y = x$, the required area is twice that of the shaded region in Fig. 7-20. The boundary of the latter region has the parametrization

$$x = \begin{cases} 3t/(1 + t^3) & (0 \le t \le 1), \\ 3(2 - t)/2 & (1 \le t \le 2), \end{cases}$$

$$y = \begin{cases} 3t^2/(1 + t^3) & (0 \le t \le 1), \\ 3(2 - t)/2 & (1 \le t \le 2). \end{cases}$$

Consequently,

$$x(t)y'(t) - y(t)x'(t) = \begin{cases} 6t^2/(1 + t^3)^2 & (0 \le t \le 1), \\ 0 & (1 \le t \le 2). \end{cases}$$

Therefore, we have

$$A = \int_0^2 [x(t)y'(t) - y(t)x'(t)]\, dt = \int_0^1 \frac{6t^2\, dt}{(1 + t^3)^2} = -\frac{2}{1 + t^3}\Big|_0^1 = 1.$$

Exercises

1. Find the area of the region bounded by the one arch of the sine curve $y = A\sin \omega x$ and the x-axis.

2. Find the area of the region inside the closed curve $y^2 = x^2 - x^4$.

3. Find the area of the region bounded by the curve $\sqrt{x} + \sqrt{y} = 1$ and the coordinate axes.

4. Find the area of the bounded region lying between the curves $y = 1/(1 + x^2)$ and $y = x^2/2$.

5. Find the area of the region bounded by the curve $x^2y = a^3$, the lines $x = 2a$, $y = 2a$, and the coordinate axes.

6. Find the area of the "triangular" region in the first quadrant bounded by the curves $y = \sin x$, $y = \cos x$, and the y-axis.

7. Find the area of the bounded region which lies between the curves $x^2 = 2y + 1$ and $y - x - 1 = 0$.

8. Prove that the area of the bounded region which lies between the parabolas $y^2 = 2ax$ and $x^2 = 2by$ $(a, b > 0)$ is $4(ab/3)$.

9. a) The *positive* and *negative parts* of a function f are defined to be
$$f^+(x) = \max \{f(x), 0\}, \qquad f^-(x) = \max \{-f(x), 0\}.$$
Interpret graphically. Show that $f = f^+ - f^-$ and $|f| = f^+ + f^-$.

 b) Assuming that both f^+ and f^- are Newton-integrable on $[a, b]$, interpret geometrically each of the integrals $\int_a^b f$ and $\int_a^b |f|$.

10. Given that f and g are continuous on $[a, b]$, find a formula for the area of the region bounded by the curves $x = f(y)$, $x = g(y)$, and the horizontal lines $y = a$, $y = b$.

11. Find the area of the bounded region which lies between the parabolas $y^2 = 4 - x$ and $y^2 = 4 - 2x$.

12. Show that the area of the region enclosed by the astroid $x^{2/3} + y^{2/3} = a^{2/3}$ is $3\pi a^2/8$. [Use the parametrization $x = a \cos^3 \theta$, $y = a \sin^3 \theta$ $(0 \le \theta \le 2\pi)$.]

13. Sketch the curve $x = a \sin 2t$, $y = a \sin t$, and determine the area of the region enclosed by one of its loops.

14. Suppose that $P_1 = (x_1, y_1)$, $P_2 = (x_2, y_2)$, $P_3 = (x_3, y_3)$ are the vertices of a triangle. Write down a parametrization for the perimeter of the triangle. Show that the area of the triangle is equal to the absolute value of
$$\frac{1}{2}\begin{vmatrix} x_1 & y_1 & 1 \\ x_2 & y_2 & 1 \\ x_3 & y_3 & 1 \end{vmatrix} = \frac{1}{2}\{(x_2y_3 - x_3y_2) + (x_3y_1 - x_1y_3) + (x_1y_2 - x_2y_1)\}.$$
Interpret geometrically the sign of this expression.

15. Suppose that as t increases from a to b, the point $P_t = (x(t), y(t))$ traverses the curve γ shown in Fig. 7-21 in a counterclockwise direction exactly once. Argue that the area of the region enclosed by γ is equal to
$$-\int_a^b y(t)x'(t)\, dt.$$

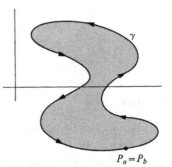

FIG. 7-21

16. State as carefully and as completely as you can the assumptions which we seem to have made in this section regarding areas of plane figures.

17. Sketch the curve $x = t^2 - 1$, $y = t^3 - t$, and find the area of the region enclosed by its loop.

7-5 Areas in polar coordinates

When calculating areas, it is sometimes convenient to use *polar coordinates*. We can associate with a given pair of real numbers r and θ a point P in $\mathbf{R}^2$ whose coordinates are

$$x = r\cos\theta, \qquad y = r\sin\theta.$$

The numbers r and θ are then called *polar coordinates* of the point; r is sometimes called the *radius vector*, and θ is called the *argument* or *polar angle*. If $r > 0$, then r is simply the distance from the origin to P, and θ is the radian measure of the directed angle whose initial side is the positive x-axis (often called the *polar axis*) and whose terminal side is the ray which emanates from 0 and passes through P, as shown in Fig. 7–22. If $r < 0$, then $(-r, \theta + \pi)$ are also polar coordinates of P. (Often only nonnegative values are allowed for r.) Unlike Cartesian coordinates, the polar coordinates of a point are not unique. Even if $r > 0$, the polar angle is unique only to within integral multiples of 2π. Furthermore, the origin has infinitely many polar coordinates. We shall use $[r; \theta]$ to denote the point in the plane whose polar coordinates are r and θ.

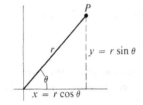

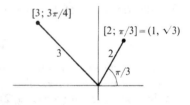

FIG. 7–22　　　　　　　　　　　　　　　　**FIG. 7–23**

Example 1. Plot the points $[2; \pi/3]$, $[2; -11\pi/3]$, $[3; 3\pi/4]$, $[1; -\pi/4]$, and $[-2; \pi/6]$. Also determine the Cartesian coordinates of these points.

To plot the point $[2; \pi/3]$, we simply go out a distance 2 on the ray emanating from 0 which makes an angle of $\pi/3$ radians (or 60°) with the positive x-axis (see Fig. 7–23). Since $\pi/3$ and $-11\pi/3$ differ by an integral multiple of 2π, $[2; \pi/3] = [2; -11\pi/3]$. The next three points can be plotted similarly. To plot the point $[-2; \pi/6]$, we go out two units on the ray emanating from 0 which makes the angle $\pi + \pi/6 = 7\pi/6$ with the positive x-axis. Observe that this point is simply the reflection of the point $[2; \pi/6]$ in the origin.

The rectangular coordinates of the point $[2; \pi/3]$ are

$$(2\cos(\pi/3), 2\sin(\pi/3)) = (1, \sqrt{3}).$$

Similarly,

$$[3; 3\pi/4] = \left(-\frac{3\sqrt{3}}{2}, \frac{3\sqrt{3}}{2}\right),$$

$$[1; -\pi/4] = \left(\frac{\sqrt{2}}{2}, -\frac{\sqrt{2}}{2}\right),$$

$$[-2; \pi/6] = (-\sqrt{3}, -1).$$

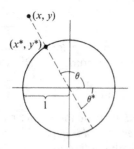

FIG. 7-24   **FIG. 7-25**

Let us consider now the problem of determining the polar coordinates of a given point (x, y). If (x, y) is not the origin, then there exist unique numbers r and θ such that $[r; \theta] = (x, y)$, $r > 0$, and $-\pi < \theta \leq \pi$. The number θ is called the *principal value* of the polar angle. The number r is easily obtained. Since $x = r \cos \theta$ and $y = r \sin \theta$, we get

$$x^2 + y^2 = r^2 \cos^2 \theta + r^2 \sin^2 \theta = r^2,$$

so that

$$r = \sqrt{x^2 + y^2}.$$

The determination of θ is somewhat more delicate. If $x \neq 0$, then

$$\frac{y}{x} = \frac{r \sin \theta}{r \cos \theta} = \tan \theta.$$

Since the tangent function has period π, this equation only determines θ to within an integral multiple of π. The tangent function is continuous and strictly increasing on the open interval $(-\pi/2, \pi/2)$; also

$$\lim_{x \to -\pi/2+} \tan x = -\infty, \qquad \lim_{x \to \pi/2-} \tan x = +\infty.$$

It follows that there is a unique number θ^* such that $-\pi/2 < \theta^* < \pi/2$ and $y/x = \tan \theta^*$. It can be easily shown that θ, the principal value of the polar angle, can be obtained from θ^* according to the rule

$$\theta = \begin{cases} \theta^* & \text{if } x > 0, \\ \pi + \theta^* & \text{if } x < 0, y \geq 0, \\ -\pi + \theta^* & \text{if } x < 0, y < 0. \end{cases}$$

(The second alternative is pictured in Fig. 7–24.) If $x = 0$, then $\theta = \pi/2$ or $-\pi/2$ according as y is positive or negative. A nicer scheme (at least theoretically) for determining θ is the following. If we set

$$x^* = x/r, \qquad y^* = y/r,$$

then (x^*, y^*) lies on the unit circle $x^2 + y^2 = 1$. If $(x^*, y^*) \neq (-1, 0)$, then $\phi = \theta/2$ is the angle indicated in Fig. 7–25 between the x-axis and the line joining $(-1, 0)$ and (x^*, y^*). Then $-\pi/2 < \phi < \pi/2$ and

$$\tan \phi = \frac{x^*}{1 + y^*};$$

the terms of the latter equation are the slope of the line determined by $(-1, 0)$ and (x^*, y^*). The values of ϕ and hence $\theta = 2\phi$ are thereby determined. If $(x^*, y^*) = (-1, 0)$, then $\theta = \pi$.

Example 2. Determine the polar coordinates of the point $(-2, 2\sqrt{3})$.

We first determine r and θ such that $(-2, 2\sqrt{3}) = [r; \theta]$, $r > 0$, and $-\pi < \theta \leq \pi$. We have

$$r = \sqrt{2^2 + (2\sqrt{3})^2} = \sqrt{16} = 4.$$

We determine θ by the two methods discussed above:

$$\tan \theta^* = \frac{2\sqrt{3}}{(-2)} = -\sqrt{3} \qquad \left(-\frac{\pi}{2} < \theta^* < \frac{\pi}{2}\right).$$

Thus $\theta^* = -\pi/3$ and $\theta = \pi - \pi/3 = 2\pi/3$. By the second method, we get

$$x^* = -\tfrac{1}{2}, \qquad y^* = \sqrt{3}/2,$$

and

$$\tan \phi = \frac{\sqrt{3}/2}{1 - \tfrac{1}{2}} = \sqrt{3} \qquad \left(-\frac{\pi}{2} < \phi < \frac{\pi}{2}\right).$$

Thus $\phi = \pi/3$ and $\theta = 2\phi = 2\pi/3$.

In general, $[4; 2\pi/3 + 2n\pi]$ and $[-4; -\pi/3 + 2n\pi]$ are polar coordinates of the point $(-2, 2\sqrt{3})$ for each $n \in \mathbf{Z}$.

If f is a real-valued function of two real variables (i.e., the domain of f is a subset of $\mathbf{R}^2$), then we can consider the set defined by

$$S = \{[r; \theta] \,|\, f(r, \theta) = 0\}$$

or, if only nonnegative values of r are considered, the set

$$S = \{[r; \theta] \,|\, r \geq 0 \text{ and } f(r, \theta) = 0\}.$$

We then say that $f(r, \theta) = 0$ is the *equation of S in polar coordinates.* The simplest polar equations are of the form

$$r = a \qquad \text{and} \qquad \theta = \alpha,$$

where a and α are constants. The first equation is that of a circle with radius a (assuming $a > 0$) and center at the origin. If only nonnegative values of r are considered, then the second is the equation of a ray emanating from the origin. (See Fig. 7–26.)

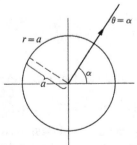

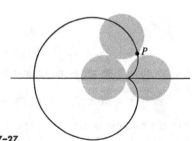

FIG. 7–26   **FIG. 7–27**

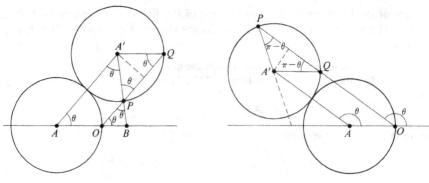

FIG. 7-28 **FIG. 7-29**

Example 3. To illustrate the handiness of polar coordinates in certain situations, we study now a *cardioid*. This curve is generated in somewhat the same fashion as a cycloid. Two circles C and C' have the same radius a. Circle C is fixed, while circle C' rolls without slipping around the outside of C. The path which is swept out by a fixed point on the circumference of C' is called a cardioid (meaning, of course, heart-shaped), and is shown in Fig. 7-27.

We introduce coordinates so that the centers of C' and C initially lie on the polar axis, C is to the left of C', and $P = 0$ is the common point of tangency. Consider now the polar coordinates $[r; \theta]$ of the point P at some subsequent time, where $0 < \theta < \pi/2$. A little geometry shows that the angles are as indicated in Fig. 7-28. Since C' moves without slipping, it follows that $\angle BAA' = \angle BA'A$. Consequently, triangle ABA' is isosceles, and therefore $d(A, B) = d(A', B)$, where $d(A, B)$ denotes the distance from A to B. Now

$$d(O, B) = d(A, B) - a = d(A', B) - a = d(P, B).$$

Thus triangle OBP is isosceles, and therefore $\theta = \angle BOP = \angle OPB$. Since ABA' and OBP are isosceles triangles with a common vertex angle, the bases AA' and OP are parallel. Consequently, $\angle BAA' = \angle BOP = \theta$. In summary,

$$\angle BAA' = \angle BA'A = \angle BOP = \angle OPB = \theta.$$

The line $A'Q$ is constructed parallel to the x-axis; Q is the point of intersection with C'. Triangle $PA'Q$ is isosceles and its vertex angle is clearly $\pi - 2\theta$. The base angles are therefore both equal to θ. It then follows that $\angle OPQ$ is a straight angle. Now, $d(P, Q) = 2a \cos \theta$, and thus

$$r = d(O, P) = d(O, Q) - d(P, Q)$$
$$= d(A, A') - d(P, Q) = 2a - 2a \cos \theta$$
$$= 2a(1 - \cos \theta).$$

One can check that the equation also holds if $\pi/2 < \theta < \pi$ (see Fig. 7-29). In this case we have

$$r = d(O, Q) + d(Q, P) = 2a + 2a \cos (\pi - \theta) = 2a(1 - \cos \theta).$$

Since the cardioid is symmetric about the polar axis, and since $\cos (-\theta) = \cos \theta$, we can conclude that $r = 2a(1 - \cos \theta)$ is the equation of the cardioid in polar coordinates.

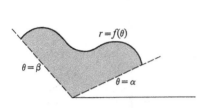

FIG. 7-30

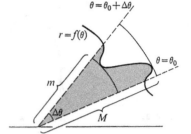

FIG. 7-31

We shall now derive a formula for the area of a region whose boundaries are given in terms of polar coordinates. Let f be a function which is nonnegative and continuous throughout the interval $[\alpha, \beta]$, where $\beta - \alpha \le 2\pi$. We consider the region bounded by the curve $r = f(\theta)$ and the rays $\theta = \alpha$ and $\theta = \beta$ (see Fig. 7-30). If $\alpha \le \theta_0 \le \beta$, then we shall let $A(\theta_0)$ denote the area of the region bounded by the curve $r = f(\theta)$ and the rays $\theta = \alpha$ and $\theta = \theta_0$. If $\Delta\theta > 0$, then $A(\theta_0 + \Delta\theta) - A(\theta_0)$ is the area of a sector which we can estimate in terms of the areas of circular sectors (see Fig. 7-31). Specifically,

$$\frac{\Delta\theta}{2\pi}(\pi m^2) \le A(\theta_0 + \Delta\theta) - A(\theta_0) \le \frac{\Delta\theta}{2\pi}(\pi M^2),$$

where

$$m = \min\ \{f(\theta) \mid \theta_0 \le \theta \le \theta_0 + \Delta\theta\},$$
$$M = \max\ \{f(\theta) \mid \theta_0 \le \theta \le \theta_0 + \Delta\theta\}.$$

Since f is continuous at θ_0, it follows that $m \to f(\theta_0)$ and $M \to f(\theta_0)$ as $\Delta\theta \to 0+$. Thus, upon dividing the inequalities above by $\Delta\theta$ and taking limits, we get

$$\lim_{\Delta\theta \to 0+} \frac{A(\theta_0 + \Delta\theta) - A(\theta_0)}{\Delta\theta} = \tfrac{1}{2}[f(\theta_0)]^2.$$

We get the same for left-sided limits. Thus

$$A'(\theta) = \tfrac{1}{2}[f(\theta)]^2,$$

so that the required area is

$$A = \tfrac{1}{2}\int_\alpha^\beta [f(\theta)]^2\ d\theta.$$

Example 4. Find the area of the region enclosed by the cardioid $r = 2a(1 - \cos\theta)$.
   We have

$$A = \tfrac{1}{2}\int_0^{2\pi} 4a^2(1 - \cos\theta)^2\ d\theta$$

$$= 2a^2\int_0^{2\pi} \left(1 - 2\cos\theta + \frac{1 + \cos 2\theta}{2}\right) d\theta$$

$$= 2a^2(\tfrac{3}{2}\theta - 2\sin\theta + \tfrac{1}{4}\sin 2\theta)\Big|_0^{2\pi}$$

$$= 6\pi a^2.$$

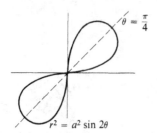

FIG. 7-32 $r^2 = a^2 \sin 2\theta$

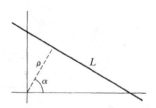

FIG. 7-33

Example 5. Find the area of the region enclosed by the *lemniscate*

$$(x^2 + y^2)^2 = 2a^2xy.$$

We first determine the equation of the lemniscate in polar coordinates by making the substitutions

$$x = r \cos \theta, \qquad y = r \sin \theta,$$

where $r \geq 0$. We get

$$r^4 = 2a^2r^2 \cos \theta \sin \theta = a^2r^2 \sin 2\theta \qquad \text{or} \qquad r^2 = a^2 \sin 2\theta.$$

The right-hand side is negative if $\pi/2 < \theta < \pi$ or $3\pi/2 < \theta < 2\pi$. Consequently, the curve lies entirely in the first and third quadrants, as shown in Fig. 7-32. Note that the curve is a figure eight made up of congruent loops. The area of the region is therefore twice the area enclosed by the loop in the first quadrant. Thus

$$A = 2 \cdot \tfrac{1}{2} \int_0^{\pi/2} r^2 \, d\theta$$

$$= \int_0^{\pi/2} a^2 \sin 2\theta \, d\theta$$

$$= -a^2 \frac{\cos 2\theta}{2} \Big|_0^{\pi/2} = a^2.$$

Exercises

1. Plot the following points and find their rectangular coordinates.
   a) $[1; \pi/4]$ b) $[1; 3\pi]$ c) $[-1; \pi]$
   d) $[2; 4\pi/3]$ e) $[-2; 13\pi/4]$ f) $[2; -5\pi/6]$
   g) $[-2; -5\pi/6]$ h) $[-3/2; -9\pi/4]$

2. Prove the rule stated in the text for obtaining θ from θ^*.

3. Find all the polar coordinates of each of the following points.
   a) $(-1, 1)$ b) $(2, 0)$ c) $(0, 3)$ d) $(\sqrt{3}, -1)$ e) $(-2, -2\sqrt{3})$

4. Plot the set of all points $[r; \theta]$ such that
   a) $1 < r \leq 2$,   b) $r = 3$,
   c) $\theta = \pi/3, \quad r > 0$,   d) $\theta = \pi/3, \quad r < 0$,
   e) $\pi/4 \leq \theta \leq 4\pi/3, \quad r > 0$,   f) $\theta = \pi/4, \quad 1 \leq r \leq 2$,
   g) $r = 2, \quad 5\pi/2 \leq \theta < 3\pi$.

5. Let L be a line which does not pass through the origin, let p be its distance from the origin, and let α be the polar angle made by the ray emanating from the origin which intersects L perpendicularly (see Fig. 7–33).

   a) Show that the polar equation of L is $r \cos (\theta - \alpha) = p$.
   b) Find the equation of L in rectangular coordinates. (The equation is known as the *normal form* of L.)

6. Let a and b be positive numbers, and let $P_1 = (-a, 0)$, $P_2 = (a, 0)$. Let C be the set of all points P such that the product of the distances from P to the points P_1 and P_2 is equal to b^2. Then C is called a *lemniscate*.

   a) Prove that the polar equation of C is $b^4 = a^4 + r^4 - 2a^2r^2 \cos 2\theta$.
   b) Sketch C for the case in which $b = 2a$. In general, if $b > a$, then C is a closed curve.
   c) Sketch C for the case in which $b = a$. Show that one can obtain the lemniscate of Example 5 by rotating C through $\pi/4$ radians.
   d) Sketch C for the case in which $2b = a$. In general, if $b < a$, then C consists of two closed curves called *ovals of Cassini*.

7. This exercise is concerned with a family of curves known as *conchoids of Nicomedes*. Let a and b be positive numbers. We define C to be the set of all points P with the property that the line drawn from O to P intersects the line $x = a$ at a point whose distance from P is b.

   a) Show that C is the union of two curves whose polar equations are

   $$r = a \sec \theta + b \quad \text{and} \quad r = a \sec \theta - b.$$

   b) Sketch the curves in each of the following cases: $a < b, a = b, a > b$.
   c) Show that the equation of C in rectangular coordinates is

   $$(x^2 + y^2)(x - a)^2 = b^2x^2.$$

8. a) Show that the polar equation of the circle of radius a and center $(a, 0)$ is $r = 2a \cos \theta$.
   b) If a and b are positive numbers, then the curve $r = a + b \cos \theta$ is called a *limaçon*. Sketch one of these curves in each of the following cases: $a < b, a = b, a > b$.

9. Sketch each of the following curves and find the area of the region which they enclose:

   a) the circle $r = 8 \cos \theta$,   b) the circle $r = 6 \sin \theta$,
   c) the lemniscate $r^2 = a^2 \cos 2\theta$,   d) the cardioid $r = a(1 - \sin \theta)$,
   e) the rose $r = a \cos 2\theta$,   f) the rose $r = a \cos 3\theta$,
   g) the rose $r = a \cos n\theta$ $(n \in \mathbf{Z}^+)$,
   h) the limaçon $r = a + b \cos \theta$ $(a > b > 0)$.

10. Find the areas enclosed by both the large and the small loops for each of the following limaçons:

    a) $r = 1 + 2 \sin \theta$,   b) $r = 1 + \sqrt{2} \cos \theta$,   c) $r = \sqrt{3} - 2 \sin \theta$.

11. Let $a > 0$. Sketch the *spiral of Archimedes*, $r = a\theta$. Find the area swept out, starting from $\theta = 0$, in (a) one revolution, (b) two revolutions, (c) n revolutions $(n \in \mathbf{Z}^+)$.

12. Show that the area of the region $\{[r; \theta] \mid 0 \le r \le 1 \text{ and } 0 \le r \le 1 + \sin \theta\}$ is

    $$\frac{\pi}{6} + \frac{1 - \sqrt{3}}{2}.$$

7-6 Volumes

We consider first volumes of revolution.

Suppose that f is continuous and nonnegative on $[a, b]$. We shall calculate the volume of the region in three-space which is obtained by revolving the set

$$S = \{(x, y) \mid a \leq x \leq b \text{ and } 0 \leq y \leq f(x)\}$$

about the x-axis [see Fig. 7–34(a)].

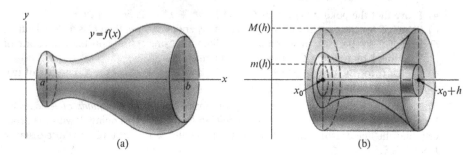

(a)   (b)

FIG. 7–34

If $a \leq x_0 \leq b$, we define $V(x_0)$ to be the volume of the region obtained by revolving the set

$$S_{x_0} = \{(x, y) \mid a \leq x \leq x_0 \text{ and } 0 \leq y \leq f(x)\}$$

about the x-axis. If $h > 0$, then $V(x_0 + h) - V(x_0)$ is the volume of a slab of the region and is less than $\pi[M(h)]^2 h$, the volume of a circumscribed right circular cylinder [Fig. 7–34(b)], where, as before,

$$M(h) = \max \{f(x) \mid x_0 \leq x \leq x_0 + h\}.$$

Similarly,

$$\pi[m(h)]^2 h \leq V(x_0 + h) - V(x_0),$$

where

$$m(h) = \min \{f(x) \mid x_0 \leq x \leq x_0 + h\}.$$

Thus

$$\pi[m(h)]^2 \leq \frac{V(x_0 + h) - V(x_0)}{h} \leq \pi[M(h)]^2.$$

By passing to limits and then considering negative values of h, we get

$$V'(x_0) = \pi[f(x_0)]^2.$$

Hence, the required volume is seen to be

$$V = \pi \int_a^b [f(x)]^2 \, dx.$$

Example 1. Find the volume of a sphere of radius r.

We can apply the formula just derived with $f(x) = \sqrt{r^2 - x^2}$, $a = -r$, and $b = r$. Thus

$$V = \pi \int_{-r}^{r} (r^2 - x^2)\, dx = \pi \left(r^2 x - \frac{x^3}{3} \right) \Big|_{-r}^{r}$$

$$= \tfrac{4}{3}\pi r^3.$$

Example 2. The center of a disk of radius a is at a distance $b > a$ from a line L. Find the volume of the solid *torus* (doughnut) obtained by revolving the disk about the line L (see Fig. 7–35).

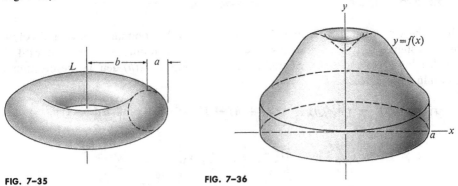

FIG. 7–35   **FIG. 7–36**

We introduce coordinates in such a way that the x-axis is L and that the semicircular boundaries of the disk have the equations

$$y = f(x) = b + \sqrt{a^2 - x^2} \qquad \text{and} \qquad y = g(x) = b - \sqrt{a^2 - x^2}.$$

The volume of the torus is then

$$V = \pi \int_{-a}^{a} [f(x)]^2\, dx - \pi \int_{-a}^{a} [g(x)]^2\, dx$$

$$= 4\pi b \int_{-a}^{a} \sqrt{a^2 - x^2}\, dx.$$

Although we cannot yet find an antiderivative of $\sqrt{a^2 - x^2}$, we can nevertheless complete the calculation by observing that $\int_{-a}^{a} \sqrt{a^2 - x^2}\, dx$ is half the area of a circle of radius a. Hence

$$V = 2\pi^2 a^2 b.$$

Let us consider next the volume of a solid of revolution obtained by revolving the graph of a function about the y-axis. Suppose that f is continuous and non-negative on the interval $[0, a]$. We shall find the volume of the solid obtained by revolving the set

$$\{(x, y) \mid 0 \le x \le a \text{ and } 0 \le y \le f(x)\}$$

about the y-axis (see Fig. 7–36).

Let $V(x_0)$ be the volume of the solid obtained by revolving the set

$$\{(x, y) \mid 0 \le x \le x_0 \text{ and } 0 \le y \le f(x)\}$$

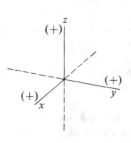

FIG. 7–37

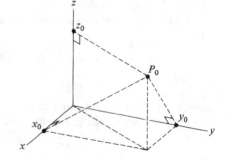

FIG. 7–38

about the y-axis. If $h > 0$, then $V(x_0 + h) - V(x_0)$ is bounded above and below by the volumes of two cylindrical shells having inner and outer radii x_0 and $x_0 + h$ and heights $M(h)$ and $m(h)$, where, as before, $M(h)$ and $m(h)$ denote the maximum and minimum values of f on $[x_0, x_0 + h]$. Thus

$$\pi[(x_0 + h)^2 - x_0^2]M(h) \leq V(x_0 + h) - V(x_0) \leq \pi[(x_0 + h)^2 - x_0^2]m(h).$$

Since

$$\lim_{h \to 0} \frac{(x_0 + h)^2 - x_0^2}{h} = \frac{d}{dx}(x^2)\Big|_{x=x_0} = 2x_0,$$

then dividing by h, passing to limits, pinching, and paying the now customary lip service to negative values of h, we get

$$V'(x_0) = 2\pi x_0 f(x_0).$$

It follows that the required volume is

$$V = 2\pi \int_0^a x f(x)\, dx.$$

Example 3. A hole of diameter a is drilled through the center of a sphere of radius a. Find the remaining volume.

We apply the preceding formula with $f(x) = \sqrt{a^2 - x^2}$. Taking advantage of symmetry, we find that the volume is

$$V = 4\pi \int_0^a x\sqrt{a^2 - x^2}\, dx - 4\pi \int_0^{a/2} x\sqrt{a^2 - x^2}\, dx$$

$$= 4\pi \int_{a/2}^a x\sqrt{a^2 - x^2}\, dx$$

$$= -\tfrac{4}{3}\pi(a^2 - x^2)^{3/2}\Big|_{x=a/2}^{x=a}$$

$$= \tfrac{1}{2}\sqrt{3}\,\pi a^3.$$

Many solids are conveniently described in terms of three-dimensional Cartesian coordinates. We can introduce such a coordinate system by first selecting three

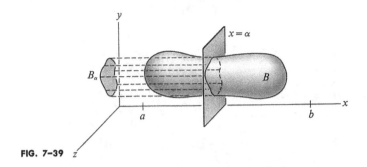

FIG. 7-39

mutually perpendicular straight lines; these lines are called the *coordinate axes*, and their common point of intersection O is called the *origin*. Using a common unit of length, we introduce coordinates onto each of the coordinate axes (commonly called the x-, y-, and z-axes). This is usually done in such a way that if the index and middle fingers of the right hand are pointed in the directions of the positive x- and y-axes, then the right thumb will point in the direction of the positive z-axis (see Fig. 7–37). In this case the coordinate system is said to be *right-handed;* otherwise it is called *left-handed.* The three planes determined by pairs of coordinate axes are called the *coordinate planes.* For instance, the plane determined by the x- and y-axes is called the xy-plane. We can then establish a one-to-one correspondence between points in space and ordered triples of real numbers. Given any point P_0 in space, we let x_0, y_0, and z_0 be the perpendicular projections of P_0 onto the x-, y-, and z-axes, respectively, as shown in Fig. 7–38. We then associate with the point P_0 the ordered triple (x_0, y_0, z_0), and thus we get the above-mentioned correspondence between points in space and elements of $\mathbf{R} \times \mathbf{R} \times \mathbf{R} = \mathbf{R}^3$, the set of all ordered triples of real numbers.

Let B be a bounded subset of $\mathbf{R}^3$. For each real number α, we let

$$B_\alpha = \{(y, z) \mid (\alpha, y, z) \in B\}.$$

Geometrically, B_α can be thought of as the projection onto the yz-plane of the cross section of B cut off by the plane $x = \alpha$, as shown in Fig. 7–39. Since B is bounded, there exist numbers a and b such that $B_x = \varnothing$ if x lies outside the interval $[a, b]$. We will show that under certain conditions the volume of B is given by the formula

$$V(B) = \int_a^b A(B_x)\, dx,$$

where $A(B_x)$ denotes the area of B_x.

We let $V(x_0)$ denote the volume of the set $\{(x, y, z) \in B \mid x \le x_0\}$. What we must find are conditions on the cross sections B_x which will enable us to place upper and lower estimates on $V(x_0 + h) - V(x_0)$. We shall say that B *expands* (in the x-direction) *throughout an interval I* iff $(\forall x, y \in I)(x \le y \Rightarrow B_x \subset B_y)$. Note that if B expands throughout the interval $[x_0, x_0 + h]$, then the slab consisting of those

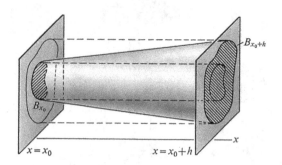

FIG. 7-40 $x = x_0$ $x = x_0 + h$

points of B which lie between the planes $x = x_0$ and $x = x_0 + h$, namely,

$$\{(x, y, z) \in B \mid x_0 \le x \le x_0 + h\},$$

is a subset of the cylinder

$$\{(x, y, z) \mid x_0 \le x \le x_0 + h \text{ and } (y, z) \in B_{x_0+h}\} = [x_0, x_0 + h] \times B_{x_0+h}$$

and contains within itself the cylinder

$$\{(x, y, z) \mid x_0 \le x \le x_0 + h \text{ and } (y, z) \in B_{x_0}\} = [x_0, x_0 + h] \times B_{x_0}.$$

(See Fig. 7–40.) Since the volume of a solid cylinder is equal to the area of its base times its height, it follows that $A(B_{x_0+h})h$ and $A(B_{x_0})h$ are upper and lower estimates of $V(x_0 + h) - V(x_0)$, the volume of the slab. Thus

$$A(B_{x_0}) \le \frac{V(x_0 + h) - V(x_0)}{h} \le A(B_{x_0+h}).$$

Furthermore, if we assume that $A(B_x)$ is a continuous function, it follows that

$$V'(x_0) = A(B_{x_0}).$$

The formula $V(B) = \int_a^b A(B_x)\, dx$ will therefore be valid if B expands in the x-direction and $A(B_x)$ is a continuous function of x. By simply reversing the direction of the x-axis, it is immediately apparent that the same will be true if B *contracts* in the x-direction. Finally, because of the additivity of volumes and integrals, it should be reasonably clear that the following is true.

Proposition. *The formula*

$$V(B) = \int_a^b A(B_x)\, dx$$

is valid if the following are satisfied:

1) The interval $[a, b]$ can be partitioned into a finite number of subintervals $[a_1, x_1], [x_1, x_2], \ldots, [x_{n-1}, b]$ on which B either expands or contracts.

2) The function $\varphi(x) = A(B_x)$ is continuous on $[a, b]$.

While this result is by no means the best possible, it has many interesting applications.

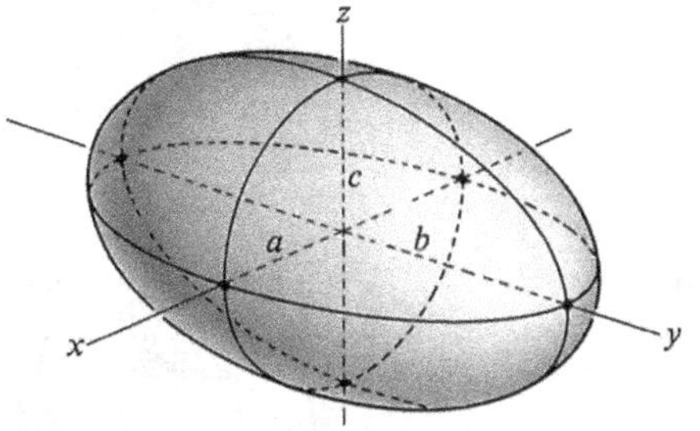

FIG. 7–41 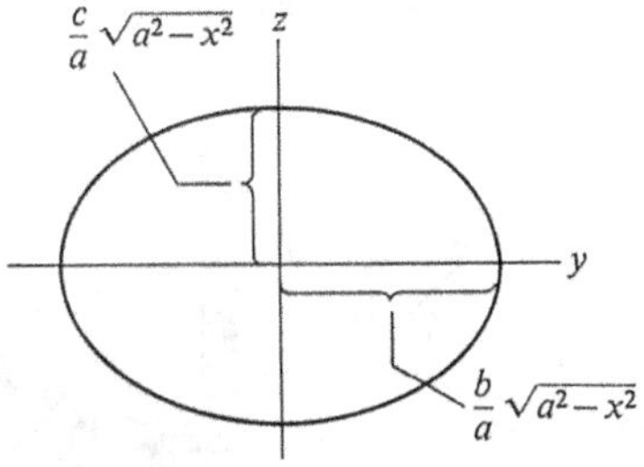   **FIG. 7–42**

Example 4. Find the volume of the region enclosed by the ellipsoid

$$\frac{x^2}{a^2} + \frac{y^2}{b^2} + \frac{z^2}{c^2} = 1.$$

(See Fig. 7–41.)

The cross section

$$B_x = \left\{ (y, z) \, \middle| \, \frac{x^2}{a^2} + \frac{y^2}{b^2} + \frac{z^2}{c^2} \le 1 \right\}$$

is the null set if $|x| > a$, and is the inside of the ellipse

$$\frac{y^2}{[(b/a)\sqrt{a^2 - x^2}]^2} + \frac{z^2}{[(c/a)\sqrt{a^2 - x^2}]^2} = 1$$

if $-a \le x \le a$ (see Fig. 7–42). The solid B expands in the interval $[-a, 0]$ and contracts in the interval $[0, a]$. Moreover,

$$A(B_x) = \pi \left(\frac{b}{a}\sqrt{a^2 - x^2} \right) \left(\frac{c}{a}\sqrt{a^2 - x^2} \right) = \frac{\pi bc}{a^2} (a^2 - x^2)$$

is continuous on the interval $[-a, a]$. Hence

$$V = \int_{-a}^{a} \frac{\pi bc}{a^2} (a^2 - x^2) \, dx = \tfrac{4}{3}\pi abc.$$

Setting $a = b = c$, we get the volume of a sphere as a special case.

Example 5. The axes of two solid right circular cylinders each of radius a intersect at right angles. Find the volume of their intersection B (see Fig. 7–43).

We introduce coordinate axes so that the cylinders are defined by the inequalities

$$x^2 + y^2 \le a^2 \quad \text{and} \quad x^2 + z^2 \le a^2.$$

Then if $-a \le x \le a$,

$$B_x = \{(y, z) \mid x^2 + y^2 \le a^2 \text{ and } x^2 + z^2 \le a^2\}$$
$$= \{(y, z) \mid |y|, |z| \le \sqrt{a^2 - x^2}\}.$$

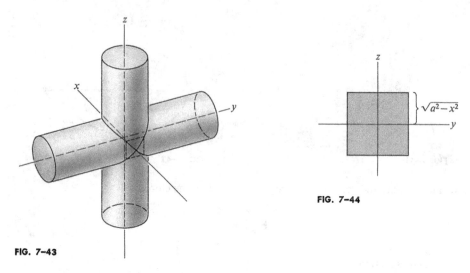

FIG. 7-44

FIG. 7-43

The latter is a square with side length $2\sqrt{a^2 - x^2}$, as shown in Fig. 7–44. As in the previous example, B expands throughout the interval $[-a, 0]$ and contracts throughout the interval $[0, a]$. Also,

$$A(B_x) = 4(a^2 - x^2)$$

is continuous on $[-a, a]$. Thus

$$V(B) = \int_{-a}^{a} 4(a^2 - x^2)\, dx = \tfrac{16}{3}a^3.$$

Exercises

1. The bounded region which lies between the parabolas $y = x^2$ and $x = y^2$ is revolved about the x-axis. Find the volume of the solid so generated.

2. The circle $x^2 + (y - a)^2 = 4a^2$ is revolved about the x-axis. Find the volume of the region which lies between the outer and inner surfaces of revolution so generated.

3. Let S be the region bounded by the curve $y = 1/(1 + x^2)^2$, the x-axis, and the lines $x = 1$, $x = 2$. Find the volume of the region generated by revolving S about the y-axis.

4. Find the volume of the region enclosed by the surface of revolution obtained by rotating the astroid $x^{2/3} + y^{2/3} = a^{2/3}$ about the x-axis.

5. Let f and g be functions such that f' and g are both nonnegative and continuous on the interval $[a, b]$. Let S be the region bounded by the curve

$$\gamma: \quad x = f(t), \quad y = g(t) \qquad (a \le t \le b),$$

the x-axis, and the vertical lines $x = f(a), x = f(b)$.

a) Derive a formula for the area of the solid generated by revolving S about the x-axis.

b) Derive a formula for the area of the solid generated by revolving S about the y-axis, assuming in this case that $f(a) \ge 0$.

c) Apply the first of these formulas when γ is the arch of the cycloid

$$x = a(t - \sin t), \quad y = a(1 - \cos t) \qquad (0 \le t \le 2\pi).$$

6. Suppose that f is nonnegative on the interval $[\alpha, \beta]$, and let S be the sector bounded by the curve $r = f(\theta)$ and the rays $\theta = \alpha$, $\theta = \beta$.

   a) If $0 \le \alpha \le \beta \le \pi$, show that the volume of the solid obtained by revolving S about the x-axis is equal to
   $$\tfrac{2}{3}\pi \int_\alpha^\beta [f(\theta)]^3 \sin \theta \, d\theta.$$

   b) Derive a similar formula for the volume of the solid obtained by revolving S about the y-axis, assuming in this case that $-\pi/2 \le \alpha \le \beta \le \pi/2$.

7. Using the previous exercise, find the volume of the solid generated by revolving the region inside the cardioid $r = 4(1 - \cos \theta)$ about the polar axis.

8. Solve the preceding problem for the curve $r = a \cos^2 \theta$.

9. Derive the formula for the volume of a right circular cone.

10. Find the volume of a solid whose base is a disk of radius 5 if all the plane sections perpendicular to a fixed diameter of the disk are equilateral triangles.

11. Solve the preceding problem for the case in which all the plane sections are isosceles triangles with altitude 3.

12. A horn is generated by a circle which moves as follows: the plane of the circle is perpendicular to the x-axis, the center of the circle lies in the xy-plane, and its diameter in the xy-plane is the line segment cut off by the curves $2y = \sqrt{x}$ and $y = \sqrt{x}$.

    a) Find the volume of the horn generated as the center of the circle moves from $x = 0$ to $x = 4$. Do this by applying the formula $V(B) = \int_a^b A(B_x) \, dx$.

    b) Observe that the hypotheses which we used to derive this formula are *not* satisfied. Show, however, that the formula is nevertheless valid in this instance.

13. Find the volume of the solid which is bounded by the surfaces
    $$\frac{x^2}{a^2} + \frac{y^2}{b^2} - \frac{z^2}{c^2} = 1, \qquad z = 0, \qquad \text{and} \qquad z = c.$$

14. Find the volume of the solid bounded by the paraboloid
    $$\frac{x^2}{a^2} + \frac{y^2}{b^2} = \frac{z}{h}.$$

and the plane $z = h$.

15. A right elliptic cylinder with base $x^2/a^2 + y^2/b^2 = 1$ is intercepted by a plane which passes through the y-axis, making an angle θ with the xy-plane. Find the volume of the wedge-shaped solid so formed (see Fig. 7–45).

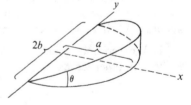

FIG. 7-45

16. Find the volume of the region bounded by the cylinder $x^2 + y^2 = 1$ and the planes $y + z = 1$, $y = 0$, $z = 0$ $(y \ge 0)$.

17. A square hole of side length 2 in. is cut through a cylindrical post of radius 2 in. The axis of the hole and the axis of the cylinder intersect at right angles.

    a) If a pair of opposite sides of the hole are perpendicular to the axis of the post, show that the volume cut out is $\tfrac{4}{3}(3\sqrt{3} + 2\pi)$.

    b) If the sides of the hole make a $45°$ angle with the axis of the post, show that the volume cut out is $\tfrac{4}{3}(10\sqrt{2} + 3\pi\sqrt{2} - 16)$.

7-7 Path length

Suppose that the path γ of a particle moving in $\mathbf{R}^2$ is described by the parametric equations

$$x = f(t), \qquad y = g(t) \qquad (a \leq t \leq b).$$

We shall be concerned with calculating the total distance traveled by the particle along γ.

First we place some restrictions on the functions f and g. If we merely assume that f and g are continuous (and certainly we shall want to assume at least this much), then the path can be very pathological. It can intersect itself infinitely often; it can even pass through every point of a square! Also, there can exist paths with no self-intersections which nevertheless have positive areas! None of these pathologies can occur, however, if f and g possess continuous derivatives on $[a, b]$ which do not vanish simultaneously. (By the derivative of f at the endpoints a and b we shall mean the one-sided derivatives

$$\lim_{h \to 0+} [f(a + h) - f(a)]/h$$

and

$$\lim_{h \to 0-} [f(b + h) - f(b)]/h,$$

respectively.) In this case we say that the path γ is smooth and that the functions f and g provide a smooth parametrization of γ. Geometrically, this condition on f and g guarantees that γ possesses a tangent at each point and that this tangent line varies continuously as we move along γ. [The tangent line to γ when $t = t_0$ has the parametric equations

$$x = f(t_0) + f'(t_0)(t - t_0), \qquad y = g(t_0) + g'(t_0)(t - t_0);$$

it is the limiting position of secant lines through the point $\big(f(t_0), g(t_0)\big)$.]

We now seek a suitable definition of the *path length* of γ, assuming that γ is smooth. We first specify certain properties of path lengths.

1) We certainly want path length to be *additive;* in other words, if

$$a = t_0 < t_1 < t_2 < \cdots < t_n = b$$

is any partition of the parametric interval, and if we let γ_k be the subpath having the parametric equations

$$x = f(t), \qquad y = g(t) \qquad (t_{k-1} \leq t \leq t_k)$$

for $k = 1, \ldots, n$, then the path length of γ should be the sum of the path lengths of $\gamma_1, \ldots, \gamma_n$ (see Fig. 7–46).

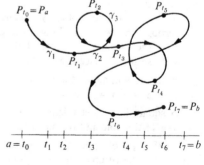

FIG. 7–46

2) Also, we ask that for points on γ which are close together, the ratio of path length to chord length should be nearly one; more precisely, if P_{t_0} denotes the point on γ corresponding to the value t_0 of the parameter, then we demand that

$$\lim_{t \to t_0} \frac{d(P_t, P_{t_0})}{\overparen{P_t P_{t_0}}} = 1,$$

where $d(P_t, P_{t_0})$ denotes the (chordal) distance between P_t and P_{t_0}, and $\overparen{P_t P_{t_0}}$ denotes the path length of the subpath determined by t_0 and t. Note that in this notation the first requirement can be written

$$\overparen{P_{t_0} P_{t_1}} + \overparen{P_{t_1} P_{t_2}} + \cdots + \overparen{P_{t_{n-1}} P_{t_n}} = \overparen{P_{t_0} P_{t_n}} = \overparen{P_a P_b}.$$

Let $s(t) = \overparen{P_a P_t}$. Let us calculate $s'(t_0)$ on the assumption that the above requirements on path length can be met. From the additivity requirement, we observe that

$$s(t_0 + h) - s(t_0) = \overparen{P_a P_{t_0+h}} - \overparen{P_a P_{t_0}}$$

$$= \begin{cases} \overparen{P_{t_0} P_{t_0+h}} & \text{if } h > 0 \\ -\overparen{P_{t_0} P_{t_0+h}} & \text{if } h < 0 \end{cases}$$

$$= (\operatorname{sgn} h)\overparen{P_{t_0} P_{t_0+h}}.$$

Hence

$$s'(t_0) = \lim_{h \to 0} \frac{s(t_0 + h) - s(t_0)}{h}$$

$$= \lim_{h \to 0} \frac{(\operatorname{sgn} h)\overparen{P_{t_0} P_{t_0+h}}}{h}$$

$$= \lim_{h \to 0} \frac{\overparen{P_{t_0} P_{t_0+h}}}{d(P_{t_0}, P_{t_0+h})} \frac{(\operatorname{sgn} h)\, d(P_{t_0}, P_{t_0+h})}{h}$$

$$= \lim_{h \to 0} \frac{d(P_{t_0}, P_{t_0+h})}{|h|},$$

where we assume momentarily that the latter limit exists and that $d(P_{t_0}, P_{t_0+h}) \neq 0$ for sufficiently small $h \neq 0$. Now

$$d(P_{t_0}, P_{t_0+h}) = \sqrt{[f(t_0 + h) - f(t_0)]^2 + [g(t_0 + h) - g(t_0)]^2},$$

and by the mean-value theorem, we can write

$$f(t_0 + h) - f(t_0) = f'(\tau)h, \qquad g(t_0 + h) - g(t_0) = g'(\tau')h$$

for suitable values of τ and τ' between t_0 and $t_0 + h$. Since f' and g' are continuous at t_0,

$$\lim_{h \to 0} \frac{d(P_{t_0}, P_{t_0+h})}{|h|} = \lim_{h \to 0} \sqrt{[f'(\tau)]^2 + [g'(\tau')]^2}$$

$$= \sqrt{[f'(t_0)]^2 + [g'(t_0)]^2}.$$

(A complete justification of the last step may be given as follows. Let $\{h_n\}$ be any sequence converging properly to zero. Let τ_n and τ_n' be the values of τ and τ' corresponding to h_n. Then $\tau_n \to t_0$ and $\tau_n' \to t_0$ as $n \to \infty$. Since f'^n and g' are continuous at t_0, it follows that $f'(\tau_n) \to f'(t_0)$ and $g'(\tau_n') \to g'(t_0)$. Finally, because of the continuity of the arithmetic operations and the square root function, we have

$$\lim_{n \to \infty} \frac{d(P_{t_0}, P_{t_0+h_n})}{|h_n|} = \lim_{n \to \infty} \sqrt{[f'(\tau_n)]^2 + [g'(\tau_n')]^2}$$

$$= \sqrt{[f'(t_0)]^2 + [g'(t_0)]^2},$$

and the desired result follows by the linking limit lemma.)

Since f' and g' are never zero simultaneously, the limit is positive and hence so is $d(P_{t_0}, P_{t_0+h})$ for sufficiently small $h \neq 0$. Thus

$$s'(t) = \sqrt{[f'(t)]^2 + [g'(t)]^2}.$$

It follows that the path length of γ must be

$$l(\gamma) = \int_a^b \sqrt{[f'(t)]^2 + [g'(t)]^2}\, dt$$

if our requirements on path length are to be met.

Definition 1. If

$$x = f(t), \qquad y = g(t) \qquad (a \leq t \leq b)$$

is a smooth parametrization of a path γ, then the **path length** of γ is defined to be

$$l(\gamma) = \int_a^b \sqrt{[f'(t)]^2 + [g'(t)]^2}\, dt = \int_a^b \sqrt{(dx/dt)^2 + (dy/dt)^2}\, dt.$$

The legitimacy of this definition depends on an important theorem which will be proved in the next chapter (and also in Chapter 13), namely, that every continuous function has an antiderivative. Assuming this, however, one can check that path length, so defined, has the properties stated above: additivity of path length follows from the additive property of definite integrals, and the proof of the second property is essentially the same as our discussion preceding the definition.

Example 1. Let us first check our formula in the case of a line segment. The parametric equations

$$x = (1 - t)x_1 + tx_2, \qquad y = (1 - t)y_1 + ty_2 \qquad (0 \leq t \leq 1)$$

describe the path γ of a particle moving with uniform speed along the line segment from $P_1 = (x_1, y_1)$ to $P_2 = (x_2, y_2)$. Since

$$dx/dt = x_2 - x_1, \qquad dy/dt = y_2 - y_1,$$

from the formula we get

$$l(\gamma) = \int_0^1 \sqrt{(x_2 - x_1)^2 + (y_2 - y_1)^2}\, dt$$

$$= \sqrt{(x_2 - x_1)^2 + (y_2 - y_1)^2}$$

$$= d(P_1, P_2).$$

Observe that if we take a second parametrization of γ, say

$$x = (1 - \tan t)x_1 + (\tan t)x_2,$$
$$y = (1 - \tan t)y_1 + (\tan t)y_2 \qquad (0 \le t \le \pi/4),$$

then the path length will be the same. Thus, since

$$dx/dt = (\sec^2 t)(x_2 - x_1), \qquad dy/dt = (\sec^2 t)(y_2 - y_1),$$

we get

$$l(\gamma) = \int_0^{\pi/4} (\sec^2 t)\sqrt{(x_2 - x_1)^2 + (y_2 - y_1)^2}\, dt$$
$$= \sqrt{(x_2 - x_1)^2 + (y_2 - y_1)^2}\, \tan t\Big|_0^{\pi/4}$$
$$= d(P_1, P_2).$$

We shall later prove that the length of a path γ is always independent of its parametrization.

Example 2. Let us also check things out for a circle. Let

$$\gamma_1: \ x = r \cos t, \ y = r \sin t \qquad (0 \le t \le 2\pi),$$
$$\gamma_2: \ x = r \cos t, \ y = r \sin t \qquad (0 \le t \le 4\pi).$$

Then

$$\sqrt{(dx/dt)^2 + (dy/dt)^2} = \sqrt{r^2 \sin^2 t + r^2 \cos^2 t} = r.$$

Hence, by the formula,

$$l(\gamma_1) = \int_0^{2\pi} r\, dt = 2\pi r, \qquad l(\gamma_2) = \int_0^{4\pi} r\, dt = 4\pi r.$$

This is as it should be. In the first case we travel around the circle once, and thus the path length is the circumference $2\pi r$. In the second case we move twice around the circle, so the path length is doubled.

This example shows that a path should not be regarded as simply a point set.

Example 3. Find the arc length of one arch of the cycloid

$$x = a(\theta - \sin \theta), \qquad y = a(1 - \cos \theta).$$

We shall interpret the problem to be that of finding the path length of the path

$$\gamma: \ x = a(\theta - \sin \theta), \ y = a(1 - \cos \theta) \qquad (0 \le \theta \le 2\pi).$$

Actually, the parametrization is not smooth, since dx/dt and dy/dt both vanish at the endpoints. Nevertheless, if $0 < \theta_1 < \theta_2 < 2\pi$, then the parametrization is smooth when θ is restricted to the interval $[\theta_1, \theta_2]$, and

$$\overparen{P_{\theta_1}P_{\theta_2}} = \int_{\theta_1}^{\theta_2} \sqrt{a^2(1 - \cos \theta)^2 + a^2 \sin^2 \theta}\, d\theta$$
$$= \int_{\theta_1}^{\theta_2} a\sqrt{2 - 2\cos \theta}\, d\theta$$
$$= 2a \int_{\theta_1}^{\theta_2} \sin(\theta/2)\, d\theta$$
$$= 4a[\cos(\theta_1/2) - \cos(\theta_2/2)].$$

Since

$$\lim_{\substack{\theta_1 \to 0 \\ \theta_2 \to 2\pi}} \overset{\frown}{P_{\theta_1}P_{\theta_2}} = 8a,$$

we take this to be the path length of γ.

The last example would seem to indicate that our definition is somewhat too restrictive. We surely ought to be able to tolerate a finite number of points at which dx/dt and dy/dt vanish simultaneously.

Definition 2. We say that a path γ is **piecewise smooth** if it has a parametrization

$$x = f(t), \qquad y = g(t) \qquad (a \le t \le b)$$

such that

1) the functions f' and g' are continuous on $[a, b]$, and

2) there are at most a finite number of points $t \in [a, b]$ for which

$$f'(t) = g'(t) = 0.$$

As before, we define the path length of γ by the formula

$$l(\gamma) = \int_a^b \sqrt{[f'(t)]^2 + [g'(t)]^2}\, dt.$$

Consider the path γ having the parametric equations

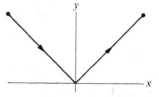

FIG. 7–47

$$x = \begin{cases} -t^2 & \text{if } -1 \le t \le 0, \\ t^2 & \text{if } 0 \le t \le 1, \end{cases} \qquad y = t^2 \qquad (-1 \le t \le 1).$$

As a point set, γ is the union of two line segments, as shown in Fig. 7–47. It is easily checked that γ is piecewise smooth. When $t = 0$, both dx/dt and dy/dt are zero. Thus γ has a *singular point* at the origin. It should be intuitively clear that the singularity of this point is due to a pathology of the path itself, and not merely to this particular parametrization. In other words, γ does not possess a completely smooth parametrization, since, if it did, the path would have a well-defined tangent line at the origin. This example shows that a piecewise smooth curve can have a finite number of sharp corners. Also, it is possible for a piecewise smooth curve to double back on itself as illustrated by the path

$$x = \cos t, \qquad y = 0 \qquad (0 \le t \le 2x).$$

(See Fig. 7–48.)

Because of the additivity of definite integrals, it follows that path length continues to be additive for this larger class of piecewise smooth curves. At singular points, however, our second property, namely,

$$\lim_{t \to t_0} \frac{d(P_t, P_{t_0})}{\overset{\frown}{P_t P_{t_0}}} = 1,$$

may fail to hold.

FIG. 7–48

We shall now consider the global relationships between path length and chord length.

Proposition 3. *"The shortest distance between two points is a straight line."* *More precisely, if P_1 and P_2 are any two points in $\mathbf{R}^2$, and if γ is any piecewise smooth path having P_1 and P_2 as its initial and terminal points, then*

$$l(\gamma) \geq d(P_1, P_2).$$

Proof. A proof using vector methods will be outlined in Chapter 12. The present proof involves two cases. We use previously established notations.

Case I. $P_a \neq P_t$ whenever $a < t \leq b$. We set

$$\varphi(t) = \widehat{P_aP_t} - d(P_a, P_t)$$
$$= \int_a^t \sqrt{[f'(t)]^2 + [g'(t)]^2}\, dt - \sqrt{[f(t) - f(a)]^2 + [g(t) - g(a)]^2}.$$

Then

$$\varphi'(t) = \sqrt{[f'(t)]^2 + [g'(t)]^2} - \frac{[f(t) - f(a)]f'(t) + [g(t) - g(a)]g'(t)}{\sqrt{[f(t) - f(a)]^2 + [g(t) - g(a)]^2}}.$$

By a special case of the Cauchy-Schwarz inequality, namely,

$$a_1b_1 + a_2b_2 \leq \sqrt{a_1^2 + a_2^2}\,\sqrt{b_1^2 + b_2^2},$$

it follows that $\varphi'(t) \geq 0$. Hence φ is a nondecreasing function, and so

$$\widehat{P_1P_2} - d(P_1, P_2) = \varphi(b) \geq \varphi(a) = 0,$$

the desired result.

Case II. There exists a number t such that $a < t \leq b$ and $P_a = P_t$. Among such t there is a largest one, call it T. (The existence of T may be established as follows. Set $h(t) = [f(t) - f(a)]^2 + [g(t) - g(a)]^2$. Then h is continuous on $[a, b]$. Put $S = \{t \mid a < t \leq b \text{ and } h(t) = 0\}$ and $T = \text{lub } S$. There exists a sequence $t_1, t_2, \ldots$ such that $t_n \in S$ for each $n \in \mathbf{Z}^+$ and $t_n \to T$. Since h is continuous at T, it follows that $h(T) = \lim h(t_n) = 0$. Clearly, T is the largest member of S.) By case I,

$$\widehat{P_1P_2} = \widehat{P_1P_T} + \widehat{P_TP_2} \geq \widehat{P_TP_2} \geq d(P_T, P_2) = d(P_1, P_2).$$

Given a path

$$\gamma:\quad x = f(t),\qquad y = g(t)\qquad (a \leq t \leq b),$$

let us consider a partition of the parametric interval $[a, b]$, where the points of subdivision are, say,

$$a = t_0 < t_1 < \cdots < t_n = b.$$

Then by the preceding result,

$$\sum_{k=1}^n d(P_{t_{k-1}}, P_{t_k}) \leq \sum_{k=1}^n \widehat{P_{t_{k-1}}P_{t_k}} = \widehat{P_{t_0}P_{t_n}} = \widehat{P_aP_b}.$$

The sum $\sum_{k=1}^{n} d(P_{t_{k-1}}, P_{t_k})$ can be thought of as the length of an *inscribed polygon*. Thus the path length of γ is an upper bound of the lengths of polygons inscribed in γ (see Fig. 7–49). Since

$$\lim_{t \to t'} \frac{d(P_t, P_{t'})}{\overparen{P_t P_{t'}}} = 1,$$

we can expect the path length of γ to be the least upper bound of the lengths of such polygons.

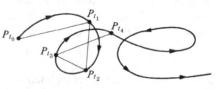

FIG. 7–49

Proposition 4. *If γ is a piecewise smooth path, then $l(\gamma)$ is the least upper bound of the lengths of the polygons inscribed in γ.*

Proof. Given an $\epsilon > 0$, we must show that the partition above of $[a, b]$ can be chosen so that

$$l(\gamma) - \sum_{k=1}^{n} d(P_{t_{k-1}}, P_{t_k}) < \epsilon.$$

Note that

$$l(\gamma) - \sum_{k=1}^{n} d(P_{t_{k-1}}, P_{t_k}) = \sum_{k=1}^{n} \{ \overparen{P_{t_{k-1}} P_{t_k}} - d(P_{t_{k-1}}, P_{t_k}) \}.$$

As before, we can write

$$d(P_{t_{k-1}}, P_{t_k}) = \sqrt{[f'(\tau_k)]^2 + [g'(\tau_k')]^2} \, (t_k - t_{k-1}),$$

where τ_k and τ_k' belong to the interval (t_{k-1}, t_k). Thus

$$\overparen{P_{t_{k-1}} P_{t_k}} - d(P_{t_{k-1}}, P_{t_k}) = \int_{t_{k-1}}^{t_k} \varphi_k(t) \, dt,$$

where

$$\varphi_k(t) = \sqrt{[f'(t)]^2 + [g'(t)]^2} - \sqrt{[f'(\tau_k)]^2 + [g'(\tau_k')]^2}.$$

It is easily shown, however, that

$$\varphi_k(t) \le \sqrt{2} \max \{ |f'(t) - f'(\tau_k)|, |g'(t) - g'(\tau_k')| \}.$$

Having done this spade work, we can now complete the proof with relative ease. Since f' and g' are uniformly continuous on $[a, b]$, we can choose a $\delta > 0$ such that

$$\left. \begin{array}{c} |f'(t) - f'(t')| \\ |g'(t) - g'(t')| \end{array} \right\} < \frac{\epsilon}{\sqrt{2}\,(b - a)}$$

whenever t and t' belong to $[a, b]$ and $|t - t'| < \delta$. We now choose the above partition of $[a, b]$ so fine that

$$(t_k - t_{k-1}) < \delta \qquad (k = 1, \ldots, n).$$

It then follows that for all $t \in [t_{k-1}, t_k]$,

$$\varphi_k(t) < \frac{\epsilon}{b - a}.$$

Thus

$$\int_{t_{k-1}}^{t_k} \varphi_k(t)\, dt < \frac{\epsilon}{b - a}(t_k - t_{k-1})$$

and

$$\widehat{P_a P_b} - \sum_{k=1}^{n} d(P_{k-1}, P_k) = \sum_{k=1}^{n} \int_{t_{k-1}}^{t_k} \varphi_k(t)\, dt < \sum_{k=1}^{n} \frac{\epsilon}{(b - a)}(t_k - t_{k-1}) = \epsilon.$$

This result can be used to define path length for a more general family of paths, namely, the so-called *rectifiable* ones. Another consequence of this result is that *path length is independent of parametrization*. In other words, if, under two different parametrizations, the same points are traversed in the same order (although with possibly different speeds), then the path length will be the same. This is clear since the class of inscribed polygons is the same for both parametrizations.

Exercises

1. Find the point on the arch of the cycloid $x = a(t - \sin t), y = a(1 - \cos t), 0 \le t \le 2$, that divides the arc into two parts such that the ratio of their arc lengths is $\frac{1}{3}$.

2. a) Find the length of the astroid $x^{2/3} + y^{2/3} = a^{2/3}$.
   *b) Find the length of the curve $(x/a)^{2/3} + (y/b)^{2/3} = 1$.

3. Find the lengths of the following paths:
   a) $x = a\cos t + at\sin t, \quad y = a\sin t - at\cos t \quad (0 \le t \le \pi/2)$.
   b) $x = t^2/2, \quad y = \frac{1}{3}(2t + 1)^{3/2} \quad (0 \le t \le 4)$.
   c) $x = 2t^2, \quad y = t^3 \quad (1 \le t \le 2)$.

4. Show that one arch of the sine curve $y = \sin x$ has the same length as the ellipse

$$x^2 + (y^2/2) = 1.$$

5. a) Let f be a function with a continuous first derivative on the interval $[a, b]$. Let γ be a path which traces out the points on the graph of f between $x = a$ and $x = b$ exactly once. Show that

$$l(\gamma) = \int_a^b \sqrt{1 + [f'(x)]^2}\, dx.$$

b) Suppose that f is nonnegative and has a continuous derivative on $[\alpha, \beta]$. Show that the length of the path $\gamma: r = f(\theta), \alpha \le \theta \le \beta$, is given by

$$l(\gamma) = \int_\alpha^\beta \sqrt{[f(\theta)]^2 + [f'(\theta)]^2}\, d\theta.$$

6. Use Exercise 5(a) to find the arc length of each of the following curves between the points indicated:

   a) $y = x^{3/2}$ from $x = 0$ to $x = 4$.

   b) $y = x^3/3 + 1/(4x)$ from $x = 1$ to $x = 3$.

7. Use Exercise 5(b) to find the length of the following paths:

   a) the cardioid $r = a(1 - \cos \theta)$, $0 \le \theta \le 2\pi$,

   b) the circle $r = 2a \cos \theta$, $0 \le \theta \le 2\pi$.

8. If g has a continuous derivative on the interval $[a, b]$, show that the path

$$\gamma: \quad \theta = g(r) \qquad (a \le r \le b),$$

   has length

$$l(\gamma) = \int_a^b \sqrt{1 + r^2[g'(r)]^2}\, dr.$$

9. Derive the equations given at the beginning of this section for tangent lines to curves defined parametrically.

10. In this exercise your job will be to develop a theory of surface area for those surfaces of revolution obtained by revolving piecewise smooth paths (about, say, the x-axis).

    a) First list properties which you want surface area to have.

    b) From these properties, show that

$$2\pi \int_a^b |g(t)| \sqrt{[f'(t)]^2 + [g'(t)]^2}\, dt$$

    must be the surface area of the surface generated by revolving the path

$$\gamma: \quad x = f(t), \quad y = g(t)$$

    about the x-axis.

    c) Prove an analog for surface area of the fact that "the shortest distance between two points is a straight line."

11. Use Exercise 10(b) to derive the formula for the surface area of a sphere.

12. Find the surface area of a torus obtained by revolving a circle of radius a about a line L, if the distance from the center of the generating circle to L is b, where $b > a$.

13. Find the area of the surface generated by revolving the arc of the parabola $y^2 = 4ax$, $0 \le x \le 3a$, about the x-axis.

14. Find the area of the surface generated by revolving one arch of the cycloid

$$x = a(t - \sin t), \qquad y = a(1 - \cos t)$$

    about the x-axis.

15. Solve the preceding problem for the arc of the lemniscate

$$r^2 = a^2 \cos 2\theta \qquad (0 \le \theta \le \pi/4).$$

16. Find the area of the surface generated by revolving the cardioid $r = a(1 - \cos \theta)$ about the x-axis.

7-8 Moments and centroids

Imagine a horizontal weightless plane π which is supported along a line L. We shall consider the tendency of the plane to rotate about L when various weighted objects are placed upon the plane. The strength of this rotational tendency is measured by the *moment about L*.

We consider first the moment which is produced by a finite number of point-masses. For simplicity, we introduce coordinates in such a way that L is parallel to the y-axis, and thus has an equation of the form $x = x_0$ (see Fig. 7–50). If a particle p of mass m and negligible diameter (i.e., a "point-mass") is placed on the plane at the point (x, y), then the moment of the particle about the line is

$$M_{x=x_0}(p) = (x - x_0)m.$$

If the moment is positive, then a clockwise rotation about L will result; a negative moment will result in a counterclockwise rotation. Note that the absolute value of the moment is the product of the mass and the distance of p from the line. If, instead of having a single point-mass, we have a system $\mathcal{S}$ of n point-masses, then the moment about L can be found by adding together the moments produced by the individual particles. Thus, if the kth particle has mass m_k and coordinates (x_k, y_k), then

$$M_{x=x_0}(\mathcal{S}) = \sum_{k=1}^{n} (x_k - x_0)m_k.$$

The system will be in equilibrium (have no tendency to rotate about L) iff

$$M_{x=x_0}(\mathcal{S}) = 0,$$

or, solving for x_0,

$$x_0 = \frac{1}{M} \sum_{k=1}^{n} x_k m_k,$$

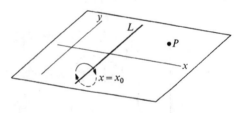

FIG. 7–50

where $M = m_1 + m_2 + \cdots + m_n$ is the total mass of the system. We denote this value of x_0 by $\bar{x}$.

Similarly, the moment of $\mathcal{S}$ about the line $y = y_0$ is

$$M_{y=y_0}(\mathcal{S}) = \sum_{k=1}^{n} (y_k - y_0)m_k.$$

The system will have no tendency to rotate about the line $y = y_0$ iff

$$y_0 = \bar{y} = \frac{1}{M} \sum_{k=1}^{n} y_k m_k.$$

The point $(\bar{x}, \bar{y})$ is called the *center of gravity* of the system. If supported at this one point, the plane will balance perfectly. We shall also show in the exercises that in many respects *the system* S *will behave as if its entire mass M were concentrated at its center of gravity.*

Throughout the remainder of the section we shall consider moments produced by continuous rather than discrete distributions of matter. As one might expect, integrals will take the place of sums. We shall consider thin plates ("plane laminas"), and for the sake of simplicity we shall suppose that they are homogeneous and have a common density. For mathematical purposes, then, we may regard such a plate as a bounded subset of the plane whose mass is equal to its area. We shall make the following assumptions about the moments of such plates. Let $x_0 \in \mathbf{R}$.

M1. Additivity. *If B is the union of two nonoverlapping plates B_1 and B_2, then*

$$M_{x=x_0}(B) = M_{x=x_0}(B_1) + M_{x=x_0}(B_2).$$

M2. Comparison Property. *If the plate B lies between the lines $x = a$ and $x = b$ (where $a < b$), then*

$$(a - x_0)A(B) \leq M_{x=x_0}(B) \leq (b - x_0)A(B).$$

(See Fig. 7–51.) The right-hand inequality says, in effect, that the moment of B does not exceed the moment which would result from the concentration of the entire mass of B along the line $x = b$.

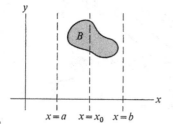

FIG. 7–51

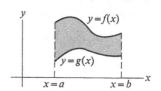

FIG. 7–52

We shall now calculate the moment about the y-axis of a plate B which is bounded by the graphs of two continuous functions f and g and the lines $x = a$ and $x = b$. We shall assume $f(x) \geq g(x)$ for all $x \in [a, b]$. This plate is pictured in Fig. 7–52. If $x_0 \in [a, b]$, we denote by $M(x)$ the moment about the y-axis of the set

$$\{(x, y) \mid a \leq x \leq x_0 \text{ and } g(x) \leq y \leq f(x)\}.$$

If $h > 0$, then the $M(x_0 + h) - M(x_0)$ is (by M1) the moment of that portion B_h of B which lies in the vertical strip determined by the lines $x = x_0$ and $x = x_0 + h$ (see Fig. 7–53). Thus, by the comparison property,

$$x_0 A(B_h) \leq M(x_0 + h) - M(x_0) = M_{x=0}(B_h)$$
$$\leq (x_0 + h)A(B_h).$$

We have shown previously that $\lim_{h\to 0+} A(B_h)/h = f(x_0) - g(x_0)$. Hence by dividing the above inequalities by h and letting $h \to 0+$, we get

$$\lim_{h\to 0+} \frac{M(x_0 + h) - M(x_0)}{h} = x_0[f(x_0) - g(x_0)].$$

Consideration of negative values of h leads to the equation

$$M'(x_0) = x_0[f(x_0) - g(x_0)].$$

Hence

$$M_{x=0}(B) = \int_a^b x[f(x) - g(x)]\, dx.$$

Proposition 1. *Suppose that f and g are continuous on $[a, b]$ and $f(x) \geq g(x)$ for all $x \in [a, b]$. If B is the region bounded by the graphs of f and g and by the lines $x = a$ and $x = b$, then*

$$M_{x=0}(B) = \int_a^b x[f(x) - g(x)]\, dx.$$

Corollary 2. *The moments of a rectangle $[a, b] \times [c, d]$ about the x- and y-axes are*

$$M_{x=0} = \tfrac{1}{2}(b^2 - a^2)(d - c), \qquad M_{y=0} = \tfrac{1}{2}(d^2 - c^2)(b - a).$$

(See Fig. 7–54.)

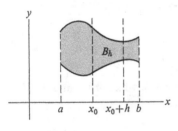

FIG. 7–53

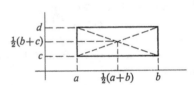

FIG. 7–54

Proof. By the theorem,

$$M_{x=0} = \int_a^b x(d - c)\, dx$$
$$= (d - c)\frac{x^2}{2}\Big|_a^b$$
$$= \tfrac{1}{2}(b^2 - a^2)(d - c).$$

We can obtain the second formula by interchanging the roles of x and y.

Note that the absolute value of $M_{x=0}$ is equal to the product of the area of the rectangle times the distance from the y-axis to the center of the rectangle (intersection point of the diagonals).

Next we shall calculate the moment of B about the x-axis. If $x_0 \in [a, b]$, we let $M(x_0)$ denote the moment about the x-axis of the set $\{(x, y) \mid a \leq x \leq x_0$ and

$g(x) \le y \le f(x)\}$. If $h > 0$, then $M(x_0 + h) - M(x_0) = M_{y=0}(B_h)$, where B_h is as above. We consider only the case in which $f(x_0) > g(x_0)$. Let $M_f(h)$ and $m_f(h)$ denote the maximum and minimum values of f on the interval $[x_0, x_0 + h]$; let $M_g(h)$ and $m_g(h)$ be similarly defined. If h is sufficiently small, then $M_g(h) < m_f(h)$. We break B_h into three regions I, II, III, as indicated in Fig. 7-55. Then

$$M_{y=0}(B_h) = M_{y=0}(\text{I}) + M_{y=0}(\text{II}) + M_{y=0}(\text{III}).$$

Since II is a rectangle, it follows that

$$M_{y=0}(\text{II}) = \tfrac{1}{2}(m_f + M_g)(m_f - M_g)h.$$

By the comparison property,

$$m_g(h)A(\text{I}) \le M_{y=0}(\text{I}) \le M_g(h)A(\text{I})$$

and

$$m_f(h)A(\text{III}) \le M_{y=0}(\text{III}) \le M_f(h)A(\text{III}).$$

Thus, the difference quotient

$$\frac{M(x_0 + h) - M(x_0)}{h} = \frac{M_{y=0}(B_h)}{h}$$

lies between

$$m_g(h)\frac{A(\text{I})}{h} + \tfrac{1}{2}(m_f^2 - M_g^2) + m_f(h)\frac{A(\text{III})}{h}$$

and

$$M_g(h)\frac{A(\text{I})}{h} + \tfrac{1}{2}(m_f^2 - M_g^2) + M_f(h)\frac{A(\text{III})}{h}.$$

As $h \to 0+$, the terms $m_g(h)$ and $M_g(h)$ tend to $g(x_0)$, while $m_f(h)$ and $M_f(h)$ tend to $f(x_0)$. Since

$$0 < \frac{A(\text{I})}{h} \le M_g(h) - m_g(h)$$

and the latter term tends to zero, we see that $A(\text{I})/h \to 0$. Similarly, $A(\text{III})/h \to 0$. By the usual pinching argument, we get

$$\lim_{h \to 0+} \frac{M(x_0 + h) - M(x_0)}{h} = \tfrac{1}{2}[f(x_0)^2 - g(x_0)^2],$$

which is only a hop from the final result:

$$M_{y=0}(B) = \tfrac{1}{2}\int_a^b [f(x)^2 - g(x)^2]\,dx.$$

Proposition 3. *Let B be the same as in Proposition 1. Then*

$$M_{y=0}(B) = \tfrac{1}{2}\int_a^b [f(x)^2 - g(x)^2]\,dx.$$

Example 1. Determine the moments of the semicircle $\{(x, y) \mid x^2 + y^2 \le a^2 \text{ and } y \ge 0\}$ about the x- and y-axes.

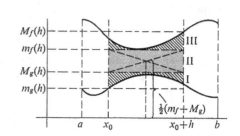

FIG. 7-55

The moment about the y-axis is

$$M_{x=0} = \int_{-a}^{a} x\sqrt{a^2 - x^2}\, dx.$$

The integral is zero, since the integrand is an odd function and the interval of integration $[-a, a]$ is symmetric about the origin. The result is what one would expect, since the set is symmetric about the y-axis. The moment about the x-axis is

$$M_{y=0} = \tfrac{1}{2}\int_{-a}^{a} (a^2 - x^2)\, dx = \tfrac{2}{3}a^3.$$

So far we have only considered the moment of a region about the x- and y-axes. Other moments can be calculated by use of the *translation formula:*

$$M_{x=x_0+h}(B) = M_{x=x_0}(B) - hA(B).$$

In proving this formula we shall abbreviate $M_{x=x_0}(B)$ to $M_{x_0}(B)$. Let $\epsilon > 0$ be arbitrary. We shall first assume that B lies in some vertical strip of width ϵ; to be specific, assume that B lies between the lines $x = a$ and $x = a + \epsilon$. By the comparison property, we then have

$$(a - (x_0 + h))A(B) \le M_{x_0+h}(B) \le (a + \epsilon - (x_0 + h))A(B)$$

and

$$(a - x_0)A(B) \le M_{x_0}(B) \le (a + \epsilon - x_0)A(B).$$

It follows that

$$-\epsilon A(B) \le M_{x_0+h}(B) - M_{x_0}(B) + hA(B) \le \epsilon A(B). \qquad (*)$$

Note that a does not appear in the latter inequality. Furthermore, this inequality is valid for any bounded set B, for in the general case B can be broken up into a finite number of nonoverlapping pieces $B_1, \ldots, B_n$, each lying in a vertical strip of width ϵ. The inequality will hold for each B_k, so addition of the inequalities and an appeal to the additivity of areas and moments will yield the inequality for the set B itself. Since, however, $\epsilon > 0$ is arbitrary, it follows from $(*)$ that

$$M_{x_0+h}(B) - M_{x_0}(B) + hA(B) = 0.$$

From the translation formula, it follows that if $A(B) > 0$, then there is exactly one number $\bar{x}$ for which

$$M_{x=\bar{x}}(B) = 0,$$

namely,

$$\bar{x} = \bar{x}(B) = x_0 + \frac{M_{x=x_0}(B)}{A(B)} = \frac{M_{x=0}(B)}{A(B)}.$$

Similarly,

$$\bar{y} = \bar{y}(B) = y_0 + \frac{M_{y=y_0}(B)}{A(B)} = \frac{M_{y=0}(B)}{A(B)}$$

is the unique number for which $M_{y=\bar{y}}(B) = 0$. As in the discrete case, the point $(\bar{x}, \bar{y})$ is called the *center of gravity* of B; if the plane is supported at this point, it will balance.

Example 2. Find the center of gravity of the semicircle described in Example 1. The area of the semicircle is $\frac{1}{2}\pi a^2$. Thus

$$\bar{x} = \frac{M_{x=0}}{\frac{1}{2}\pi a^2} = 0, \qquad \bar{y} = \frac{M_{y=0}}{\frac{1}{2}\pi a^2} = \frac{\frac{2}{3}a^3}{\frac{1}{2}\pi a^2} = \frac{4a}{3\pi}.$$

The reader will note that for the region B of Propositions 1 and 2 the coordinates of the center of gravity have the formulas

$$\bar{x} = \frac{\int_a^b x[f(x) - g(x)]\,dx}{\int_a^b [f(x) - g(x)]\,dx}, \qquad \bar{y} = \frac{\frac{1}{2}\int_a^b [f(x)^2 - g(x)^2]\,dx}{\int_a^b [f(x) - g(x)]\,dx}.$$

In certain respects, a plate B will behave as though its entire mass were concentrated at its center of gravity. In particular, if a plate is broken up into a finite number of nonoverlapping pieces, then the center of gravity of the plate can be found from the centers of gravity of the pieces, in accordance with this principle.

Proposition 4. *If a plate B is broken into nonoverlapping pieces $B_1, \ldots, B_n$, then*

$$\bar{x}(B) = \frac{1}{A(B)} \sum_{k=1}^{n} \bar{x}(B_k)A(B_k), \qquad \bar{y}(B) = \frac{1}{A(B)} \sum_{k=1}^{n} \bar{y}(B_k)A(B_k).$$

In other words, we can compute the center of gravity of B by replacing each piece B_k with a point-mass located at the center of gravity of B_k whose mass is that of B_k, that is, $A(B_k)$.

Proof

$$\bar{x}(B)A(B) = M_{x=0}(B) = \sum_{k=1}^{n} M_{x=0}(B_k) = \sum_{k=1}^{n} \bar{x}(B_k)A(B_k).$$

Example 3. Determine the center of gravity of the shaded region B shown in Fig. 7-56. Also determine the amount of force F which would have to be applied at the point $(-2a, 0)$ to keep the figure in equilibrium if it is supported along the line $x = -a$.

The center of gravity of the larger semicircle is $(0, 8a/3\pi)$; the center of gravity of the smaller one is $(a, 4a/3\pi)$. Thus by the theorem,

$$\frac{8a}{3\pi} \cdot 2\pi a^2 = \frac{4a}{3\pi} \cdot \frac{1}{2}\pi a^2 + \bar{y}(B)\frac{3}{2}\pi a^2,$$

$$0 = a \cdot \frac{1}{2}\pi a^2 + \bar{x}(B)\pi \frac{3}{2}a^2.$$

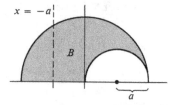

Therefore,

$$\bar{x}(B) = -\frac{a}{3}, \qquad \bar{y}(B) = \frac{28}{9\pi}a.$$

FIG. 7-56

By the translation formula,

$$M_{x=-a}(B) = M_{x=0}(B) + aA(B) = \bar{x}(B)A(B) + aA(B) = \pi a^3.$$

Thus, for the figure to balance, we must have $aF = \pi a^3$, or $F = \pi a^2$.

Exercises

1. Find the center of gravity of the lamina bounded by the curve $\sqrt{x} + \sqrt{y} = 1$ and the coordinate axes. Also find the moment of the lamina about the line $x = -1$.

2. a) Find the center of gravity of the lamina lying in the first quadrant which is bounded by the ellipse $(x/a)^2 + (y/b)^2 = 1$ and the coordinate axes.
   b) If $b < a$, find the center of gravity of the relative complement of the disk
   $$x^2 + y^2 \le a^2$$
   and the lamina in (a).

3. Find the moment about the x-axis of the lamina bounded by the curves $y = 2/(1 + x^2)$ and $y = x^2$.

4. Find the center of gravity of the lamina bounded by the closed curve $y^2 = ax^3 - x^4$.

5. Find the center of gravity of the lamina bounded by one arch of the cycloid
   $$x = a(t - \sin t), \qquad y = a(1 - \cos t),$$
   and the x-axis.

6. Find the center of gravity of the lamina lying in the first quadrant which is bounded by the astroid $x^{2/3} + y^{2/3} = a^{2/3}$ and the coordinate axes.

7. Suppose that
   $$\gamma: \quad x = x(t), \quad y = y(t) \qquad (a \le t \le b)$$
   is a simple loop whose points are traced out once in a counterclockwise direction. If B is the lamina bounded by the loop, prove that
   $$M_{x=0}(B) = \tfrac{1}{2}\int_a^b [x(t)]^2 y'(t)\, dt$$
   $$= -\int_a^b x(t)x'(t)y(t)\, dt.$$
   Similarly,
   $$M_{y=0}(B) = -\tfrac{1}{2}\int_a^b [y(t)]^2 x'(t)\, dt = \int_a^b y(t)y'(t)x(t)\, dt.$$

8. Using the preceding problem, show that the center of gravity of a triangle is the intersection of its medians.

9. Find the center of gravity of the lamina bounded by the loop
   $$x = a\sin 2t, \qquad y = a\sin t \qquad (0 \le t \le \pi).$$

10. Deduce the formulas for moments given in the text from the formulas given in Exercise 7.

11. Let f be continuous and positive throughout the interval $[\alpha, \beta]$, where $\beta - \alpha \le 2\pi$. Let B be the lamina bounded by the curve $r = f(\theta)$ and the rays $\theta = \alpha, \theta = \beta$. Prove that
    $$M_{y=0}(B) = \tfrac{1}{3}\int_\alpha^\beta [f(\theta)]^3 \cos\theta\, d\theta,$$
    $$M_{x=0}(B) = \tfrac{1}{3}\int_\alpha^\beta [f(\theta)]^3 \sin\theta\, d\theta.$$

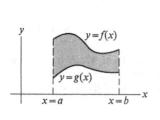

FIG. 7-57   **FIG. 7-58**

12. Find the center of gravity of the lamina bounded by the cardioid $r = a(1 + \cos \theta)$.

13. Find the center of gravity of the lamina bounded by the right-hand loop of the lemniscate of Bernoulli: $r^2 = a^2 \cos 2\theta$.

14. *Pappus' theorem* states that if a lamina is revolved about a line which lies to one side of the lamina, then the volume of the solid so generated is equal to the product of the area of the lamina and the distance traveled by the center of gravity of the lamina. Prove Pappus' theorem for the cases in which the lamina indicated in Fig. 7-57 is revolved about the two coordinate axes.

15. Use Pappus' theorem to find the volume of a torus.

16. Use Pappus' theorem to find the center of gravity of the lamina bounded by the semicircle $y = \sqrt{a^2 - x^2}$ and the x-axis.

17. Suppose that the lamina B is part of a vertical dam wall, where the water comes to the top of the dam (see Fig. 7-58). Prove that the force exerted by the water against B is equal to $wM_{y=0}(B)$, where w is the weight of water per unit volume and the coordinates are introduced in such a way that the dam wall is part of the upper xy-plane and the x-axis is located at the surface of the water.

7-9 Miscellaneous applications to physics

Application 1. *Escape Velocity.* Neglecting air resistance, we shall find the initial velocity required for a particle projected vertically upward from the earth's surface to "escape" the earth's gravitational field.

Consider a particle of mass m projected upward from the earth's surface with an initial velocity v_0. We denote by R the radius of the earth, by g the gravitational constant (the acceleration caused by gravity at the surface of the earth), and by s the distance of the particle from the center of the earth at time t. Physical intuition suggests that if v_0 is quite small, then the particle will move steadily upward for a while and will then fall back to earth. On the other hand, we would expect that if v_0 is sufficiently large, then the particle will move away from the earth indefinitely.

By Newton's law of gravitation, the force acting on the particle is $F = -mgR^2s^{-2}$. (Why the minus sign?) Newton's second law of motion tells us, however, that

$$F = ma = m\frac{d^2s}{dt^2} = m\frac{dv}{dt} = m\frac{dv}{ds}\cdot\frac{ds}{dt} = mv\frac{dv}{ds}.$$

(Are we justified in regarding s as the independent variable?) We therefore get the equation

$$mv\frac{dv}{ds} = -\frac{mgR^2}{s^2}.$$

We cancel the m's and take antiderivatives of each side, thus obtaining

$$\tfrac{1}{2}v^2 = \frac{gR^2}{s} + C.$$

When $s = R$, we have

$$\tfrac{1}{2}v_0^2 = gR + C,$$

so that $C = \tfrac{1}{2}v_0^2 - gR$. Hence

$$\left(\frac{ds}{dt}\right)^2 = v^2 = (v_0^2 - 2gR) + \frac{2gR^2}{s}.$$

If the particle falls back to earth, then at the moment when it is at its maximum height, $ds/dt = 0$. From our last displayed equation we see that this can occur only if $v_0^2 - 2gR < 0$, or equivalently, $v_0 < \sqrt{2gR}$. If, on the other hand, $v_0 = \sqrt{2gR}$, then

$$\sqrt{s}\,\frac{ds}{dt} = v_0\sqrt{R}.$$

From this we get

$$\tfrac{2}{3}(s^{2/3} - R^{2/3}) = \int_0^t \sqrt{s}\,\frac{ds}{dt}\,dt = \int_0^t v_0\sqrt{R}\,dt = v_0\sqrt{R}\,t.$$

Hence

$$s = R\left(1 + \frac{3v_0 t}{2R}\right)^{2/3}.$$

As $t \to +\infty$, $s \to +\infty$, that is, the particle moves away from the earth indefinitely.

Application 2. *Flow Through an Orifice.* Find the time required to empty a water-filled cylindrical tank through a hole punched in the bottom of the tank (see Fig. 7–59).

We introduce the following symbols:

H = height of tank,

h = height of water at time t,

A = cross-sectional area of the tank,

a = area of the hole.

To solve the problem, we must use *Torricelli's law.* According to this law, the velocity of the water leaving the tank is $k\sqrt{2gh}$, where k is a constant. The volume of water which leaves the tank during the time interval from t to $t + \Delta t$ is approximately

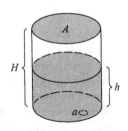

FIG. 7–59

$ak\sqrt{2gh}\,\Delta t$. On the other hand, the same volume is exactly equal to $-A\,\Delta h$, where $\Delta h = h(t + \Delta t) - h(t)$. Thus

$$ak\sqrt{2gh}\,\Delta t = -A\,\Delta h.$$

A pinching argument shows that

$$\frac{dh}{dt} = \lim_{\Delta t \to 0} \frac{\Delta h}{\Delta t} = -C\sqrt{h},$$

where $C = ak\sqrt{2g}/A$. If T is the time required to drain the tank, then

$$2\sqrt{H} = 2\sqrt{h(0)} - 2\sqrt{h(T)}$$

$$= \int_T^0 \frac{1}{\sqrt{h(t)}} \cdot \frac{dh}{dt}\, dt,$$

$$= \int_T^0 (-C)\, dt = CT.$$

Hence

$$T = \frac{2\sqrt{H}}{C} = \frac{A}{ka}\sqrt{\frac{2H}{g}}.$$

One often sees differentials used in solving such problems. In place of the discussion above, one might see the following:

The volume of water leaving the tank during the time interval from t to $t + dt$ is, on the one hand, $ak\sqrt{2gh}\,dt$, and on the other, $-A\,dh$. Thus

$$ak\sqrt{2gh}\,dt = -A\,dh \qquad \text{or} \qquad C\,dt = -\frac{dh}{\sqrt{h}},$$

where $C = ak\sqrt{2g}/A$. Let T be the time required to drain the tank. Then as t varies from 0 to T, h varies from H to 0. Hence

$$CT = \int_{t=0}^{t=T} C\,dt = -\int_{h=H}^{h=0} h^{-1/2}\, dh = -2h^{1/2}\Big|_H^0 = 2\sqrt{H}.$$

Therefore,

$$T = \frac{2\sqrt{H}}{C} = \frac{A}{ka}\sqrt{\frac{2H}{g}}.$$

Although this formalism obscures such matters as the passage to limits, it has its place. It often leads to the answer more quickly, though somewhat less safely. Furthermore, while one is writing down these admittedly meaningless equations, one can check mentally that the argument could be transformed into more respectable mathematics by the use of the pinching theorem, chain rule, etc. The prudery of valuing rigor above all else has this danger: it can cut off the dialogue between mathematics and the physical sciences, a dialogue which in the past has proved fruitful to both parties.

Application 3. *Elongation of a Wire.* A wire acted upon a by force F will elongate by the amount

$$\Delta l = \frac{Fl}{Ea},$$

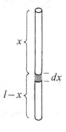

where

l = length of the wire,

a = cross-sectional area of the wire,

E = modulus of elasticity.

FIG. 7-60

Find the elongation of a wire produced by its own weight if the wire hangs vertically.

We shall use differentials in solving this problem. We consider a segment of the wire with length dx at a distance x from the top (see Fig. 7-60). The force acting on this infinitesimal segment is $(l - x)a$, where μ is the density (mass per unit volume) of the wire. (We shall suppose that μ is constant.) Consequently, this infinitesimal segment is elongated by the amount

$$\frac{(l - x)a\mu\, dx}{aE} = \frac{\mu}{E}(l - x)\, dx.$$

Hence for the entire wire the elongation is

$$\Delta l = \int_0^l \frac{\mu}{E}(l - x)\, dx = \frac{\mu l^2}{2E}.$$

The student is urged to derive the same result by a more careful argument.

Application 4. *Work Done in a Gravitational Force Field.* The problem here is to calculate the work required to move a particle of mass m from one point to another in the gravitational force field produced by a particle of mass M.

FIG. 7-61   0 $x(t)$

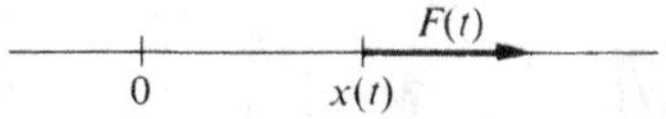

First, it is necessary to agree on the meaning of the term "work." We begin by considering the simple case of linear motion. Suppose a particle moves along a straight line onto which coordinates have been introduced; we denote by $x(t)$ the coordinate of the particle at time t. Suppose also that at time t a force $F(t)$ acts on the particle, the direction of the force being that of the line (see Fig. 7-61). [In absolute value, $F(t)$ is equal to the magnitude of the force; $F(t)$ is positive or negative according as it tends to produce motion in the positive or negative direction.] Then the work done on the particle during the time interval (t_1, t_2) is defined to be

$$W = \int_{t_1}^{t_2} F(t) \frac{dx}{dt}\, dt.$$

Observe that if $F(t)$ is a constant, say $F(t) = F$, then the work

$$W = \int_{t_1}^{t_2} F \frac{dx}{dt}\, dt = F[x(t_2) - x(t_1)]$$

is the product of the force and the change in displacement.

If, instead of moving along a straight line, the particle moves along a smooth path

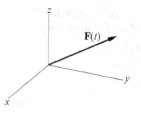

FIG. 7-62

$$\gamma: \quad x = x(t), \quad y = y(t), \quad z = x(t) \qquad (t_1 \leq t \leq t_2)$$

in three-space, the expression for the work done on the particle is more complicated. Let the force acting on the particle at time t be specified by the vector $F(t)$, whose length is the magnitude of force, whose direction is that of the force, and whose initial point is the origin (see Fig. 7-62). We then denote by $F_x(t)$, $F_y(t)$, and $F_z(t)$ the x-, y-, and z-coordinates, respectively, of the terminal point of $F(t)$. These numbers are called the *components* of $F(t)$. The work done on the particle is then defined to be

$$W = \int_{t_1}^{t_2} \left[F_x(t) \frac{dx}{dt} + F_y(t) \frac{dy}{dt} + F_z(t) \frac{dz}{dt} \right] dt.$$

It can be shown that W does not depend on the choice of the coordinate system.

We turn now to the problem stated. We first introduce coordinates so that the particle of mass M is located at the origin. Let

$$\gamma: \quad x = x(t), \quad y = y(t), \quad z = z(t) \qquad (t_1 \leq t \leq t_2)$$

be the path over which the particle P of mass m is moved. We denote by $r(t)$ the distance from P to O at time t. Thus

$$r(t) = \sqrt{x(t)^2 + y(t)^2 + z(t)^2}.$$

According to the law of gravitation, the force $F(t)$ acting on P at time t has magnitude $MmG/r(t)^2$, and is directed toward the origin. It follows that the components of $F(t)$ are

$$-\frac{MmGx(t)}{r(t)^3}, \qquad -\frac{MmGy(t)}{r(t)^3}, \qquad -\frac{MmGz(t)}{r(t)^3}.$$

Hence the work done on P by the field is

$$W = -MmG \int_{t_1}^{t_2} \frac{x(t)x'(t) + y(t)y'(t) + z(t)z'(t)}{[x(t)^2 + y(t)^2 + z(t)^2]^{3/2}}\, dt$$

$$= MmG[x(t)^2 + y(t)^2 + z(t)^2]^{-1/2}\Big|_{t_1}^{t_2}$$

$$= \frac{MmG}{r(t)}\Big|_{t_1}^{t_2} = MmG\left[\frac{1}{r(t_2)} - \frac{1}{r(t_1)}\right].$$

Since the work done by the force field depends only on the endpoints of γ, we say that the field is *conservative*.

Application 5. *Kinetic Energy Formula.* Let $\mathbf{F}(t)$ represent the sum of all the forces acting on a particle P of mass m at time t. Find the work done by the resultant force $\mathbf{F}(t)$ during the time interval (t_1, t_2).

According to Newton's second law of motion, the particle will travel along a path

$$\gamma: \quad x = x(t), \quad y = y(t), \quad z = z(t) \qquad (t_1 \leq t \leq t_2),$$

where

$$F_x(t) = m\frac{d^2x}{dt^2}, \qquad F_y(t) = m\frac{d^2y}{dt^2}, \qquad F_z(t) = m\frac{d^2z}{dt^2};$$

here F_x, F_y, and F_z are the components of $\mathbf{F}$. Hence the work done is

$$W = \int_{t_1}^{t_2} m[x''(t)x'(t) + y''(t)y'(t) + z''(t)z'(t)]\, dt$$

$$= \tfrac{1}{2}m\left[\left(\frac{dx}{dt}\right)^2 + \left(\frac{dy}{dt}\right)^2 + \left(\frac{dz}{dt}\right)^2\right]\Big|_{t_1}^{t_2}.$$

However,

$$\left(\frac{dx}{dt}\right)^2 + \left(\frac{dy}{dt}\right)^2 + \left(\frac{dz}{dt}\right)^2 = \left(\frac{ds}{dt}\right)^2,$$

where s denotes path length along γ. Thus

$$W = \tfrac{1}{2}mv_2^2 - \tfrac{1}{2}mv_1^2,$$

where v_1 and v_2 represent the speed of P at times t_1 and t_2. If v is the speed of P at time t, then

$$\text{KE} = \tfrac{1}{2}mv^2$$

is called the *kinetic energy* of the particle. We have shown, then, that *the work done by the resultant force on a particle is equal to its change in kinetic energy.*

Application 6. *Gravitational Force Produced by a Rod.* Find the gravitational force produced on a particle of mass m by a thin bar of length $2a$ if the particle is at a distance a from the bar and equidistant from the endpoints of the bar, as shown in Fig. 7–63.

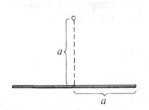

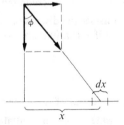

FIG. 7-63   **FIG. 7-64**

We introduce coordinates so that the coordinates of the ends of the bar are $(-a, 0)$ and $(a, 0)$, while the particle itself is located at the point $(0, a)$. We consider an infinitesimal segment of the bar with coordinate x and length dx (see Fig. 7–64).

The force produced by this segment has magnitude

$$\frac{Gm\mu \, dx}{a^2 + x^2},$$

where μ is the linear density of the rod. The components of this force are

$$\sin \phi \, \frac{Gm\mu \, dx}{a^2 + x^2} = \frac{Gm\mu x \, dx}{(a^2 + x^2)^{3/2}}, \qquad -\cos \phi \, \frac{Gm\mu \, dx}{a^2 + x^2} = -\frac{aGm\mu \, dx}{(a^2 + x^2)^{3/2}}.$$

The horizontal components cancel because of symmetry. Hence the force acting on the particle will be the "sum" of the vertical components, namely,

$$F = -\int_{-a}^{a} \frac{aGm\mu \, dx}{(a^2 + x^2)^{3/2}} = -aGm\mu \int_{-a}^{a} \frac{dx}{(a^2 + x^2)^{3/2}}.$$

We evaluate the latter integral by making the substitution $x = a \tan \theta$. We get

$$\int_{-a}^{a} \frac{dx}{(a^2 + x^2)^{3/2}} = \frac{1}{a^2} \int_{-\pi/4}^{\pi/4} \sin \theta \, d\theta = \frac{\sqrt{2}}{a^2}.$$

Hence

$$F = -\frac{\sqrt{2} \, Gm\mu}{a}.$$

Exercises

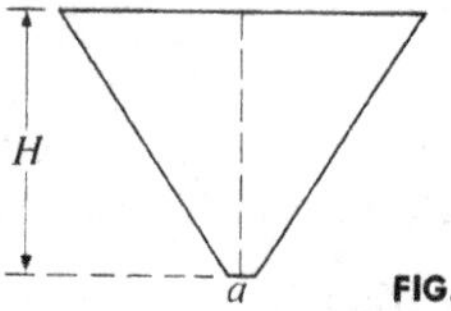

1. Solve the problem about draining a tank for the case in which the tank is an inverted cone, as shown in Fig. 7-65.

FIG. 7-65

2. Give a careful solution to the problem concerning the elongation of the wire.

3. The pyramid of Cheops has a square base with side length 200 m; its height is 140 m, and the density of the stone is about 2.5 g/cm³. Find the work done during its construction in overcoming the force of gravity.

4. Consider the discussion of work in Application 4 for the case of linear motion. Observe that if F is a function solely of displacement, say $F(t) = G(x(t))$, then

$$W = \int_{x(t_1)}^{x(t_2)} G(x) \, dx,$$

the integral of the force with respect to displacement.

5. A water drop with initial mass M falls under the force of gravity while, in the meantime, evaporating at the rate of m units of mass per second. What is the amount of work done by gravity from the initial moment until the drop has completely evaporated? (Neglect air resistance.)

6. Air at atmospheric pressure is contained in a cylinder with initial volume $V_0 = 0.1\ m^3$. A piston at one end of the cylinder is driven rapidly so as to compress the air to a volume $V = 0.03\ m^3$. Assuming that there is no transmission of heat in the process, find the work done by the piston. (Such a process is called *adiabatic*, and in this case the pressure and volume of the gas are related by $PV^\gamma = P_0 V_0^\gamma$, where γ is a constant. In the case of air, $\gamma = 1.40$, and the atmospheric pressure is $1.033\ \text{kg/cm}^2$.)

7. a) A particle of mass m is twirled about on the end of a weightless string of length r at an angular velocity of ω rad/sec (see Fig. 7–66). Hence by a suitable choice of coordinates, the equations of motion are

$$x = r \cos \omega t, \qquad y = r \sin \omega t.$$

FIG. 7–66

   Apply Newton's second law to show that the force $\mathbf{F}$ required to maintain this motion is directed toward the center of the circle and has magnitude $mr\omega^2$.

   b) Suppose that a rod of length r rotates about one of its ends with an angular velocity ω rad/sec. Given that the linear density of the rod is γ (mass per unit length), find the force acting on the fixed end.

8. At normal temperatures the resistance R of a metallic conductor is directly proportional to the temperature. For most pure metals we have

$$R = R_0(1 + 0.004\theta),$$

where R_0 is the resistance at $0°C$ and θ is the temperature in degrees centigrade. If $R_0 = 10$ ohms and the conductor is heated at a uniform rate from $20°C$ to $200°C$ in the course of 10 min, find the amount of electricity (in coulombs) which flows through the circuit if a voltage of $120V$ is maintained. [Use *Ohm's law*, $E = IR$. Here E is the electromotive force (in volts), R is the resistance (in ohms), and I is the current (in amperes = coul/sec).]

9. A rod of length l and mass M attracts a particle of mass m which lies along the continuation of the rod at a distance s from one of the ends. Find the force of attraction between the two objects.

10. a) A wire ring has mass M and radius r. Find the force with which it attracts a particle of mass m located on the line through the center of the ring and perpendicular to its plane.

    b) Calculate the work required to move the particle from one point to another along the line.

    c) Find the force of attraction exerted by a disk on a particle located along the line perpendicular to the disk and passing through its center.

11. An infinite straight wire conductor running along the z-axis carries a steady current. As a result, it produces an electromagnetic field. Assume that the force $\mathbf{F}$ which it produces on a magnetic pole located at the point (x, y, z) has components

$$F_x = -\frac{y}{x^2 + y^2}, \qquad F_y = \frac{x}{x^2 + y^2}, \qquad F_z = 0.$$

Find the work required to move the pole from the point $(1, 0, 0)$ to the point $(0, 1, 1)$ along each of the following paths:

$$\gamma_1: \quad x = 1 - t, \quad y = t, \quad z = t \qquad\qquad (0 \leq t \leq 1),$$
$$\gamma_2: \quad x = \cos t, \quad y = \sin t, \quad z = 2t/5\pi \qquad (0 \leq t \leq 5\pi/2),$$
$$\gamma_3: \quad x = \cos t, \quad y = \sin t, \quad z = \sin t \qquad (0 \leq t \leq 8\pi),$$

$$\gamma_4: \quad x = \begin{cases} \cos (7\pi t/2) & (0 \leq t \leq 1), \\ 0 & (1 \leq t \leq 2), \end{cases}$$

$$y = \begin{cases} -\sin (7\pi t/2) & (0 \leq t \leq 1), \\ 1 & (1 \leq t \leq 2), \end{cases}$$

$$z = \begin{cases} 0 & (0 \leq t \leq 1), \\ t - 1 & (1 \leq t \leq 2). \end{cases}$$

Describe each of the paths. Is the force field conservative? If γ is *any* path having $(1, 0, 0)$ and $(0, 1, 1)$ as initial and terminal points, respectively, what do you suppose can be said regarding the work required to move the pole along γ?

*8 □ THE RIEMANN INTEGRAL

An important theoretical question raised, but not settled, in the preceding chapter is this: Does every continuous function have an antiderivative? We shall give an affirmative answer to this question and, in the process, we shall develop an entirely different approach to the general subject of integration. This approach bears the name of its inventor—Riemann. For the past century the Riemann integral has performed a yeoman's service, although the Bourbaki school is no doubt right in maintaining that its importance has been exaggerated. (The Bourbaki approach will be taken up in the chapter on infinite series.) The aspersions cast upon the lowly Riemann integral need not be taken too seriously, however. For while the Riemann theory is admittedly not so elegant as, say, that of Lebesgue, the Riemann integral is more than adequate for the purposes of elementary analysis, and it serves as a good stepping stone toward the study of the more sophisticated Lebesgue integral. It is an excellent testing ground for anyone with an interest in analysis.

8–1 Definite integrals and Riemann integrability

Throughout this section we shall confine our attention to real-valued functions which are defined and bounded on some fixed closed bounded interval $[a, b]$. We denote the class of all such functions by $\mathcal{B}([a, b])$, or, when there is no danger of ambiguity, by simply $\mathcal{B}$. It is easily checked that $\mathcal{B}$ is closed under addition, subtraction, multiplication, and lattice operations. (The latter will be studied in the third section of this chapter.)

We begin with an abstract definition of a definite integral.

Definition 1. By a **definite integral** on $[a, b]$ we shall mean a real-valued function I with the following properties:

1) The domain of I is a subset of $\mathcal{B}$ which is closed under addition and multiplication by constants.

2) I is linear. That is, if $f, g \in \mathcal{D}_I$ and c is a constant, then

$$I(f + g) = I(f) + I(g), \qquad I(cf) = cI(f).$$

3) I is positive. That is, if $f \in \mathcal{D}_I$ and if f is nonnegative on $[a, b]$, then

$$I(f) \geq 0.$$

4) If J is any subinterval of $[a, b]$ with endpoints c and d $(c \leq d)$, then $\varphi_J \in \mathcal{D}_I$ and

$$I(\varphi_J) = d - c.$$

[Recall that $\varphi_J(x)$ is one or zero according as $x \in J$ or $x \notin J$.]

Example 1. Let $\mathcal{D}_I$ consist of all functions in $\mathcal{B}$ which have antiderivatives on $[a, b]$. For any such function f, let $I(f)$ be the Newton integral $\int_a^b f(x)\, dx$. Then by the properties previously established for Newton integrals, I is a definite integral in the abstract sense just defined.

Example 2. One can obtain the very simplest definite integral I_0 from the Newton integral by restricting its domain. The domain of I_0 consists of functions f of the form

$$f = \sum_{k=1}^{n} a_k \varphi_{J_k},$$

where $a_1, \ldots, a_n$ are real constants and $J_1, \ldots, J_n$ are subintervals of $[a, b]$. Such functions are called *step functions* for the good reason that their graphs resemble series of steps. For, note that if the endpoints of $J_1, \ldots, J_n$ are included among the numbers

$$a = x_0 < x_1 < \cdots < x_r = b,$$

then f is constant on each of the open intervals $(x_0, x_1), (x_1, x_2), \ldots, (x_{r-1}, x_r)$, as shown in Fig. 8-1.

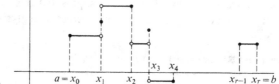

FIG. 8-1

The class $\mathcal{S} = \mathcal{S}([a, b])$ of all step functions on $[a, b]$ is clearly closed under addition and multiplication by constants. For each $f \in \mathcal{S}$, we define $I_0(f)$ to be, just as in Example 1, the Newton integral $\int_a^b f(x)\, dx$. Because of the linearity of the Newton integral, we find that

$$I_0(f) = \int_a^b f(x)\, dx = \sum_{k=1}^{n} a_k \int_a^b \varphi_{J_k}(x)\, dx = \sum_{k=1}^{n} a_k l(J_k),$$

where $l(J_k)$ denotes the length of the interval J_k. Since the Newton integral is linear and positive, I_0 has these properties *a fortiori*. Thus I_0 is a definite integral.

Proposition 2. *Let I be any definite integral. Then I is an extension of the integral I_0 of Example 2 [that is, $\mathcal{S} = \mathcal{D}_{I_0} \subset \mathcal{D}_I$ and for each $f \in \mathcal{S}$, we note that $I(f) = I_0(f)$]. Also, I is monotone, and when both f and $|f|$ belong to $\mathcal{D}_I$, we have*

$$|I(f)| \leq I(|f|).$$

Proof. The fact that $\mathfrak{D}_I$ includes the step functions follows from induction on properties (1) and (4). If f is a step function, say,

$$f = \sum_{k=1}^{n} a_k \varphi_{J_k},$$

then it also follows by induction on (2) and (4) that

$$I(f) = \sum_{k=1}^{n} I(a_k \varphi_{J_k}) = \sum_{k=1}^{n} a_k I(\varphi_{J_k}) = \sum_{k=1}^{n} a_k l(J_k),$$

where $l(J_k)$ denotes the length of the interval J_k.

The proofs of the remaining properties of I are exactly the same as those for the Newton integral.

Let I continue to denote an abstract definite integral on $[a, b]$. We shall attempt to get estimates for the number $I(f)$ if f is an arbitrary member of $\mathfrak{D}_I$. If f is a step function, then we know that there can be no question as to the value of $I(f)$; it is simply $\int_a^b f(x)\,dx$. If f is not a step function, we can still get some upper and lower estimates on $I(f)$. Suppose g is a step function such that $g \geq f$ on $[a, b]$. (We mean by this that $g(x) \geq f(x)$ for all $x \in [a, b]$.) Such a step function exists, since f is bounded on $[a, b]$. Then by the monotonicity of I, it follows that

$$I(f) \leq I(g) = \int_a^b g.$$

Thus $I(f)$ is a lower bound of the set $\{\int_a^b g \mid g \in \mathcal{S} \text{ and } g \geq f\}$. If we let $\overline{I}(f)$ denote the greatest lower bound of this set, then we have $I(f) \leq \overline{I}(f)$. Similarly,

$$I(f) \geq \underline{I}(f) = \mathrm{lub}\left\{\int_a^b g \mid g \in \mathcal{S} \text{ and } g \leq f\right\}.$$

Definition 3. Let f be any bounded function on $[a, b]$. Then the numbers

$$\overline{I}(f) = \mathrm{glb}\left\{\int_a^b g \mid g \in \mathcal{S} \text{ and } f \leq g\right\},$$

$$\underline{I}(f) = \mathrm{lub}\left\{\int_a^b g \mid g \in \mathcal{S} \text{ and } g \leq f\right\}$$

are called the **upper** and **lower Riemann integrals of** f **on** $[a, b]$.

Using these definitions, we can now summarize the preceding discussion.

Proposition 4. *If I is any definite integral on $[a, b]$ and $f \in \mathfrak{D}_I$, then*

$$\underline{I}(f) \leq I(f) \leq \overline{I}(f).$$

Neither $\underline{I}$ nor $\overline{I}$ is a definite integral. How close each comes to being an integral can be judged by the next result.

Proposition 5. *Properties of Upper and Lower Integrals. Let f and g be bounded functions on [a, b], and let c ∈ R. The following hold:*

1) $\underline{L}(f) \le \overline{I}(f)$ *and equality holds if f is a step function.*
2) $\underline{L}$ *and* $\overline{I}$ *are both positive and monotone.*
3) $\overline{I}(f + g) \le \overline{I}(f) + \overline{I}(g), \underline{L}(f + g) \ge \underline{L}(f) + \underline{L}(g).$
4) $\overline{I}(cf) = c\overline{I}(f)$ *and* $\underline{L}(cf) = c\underline{L}(f)$ *if* $c > 0$, *while*
   $\overline{I}(cf) = c\underline{L}(f)$ *and* $\underline{L}(cf) = c\overline{I}(f)$ *if* $c < 0$.

Proof of (1). Because I_0 is monotone, each member of $S = \{\int_a^b g \mid g \in \mathsf{S}$ and $g \le f\}$ is a lower bound of $T = \{\int_a^b g \mid g \in \mathsf{S}$ and $f \le g\}$. Thus for each $g \in \mathsf{S}$ such that $g \le f$, we have

$$\int_a^b g \le \text{glb } T = \overline{I}(f).$$

Since $\overline{I}(f)$ is therefore an upper bound of S, it follows that

$$\overline{I}(f) \ge \text{lub } S = \underline{L}(f).$$

If $f \in \mathsf{S}$, then, because of the monotonicity of I_0, $\int_a^b f$ will be the smallest member of T and the largest member of S, so that $\overline{I}(f) = \underline{L}(f) = \int_a^b f$.

Proof of (2). Suppose $f \le g$. Then $S = \{\int_a^b h \mid h \in \mathsf{S}$ and $h \le f\}$ is a subset of $T = \{\int_a^b h \mid h \in \mathsf{S}$ and $h \le g\}$, and thus

$$\underline{L}(f) = \text{lub } S \le \text{lub } T = \underline{L}(g).$$

Similarly, $\overline{I}(f) \le \overline{I}(g)$.

Proof of (3). Let ϵ be any positive number. Then there exist step functions f^* and g^* such that

$$f \le f^* \quad \text{and} \quad \int_a^b f^* < \overline{I}(f) + \epsilon/2,$$

$$g \le g^* \quad \text{and} \quad \int_a^b g^* < \overline{I}(g) + \epsilon/2.$$

It follows that $f + g \le f^* + g^*$ and

$$\overline{I}(f + g) \le \int_a^b (f^* + g^*) = \int_a^b f^* + \int_a^b g^* \le \overline{I}(f) + \overline{I}(g) + \epsilon.$$

Since $\epsilon > 0$ is arbitrary, we have $\overline{I}(f + g) \le \overline{I}(f) + \overline{I}(g)$.
    The proof that $\underline{L}(f + g) \ge \underline{L}(f) + \underline{L}(g)$ is similar.

    The proof of (4) is left to the reader.

    Of special interest are those functions f for which $\underline{L}(f) = \overline{I}(f)$, for by Proposition 4, the values of $I(f)$ must then be the same for *all* definite integrals I whose domains include f.

Definition 6. If $L(f) = \overline{I}(f)$, we say that the function f is **Riemann-integrable** on $[a, b]$, and write $f \in \mathfrak{R}$, [or, more explicitly, $\mathfrak{R}([a, b])$]. We define I_R to be the function with domain $\mathfrak{R}$ which assigns to each $f \in \mathfrak{R}$ the number

$$I_R(f) = L(f) = \overline{I}(f).$$

Theorem 7. I_R *is an integral on* $[a, b]$.

Proof. Let $f, g \in \mathfrak{R}$. Then

$$L(f + g) \geq L(f) + L(g) = \overline{I}(f) + \overline{I}(g) \geq \overline{I}(f + g).$$

But since $L(f + g) \leq \overline{I}(f + g)$, only equalities can hold in the above display. It follows that $(f + g) \in \mathfrak{R}$ and that

$$I_R(f + g) = I_R(f) + I_R(g).$$

We shall leave to the reader the proof that $(cf) \in \mathfrak{R}$ and that

$$I_R(cf) = cI_R(f),$$

where c is a constant.

The positivity of I_R follows from that of L and $\overline{I}$. Since every step function is Riemann-integrable, property (4) also holds. Hence we have verified that I_R is an integral.

Definition 8. I_R is called the **Riemann integral** on $[a, b]$.

We have then two definite integrals, the Newton integral and the Riemann integral. It is natural to ask whether these two integrals are in fact equal, and, if not, which of the two is to be preferred. The answer to the first question is negative. The domains of definition of the two integrals are different, and neither domain is a subset of the other. Nevertheless, the two integrals do agree on the intersection of their domains.

Theorem 9. Second Form of the Fundamental Theorem of Calculus. *If I is a definite integral on* $[a, b]$, *and if* $f \in \mathfrak{D}_I \cap \mathfrak{R}$, *then* $I_R(f) = I(f)$. *In particular, if f is both Riemann- and Newton-integrable, then*

$$I_R(f) = \int_a^b f(x)\, dx.$$

Proof. By Proposition 4, $L(f) \leq I(f) \leq \overline{I}(f)$. Since $L(f) = \overline{I}(f) = I_R(f)$, we have the result.

Henceforth, if f is Riemann-integrable on $[a, b]$, we shall denote the Riemann integral of f by $\int_a^b f(x)\, dx$. (Because of the result just proved, there is no conflict with our previous use of the notation for the Newton integral.)

How large is the class of Riemann-integrable functions? All we know at present is that $\mathcal{S} \subset \mathfrak{R} \subset \mathfrak{B}$. We shall show that both inclusions are proper and that $\mathfrak{R}$ contains most functions of interest in analysis.

Proposition 10. *Let $f \in \mathcal{B}$. The following are equivalent:*

1) $f \in \mathcal{R}$.

2) Riemann Condition. For each $\epsilon > 0$, there exist step functions g and h such that $g \le f \le h$ on $[a, b]$ and

$$\int_a^b (h - g) < \epsilon.$$

3) There exist sequences of step functions $\{g_n\}$ and $\{h_n\}$ such that for each $n \in \mathbf{Z}^+$ we have $g_n \le f \le h_n$ on $[a, b]$ and

$$\lim_{n \to \infty} \int_a^b g_n = \lim_{n \to \infty} \int_a^b h_n,$$

in which case the common limit is $\int_a^b f$.

This is a trivial exercise in handling lub's and glb's.

Theorem 11. *If f is continuous on $[a, b]$, then f is Riemann-integrable on $[a, b]$.*

Proof. Let $\epsilon > 0$ be given. Since f is uniformly continuous on $[a, b]$, there exists a $\delta > 0$ such that for all $x, y \in [a, b]$,

$$|x - y| < \delta \Rightarrow |f(x) - f(y)| < \epsilon/(b - a).$$

We next partition the interval $[a, b]$ into subintervals of length less than δ by means of the points of subdivision

$$a = x_0 < x_1 < x_2 < \cdots < x_n = b.$$

(Thus, $x_k - x_{k-1}$, the length of the kth subinterval, is less than δ for $k = 1, \ldots, n$.) If we set

$$m_k = \min \{f(x) \mid x_{k-1} \le x \le x_k\}$$

and

$$M_k = \max \{f(x) \mid x_{k-1} \le x \le x_k\},$$

then it will follow that $M_k - m_k < \epsilon/(b - a)$ for $k = 1, \ldots, n$. We then set

$$g = \sum_{k=1}^{n-1} m_k \varphi_{[x_{k-1}, x_k)} + m_n \varphi_{[x_{n-1}, x_n]}$$

and

$$h = \sum_{k=1}^{n-1} M_k \varphi_{[x_{k-1}, x_k)} + M_n \varphi_{[x_{n-1}, x_n]}.$$

Then g and h are step functions, $g \le f \le h$ on $[a, b]$, and

$$\int_a^b (h - g) = \sum_{k=1}^n (M_k - m_k)(x_k - x_{k-1}) < \sum_{k=1}^n \frac{\epsilon}{b - a} (x_k - x_{k-1}) = \epsilon$$

(see Fig. 8–2). Hence by the Riemann condition, f is Riemann-integrable on $[a, b]$.

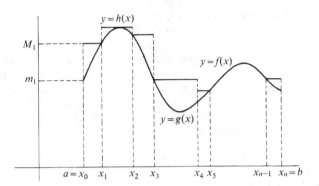

FIG. 8-2

This theorem proves that the inclusion $S \subset \mathfrak{R}$ is proper. It is also the first step toward proving that continuous functions have antiderivatives. The fact that the inclusion $\mathfrak{R} \subset \mathfrak{B}$ is also proper follows from the next example.

Example 3. Determine the upper and lower Riemann integrals of the function

$$f(x) = \begin{cases} 1 & \text{if } x \text{ is rational,} \\ 0 & \text{if } x \text{ is irrational} \end{cases}$$

on the interval $[0, 1]$.

Let g be any step function such that $g \geq f$ on the interval. We can then partition the interval $[0, 1]$ by means of the points of division

$$0 = x_0 < x_1 < x_2 < \cdots < x_n = 1$$

in such a way that g is constant on each of the open intervals (x_0, x_1), (x_1, x_2), $\ldots$, (x_{n-1}, x_n). Since each of these intervals contains rational numbers, it follows that $g \geq g^*$ on $[0, 1]$, where g^* is the step function $\sum_{k=1}^{n} \varphi_{(x_{k-1}, x_k)}$. Thus

$$\int_0^1 g \geq \int_0^1 g^* = \sum_{k=1}^{n} (x_k - x_{k-1}) = 1.$$

Hence

$$\bar{I}(f) = \text{glb} \left\{ \int_0^1 g \mid g \in S \text{ and } g \geq f \right\} \geq 1.$$

On the other hand, the function $g(x) = 1$ is a step function, $g \geq f$ on $[0, 1]$, and $\int_0^1 g = 1$. Hence $\bar{I}(f) = 1$.

Similarly, one can argue (from the fact that the irrational numbers are dense) that $\underline{I}(f) = 0$. It follows that the function f is not Riemann-integrable on $[0, 1]$.

More can be learned from this example. If we set $g(x) = 1 - f(x)$, then the same sort of argument as above will show that $\bar{I}(g) = 1$ and $\underline{I}(g) = 0$. Since

$$f(x) + g(x) = 1,$$

we get

$$1 = \bar{I}(f + g) < \bar{I}(f) + \bar{I}(g) = 2 \quad \text{and} \quad 1 = \underline{I}(f + g) > \underline{I}(f) + \underline{I}(g) = 0.$$

This proves the earlier assertion that $\bar{I}$ and $\underline{I}$ are not definite integrals.

Example 4. If $a > 1$, show that $\lim_{n\to\infty} n(\sqrt[n]{a} - 1)$ exists and is equal to $\int_1^a x^{-1}\,dx$.

The proof of the assertion above follows rather naturally from a particular attack on the problem of calculating the Riemann integral $\int_1^a x^{-1}\,dx$. (Since the integrand x^{-1} is continuous on the interval $[1, a]$, we know that the integral exists.) Since we have no function in our present repertoire whose derivative is x^{-1}, we cannot calculate the integral via the second fundamental theorem of calculus. We can calculate by this theorem, however, the integral $\int_1^a x^{\mathfrak{p}}\,dx$ for every rational number $\mathfrak{p}$ *except* -1, and we might hope to be able to obtain $\int_1^a x^{-1}\,dx$ as a limit of such integrals as $\mathfrak{p} \to -1$. If we take $\mathfrak{p}$ to be of the form $-1 + 1/n$, where $n \in \mathbf{Z}^+$, then we get

$$\int_1^a x^{-1+1/n}\,dx = nx^{1/n}\Big|_1^a = n(\sqrt[n]{a} - 1).$$

Hence it suffices to show that

$$n(\sqrt[n]{a} - 1) - \int_1^a x^{-1}\,dx = \int_1^a x^{-1}(\sqrt[n]{x} - 1)\,dx$$

tends to zero as $n \to \infty$. If $x \in [1, a]$, then $0 \le x^{-1} \le 1$ and $0 \le \sqrt[n]{x} - 1 \le \sqrt[n]{a} - 1$, so that

$$0 \le x^{-1}(\sqrt[n]{x} - 1) \le \sqrt[n]{a} - 1.$$

By the monotone property of definite integrals, we get

$$0 \le \int_1^a x^{-1}(\sqrt[n]{x} - 1)\,dx$$
$$\le \int_1^a (\sqrt[n]{a} - 1)\,dx$$
$$= (a - 1)(\sqrt[n]{a} - 1).$$

Since $\lim_{n\to\infty} \sqrt[n]{a} = 1$, it follows that the expression $(a - 1)(\sqrt[n]{a} - 1)$ tends to zero as $n \to \infty$. By the pinching theorem, the expression

$$n(\sqrt[n]{a} - 1) - \int_1^a x^{-1}\,dx$$

must also tend to zero.

Exercises

1. In this problem we shall take $[a, b]$ to be the interval $[-8, 7]$.

   a) Let f and f^* be step functions defined as follows:
   $$f = 3\varphi_{[-5,3)} + 2\varphi_{[1,5]} - \varphi_{(0,4)},$$
   $$f^* = 3\varphi_{[-5,1)} - \varphi_{(0,5]} + 5\varphi_{[1,4)} - 3\varphi_{[3,5]} + 6\varphi_{[4,5]}.$$
   Show that $f = f^*$, draw the graph of the function, and calculate $\int_{-8}^7 f$ using each of the representations of f as a step function.

   b) Let $g = 4\varphi_{[0,0]} - \varphi_{[-2,3]} + \frac{3}{2}\varphi_{(3,6]}$. Determine numbers
   $$-8 = x_0 < x_1 < x_2 < \cdots < x_n = 7$$
   such that the functions $f + g$ and fg are constant on each of the open intervals $(x_0, x_1), (x_1, x_2), \ldots, (x_{n-1}, x_n)$. Find $\int_{-8}^7 (f+g)$ and $\int_{-8}^7 fg$.

2. Prove that the family $\mathcal{S}$ of all step functions on $[a, b]$ is closed under multiplication.

3. What can you say about a step function f on $[a, b]$ such that $\int_a^b |f| = 0$?

4. If f is discontinuous at a point x_0 but the limits $\lim_{x \to x_0+} f(x)$ and $\lim_{x \to x_0-} f(x)$ nevertheless exist, we say that f has a *jump discontinuity at* x_0. Observe that a step function can have only jump discontinuities.

   a) Let $f \in \mathcal{B}([a, b])$. If f has a jump discontinuity at x_0, show that $f - g$ is continuous at x_0 if g is the step function

   $$g = (B - A)\varphi_{[a, x_0)} + [B - f(x_0)]\varphi_{[x_0, x_0]},$$

   where $A = \lim_{x \to x_0-} f(x)$ and $B = \lim_{x \to x_0+} f(x)$. Interpret this graphically.

   b) Suppose that $f \in \mathcal{B}$ and that f is continuous on $[a, b]$ except for a finite number of jump discontinuities. Prove that f can be written as $f^* + g$, where f^* is continuous on $[a, b]$ and g is a step function. Infer that f is Riemann-integrable on $[a, b]$.

5. a) Prove that if f is nondecreasing on $[a, b]$, then f is Riemann-integrable on $[a, b]$. [Divide the interval $[a, b]$ into n subintervals of equal length. The endpoints of these subintervals will be $x_0, x_1, \ldots, x_n$, where $x_k = a + k(b - a)/n$ ($k = 0, \ldots, n$). Let

   $$g = \sum_{k=1}^{n-1} f(x_{k-1})\varphi_{[x_{k-1}, x_k)} + f(x_{n-1})\varphi_{[x_{n-1}, x_n]},$$

   $$h = \sum_{k=1}^{n-1} f(x_k)\varphi_{[x_{k-1}, x_k)} + f(x_n)\varphi_{[x_{n-1}, x_n]}.$$

   Show that $g \le f \le h$ on $[a, b]$ and that

   $$\int_a^b (h - g) = \frac{b - a}{n}[f(b) - f(a)].$$

   Then use the Riemann condition.]

   b) Deduce from (a) that if f is nonincreasing on $[a, b]$, then f is Riemann-integrable on $[a, b]$.

6. a) If $f \in \mathcal{B}$, show that

   $$\bar{I}(f) = \text{glb} \left\{ \int_a^b g \mid g \in \mathcal{R} \text{ and } g \ge f \text{ on } [a, b] \right\},$$

   $$\underline{I}(f) = \text{lub} \left\{ \int_a^b g \mid g \in \mathcal{R} \text{ and } g \le f \text{ on } [a, b] \right\}.$$

   b) If $f \in \mathcal{R}$ and $g \in \mathcal{B}$, show that

   $$\bar{I}(f + g) = \int_a^b f + \bar{I}(g),$$

   $$\underline{I}(f + g) = \int_a^b f + \underline{I}(g).$$

7. a) Prove that if f is Riemann-integrable on $[a, b]$, then f is Riemann-integrable on every closed bounded subinterval of $[a, b]$.

   b) Prove that if f is bounded on $[a, b]$ and Riemann-integrable on every subinterval $[c, d]$, where $a < c < d < b$, then f is Riemann-integrable on $[a, b]$. In particular, it follows that if f is bounded on $[a, b]$ and continuous on the open interval (a, b), then f is Riemann-integrable on $[a, b]$.

8. Prove that

$$\bar{I}(cf) = \begin{cases} c\bar{I}(f) & \text{if } c \geq 0, \\ c\underline{I}(f) & \text{if } c < 0, \end{cases}$$

where $f \in \mathcal{B}$ and c is a constant. Deduce from this an analogous result for $\underline{I}(cf)$.

9. Given that f and g are bounded real-valued functions on $[a, b]$, we have used $f \geq g$ to mean that $f(x) \geq g(x)$ for all $x \in [a, b]$. Show that the relation "$\geq$" has the following properties:

a) *Transitivity*. If $f \geq g$ and $g \geq h$, then $f \geq h$.
b) *Antireflexivity*. If $f \geq g$ and $g \geq f$, then $f = g$.

Such a relation is called a *partial ordering*. Show, however, that it is possible to have functions $f, g \in \mathcal{B}$ such that neither $f \geq g$ nor $g \geq f$ is true.

8-2 The Riemann integral as a limit of sums

As in the preceding section we shall work over a fixed closed bounded interval $[a, b]$. If $f \in \mathcal{B}$, then we can associate with f certain step functions, as follows. First we partition $[a, b]$ into subintervals by selecting certain points of division, say,

$$a = x_0 < x_1 < x_2 < \cdots < x_n = b.$$

From each of the subintervals thereby determined, namely, $[x_0, x_1]$, $[x_1, x_2]$, $\ldots$, $[x_{n-1}, x_n]$, we select a point at random; let these points be $\xi_1, \xi_2, \ldots, \xi_n$, respectively. We then form a step function f^* whose value on the kth open interval (x_{k-1}, x_k) is $f(\xi_k)$, for $k = 1, \ldots, n$. The values of f^* at $x_0, x_1, \ldots, x_n$ can be anything at all. Then

$$\int_a^b f^* = \sum_{k=1}^n f(\xi_k)(x_k - x_{k-1})$$

is called a *Riemann sum*. We shall prove that if f is Riemann-integrable, then such a sum will be a good approximation to $\int_a^b f$, provided our partition of $[a, b]$ is "sufficiently fine."

Definition 1. By a **partition** of $[a, b]$ we shall mean any $(n + 1)$-tuple

$$\pi = (x_0, x_1, \ldots, x_n)$$

such that

$$a = x_0 < x_1 < \cdots < x_n = b.$$

We define the **norm** of π to be

$$\|\pi\| = \max \{(x_k - x_{k-1}) \mid k = 1, \ldots, n\},$$

the length of the longest subinterval determined by π. By a **selector** for π we shall mean an n-tuple $\sigma = (\xi_1, \ldots, \xi_n)$ such that

$$x_{k-1} \leq \xi_k \leq x_k,$$

for $k = 1, \ldots, n$.

The norm of π will serve as a measure of its "fineness."

If $\pi = (x_0, x_1, \ldots, x_n)$ is any partition of $[a, b]$, and if $\sigma = (\xi_1, \ldots, \xi_n)$ is any selector for π, then we can define a function $I_{\pi, \sigma}$ on $\mathfrak{B} = \mathfrak{B}([a, b])$ as follows. For each $f \in \mathfrak{B}$, we define

$$I_{\pi, \sigma}(f) = \sum_{k=1}^{n} f(\xi_k)(x_k - x_{k-1}).$$

Then $I_{\pi, \sigma}$ comes close to being a definite integral: it is linear, positive, and monotone. The only property of definite integrals which it lacks is property (4); that is, if J is a subinterval of $[a, b]$, then $I_{\pi, \sigma}(\varphi_J)$ is not necessarily equal to the length of J. One can show, nevertheless, that if c and d are the endpoints of J, then

$$|I_{\pi, \sigma}(\varphi_J) - (d - c)| \leq 4\|\pi\|.$$

A proof may be given as follows. We can choose integers j and k such that

$$x_{j-1} < c < x_{j+1} \quad \text{and} \quad x_{k-1} < d < x_{k+1}.$$

Then $\varphi_J(\xi_i) = 0$ if either $i < j - 1$ or $i > k + 1$, and $\varphi_J(\xi_i) = 1$ if $j + 1 < i < k - 1$. Assuming that there are integers between $j + 1$ and $k - 1$, we may write

$$x_{k-1} - x_{j+1} = \sum_{i=j+2}^{k-1} (x_i - x_{i-1})$$

$$\leq I_{\pi, \sigma}(\varphi_J) = \sum_{i=j}^{k+1} \varphi(\xi_i)(x_i - x_{i-1})$$

$$\leq \sum_{i=j}^{k+1} (x_i - x_{i-1}) = x_{k+1} - x_{j-1}.$$

If there are no integers between $j + 1$ and $k - 1$, then the inequality $I_{\pi, \sigma}(\varphi_J) \leq x_{k+1} - x_{j-1}$ is still valid (for the same reasons) and $x_{k-1} - x_{j+1} \leq 0 \leq I_{\pi, \sigma}(\varphi_J)$. Thus, in all cases

$$x_{k-1} - x_{j+1} \leq I_{\pi, \sigma}(\varphi_J) \leq x_{k+1} - x_{j-1}.$$

But from the way j and k were chosen, we also have

$$x_{k-1} - x_{j+1} \leq d - c \leq x_{k+1} - x_{j-1}.$$

Since $I_{\pi, \sigma}(\varphi_J)$ and $d - c$ both lie in the interval $[x_{k-1} - x_{j+1}, x_{k+1} - x_{j-1}]$, the distance between them is less than or equal to the length of the interval. Thus

$$\begin{aligned}|I_{\pi, \sigma}(\varphi_J) - (d - c)| &\leq (x_{k+1} - x_{j-1}) - (x_{k-1} - x_{j+1}) \\ &= (x_{k+1} - x_k) + (x_k - x_{k-1}) + (x_{j+1} - x_j) + (x_j - x_{j-1}) \\ &\leq 4\|\pi\|.\end{aligned}$$

If one is willing to go to the bother, one can get a sharper result, namely,

$$|I_{\pi, \sigma}(\varphi_J) - (d - c)| \leq 2\|\pi\|.$$

Thus, the finer π is, the closer $I_{\pi, \sigma}$ is to being a definite integral. This suggests that we can obtain a definite integral from $I_{\pi, \sigma}$ by somehow taking a limit as $\|\pi\| \to 0$.

We now consider an entire sequence Π of partitions $\pi_1, \pi_2, \ldots$ of $[a, b]$ such that $\lim_{n \to \infty} \|\pi_n\| = 0$. We also consider an associated sequence Σ of selectors $\sigma_1, \sigma_2, \ldots,$ where for each $n \in \mathbf{Z}^+$, σ_n is a selector for π_n. If $f \in \mathfrak{B}$, we then define

$$I_{\Pi,\Sigma}(f) = \lim_{n \to \infty} I_{\pi_n, \sigma_n}(f)$$

whenever the limit exists.

Proposition 2. $I_{\Pi,\Sigma}$ *is a definite integral on* $[a, b]$.

Proof. We shall use the abbreviation $I_n = I_{\pi_n, \sigma_n}$. Suppose f and g belong to the domain of $I_{\Pi,\Sigma}$. Then

$$I_{\Pi,\Sigma}(f) = \lim_{n \to \infty} I_n(f),$$

$$I_{\Pi,\Sigma}(g) = \lim_{n \to \infty} I_n(g).$$

Since I_n is linear, we have $I_n(f + g) = I_n(f) + I_n(g)$ for each $n \in \mathbf{Z}^+$. Thus

$$\lim_{n \to \infty} I_n(f + g) = I_{\Pi,\Sigma}(f) + I_{\Pi,\Sigma}(g),$$

from which it follows that $f + g$ belongs to the domain of $I_{\Pi,\Sigma}$ and that

$$I_{\Pi,\Sigma}(f + g) = I_{\Pi,\Sigma}(f) + I_{\Pi,\Sigma}(g).$$

Also, if c is any constant, one can show similarly that cf belongs to the domain of $I_{\Pi,\Sigma}$ and that $I_{\Pi,\Sigma}(cf) = cI_{\Pi,\Sigma}(f)$. If $f \geq 0$ on $[a, b]$, then $I_n(f) \geq 0$ for each $n \in \mathbf{Z}^+$, so that $I_{\Pi,\Sigma}(f) = \lim_{n \to \infty} I_n(f) \geq 0$.

If J is any subinterval of $[a, b]$ having endpoints c and d (with $c \leq d$), then

$$(d - c) - 4\|\pi_n\| \leq I_n(\varphi_J) \leq (d - c) + 4\|\pi_n\|.$$

Since $\lim_{n \to \infty} \|\pi_n\| = 0$, we get, via the pinching theorem,

$$I_{\Pi,\Sigma}(\varphi_J) = \lim_{n \to \infty} I_n(\varphi_J) = d - c.$$

This proves that $I_{\Pi,\Sigma}$ is a definite integral.

As a corollary, it follows that if f belongs to the domain of $I_{\Pi,\Sigma}$ and is also Riemann-integrable, then $I_{\Pi,\Sigma}(f) = \int_a^b f$. We shall now prove a stronger result, namely, that if f is Riemann-integrable, then f will automatically belong to the domain of $I_{\Pi,\Sigma}$.

Theorem 3. *Let* Π *and* Σ *be as above. Then* $\mathfrak{R}$, *the class of Riemann-integrable functions on* $[a, b]$, *is a subset of the domain of* $I_{\Pi,\Sigma}$, *and for each* $f \in \mathfrak{R}$,

$$I_{\Pi,\Sigma}(f) = \int_a^b f.$$

Proof. Let $f \in \mathfrak{R}$. We shall prove that

$$\lim_{n \to \infty} I_n(f) = \int_a^b f.$$

Let $\epsilon > 0$ be given. By the Riemann condition, there exist step functions g and h such that $g \leq f \leq h$ on $[a, b]$ and

$$\int_a^b (h - g) < \epsilon/3.$$

For each $n \in \mathbf{Z}^+$, we have

$$I_n(g) \leq I_n(f) \leq I_n(h).$$

Moreover, since g and h are step functions, they belong to the domain of $I_{\Pi,\Sigma}$ and

$$\int_a^b g = I_{\Pi,\Sigma}(g) = \lim_{n \to \infty} I_n(g),$$

$$\int_a^b h = I_{\Pi,\Sigma}(h) = \lim_{n \to \infty} I_n(h).$$

Thus for n sufficiently large, say $n > N$, we have

$$I_n(h) < \int_a^b h + \epsilon/3, \qquad I_n(g) > \int_a^b g - \epsilon/3,$$

from which we get

$$\int_a^b g - \epsilon/3 \leq I_n(f) \leq \int_a^b h + \epsilon/3.$$

Since the inequality

$$\int_a^b g \leq \int_a^b f \leq \int_a^b h$$

also holds, we get at once

$$\left| I_n(f) - \int_a^b f \right| < \epsilon$$

for all $n > N$. Thus $\lim_{n \to \infty} I_n(f) = \int_a^b f$, as asserted.

Corollary 4. *If f is Riemann-integrable on the interval $[a, b]$, then*

$$\lim_{n \to \infty} \frac{b-a}{n} \sum_{k=1}^n f\left(a + \frac{k(b-a)}{n}\right) = \int_a^b f.$$

Proof. We find that

$$\frac{b-a}{n} \sum_{k=1}^n f\left(a + \frac{k(b-a)}{n}\right) = I_{\pi_n, \sigma_n}(f),$$

where

$$\pi_n = (a, a + h, a + 2h, \ldots, a + nh),$$
$$\sigma_n = (a + h, a + 2h, \ldots, a + nh),$$
$$h = (b - a)/n.$$

Since $\|\pi_n\| = (b - a)/n \to 0$ as $n \to \infty$, the result follows.

By specializing f, a, and b, we can come up with some rather remarkable limits.

Example 1. If r is any positive rational number, discuss the behavior of the sum

$$S_r(n) = \sum_{k=1}^{n} k^r$$

as $n \to \infty$.

Since

$$\frac{1}{n^{r+1}} S_r(n) = \frac{1}{n} \sum_{k=1}^{n} \left(\frac{k}{n}\right)^r,$$

we see from the corollary above that

$$\lim_{n \to \infty} \frac{1}{n^{r+1}} S_r(n) = \int_0^1 x^r \, dx,$$

inasmuch as the function x^r is continuous and therefore Riemann-integrable on $[0, 1]$. Since x^r has an antiderivative, we have, by the second form of the fundamental theorem,

$$\int_0^1 x^r \, dx = \frac{x^{r+1}}{r+1}\Big|_0^1 = \frac{1}{r+1}.$$

Thus

$$\lim_{n \to \infty} \frac{1}{n^{r+1}} S_r(n) = \frac{1}{r+1}.$$

We say in this case that $S_r(n)$ is *asymptotic* to $n^{r+1}/(r+1)$ as $n \to \infty$. More generally, two sequences $\{a_n\}$ and $\{b_n\}$ are said to be asymptotic, and we write $a_n \sim b_n$, iff $\lim_{n \to \infty} a_n/b_n = 1$.

Example 2. Discuss the behavior of the sum

$$S_n = \sum_{k=1}^{n} \frac{k}{\sqrt{n^2 + k^2}}$$

as $n \to \infty$.

Since

$$\frac{1}{n} S_n = \frac{1}{n} \sum_{k=1}^{n} \frac{k/n}{\sqrt{1 + (k/n)^2}},$$

we find that

$$\lim_{n \to \infty} \frac{1}{n} S_n = \int_0^1 \frac{x \, dx}{\sqrt{1 + x^2}} = \sqrt{1 + x^2}\Big|_0^1 = \sqrt{2} - 1.$$

Thus $S_n \sim (\sqrt{2} - 1)n$.

Example 3. If $a > 1$, show once again that

$$\lim_{n \to \infty} n(\sqrt[n]{a} - 1) = \int_1^a x^{-1} \, dx.$$

We partition $[1, a]$ into n subintervals in such a way that the points of subdivision form a geometric progression. Let

$$\pi_n = (1, a^{1/n}, a^{2/n}, \ldots, a^{n/n}).$$

If we let

$$\sigma_n = (1, a^{1/n}, \ldots, a^{(n-1)/n}),$$

then

$$I_{\pi_n, \sigma_n}(f) = \sum_{k=1}^{n} a^{-(k-1)/n}(a^{k/n} - a^{(k-1)/n})$$

$$= \sum_{k=1}^{n} (a^{1/n} - 1)$$

$$= n(\sqrt[n]{a} - 1),$$

where $f(x) = x^{-1}$. Now

$$\|\pi_n\| = \max \{a^{k/n} - a^{(k-1)/n} \mid k = 1, \ldots, n\}$$

$$= \max \{a^{k/n}(1 - a^{-1/n}) \mid k = 1, \ldots, n\}$$

$$= a(1 - a^{-1/n}) = a(1 - 1/\sqrt[n]{a}),$$

and since $\lim_{n \to \infty} \sqrt[n]{a} = 1$, we see that $\lim_{n \to \infty} \|\pi_n\| = 0$. Thus

$$\lim_{n \to \infty} n(\sqrt[n]{a} - 1) = \lim_{n \to \infty} I_{\pi_n, \sigma_n}(f) = \int_1^a x^{-1} \, dx.$$

In the exercises the reader will be asked to show that if $f \in \mathcal{B}$, then there exists a sequence of partitions Π and a corresponding sequence of selectors Σ (both depending on f) such that $I_{\Pi, \Sigma}(f) = \overline{I}(f)$. Similarly, Π^* and Σ^* can be chosen so that $I_{\Pi^*, \Sigma^*}(f) = \underline{I}(f)$. This leads to a strengthening of the previous theorem.

Theorem 5. *Let $f \in \mathcal{B}$. The following are logically equivalent:*

1) *$f \in \mathcal{R}$ and $J = \int_a^b f$.*

2) *f belongs to the domain of each of the integrals $I_{\Pi, \Sigma}$, and in all cases $I_{\Pi, \Sigma}(f) = J$.*

3) *For each $\epsilon > 0$, there exists a $\delta > 0$ such that for each partition π of $[a, b]$ and for each selector σ for π,*

$$\|\pi\| < \delta \Rightarrow |I_{\pi, \sigma}(f) - J| < \epsilon.$$

Properties (2) and (3) are frequently taken as definitions of the Riemann integral. A shorthand way of writing (3) is

$$\lim_{\|\pi\| \to 0} I_{\pi, \sigma}(f) = J.$$

Proof. We have already established that (1) $\Rightarrow$ (2). To prove that (2) $\Rightarrow$ (1), we prove the contrapositive, namely $\sim$(1) $\Rightarrow$ $\sim$(2). If f is not Riemann-integrable, then we can choose Π, Σ, Π^*, Σ^* such that

$$I_{\Pi, \Sigma}(f) = \overline{I}(f) \neq \underline{I}(f) = I_{\Pi^*, \Sigma^*}(f),$$

thereby showing $\sim$(2).

The proof that (2) $\Leftrightarrow$ (3) is essentially the same as the proof of the linking limit lemma.

Another consequence of the remarks preceding the theorem is the fact that $\mathcal{R}$ *is the largest class of bounded functions for which definite integrals on $[a, b]$ agree.*

Exercises

1. Let $f(x) = x^2$ and $[a, b] = [1, 3]$.

   a) Let $\pi = (1, 1.5, 2.1, 2.6, 3)$ and $\sigma = (1, 2, 2.5, 2.7)$. Calculate $I_{\pi,\sigma}(f)$ and $\|\pi\|$. Also calculate $I_{\pi,\sigma}(\varphi_J)$, where $J = (2, 2.6]$.

   b) Let $\pi_n = (1, 1 + 2/n, 1 + 4/n, 1 + 6/n, \ldots, 3)$ and $\sigma_n = (1 + 2/n, 1 + 4/n, \ldots, 3)$. Calculate $I_n(f) = I_{\pi_n,\sigma_n}(f)$, and show directly that as $n \to \infty$, $\|\pi_n\| \to 0$ and $I_n(f) \to \int_1^3 x^2\, dx$, where the integral is interpreted in the Newtonian sense.

   c) Let $\pi_n = (1, 3^{1/n}, 3^{2/n}, 3^{3/n}, \ldots, 3)$ and $\sigma_n = (1, 3^{1/n}, 3^{2/n}, \ldots, 3^{(n-1)/n})$. Compute $\|\pi_n\|$ and $I_n(f) = I_{\pi_n,\sigma_n}(f)$, and show that $\|\pi_n\| \to 0$ and $I_n(f) \to \int_1^3 x^2\, dx$.

   d) For each $n \in \mathbf{Z}^+$, let

   $$\pi_n = (1, 2, 2\tfrac{1}{2}, 2\tfrac{3}{4}, 2\tfrac{7}{8}, \ldots, 3 - 1/2^n, 3)$$

   and

   $$\sigma_n = (1, 2, 2\tfrac{1}{2}, \ldots, 3 - 1/2^n).$$

   What is $\|\pi_n\|$? If $f \in \mathfrak{B}([1, 3])$, we define

   $$I(f) = \lim_{n \to \infty} I_{\pi_n,\sigma_n}(f)$$

   whenever the latter limit exists. Prove that I is not a definite integral on $[1, 3]$. What properties does I share with definite integrals?

2. If $\pi = (x_0, x_1, \ldots, x_n)$ is a partition of $[a, b]$, and if g is a step function on $[a, b]$, we say that π *fits* g iff g is constant on each of the open intervals $(x_0, x_1), \ldots, (x_{n-1}, x_n)$.

   a) Let $f \in \mathfrak{B}$, and let $\pi = (x_0, x_1, \ldots, x_n)$ be a partition of $[a, b]$. Among all step functions g such that $f \leq g$ on $[a, b]$ and π fits g, show that there is a smallest such function g_π. [The value of g_π on the open interval (x_0, x_1) will be the least upper bound of the values of f on the same interval, etc.] Show that $\int_a^b g_\pi$ is the least upper bound of all Riemann sums $I_{\pi,\sigma}(f)$, where $\sigma = (\xi_1, \ldots, \xi_n)$ is a selector for π such that ξ_k belongs to the *open* interval (x_{k-1}, x_k) for $k = 1, \ldots, n$.

   b) We say that a partition $\pi' = (x_0, x_1, \ldots, x_n)$ is a *refinement* of $\pi = (y_0, y_1, \ldots, y_m)$ iff $\{y_0, y_1, \ldots, y_m\}$ is a subset $\{x_0, x_1, \ldots, x_n\}$. Equivalently, each open interval of the form (x_{j-1}, x_j) is contained in an interval of the form (y_{k-1}, y_k). Observe that if π fits g and if π' is a refinement of π, then π' fits g. Deduce that if π' is a refinement of π, then $g_{\pi'} \leq g_\pi$.

   c) Show that there exists an integral of the form $I_{\Pi,\Sigma}$ such that $I_{\Pi,\Sigma}(f) = \overline{I}(f)$. [Given $n \in \mathbf{Z}^+$, show that there exists a step function g_n such that $f \leq g_n$ on $[a, b]$ and $\int_a^b g_n - \overline{I}(f) < 1/n$. Show that there exists a partition π_n of $[a, b]$ such that $\|\pi_n\| < 1/n$ and π_n fits g_n. Then $f \leq g_{\pi_n} \leq g_n$, so that

   $$\overline{I}(f) \leq \overline{I}(g_{\pi_n}) = \int_a^b g_{\pi_n} \leq \int_a^b g_n \leq \overline{I}(f) + 1/n.$$

   Show that there exists a selector σ_n for π_n such that

   $$\int_a^b g_{\pi_n} \geq I_{\pi_n,\sigma_n}(f) \geq \int_a^b g_{\pi_n} - 1/n.$$

   Show that $\|\pi_n\| \to 0$ and $I_{\pi_n,\sigma_n}(f) \to \overline{I}(f)$ as $n \to \infty$.] One can also show that there exists an integral of the form I_{Π^*,Σ^*} such that $I_{\Pi^*,\Sigma^*}(f) = \underline{I}(f)$.

3. a) Let I_1 and I_2 be any two definite integrals on $[a, b]$, and let $0 \leq \lambda \leq 1$. We define I to be the function with domain $\mathcal{D}_{I_1} \cap \mathcal{D}_{I_2}$ such that

$$I(f) = (1 - \lambda)I_1(f) + \lambda I_2(f)$$

for each $f \in \mathcal{D}_{I_1} \cap \mathcal{D}_{I_2}$. Prove that I is a definite integral on $[a, b]$.

    b) Let $f \in \mathcal{B}([a, b])$, and let S be the set of all numbers of the form $I(f)$ such that I is a definite integral on $[a, b]$ and $f \in \mathcal{D}_I$. Prove that S is the closed bounded interval $[\underline{I}(f), \overline{I}(f)]$. [Use (a), Exercise 2(c), and Proposition 4, Section 8-1.]

8-3 Further properties of Riemann integrals

The Riemann integral, like the Newton integral, is additive.

Proposition 1. Additivity. *If f is $\mathcal{R}$-integrable on $[a, b]$ and $[b, c]$, then f is $\mathcal{R}$-integrable on $[a, c]$ and*

$$\int_a^b f + \int_b^c f = \int_a^c f.$$

Proof. We know that the above equation is true for all step functions, and from this we obtain the general result by passing to limits. Since f is $\mathcal{R}$-integrable on $[a, b]$, there exist sequences of step functions $\{g_n\}$ and $\{h_n\}$ on $[a, b]$ such that $g_n \leq f \leq h_n$ on $[a, b]$ for all $n \in \mathbf{Z}^+$ and

$$\lim_{n \to \infty} \int_a^b g_n = \lim_{n \to \infty} \int_a^b h_n = \int_a^b f.$$

Similarly, there exist step functions $\{g_n^*\}$ and $\{h_n^*\}$ such that $g_n^* \leq f \leq h_n^*$ on $[b, c]$ and

$$\lim_{n \to \infty} \int_b^c g_n^* = \lim_{n \to \infty} \int_b^c h_n^* = \int_b^c f.$$

We define

$$G_n(x) = \begin{cases} g_n(x) & \text{if} \quad a \leq x < b, \\ g_n^*(x) & \text{if} \quad b \leq x \leq c, \end{cases}$$

$$H_n(x) = \begin{cases} h_n(x) & \text{if} \quad a \leq x < b, \\ h_n^*(x) & \text{if} \quad b \leq x \leq c. \end{cases}$$

Then for each $n \in \mathbf{Z}^+$, we find that G_n and H_n are step functions, $G_n \leq f \leq H_n$ on $[a, c]$, and

$$\lim_{n \to \infty} \int_a^c G_n = \lim_{n \to \infty} \left(\int_a^b g_n + \int_b^c g_n^* \right)$$

$$= \int_a^b f + \int_b^c f = \lim_{n \to \infty} \left(\int_a^b h_n + \int_b^c h_n^* \right)$$

$$= \lim_{n \to \infty} \int_a^c H_n.$$

Thus, f is $\mathcal{R}$-integrable on $[a, c]$, and

$$\int_a^c f = \lim_{n \to \infty} \int_a^c G_n = \int_a^b f + \int_b^c f.$$

The Riemann integral is superior to the Newton integral in at least one respect, namely, the class of $\mathcal{R}$-integrable functions is closed under multiplication and the lattice operations. Recall the definition of the lattice operations. If f and g are real-valued functions defined on a set S, then $f \vee g$ and $f \wedge g$ are functions defined on S such that

$$(f \vee g)(x) = \max \{f(x), g(x)\}, \qquad (f \wedge g)(x) = \min \{f(x), g(x)\}$$

for all $x \in S$. We also define the *positive* and *negative* parts of a function to be

$$f^+ = f \vee 0, \qquad f^- = (-f) \vee 0.$$

Then it is easy to check that

$$f = f^+ - f^- \qquad \text{and} \qquad |f| = f^+ + f^-.$$

Proposition 2. Closure under Lattice Operations. *If f and g are $\mathcal{R}$-integrable on $[a, b]$, then so are $f^+, f^-, |f|, f \vee g$, and $f \wedge g$.*

Proof. Let $\epsilon > 0$ be given. Since f is $\mathcal{R}$-integrable on $[a, b]$, there exist step functions g and h such that $g \leq f \leq h$ on $[a, b]$ and

$$\int_a^b (h - g) < \epsilon.$$

It is easily verified that g^+ and h^+ are step functions, that $g^+ \leq f^+ \leq h^+$, and that $h^+ - g^+ \leq h - g$ on $[a, b]$. Consequently, $\int_a^b (h^+ - g^+) < \epsilon$. Thus f^+ is $\mathcal{R}$-integrable on $[a, b]$.

The integrability of the remaining functions follows then from the identities

$$f^- = f^+ - f,$$
$$|f| = f^+ + f^-,$$
$$f \vee g = \tfrac{1}{2}[f + g + |f - g|],$$

and

$$f \wedge g = \tfrac{1}{2}[f + g - |f - g|].$$

Proposition 3. Closure under Multiplication. *If f and g are $\mathcal{R}$-integrable on $[a, b]$, then go is fg.*

Proof. Since $(fg)^+ = f^+g^+ + f^-g^-$ and $(fg)^- = f^+g^- + f^-g^+$, it suffices to consider the case in which f and g are nonnegative.

Let $\epsilon > 0$ be given. Since f and g are $\mathcal{R}$-integrable, we can find step functions f_*, f^*, g_*, g^* such that $0 \leq f_* \leq f \leq f^* \leq M_f$ and $0 \leq g_* \leq g \leq g^* \leq M_g$ on $[a, b]$, and

$$\int_a^b (f^* - f_*) < \frac{\epsilon}{2M_g}, \qquad \int_a^b (g^* - g_*) < \frac{\epsilon}{2M_f},$$

where $M_f > \text{lub} \{f(x) \mid a \leq x \leq b\}$ and M_g is similarly defined. Then

$$f_* g_* \leq fg \leq f^* g^*,$$

and

$$\int_a^b (f^*g^* - f_*g_*) = \int_a^b f^*(g^* - g_*) + \int_a^b g_*(f^* - f_*)$$

$$\leq \int_a^b M_f(g^* - g_*) + \int_a^b M_g(f^* - f_*)$$

$$< M_f \frac{\epsilon}{2M_f} + M_g \frac{\epsilon}{2M_g} = \epsilon.$$

Thus fg is $\mathcal{R}$-integrable.

We conclude this section with a mean-value theorem for integrals.

Theorem 4. Generalized Mean-Value Theorem for Integrals. *Suppose that*

1) f *is continuous on* $[a, b]$,

2) g *is Riemann-integrable on* $[a, b]$ *and does not change sign on the interval.*
Then there exists a number ξ *such that* $a \leq \xi \leq b$ *and*

$$\int_a^b f(x)\, g(x)\, dx = f(\xi) \int_a^b g(x)\, dx.$$

Proof. We may assume that $g \geq 0$ on $[a, b]$. We let M and m respectively denote the maximum and minimum values of f on $[a, b]$. Then

$$mg(x) \leq f(x) g(x) \leq Mg(x)$$

for all $x \in [a, b]$. Since the three functions involved are Riemann-integrable, we get

$$m \int_a^b g(x)\, dx \leq \int_a^b f(x)\, g(x)\, dx \leq M \int_a^b g(x)\, dx.$$

If $\int_a^b g(x)\, dx = 0$, then $\int_a^b f(x) g(x)\, dx = 0$ and any $\xi \in [a, b]$ will have the required property. If $\int_a^b g(x)\, dx > 0$, then we have

$$m \leq \frac{\int_a^b f(x)\, g(x)\, dx}{\int_a^b g(x)\, dx} \leq M.$$

By the intermediate-value theorem, the function f assumes all values between m and M. In particular, the equation

$$f(\xi) = \frac{\int_a^b f(x)\, g(x)\, dx}{\int_a^b g(x)\, dx}$$

must hold for some $\xi \in [a, b]$.

Corollary 5. Mean-Value Theorem for Integrals. *If f is continuous on the interval* $[a, b]$, *then there exists a* $\xi \in [a, b]$ *such that*

$$\int_a^b f(x)\, dx = (b - a)f(\xi).$$

Proof. Set $g(x) \equiv 1$ and apply the generalized mean-value theorem.

Exercises

1. We shall use $\overline{\int}_a^b f$ to denote the upper Riemann integral of f on $[a, b]$.
   a) If f is bounded on both $[a, b]$ and $[b, c]$, show that

   $$\overline{\int}_a^b f + \overline{\int}_b^c f = \overline{\int}_a^c f.$$

   b) State and deduce from (a) an analogous result for lower integrals.
   c) Using (a) and (b), give an alternative proof of the additivity of Riemann integrals.

2. Using the additive property of Riemann integrals and Exercise 7, Section 8–1, prove that if f is bounded and continuous at all but a finite number of points of $[a, b]$, then f is Riemann-integrable on $[a, b]$.

3. We say that a function f is *bounded away from zero* on a set S iff there exists an $\epsilon > 0$ such that $|f(x)| > \epsilon$ for all $x \in S$. Prove that if f is Riemann-integrable and bounded away from zero on $[a, b]$, then $1/f$ is Riemann-integrable on $[a, b]$.

4. If f and g are bounded on $[a, b]$, prove that

   $$\overline{I}(f \vee g) + \overline{I}(f \wedge g) \leq \overline{I}(f) + \overline{I}(g),$$
   $$\underline{I}(f \vee g) + \underline{I}(f \wedge g) \geq \underline{I}(f) + \underline{I}(g).$$

   Use these inequalities to give an alternative proof of the closure of Riemann-integrable functions under lattice operations.

5. A function f defined on $\mathbf{R}$ is said to have *compact support* iff there exists a closed bounded interval $[a, b]$ outside of which the function f is identically zero [that is, $x \notin [a, b] \Rightarrow f(x) = 0$].

   a) Prove that the bounded real-valued functions having compact support are closed under addition, multiplication, and lattice operations.
   b) Suppose that f has compact support, that f vanishes outside the interval $[a, b]$, and that f is Riemann-integrable on $[a, b]$. We then say that f is *Riemann-integrable on* $\mathbf{R}$, and we define

   $$\int_{-\infty}^{\infty} f(x)\, dx = \int_a^b f(x)\, dx.$$

   Argue convincingly that these definitions do not depend on a particular choice of the interval $[a, b]$. (Prove, in other words, that if f also vanishes outside $[a^*, b^*]$, then f is Riemann-integrable on $[a^*, b^*]$ and $\int_a^b f = \int_{a^*}^{b^*} f$.) Also define appropriately the upper and lower integrals for a bounded function having compact support.
   c) If f and g are Riemann-integrable on $\mathbf{R}$, show that $f + g$ is also and that

   $$\int_{-\infty}^{\infty} (f + g) = \int_{-\infty}^{\infty} f + \int_{-\infty}^{\infty} g.$$

   Under what other operations is the family of Riemann-integrable functions closed?
   d) State and prove properties of upper and lower Riemann integrals on $\mathbf{R}$.

6. If A is any subset of $\mathbf{R}$, we define the *characteristic* (or *indicator*) *function* of A as follows:

   $$\varphi_A(x) = \begin{cases} 1 & \text{if } x \in A, \\ 0 & \text{if } x \notin A. \end{cases}$$

Prove the following properties of characteristic functions:

a) $A = B$ iff $\varphi_A = \varphi_B$,

b) $A \subset B$ iff $\varphi_A \leq \varphi_B$,

c) $\varphi_{A \cap B} = \varphi_A \wedge \varphi_B = \varphi_A \varphi_B$,

d) $\varphi_{A \cup B} = \varphi_A \vee \varphi_B = \varphi_A + \varphi_B - \varphi_A \varphi_B$,

e) $\varphi_{A'} = 1 - \varphi_A$,

f) $\varphi_{A \backslash B} = \varphi_A - \varphi_A \varphi_B$.

7. A set $A \subset \mathbf{R}$ is said to be *bounded* iff A is a subset of some closed bounded interval, or equivalently, φ_A has compact support. We define the *outer* and *inner Jordan contents* of a bounded set $A \subset \mathbf{R}$ to be respectively

$$I^*(A) = \overline{\int_{-\infty}^{\infty}} \varphi_A \quad \text{and} \quad I_*(A) = \underline{\int_{-\infty}^{\infty}} \varphi_A.$$

If $I^*(A) = I_*(A)$ [or equivalently, φ_A is Riemann-integrable on $\mathbf{R}$], we say that A is *Jordan-measurable*, and $I(A) = I^*(A) = I_*(A)$ is called the *Jordan content* of A. [Intuitively, $I(A)$ is the length of A.]

a) Given that J is any bounded interval, observe that J is Jordan-measurable and that $I(J)$ is the length of J. If $A \subset \mathbf{R}$ is bounded, show that $I^*(A)$ is the greatest lower bound of all numbers of the form $\sum_{k=1}^{n} I(J_k)$, where $J_1, \ldots, J_n$ are disjoint bounded intervals and $A \subset J_1 \cup \cdots \cup J_n$. State a similar result for $I_*(A)$.

b) Show that the family $\mathcal{J}$ of Jordan-measurable subsets of $\mathbf{R}$ is closed under the formation of finite unions, intersections, and relative complements. (One often summarizes this by saying that $\mathcal{J}$ is a *ring of sets.*) If A and B are any two Jordan-measurable sets, prove that

$$I(A \cup B) + I(A \cap B) = I(A) + I(B).$$

In particular, if A and B are disjoint, then

$$I(A \cup B) = I(A) + I(B).$$

(Because of this property, we say that I is *finitely additive.*)

8. a) Let f be continuous on $[a, b]$, and let π be a partition of $[a, b]$. Using the mean-value theorem for integrals, show that there exists a selector σ for π such that

$$\int_a^b f = I_{\pi,\sigma}(f).$$

b) Suppose that f has a bounded derivative on $[a, b]$, say $|f'(x)| \leq M$ for all $x \in [a, b]$. Using (a) and the mean-value theorem, prove that

$$\left| I_{\pi,\sigma}(f) - \int_a^b f \right| \leq (b - a)M\|\pi\|$$

for every partition π of $[a, b]$ and every selector σ for π. Also observe that if $\pi = (x_0, x_1, \ldots, x_n)$, and the selector $\sigma = (\xi_1, \ldots, \xi_n)$ consists of the midpoints of π [i.e., $\xi_k = (x_{k-1} + x_k)/2$, for $k = 1, \ldots, n$], then we have the sharper inequality

$$\left| I_{\pi,\sigma}(f) - \int_a^b f \right| \leq \tfrac{1}{2}(b - a)M\|\pi\|.$$

c) Suppose that you are asked to calculate the integral $\int_1^2 x^{-1}\, dx$ to four decimal places by means of Riemann sums. Plan such a calculation so that a minimum amount of arithmetic is involved. (Be explicit as to which Riemann sum is to be computed.)

9. The Cauchy-Schwarz inequality for integrals states that if f and g are Riemann-integrable on $[a, b]$, then

$$\left(\int_a^b fg\right)^2 \le \left(\int_a^b f^2\right)\left(\int_a^b g^2\right).$$

a) Prove the inequality for integrals by mimicking the proof outlined for sums. (See Exercise 17, Section 1–9.)
b) Given that f and g are continuous, prove that equality holds iff one of the functions is a constant times the other.
c) Using the Cauchy-Schwarz inequality, get an upper estimate on the integral

$$\int_0^{\pi/2} \sqrt{x \sin x}\, dx.$$

8-4 The fundamental theorem of calculus

We shall now examine the behavior of a Riemann integral $\int_a^b f$ when the endpoints a and b are allowed to vary. In doing so, we eventually get to the point where it is troublesome always to require that $a < b$. From our experience with the Newton integral, it should be clear how $\int_a^b f$ is to be defined if $a \ge b$.

Definition 1. If $a > b$, we define

$$\int_a^b f = -\int_b^a f,$$

provided the latter integral exists. We define $\int_{-a}^a f = 0$.

By considering all possible orderings for a, b, and c, one can apply previous results to prove that

$$\int_a^b f + \int_b^c f = \int_a^c f,$$

regardless of the relative order of the numbers a, b, and c; the existence of any two of the integrals implies the existence of the third.

Theorem 2. The First Form of the Fundamental Theorem of Calculus. *Let f be a real-valued function defined on an interval J which is Riemann-integrable on every closed bounded subinterval of J. Given a number $a \in J$, we define a second function F on J as follows:*

$$F(x) = \int_a^x f$$

for all $x \in J$. Then F is continuous on J and at each point $x_0 \in J$ where f is continuous, F is differentiable and

$$F'(x_0) = f(x_0).$$

Proof. It suffices, in proving the first assertion, to show that F is uniformly continuous on each closed bounded interval $[c, d] \subset J$. Since f is Riemann-integrable on $[c, d]$, it is, in particular, bounded on $[c, d]$. Hence we can find a number M such that $|f(x)| \leq M$ for all $x \in [c, d]$. Let x and y be any two points in $[c, d]$ with, say, $x \leq y$. Then

$$F(y) - F(x) = \int_a^y f - \int_a^x f = \int_x^y f,$$

so that

$$|F(y) - F(x)| = \left| \int_x^y f \right| \leq \int_x^y |f|$$

$$\leq \int_x^y M = M(y - x).$$

From this inequality it is obvious that F is uniformly continuous on $[c, d]$. (Given $\epsilon > 0$, one can take $\delta = \epsilon/M$.)

Suppose that f is right-continuous at x_0 where $x_0 \in J$. Consider $x \in J$ with $x > x_0$. Then

$$F(x) - F(x_0) = \int_{x_0}^x f(t)\, dt$$

and

$$f(x_0) = \frac{1}{x - x_0} \int_{x_0}^x f(x_0)\, dt.$$

From these two equations we get

$$\frac{F(x) - F(x_0)}{x - x_0} - f(x_0) = \frac{1}{x - x_0} \int_{x_0}^x [f(t) - f(x_0)]\, dt,$$

and thus

$$\left| \frac{F(x) - F(x_0)}{x - x_0} - f(x_0) \right| \leq \frac{1}{x - x_0} \int_{x_0}^x |f(t) - f(x_0)|\, dt.$$

Let $\epsilon > 0$ be given. Since f is right-continuous at x_0, there exists a $\delta > 0$ such that for all t,

$$x_0 < t < x_0 + \delta \Rightarrow |f(t) - f(x_0)| < \epsilon.$$

Thus, if $x_0 < x < x_0 + \delta$, then

$$\int_{x_0}^x |f(t) - f(x_0)|\, dt \leq \int_{x_0}^x \epsilon\, dt = \epsilon(x - x_0),$$

so that

$$\left| \frac{F(x) - F(x_0)}{x - x_0} - f(x_0) \right| \leq \epsilon.$$

This proves that

$$\lim_{x \to x_0+} \frac{F(x) - F(x_0)}{x - x_0} = f(x_0).$$

In other words, F is right-differentiable at x_0 and $F'(x_0+) = f(x_0)$.

Similarly, if f is left-continuous at x_0, then it can be shown that $F'(x_0-) = f(x_0)$. It follows that if f is continuous at x_0 in the usual two-sided sense, then F is differentiable at x_0 in the usual two-sided sense and $F'(x_0) = f(x_0)$. ·

Corollary 3. *If f is continuous on an interval J (or continuous at all but a finite number of points of J), then f has an antiderivative on J.*

Although this result is an immediate consequence of Theorem 2, we shall sketch a separate proof for the case in which f is continuous on J.

Proof. Let F be as above. If x and x_0 belong to J, then by the mean-value theorem for integrals, we may write

$$F(x) - F(x_0) = \int_{x_0}^{x} f(t)\, dt = (x - x_0)f(\xi)$$

for some ξ between x_0 and x. Since f is continuous at x_0, it follows that $f(\xi) \to f(x_0)$ as $x \to x_0$. Thus

$$F'(x_0) = \lim_{x \to x_0} \frac{F(x) - F(x_0)}{x - x_0} = f(x_0).$$

The fundamental theorem vindicates much of our earlier work. The assumption that continuous functions have antiderivatives was made, for instance, in the discussion of arc length. Also, in calculating such integrals as $\int_{-1/2}^{\sqrt{3}/2} \sqrt{1 - x^2}\, dx$, we have had to assume that the integrands possess antiderivatives, even though we could produce none.

The fundamental theorem is concerned with the behavior of a Riemann integral as its upper limit of integration is allowed to vary. The dependence upon the lower limit of integration can also be studied. Consider, for instance,

$$F(x) = \int_{x}^{b} f(t)\, dt,$$

where f is continuous on **R**. Then

$$F'(x) = \frac{d}{dx}\left[-\int_{b}^{x} f(t)\, dt \right] = -f(x),$$

by the fundamental theorem. Consider also the problem of calculating dy/dx if

$$y = \int_{x^2}^{\sec x} dt/t \qquad (0 < x < \pi/2).$$

If we let $F(x) = \int_{1}^{x} dt/t$, then by the fundamental theorem,

$$F'(x) = 1/x.$$

Hence, by the second form of the fundamental theorem of calculus,

$$y = F(\sec x) - F(x^2).$$

We next apply the chain rule, and obtain

$$\frac{dy}{dx} = F'(\sec x)\frac{d}{dx}(\sec x) - F'(x^2)\frac{d}{dx}(x^2)$$

$$= \frac{1}{\sec x}(\sec x \tan x) - \frac{1}{x^2}(2x)$$

$$= \tan x - \frac{2}{x}.$$

The fundamental theorem can also be used to prove the existence of functions with certain properties. Let us prove, for example, that *to each real number x there corresponds a unique number* $y = \text{Tan}^{-1} x$ *such that* $-\pi/2 < y < \pi/2$ *and* $\tan y = x$. *Moreover, the function* Tan^{-1} *so obtained possesses continuous derivatives of all orders.* The tangent function is continuous on the interval $(-\pi/2, \pi/2)$ and

$$\lim_{x \to \pi/2-} \tan x = +\infty, \qquad \lim_{x \to -\pi/2+} \tan x = -\infty.$$

It follows from the intermediate-value theorem that to each real number x there corresponds *at least* one number y such that $-\pi/2 < y < \pi/2$ and $x = \tan y$. Since

$$\frac{d}{dy} (\tan y) = \sec^2 y > 0 \qquad (-\pi/2 < y < \pi/2),$$

the tangent function is strictly increasing on the interval $(-\pi/2, \pi/2)$, and thus the number $y = \text{Tan}^{-1} x$ mentioned above is unique. Rather than attempting to prove directly that the function Tan^{-1} is differentiable, we shall prove that it is equal to a function which is known to be differentiable. To determine the latter function, we *assume* provisionally that Tan^{-1} is differentiable. Set

$$y = \text{Tan}^{-1} x.$$

Then $x = \tan y$, and the chain rule yields

$$1 = \frac{d}{dx} (x) = (\sec^2 y) \frac{dy}{dx}.$$

Hence

$$\frac{dy}{dx} = \frac{1}{\sec^2 y} = \frac{1}{1 + \tan^2 y} = \frac{1}{1 + x^2}.$$

This suggests that $\text{Tan}^{-1} x$ is an antiderivative of $1/(1 + x^2)$. We also note that $\text{Tan}^{-1} 0 = 0$. However, we know by the fundamental theorem that the function

$$F(x) = \int_0^x \frac{dt}{1 + t^2}$$

also has these two properties. If we can now show that

$$F(x) = \text{Tan}^{-1} x$$

for all $x \in \mathbf{R}$, we shall be finished. But, if $-\pi/2 < y < \pi/2$, then

$$F(\tan y) = \int_0^{\tan y} \frac{dt}{1 + t^2} = y.$$

(See Example 6, Section 7–3.) By definition, $y = \text{Tan}^{-1} x$ iff $-\pi/2 < y < \pi/2$ and $x = \tan y$. Thus

$$F(x) = F(\tan y) = y = \text{Tan}^{-1} x$$

for all $x \in \mathbf{R}$.

Exercises

1. The purpose of this exercise is to present an alternative proof of the corollary to the fundamental theorem, a proof which does not make use of the Riemann integral. Suppose that f is a function continuous on an interval J. Prove that f has an antiderivative on J.

   a) Show that it suffices to consider the case in which J is a closed bounded interval $[a, b]$. Be as explicit as possible about how such a reduction can be accomplished when J is either the interval $[0, +\infty)$ or the interval $(1, 2]$. Then comment on the general case.

   b) Prove that for each $n \in \mathbf{Z}^+$ there exists a step function f_n on $[a, b]$ such that

$$|f(x) - f_n(x)| < 1/n$$

   for all $x \in [a, b]$. (Use the fact that f is uniformly continuous on $[a, b]$.)

   c) For each $x \in [a, b]$ and $n \in \mathbf{Z}^+$, define

$$F_n(x) = \int_a^x f_n(t)\, dt,$$

   where the integral is interpreted in the Newtonian sense. Show that

$$|F_n(x) - F_m(x)| \le \left(\frac{1}{n} + \frac{1}{m}\right)(x - a)$$

$$\le \left(\frac{1}{n} + \frac{1}{m}\right)(b - a),$$

   and deduce from this that the sequence $\{F_n(x)\}$ is Cauchy. Set

$$F(x) = \lim_{n \to \infty} F_n(x) = \lim_{n \to \infty} \int_a^x f_n(t)\, dt,$$

   and prove that the function F so obtained is an antiderivative of f. [Suppose that $a < x_0 < b$, and let $h > 0$ be sufficiently small so that $x_0 + h \le b$. Show that

$$e(h) = \frac{F(x_0 + h) - F(x_0)}{h} - f(x_0)$$

$$= \lim_{n \to \infty} \frac{1}{h} \int_{x_0}^{x_0+h} [f_n(t) - f(x_0)]\, dt.$$

   From the fact that

$$|f_n(t) - f(x_0)| \le 1/n + |f(t) - f(x_0)|,$$

   deduce that

$$|e(h)| \le \max\{|f(t) - f(x_0)| \mid x_0 \le t \le x_0 + h\}$$

   and hence that

$$\lim_{h \to 0+} e(h) = 0.$$

   Treat left-hand limits similarly. Show finally that F is continuous at the endpoints a and b.]

2. Find dy/dx in each of the following cases.

a) $y = \int_x^5 \sqrt{25 - t^2}\, dt$

b) $y = \int_1^x t^{-1}\, dt$

c) $y = \int_1^{3x} u^{-1}\, du$

d) $y = \int_{1/x}^{47} v^{-1}\, dv$

e) $y = \int_{x^3}^{\cos x} \sin (t^2)\, dt$

3. Assuming that the equation below defines implicitly a function y of x, find dy/dx:

$$x \sin xy + \int_{\sqrt{y}}^{x^2 \sin y} \frac{\sin t}{t}\, dt = 1.$$

4. Recall that f is said to be *periodic* with *period* T iff

$$f(x + T) = f(x)$$

for all x in **R**. If f is a continuous function which is periodic with period T, show that

$$\int_0^T f(t)\, dt = \int_a^{a+T} f(t)\, dt$$

for all $a \in \mathbf{R}$. [Let $G(x) = \int_x^{x+T} f(t)\, dt$, and use the fundamental theorem to show that $G'(x)$ is identically zero.] Interpret this result geometrically.

*5. Let f be continuous on the entire line.

a) Prove the translation-invariance formula:

$$\int_a^b f(x)\, dx = \int_{a+c}^{b+c} f(x - c)\, dx.$$

b) For any fixed $\delta > 0$, we define $f_\delta(x)$ by the equation

$$F_\delta(x) = \frac{1}{2\delta} \int_{-\delta}^{+\delta} f(x + t)\, dt.$$

Show that F_δ is differentiable on the entire line.

c) We consider now a fixed closed bounded interval $[a, b]$. Prove that for each $\epsilon > 0$ there exists a $\delta > 0$ such that

$$|F_\delta(x) - f(x)| < \epsilon$$

for all x in $[a, b]$. This shows, in effect, that any continuous function can be approximated reasonably well by differentiable functions.

6. Use the two forms of the fundamental theorem of calculus and the mean-value theorem to prove the following strengthened version of the mean-value theorem for integrals. *If f is continuous on $[a, b]$, then there exists a number ξ such that $a < \xi < b$ and*

$$\int_a^b f(x)\, dx = (b - a)f(\xi).$$

**8–5 A deeper look at areas

In our earlier treatment of areas and volumes we took much for granted. We shall now enter into a deeper study of these notions.

Much of our previous work can be justified by the general principle that "the whole of a thing is equal to the sum of its parts." In a less sophisticated era this statement was taken to be a "self-evident truth," and even today few people are inclined to take issue with it. The statement is, however, far too vague to be used in serious mathematics, and besides, there are reasonable interpretations of the terms "whole" and "sum of its parts" which make the statement false. Nevertheless, some half-truths are useful, and it is unwise to toss them aside without first searching for a better replacement. In this particular instance, there is an entire branch of mathematics called "measure theory" which is devoted to determining when and in what sense the whole of a thing is equal to the sum of its parts.

Students of measure theory attempt, among other things, to study axiomatically such concepts as length, area, and volume. Let us consider first the notion of volume. One might be tempted to set down the following "axioms."

V1. Every bounded set S has a volume V(S). (We say that a set in three-space is bounded if all its points are contained within some rectangular parallelepiped.)

V2. If a bounded set S is the union of a finite number of nonoverlapping sets $S_1, \ldots, S_n$, then $V(S) = V(S_1) + \cdots + V(S_n)$.

V3. The volumes of any two bounded and congruent sets are the same.

V4. The volume of any rectangular parallelepiped is equal to the product of its length, breadth, and height.

Axiom V2 is, of course, a precise formulation of the statement that the whole of a thing is equal to the sum of its parts. Although these axioms might appear reasonable, they have one very serious defect: they are inconsistent. Their inconsistency is most strikingly illustrated by a theorem which so staggers the imagination that one is likely to dismiss it as fantasy. This theorem is usually called the *Banach-Tarski paradox.* Suppose that we are given two spheres S and S^*, where, say, S is the size of a pea and S^* is the size of the sun. Then, according to Tarski's paradox, it is possible to find an integer n and sets $S_1, \ldots, S_n$ and $S_1^*, \ldots, S_n^*$ such that

1) the sets $S_1, \ldots, S_n^*$ are nonoverlapping and S is their union;

2) the sets $S_1^*, \ldots, S_n^*$ are nonoverlapping and S^* is their union;

3) the starred and unstarred sets are pairwise congruent, that is, $S_1 \simeq S_1^*$, $S_2 \simeq S_2^*, \ldots, S_n \simeq S_n^*$.

Thus, according to Tarski's paradox, one can take a sphere the size of the sun, split it into a finite number of pieces, and by merely reassembling these pieces (moving them rigidly through space), form a sphere which can be tucked into one's pocket! We shall make no attempt to prove this remarkable theorem. Assuming that it is true, however, it is immediately clear that the axioms for volume listed above are inconsistent. For by V2 and V3, we would then have

$$V(S) = V(S_1) + \cdots + V(S_n)$$
$$= V(S_1^*) + \cdots + V(S_n^*) = V(S^*).$$

In other words, the volumes of any two spheres would have to be the same. One can easily deduce from the axioms that this is not possible.

The moral of Tarski's paradox is perfectly clear: our intuitive notions of volume can lead us astray, and they should therefore be examined critically. This is the task of measure theory. In developing a reasonable theory of volume, we must first decide what we might like to be true. Clearly, we would like Axioms V2, V3, and V4 to hold, even though it is clear from the preceding discussion that we cannot expect them to hold for *all* bounded sets. One can prove that there is quite a large class $\mathfrak{M}$ of bounded sets for which Axioms V2, V3, and V4 hold. This class $\mathfrak{M}$ includes all the bounded sets usually studied in solid geometry, and, in particular, it includes all rectangular parallelepipeds. It does not include, however, the perverse sets whose existence is asserted in Tarski's paradox (since, if it did, we would be forced to conclude that any two spheres have the same volume).

Our main concern is with area rather than volume. We start, however, with an even simpler concept—length. First we review some results from previous exercises.

A function f defined on **R** has *compact support* if it takes on the value zero throughout the complement of some closed bounded interval $[a, b]$. If f is also bounded, we define the *upper and lower Riemann integrals of f over* **R** to be

$$\overline{\int_{-\infty}^{\infty}} f(x)\, dx = \overline{\int_a^b} f(x)\, dx$$

and

$$\underline{\int_{-\infty}^{\infty}} f(x)\, dx = \underline{\int_a^b} f(x)\, dx,$$

respectively. It can be checked that the value of these integrals is independent of a particular choice of a and b. Furthermore, these upper and lower integrals are positive and monotone and

$$\underline{\int_{-\infty}^{\infty}} f \leq \overline{\int_{-\infty}^{\infty}} f.$$

If the upper and lower integrals of f are equal, we say that f is *Riemann-integrable over* **R**, and we use $\int_{-\infty}^{\infty} f$ to denote the common value of these integrals. It follows easily from our previous work that the family $\mathfrak{R}(-\infty, \infty)$ of all functions Riemann-integrable over **R** is closed under addition, multiplication, and the lattice operations. We also have the usual properties of definite integrals (linearity, positivity, monotonicity).

If A is a bounded subset of **R**, then the characteristic function of A,

$$\varphi_A(x) = \begin{cases} 1 & \text{if } x \in A, \\ 0 & \text{if } x \notin A, \end{cases}$$

is bounded and has compact support. We may therefore define the *inner and outer* (Jordan) *length* of A to be

$$l_*(A) = \int_{-\infty}^{\infty} \varphi_A$$

and

$$l^*(A) = \overline{\int_{-\infty}^{\infty}} \varphi_A,$$

respectively. Should these two numbers be equal, we say that A is Jordan-measurable, and $l(A) = l_*(A) = l^*(A)$ is called the *length* (or linear Jordan content) of A. (Note, then, that A is Jordan-measurable iff φ_A is Riemann-integrable.)

Certain closure properties of the family $\mathcal{J}(\mathbf{R})$ of all Jordan-measurable subsets of $\mathbf{R}$ follow from the closure properties of $\mathfrak{R}(-\infty, \infty)$. In particular, the union of two disjoint Jordan-measurable sets is again Jordan-measurable. (Two sets A and B are *disjoint* iff $A \cap B = \varnothing$. A family of sets, say, $\{A_1, A_2, \ldots, A_n\}$, is disjoint iff any two of the sets in the family are disjoint.) Also, if A and B are Jordan-measurable sets, then the relative complement $A \backslash B = A \cap B'$ is again a Jordan-measurable set. A family of sets with these properties is called a *ring* (or *Boolean algebra*).

Definition 1. Let S be a set. By a **ring** of subsets of S we shall mean a nonempty family $\mathfrak{R}$ of subsets of S such that

1) if A and B are disjoint sets belonging to $\mathfrak{R}$, then $A \cup B$ is also a member of $\mathfrak{R}$;

2) if $A \in \mathfrak{R}$ and $B \in \mathfrak{R}$, then $(A \backslash B) \in \mathfrak{R}$.

In other words, a ring is a nonvoid family of sets which is closed under the formation of disjoint unions and relative complements.

It also can be shown that the set function l has these properties:

1) If A and B are disjoint measurable sets, then

$$l(A \cup B) = l(A) + l(B).$$

2) $l(A) \geq 0$ for all $A \in \mathcal{J}(\mathbf{R})$.

3) If I is a bounded interval, then I is Jordan-measurable and $l(I)$ is the difference of its endpoints.

Definition 2. By a **content** we shall mean a set function μ with the following properties:

1) $\mathfrak{D}_\mu$ is a ring of sets.

2) *Positivity.* For all $A \in \mathfrak{D}_\mu$, we have $\mu(A) \geq 0$.

3) *Finite Additivity.* If A and B are disjoint members of $\mathfrak{D}_\mu$, then

$$\mu(A \cup B) = \mu(A) + \mu(B).$$

We shall now give examples of rings and contents.

Example 1. It is easily shown that the family $\mathfrak{B}$ of all bounded subsets of $\mathbf{R}$ is a ring. Although the functions l_* and l^*, defined on $\mathfrak{B}$, are positive, they are not finitely additive, and therefore are not contents. Let A be the set of all rational numbers in the interval

$[0, 1]$, and let $B = [0, 1] \backslash A$. Then, by Example 3, Section 8-1, $l_*(A) = l_*(B) = 0$ and $l^*(A) = l^*(B) = 1$, whereas

$$1 = l^*(A \cup B) \neq l^*(A) + l^*(B),$$
$$1 = l_*(A \cup B) \neq l_*(A) + l_*(B).$$

Example 2. Let $S = \{1, 2, 3\}$, and let $\mathcal{F}$ be the family consisting of the following subsets of S:

$$A_1 = \{1\}, \qquad A_2 = \{2, 3\}, \qquad A_3 = \{1, 3\}.$$

Then $\mathcal{F}$ is not a ring of subsets of S. For one thing, $\mathcal{F}$ is not closed under the formation of disjoint unions. $(A_1 \cup A_2 = S \notin \mathcal{F}.)$ In addition, $\mathcal{F}$ is not closed under the formation of relative complements. $(A_3 \backslash A_1 = \{3\} \notin \mathcal{F}.)$

Example 3. Let S be any set, and let $\mathcal{F}$ consist of all subsets of S (including $\varnothing$) which contain only a finite number of elements. For instance, if S were the set of positive integers, then the set $\{1, 5, 20\}$ would belong to $\mathcal{F}$, whereas the set of all even integers would not. It should be fairly clear that $\mathcal{F}$ is a ring. (Suppose that A and B are disjoint members of $\mathcal{F}$. Then A has a finite number of elements, say m, and B has a finite number of elements n. Since A and B are disjoint, $A \cup B$ will contain $m + n$ elements, and thus $A \cup B$, being a finite set, must belong to $\mathcal{F}$. Hence, $\mathcal{F}$ is closed under the formation of disjoint unions. The requirement that $\mathcal{F}$ be closed under the formation of relative complements is satisfied, since any subset of a finite set is finite.) Moreover, if we let $\mu(A)$ be the number of elements of A ($A \in \mathcal{F}$), then it is clear that μ is a content.

We shall prove next some basic facts about rings and contents.

Proposition 3. *Let $\mathcal{R}$ be a ring of subsets of a set S. Then $\mathcal{R}$ contains the null set and is closed under the formation of finite unions and intersections.*

Proof. Let $A \in \mathcal{R}$. (Since, by assumption, $\mathcal{R}$ is a *nonempty* family of sets, such an A exists.) Since $\mathcal{R}$ is closed under the formation of relative complements, $\varnothing = A \backslash A$ belongs to $\mathcal{R}$.

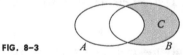

FIG. 8-3

Suppose that $A \in \mathcal{R}$ and $B \in \mathcal{R}$. Then $C = B \backslash A$ is also in $\mathcal{R}$ (since $\mathcal{R}$ is closed under the formation of relative complements). Consequently, $A \cap B = B \backslash C$ is in $\mathcal{R}$. Also, since A and C are disjoint sets in $\mathcal{R}$, it follows that $A \cup B = A \cup C$ is in $\mathcal{R}$ (see Fig. 8-3). (By definition, $\mathcal{R}$ is closed under the formation of the union of two disjoint sets.) Thus we have proved that the union or intersection of any two sets in $\mathcal{R}$ is also in $\mathcal{R}$. It follows by a simple induction argument that if $A_1, \ldots, A_n$ are sets in $\mathcal{R}$, then $A_1 \cup \cdots \cup A_n$ and $A_1 \cap \cdots \cap A_n$ also belong to $\mathcal{R}$.

Proposition 4. *Let μ be a content defined on a ring $\mathcal{R}$ of subsets of a set S. Then μ has the following properties:*

*1) μ is **monotone** in the sense that $\mu(A) \leq \mu(B)$ whenever A and B are members of $\mathcal{R}$ and $A \subset B$.*

2) *If $A_1, \ldots, A_n$ are disjoint sets in $\mathfrak{R}$, then*

$$\mu\left(\bigcup_{k=1}^{n} A_k\right) = \sum_{k=1}^{n} \mu(A_k).$$

3) *μ is **finitely subadditive**; that is, if $A_1, \ldots, A_n$ belong to $\mathfrak{R}$ (but are not necessarily disjoint), then*

$$\mu\left(\bigcup_{k=1}^{n} A_k\right) \leq \sum_{k=1}^{n} \mu(A_k).$$

4) *If $A \in \mathfrak{R}$ and $B \in \mathfrak{R}$, then*

$$\mu(A \cup B) + \mu(A \cap B) = \mu(A) + \mu(B).$$

Proof. 1) Suppose that $A \in \mathfrak{R}$, $B \in \mathfrak{R}$, and $A \subset B$. Then $C = B \backslash A$ also belongs to $\mathfrak{R}$, $A \cap C = \varnothing$, and $A \cup C = B$ (see Fig. 8–4). Consequently,

$$\mu(B) = \mu(A) + \mu(C) \geq \mu(A) + 0 = \mu(A).$$

Property (2) can be proved by a simple induction argument.

FIG. 8–4 B **FIG. 8–5** A B

We prove next property (4). Suppose that A and B belong to $\mathfrak{R}$. Since $\mathfrak{R}$ is a ring, it contains the sets $A \cap B$, $A \cup B$, and $C = B \backslash A$. Since C and $A \cap B$ are disjoint (see Fig. 8–5),

$$\mu(C) + \mu(A \cap B) = \mu\big(C \cup (A \cap B)\big) = \mu(B).$$

Since A and C are disjoint,

$$\mu(A) + \mu(C) = \mu(A \cup C) = \mu(A \cup B).$$

Combining the two equations, we get

$$\mu(A \cup B) + \mu(A \cap B) = \mu(A) + \mu(C) + \mu(A \cap B)$$
$$= \mu(A) + \mu(B).$$

It follows from (4) that

$$\mu(A \cup B) \leq \mu(A) + \mu(B)$$

for any two sets A and B in $\mathfrak{R}$. From this inequality, the finite subadditivity of μ follows by induction.

It is possible for a set function to be monotone without being a content. For instance, I_* and I^* are both monotone.

We shall develop the Jordan theory of area. It parallels the theory of the Riemann integral to a considerable extent. We shall have the following correspondences:

functions	sets
definite integral	area content
lower Riemann integral	inner Jordan content
upper Riemann integral	outer Jordan content
Riemann-integrable function	Jordan-measurable set
Riemann integral	Jordan content
step function	block

Definition 5. By an **area content** we shall mean any content μ such that

1) $\mathfrak{D}_\mu$ is a family of bounded subsets of $\mathbf{R}^2$,
2) if I and J are bounded intervals of $\mathbf{R}$, then $I \times J$ belongs to $\mathfrak{D}_\mu$ and

$$\mu(I \times J) = l(I)l(J)$$

(see Fig. 8–6).

Sets of the form $I \times J$, where I and J are intervals, are called *two-dimensional intervals*. A bounded two-dimensional interval is just a rectangle with sides parallel to the coordinate axes. We shall call such rectangles *standard rectangles*. An area content assigns to such a rectangle the product of its base and height.

Let us assume momentarily that an area content μ exists. If B is the union of a finite number of disjoint standard rectangles, then B must belong to $\mathfrak{D}_\mu$. Such a set B will be called a *block*. The family $\mathfrak{B}_0$ of all blocks forms a ring, as will be shown in the exercises.

Proposition 6. *There exists a unique area content m whose domain is $\mathfrak{B}_0$. Moreover, every area content is an extension of m.*

We shall prove only the existence of m. If B is a block, then for each $x \in \mathbf{R}$, the vertical cross section $B_x = \{y \mid (x, y) \in B\}$ is a finite disjoint union of bounded

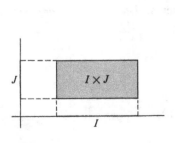

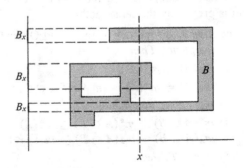

FIG. 8–6   **FIG. 8–7**

intervals and hence belongs to $\mathscr{J}(\mathbf{R})$. Moreover, the function $f(x) = l(B_x)$ has compact support and is a step function (see Fig. 8–7). Thus we may define

$$m(B) = \int_{-\infty}^{\infty} l(B_x)\, dx.$$

If B and C are disjoint blocks, then for each $x \in \mathbf{R}$, the cross sections B_x and C_x are disjoint sets in $\mathscr{J}(\mathbf{R})$, $(B \cup C)_x = B_x \cup C_x$, and thus

$$l((B \cup C)_x) = l(B_x) + l(C_x).$$

The identity

$$m(B \cup C) = m(B) + m(C)$$

follows upon integration from the linearity property of Riemann integrals. Thus m is finitely additive. It is obvious, on the other hand, that m is nonnegative.

If $B = I \times J$ is a standard rectangle, then

$$B_x = \begin{cases} J & \text{if } x \in I, \\ \varnothing & \text{if } x \notin I, \end{cases}$$

and $l(B_x) = l(J)\varphi_I(x)$, so that

$$m(B) = \int_{-\infty}^{\infty} l(J)\varphi_I(x)\, dx = l(J)l(I).$$

Assuming, then, that the family $\mathscr{B}_0$ of blocks is a ring, we have shown that m is an area content on $\mathscr{B}$.

Definition 7. Let B be any bounded subset of $\mathbf{R}^2$. The numbers

$$m_*(B) = \text{lub } \{m(C) \mid C \in \mathscr{B}_0 \text{ and } C \subset B\}$$

and

$$m^*(B) = \text{glb } \{m(C) \mid C \in \mathscr{B}_0 \text{ and } B \subset C\},$$

respectively, are called the **inner** and **outer Jordan content** of B.

Proposition 8. *Suppose that μ is an area content, or more generally, a monotone set function which extends m. Then for each bounded set $B \in \mathscr{D}_\mu$,*

$$m_*(B) \leq \mu(B) \leq m^*(B).$$

This is analogous to a theorem concerning upper and lower Riemann integrals and is proved by the same method.

Proposition 9. Properties of Inner and Outer Contents. *Let A and B be any bounded sets in the plane. Then*

1) $m_*(A) \leq m^*(A)$;
2) $m_*(A) = m^*(A) = m(A)$, if $A \in \mathscr{B}_0$;
3) m_* and m^* are monotone;
4) $m^*(A \cup B) + m^*(A \cap B) \leq m^*(A) + m^*(B)$,
   $m_*(A \cup B) + m_*(A \cap B) \geq m_*(A) + m_*(B)$;
5) *if $A \subset B$, then*
$$m^*(B \backslash A) \leq m^*(B) - m_*(A),$$
$$m_*(B \backslash A) \geq m_*(B) - m^*(A).$$

Proof. We shall prove only the first inequalities in (4) and (5).

4) Let ϵ be any positive number. We can find blocks A^* and B^* such that $A \subset A^*$, $B \subset B^*$,

$$m^*(A) + \epsilon > m(A^*), \quad \text{and} \quad m^*(B) + \epsilon > m(B^*).$$

(Why?) Clearly, $A \cap B \subset A^* \cap B^*$ and $A \cup B \subset A^* \cup B^*$. Hence

$$m^*(A \cup B) + m^*(A \cap B) \leq m(A^* \cup B^*) + m(A^* \cap B^*)$$
$$= m(A^*) + m(B^*) \leq m^*(A) + m^*(B) + 2\epsilon.$$

Since $\epsilon > 0$ is arbitrary, we must have

$$m^*(A \cup B) + m^*(A \cap B) \leq m^*(A) + m^*(B)$$

(see Fig. 8–8).

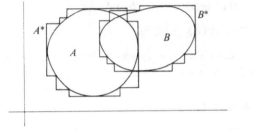

FIG. 8–8   **FIG. 8–9**

5) Let $A \subset B$. Suppose that B^* and A_* are blocks such that $B \subset B^*$ and $A_* \subset A$. Then $A_* \subset B^*$ and $B^* \backslash A_*$ is a block containing $B \backslash A$, as illustrated in Fig. 8–9. Thus

$$m^*(B \backslash A) \leq m(B^* \backslash A_*) = m(B^*) - m(A_*)$$

or

$$m(A_*) + m^*(B \backslash A) \leq m(B^*).$$

It follows that for each block A_* such that $A_* \subset A$, the number $m(A_*) + m^*(B \backslash A)$ is a lower bound of the set $\{m(B^*) \mid B^* \in \mathcal{B}_0 \text{ and } B \subset B^*\}$. Hence

$$m(A_*) + m^*(B \backslash A) \leq \text{glb } \{m(B^*) \mid B^* \in \mathcal{B}_0 \text{ and } B \subset B^*\} = m^*(B).$$

Since $m^*(B) - m^*(B \backslash A)$ is an upper bound of the set $\{m(A_*) \mid A_* \in \mathcal{B}_* \text{ and } A_* \subset A\}$, we get

$$m^*(B) - m^*(B \backslash A) \geq m_*(A).$$

Thus $m^*(B \backslash A) \leq m^*(B) - m_*(A)$.

Definition 10. We say that a bounded subset A of the plane is **Jordan-measurable** iff $m_*(A) = m^*(A)$, in which case the **Jordan content** of A is defined to be

$$m(A) = m_*(A) = m^*(A).$$

Observe that this does not conflict with our previous use of the letter m, since if A is a block, then $m_*(A) = m^*(A) = $ area of A.

Theorem 11. *m is an area content.*

Proof. Suppose that A and B are Jordan-measurable. Then

$$m^*(A \cup B) + m^*(A \cap B) \leq m^*(A) + m^*(B)$$
$$= m_*(A) + m_*(B) \leq m_*(A \cup B) + m_*(A \cap B).$$

Since $m_*(A \cup B) \leq m^*(A \cup B)$ and $m_*(A \cap B) \leq m^*(A \cap B)$, it follows that

$$m_*(A \cup B) = m^*(A \cup B) \quad \text{and} \quad m_*(A \cap B) = m^*(A \cap B).$$

Thus $A \cup B$ and $A \cap B$ are Jordan-measurable. Furthermore, we also have

$$m(A \cup B) + m(A \cap B) = m(A) + m(B).$$

In particular, if A and B are disjoint, then $m(A \cap B) = 0$ and

$$m(A \cup B) = m(A) + m(B).$$

To complete the proof, we must show that $\mathcal{J}$ is closed under the formation of relative complements. Let A and B be Jordan-measurable. We will prove that $A \backslash B$ is Jordan-measurable. Since

$$A \backslash B = A \backslash (A \cap B)$$

and since $A \cap B \in \mathcal{J}$, it suffices to consider the case in which $B \subset A$. Now

$$m^*(A \backslash B) \leq m^*(A) - m_*(B) = m_*(A) - m^*(B) \leq m_*(A \backslash B).$$

Since $m_*(A \backslash B) \leq m^*(A \backslash B)$, we have $m^*(A \backslash B) = m_*(A \backslash B)$. Thus $A \backslash B$ is Jordan-measurable.

Before we accept Jordan content as an explication of the term "area," we must check whether the family $\mathcal{J}$ is sufficiently large, that is, whether it includes most of the sets we might be interested in. Also we need an effective means of calculating Jordan contents.

Theorem 12. Fubini Theorem for Jordan-Measurable Sets. *Let B be a Jordan-measurable subset of $\mathbf{R}^2$. Then the functions*

$$f(x) = l_*(B_x) \quad \text{and} \quad g(x) = l^*(B_x)$$

are both Riemann-integrable over $\mathbf{R}$, and

$$m(B) = \int_{-\infty}^{\infty} l_*(B_x)\, dx = \int_{-\infty}^{\infty} l^*(B_x)\, dx.$$

Proof. For each bounded set $A \subset \mathbf{R}^2$, we define

$$\mu(A) = \underline{\int_{-\infty}^{\infty}} l_*(B_x)\, dx.$$

Then it can be verified that μ is a monotone set function which extends the area content on $\mathcal{B}_0$. Consequently, we have (by Proposition 8),

$$m_*(B) \le \mu(B) = \int_{-\infty}^{\infty} l_*(B_x)\, dx.$$

Similarly,

$$\overline{\int}_{-\infty}^{\infty} l_*(B_x)\, dx \le m^*(B).$$

Thus

$$m_*(B) \le \underline{\int}_{-\infty}^{\infty} l_*(B_x)\, dx \le \overline{\int}_{-\infty}^{\infty} l_*(B_x)\, dx \le m^*(B).$$

But B is Jordan-measurable, and thus $m_*(B) = m^*(B)$. It follows that $f(x) = l_*(B_x)$ is Riemann-integrable and that

$$m(B) = \int_{-\infty}^{\infty} l_*(B_x)\, dx.$$

A similar proof works for l^*.

The problem of finding the "area beneath a curve" is often used to motivate the definition of the Riemann integral. It can be seen from the next result that Jordan content is the appropriate explication of area in this context.

Corollary 13. *Let f be a nonnegative function on the interval $[a, b]$, and let O_f be the "ordinate set"* $\{(x, y) \mid a \le x \le b \text{ and } 0 \le y \le f(x)\}$ *(see* Fig. 8–10*). Then O_f is Jordan-measurable if and only if f is Riemann-integrable on $[a, b]$, in which case*

$$m(O_f) = \int_a^b f(x)\, dx.$$

Proof. If O_f is Jordan-measurable, then

$$m(O_f) = \int_{-\infty}^{\infty} l_*((O_f)_x)\, dx.$$

But

$$l_*((O_f)_x) = \begin{cases} f(x) & \text{if } a \le x \le b, \\ 0 & \text{otherwise,} \end{cases}$$

so that

$$m(O_f) = \int_a^b f(x)\, dx.$$

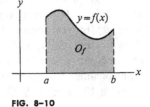

FIG. 8–10

Suppose, conversely, that f is Riemann-integrable on $[a, b]$. Given $\epsilon > 0$, there exist step functions g and h such that $g \le f \le h$ on $[a, b]$ and $\int_a^b (h - g) < \epsilon$. But O_g and O_h are blocks, $O_g \subset O_f \subset O_h$, and

$$m(O_h) - m(O_g) = \int_a^b h - \int_a^b g = \int_a^b (h - g) < \epsilon.$$

Thus O_f is Jordan-measurable. (We are assuming an easily proved criterion for Jordan measurability analogous to the Riemann condition.)

We showed earlier that if the functions f and g are continuous, then $\int_a^b (f - g)$ represents the area of a region bounded by the graphs of the functions. In doing so, we took for granted the existence of areas of various sets and the monotonicity of area. We can now deliver a definitive statement concerning such matters.

Corollary 14. *Let f and g be defined on $[a, b]$, and let B be the region bounded by the graphs of f and g and the lines $x = a$ and $x = b$. If B is Jordan-measurable, then $|f - g|$ is Riemann-integrable on $[a, b]$ and*

$$m(B) = \int_a^b |f - g|.$$

Conversely, if f and g are Riemann-integrable on $[a, b]$, then B is Jordan-measurable.

Proof. The first assertion follows from the Fubini theorem and the observation that

$$|f(x) - g(x)| = l_*(B_x) = l^*(B_x)$$

if $a \leq x \leq b$.

The second part can be handled in two stages.

Case I. The functions f and g are both nonnegative on the interval $[a, b]$. If f and g are Riemann-integrable on $[a, b]$, then so are $f \wedge g$ and $f \vee g$. Therefore, $O_{f \wedge g}$ and $O_{f \vee g}$ are Jordan-measurable, and hence so is $B = O_{f \vee g} \backslash O_{f \wedge g}$.

Case II. If f and g are Riemann-integrable, they are bounded. A constant C can therefore be chosen so that $f^* = f + C$ and $g^* = g + C$ are both nonnegative on $[a, b]$. By case I, the region B^* bounded by the graphs of f^* and g^* is Jordan-measurable. However, one can obtain B from B^* by translating B^* vertically, as shown in Fig. 8–11, and it is easily shown that such an operation does not alter the outer or inner Jordan content of a set (see Exercise 17, Section 8–5). Thus

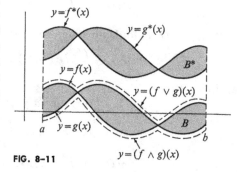

$$m_*(B) = m_*(B^*) = m^*(B^*) = m^*(B),$$

that is, B is Jordan-measurable.

FIG. 8–11

In this last proof we have used the fact that m is translation invariant. A stronger result is true, namely, that m is invariant under congruence. If A and B are congruent subsets of $\mathbf{R}^2$, then $m_*(A) = m_*(B)$ and $m^*(A) = m^*(B)$, so that if one set is Jordan-measurable, the other is also, and the Jordan contents are equal. We shall defer proofs of these assertions until a later chapter.

What we have done on the line and in the plane with Jordan content can be extended to higher-dimensional spaces by induction. Some of this is carried out in the exercises.

Let us close with a backward look at what we have done. The real point at issue in this section is the *consistency* of our intuitive notions of areas (additivity, invariance under congruence, etc.). While it might not occur to most people that this is a serious and nontrivial problem, the Banach-Tarski theorem strikingly demonstrates that it is. Certainly, one cannot take refuge in the comfortable statement, "the whole of a thing is equal to the sum of its parts." Far from being the "self-evident truth" it was once thought to be, it is an utterance quite devoid of real meaning. One can test its validity only in very limited situations where an exact procedure for *measuring* the "whole" and the "parts" has been specified. In the case of areas, for instance, the statement is false if we take the outer Jordan content as a measure of the area of a bounded set. What we have shown, however, is that if we restrict our attention to Jordan-measurable sets, the statement is true when Jordan content is taken as the measure of area. Fortunately, the class of Jordan-measurable sets is large enough to satisfy most needs in elementary analysis, and one cannot get outside the family by the usual finite set operations. Furthermore, Jordan content goes hand in glove with the Riemann integral. The Fubini theorem tells us, for instance, that the content of a Jordan-measurable set is the Riemann integral of the lengths of vertical cross sections. Also, we have seen that regions bounded by the graphs of functions are Jordan-measurable if the functions are Riemann-integrable. We have not, to be sure, given the *only* possible explication of area, nor ultimately the best one. We have given, however, the interpretation of "area" most appropriate at this stage. Perhaps the time will come when the reader will need to learn about the Lebesgue measure of area and the Lebesgue integral, but that is another story.

Exercises

1. Consider the following two-dimensional intervals:

$$B_1 = [0, 4] \times [0, 3], \qquad B_2 = [1, 3) \times (2, 3),$$
$$B_3 = (2, 5] \times [2, 5), \qquad B_4 = [0, 3) \times [4, 5).$$

It follows from the general theory that each of the sets $B_1 \cup B_2 \cup B_3 \cup B_4$, $B_1 \backslash B_2$, $(B_3 - B_2) \cap B_1$ is a block. Prove that this is so by exhibiting in each case a decomposition of the set into disjoint elementary blocks. Find the area of each block.

2. The purpose of this exercise is to prove that the family $\mathfrak{B}_0$ of blocks is a ring. It is obvious that the union of two disjoint blocks is again a block. We show that $\mathfrak{B}_0$ is closed under the formation of relative complements by a series of steps.

   a) Prove that the intersection of two two-dimensional intervals is again a two-dimensional interval which is bounded if either of the original intervals is bounded.

   b) Prove that the complement of a bounded two-dimensional interval can be expressed as a finite union of unbounded two-dimensional intervals. What is the minimum number of intervals in such a decomposition?

   c) Show that the relative complement of two bounded two-dimensional intervals is a block. Draw illustrations of the types of blocks that can arise in this way. (Let

A and B be the two-dimensional intervals. If $B' = C_1 \cup \cdots \cup C_m$, where $C_1, \ldots,$ C_m are disjoint unbounded intervals, show that $A \cap C_1, \ldots, A \cap C_m$ are disjoint bounded intervals and that A is their union.)

d) Verify the set identity

$$\left(\bigcup_{k=1}^{n} A_k \right) \setminus B = \bigcup_{k=1}^{n} (A_k \setminus B),$$

and use it to show that if A is a block and B is a bounded interval, then $A \setminus B$ is a block.

e) Verify the set identity

$$A \setminus (B \cup C) = (A \setminus B) \setminus C,$$

and use it to prove by induction that $\mathfrak{B}_0$ is closed under the formation of relative complements.

3. Let μ be a content on a ring of sets $\mathfrak{R}$.

a) Prove that $\mu(\varnothing) = 0$.

b) If A, B, and C are in $\mathfrak{R}$, prove that

$$\mu(A \cup B \cup C) = \mu(A) + \mu(B) + \mu(C) - \mu(A \cap B)$$
$$- \mu(A \cap C) - \mu(B \cap C) + \mu(A \cap B \cap C).$$

*c) If $A_1, \ldots, A_n$ belong to $\mathfrak{R}$, prove that

$$\mu \left(\sum_{k=1}^{n} A_k \right) = S_1 - S_2 + S_3 - \cdots + (-1)^{n-1} S_n,$$

where

$$S_k = \sum \mu(A_{i_1} \cap A_{i_2} \cap \cdots \cap A_{i_k});$$

the summation is taken over all sets of indices $(i_1, \ldots, i_k)$ for which

$$1 \leq i_1 < i_2 < \cdots < i_k \leq n.$$

4. a) Let $\mathfrak{F}$ be the ring of all finite subsets of $\mathbf{Z}^+$. Let μ be the function which assigns to each set in $\mathfrak{F}$ the sum of the numbers in the set. Thus $\mu(\{1, 3, 8\}) = 12$. Show that μ is a content.

b) More generally, let $\mathfrak{F}$ be the ring of finite subsets of a set S, and let f be a nonnegative real-valued function defined on S. We define

$$\mu_f(A) = \sum_{p \in A} f(p), \qquad \mu_f(\varnothing) = 0 \qquad (\varnothing \neq A \in \mathfrak{F}).$$

Verify that μ_f is a content. Prove, moreover, that every content μ on $\mathfrak{F}$ is of this form.

5. Let $\mathfrak{B}^*$ be the class of all bounded subsets of the plane. Prove in detail that $\mathfrak{B}^*$ is a ring of sets. For each nonvoid set A in $\mathfrak{B}^*$, we define the diameter of A to be the least upper bound of the set of all distances between points in A; thus

$$d(A) = \text{lub } \{|PQ| \mid P \in A \text{ and } Q \in A\}.$$

We define $d(\varnothing) = 0$. Show that d is not a content although it is monotone. Is d subadditive?

6. Let $\mathfrak{R}$ consist of all subsets of the real line which can be expressed as a finite disjoint union of bounded intervals of the form $[a, b)$. Convince yourself that $\mathfrak{R}$ is a ring. Let f be any nondecreasing real-valued function defined on $\mathbf{R}$. Show that there is a unique content μ_f on $\mathfrak{R}$ with the property that

$$\mu_f([a, b)) = f(b) - f(a).$$

(A detailed verification would be quite lengthy. Try to indicate at least the crucial steps in the verification, and indicate what technical difficulties there are.) Show conversely that every content on $\mathfrak{R}$ arises in this way.

7. Let $\mathfrak{R}$ be a ring of sets. Define

$$A + B = (A \backslash B) \cup (B \backslash A), \qquad AB = A \cap B$$

for all $A \in \mathfrak{R}$ and $B \in \mathfrak{R}$. Show that with these operations, $\mathfrak{R}$ satisfies Axioms A1, A2, A3, A4, A5, M1, M2, M5, and AM for a field. Such an algebraic system is called a *commutative ring*. Note that this ring $\mathfrak{R}$ has two somewhat peculiar properties, namely,

$$A + A = 0 \qquad \text{and} \qquad AA = A$$

for all $A \in \mathfrak{R}$.

8. Let $\mathfrak{F}$ be a ring of subsets of a set S. If S itself is a member of $\mathfrak{F}$, we say that $\mathfrak{F}$ is a *field* of subsets of S. Show that a field $\mathfrak{F}$ is closed under the formation of complements. It follows that any set which can be built up from a finite number of sets in $\mathfrak{F}$ by the usual set operations ($\cup, \cap, '$) also belongs to $\mathfrak{F}$. Also observe that with the operations of the previous problem, $\mathfrak{F}$ also satisfies Axiom M3. Such an algebraic structure is called a *commutative ring with identity*.

   Let $\mathfrak{R}$ be a ring of subsets of a set S, and let $\mathfrak{F}$ consist of all sets A such that

$$A \in \mathfrak{R} \qquad \text{or} \qquad A' \in \mathfrak{R}.$$

Prove that $\mathfrak{F}$ is a field of subsets of S. Deduce that if $\mathfrak{F}$ consists of all subsets of S which either are finite or have a finite complement, then $\mathfrak{F}$ is a field.

9. State and prove a criterion for Jordan measurability (in the plane) analogous to the Riemann condition for Riemann integrability.

10. If A and B are Jordan-measurable subsets of $\mathbf{R}$, prove that $A \times B$ is a Jordan-measurable subset of $\mathbf{R}^2$ and that

$$m(A \times B) = l(A)l(B).$$

Prove conversely that if $A \times B$ is Jordan-measurable, then A and B are Jordan-measurable. (Use Exercise 9 to prove the first assertion. Use the Fubini theorem to prove the second.)

11. Prove the inequalities

$$m_*(A \cup B) + m_*(A \cap B) \geq m_*(A) + m_*(B), \qquad m_*(B - A) \geq m_*(B) = m^*(A),$$

where A and B are bounded sets, and in the second case, $A \subset B$.

*12. Prove that a bounded set $A \subset \mathbf{R}^2$ is Jordan-measurable iff for every bounded set $B \subset \mathbf{R}^2$,

$$m^*(B) = m^*(B \cap A) + m^*(B \backslash A).$$

[To prove "$\Rightarrow$," consider any $C \in \mathcal{B}$ with $B \subset C$. Show that

$$m(C) \geq m^*(B \cap A) + m^*(B \backslash A),$$

etc. To prove "$\Leftarrow$," consider any $\epsilon > 0$. Choose blocks C and D such that $C \subset A \subset D$, $m(D) - m^*(A) < \epsilon$, and $m^*(A) - m(C) < \epsilon$. Apply the above identity to $B = D \backslash C$ to get $m^*(A) - m_*(A) < 2\epsilon$, etc.]

13. If $A \subset B$ and B is Jordan-measurable, prove that

$$m^*(A) = m(B) - m_*(B \backslash A), \qquad m_*(A) = m(B) - m^*(B \backslash A).$$

14. We shall say that a set $B \subset \mathbf{R}^2$ is *Fubini-measurable* iff the integrals $\int_{-\infty}^{\infty} l_*(B_x)\,dx$ and $\int_{-\infty}^{\infty} l^*(B_x)\,dx$ exist and are equal, in which case we define the *Fubini content* of B to be

$$m_F(B) = \int_{-\infty}^{\infty} l_*(B_x)\,dx = \int_{-\infty}^{\infty} l^*(B_x)\,dx.$$

Prove that m_F is a monotone set function extending the content on $\mathcal{B}$. Prove that the union of two disjoint Fubini-measurable sets A and B is Fubini-measurable and that $m_F(A \cup B) = m_F(A) + m_F(B)$. Prove, however, that m_F is not a content since the family of Fubini-measurable sets is not a ring. (Let A be the unit square $[0, 1] \times [0, 1]$. Let B be the set whose vertical cross sections are as follows:

$$B_x = \begin{cases} [0, 1] & \text{if } 0 \leq x \leq 1 \quad \text{and} \quad x \text{ is rational}, \\ [1, 2] & \text{if } 0 < x < 1 \quad \text{and} \quad x \text{ is irrational}, \\ \varnothing & \text{otherwise}. \end{cases}$$

Show that A and B are Fubini-measurable, whereas neither $A \backslash B$ nor $B \backslash A$ is.)

15. State and prove a version of the Fubini theorem involving horizontal rather than vertical cross sections.

16. If B is a subset of $\mathbf{R}^2$ and $r > 0$, we denote by rB the set $\{(rx, ry) \mid (x, y) \in B\}$.

   a) For each bounded subset $B \subset \mathbf{R}^2$, we let $\mu_r(B)$ denote the number of lattice points in rB divided by r^2. (By a lattice point in $\mathbf{R}^2$ is meant an ordered pair of integers.) For a fixed r, show that μ_r is a content, but not an area content.

   *b) We say that a bounded set $A \subset \mathbf{R}^2$ is *lattice-measurable* iff $\lim_{r \to +\infty} \mu_r(A)$ exists, in which case the limit is denoted by $\mu(A)$ and is called the *lattice content* of A. Prove that μ is an area content. Deduce that if A is both Jordan- and lattice-measurable, then $\mu(A) = m(A)$.

   *c) Prove that every Jordan-measurable set is lattice-measurable.

17. a) If $\alpha = (a_1, a_2)$ is a fixed point in $\mathbf{R}^2$, and if $B \subset \mathbf{R}^2$, we define the *translate of B by α* to be the set

$$B + \alpha = \{(x + a_1, y + a_2) \mid (x, y) \in B\}.$$

Interpret the definition geometrically through examples. If B is a bounded subset of $\mathbf{R}^2$, show that

$$m_*(B + \alpha) = m_*(B), \qquad m^*(B + \alpha) = m^*(B).$$

(Check this first for two-dimensional intervals, then for blocks, and finally for bounded sets.) Deduce that B is Jordan-measurable iff $B + \alpha$ is, in which case

$$m(B + \alpha) = m(B).$$

b) If $B \subset \mathbf{R}^2$, we define
$$B^c = \{(x, y) \mid (y, x) \in B\}.$$

Show that B^c can be thought of as the reflection of B in the line $y = x$. If B is bounded, prove that $m_*(B) = m_*(B^c)$ and $m^*(B) = m^*(B^c)$.

18. By a *disk* centered at a point (x_0, y_0) we shall mean any set of the form
$$\{(x, y) \mid (x - x_0)^2 + (y - y_0)^2 < r\},$$

where $r > 0$. If A is a subset of $\mathbf{R}^2$, we say that a point P is an *interior point of A* iff *there exists* a disk D with center P such that $D \subset A$. We say that P is a *closure point* of A iff *every* disk D with center P contains a point which belongs to A (see Fig. 8–12). The *interior* of A, which we denote by A^0, is the set consisting of all interior points of A. The *closure* of A, which we denote by $\overline{A}$, is the set consisting of all closure points of A. For instance, if $A = (1, 2] \times [0, 2)$, one can easily convince oneself that

$$A^0 = (1, 2) \times (0, 2) \quad \text{and} \quad \overline{A} = [1, 2] \times [0, 2].$$

It is clear that $A^0 \subset A \subset \overline{A}$. In addition, the operations of "interior" and "closure" are monotone: if $A \subset B$, then $A^0 \subset B^0$ and $\overline{A} \subset \overline{B}$.

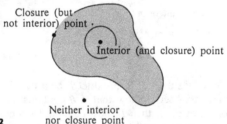

Closure (but not interior) point

Interior (and closure) point

Neither interior nor closure point

FIG. 8–12

a) Find the interior and closure of each of the following sets. So far as it is possible, draw a picture of each of the sets.

$$\{(x, y) \mid 1 < x^2 + y^2 \leq 4\}$$
$$\{(x, y) \mid 0 \leq y < 2x\}$$
$$\{(x, y) \mid x \text{ and } y \text{ are integers}\}$$
$$\{(1/m, 1/n) \mid m, n \in \mathbf{Z}^+\}$$
$$\{(x, y) \mid x \text{ and } y \text{ are rational numbers in } [0, 1]\}$$
$$\{(x, y) \mid x \text{ and } y \text{ are irrational numbers in } [0, 1]\}$$

b) If B is a block, convince yourself that B^0 and $\overline{B}$ are blocks and that
$$m(B^0) = m(B) = m(\overline{B}).$$
Prove then that
$$m_*(A) = m_*(A^0) \quad \text{and} \quad m^*(A) = m^*(\overline{A})$$
for every bounded set A. [Let B be any block such that $A \subset B$. Argue that
$$m^*(\overline{A}) \leq m^*(\overline{B}) = m(B).$$
Deduce that $m^*(\overline{A}) \leq m^*(A)$, etc.]

   c) Use (a) and (b) to determine the outer and inner Jordan contents of the last two sets listed in (a).

   d) Prove that a set A is Jordan-measurable iff A^0 and $\overline{A}$ are both Jordan-measurable and $m(A^0) = m(\overline{A})$.

19. (Continuation of Exercise 18.) The *boundary* of a set $A \subset \mathbf{R}^2$ is the set $\partial A = \overline{A} \backslash A^0$. Any point in ∂A is called a *boundary point* of A.

   a) Find the boundary of each of the sets in part (a) of Exercise 18.

   b) Show that P is a boundary point of A iff every disk with center P contains at least one point in A and at least one point not in A. Deduce that $\partial A = \overline{A} \cap \overline{(A')}$.

   c) P is an *exterior point* of A iff $P \notin \overline{A}$. Prove that P is an exterior point of A iff P is an interior point of A'. (See Fig. 8–13.)

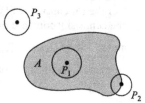

P_1: interior point of A

P_2: boundary point of A

FIG. 8–13 P_3: exterior point of A

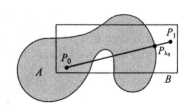

FIG. 8–14

   d) Let A be an arbitrary subset of $\mathbf{R}^2$, and let B be a two-dimensional interval (or, if you prefer, any convex set). Show that if B contains both an interior point of A and an exterior point of A, then B must contain a boundary point of A (see Fig. 8–14). [Suppose that $P_0 \in A^0 \cap B$ and that $P_1 \in (A')^0 \cap B$. Set $P_0 = (x_0, y_0)$ and $P_1 = (x_1, y_1)$. The line segment joining P_0 and P_1 lies in B. If $0 \le \lambda \le 1$, we let P_λ be the point on this line segment having the coordinates

$$x = (1 - \lambda)x_0 + \lambda x_1, \qquad y = (1 - \lambda)y_0 + \lambda y_1.$$

Let $\lambda_0 = \operatorname{lub} \{\lambda \mid 0 \le \lambda \le 1 \text{ and } P_\lambda \in A^0\}$. Prove that P_{λ_0} is a boundary point of A.]

   e) If A is a bounded subset of $\mathbf{R}^2$, prove that

$$m^*(\partial A) = m^*(A) - m_*(A).$$

[Using Exercise 18, show that $m^*(\partial A) \le m^*(A) - m_*(A)$. To establish the reverse inequality, consider any block B such that $\partial A \subset B$. Choose a two-dimensional interval C such that $A \cup B \subset C^0$. Then $C\backslash B$ is a block, and hence can be written as a finite disjoint union of bounded two-dimensional intervals (see Fig. 8–15). Since none of these rectangles contains boundary points of A, it follows from (d) that each of these rectangles consists entirely of interior points of A or else entirely of exterior points of A. Let D be the union of those rectangles consisting of interior points of A. Show that

$$m(B) = m(B \cup D) - m(D) \ge m^*(\overline{A}) - m_*(A^0) = m^*(A) - m_*(A).$$

Deduce that $m^*(\partial A) \ge m^*(A) - m_*(A)$.]

   f) Show that A is Jordan-measurable iff $m^*(\partial A) = 0$.

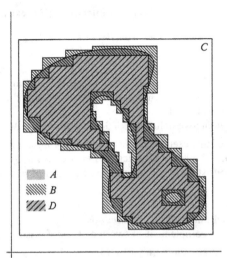

FIG. 8-15

20. *Higher-Dimensional Jordan Contents.* We have treated Jordan content on the line **R** and in the plane **R**^{2}. The next step would be to treat Jordan content in three-space **R**^{3}. Such a content would correspond to our intuitive notion of volume. From there we can climb upwards to **R**^{4}, **R**^{5}, etc. We shall outline an inductive procedure which will do just this.

a) We start with some general observations. Let S be a set, $\mathcal{R}_0$ a ring of subsets of S, and μ_0 a content with domain $\mathcal{R}_0$. We shall say that a set $A \subset S$ is $\mathcal{R}_0$-*bounded* iff there exists a set $B \in \mathcal{R}_0$ such that $A \subset B$. Prove that the family $\mathcal{R}^*$ of all $\mathcal{R}_0$-bounded subsets of S is a ring. We then define for each $A \in \mathcal{R}^*$

$$\mu^*(A) = \text{glb } \{\mu_0(B) \mid B \in \mathcal{R}_0 \text{ and } A \subset B\},$$
$$\mu_*(A) = \text{lub } \{\mu_0(B) \mid B \in \mathcal{R}_0 \text{ and } B \subset A\}.$$

The set functions μ^* and μ_* are called the *outer* and *inner contents induced by* μ_0. One can verify that they have all the properties listed for outer and inner Jordan contents. We say that a set $A \in \mathcal{R}^*$ is μ_0-*measurable* iff $\mu_*(A) = \mu^*(A)$, and we then define $\mu(A) = \mu_*(A) = \mu^*(A)$. It can then be shown (by *exactly* the same proofs as before) that the family $\mathcal{R}$ of μ_0-measurable sets is a ring and that μ is a content on $\mathcal{R}$ which extends μ_0. We call μ the *content induced* by μ_0.

In particular, we can let $S = \mathbf{R}^2$; we can let $\mathcal{R}_0$ be the family of blocks, and μ_0 the unique area content on $\mathcal{R}_0$. Then $\mathcal{R}^*$ is the family of all bounded subsets of $\mathbf{R}^2$, μ_* and μ^* are the inner and outer Jordan contents, $\mathcal{R}$ is the family of Jordan-measurable subsets, and μ is the Jordan content.

b) The three-dimensional analog of a standard rectangle is a Cartesian product of three bounded intervals, that is, a rectangular parallelepiped with edges parallel to the coordinate axes. By a three-dimensional block we shall mean a finite disjoint union of such things. The family $\mathcal{B}_0$ of all three-dimensional blocks is a ring. (See the following exercise.) Prove that there exists a unique content function μ_0 with domain $\mathcal{B}_0$ such that

$$\mu_0(I \times J \times K) = l(I)l(J)l(K)$$

whenever I, J, and K are bounded intervals. [For existence, define

$$\mu_0(B) = \int_{-\infty}^{\infty} m(B_x)\, dx,$$

for each $B \in \mathcal{B}_0$, where $B_x = \{(y, z) \mid (x, y, z) \in B\}$, after justifying the existence of the integral.] We then define the three-dimensional Jordan content to be the content μ induced by μ_0. State and prove a Fubini theorem for three-dimensional Jordan-measurable sets.

c) Outline an inductive definition of Jordan content in $\mathbf{R}^n$.

*21. One objective of this exercise is to show that the family of blocks in $\mathbf{R}^3$ is a ring. [See Exercise 20(b).] We begin abstractly. Let S be a set, $\mathcal{I}$ a family of subsets of S, and $\mathcal{F}$ the family consisting of all finite disjoint unions of sets in $\mathcal{I}$. We say that $\mathcal{I}$ is a *pre-ring* iff

1) $\varnothing \in \mathcal{I}$,
2) $\mathcal{I}$ is closed under finite intersections,
3) the relative complement of two members of $\mathcal{I}$ is a member of $\mathcal{F}$.

a) If $S = \mathbf{R}$ and $\mathcal{I}$ is the family of all bounded intervals, show that $\mathcal{I}$ is a pre-ring.
b) Given that $\mathcal{I}$ is a pre-ring, prove that $\mathcal{F}$ (as defined above) is a ring. Deduce from (a) that the family of all finite disjoint unions of bounded intervals of $\mathbf{R}$ is a ring.
c) Let $\mathcal{I}_k$ be a pre-ring of subsets of S_k, for $k = 1, 2$. Prove that

$$\mathcal{I} = \{I_1 \times I_2 \mid I_1 \in \mathcal{I}_1 \text{ and } I_2 \in \mathcal{I}_2\}$$

is a pre-ring of subsets of the set $S_1 \times S_2$.
d) Deduce that the blocks in $\mathbf{R}^2$ and $\mathbf{R}^3$ form rings. Prove by induction that the blocks in $\mathbf{R}^n$ form a ring.

22. In our earlier discussions of moments of plane laminas we assumed that it was possible to assign moments to most bounded sets in the plane, but we did not explicitly single out such a family of sets. The purpose of this exercise is to show that moments can be defined for Jordan-measurable sets in such a way that the previously stated axioms for moments are satisfied.

a) If f and g are Riemann-integrable on $[a, b]$ and if $\int_a^b |g(x)|\, dx = 0$, show that $\int_a^b f(x)g(x)\, dx = 0$. (Examine the proof of the generalized mean-value theorem for integrals.)
b) Let B be a Jordan-measurable subset of $\mathbf{R}^2$ and let x_0 be any real number. Show that the integrals

$$\int_{-\infty}^{\infty} (x - x_0) l_*(B_x)\, dx$$

and

$$\int_{-\infty}^{\infty} (x - x_0) l^*(B_x)\, dx$$

exist and are equal. [Use the Fubini theorem for Jordan-measurable sets and (a).]
c) If B is Jordan-measurable, we define

$$M_{x=x_0}(B) = \int_{-\infty}^{\infty} (x - x_0) l_*(B_x)\, dx = \int_{-\infty}^{\infty} (x - x_0) l^*(B_x)\, dx.$$

Prove that the axioms for moments are satisfied.

*8–6 Necessary and sufficient conditions for Riemann integrability

We have proved that continuous functions are Riemann-integrable. While continuity is an important sufficient condition for Riemann integrability, it certainly is not necessary. Monotone functions are also Riemann-integrable, even though they can have an infinite number of discontinuities. In this section we shall settle the following question: Just how discontinuous can a function be and still be Riemann-integrable?

If f is defined on an interval I, then we define the *oscillation* of f on I to be

$$\Omega(f, I) = \text{lub } \{f(x) - f(y) \mid x \in I \text{ and } y \in I\}.$$

If I is a closed bounded interval, and if f is continuous on I, then $\Omega(f, I)$ is simply the difference between the maximum and minimum values of f on I. In any case, $\Omega(f, I)$ is nonnegative; if f is unbounded on I, then $\Omega(f, I) = +\infty$. If $J \subset I$, then clearly $\Omega(f, J) \leq \Omega(f, I)$. Consequently, if f is defined throughout a neighborhood of x_0, then as $\delta \to 0+$, the number $\Omega(f, [x_0 - \delta, x_0 + \delta])$ is nondecreasing and bounded below by zero. Thus

$$\omega_f(x_0) = \lim_{\delta \to 0+} \Omega(f, [x_0 - \delta, x_0 + \delta])$$

exists (although $+\infty$ is a possible value of the limit). The number $\omega_f(x_0)$ is called the *oscillation* of f at x_0.

Example 1. Let

$$f(x) = \begin{cases} \sin(1/x) & \text{if } x \neq 0, \\ 0 & \text{if } x = 0. \end{cases}$$

Then for each $\delta > 0$, we have $\Omega(f, [-\delta, \delta]) = 2$. Thus $\omega_f(0) = 2$.

Example 2. Let

$$f(x) = \begin{cases} 3x & \text{if } x < 1, \\ 2x + 2 & \text{if } x \geq 1. \end{cases}$$

Since f is strictly increasing,

$$\Omega(f, [1 - \delta, 1 + \delta]) = f(1 + \delta) - f(1 - \delta) = 5\delta + 1$$

for all $\delta > 0$. Thus $\omega_f(1) = 1$.

Proposition 1. $\omega_f(x_0) = 0$ iff f is continuous at x_0.

The proof is an easy exercise.

The next result is a generalization of our theorem concerning uniform continuity. If f is defined on the interval $[a, b]$, we define the oscillation of f at the endpoints by

$$\omega_f(a+) = \lim_{\delta \to 0+} \Omega(f, [a, a + \delta]),$$

$$\omega_f(b-) = \lim_{\delta \to 0+} \Omega(f, [b - \delta, b]).$$

Proposition 2. *Suppose that f is defined on the interval [a, b] and that $\omega_f(x) < \epsilon$ for all $x \in [a, b]$. Then there exists a $\delta > 0$ such that for all x, y $\in [a, b]$,*

$$|x - y| < \delta \Rightarrow |f(x) - f(y)| < \epsilon.$$

Proof. Suppose this to be false. Then for each $n \in \mathbf{Z}^+$ there exist numbers x_n and y_n in [a, b] such that

$$|x_n - y_n| < 1/n \quad \text{and} \quad |f(x_n) - f(y_n)| \geq \epsilon.$$

By the sequential compactness property, $\{x_n\}$ contains a convergent subsequence, say, $x_{n_m} \to x_0$ as $m \to \infty$. Then $y_{n_m} \to x_0$ also. Since

$$\epsilon > \omega_f(x_0) = \lim_{\delta \to 0+} \Omega(f, [x_0 - \delta, x_0 + \delta]),$$

there exists a $\delta > 0$ such that

$$\Omega(f, [x_0 - \delta, x_0 + \delta]) < \epsilon.$$

If m is sufficiently large, then both x_{n_m} and y_{n_m} belong to $[x_0 - \delta, x_0 + \delta]$, and thus

$$|f(x_{n_m}) - f(y_{n_m})| \leq \Omega(f, [x_0 - \delta, x_0 + \delta]) < \epsilon,$$

a contradiction. (We have assumed that $a < x_0 < b$. It should be clear what modifications we would need if x_0 were either a or b.)

Corollary 3. *If f is as above, then there exist step functions g and h on [a, b] such that $g \leq f \leq h$ and*

$$\int_a^b (h - g) \leq \epsilon(b - a).$$

We shall assume henceforth that f is bounded on the interval [a, b]. If $\epsilon > 0$, we let

$$S(\epsilon) = \{x \mid a \leq x \leq b \text{ and } \omega_f(x) \geq \epsilon\}.$$

We shall also let M be an upper bound of the set $\{|f(x)| \mid a \leq x \leq b\}$.

Proposition 4. *The function f is Riemann-integrable on [a, b] iff for each $\epsilon > 0$ the set $S(\epsilon)$ has Jordan content zero.*

Proof. Let us suppose first that for some $\epsilon > 0$, $I^*(S(\epsilon)) > 0$. If g and h are step functions such that $g \leq f \leq h$ on [a, b], then

$$h(x) - g(x) \geq \epsilon \varphi_{S(\epsilon)}(x)$$

for all $x \in [a, b]$, with at most a finite number of exceptions. [Choose

$$a = x_0 < x_1 < \cdots < x_n = b$$

such that g and h are constant on each of the open intervals $(x_0, x_1), (x_1, x_2), \ldots, (x_m, x_n)$. Consider any $x \in [a, b]$ other than one of these x_k's, say, $x_{k-1} < x < x_k$. If $x \notin S(\epsilon)$, then

$$\epsilon \varphi_{S(\epsilon)}(x) = 0 \leq h(x) - g(x).$$

If $x \in S(\epsilon)$, then

$$\epsilon \varphi_{S(\epsilon)}(x) = \epsilon \leq \omega_f(x) \leq \Omega(f, (x_{k-1}, x_k)) \leq h(x) - g(x).$$

Thus

$$\int_a^b (h - g) \geq \overline{\int_a^b} \epsilon \varphi_{S(\epsilon)}(x) \, dx = \epsilon l^*(S(\epsilon)),$$

as asserted.] By the Riemann condition, f is not Riemann-integrable on $[a, b]$.

Suppose next that $l^*(S(\epsilon)) = 0$ for each $\epsilon > 0$. Let $\epsilon > 0$ be given. Set

$$\epsilon' = \epsilon/2(b - a).$$

Since $l^*(S(\epsilon')) = 0$, we can choose a partition

$$a = x_0 < x_1 < x_2 < \cdots < x_{2n+1} = b$$

such that $S(\epsilon')$ is contained in the union of the intervals $[x_0, x_1), (x_2, x_3), (x_4, x_5),$ $\ldots, (x_{2n}, x_{2n+1}]$ and such that the sum of the lengths of these intervals is as small as we please, say,

$$\sum_{k=0}^n (x_{2k+1} - x_{2k}) < \frac{\epsilon}{4M}.$$

On each of the closed intervals $[x_1, x_2], [x_3, x_4], \ldots, [x_{2n-1}, x_{2n}]$, the oscillation of f is less than ϵ', so we may choose step functions $g_1, \ldots, g_n, h_1, \ldots, h_n$ such that for each $k = 1, \ldots, n$, g_k and h_k are step functions on $[x_{2k-1}, x_{2k}], g_k \leq f \leq h_k$ on the same interval, and

$$\int_{x_{2k-1}}^{x_{2k}} (h_k - g_k) \leq \epsilon'(x_{2k} - x_{2k-1}).$$

We now define step functions g and h on $[a, b]$ as follows. On each of the intervals $[x_0, x_1), (x_2, x_3), \ldots, (x_{2n}, x_{2n+1}]$, g and h have the values $-M$ and M, respectively; on an interval of the form $[x_{2k-1}, x_{2k}]$, g and h are equal to g_k and h_k, respectively. Then $g \leq f \leq h$ on $[a, b]$ and

$$\int_a^b (h - g) = \sum_{k=1}^{2n} \int_{x_{k-1}}^{x_k} (h - g)$$

$$= \sum_{k=0}^n 2M(x_{2k+1} - x_{2k}) + \sum_{k=1}^n \int_{x_{2k-1}}^{x_{2k}} (h_k - g_k)$$

$$\leq 2M \sum_{k=0}^n (x_{2k+1} - x_{2k}) + \epsilon' \sum_{k=1}^n (x_{2k} - x_{2k-1})$$

$$< 2M \frac{\epsilon}{4M} + \epsilon'(b - a) = \epsilon.$$

Thus f is Riemann-integrable.

Definition 5. A set $E \subset \mathbf{R}$ has Lebesgue **measure zero** iff for each $\epsilon > 0$ there exists a sequence $I_1, I_2, I_3, \ldots$ of open intervals such that

$$E \subset \bigcup_{k=1}^{\infty} I_k$$

and for each $n \in \mathbf{Z}^+$,

$$\sum_{k=1}^{n} l(I_k) < \epsilon.$$

The inclusion $E \subset \bigcup_{k=1}^{\infty} I_k$ states that each member of E belongs to at least one of the intervals I_k. We often paraphrase this by saying that the family of intervals $\{I_1, I_2, \ldots\}$ *covers* E. If we set

$$s_n = \sum_{k=1}^{n} l(I_k),$$

then s_n is a nondecreasing sequence bounded above by ϵ; it therefore converges and

$$\lim_{n \to \infty} s_n \le \epsilon.$$

This limit is denoted by $\sum_{n=1}^{\infty} l(I_n)$ and can be thought of as the sum of the lengths of the intervals $I_1, I_2, \ldots$ We may therefore paraphrase our definition as follows. $E \subset \mathbf{R}$ has Lebesgue measure zero iff for each $\epsilon > 0$, E can be covered by a sequence of open intervals, the sum of whose lengths is less than or equal to ϵ.

Example 3. The rational numbers have Lebesgue measure zero.

We observed earlier that the rational numbers can be enumerated in a sequence $r_1, r_2,$ $\ldots$ Given $\epsilon > 0$, set $I_k = (r_k - \epsilon 2^{-k-1}, r_k + \epsilon 2^{-k-1})$. Then $r_k \in I_k$ and $l(I_k) = \epsilon 2^{-k}$. Consequently, the intervals $I_1, I_2, \ldots$ cover the set of rationals, and

$$\sum_{k=1}^{\infty} l(I_k) = \lim_{n \to \infty} \sum_{k=1}^{n} \epsilon 2^{-k} = \lim_{n \to \infty} \epsilon(1 - 2^{-n}) = \epsilon.$$

From the proof we see more generally that if the elements of a set can be arranged in a sequence, then the set has Lebesgue measure zero. Such a set is called *countable*.

Proposition 6. *Every set having Jordan content zero has Lebesgue measure zero. The family of sets with Lebesgue measure zero is a ring. Every subset of a set of Lebesgue measure zero also has Lebesgue measure zero. Moreover, if $E_1, E_2, \ldots$ is a sequence of sets each having Lebesgue measure zero, then their union $\bigcup_{n=1}^{\infty} E_n$ also has Lebesgue measure zero.*

Proof. All assertions but the last are trivial. Suppose that $E_1, E_2, \ldots$ is a sequence of sets having Lebesgue measure zero. Let $\epsilon > 0$ be given. Since E_k has Lebesgue measure zero, there exists a sequence of open intervals $I_{k1}, I_{k2}, I_{k3}, \ldots$ covering E_k such that

$$\sum_{n=1}^{\infty} l(I_{kn}) \le \frac{\epsilon}{2^k}.$$

The doubly indexed family of open intervals $\{I_{mn} \mid m, n \in \mathbf{Z}^+\}$ covers $E = \bigcup_{k=1}^{\infty} E_k$, and can be arranged in a sequence (for instance, $I_{11}, I_{21}, I_{12}, I_{31}, I_{22}, I_{13}, I_{41}, I_{32},$ ...). Also, the sum of the lengths of these intervals is less than or equal to

$$\sum_{k=1}^{\infty} \frac{\epsilon}{2^k} = \epsilon.$$

This proves that $E = \bigcup_{k=1}^{\infty} E_k$ has Lebesgue measure zero.

Observe that a set which has Lebesgue measure zero need not have Jordan content zero. For example, the set of rational numbers in $[0, 1]$ has Lebesgue measure zero but outer Jordan content one. There is, nevertheless, an important instance in which Lebesgue measure zero does imply Jordan content zero.

Definition 7. A set $E \subset \mathbf{R}$ is **compact** iff to each sequence $\{x_n\}$ of points in E there corresponds a subsequence $\{x_{n_m}\}$ and an element $x_0 \in E$ such that

$$\lim_{m \to \infty} x_{n_m} = x_0.$$

Example 4. By the sequential compactness property, it follows easily that each closed bounded interval $[a, b]$ is compact.

Example 5. Each set of the form $S(\epsilon)$ is compact.

Let $\{x_n\}$ be a sequence of points in $S(\epsilon)$. Since $\{x_n\} \subset [a, b]$, it follows from the sequential compactness property that $\{x_n\}$ has a subsequence $\{x_{n_m}\}$ which converges to some $x_0 \in [a, b]$. We must show that $x_0 \in S(\epsilon)$. Suppose that it is not. Then $\omega_f(x_0) < \epsilon$. Hence there is a neighborhood V of x_0 such that $\Omega(f, V) < \epsilon$. If m is sufficiently large, then $x_{n_m} \in V$, and

$$\omega_f(x_{n_m}) \leq \Omega(f, V) < \epsilon.$$

But then $x_{n_m} \notin S(\epsilon)$, a contradiction. Thus $S(\epsilon)$ is compact.

Proposition 8. *A compact set E with Lebesgue measure zero also has Jordan content zero.*

Proof. Let $\epsilon > 0$ be given. Since E has Lebesgue measure zero, there exists a sequence of open intervals $I_1, I_2, \ldots$ such that

$$E \subset \bigcup_{n=1}^{\infty} I_n \quad \text{and} \quad \sum_{n=1}^{\infty} l(I_n) \leq \epsilon.$$

It suffices to prove that there exists an n such that $E \subset \bigcup_{k=1}^{n} I_k$. (Why?) Suppose this is false. Then for each $n \in \mathbf{Z}^+$ we can choose an

$$x_n \in \left(E \setminus \bigcup_{k=1}^{n} I_k \right).$$

Since E is compact, the resulting sequence $\{x_n\}$ contains a subsequence $\{x_{n_m}\}$ which converges to some $x_0 \in E$. Choose $N \in \mathbf{Z}^+$ such that $x_0 \in I_N$. Then for m sufficiently large,

$$x_{n_m} \in I_N \subset \bigcup_{k=1}^{N} I_k,$$

which is impossible.

We can now prove the main result.

Theorem 9. *Let f be a bounded function on* $[a, b]$. *Then f is Riemann-integrable on* $[a, b]$ *iff the set of points in* $[a, b]$ *at which f is discontinuous has Lebesgue measure zero.*

Proof. Let D be the set of points in $[a, b]$ at which f is discontinuous. Then

$$D = \{x \in [a, b] \mid \omega_f(x) > 0\}$$

$$= \bigcup_{n=1}^{\infty} \{x \in [a, b] \mid \omega_f(x) \geq 1/n\}$$

$$= \bigcup_{n=1}^{\infty} S(1/n).$$

Suppose f is Riemann-integrable on $[a, b]$. Then each set $S(1/n)$ has Jordan content zero and hence Lebesgue measure zero. Thus D also has Lebesgue measure zero.

Suppose, on the other hand, that D has Lebesgue measure zero. Then each of the sets $S(1/n)$ has Lebesgue measure zero. Given $\epsilon > 0$, we can choose $n \in \mathbf{Z}^+$ such that $1/n < \epsilon$. Then $S(\epsilon) \subset S(1/n)$, so $S(\epsilon)$ has Lebesgue measure zero. Since $S(\epsilon)$ is compact, however, its Jordan content is also zero. It then follows that f is Riemann-integrable.

Corollary 10. *The Riemann-integrable functions are closed under addition, multiplication, and the lattice operations. If f is Riemann-integrable and bounded away from zero on* $[a, b]$, *then* $1/f$ *is Riemann-integrable on* $[a, b]$; *if f is continuous on an interval containing the range of g and if g is Riemann-integrable on* $[a, b]$, *then* $f \circ g$ *is Riemann-integrable on* $[a, b]$.

Exercises

1. Use the theorems of this section to show that the ruler function is Riemann-integrable on every closed bounded interval.

*2. The *Cantor middle third set* is defined as follows. From the unit segment $[0, 1]$ we first remove the open middle third $(\frac{1}{3}, \frac{2}{3})$, and we let C_1 be the set remaining. (Thus $C_1 = [0, \frac{1}{3}] \cup [\frac{2}{3}, 1]$.) From each of the segments in C_1 we remove the open middle third, and let C_2 be the set of points remaining. (Thus $C_2 = [0, \frac{1}{9}] \cup [\frac{2}{9}, \frac{1}{3}] \cup [\frac{2}{3}, \frac{7}{9}] \cup [\frac{8}{9}, 1]$.) Continuing in this way, we get a sequence of sets $C_1 \supset C_2 \supset C_3 \supset \cdots$ We define the Cantor set C to be their intersection.

   a) Show that C has Lebesgue measure zero.
   b) Show that C can be identified with the set of all numbers in the interval $[0, 1]$ which have no 1's appearing in their expansions to base 3.
   c) Show that C is uncountable.
   d) Show that the characteristic function of C, φ_C, has C as its set of points of discontinuity. Infer that φ_C is Riemann-integrable.

3. Prove Corollary 10.

*4. a) Let f and g be continuous on $[a, b]$. If $\pi = (x_0, x_1, \ldots, x_n)$ is a partition of $[a, b]$, and if $\sigma = (\xi_1, \ldots, \xi_n)$ and $\sigma^* = (\xi_1^*, \ldots, \xi_n^*)$ are selectors for π, we define

$$I_{\pi,\sigma,\sigma^*}(f, g) = \sum_{k=1}^{n} f(\xi_k) \, g(\xi_k^*)(x_k - x_{k-1}).$$

Suppose that $\{\pi_n\}$ is a sequence of partitions of $[a, b]$ such that $\|\pi_n\| \to 0$. For each $n \in \mathbf{Z}^+$, let σ_n and σ_n^* be selectors for π_n. Show that

$$\lim_{n \to \infty} I_{\pi_n,\sigma_n,\sigma_n^*}(f, g) = \int_a^b f(x) \, g(x) \, dx$$

(*Bliss' formula*).

b) Show that Bliss' formula is valid if one of the functions is continuous and the other is Riemann-integrable.

5. Show that the composition product of two Riemann-integrable functions need not be Riemann-integrable.

9 □ TRANSCENDENTAL FUNCTIONS

A function f is said to be *algebraic* if there exists a polynomial in two variables $P(x, y)$ such that $P(x, f(x)) = 0$ for all $x \in \mathfrak{D}_f$. For example, the function

$$f(x) = x - \sqrt{1 - x^2}$$

is algebraic, since the polynomial

$$P(x, y) = y^2 - 2xy + 2x^2 - 1$$

has the required property. A function which is not algebraic is said to be *transcendental*. Among the functions which we have already studied, $\sin x$ is an example of such a function. (An algebraic function cannot have an infinite number of zeros without being equal to zero. Consequently, the sine function, which has zeros at all integral multiples of π, is not algebraic.) The dichotomy of algebraic versus transcendental functions is of limited usefulness. It is roughly comparable to the classification of objects as jelly beans and nonjelly beans. Just as it is difficult to make many profound statements about the class of all nonjelly beans, one should not expect to find a deep general theory concerning transcendental functions. However, although we shall have very little to say about transcendental functions in general, we shall have a great deal to say about certain specific ones.

Transcendental functions are thrust upon us by integration. For while the algebraic functions are closed under many operations (addition, multiplication, composition, differentiation), there are many simple algebraic functions, such as $1/x$, $1/(1 + x^2)$, $\sqrt{1 + x^2}$, whose antiderivatives are transcendental.

9-1 General theory of inverse functions

Let f be a function mapping a set A into a set B. For the moment we shall think of f as a process which converts elements of A into elements of B. If the process is a reversible one, we denote the reverse process by f^{-1}, and we call it the *inverse* of the function f. Thus if $f(4) = 9$, then $f^{-1}(9) = 4$; more generally, if $y = f(x)$, then $f^{-1}(y) = x$. Suppose, for example, that f is the function that converts each number into its cube.

Thus $f(x) = x^3$. Then f^{-1}, the deconverter, is the function that converts each number into its cube root.

Definition 1. Let f be a function. We say that f is **invertible** if there exists a function g such that the equations

$$y = f(x) \quad \text{and} \quad x = g(y)$$

are logically equivalent, in which case the function g (which is clearly unique) is called the **inverse** of f, and is denoted by f^{-1}.

Example 1. Suppose $f(x) = 2x - 1$. Since the equation $y = 2x - 1$ is equivalent to the equation $x = \frac{1}{2}(y + 1)$, it is clear that f is invertible and that

$$f^{-1}(y) = \tfrac{1}{2}(y + 1).$$

The graphs of f and f^{-1} are shown in Fig. 9-1.

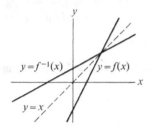

FIG. 9-1

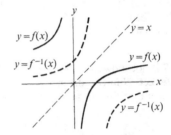

FIG. 9-2

Example 2. Let $f(x) = 1 - 1/x$. Solving the equation $y = 1 - 1/x$ for x gives

$$x = 1/(1 - y).$$

Hence f is invertible and

$$f^{-1}(u) = 1/(1 - u).$$

The graphs of f and f^{-1} are shown in Fig. 9-2. Note that the domain of f and the range of f^{-1} are the line with the number zero deleted, while the range of f and the domain of f^{-1} are the line with the number one deleted.

Before giving additional examples, we make the following observations.

Proposition 2. *Let f be an invertible function. Then*

1) $\mathfrak{D}_{f^{-1}} = \mathfrak{R}_f$ *and* $\mathfrak{R}_{f^{-1}} = \mathfrak{D}_f$;

2) $f\big(f^{-1}(x)\big) = x$ *for all* $x \in \mathfrak{R}_f$, $\quad f^{-1}\big(f(x)\big) = x$ *for all* $x \in \mathfrak{D}_f$;

3) f^{-1} *is invertible and* $(f^{-1})^{-1} = f$;

4) If f is a real-valued function of a real variable, then the graph of f^{-1} is the reflection of the graph of f in the line $y = x$.

The statements are all trivial consequences of the definition. We shall comment only on (4). Suppose that (u, v) is a point on the graph of f. Then $v = f(u)$, or,

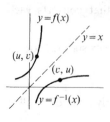

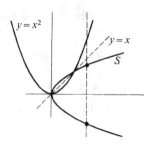

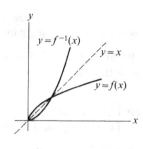

FIG. 9–3   **FIG. 9–4**   **FIG. 9–5**

equivalently, $u = f^{-1}(v)$. In other words, (v, u), which is the reflection of (u, v) in the line $y = x$, is on the graph of f^{-1}, as shown in Fig. 9–3.

Example 3. Let $f(x) = x^2$. Then f is not invertible. Suppose, to the contrary, that f were invertible. Then the graph of f^{-1} would be the reflection of the graph of f in the line $y = x$, namely, the set

$$S = \{(x, y) \mid x = y^2\}.$$

The set S is not, however, the graph of a function, since each vertical line in the right half-plane intersects S in two distinct points (see Fig. 9–4).

The fact that f is not invertible is also evidenced by the fact that the equation

$$y = f(x) = x^2$$

is equivalent to the equation

$$x = \pm\sqrt{y},$$

and the right-hand side of the latter equation is not a function of y.

Example 4. Let $f(x) = \sqrt{x}$. Then f is invertible and f^{-1} is the function with domain $[0, \infty)$ such that $f^{-1}(x) = x^2$ (see Fig. 9–5).

Closely associated with the notion of inverse functions is the notion of one-to-oneness.

Definition 3. A function f is said to be **one-to-one** (or **injective**) iff for any two points x_1 and x_2 in the domain of f,

$$x_1 \neq x_2 \Rightarrow f(x_1) \neq f(x_2),$$

or, equivalently,

$$f(x_1) = f(x_2) \Rightarrow x_1 = x_2.$$

In other words, a one-to-one function is one which maps distinct points into distinct points.

Example 5. The function $f(x) = x^3$ is one-to-one. More generally, any strictly increasing (or strictly decreasing) function is one-to-one. The function $f(x) = x^2$, on the other hand, is not one-to-one, since, for instance, $f(-3) = 9 = f(3)$. The antithesis of a one-to-one function is a constant function (whose domain contains more than one point).

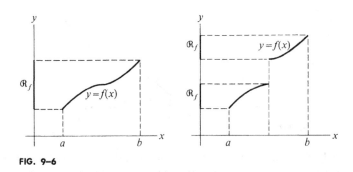

FIG. 9–6

Proposition 4. *A function f is invertible iff it is one-to-one.*

Proof. Suppose that f is invertible. Then

$$f(x_1) = f(x_2) \Rightarrow x_1 = f^{-1}(f(x_1)) = f^{-1}(f(x_2)) = x_2.$$

Thus f is one-to-one. Suppose, conversely, that f is one-to-one. Then for each element y in the range of f there is exactly one element $x = g(y)$ such that $y = f(x)$. From this it is clear that f is invertible and that the function g so obtained is the inverse of f.

Corollary 5. *A strictly increasing (or strictly decreasing) function is invertible.*

If f is an invertible function with certain properties, then it is natural to ask which of these properties are inherited by f^{-1}. We will prove that in certain cases the properties of continuity and differentiability are so inherited.

Proposition 6. *Let f be a strictly increasing function whose domain is a closed bounded interval* $[a, b]$. *Then f is continuous iff* $\mathcal{R}_f = [f(a), f(b)]$.

A glance at Fig. 9–6 should help to convince the reader that the theorem is true.

Proof. Observe, first of all, that $\mathcal{R}_f \subset [f(a), f(b)]$, whether or not f is continuous. This follows easily from the fact that f is strictly increasing. (Suppose $y \in \mathcal{R}_f$. Then y is of the form $f(x)$ for some x between a and b. Since f is increasing,

$$f(a) \leq f(x) = y \leq f(b).$$

Thus $y \in [f(a), f(b)]$.)

If f is continuous, then the reverse inclusion, namely, $[f(a), f(b)] \subset \mathcal{R}_f$, is an immediate consequence of the intermediate-value theorem.

Let us suppose now that $\mathcal{R}_f = [f(a), f(b)]$. We extend f to the entire line by defining it outside the interval $[a, b]$ as follows:

$$f(x) = \begin{cases} x - b + f(b) & \text{if } x > b, \\ x - a + f(a) & \text{if } x < a. \end{cases}$$

Geometrically, this amounts to attaching rays at the endpoints of the graph of f, as shown in Fig. 9–7.

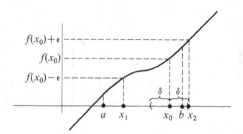

FIG. 9–7

It is easily checked that the function f, so extended, is a strictly increasing function whose range is the entire real line. We shall prove that f is everywhere continuous. Given any number x_0 and any $\epsilon > 0$, there exist numbers x_1 and x_2 such that

$$f(x_1) = f(x_0) - \epsilon \quad \text{and} \quad f(x_2) = f(x_0) + \epsilon.$$

Since f is an increasing function, $x_1 < x_0 < x_2$. We let δ be the distance from x_0 to the closer of the two points x_1 and x_2. [Thus $\delta = \min (x_0 - x_1, x_2 - x_0)$.] If $|x - x_0| < \delta$, then, by the way δ was chosen, $x_1 < x < x_2$. Since f is strictly increasing, this in turn implies that

$$f(x_0) - \epsilon = f(x_1) < f(x) < f(x_2) = f(x_0) + \epsilon$$

of $|f(x) - f(x_0)| < \epsilon$. This proves, of course, that f is continuous at x_0.

Theorem 7. *Let f be a function with the following properties:*

  1) The domain of f is a closed bounded interval;

  2) f is strictly increasing;

  3) f is continuous.

Then f^{-1} exists and has the same three properties.

Proof. Since f is strictly increasing, f^{-1} exists and is strictly increasing. Let $\mathfrak{D}_f = [a, b]$. Since f is continuous,

$$\mathfrak{D}_{f^{-1}} = \mathfrak{R}_f = [f(a), f(b)],$$

by the previous theorem. Thus f^{-1} has property (1). Since f^{-1} is strictly increasing, and since

$$\mathfrak{R}_{f^{-1}} = \mathfrak{D}_f = [a, b] = [f^{-1}(f(a)), f^{-1}(f(b))],$$

it also follows from the preceding theorem that f^{-1} has property (3).

With a little more work, one can prove a refinement of the theorem, namely, that *condition (1) can be replaced by*

  1') The domain of f is an interval.

We shall use this stronger result in subsequent sections.

We shall next be concerned with differentiability of inverse functions. Suppose that f is invertible and that both f and f^{-1} are differentiable. Then the derivative of f^{-1} can easily be related to the derivative of f. To simplify notation, let $g = f^{-1}$. Then

$$f(g(y)) = y$$

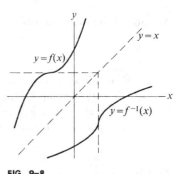

FIG. 9–8

for all $y \in \mathcal{R}_f = \mathcal{D}_g$. Using the chain rule, we get, upon differentiating both sides of the equation,

$$f'(g(y))g'(y) = 1$$

or

$$g'(y) = \frac{1}{f'(g(y))}.$$

Note that the validity of this calculation hinges upon the *assumption* that f and f^{-1} are both differentiable. Our next theorem will guarantee the differentiability of f^{-1} in certain cases. Note also that trouble will surely occur if f' vanishes. If, for instance, $f'(x_0) = 0$ and $y_0 = f(x_0)$, then $g = f^{-1}$ is not differentiable at y_0, since if it were, we would have

$$0 = f'(x_0)g'(y_0) = 1,$$

which is impossible. Geometrically, this corresponds to the fact that if a curve is reflected in the line $y = x$, then every horizontal tangent line is mapped onto a vertical tangent line, as illustrated in Fig. 9–8.

Before proceeding to the next theorem, we need a definition.

Definition 8. Let S be a set of real numbers. We say that x_0 is an **interior point** of S iff there exists a $\delta > 0$ such that the open interval $(x_0 - \delta, x_0 + \delta)$ is a subset of S. The set of all interior points of S is called the **interior** of S.

For example, the interior of the interval $[4, 7)$ is the interval $(4, 7)$. The interior of the set of all rational numbers is the null set.

Theorem 9. *Let f be a function with the following properties:*

1) The domain of f is an interval;

2) f is continuous;

3) f' exists and is positive throughout the interior of the domain of f.

Then $f^{-1} = g$ exists and has the same three properties. Moreover,

$$g'(y) = \frac{1}{f'(g(y))}$$

for all y in the interior of the domain of $g = f^{-1}$.

Proof. The stated properties of f imply that f is strictly increasing. Hence, by the previous theorem, $g = f^{-1}$ exists and has properties (1) and (2).

Thus it suffices to show that g is differentiable at interior points of its domain. The ideas involved are quite simple. In less precise notation, we wish to show that

$$\frac{dx}{dy} = \frac{1}{dy/dx}.$$

We do this by taking limits of both sides of the equation:

$$\frac{\Delta x}{\Delta y} = \frac{1}{\Delta y / \Delta x}.$$

Essential to this step is the fact that as $\Delta y \to 0$, $\Delta x \to 0$ *properly*. The convergence of Δx to 0 is guaranteed by the continuity of g; propriety of the convergence comes from the fact that g is injective. Guided by these remarks, one can readily construct a detailed proof of property (3).

Exercises

1. Determine in each case whether f is invertible. When f is invertible, calculate f^{-1} and graph both f and f^{-1}.

a) $f(x) = \sin x$ b) $f(x) = 3x^3 - 2$
c) $f(x) = 5$ d) $f(x) = 1 - x$
e) $f(x) = 1/x$ f) $f(x) = x + [x]$
g) $f(x) = x|x|$ h) $f(x) = \sqrt{1 - x^{-1}}$

2. Prove the existence, continuity, and differentiability of nth roots by applying the theorems in this section to the function x^n ($x \geq 0$). (Note that this makes unnecessary all the hard work in Exercise 1, Section 3–6.)

3. Let $f(x) = (ax + b)/(cx + d)$, where a, b, c, and d are constants such that c and d are not both zero. Show that f is invertible iff $ad - bc \neq 0$. Also, calculate f^{-1} for the case in which f is invertible.

4. Given that f is a strictly increasing convex function, show that f^{-1} is concave.

5. Given that f and g are invertible functions, prove that $f \circ g$ is invertible and

$$(f \circ g)^{-1} = g^{-1} \circ f^{-1}.$$

6. Suppose that f is a continuous function whose domain is an interval. Show that f is invertible iff f is strictly monotone (that is, either strictly increasing or strictly decreasing).

7. State sufficient conditions for a function to have a twice differentiable inverse function. Obtain a general formula for the second derivative of an inverse function.

8. Suppose that the derivative of a function f exists and is monotone throughout some closed bounded interval I. Using Darboux's theorem (page 202) and the theorems in this section, deduce that f' is continuous throughout I.

9. Let $f(x) = x + \sin x$. Show that f is an invertible function whose domain and range is the entire real line. What can you say concerning the differentiability of f^{-1}.

10. Let $f(x)$ be continuous and strictly increasing on the interval $[0, c]$. Suppose further that $f(0) = 0$, $0 \leq a \leq c$, and $0 \leq b \leq f(c)$. Prove *Young's inequality:*

$$\int_0^a f(x)\, dx + \int_0^b f^{-1}(x)\, dx \geq ab.$$

(Show that the difference between the left- and right-hand sides of the inequality is equal to the area of the shaded regions shown in Fig. 9–9.) When does equality hold?

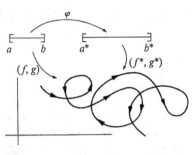

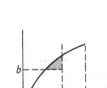 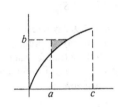

FIG. 9–9   **FIG. 9–10**

11. In Chapter 7 we argued that the path length of a piecewise smooth path was independent of its parametrization. We could not give a rigorous proof of this fact, since we could not formulate a precise meaning for equivalent parametrizations. We do so now.

a) We say that

$$x = f(t), \qquad y = g(t) \qquad (a \leq t \leq b)$$

and

$$x = f^*(t), \qquad y = g^*(t) \qquad (a^* \leq t \leq b^*)$$

are *equivalent parametrizations* of a path γ iff there exists a continuous strictly increasing function φ mapping $[a, b]$ onto $[a^*, b^*]$ such that $f = f^* \circ \varphi$ and $g = g^* \circ \varphi$ (see Fig. 9–10). We shall indicate this state of affairs by writing $(f, g) \sim (f^*, g^*)$. Although this definition is not *symmetric* with respect to the functions involved, apply the theorems in this section to show that this is actually the case. [Prove, in other words, that if $(f, g) \sim (f^*, g^*)$, then $(f^*, g^*) \sim (f, g)$.] Observe also that this relation is *reflexive* [that is, $(f, g) \sim (f, g)$] and *transitive* [that is, $(f, g) \sim (f^*, g^*)$ and $(f^*, g^*) \sim (f^{**}, g^{**})$ imply $(f, g) \sim (f^{**}, g^{**})$]. Consequently, the use of the term "equivalent" is justified.

b) Show that

$$x = \frac{1 - t^2}{1 + t^2}, \qquad y = \frac{2t}{1 + t^2} \qquad (-1 \leq t \leq 1)$$

and

$$x = \cos \theta, \qquad y = \sin \theta \qquad (-\pi/2 \leq \theta \leq \pi/2)$$

are equivalent parametrizations of a semicircle.

c) Prove that the path length of a piecewise smooth path is independent of its parametrization.

d) Prove that a piecewise smooth path γ has a unique parametrization

$$x = f(s), \qquad y = g(s) \qquad (0 \leq s \leq b)$$

such that for all $s \in [a, b]$, we have $s = \overarc{P_0 P_s}$. We say in this case that γ is parametrized by arc length.

e) If the function φ described in (a) also has a continuous positive derivative throughout (a, b), we say that the parametrizations are *smoothly equivalent*. Prove that this relation is symmetric, reflexive, and transitive.

*12. Suppose that φ is continuous on $[a, b]$ and that φ' exists and is positive and continuous on (a, b). If $S \subset [a, b]$ has Lebesgue measure zero, prove that

$$\varphi(S) = \{\varphi(x) \mid x \in S\}$$

also has Lebesgue measure zero. [Let $S_n = S \cap I_n$, where $I_n = [a + 1/n, b - 1/n]$. Let M_n be the maximum value of φ' on I_n. If J is any subinterval of I_n, show that $\varphi(J)$ is an interval whose length is $M_n l(J)$ at most. Deduce that for each n, $\varphi(S_n)$ has Lebesgue measure zero. Show that the sets $\varphi(S)$ and $\bigcup_{n=1}^{\infty} \varphi(S_n)$ differ by two points at most. Conclude finally that $\varphi(S)$ has Lebesgue measure zero.]

*13. Prove *the substitution rule for Riemann integrals:*

$$\int_a^b f(\varphi(x))\varphi'(x)\,dx = \int_{\varphi(a)}^{\varphi(b)} f(u)\,du,$$

provided the following are satisfied:

1) φ' is continuous on $[a, b]$;

2) $\varphi'(x) \neq 0$ for all $x \in [a, b]$, with at most a finite number of exceptions;

3) f is Riemann-integrable on an interval I containing the image of $[a, b]$ under φ (that is, $\varphi(x) \in I$ for all $x \in [a, b]$).

[Using the additive property of integrals, show that it suffices to consider the situation in which $\varphi'(x) \neq 0$ except possibly at the endpoints a and b. For definiteness, assume that φ' is positive throughout the open interval (a, b). Apply the inverse function theorems, the preceding problem, and the theorem on Riemann integrability (Theorem 9, Section 8–6) to show that if f is Riemann-integrable on $[u(a), u(b)]$, then $f \circ \varphi$ is Riemann-integrable on $[a, b]$. For each such f, define

$$I(f) = \int_a^b f(\varphi(x))\varphi'(x)\,dx.$$

Prove that I is a definite integral (in the abstract sense of Chapter 8) on the interval $[u(a), u(b)]$. The substitution rule for Newton integrals should prove helpful in proving that I behaves properly for characteristic functions of subintervals of $[u(a), u(b)]$. Then apply the second form of the fundamental theorem of calculus.]

*14. Construct a second proof of the above substitution rule along the following lines. As before, it suffices to consider the case in which $\varphi'(x) > 0$ for all $x \in (a, b)$. If $\pi = (y_0, y_1, \ldots, y_r)$ is a partition of $[\varphi(a), \varphi(b)]$, we let $\varphi^{-1}(\pi)$ be the partition $(\varphi^{-1}(y_0), \varphi^{-1}(y_1), \ldots, \varphi^{-1}(y_r))$ of $[a, b]$. Consider now a sequence of partitions $\pi_1, \pi_2, \ldots$ of $[\varphi(a), \varphi(b)]$ such that $\|\pi_n\| \to 0$. Also let $\sigma_1, \sigma_2, \ldots$ be a corresponding sequence of selectors. Show that $\|\varphi^{-1}(\pi_n)\| \to 0$. (Use the uniform continuity of the function φ^{-1}.) Show that the Riemann sum $I_{\pi_n, \sigma_n}(f)$ can, by use of the mean-value theorem, be written as

$$I_{\varphi^{-1}(\pi_n), \; \varphi^{-1}(\sigma_n), \; \sigma_n'}(f \circ \varphi, \varphi')$$

for a suitable selector σ_n' corresponding to the partition $\varphi^{-1}(\pi_n)$ of $[a, b]$. Then apply Bliss' formula (see Exercise 4, Section 8–6).

15. Let $x = x(t), y = y(t), a \leq t \leq b$, be parametric equations of a piecewise smooth path γ. Suppose that $a < t_0 < b$ and that the corresponding point on γ,

$$(x_0, y_0) = (x(t_0), y(t_0)),$$

is a regular point. [Recall that this means that $x'(t_0)$ and $y'(t_0)$ are not both zero.] If $x'(t_0) \neq 0$, show that for values of t close to t_0, the points of γ lie on a curve of the form $y = f(x)$, where f possesses a continuous derivative throughout some neighborhood of x_0. Deduce that the tangent line to γ where $t = t_0$ is given by either

$$x'(t_0)(x - x_0) + y'(t_0)(y - y_0) = 0$$

or

$$x = x_0 + x'(0)(t - t_0), \qquad y = y_0 + y'(0)(t - t_0).$$

[More intuitively, we can treat y as a function of x, and we then have

$$\frac{dy}{dx} = \frac{dy/dt}{dx/dt}.$$

The ideas involved here traditionally come under the heading of *parametric differentiation*.] What is the situation if $x'(t_0) = 0$?

9-2 The inverse trigonometric functions

Although a given function need not be invertible, we can make it so by cutting down the size of its domain. Recall an earlier definition.

Definition 1. Let f be a function, and let A be a set. By the **restriction** of f to A we shall mean the function g with domain $\mathfrak{D}_f \cap A$ such that

$$g(x) = f(x)$$

for all $x \in \mathfrak{D}_f \cap A$. We write $g = f \,|\, A$.

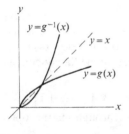

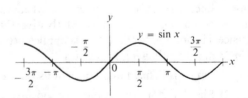

FIG. 9–11   FIG. 9–12

Example 1. Let $f(x) = x^2$, and let g be the restriction of f to the set of nonnegative numbers. The graph of g is shown in Fig. 9–11. Note that the function g, unlike f, is invertible. In fact, $g^{-1}(x) = \sqrt{x}$.

Let us consider now the sine function, whose graph is shown in Fig. 9–12. Although it is rather badly many-to-one, it is monotone (and hence one-to-one) on each interval of the form $[(2n - 1)(\pi/2), (2n + 1)(\pi/2)]$, where n is an integer. Hence by restricting the function to any one of these intervals, we obtain an invertible

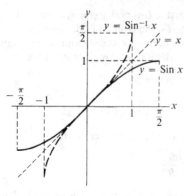

FIG. 9–13   FIG. 9–14

function. The inverse functions so obtained are called (for historical reasons dating back to the stone age of analysis) *branches of the arc sine.* We shall arbitrarily confine our attention to that branch obtained by restricting the sine function to the interval $[-\pi/2, \pi/2]$.

Definition 2. Let Sin denote the restriction of the sine function to the interval $[-\pi/2, \pi/2]$. Then Sin^{-1} is called the **principle branch of the arc sine**.

Note, then, that $y = \text{Sin}^{-1}x$ iff $x = \text{Sin } y$ and $-\pi/2 \leq y \leq \pi/2$. For example,

$$\text{Sin}^{-1}\frac{1}{2} = \frac{\pi}{6} \quad \text{and} \quad \text{Sin}^{-1}\left(-\frac{\sqrt{2}}{2}\right) = -\frac{\pi}{4}.$$

The graphs of Sin and Sin^{-1} are shown in Fig. 9–13.

Observe that Sin is continuous because it is the restriction of a continuous function. Since $d(\sin x)/dx = \cos x > 0$ whenever $-\pi/2 < x < \pi/2$, the function Sin has a continuous and positive derivative throughout the interior of its domain. Hence Theorem 9, Section 9–1, is applicable, and we can immediately conclude that Sin^{-1} has the following properties:

1) Sin^{-1} is defined and continuous on the closed interval $[-1, +1]$;

2) Sin^{-1} has a continuous and positive derivative throughout the open interval $(-1, +1)$.

We shall now compute the derivative of Sin^{-1}. Let

$$y = \text{Sin}^{-1} x.$$

Then $x = \text{Sin } y$, and since Sin^{-1} is differentiable, we can use the chain rule to differentiate the latter equation with respect to x. We get

$$1 = (\cos y)\frac{dy}{dx} \quad \text{or} \quad \frac{dy}{dx} = \frac{1}{\cos y}.$$

Our calculation is legitimate only for x in the open interval $(-1, +1)$. However,

$$-1 < x < +1 \Rightarrow -\pi/2 < y < \pi/2 \Rightarrow \cos y > 0$$
$$\Rightarrow \cos y = \sqrt{1 - \sin^2 y} = \sqrt{1 - x^2}.$$

Thus we have the following differentiation formula:

$$\frac{d}{dx}[\operatorname{Sin}^{-1} x] = \frac{1}{\sqrt{1 - x^2}} \qquad (|x| < 1).$$

Of somewhat less importance is the principle branch of the arc cosine.

Definition 3. Let Cos be the restriction of the cosine function to the interval $[0, \pi]$. Then Cos^{-1} is called the **principle branch of the arc cosine.**

Since the cosine function is strictly decreasing throughout the interval $[0, \pi]$, it is clear that Cos is invertible. The graphs of Cos and Cos^{-1} are given in **Fig. 9–14.**

As in the case of the arc sine, one can apply the general theory of inverse functions to show that Cos^{-1} is differentiable throughout the open interval $(-1, +1)$, and an easy calculation yields

$$\frac{d}{dx}[\operatorname{Cos}^{-1} x] = -\frac{1}{\sqrt{1 - x^2}} \qquad (|x| < 1).$$

Since Cos^{-1} and $-\operatorname{Sin}^{-1}$ have the same derivative throughout the open interval $(-1, +1)$, they differ by a constant in this interval. Thus

$$\operatorname{Cos}^{-1} x = C - \operatorname{Sin}^{-1} x$$

if $-1 < x < 1$. By continuity, the equation also holds if $x = 1$ or $x = -1$. The constant C can be determined by setting $x = 0$. Thus

$$\pi/2 = \operatorname{Cos}^{-1} 0 = C - \operatorname{Sin}^{-1} 0 = C.$$

Therefore,

$$\operatorname{Cos}^{-1} x = \pi/2 - \operatorname{Sin}^{-1} x.$$

Geometrically, this corresponds to the fact that the base angles of a right triangle are complementary (see Fig. 9–15).

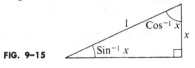

FIG. 9–15

Definition 4. Let Tan be the restriction of the tangent function to the interval $(-\pi/2, \pi/2)$. Then Tan^{-1} is called the **principle branch of the arc tangent.**

Since the derivative of the tangent function is always positive, it is clear that Tan^{-1} exists. It follows from the general theory that the domain of Tan^{-1} is an interval, and since

$$\lim_{x \to -\pi/2+} \tan x = -\infty \qquad \text{and} \qquad \lim_{x \to \pi/2-} \tan x = +\infty,$$

it is obvious that this interval (which is the range of Tan) must be the entire real line (see Fig. 9–16). It also follows from the general theory that Tan^{-1} has a continuous derivative everywhere. We shall now calculate the derivative. Let $y = \text{Tan}^{-1} x$. Then $x = \text{Tan}\, y$ and

$$1 = \frac{d}{dx}(x) = \frac{d}{dx}(\tan y) = \sec^2 y \frac{dy}{dx}.$$

Hence

$$\frac{dy}{dx} = \frac{1}{\sec^2 y} = \frac{1}{1 + \tan^2 y} = \frac{1}{1 + x^2}.$$

This proves the differentiation formula

$$\frac{d}{dx}[\text{Tan}^{-1} x] = \frac{1}{1 + x^2}.$$

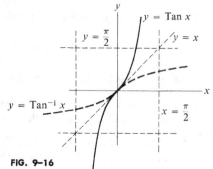

FIG. 9–16

Inverse functions corresponding to the remaining three basic trigonometric functions can also be defined. This is left to the reader.

Example 2. A certain amount of care must be exercised in handling the inverse trigonometric functions. One might be tempted to conclude, for instance, that the expressions $\sin(\text{Sin}^{-1} x)$ and $\text{Sin}^{-1}(\sin x)$ both simplify to x. This is not true, however.

Consider first the expression $\sin(\text{Sin}^{-1} x)$. If $-1 \le x \le 1$, then

$$\sin(\text{Sin}^{-1} x) = \text{Sin}(\text{Sin}^{-1} x) = x.$$

The expression is meaningless, however, if $|x| > 1$. In other words, $\sin \circ \text{Sin}^{-1}$ is the restriction of the identity mapping to $[-1, +1]$.

Consider next the expression $\text{Sin}^{-1}(\sin x)$. If $-\pi/2 \le x \le \pi/2$, then

$$\text{Sin}^{-1}(\sin x) = \text{Sin}^{-1}(\text{Sin}\, x) = x.$$

Note, however, that $\text{Sin}^{-1}(\sin \pi) = \text{Sin}^{-1} 0 = 0$. Thus $\text{Sin}^{-1}(\sin x)$ is not identically x. Suppose next that $\pi/2 < x < 3\pi/2$. Then

$$\frac{d}{dx}[\text{Sin}^{-1}(\sin x)] = \frac{1}{\sqrt{1 - \sin^2 x}} \cos x.$$

Since $\cos x < 0$, we see that $\sqrt{1 - \sin^2 x} = -\cos x$. Thus, throughout the interval $(\pi/2, 3\pi/2)$,

$$\frac{d}{dx}[\text{Sin}^{-1}(\sin x)] = -1 = \frac{d}{dx}(-x).$$

It follows that there exists a constant C such that

$$\text{Sin}^{-1}(\sin x) = C - x$$

for all x in $(\pi/2, 3\pi/2)$. Setting $x = \pi$, we get $0 = C - \pi$. Hence

$$\text{Sin}^{-1}(\sin x) = \pi - x$$

throughout the interval $(\pi/2, 3\pi/2)$. Since $\mathrm{Sin}^{-1}(\sin x)$ is clearly periodic with period 2π, the values of $\mathrm{Sin}^{-1}(\sin x)$ are determined by the values which we have already calculated (see Fig. 9–17). Note that the function $\mathrm{Sin}^{-1} \circ \sin$ is not differentiable at points of the form $(2n + 1)(\pi/2)$, where n is an integer.

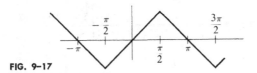

FIG. 9–17

Example 3. There are relations for the inverse functions corresponding to most of the trigonometric identities; the inverse relations, however, are frequently more complicated. For instance, corresponding to the identity

$$\tan\left(\frac{\pi}{2} - \phi\right) = \frac{1}{\tan \phi} \qquad (*)$$

(where ϕ is not an integral multiple of $\pi/2$), we have the somewhat more complicated relation

$$\mathrm{Tan}^{-1}\frac{1}{x} = \begin{cases} \pi/2 - \mathrm{Tan}^{-1} x & \text{if } x > 0, \\ -\pi/2 - \mathrm{Tan}^{-1} x & \text{if } x < 0. \end{cases} \qquad (**)$$

Several proofs of this relation are possible. The one which we give here involves the direct use of the identity $(*)$.

Suppose that $x > 0$, and let $\phi = \mathrm{Tan}^{-1} x$. Then $0 < \phi < \pi/2, 0 < \pi/2 - \phi < \pi/2$, and $x = \tan \phi$. Thus

$$\frac{1}{x} = \frac{1}{\tan \phi} = \tan\left(\frac{\pi}{2} - \phi\right) = \mathrm{Tan}\left(\frac{\pi}{2} - \phi\right).$$

Hence

$$\mathrm{Tan}^{-1}\frac{1}{x} = \mathrm{Tan}^{-1}\left[\mathrm{Tan}\left(\frac{\pi}{2} - \phi\right)\right] = \frac{\pi}{2} - \phi = \frac{\pi}{2} - \mathrm{Tan}^{-1} x.$$

Since Tan is an odd function, so is Tan^{-1}. Thus, if $x < 0$, we have, by the previous case,

$$\mathrm{Tan}^{-1}\frac{1}{x} = -\mathrm{Tan}^{-1}\left(-\frac{1}{x}\right) = -\frac{\pi}{2} + \mathrm{Tan}^{-1}(-x)$$

$$= -\frac{\pi}{2} - \mathrm{Tan}^{-1} x.$$

Example 4. We shall now prove a generalization of the preceding result. If $xy \neq 1$, then

$$\mathrm{Tan}^{-1} x + \mathrm{Tan}^{-1} y = \mathrm{Tan}^{-1}\frac{x + y}{1 - xy} + \pi\omega(x, y),$$

where

$$\omega(x, y) = \begin{cases} 0 & \text{if } xy < 1, \\ +1 & \text{if } xy > 1 \text{ and } x > 0, \\ -1 & \text{if } xy > 1 \text{ and } x < 0. \end{cases}$$

We shall prove the identity by regarding y as a constant and x as a variable. First we consider the case in which $y > 0$. Let

$$f(x) = \operatorname{Tan}^{-1} \frac{x + y}{1 - xy} \qquad (x \neq y^{-1}).$$

Then

$$f'(x) = \frac{1}{1 + \left(\dfrac{x + y}{1 - xy}\right)^2} \cdot \frac{(1 - xy) - y(x + y)}{(1 - xy)^2}$$

$$= \frac{1 + y^2}{(1 - xy)^2 + (x + y)^2} = \frac{1}{1 + x^2} \cdot$$

Hence, on each of the intervals $(-\infty, 1/y)$ and $(1/y, +\infty)$, the functions f and Tan^{-1} differ by a constant. Thus we can write

$$f(x) = \begin{cases} C_1 + \operatorname{Tan}^{-1} x & \text{if } x < 1/y, \\ C_2 + \operatorname{Tan}^{-1} x & \text{if } x > 1/y. \end{cases}$$

The constant C_1 may be found by setting $x = 0$. This gives

$$\operatorname{Tan}^{-1} y = C_1.$$

The constant C_2 may be found by taking limits as $x \to +\infty$. This gives

$$C_2 + \frac{\pi}{2} = \operatorname{Tan}^{-1}\left(-\frac{1}{y}\right) = -\operatorname{Tan}^{-1}\frac{1}{y} = -\frac{\pi}{2} + \operatorname{Tan}^{-1} y.$$

Thus $C_2 = \operatorname{Tan}^{-1} y - \pi$. We have proved that

$$\operatorname{Tan}^{-1} x + \operatorname{Tan}^{-1} y = \begin{cases} \operatorname{Tan}^{-1} \dfrac{x + y}{1 - xy} & \text{if } y > 0 \text{ and } xy < 1, \\[2ex] \operatorname{Tan}^{-1} \dfrac{x + y}{1 - xy} + \pi & \text{if } y > 0 \text{ and } xy > 1. \end{cases}$$

The other cases may be treated similarly.

Throughout the rest of this section we shall be concerned with a series expansion for $\operatorname{Tan}^{-1} x$. We have shown that

$$\frac{d}{dx}[\operatorname{Tan}^{-1} x] = \frac{1}{1 + x^2} \cdot$$

We can expand the right-hand side by performing "long division," thus obtaining

$$
\begin{array}{r}
1 - x^2 + x^4 - x^6 + \cdots \\
1 + x^2 \,\overline{\big)\, 1 } \\
1 + x^2 \\
\hline
- x^2 \\
- x^2 - x^4 \\
\hline
+ x^4 \\
x^4 + x^6 \\
\hline
- x^6
\end{array}
$$

Thus

$$\frac{1}{1+x^2} = 1 - x^2 + x^4 - x^6 + \cdots$$

We now integrate both sides of this equation and get

$$\text{Tan}^{-1} x = \int_0^x \frac{dx}{1+x^2} = \int_0^x dx - \int_0^x x^2 \, dx + \int_0^x x^4 \, dx - \int_0^x x^6 \, dx + \cdots$$

$$= x - \frac{x^3}{3} + \frac{x^5}{5} - \frac{x^7}{7} + \cdots$$

In particular, by setting $x = 1$, we get a strikingly beautiful expression for π, namely,

$$\pi/4 = 1 - \tfrac{1}{3} + \tfrac{1}{5} - \tfrac{1}{7} + \cdots$$

In case the reader should be in doubt, let it be stressed that the preceding paragraph is not offered as a model of mathematical rigor. Its only purpose is to suggest what might be true. What we shall now prove is that the series expansion

$$\text{Tan}^{-1} x = x - \frac{x^3}{3} + \frac{x^5}{5} - \frac{x^7}{7} + \cdots$$

is valid whenever $|x| \leq 1$. Before we can even think of constructing a proof, however, we need to give a precise meaning to the polynomial-run-wild which appears on the right-hand side of the equation.

Definition 5. Let $a_1, a_2, a_3, \ldots$ be a sequence of real numbers. For each positive integer n, we let s_n, the **nth partial sum,** be the sum of the first n terms of the sequence; thus

$$s_n = \sum_{k=1}^{n} a_k.$$

If $S = \lim_{n \to \infty} s_n$ exists, we say that the **infinite series**

$$\sum_{k=1}^{\infty} a_k = a_1 + a_2 + a_3 + \cdots$$

converges and has **sum** S. Hence we write

$$\sum_{k=1}^{\infty} a_k = S.$$

Otherwise, the infinite series is said to **diverge.**

Example 5. Consider the infinite series

$$\sum_{k=1}^{\infty} \frac{1}{k(k+1)}.$$

By a formula from the first chapter, the nth partial sum of the series is

$$s_n = \sum_{k=1}^{n} \frac{1}{k(k+1)} = \frac{n}{n+1}.$$

Since $\lim_{n \to \infty} s_n = 1$, the infinite series converges and

$$\sum_{k=1}^{\infty} \frac{1}{k(k+1)} = 1.$$

Example 6. Consider the geometric series

$$1 + x + x^2 + \cdots = \sum_{k=0}^{\infty} x^k.$$

If $x \neq 1$, the nth partial sum is given by

$$s_n(x) = \sum_{k=1}^{n} x^{k-1} = \frac{1 - x^n}{1 - x}.$$

If $|x| < 1$, $\lim_{n \to \infty} x^n = 0$ and

$$\lim_{n \to \infty} s_n(x) = \frac{1}{1 - x}.$$

If $|x| > 1$, the sequence $\{s_n(x)\}$ is unbounded and hence has no limit. If $x = -1$, the partial sums alternate between 1 and 0, and hence have no limit. If $x = 1$, $s_n(x) = n$, so that the series once again diverges.

In conclusion, the series $\sum_{k=0}^{\infty} x^k$ converges iff $|x| < 1$, in which case

$$\sum_{k=0}^{\infty} x^k = \frac{1}{1 - x}.$$

We return now to our discussion of the arc tangent. Replacing x by $-t^2$ in the equation

$$\frac{1}{1 - x} = 1 + x + x^2 + \cdots + x^{n-1} + \frac{x^n}{1 - x},$$

we get

$$\frac{1}{1 + t^2} = 1 - t^2 + t^4 - \cdots + (-1)^{n-1}t^{2n-2} + \frac{(-1)^n t^{2n}}{1 + t^2}.$$

We next integrate both sides from 0 to x. This yields

$$\operatorname{Tan}^{-1} x = \int_0^x \frac{dt}{1 + t^2} = x - \frac{x^3}{3} + \frac{x^5}{5} - \cdots + (-1)^{n-1}\frac{x^{2n-1}}{2n - 1} + \int_0^x \frac{(-1)^n t^{2n}}{1 + t^2}\, dt$$

$$= S_n(x) + R_n(x),$$

where

$$S_n(x) = x - \frac{x^3}{3} + \frac{x^5}{5} - \cdots + (-1)^{n-1}\frac{x^{2n-1}}{2n - 1}$$

and

$$R_n(x) = (-1)^n \int_0^x \frac{t^{2n}}{1 + t^2}\, dt.$$

If $0 \leq x \leq 1$, we have, by the comparison property of integrals,

$$|R_n(x)| = \int_0^x \frac{t^{2n}}{1 + t^2}\, dt \leq \int_0^x t^{2n}\, dt$$

$$= \frac{x^{2n+1}}{2n + 1} \leq \frac{1}{2n + 1}.$$

Since the latter tends to zero as $n \to \infty$, it follows that $\lim_{n \to \infty} R_n(x) = 0$ and thus

$$\lim_{n \to \infty} S_n(x) = \lim_{n \to \infty} [\text{Tan}^{-1} x - R_n(x)] = \text{Tan}^{-1} x.$$

But $S_n(x)$ is just the nth partial sum of the infinite series $x - x^3/3 + x^5/5 - \cdots$ This proves that the latter series converges and

$$\text{Tan}^{-1} x = x - \frac{x^3}{3} + \frac{x^5}{5} - \frac{x^7}{7} + \cdots,$$

provided $0 \leq x \leq 1$. Since the replacement of x by $-x$ simply results in a sign change for each member of the equation, we conclude that the series expansion is valid whenever $|x| \leq 1$. In particular,

$$\frac{\pi}{4} = 1 - \tfrac{1}{3} + \tfrac{1}{5} - \tfrac{1}{7} + \cdots$$

Exercises

1. Find the following.
   a) $\text{Sin}^{-1} \sqrt{3}/2$ b) $\text{Cos}^{-1} \tfrac{1}{2}$
   c) $\text{Tan}^{-1} \sqrt{3}$ d) $\text{Tan}^{-1} (-1/\sqrt{3})$
   e) $\text{Tan}^{-1} (-1)$ f) $\sin (\text{Tan}^{-1} 5)$
   g) $\tan (\text{Sin}^{-1} \tfrac{1}{3})$

2. Find dy/dx in each of the following cases. (You can occasionally shorten the computation by first simplifying the function.)
   a) $\text{Sin}^{-1} \sqrt{x}$ b) $\text{Sin}^{-1} \sqrt{1 - x^2}$
   c) $x\, \text{Sin}^{-1} x + \sqrt{1 - x^2}$ d) $\text{Sin}^{-1} x - x\sqrt{1 - x^2}$
   e) $\text{Tan}^{-1} (x + 1)$ f) $\text{Tan}^{-1} (x + \sqrt{1 + x^2})$
   g) $\text{Sin}^{-1} (\sin x - \cos x)$ h) $\sqrt{\text{Sin}^{-1} 3x}$
   i) $\text{Tan}^{-1} \left(\dfrac{x}{a}\, \text{Tan}^{-1} \dfrac{x}{a}\right)$ j) $\text{Tan}^{-1} \dfrac{x^{1/3} + a^{1/3}}{1 - a^{1/3}x^{1/3}}$

3. Verify the following integration formulas.

   a) $\displaystyle \int \frac{dx}{a^2 + x^2} = \frac{1}{a} \text{Tan}^{-1} \frac{x}{a} + C$ b) $\displaystyle \int \frac{dx}{\sqrt{a^2 - x^2}} = \text{Sin}^{-1} \frac{x}{a} + C$

4. Calculate each of the following integrals.

a) $\displaystyle\int_{-1/2}^{\sqrt{3}/2} \frac{dx}{\sqrt{1-x^2}}$

b) $\displaystyle\int_{-\sqrt{3}}^{1} \frac{1}{1+x^2}$

c) $\displaystyle\int_{0}^{3} \frac{dx}{x^2+9}$

d) $\displaystyle\int_{5\sqrt{2}/2}^{-5\sqrt{3}/2} \frac{dx}{\sqrt{25-x^2}}$

e) $\displaystyle\int \frac{\mathrm{Cos}^{-1}x}{\sqrt{1-x^2}}\,dx$

f) $\displaystyle\int \frac{\cos x}{1+\sin^2 x}\,dx$

g) $\displaystyle\int \frac{dx}{\sqrt{x(1-x)}}$

5. Prove by at least two methods that

$$\mathrm{Tan}^{-1}\frac{x}{\sqrt{1-x^2}} = \mathrm{Sin}^{-1}x$$

if $|x| < 1$.

6. Prove that

$$\mathrm{Sin}^{-1}x + \mathrm{Sin}^{-1}y = (-1)^{\omega}\,\mathrm{Sin}^{-1}(x\sqrt{1-y^2}+y\sqrt{1-x^2}) + \pi\omega,$$

where

$$\omega = \begin{cases} 0 & \text{if } xy \le 0 \text{ or } x^2+y^2 \le 1, \\ 1 & \text{if } xy > 0,\, x > 0, \text{ and } x^2+y^2 > 1, \\ -1 & \text{if } xy > 0,\, x < 0, \text{ and } x^2+y^2 < 1. \end{cases}$$

7. Plot carefully the graph of $\mathrm{Sin}^{-1}(\cos x)$.

8. Let f be the inverse of the sine function restricted to the interval $[\pi/2, 3\pi/2]$. Show that

$$f'(x) = -\frac{1}{\sqrt{1-x^2}}.$$

9. Let Sec be the restriction of the secant function to the set $[0, \pi/2) \cup (\pi/2, \pi]$. Show that Sec is invertible, and graph both the function and its inverse. Show that

$$\frac{d}{dx}[\mathrm{Sec}^{-1}x] = \frac{1}{|x|\sqrt{x^2-1}} \qquad (|x| > 1),$$

$$\mathrm{Sec}^{-1}x = \mathrm{Cos}^{-1}(1/x).$$

10. Show that

$$\pi/4 = \mathrm{Tan}^{-1}(120/119) - \mathrm{Tan}^{-1}(1/239)$$
$$= 4\,\mathrm{Tan}^{-1}(1/5) - \mathrm{Tan}^{-1}(1/239).$$

Using the latter expression, together with the series expansion

$$\mathrm{Tan}^{-1}x = x - \frac{x^3}{3} + \frac{x^5}{5} - \frac{x^7}{7} + \cdots,$$

compute π to four decimal places.

11. *The Chebyshev Polynomials.* For each nonnegative integer n, we define the function T_n as follows:

$$T_n(x) = \cos(n \operatorname{Cos}^{-1} x) \qquad (|x| \leq 1).$$

a) Show that $T_0(x), T_1(x),$ and $T_2(x)$ are the restrictions of the polynomials $1, x,$ and $2x^2 - 1$, respectively, to the interval $[-1, +1]$.

b) Show that if $m \leq n$, then

$$2T_m(x)T_n(x) = T_{n+m}(x) + T_{n-m}(x).$$

In particular, deduce the recursion formula

$$T_{n+1}(x) = 2xT_n(x) - T_{n-1}(x).$$

Use this formula to calculate T_3 and T_4.

c) Using the recursion formula above, prove by induction that T_n is (the restriction of) a polynomial of degree n with integer coefficients whose leading term is $2^{n-1}x^n$. Show, moreover, that T_n is (the restriction of) an even function if n is even and an odd function if n is odd. The function T_n is called the nth *Chebyshev polynomial*.

*d) Prove that

$$T_n(x) = \frac{(-2)^n n!}{(2n)!} \sqrt{1 - x^2} \, \frac{d^n}{dx^n} (1 - x^2)^{n-1/2}.$$

(Verify the equation directly if $n = 0$ and $n = 1$. Then show that the expression on the right-hand side of the equation satisfies the same recursion formula as T_n.)

e) If $n > 0$, show that

$$\int_{-1}^{+1} \frac{T_n(x)}{\sqrt{1 - x^2}} \, dx = 0.$$

[*Hint:* Differentiate $\sin(n \operatorname{Cos}^{-1} x)$.] Using (b), deduce the *orthogonality relation*

$$\int_{-1}^{+1} \frac{T_m(x)T_n(x)}{\sqrt{1 - x^2}} \, dx = \begin{cases} 0 & \text{if } m \neq n, \\ \pi/2 & \text{if } m = n \neq 0, \\ \pi & \text{if } m = n = 0. \end{cases}$$

f) Show that $y = T_n(x)$ is a solution of the differential equation

$$(1 - x^2)y'' - xy' + n^2 y = 0.$$

g) Let $x = \cos 1°$. Show that x is an algebraic number by observing that $T_{90}(x) = 0$.

h) For a fixed n, show that T_n has exactly n real zeros and that they lie in the open interval $(-1, +1)$. In other words, show that the numbers $r_1, r_2, \ldots, r_n$ are zeros of T_n, where

$$r_k = \cos \frac{(2k - 1)\pi}{2n}.$$

Show that the extreme values of T_n on the interval $[-1, +1]$ are plus and minus one. More specifically, let

$$e_k = \cos \frac{k\pi}{2n} \qquad (k = 0, \ldots, n),$$

and show that

$$T_n(e_k) = (-1)^k.$$

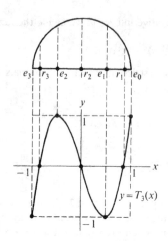

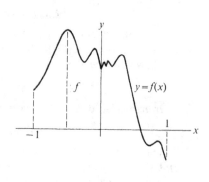

FIG. 9–18   FIG. 9–19

The case in which $n = 3$ is illustrated in Fig. 9–18. Plot in a similar manner the graphs of T_4 and T_5.

i) A *monic* polynomial is a polynomial whose highest-order term has coefficient one. Thus, $x^3 - 5x + 1$ and $x^7 - 18x^3 + 4$ are monic polynomials, whereas $3x^2 - 1$ is not. Given any function f continuous on the interval $[-1, +1]$, we define

$$\|f\| = \max \{|f(x)| \mid -1 \leq x \leq 1\}$$

(see Fig. 9–19). In function theory $\|f\|$ is usually called the *uniform norm* of f on the interval $[-1, +1]$. Some numerical analysts, however, prefer to call $\|f\|$ the *merit figure* of f. Prove that among all monic polynomials of degree n, the polynomial $2^{-n+1}T_n$ has least norm. Prove, in other words, that if P is any monic polynomial of degree n, then

$$\|P\| \geq \|2^{-n+1}T_n\| = 2^{-n+1}.$$

(Suppose that P is a monic polynomial of degree n such that $\|P\| < 2^{-n+1}$. By considering the values of P at $e_0, e_1, \ldots, e_n$, deduce that $2^{-n+1}T_n - P$ is a polynomial of degree $n - 1$ having $n + 1$ changes of sign. Show that this is impossible.) Actually, one can prove that $2^{-n+1}T_n$ is the only monic polynomial of degree n whose norm is 2^{-n+1}. This is called the *minimax property* of T_n.

12. Prove that

$$\text{Cos}^{-1}\left(\frac{a \cos x + b}{a + b \cos x}\right) - 2 \, \text{Tan}^{-1}\left(\sqrt{\frac{a - b}{a + b}} \tan \frac{x}{2}\right)$$

$(0 < b \leq a, 0 \leq x < \pi)$ is a constant, and find the constant.

9–3 Definitions and basic properties of the exponential and logarithmic functions

The definition of a logarithm found in most high school algebra texts must be regarded as suspect. Consider for a moment $\log_{10} 13$. According to the usual definition, $\log_{10} 13$ is that power to which 10 must be raised in order to get 13. In other words, $x = \log_{10} 13$ is a solution of the equation $10^x = 13$. But how do we know that a solution exists? Certainly no rational number satisfies the equation. (Suppose to the contrary that $10^{m/n} = 13$, where m and n are positive integers. We would then have $10^m = 13^n$. The latter is impossible, since the left-hand side would be an even number, whereas the right-hand side would be odd.) Noting that $10^1 = 10$ and $10^2 = 100$, one might be tempted to invoke the intermediate-value theorem to deduce that there is a real number x between 1 and 2 such that $10^x = 13$. This argument is fine *provided* we can prove that 10^x is a continuous function. However, before we can do that we must agree on the proper meaning of 10^x for irrational values of x, and this is the real issue. Contemplate the following equations:

$$10^{1.7} = \sqrt[10]{10^{17}} \quad \text{and} \quad 10^{4.579} = \sqrt[1000]{10^{4579}}.$$

Then ask yourself what it means to raise 10 to the πth power.

Although there are more direct solutions to the problems raised in the preceding paragraph, there is none more elegant or appropriate to our purposes than the one which we now present.

Definition 1. If $x > 0$, we define the **natural logarithm** of x to be

$$\ln x = \int_1^x \frac{dt}{t}.$$

The integrand $1/t$ is both monotone and continuous on the interval $(0, +\infty)$, and thus by either Theorem 11, Section 8-1, or Exercise 5, Section 8-1, the function is integrable on any closed bounded subinterval of $(0, +\infty)$.

From the fundamental theorem we get immediately one of the important properties of the natural logarithm, namely, that it is a differentiable function and that

$$\frac{d}{dx}(\ln x) = \frac{1}{x}.$$

Since the derivative is always positive, the function is strictly increasing; since the second derivative, namely, $-1/x^2$, is always negative, the function is strictly concave, as shown in Fig. 9–20. Also, note that

$$\ln 1 = \int_1^1 t^{-1} \, dt = 0.$$

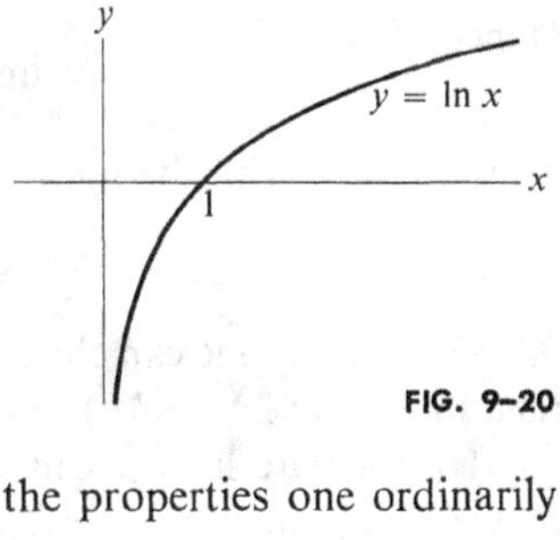

FIG. 9–20

We will now show that the natural logarithm has the properties one ordinarily expects of logarithms.

Theorem 2. *Let a and b be any two positive numbers, and let r be any rational number. Then*

1) $\ln ab = \ln a + \ln b$,   2) $\ln a/b = \ln a - \ln b$,   3) $\ln a^r = r \ln a$.

Proof. By the chain rule,

$$\frac{d}{dx}(\ln ax) = \frac{1}{ax}\frac{d}{dx}(ax) = \frac{a}{ax} = \frac{1}{x} = \frac{d}{dx}(\ln x).$$

Hence there exists a constant C such that

$$\ln ax = C + \ln x$$

for all $x > 0$. Setting $x = 1$, we get

$$\ln a = C + \ln 1 = C.$$

Hence

$$\ln ax = \ln a + \ln x$$

for all positive numbers a and x.

Similarly,

$$\frac{d}{dx}(\ln x^r) = \frac{1}{x^r}\frac{d}{dx}(x^r) = \frac{1}{x^r}(rx^{r-1}) = \frac{r}{x} = \frac{d}{dx}(r \ln x).$$

Hence there exists a constant C such that

$$\ln x^r = C + r \ln x$$

for all $x > 0$. Setting $x = 1$, we get $0 = \ln 1 = C$. Thus

$$\ln x^r = r \ln x$$

for all $x > 0$.

Property (2), of course, follows immediately from (1) and (3):

$$\ln a/b = \ln ab^{-1} = \ln a + \ln b^{-1} = \ln a + (-1)\ln b = \ln a - \ln b.$$

Theorem 3
$$\lim_{x\to+\infty} \ln x = +\infty \qquad and \qquad \lim_{x\to 0+} \ln x = -\infty.$$

Proof. Since the natural logarithm is a strictly increasing function, $\ln 2 > \ln 1 = 0$. Hence

$$\lim_{n\to\infty} \ln 2^n = \lim_{n\to\infty} n \ln 2 = +\infty.$$

This, coupled with the monotonicity of $\ln x$, implies that

$$\lim_{x\to+\infty} \ln x = +\infty.$$

(Given $M > 0$, one can choose a positive integer N such that $\ln 2^N > M$. If $x > 2^N$, then $\ln x > \ln 2^N > M$.)

To calculate $\lim_{x\to 0+} \ln x$, we make the substitution $y = 1/x$. Since $y \to +\infty$ as $x \to 0+$, and conversely, we have

$$\lim_{x\to 0+} \ln x = \lim_{y\to+\infty} \ln y^{-1} = \lim_{y\to+\infty} -\ln y = -\infty.$$

Corollary 4. *The range of the natural logarithm is the entire real line.*

Proof. Given any real number y, we can find a real number a (sufficiently close to zero) and a real number b (sufficiently large) such that

$$\ln a < y < \ln b.$$

By the intermediate-value theorem, there exists a number x between a and b such that $y = \ln x$.

Because it is a strictly increasing function, the natural logarithm is invertible. Its inverse is the most important function in analysis.

Definition 5. We denote by exp the inverse of the function ln. This function is known as the **exponential function**.

It follows at once from the general theory of inverse functions that exp is an everywhere differentiable function whose domain is the entire line and whose range is the set of all positive numbers. Also, note that we have the identities

$$\ln (\exp x) = x, \qquad \exp (\ln x) = x;$$

the first holds for all real numbers, the second holds for all positive numbers. In particular,

$$\exp 0 = \exp (\ln 1) = 1.$$

The graph of this function is shown in Fig. 9–21.

Let us calculate the derivative of the exponential function. Let

$$y = \exp x.$$

Then $x = \ln y$ and

$$1 = \frac{d}{dx} (x) = \frac{d}{dx} (\ln y) = \frac{1}{y} \frac{dy}{dx}.$$

Thus

$$\frac{dy}{dx} = y = \exp x.$$

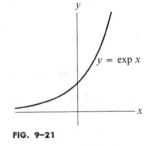

FIG. 9–21

In other words, the exponential function has the remarkable property of being its own derivative. We shall later prove that (modulo constant factors) the exponential function is unique in this respect. It follows, of course, that the exponential function has continuous derivatives of all orders, all of which are positive. In particular, the function is strictly increasing and strictly convex. We now summarize our discussion up to this point.

Theorem 6. *The exponential function is a strictly increasing and strictly convex function whose domain is the entire real line and whose range is the set of positive numbers. It is everywhere differentiable and*

$$\frac{d}{dx} (\exp x) = \exp x.$$

To each property of the natural logarithmic function there corresponds a property of the exponential function. We consider some of these now.

Theorem 7. *Let x and y be any two real numbers, and let r be any rational number.* Then

   1) $\exp(x + y) = (\exp x)(\exp y)$,

   2) $\exp(x - y) = (\exp x)/(\exp y)$,

   3) $(\exp x)^r = \exp(rx)$,

   4) $\lim_{x \to +\infty} \exp x = +\infty$ *and* $\lim_{x \to -\infty} \exp x = 0$.

Proof. Let $a = \exp x$ and $b = \exp y$. Then $x = \ln a$ and $y = \ln b$. Hence

$$x + y = \ln a + \ln b = \ln ab$$

or, equivalently,

$$\exp(x + y) = ab = (\exp x)(\exp y),$$

thus proving property (1).

   The other properties have similar proofs.

   Let $a > 0$, and let r be any rational number. Then

$$a^r = [\exp(\ln a)]^r = \exp(r \ln a)$$

by property (3) above. This suggests that we *define* a^x for irrational x to be $\exp(x \ln a)$.

Definition 8. Let $a > 0$, and let x be any real number. We define

$$a^x = \exp(x \ln a).$$

We shall prove first the general laws of exponents.

Theorem 9. General Exponent Laws. *Let a and b be positive real numbers. The identities*

   1) $a^x a^y = a^{x+y}$,

   2) $a^x/a^y = a^{x-y}$,

   3) $(ab)^x = a^x b^x$,

   4) $(a^x)^y = a^{xy}$

hold for all real numbers x and y.

Proof of (1)

$$a^x a^y = \exp(x \ln a) \exp(y \ln a)$$
$$= \exp(x \ln a + y \ln a) = \exp[(x + y) \ln a] = a^{x+y}.$$

Proof of (4). Let $z = a^x = \exp(x \ln a)$. Then $\ln z = x \ln a$. Hence

$$(a^x)^y = \exp(y \ln z) = \exp(yx \ln a) = a^{xy}.$$

The proofs of (2) and (3) are left to the reader.

   Note that if we let $e = \exp 1$, then $\ln e = 1$ and

$$e^x = \exp(x \ln e) = \exp x$$

for all real numbers x. This number e is unquestionably one of the most important numbers in all mathematics.

Definition 10. The number $e = \exp 1$ is called the **base of the natural (or Napierian) logarithms.**

In the next section we will prove that e is an irrational number whose value to 20 places is

$$e = 2.71828182845904523536.$$

Harder to prove, but nevertheless true, is the fact that e, like π, is transcendental. The reason for calling e the base of the natural logarithms will soon be clear.

Graphs of the function a^x for different values of a are shown in **Fig. 9–22.** By definition,

$$a^x = \exp(x \ln a) = e^{\lambda x},$$

where $\lambda = \ln a$. If $a > 1$, it follows that $\lambda > 0$, and the function a^x is strictly increasing, since its derivative,

$$\frac{d}{dx}(a^x) = \lambda e^{\lambda x},$$

is always positive. If $0 < a < 1$, the function is strictly decreasing, since the derivative is always negative. If $a = 1$, the function is equal to one. Consequently, if a is any positive number other than one, the function a^x is invertible.

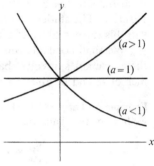

FIG. 9–22

Definition 11. If a is any positive integer other than one, we define $\log_a$ to be the inverse of the function f, where $f(x) = a^x$. For each $x > 0$, the number $\log_a x$ is called the **logarithm of x to the base a.**

Examples. $\log_{1/2} 4 = -2$, $\log_4 2 = \frac{1}{2}$.

Note that if $a = e$, then $e^x = \exp x$ and, consequently, $\log_e$, the inverse function of e^x, must be $\ln$; that is,

$$\ln x = \log_e x$$

for all positive x. This explains why e is called the base of the natural logarithms.

Theorem 12. Let a and b be two positive numbers other than one. Then

    1) $\log_a xy = \log_a x + \log_a y$,

    2) $\log_a x/y = \log_a x - \log_a y$,

    3) $\log_a x^z = z \log_a x$,

    4) $\log_b x = (\log_b a)(\log_a x)$ (change of base rule)

for all positive numbers x and y and for all real numbers z.

Proof. Properties (1), (2), and (3) are equivalent to the exponent laws. The change of base rule follows from (3). Let $z = \log_a x$. Then

$$\log_b x = \log_b a^z = z \log_b a = (\log_b a)(\log_a x).$$

The change of base rule is easily remembered, since it involves simple cancellation, as in the case of fractions.

Because of the change of base rule, one can convert logarithms from one base to another merely by multiplying by a constant factor (sometimes called the *modulus of conversion*). The only two bases of any real importance are e and 10. The importance of logarithms to the base 10 stems solely from the use of decimal representations. This is due to the fact that if x and y differ only in the location of the decimal point, then their logarithms to the base 10 differ by an integer. Hence it suffices to tabulate logarithms to the base 10 only for numbers between 10 and 100 (or between 1 and 10, etc.). The student's awareness of the importance of natural logarithms will grow with his further study of analysis.

We shall conclude this section with some examples of differentiation. The first three examples involve basic differentiation formulas which the student should memorize and be able to derive on command.

Example 1. Let $y = a^x$, where a is positive and different from one. Since $y = a^x = e^{x \ln a}$, we have by the chain rule

$$\frac{dy}{dx} = e^{x \ln a} \frac{d}{dx}(x \ln a) = (\ln a)e^{x \ln a} = (\ln a)a^x.$$

Example 2. Let $y = \log_a x$, where a is a positive number not equal to one. Since

$$y = \log_a x = (\log_a e) \log_e x = (\log_a e) \ln x,$$

we get

$$\frac{dy}{dx} = \frac{\log_a e}{x}.$$

Alternatively, we have $x = a^y$, so that

$$1 = \frac{d}{dx}(x) = \frac{d}{dx}(a^y) = (\ln a)a^y \frac{dy}{dx}$$

by Example 1. Hence

$$\frac{dy}{dx} = \frac{1}{(\ln a)a^y} = \frac{1}{(\ln a)x}.$$

The two expressions which we have obtained for dy/dx are the same, since by the change of base formula

$$(\ln a)(\log_a e) = (\log_e a)(\log_a e) = \log_e e = 1.$$

Example 3. We consider next the differentiation of an arbitrary power. Let $y = x^a$ ($x > 0$), where a is *any* real number. Since, by definition,

$$y = x^a = e^{a \ln x},$$

we have

$$\frac{dy}{dx} = e^{a \ln x} \frac{d}{dx} (a \ln x) = x^a \cdot \frac{a}{x} = ax^{a-1}.$$

Thus the differentiation formula for powers is exactly the same as before. (Previously, of course, we were only able to prove this formula—with considerable difficulty—for the case of rational exponents.)

Example 4. Let $y = x^x$ $(x > 0)$. Since

$$y = x^x = e^{x \ln x},$$

it follows that

$$\frac{dy}{dx} = e^{x \ln x} \frac{d}{dx} (x \ln x) = x^x \left(\ln x + x \cdot \frac{1}{x} \right) = (1 + \ln x) x^x.$$

Essentially the same calculation results from *logarithmic differentiation*. We first take the natural logarithm of each side of the equation $y = x^x$. This gives

$$\ln y = x \ln x.$$

Differentiating, we get

$$\frac{1}{y} \frac{dy}{dx} = 1 + \ln x,$$

so that

$$\frac{dy}{dx} = y(1 + \ln x) = (1 + \ln x) x^x.$$

Example 5. Let

$$y = \sqrt{\frac{x}{x + 1}} x^{1/x} \qquad (x > 0).$$

This expression all but cries out for logarithmic differentiation:

$$\ln y = \frac{1}{2} \ln x - \frac{1}{2} \ln (x + 1) + \frac{1}{x} \ln x.$$

Thus

$$\frac{1}{y} \frac{dy}{dx} = \frac{1}{2x} - \frac{1}{2(x + 1)} + \frac{1}{x^2} - \frac{1}{x^2} \ln x,$$

and

$$\frac{dy}{dx} = \sqrt{\frac{x}{x + 1}} x^{1/x} \left[\frac{1}{x(x + 1)} + \frac{1}{x^2} (1 - \ln x) \right].$$

We collect some of the differentiation formulas established in this section:

$$\frac{d}{dx} (\ln x) = \frac{d}{dx} (\log_e x) = \frac{1}{x},$$

$$\frac{d}{dx} (\log_a x) = \frac{\log_a e}{x} = \frac{1}{(\ln a)x},$$

$$\frac{d}{dx} (e^x) = e^x, \qquad \frac{d}{dx} (a^x) = (\ln a)a^x, \qquad \frac{d}{dx} (x^a) = ax^{a-1}.$$

Exercises

1. Calculate the derivatives of each of the following functions.

a) $(\ln x)^2$

b) $\ln (\ln x)$

c) $x \ln x - x$

d) $\ln (x + \sqrt{x^2 - 1})$

e) $\ln (\sec x + \tan x)$

f) $e^{\sin x}$

g) $2/(e^x + e^{-x})$

h) $\text{Tan}^{-1} (e^x)$

i) $\ln (e^x + \sqrt{1 + e^{2x}})$

j) 2^{-x}

k) $x^\pi \pi^x$

l) $\log_{10} (\text{Sin}^{-1} x^2)$

m) 2^{x^2}

n) e^{e^x}

o) $x^{\sqrt{x}}$

p) $x \, \text{Tan}^{-1} x - \ln \sqrt{1 + x^2}$

q) x^{x^x}

r) $\log_x e$

s) $(\sin x)^{\cos x} + (\cos x)^{\sin x}$

t) x^{a^x}

u) $\dfrac{(\ln x)^x}{x^{\ln x}}$

v) $\sqrt[3]{\dfrac{\cos x}{\ln x}} \, x^{\tan x}$

2. If $\text{Tan}^{-1} (x/y) + \ln \sqrt{x^2 + y^2} = 0$, show that

$$\frac{dy}{dx} = \frac{x + y}{x - y}.$$

3. Using the mean-value theorem, show that the inequality

$$\frac{h}{1 + h} < \ln (1 + h) < h$$

holds if $-1 < h < 0$ or $0 < h$. Also show that

$$\frac{a - b}{a} \le \ln \frac{a}{b} \le \frac{a - b}{b} \qquad (0 < b \le a).$$

4. Verify the following identities, in which the bases of the logarithms are arbitrary.

a) $\log (x + \sqrt{x^2 - 1}) = -\log (x - \sqrt{x^2 - 1})$

b) $\log \tan (\phi/2) = \log \sin \phi - \log (1 + \cos \phi) \quad (0 < \phi < \pi)$

5. Calculate the following integrals.

a) $\displaystyle\int 2^{-x} \, dx$

b) $\displaystyle\int x^{-1}(\ln x)^3 \, dx$

c) $\displaystyle\int \frac{dx}{x \ln x}$

d) $\displaystyle\int \frac{1}{\sqrt{x}} e^{\sqrt{x}} \, dx$

e) $\displaystyle\int \frac{\cos x}{1 + \sin^2 x} \, dx$

f) $\displaystyle\int \text{ctn} \, x \, dx$

g) $\displaystyle\int (1 - x^2)^{-1/2} \exp (\text{Sin}^{-1} x) \, dx$

6. Calculate the following limit:

$$\lim_{n \to \infty} [(n + 1)^{-1} + (n + 2)^{-1} + \cdots + (2n)^{-1}].$$

7. a) Integrate the geometric series

$$\frac{1}{1 + x} = 1 - x + x^2 - x^3 + \cdots$$

term by term to obtain formally the series expansion

$$\ln (1 + x) = x - \frac{x^2}{2} + \frac{x^3}{3} - \cdots$$

b) Prove that the series expansion is valid if $-1 < x \leq 1$. (Use the same technique we used in dealing with the series for the arc tangent. You will have to treat separately, however, the cases in which $x > 0$ and $x < 0$.) Note, in particular, that

$$\ln 2 = 1 - \tfrac{1}{2} + \tfrac{1}{3} - \tfrac{1}{4} + \cdots$$

8. Let $y = (\mathrm{Sin}^{-1} x)^2$. Show first that

$$(1 - x^2)(y')^2 = 4y,$$

and by differentiating once again, show that y satisfies the differential equation

$$(1 - x^2)y'' - xy' - 2 = 0.$$

Using Leibnitz's rule, show next that

$$(1 - x^2)y^{(n+2)} - (2n + 1)xy^{(n+1)} - n^2 y^{(n)} = 0.$$

Deduce that

$$y^{(n+2)}(0) = n^2 y^{(n)}(0) \qquad (n \geq 2),$$

and then find an explicit formula for $y^{(n)}(0)$.

9. Show that

$$\frac{d^n}{dx^n}\left(\frac{\ln x}{x}\right) = (-1)^n \frac{n!}{x^{n+1}}\left(\ln x - 1 - \frac{1}{2} - \cdots - \frac{1}{n}\right).$$

*10. Show that

$$\frac{d^n}{dx^n}\left(e^{x^2/2}\right) = u_n(x)e^{x^2/2},$$

where u_n is a monic polynomial of degree n. Also prove the recursion formulas

$$u_{n+1} = xu_n + u'_n \qquad \text{and} \qquad u_{n+1} = xu_n + nu_{n-1}.$$

By combining these two recursion formulas, show that u_n satisfies the differential equation

$$u''_n + xu'_n - nu_n = 0.$$

Let

$$u_n(x) = x^n + a_1 x^{n-1} + \cdots + a_n,$$

and by substituting into the above differential equation, determine the coefficients $a_1, \ldots, a_n$.

11. Fill in the details of the following proof that e^x is a transcendental function. (A simpler proof will be considered later.) Suppose to the contrary that e^x is algebraic. Then there exist polynomials $P_0, \ldots, P_n$ such that

$$P_0(x)e^{nx} + P_1(x)e^{(n-1)x} + \cdots + P_{n-1}(x)e^x + P_n(x) = 0$$

for all x. Among such equations there is one for which n is least, and we shall assume that this is true of the equation above. Then $P_n(x)$ is a nonzero polynomial. Let r be the degree of P_n and by differentiating the above equation $r + 1$ times, show that

$$Q_0(x)e^{nx} + Q_1(x)e^{(n-1)x} + \cdots + Q_{n-1}(x)e^x = 0,$$

where $Q_0, \ldots, Q_{n-1}$ are polynomials whose degrees are equal to those of $P_0, \ldots, P_{n-1}$, respectively. Observe that

$$Q_0(x)e^{(n-1)x} + Q_1(x)e^{(n-2)x} + \cdots + Q_{n-1}(x) = 0,$$

and deduce a contradiction.

9-4 Further study of the exponential function

We have observed that the exponential function is its own derivative. More generally, if $y = Ce^{\lambda x}$, where C and λ are constants, then dy/dx is proportional to y, since

$$\frac{dy}{dx} = \lambda Ce^{\lambda x} = \lambda y.$$

The differential equation

$$\frac{dy}{dx} = \lambda y$$

is of considerable importance in applications, since there are many phenomena in which the instantaneous rate of growth or decay of some quantity is directly proportional to the amount of the quantity present. We shall begin with an existence and uniqueness theorem for this differential equation.

Theorem 1. *If*

$$y = Ce^{\lambda x},$$

where C and λ are constants, then y satisfies the initial-value problem

$$\frac{dy}{dx} = \lambda y, \qquad y(0) = C. \tag{*}$$

Conversely, if y satisfies the initial-value problem (*), then $y = Ce^{\lambda x}$.

Proof. If $y = Ce^{\lambda x}$, then we have previously shown that $dy/dx = \lambda y$. Setting $x = 0$, we get $y(0) = Ce^0 = C$. Thus y is a solution of the initial-value problem (*). Suppose now that $y(x)$ is a solution of (*). Then

$$\frac{d}{dx}[e^{-\lambda x}y(x)] = e^{-\lambda x}\frac{dy}{dx} - \lambda e^{-\lambda x}y = \left(\frac{dy}{dx} - \lambda y\right)e^{-\lambda x} = 0.$$

Hence the function $e^{-\lambda x}y(x)$ must be constant, say,

$$e^{-\lambda x}y(x) = K.$$

Multiplying this equation by $e^{\lambda x}$ gives

$$y(x) = Ke^{\lambda x},$$

and setting $x = 0$, we get

$$C = y(0) = K.$$

Thus

$$y(x) = Ce^{\lambda x},$$

and the proof is complete.

Corollary 2. *The exponential function is characterized by the fact that it is equal to its own derivative and that its value at zero is one.*

We shall now consider a variety of applications of this theorem.

Application 1. At one time it was thought that the speed of a freely falling body is directly proportional to its displacement from rest position. Without knowing any physics at all (except that objects *can* fall!), we can incontrovertibly refute this superstition. Suppose that it were so. Let $s(t)$ denote the displacement from rest position of a freely falling body at time t. Then

$$\frac{ds}{dt} = ks$$

for some constant k. Hence, by the theorem,

$$s(t) = Ce^{kt},$$

where C is a constant. Setting $t = 0$, we get $0 = s(0) = C$. It follows that $s(t)$ is identically zero. In other words, it follows from this belief that a freely falling body, once at rest, is always at rest, which is but to say that freely falling bodies can't fall!

Application 2. *Newton's Law of Cooling.* Suppose that a hot skillet is lowered into a large bath of cold water. We shall attempt to describe how the temperature of the skillet varies with time. In doing so, we shall assume that the temperature of the water is not appreciably raised by the skillet (which is reasonable if there is a lot of water kept in constant motion). We shall also assume (whether or not it is good physics) the validity of Newton's law of cooling. This law states that the rate of cooling of a body is directly proportional to the difference between the temperature of a body and that of its surroundings.

Let $T(t)$ be the temperature of the skillet at time t, and let T_w be the temperature of the water. Then by Newton's law of cooling,

$$\frac{dT}{dt} = -k(T - T_w),$$

where k is a positive constant. If we make the substitution $y = T - T_w$, we get

$$\frac{dy}{dt} = -ky.$$

Hence, by the theorem,

$$T(t) - T_w = y = Ce^{-kt},$$

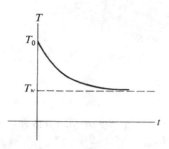

FIG. 9-23   FIG. 9-24

where C is a constant. Setting $t = 0$ yields

$$T_0 - T_w = C,$$

where T_0 is the initial temperature of the skillet. Thus

$$T = T_w + (T_0 - T_w)e^{-kt}$$

(see Fig. 9-23).

Application 3. *Growth of a Current.* Consider the circuit shown in Fig. 9-24, consisting of a battery, a resistor, and an inductance coil. Suppose that at time $t = 0$ the circuit is closed. We shall determine the current I at time t.

As any good engineer knows, the voltage drop across the resistance coil (which by Ohm's law is IR) is equal to the voltage of the battery E minus the "back emf (electromotive force)" produced by the inductance coil. The latter is directly proportional to the rate of change of the current. Thus we get the differential equation

$$IR = E - L\frac{dI}{dt},$$

where the constant L is known as the *self-inductance* of the coil. To apply the theorem, we first rewrite the equation as

$$\frac{dI}{dt} = -\frac{R}{L}\left(I - \frac{E}{R}\right).$$

If we let $y = I - E/R$, we get

$$\frac{dy}{dt} = -\frac{R}{L}y.$$

Hence, by the theorem,

$$I - \frac{E}{R} = y = Ce^{-(Rt)/L}.$$

Since there is no current at the moment the circuit is closed, setting $t = 0$, we get $-E/R = C$. Hence

$$I = \frac{E}{R}(1 - e^{-(Rt)/L})$$

(see Fig. 9-25).

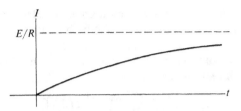

FIG. 9-25

Application 4. Observe that if $f(x) = e^{\lambda x}$, then by the exponent laws,

$$f(x + y) = f(x)f(y).$$

We will now show that this *functional equation* (plus smoothness conditions) characterizes these functions.

Proposition 3. *Suppose that $f(x)$ is a differentiable function which satisfies the functional equation*

$$f(x + y) = f(x)f(y)$$

for all real numbers x and y. Then either $f(x) = e^{\lambda x}$ for some constant λ, or else f is identically zero.

Proof. Let $C = f(0)$. Then

$$C^2 = f(0)f(0) = f(0 + 0) = f(0) = C.$$

Consequently, either $C = 1$ or $C = 0$.

Case I. $C = f(0) = 0$. Then for any real number x,

$$f(x) = f(x + 0) = f(x)f(0) = 0.$$

That is, f is identically zero.

Case II. $C = f(0) = 1$. Then

$$f'(x) = \lim_{h \to 0} \frac{f(x + h) - f(x)}{h} = \lim_{h \to 0} \frac{f(x)f(h) - f(x)}{h}$$

$$= f(x) \lim_{h \to 0} \frac{f(h) - f(0)}{h} = \lambda f(x),$$

where $\lambda = f'(0)$. Hence, by the theorem,

$$f(x) = K e^{\lambda x}$$

for some constant K. By setting $x = 0$, we see that $K = 1$, and thus that $f(x) = e^{\lambda x}$.

Although the preceding proposition is a nice application of Theorem 1, it does not begin to tell the whole story.* For one thing, the differentiability assumption on f

---

* On first reading, it might be best for the student to skip this section, glance at the statement of Theorem 5, and then continue with Application 5.

can be weakened to that of continuity or monotonicity. Because of the importance of the functional equation $f(x + y) = f(x)f(y)$ both in algebra and analysis, it fully merits the deeper study which we now undertake.

Suppose that f is a solution of the functional equation. Then f is always non-negative, since for each real number x,

$$f(x) = f\left(\frac{x}{2} + \frac{x}{2}\right) = \left[f\left(\frac{x}{2}\right)\right]^2.$$

Furthermore, if f is zero at one point, then it is identically zero; for if $f(x_0) = 0$, then

$$f(x) = f((x - x_0) + x_0) = f(x - x_0)f(x_0) = 0$$

for all x. Hence, either f is identically zero, in which case we say that f is the *trivial solution* of the functional equation, or else f is everywhere positive.

Let us suppose now that f is a nontrivial solution of the functional equation. Since f is everywhere positive, we can form a new function F as follows:

$$F(x) = \ln f(x).$$

Then

$$F(x + y) = \ln f(x + y) = \ln f(x)f(y)$$
$$= \ln f(x) + \ln f(y) = F(x) + F(y).$$

Hence the new function F satisfies the functional equation

$$F(x + y) = F(x) + F(y). \tag{*}$$

Such a function is said to be *additive*. Moreover, since the natural logarithm is continuous and strictly increasing, it follows that f and F have the same points of continuity, and that monotonicity of one of these functions implies monotonicity of the other. Conversely, if F is a solution of (*), then $f(x) = \exp F(x)$ is a nontrivial solution of our original functional equation, namely,

$$f(x + y) = f(x)f(y). \tag{**}$$

Hence, the problem of determining nontrivial solutions of (**) reduces to the problem of finding additive functions.

Theorem 4. *Let F be an additive function. Then for all rational numbers r and all real numbers x,*

$$F(rx) = r F(x).$$

Moreover, the following are logically equivalent:

    1) F is continuous at one point;

    2) F is continuous at all points;

    3) F is monotone;

    4) $F(x) = \lambda x$ for some constant $\lambda \in \mathbf{R}$.

Proof. One can prove by induction that

$$F(nx) = n F(x)$$

for all positive integers n. Since

$$F(0) = F(0 + 0) = 2 F(0),$$

we have $F(0) = 0$. If n is a positive integer,

$$F(nx) + F(-nx) = F(nx + (-nx)) = F(0) = 0,$$

so that

$$F(-nx) = -F(nx) = -n F(x).$$

Hence the identity $F(rx) = r F(x)$ holds for all integers r (positive, negative, and zero) and all real numbers x. Thus, if m and n are integers and $n \neq 0$,

$$n F\left(\frac{m}{n} x\right) = F\left(n \cdot \frac{m}{n} x\right) = F(mx) = m F(x),$$

so that

$$F\left(\frac{m}{n} x\right) = \frac{m}{n} F(x).$$

This proves that the identity $F(rx) = r F(x)$ holds for all rational numbers r and real numbers x.

Observe that if F is of the form $F(x) = \lambda x$, then F is additive and has properties (1), (2), and (3). Hence it suffices to prove that (1) $\Rightarrow$ (2), (2) $\Rightarrow$ (4), and (3) $\Rightarrow$ (4). [Observe that (2) $\Rightarrow$ (1) trivially.]

Proof of (1) $\Rightarrow$ (2). Suppose that F is continuous at x_0. We shall prove that F is continuous at any second point x_1. Let $\epsilon > 0$ be given. Since F is continuous at x_0, there exists a $\delta > 0$ such that

$$|x - x_0| < \delta \Rightarrow |F(x) - F(x_0)| < \epsilon.$$

If $|x - x_1| < \delta$, then $|(x - x_1 + x_0) - x_0| < \delta$, so that

$$|F(x - x_1 + x_0) - F(x_0)| < \epsilon.$$

But

$$|F(x - x_1 + x_0) - F(x_0)| = |F(x) - F(x_1) + F(x_0) - F(x_0)|$$
$$= |F(x) - F(x_1)|.$$

Hence

$$|x - x_1| < \delta \Rightarrow |F(x) - F(x_1)| < \epsilon.$$

This proves that F is continuous at x_1. (This argument actually proves that F is uniformly continuous on the entire line.)

Proof of (2) $\Rightarrow$ (4). We assume now that F is continuous. For all rational numbers r, we have

$$F(r) = r\lambda,$$

where $\lambda = f(1)$. Let x be any irrational number. Then we can find a sequence of rational numbers $r_1, r_2, \ldots$ which converge to x. (For instance, we could let r_n be the decimal expansion of x rounded to the nth decimal place.) Since F is continuous at x, we have

$$F(x) = \lim_{n \to \infty} F(r_n) = \lim_{n \to \infty} \lambda r_n = \lambda_x.$$

Thus $(2) \Rightarrow (4)$.

Proof of (3) $\Rightarrow$ (4). Suppose that F is monotone, and let $\lambda = F(1)$. If $\lambda = 0$, then f is zero at all the rationals. Hence the only way that F can be monotone is for F to be identically zero. Suppose next that $\lambda > 0$. If x is any real number, we have

$$F(x) \leq \text{glb } \{F(r) \mid r > x \text{ and } r \text{ is rational}\}$$

and

$$F(x) \geq \text{lub } \{F(r) \mid r < x \text{ and } r \text{ is rational}\}.$$

However, $F(r) = \lambda r$ for all rational numbers, and it is very easily verified that the greatest lower bound and the least upper bound indicated above must both be equal to λx. Thus $F(x) = \lambda x$ for any real number x. The case in which $\lambda < 0$ can be treated similarly.

The student may wonder whether there is any additive function which is not of the form $F(x) = \lambda x$, where λ is a constant. The answer is that there are such functions, that they do in fact vastly outnumber the continuous additive functions, but that their construction seems to depend on the use of *transfinite induction* and is therefore outside the scope of this book. Such a function is clearly pathological, since, according to the theorem, it must be everywhere discontinuous. In fact, one can show without too much difficulty that the closure of the graph of such a function must be the entire plane. In other words, every disk in the plane (no matter where it is located or how small its radius is) must contain infinitely many points of the graph of the function!

From the remarks preceding the proof of Theorem 4, we get the following characterization of exponential functions.

Theorem 5. *Suppose that f satisfies the functional equation $f(x + y) = f(x)f(y)$. Then f is either identically zero or everywhere positive. In the latter case*

$$f(rx) = [f(x)]^r$$

for all rational numbers r and all real numbers x. Also, the following are equivalent:

1) f is continuous at one point;

2) f is continuous at all points;

3) f is monotone;

4) f is of the form $f(x) = e^{\lambda x}$ for some fixed real number λ.

Application 5. *Compound Interest.* We have seen that if a quantity grows continuously in such a way that its instantaneous rate of growth is directly proportional to its size, then the growth of this quantity can be described by an exponential function. An analogous discrete phenomonon is offered by the compounding of interest. If an amount of money is deposited in a bank, then at the times when interest is paid, the amount by which the account grows is directly proportional to the amount of money in the account. The growth of the account is, of course, discrete rather than continuous. (We are using the terms "discrete" and "continuous" in a nontechnical sense.) We shall see, however, that the exponential function arises as a limiting case of this discrete phenomenon.

Suppose that P dollars is deposited into a bank account where it is allowed to draw interest at the rate of $100r$ percent per annum. No further deposits are then made. Let us consider the amount A which is in the account at the end of one year. This amount will depend on the number of times that interest is compounded. If simple interest is paid, then

$$A = P + rP = (1 + r)P.$$

If interest is compounded semiannually, then at the end of half a year the account will contain $[1 + (r/2)]P$ dollars, so that the amount at the end of the year will be

$$A = \left(1 + \frac{r}{2}\right)\left[\left(1 + \frac{r}{2}\right)P\right] = \left(1 + \frac{r}{2}\right)^2 P.$$

If compounded triannually, the amount will be

$$A = \left(1 + \frac{r}{3}\right)^3 P.$$

More generally, if interest is compounded n times a year, then

$$A = \left(1 + \frac{r}{n}\right)^n P.$$

It is reasonable to ask at this point how A behaves as n becomes very large. It is common knowledge among people interested in making money (and that includes most of us) that A increases as n increases; that is, the more often interest is compounded, the better the deal. Less well-known, however, is the upper limit on the amount which can be made by compounding interest more and more frequently. This is what we shall consider now.

Intuitively, it would seem that as n increases, we approach a state of continuous growth in which the instantaneous rate of growth of the account is directly proportional to its size. In this limiting or idealized state we have

$$\frac{dA}{dt} = rA,$$

and thus, by our theorem, the amount present in the account at time t is given by

$$A = Pe^{rt}.$$

In particular, the amount at the end of one year will be Pe^r. We shall justify this heuristic argument by proving that

$$\lim_{n \to \infty} \left(1 + \frac{r}{n}\right)^n P = Pe^r.$$

Theorem 6. *For every real number x,*

$$e^x = \lim_{n \to \infty} \left(1 + \frac{x}{n}\right)^n = \lim_{n \to \infty} \left(1 - \frac{x}{n}\right)^{-n}.$$

Proof

$$\lim_{h \to 0} \ln (1 + hx)^{1/h} = \lim_{h \to 0} \frac{\ln (1 + hx) - \ln (1 + 0 \cdot x)}{h}$$

$$= \frac{d}{dt} \ln (1 + xt) \Big|_{t=0} = \frac{x}{1 + xt} \Big|_{t=0} = x.$$

Since the exponential function is continuous,

$$\lim_{h \to 0} (1 + hx)^{1/h} = \lim_{h \to 0} \exp \left[\ln (1 + hx)^{1/h}\right]$$

$$= \exp x = e^x.$$

Thus, if $\{h_n\}$ is any sequence of distinct points converging to zero, we have $\lim_{n \to \infty} (1 + h_n x)^{1/h_n} = e^x$. In particular, by selecting the sequences $\{1/n\}$ and $\{-1/n\}$, we get the desired result.

Corollary 7. $e = \lim_{n \to \infty} (1 + 1/n)^n$.

In some treatments of the exponential function the above limit is taken as the definition of e.

We shall now derive an infinite series expansion for e^x by somewhat shady methods. By the previous theorem,

$$e^x = \lim_{n \to \infty} \left(1 + \frac{x}{n}\right)^n.$$

The expression inside the limit sign is a polynomial of degree n which can be written in standard form by use of the binomial theorem. Thus

$$\left(1 + \frac{x}{n}\right)^n = 1 + n\frac{x}{n} + \frac{n(n-1)}{2!} \left(\frac{x}{n}\right)^2 + \frac{n(n-1)(n-2)}{3!} \left(\frac{x}{n}\right)^3 + \cdots + \left(\frac{x}{n}\right)^n$$

$$= 1 + x + \frac{x^2}{2!} \left(1 - \frac{1}{n}\right) + \frac{x^3}{3!} \left(1 - \frac{1}{n}\right) \left(1 - \frac{2}{n}\right) + \cdots$$

We now "take the limit" of both sides as n tends to infinity and get

$$e^x = 1 + x + \frac{x^2}{2!} + \frac{x^3}{3!} + \frac{x^4}{4!} + \cdots$$

At first it might seem that in calculating this limit we have simply used the fact that the "limit of a sum is equal to the sum of the limits." If there were only a *fixed* number of terms in the expansion for $(1 + x/n)^n$, there would be no difficulty, but as it is, the number of terms becomes infinite as n increases. What is really at issue is the interchangeability of two limit operations. Such matters are always delicate, but, fortunately, they need not concern us at present, since we can establish the validity of the series expansion by a different technique.

Let us consider the general term in the series expansion for e^x, namely,

$$\frac{x^n}{n!} = \frac{x}{1} \cdot \frac{x}{2} \cdot \frac{x}{3} \cdots \frac{x}{n}.$$

If $|x| \leq 1$, the term obviously tends to zero as $n \to \infty$. If $|x| > 1$, however, there is a power struggle between the numerator and the denominator as they both race to infinity. If $|x|$ is extremely large, the numerator will take a commanding lead early in the race. Eventually, however, the ratio $|x|/n$ will be less than $\frac{1}{2}$, and once this occurs, each term will then be less than half the size of the preceding one, since

$$\frac{x^{k+1}}{(k+1)!} = \frac{x^k}{k!} \cdot \frac{x}{k+1}.$$

Thus, no matter how large a lead the numerator may gain early in the race, it's lead is quickly whittled away from this point on.

Lemma 8. *If x is any real number, then*

$$\lim_{n \to \infty} \frac{x^n}{n!} = 0.$$

Proof. Since $|x^n/n!| = |x|^n/n!$, it suffices to consider only nonnegative values of x. Let N be any positive integer such that $x/N < \frac{1}{2}$. If $n > N$, we have

$$0 \leq \frac{x^n}{n!} = \frac{x^N}{N!} \cdot \frac{x}{N+1} \cdot \frac{x}{N+2} \cdots \frac{x}{n}$$

$$\leq \frac{x^N}{N!} \cdot \underbrace{\frac{1}{2} \cdot \frac{1}{2} \cdots \frac{1}{2}}_{n-N \text{ factors}}$$

$$= \frac{x^N}{N!} \left(\frac{1}{2}\right)^{n-N} = K2^{-n},$$

where $K = (2x)^N/N!$. Since $2^{-n} \to 0$ as $n \to \infty$, the desired result follows from the pinching theorem.

Theorem 9. *For every real number x,*

$$e^x = 1 + x + \frac{x^2}{2!} + \frac{x^3}{3!} + \frac{x^4}{4!} + \cdots$$

Proof. Let

$$S_n(x) = 1 + x + \frac{x^2}{2!} + \frac{x^3}{3!} + \cdots + \frac{x^n}{n!}.$$

Since e^x is its own derivative, we can expect this property to be reflected somehow in the partial sums. This is the case, since

$$\frac{d}{dx} S_n(x) = 0 + 1 + x + \frac{x^2}{2!} + \cdots + \frac{x^{n-1}}{(n-1)!} = S_{n-1}(x).$$

We must prove that $\lim_{n\to\infty} S_n(x) = e^x$, or, equivalently, that $\lim_{n\to\infty} e^{-x} S_n(x) = 1$. Now

$$\frac{d}{dx}[e^{-x}S_n(x)] = e^{-x}\left[\frac{d}{dx} S_n(x) - S_n(x)\right]$$
$$= e^{-x}[S_{n-1}(x) - S_n(x)] = -e^{-x}\frac{x^n}{n!}.$$

Thus

$$e^{-x}S_n(x) - 1 = \int_0^x \frac{d}{dt}[e^{-t}S_n(t)]\,dt$$
$$= -\int_0^x e^{-t}\frac{t^n}{n!}\,dt$$

or

$$e^{-x}S_n(x) = 1 - R_n(x),$$

where

$$R_n(x) = \int_0^x e^{-t}\frac{t^n}{n!}\,dt.$$

It suffices to prove that

$$\lim_{n\to\infty} |R_n(x)| = 0.$$

Case I. $x \geq 0$. Since e^{-t} is a decreasing function, we have, by the comparison property of definite integrals,

$$0 \leq R_n(x) = \int_0^x e^{-t}\frac{t^n}{n!}\,dt \leq \int_0^x \frac{t^n}{n!}\,dt = \frac{x^{n+1}}{(n+1)!}.$$

The latter term tends to zero as $n \to \infty$, by the lemma, and thus $R_n(x) \to 0$, by the pinching theorem.

Case II. $x < 0$. By a change of variables, we have

$$R_n(x) = \int_0^{-x} e^u \frac{(-u)^n}{n!}\,du.$$

Thus

$$|R_n(x)| = \int_0^{-x} e^u \frac{u^n}{n!}\,du \leq \int_0^{-x} e^{-x}\frac{u^n}{n!}\,du = e^{-x}\frac{x^{n+1}}{(n+1)!}.$$

Once again, by the lemma and the pinching theorem, it follows that $\lim_{n\to\infty} |R_n(x)| = 0$.

By reexamining the estimates on $R_n(x)$, we gain the following additional information from the proof. *If the polynomial*

$$S_n(x) = 1 + x + \frac{x^2}{2!} + \cdots + \frac{x^n}{n!}$$

is used as an approximation to e^x, then the error involved is less than

$$e^x \frac{x^{n+1}}{(n+1)!} \quad \text{if } x > 0$$

and

$$\frac{(-x)^{n+1}}{(n+1)!} \quad \text{if } x < 0.$$

From the series expansion we can compute e to any required accuracy. We observe first of all that $e < 3$, since

$$e = 1 + \frac{1}{1!} + \frac{1}{2!} + \frac{1}{3!} + \frac{1}{4!} + \cdots$$
$$< 1 + 1 + \tfrac{1}{2} + (\tfrac{1}{2})^2 + (\tfrac{1}{2})^3 + \cdots$$
$$= 1 + \frac{1}{1 - \tfrac{1}{2}} = 3.$$

Suppose that we wish to compute e to ten decimal places. If we use

$$e_n = 1 + \frac{1}{1!} + \frac{1}{2!} + \cdots + \frac{1}{n!}$$

as an approximation to e, then the error involved will be less than

$$\frac{e}{(n+1)!} < \frac{3}{(n+1)!}.$$

If we demand that

$$\frac{3}{(n+1)!} < 5 \times 10^{-11},$$

we will get the required accuracy. The smallest value of n for which the inequality holds is 13. By computing e_{13}, we get the desired result.

Proposition 10. *e is irrational.*

Proof. Suppose that e were rational. Then we could express e as m/n, where m and n are positive integers both divisible by 3. But

$$\frac{m}{n} = e = 1 + \frac{1}{1!} + \frac{1}{2!} + \cdots + \frac{1}{(n-1)!} + \frac{r_n}{n!},$$

where $0 < r_n < e < 3$. Solving for r_n, we get

$$r_n = m(n-1)! - 2(n!) - [n(n-1) \cdots 3] - \cdots - n.$$

The first term is divisible by 3, since it has m as a factor. The remaining terms are integers having n as a common factor, and hence they are also divisible by 3. It follows that r_n must be an integer divisible by 3. This is impossible, however, since $0 < r_n < 3$.

Exercises

1. The following argument may be found in many calculus texts. Suppose $dy/dx = \lambda y$. Then

$$\frac{1}{y}\frac{dy}{dx} = \lambda.$$

Hence

$$\ln\frac{y(x)}{y(0)} = \int_0^x \frac{1}{y}\frac{dy}{dx}\,dx = \int_0^x \lambda\,dx = \lambda x.$$

Thus, by exponentiation,

$$y(x) = y(0)e^{\lambda x}.$$

Explain the deficiencies of this argument.

2. Suppose that f is a function such that

$$f(x) = \int_0^x f(t)\,dt$$

for all real numbers x. Prove that f must be identically zero.

3. The rate at which a quantity of radium diminishes in mass is at each instant proportional to the mass of the radium present. Show that the time required for any given amount of radium to diminish to half its original weight is a constant τ; this constant is known as the *half-life* of radium. Show that if M_0 is the mass of a quantity of radium at time $t = 0$, then the mass at time t is given by

$$M = M_0 2^{-(t/\tau)}.$$

Given that 1 gram of radium experiences a weight loss of 0.003 grams over a period of 10 yrs, determine the half-life of radium. Also determine the weight loss over a period of 34 yrs.

4. The bacterial content of milk increases 1000 times in 12 hrs when the temperature is 80°F, and this suffices to turn it sour. The rate at which the bacteria multiply is halved for each 10°F drop in temperature. Determine how long it takes milk to go sour at 70°F and at 60°F. (Although the multiplication of bacteria is a discrete phenomenon, we can, with good results, regard it as continuous when such a great number of bacteria are involved. Give a plausible physical argument to show that the rate of multiplication of the bacteria should be proportional to the amount of bacteria present.)

5. What can you say about solutions of the functional equation $f(xy) = f(x) + f(y)$?

6. In our discussion of compound interest it was asserted that for a fixed $r > 0$ the sequence $(1 + r/n)^n$ is increasing. Prove that this is so by showing that the function

$$f(x) = \ln\left(1 + \frac{r}{x}\right)^x$$

is strictly increasing throughout the interval $(0, +\infty)$. [You will need the inequality

$$\ln(1+h) > \frac{h}{1+h} \qquad (h > 0)$$

established in a previous exercise.] Note, then, that the inequality

$$\left(1 + \frac{x}{n}\right)^n < e^x < \left(1 - \frac{x}{n}\right)^{-n} \qquad (x > 0)$$

holds for all positive integers n.

7. Suppose that interest is compounded n times a year at the rate of $100r$ percent per annum. Show that if P dollars is deposited at time $t = 0$, then the amount in the account at time t (in years) is

$$A_n(t) = \left(1 + \frac{r}{n}\right)^{[nt]} P.$$

Show that

$$\lim_{n \to \infty} A_n(t) = e^{rt}P.$$

8. a) Show that

$$\frac{d}{dx}\left[-e^{-x}\left(1 + \frac{x}{n}\right)^n\right] \le \frac{x}{n} \qquad (x > 0).$$

By integrating, deduce that

$$1 - e^{-x}\left(1 + \frac{x}{n}\right)^n \le \frac{x^2}{2n} \qquad \text{or} \qquad e^x - \left(1 + \frac{x}{n}\right)^n \le \frac{x^2}{2n}e^x.$$

b) Let A be the amount to which one dollar will grow in 7 yrs if it draws interest at the rate of 4 percent per annum, and the interest is compounded semiannually. Show that A is approximately equal to $e^{0.28}$ and that the error involved is less than 0.004. [Use (a).]

9. Recall that if a is any real number,

$$\binom{a}{n} = \frac{a(a-1)\cdots(a-n+1)}{n!}$$

for all positive integers n. Also

$$\binom{a}{0} = 1.$$

The purpose of this exercise is to prove that the series expansion

$$(1+x)^a = \sum_{n=0}^{\infty} \binom{a}{n} x^n$$

is valid whenever $|x| < 1$.

a) Prove the Pascal triangle property:

$$\binom{a}{n} + \binom{a}{n+1} = \binom{a+1}{n+1}.$$

b) We let

$$S_{a,n}(x) = \sum_{k=0}^{n} \binom{a}{k} x^k.$$

Show that

$$\frac{d}{dx} S_{a,n}(x) = a S_{a-1,n-1}(x)$$

and

$$\frac{d}{dx} [(1 + x)^{-a} S_{a,n}(x)] = -a(1 + x)^{-a-1} [S_{a,n}(x) - (1 + x) S_{a-1,n-1}(x)]$$

$$= a(1 + x)^{-a-1} \binom{a-1}{n} x^n.$$

c) If $|x| < 1$, let M_x be the maximum value of $|(1 + t)^{-a-1}|$ on the interval from 0 to x. Show that

$$|1 - (1 + x)^{-a} S_{a,n}(x)| \le M_x \left| \binom{a}{n+1} x^{n+1} \right|.$$

Let b be an integer larger than $|a|$, and let $|x| < r < 1$. Since

$$\lim_{k \to \infty} \frac{b + k}{1 + k} = 1,$$

we can choose N so large that

$$\frac{b + k}{1 + k} |x| < r,$$

if $k > N$. If $n > N$, show that

$$\left| \binom{a}{n+1} x^{n+1} \right| \le K r^{n-N},$$

where

$$K = \frac{b(b + 1) \cdots (b + N)}{1 \cdot 2 \cdots (1 + N)} |x|^{N+1}.$$

Deduce finally that

$$\lim_{n \to \infty} |1 - (1 + x)^{-a} S_{a,n}(x)| = 0,$$

and hence that

$$\lim_{n \to \infty} S_{a,n}(x) = (1 + x)^a.$$

10. a) Let $x > 0$. Use Exercise 4, Section 6–3, to show that

$$\sin x - \frac{x^{4n+1}}{(4n + 1)!} \le S_{2n}(x) \le \sin x$$

and

$$\sin x \le S_{2n-1}(x) \le \sin x + \frac{x^{4n-1}}{(4n - 1)!},$$

where $S_n(x)$ is the nth partial sum of the infinite series $x - x^3/3! + x^5/5! - \cdots$

b) Deduce that

$$\sin x = x - \frac{x^3}{3!} + \frac{x^5}{5!} - \frac{x^7}{7!} + \cdots$$

if $x > 0$. (Use Lemma 8.) Show next that the series expansion is also valid if $x \leq 0$. Similarly, one can show that the series expansion

$$\cos x = 1 - \frac{x^2}{2!} + \frac{x^4}{4!} - \frac{x^6}{6!} - \cdots$$

is valid for every real number x.

11. In this exercise we shall give another proof of the validity of the series expansion for e^x.

a) Show that for all $x > 0$,

$$e^x > 1, \qquad\qquad e^{-x} < 1,$$
$$e^x > 1 + x, \qquad\qquad e^{-x} > 1 - x,$$
$$e^x > 1 + x + \frac{x^2}{2!}, \qquad e^{-x} < 1 - x + \frac{x^2}{2!},$$

by deducing each inequality from the preceding one. Prove by induction that for each $n \in \mathbf{Z}^+$,

$$e^x > T_n(x), \qquad S_{2n}(x) < e^{-x} < S_{2n+1}(x),$$

where $T_n(x)$ and $S_n(x)$ denote the nth partial sums of the infinite series

$$1 - x + \frac{x^2}{2!} - \frac{x^3}{3!} + \cdots \qquad \text{and} \qquad 1 + x + \frac{x^2}{2!} + \cdots,$$

respectively.

b) Using Lemma 8 and part (a), prove that $e^{-x} = \lim_{n \to \infty} S_n(x)$ for $x \geq 0$.

c) Show that $S_{2n+1}(x)T_{2n+1}(x)$ is an increasing function for $x \geq 0$. Deduce that

$$S_{2n+1}(x)T_{2n+1}(x) \geq S_{2n+1}(0)T_{2n+1}(0) = 1 \qquad (x \geq 0).$$

d) Using (a) and the monotone sequence property, show that $T(x) = \lim_{n \to \infty} T_n(x)$ exists and is less than or equal to e^x for each $x \geq 0$.

e) Observe that

$$e^{-x}T(x) = \lim_{n \to \infty} S_{2n+1}(x)T_{2n+1}(x) \geq 1,$$

so that $T(x) \geq e^x$ ($x \geq 0$). Deduce from (b), (c), and (d) that the power series expansions

$$e^x = 1 + x + \frac{x^2}{2!} + \frac{x^3}{3!} + \cdots \qquad \text{and} \qquad e^{-x} = 1 - x + \frac{x^2}{2!} - \frac{x^3}{3!} + \cdots$$

are valid for all $x \in \mathbf{R}$.

12. Obtain an expression for the pressure of the earth's atmosphere in terms of height above sea level, assuming that temperature is constant. (You will need the following facts. First, the pressure at any point is equal to the weight of the column of air directly above a surface of unit area. Second, $p = a\sigma$, where p is pressure, σ is the density of air, and a (in the case of constant temperature) is a constant. This is known as *Boyle's law*.)

13. Boyle's law may also be formulated in the following way. If an enclosed "ideal" gas is kept at a constant temperature, then the product of its pressure and volume is a constant. Suppose that initially an ideal gas is enclosed in a cylinder with volume V_0 under a pressure P_0. The gas is then compressed to a volume V by a piston located at one end of the cylinder. Assuming that the temperature is kept constant, calculate the work required to accomplish this compression.

9–5 The hyperbolic functions

Certain combinations of e^x and e^{-x} occur so frequently that they have been studied as functions in their own right. These functions, known as the *hyperbolic functions*, are analogous (and closely related) to the trigonometric functions. As we shall later show, they are to the geometry of the hyperbola what the trigonometric functions are to the geometry of the circle.

The *hyperbolic cosine* and the *hyperbolic sine* are defined, respectively, to be the even and odd parts of the exponential function:

$$\cosh x = \frac{e^x + e^{-x}}{2},$$

$$\sinh x = \frac{e^x - e^{-x}}{2}.$$

In terms of these functions, the remaining hyperbolic functions are defined by analogy to the trigonometric functions:

$$\tanh x = \frac{\sinh x}{\cosh x} = \frac{e^x - e^{-x}}{e^x + e^{-x}},$$

$$\operatorname{sech} x = \frac{1}{\cosh x} = \frac{2}{e^x + e^{-x}},$$

$$\operatorname{csch} x = \frac{1}{\sinh x} = \frac{2}{e^x - e^{-x}},$$

$$\operatorname{ctnh} x = \frac{1}{\tanh x} = \frac{e^x + e^{-x}}{e^x - e^{-x}}.$$

Identities analogous to those for the trigonometric functions usually hold, with an occasional change of sign. The following are samples:

$$\frac{d}{dx}(\sinh x) = \cosh x,$$

$$\frac{d}{dx}(\cosh x) = \sinh x,$$

$$\frac{d}{dx}(\tanh x) = \operatorname{sech}^2 x,$$

$$\cosh^2 x - \sinh^2 x = 1,$$

$$\sinh(x + y) = \sinh x \cosh y + \cosh x \sinh y,$$

$$\cosh(x + y) = \cosh x \cosh y + \sinh x \sinh y.$$

The proofs of these identities and others like them involve at worst some repulsively lengthy, but trivial, algebra. Consider, for instance, the analog of the Pythagorean identity:

$$\cosh^2 x - \sinh^2 x = \left(\frac{e^x + e^{-x}}{2}\right)^2 - \left(\frac{e^x - e^{-x}}{2}\right)^2$$

$$= \frac{e^{2x} + 2 + e^{-2x}}{4} - \frac{e^{2x} - 2 + e^{-2x}}{4} = 1.$$

A more complete listing of such identities may be found in the exercises.

The hyperbolic sine is clearly an odd function (the term is used here in its technical sense). Since its derivative, namely, the hyperbolic cosine, is always positive, the function is strictly increasing and, in particular, invertible. Since the function is continuous, and since

$$\lim_{x \to +\infty} \sinh x = +\infty \qquad \text{and} \qquad \lim_{x \to -\infty} \sinh x = -\infty,$$

the range of the hyperbolic sine is the entire line. Its graph is shown in Fig. 9–26. The function is invertible and its inverse is readily calculated. Let

$$y = \sinh x = \frac{e^x - e^{-x}}{2}.$$

Then

$$2ye^x = e^{2x} - 1 \qquad \text{or} \qquad (e^x)^2 - 2ye^x - 1 = 0.$$

We solve for e^x by using the quadratic formula. Thus we get

$$e^x = y \pm \sqrt{y^2 + 1}.$$

Since $e^x > 0$, only the plus sign is appropriate. Thus

$$e^x = y + \sqrt{y^2 + 1},$$

so that

$$\sinh^{-1} y = x = \ln (y + \sqrt{y^2 + 1}).$$

The derivative of $\sinh^{-1}$ may be calculated several different ways. Let

$$y = \sinh^{-1} x = \ln (x + \sqrt{x^2 + 1}).$$

Then

$$\frac{dy}{dx} = \frac{1}{x + \sqrt{x^2 + 1}} \frac{d}{dx} (x + \sqrt{x^2 + 1})$$

$$= \frac{1}{x + \sqrt{x^2 + 1}} \left(1 + \frac{x}{\sqrt{x^2 + 1}}\right)$$

$$= \frac{1}{\sqrt{x^2 + 1}}.$$

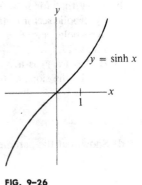

FIG. 9–26

We can also calculate dy/dx by mimicking our derivation of the derivative of the arc sine. Since

$$y = \sinh^{-1} x, \qquad x = \sinh y.$$

Thus

$$1 = \frac{d}{dx}(x) = \frac{d}{dx}(\sinh y)$$

$$= (\cosh y)\frac{dy}{dx} = \sqrt{1 + \sinh^2 y}\,\frac{dy}{dx}$$

$$= \sqrt{1 + x^2}\,\frac{dy}{dx}\,.$$

Once again we have

$$\frac{dy}{dx} = \frac{1}{\sqrt{x^2 + 1}}\,.$$

Further study of the hyperbolic functions is pursued in the exercises.

Exercises

1. Prove the following identities.
   a) $\sinh(x + y) = \sinh x \cosh y + \cosh x \sinh y$
   b) $\cosh(x + y) = \cosh x \cosh y + \sinh x \sinh y$
   c) $(\cosh x + \sinh x)^n = \cosh nx + \sinh nx$
   d) $1 - \tanh^2 x = \operatorname{sech}^2 x$ e) $\operatorname{ctnh}^2 x - 1 = \operatorname{csch}^2 x$

   f) $\dfrac{d}{dx}(\tanh x) = \operatorname{sech}^2 x$ g) $\dfrac{d}{dx}(\operatorname{ctnh} x) = -\operatorname{csch}^2 x$

   h) $\dfrac{d}{dx}(\operatorname{sech} x) = -\operatorname{sech} x \tanh x$ i) $\dfrac{d}{dx}(\operatorname{csch} x) = -\operatorname{csch} x \operatorname{ctnh} x$

2. Find double- and half-angle formulas for hyperbolic sines and cosines.

3. a) Show that $x = \cosh u$, $y = \sinh u$ is a parametrization for the right-hand branch of the hyperbola $x^2 - y^2 = 1$.
   b) Show that for $u > 0$, the parameter u can be interpreted as twice the area of the hyperbolic sector OAP, where $A = (1, 0)$ is the vertex of the right-hand branch of the hyperbola, and $P = (\cosh u, \sinh u)$ is the point corresponding to u, as shown in Fig. 9-27.
   c) If $P = (x, y)$ is a point on the right-hand branch of the hyperbola $x^2 - y^2 = 1$, let t be the slope of the line joining P and $(-1, 0)$. Show that

$$x = \frac{1 + t^2}{1 - t^2}, \qquad y = \frac{2t}{1 - t^2}\,.$$

   d) Show that the parameters t and u are related by the equation

$$t = \tanh(u/2).$$

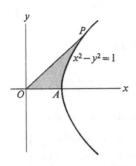

FIG. 9-27   FIG. 9-28

4. a) Show that the hyperbolic tangent is a strictly increasing function and that

$$\lim_{x\to+\infty} \tanh x = 1- \quad \text{and} \quad \lim_{x\to-\infty} \tanh x = -1+.$$

What is the range of the function? Sketch its graph.

b) Show that

$$\tanh^{-1} x = \tfrac{1}{2} \ln \frac{1+x}{1-x}.$$

c) Show by two methods that

$$\frac{d}{dx}(\tanh^{-1} x) = \frac{1}{1-x^2}.$$

5. a) Let Cosh denote the restriction of cosh to the interval $[0, +\infty)$. Show that Cosh is invertible and that

$$\text{Cosh}^{-1} x = \ln(x + \sqrt{x^2-1}) \qquad (x \geq 1).$$

b) Show by two different methods that

$$\frac{d}{dx}(\text{Cosh}^{-1} x) = \frac{1}{\sqrt{x^2-1}}.$$

6. Calculate dy/dx for each of the following. (Take advantage of the preceding exercises.)

a) $y = \text{ctnh}(\sin x)$

b) $y = \dfrac{1}{e^x + e^{-x}}$

c) $y = \left(\dfrac{e^x - e^{-x}}{e^x + e^{-x}}\right)^3$

d) $y = \dfrac{1}{2^x - 2^{-x}}$

e) $y = \left(\dfrac{e^{2x} + 1}{e^{2x} - 1}\right)^2$

f) $y = \dfrac{3^x}{3^{2x} - 1}$

7. The curve $y = a \cosh(x/a)$ is called a *catenary*. We will show in this exercise that it is the shape of a cable hanging by virtue of its own weight. (The root of "catenary" is the Latin *catena*, meaning "chain.")

a) We consider a hanging cable supported at two ends. We introduce coordinates so that the lowest point V on the cable lies along the positive y-axis (see Fig. 9-28). We consider an arbitrary point $P = (x, y)$, with $x > 0$, on the cable, and introduce the following notation:

$$s = \text{length of cable between } V \text{ and } P,$$
$$H = \text{tension of the cable at } V,$$
$$\mathbf{T} = \text{tension of the cable at } P,$$
$$\mu = \text{linear density of the cable}$$
$$= \text{weight per unit length},$$
$$\phi = \text{angle of inclination of the tangent line at } P.$$

In order for the segment of the cable between V and P to be in equilibrium, the net force acting on it must be zero. The relevant forces are the tension $\mathbf{T}$ at P, which acts in the direction of the tangent line, the horizontal tension at V, whose magnitude we denote by H, and the downward force of gravity, which in magnitude is μs. Observe that the horizontal and vertical components of $\mathbf{T}$ are $T \cos \phi$ and $T \sin \phi$, where T denotes the magnitude of $\mathbf{T}$. Deduce the equations of equilibrium

$$T \cos \phi = H \quad \text{and} \quad T \sin \phi = \mu s.$$

By division, deduce that

$$\frac{dy}{dx} = \frac{\mu}{H} s,$$

and hence that

$$\frac{d^2 y}{dx^2} = \frac{\mu}{H} \sqrt{1 + \left(\frac{dy}{dx}\right)^2}.$$

b) Show that

$$\sinh^{-1} \frac{dy}{dx} = \int_0^x \frac{y'' \, dx}{\sqrt{1 + (y')^2}} = \int_0^x \frac{\mu}{H} \, dx = \frac{\mu}{H} x,$$

and hence that

$$\frac{dy}{dx} = \sinh \frac{\mu}{H} x.$$

c) Show finally that

$$y = \frac{H}{\mu} \cosh \frac{\mu}{H} x + y_0 - \frac{H}{\mu},$$

where y_0 is the y-coordinate of V. By relocating the x-axis, if necessary, we may assume that $y_0 - H/\mu = 0$. Thus, we get

$$y = a \cosh (x/a).$$

d) Show that $T = y$. Interpret the physical significance of this fact.

8. Let M be a point on the catenary $y = a \cosh (x/a)$, and let N be its projection onto the x-axis. A semicircle with MN as diameter is constructed as indicated in Fig. 9–29, and a point P is chosen on the semicircle such that $d(N, P) = a$. Show that the line through M and P is tangent to the catenary.

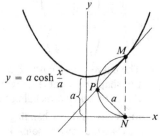

FIG. 9–29

9-6 Some important limits

As $x \to +\infty$, each of the expressions e^x, x^2, and $\ln x$ becomes infinite, but they do so at very different rates. We shall see that e^x grows at a faster rate than x^2, and that x^2, in turn, grows faster than $\ln x$.

Proposition 1. *Let $a > 1$ and $b > 0$. Then*

$$\lim_{x \to +\infty} \frac{x^b}{a^x} = 0 \quad and \quad \lim_{x \to +\infty} \frac{\log_a x}{x^b} = 0.$$

Less accurately stated, exponential functions grow more rapidly than powers, and powers, in turn, grow more rapidly than logarithms. We present two proofs.

Proof 1. Let n be any positive integer such that $n > b + 1$. It follows that if $x > 1$, then

$$\frac{x^b}{x^n} < \frac{1}{x}.$$

Let $\lambda = \ln a$. Then

$$a^x = e^{\lambda x} = 1 + \lambda x + \frac{(\lambda x)^2}{2!} + \cdots + \frac{(\lambda x)^n}{n!} + \cdots > \frac{(\lambda x)^n}{n!}$$

for all $x > 0$. Thus

$$0 < \frac{x^b}{a^x} < \frac{n!}{\lambda^n} \frac{x^b}{x^n} < \frac{n!}{\lambda^n} \frac{1}{x}$$

if $x > 1$. Since $1/x \to 0$ as $x \to +\infty$, we have, via the pinching theorem,

$$\lim_{x \to +\infty} \frac{x^b}{a^x} = 0.$$

To handle the second limit, we can make the substitution $y = \log_a x$. Then $x = a^y$ and $x^b = c^y$, where $c = a^b > 0$. Since $y \to +\infty$ as $x \to +\infty$, and conversely, we have

$$\lim_{x \to +\infty} \frac{\log_a x}{x^b} = \lim_{y \to +\infty} \frac{y}{e^y} = 0,$$

by the previous result.

Proof 2. Let $\epsilon > 0$. Then if $t \geq 1$, it follows that

$$t \geq t^{1-\epsilon} \quad or \quad t^{-1} \leq t^{-(1-\epsilon)}.$$

Consequently, if $x > 1$, then

$$\ln x = \int_1^x t^{-1} \, dt \leq \int_1^x t^{-(1-\epsilon)} \, dt = (1/\epsilon)(x^\epsilon - 1).$$

In particular, if $\epsilon < b$, then

$$0 < \frac{\ln x}{x^b} < \frac{1}{\epsilon}(x^{-(b-\epsilon)} - x^{-b}).$$

The latter expression tends to zero as $x \to +\infty$, and thus, by the pinching theorem,

$$\lim_{x \to +\infty} \frac{\ln x}{x^b} = 0.$$

Since $\log_a x = (\log_a e) \ln x$, we get immediately the more general result

$$\lim_{x \to +\infty} \frac{\log_a x}{x^b} = 0.$$

We can then obtain the other limit by making the substitution $x = b(\ln y / \ln a)$:

$$\lim_{x \to +\infty} \frac{x^b}{a^x} = \lim_{y \to +\infty} K\left(\frac{\ln y}{y}\right)^b = 0,$$

where $K = (b/\ln a)^b$.

Throughout the rest of this section we shall consider a number of rather loosely related examples.

Example 1. Note that we have as special cases of the above theorem the following:

$$\lim_{x \to +\infty} \frac{x^2}{e^x} = 0, \qquad \lim_{x \to +\infty} \frac{\ln x}{x^2} = 0.$$

In other words, as $x \to +\infty$, e^x grows more rapidly than x^2, and x^2 grows more rapidly than $\ln x$.

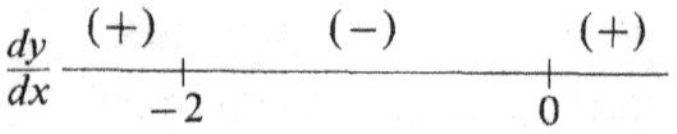

FIG. 9–30

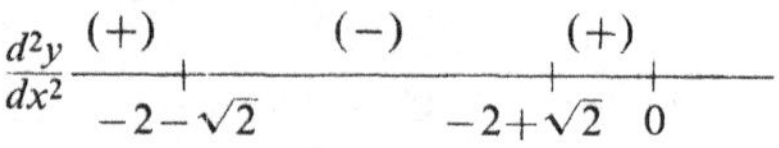

FIG. 9–31

Example 2. Sketch the graph of $y = x^2 e^x$.

As $x \to +\infty$, $x^2 e^x$ obviously becomes infinite. Consider next the behavior of the function as $x \to -\infty$.

$$\lim_{x \to -\infty} x^2 e^x = \lim_{y \to +\infty} (-y)^2 e^{-y} = \lim_{y \to +\infty} y^2/e^y = 0.$$

Hence, the x-axis is an asymptote of the graph. The derivative is

$$\frac{dy}{dx} = (x^2 + 2x)e^x = x(x + 2)e^x,$$

and it is easily seen from the latter expression that the derivative has changes of sign at $x = 0$ and $x = -2$ (Fig. 9–30). By the first-derivative test, the function has a relative minimum at $x = 0$ and a relative maximum at $x = -2$. The second derivative, namely,

$$\frac{d^2y}{dx^2} = (x^2 + 4x + 2)e^x,$$

has sign changes at $x = -2 + \sqrt{2}$ and $x = -2 - \sqrt{2}$ (Fig. 9–31). Thus the corresponding points on the graph are inflection points. The graph is shown in Fig. 9–32.

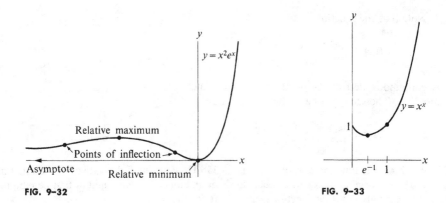

FIG. 9–32   **FIG. 9–33**

Example 3. Discuss the behavior of x^x for small positive values of x.

By definition, $x^x = e^{x \ln x}$. Now

$$\lim_{x \to 0+} x \ln x = \lim_{t \to +\infty} \frac{1}{t} \ln \frac{1}{t} = \lim_{t \to +\infty} -\frac{\ln t}{t} = 0.$$

Hence

$$\lim_{x \to 0+} x^x = e^0 = 1.$$

The derivative

$$\frac{d}{dx} (x^x) = x^x (\ln x + 1)$$

is negative if $x < e^{-1}$ and positive if $x > e^{-1}$. Moreover, as $x \to 0+$, the derivative obviously tends to minus infinity; it follows that the graph has a vertical (one-sided) tangent at $x = 0$ (see Fig. 9–33).

Example 4. Describe the behavior of the function $f(x) = e^{1/x}$.

As $x \to \pm\infty$, $1/x \to 0$, so that $e^{1/x} \to 1$. Hence the line $y = 1$ is a horizontal asymptote of the graph. At zero, the function has a remarkable discontinuity. As $x \to 0+$, $1/x \to +\infty$, and thus $e^{1/x} \to +\infty$. However, as $x \to 0-$, $1/x \to -\infty$ and $e^{1/x} \to 0$. Thus, if we define $f(0) = 0$, then f will be left-continuous at zero. Furthermore, all the left-hand derivatives will exist and be equal to zero at the point. Let us check this for the first derivative:

$$f'(x) = \frac{d}{dx} (e^{1/x}) = -\frac{1}{x^2} e^{1/x} \qquad (x \neq 0).$$

Now

$$\lim_{x \to 0-} f'(x) = \lim_{t \to +\infty} -t^2 e^{-t} = 0.$$

Since f is left-continuous at zero, it follows from the law of the mean (cf. Exercise 6, Section 6–2) that f has a zero left-hand derivative at zero. The higher-order derivatives are handled similarly. One can show that for $x \neq 0$, the nth derivative of f is equal to $e^{1/x}$ times some polynomial in $1/x$. As $x \to 0-$, the nth derivative must therefore tend to zero. The graph of $y = e^{1/x}$ is shown in Fig. 9–34.

Note that the function

$$f(x) = \begin{cases} e^{1/x} & \text{if } x < 0, \\ 0 & \text{if } x \geq 0 \end{cases}$$

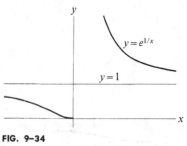

has continuous derivatives of all orders over the
entire line. The same can be said of the function

$$f(x) = \begin{cases} e^{-1/x^2} & \text{if } x \neq 0, \\ 0 & \text{if } x = 0. \end{cases}$$

FIG. 9–34

The graph of this function is shown in Fig. 9–35. Here again all derivatives of the function
are zero at the origin. This sort of behavior is quite unusual. One might at first be tempted
to think that only constant functions can behave in this way.

Example 5. In certain branches of mathematics (notably the Schwartz theory of distributions) a certain amount of proficiency in handling functions like e^{-1/x^2} is needed. The following is a typical problem. Given two disjoint sets A and B, find (if possible) a function f which
has continuous derivatives of all orders at all points, and which assumes the values 0 and 1
on the sets A and B, respectively. If, for instance, $A = (-\infty, 0]$ and $B = [1, +\infty)$, then
the function

$$f(x) = \begin{cases} 0 & \text{if } x \leq 0, \\ \exp\left[-\dfrac{1}{x^2} \exp\left(\dfrac{-1}{(x-1)^2} \right) \right], \\ 1 & \text{if } x \geq 1 \end{cases}$$

does the trick (see Fig. 9–36).

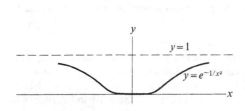

FIG. 9–35   FIG. 9–36

Example 6. Find $\lim_{x \to 0} (a^x - 1)/x$, where $a > 0$.
    We can easily evaluate this limit by interpreting it as a derivative. Thus

$$\lim_{x \to 0} \frac{a^x - 1}{x} = \lim_{h \to 0} \frac{a^{0+h} - a^0}{h} = \frac{d}{dx} (a^x) \Big|_{x=0} = (\ln a) a^x \Big|_{x=0} = \ln a.$$

Example 7. Find $\lim_{x \to 0} x^{\sin x}$.
    Since, as $x \to 0$, $\sin x$ behaves much like x itself, we would expect that

$$\lim_{x \to 0} x^{\sin x} = \lim_{x \to 0} x^x = 1.$$

This is not difficult to justify:

$$x^{\sin x} = e^{\sin x \ln x},$$

$$\lim_{x \to 0} \sin x \ln x = \lim_{x \to 0} \frac{\sin x}{x} x \ln x = 1 \cdot 0 = 0,$$

$$\lim_{x \to 0} x^{\sin x} = e^0 = 1.$$

Example 8. Find $\lim_{n \to \infty} (1/n)\sqrt[n]{(2n)!/n!}$.

In dealing with complicated products or exponents, the use of logarithms is usually helpful. We have

$$\frac{1}{n} \sqrt[n]{\frac{(2n)!}{n!}} = \sqrt[n]{\frac{(2n)(2n-1)(2n-2)\cdots(n+1)}{n^n}}$$

$$= \sqrt[n]{\frac{n+1}{n} \cdot \frac{n+2}{n} \cdots \frac{2n}{n}}$$

or

$$\ln \frac{1}{n} \sqrt[n]{\frac{(2n)!}{n!}} = \frac{1}{n} \sum_{k=1}^{n} \ln\left(1 + \frac{k}{n}\right).$$

The latter expression is a Riemann sum for the integral

$$\int_1^2 \ln x \, dx = (x \ln x - x)\Big|_1^2 = 2 \ln 2 - 1.$$

Hence

$$\lim_{n \to \infty} \ln \frac{1}{n} \sqrt[n]{\frac{(2n)!}{n!}} = 2 \ln 2 - 1,$$

so that

$$\lim_{n \to \infty} \frac{1}{n} \sqrt[n]{\frac{(2n)!}{n!}} = e^{(2 \ln 2 - 1)} = \frac{4}{e}.$$

Example 9. A similar, but more interesting and more delicate, problem is that of calculating $\lim_{n \to \infty} \sqrt[n]{n!}/n$.

Taking logarithms, we get

$$\ln \frac{\sqrt[n]{n!}}{n} = \frac{1}{n} \sum_{k=1}^{n} \ln \frac{k}{n}.$$

One might be tempted to call this a Riemann sum for the integral $\int_0^1 \ln x \, dx$. However, $\ln x$ is unbounded and hence not Riemann-integrable on the interval $[0, 1]$. Nevertheless, $\int_0^1 \ln x \, dx$ does exist as an *improper integer*, which is but to say that $\lim_{\epsilon \to 0+} \int_\epsilon^1 \ln x \, dx$ exists. In fact,

$$\lim_{\epsilon \to 0+} \int_\epsilon^1 \ln x \, dx = \lim_{\epsilon \to 0+} (-1 - \epsilon \ln \epsilon + \epsilon) = -1.$$

This suggests that

$$\lim_{n \to \infty} \frac{\sqrt[n]{n!}}{n} = e^{-1}.$$

To prove this, we set

$$a_n = \frac{1}{n}\sum_{k=1}^{n}\ln\frac{k}{n} = \frac{1}{n}\sum_{k=1}^{n-1}\ln\frac{k}{n} \quad\text{and}\quad b_n = \int_{1/n}^{1}\ln x\,dx.$$

Then $\lim_{n\to\infty} b_n = -1$, by the above. To show that a_n also has -1 as a limit, we use a pinching argument. Since $\ln x$ is monotone, it follows that

$$\frac{1}{n}\ln\frac{k}{n} = \int_{k/n}^{(k+1)/n}\ln\frac{k}{n}\,dx \le \int_{k/n}^{(k+1)/n}\ln x\,dx$$

$$\le \int_{k/n}^{(k+1)/n}\ln\frac{k+1}{n}\,dx \le \frac{1}{n}\ln\frac{k+1}{n}$$

$(k = 1,\ldots,n-1)$. Summing, we get

$$a_n \le b_n \le a_n - \frac{1}{n}\ln\frac{1}{n} \quad\text{or}\quad b_n + \frac{1}{n}\ln\frac{1}{n} \le a_n \le b_n.$$

Since $\lim_{x\to 0+} x\ln x = 0$, we have, in particular,

$$\lim_{n\to\infty}\frac{1}{n}\ln\frac{1}{n} = 0.$$

The desired result now follows from the pinching theorem.

Example 10. Prove that e^x is a transcendental function.

Suppose to the contrary that e^x is algebraic. Then there exists a polynomial $P(x, y)$ such that $P(x, e^x) = 0$ for all real numbers x. By collecting like powers of e^x, the latter equation is easily seen to be of the form

$$P_0(x)e^{nx} + P_1(x)e^{(n-1)x} + \cdots + P_n(x) = 0,$$

where $P_0,\ldots,P_n$ are polynomials and P_0 is not identically zero. Multiplying by e^{-nx} and rearranging terms gives

$$P_0(x) = -P_1(x)e^{-x} - \cdots - P_n(x)e^{-nx}.$$

As $x \to +\infty$, the right-hand side tends to zero. Since the left-hand side is a polynomial, this can occur only if P_0 is the polynomial which is identically zero. But this is a contradiction.

Exercises

1. Find the following limits. Justify your answers.

a) $\lim\limits_{x\to-\infty} \sqrt[3]{x}\,2^x$

b) $\lim\limits_{x\to+\infty} \dfrac{x\ln\ln x}{(\ln x)^2}$

c) $\lim\limits_{x\to+\infty} \dfrac{\cosh x + \sqrt{x}}{e^x + \ln x}$

d) $\lim\limits_{x\to+\infty} (\cosh x - \sinh x)$

e) $\lim\limits_{x\to-\infty} (\cosh x - \sinh x)$

f) $\lim\limits_{x\to+\infty} \left(\dfrac{x}{1+x}\right)^x$

g) $\lim_{x \to +\infty} \left(1 + \dfrac{1}{x^2}\right)^x$

h) $\lim_{x \to 0} (1 + \tan^2 \sqrt{x})^{1/2x}$

i) $\lim_{x \to 0} (1 + \sin x)^{\csc x}$

j) $\lim_{x \to 0} \dfrac{e^{x^2} - \cos x}{x^2}$

k) $\lim_{x \to +\infty} x(e^{1/x} - 1)$

l) $\lim_{x \to 0} (e^x + x)^{1/x}$

2. Investigate the continuity and differentiability of the function

$$f(x) = \begin{cases} \dfrac{x}{1 + e^{1/x}} & \text{if } x \neq 0, \\ 0 & \text{if } x = 0 \end{cases}$$

at zero. Sketch the graph of the function.

3. Sketch the graphs of the following curves.

a) $y = x^2 \ln x$

b) $y = x + (\ln x)/x$

c) $y = \ln (1 + x^2)$

d) $y = x - \ln x$

e) $y = x \sin (\ln x)$

f) $y = xe^{-x^2/2}$

g) $y = xe^{x^2/2}$

h) $y = xe^{1/x^2}$

4. Compare the growth rates of x^{x^2}, x^{2^x}, and 2^{x^x} as $x \to +\infty$.

5. Let $a > 0$. Compare the growth rates of $a^x x^a$ and x^x as $x \to +\infty$.

6. We have established the following inequalities:

$$E_n(x) < e^x < E_n(x) + e^x \frac{x^{n+1}}{(n+1)!} \qquad (x > 0),$$

$$S_{2n}(x) < \sin x < S_{2n-1}(x) \qquad (x > 0),$$

$$C_{2n}(x) < \cos x < C_{2n-1}(x) \qquad (x > 0),$$

where $E_n(x)$, $S_n(x)$, and $C_n(x)$, respectively, are the nth partial sums of the infinite series

$$1 + x + \frac{x^2}{2!} + \frac{x^3}{3!} + \cdots,$$

$$x - \frac{x^3}{3!} + \frac{x^5}{5!} - \cdots,$$

$$1 - \frac{x^2}{2!} + \frac{x^4}{4!} - \cdots$$

State similar inequalities for negative values of x. Using these inequalities and pinching arguments, calculate the following limits.

a) $\lim_{x \to 0} \dfrac{e^x - 1 - x}{x^2}$

b) $\lim_{x \to 0} \dfrac{e^x - 1 - \sin x}{x^2}$

c) $\lim_{x \to 0} \dfrac{e^{x^2} - 1}{\cos x - 1}$

d) $\lim_{x \to 0} \dfrac{e^x - e^{-x}}{\sin x \cos x}$

e) $\lim_{x \to 0} \dfrac{e^x - e^{-x} - 2x}{x - \sin x}$

f) $\lim_{x \to 0} \dfrac{e^{\tan x} - e^x}{\tan x - x}$

g) $\lim_{x \to 0} \dfrac{e^{ax} - \cos ax}{e^{bx} - \cos bx}$

7. Let $\mathfrak{F}_\infty$ denote the family of all functions f with the property that $\lim_{x\to+\infty} f(x) = +\infty$. If f and g belong to $\mathfrak{F}_\infty$, we say that f is of *lower order of magnitude than* g, and we write $f \prec g$ iff

$$\lim_{x\to+\infty} \frac{f(x)}{g(x)} = 0.$$

a) Prove that the relation $\prec$ is transitive. Show, in other words, that if $f \prec g$ and $g \prec f$, then $f \prec h$.

b) Let

$$f(x) = x^2(\sin x)^2 + x + 1 \quad \text{and} \quad g(x) = x^2(\cos x)^2 + x.$$

Show that these functions are not comparable, that is, neither $f \prec g$ nor $g \prec f$, nor does $f(x)/g(x)$ tend to a nonzero limit as $x \to +\infty$.

c) Show that

$$x \prec x \ln x \prec x^{1+\epsilon}$$

for every $\epsilon > 0$. Does the family of functions $\{x^{1+\epsilon} \mid \epsilon > 0\}$ have a greatest lower bound with respect to the order relation $\prec$?

*9-7 Some inequalities

Having completed our repertoire of elementary functions, we are in a position to do some fancy things with inequalities. One of the tools which we shall use is Jensen's inequality. It states, you will recall, that if f is convex on an interval I, then

$$f(\lambda_1 x_1 + \lambda_2 x_2 + \cdots + \lambda_a x_n) \le \lambda_1 f(x_1) + \cdots + \lambda_n f(x_n) \tag{*}$$

holds whenever $x_1, \ldots, x_n$ belong to I and $\lambda_1, \ldots, \lambda_n$ are positive numbers whose sum is one. If f is strictly convex on I, then strict inequality holds unless

$$x_1 = x_2 = \cdots = x_n.$$

Moreover, a sufficient condition for f to be strictly convex on I is that f'' be positive throughout I. If f is concave on I, then the inequality sign in $(*)$ is reversed.

Example 1. Prove that the geometric mean is less than or equal to the arithmetic mean.

Since $d^2(\ln x)/dx^2 = -1/x^2 < 0$, it follows that $\ln x$ is strictly concave on the interval $(0, +\infty)$. Hence if $x_1, \ldots, x_n$ are positive numbers, we have by Jensen's inequality

$$\ln \frac{1}{n}(x_1 + x_2 + \cdots + x_n) \ge \frac{1}{n}(\ln x_1 + \ln x_2 + \cdots + \ln x_n)$$

$$= \ln \sqrt[n]{x_1 x_2 \cdots x_n};$$

equality holds iff $x_1 = x_2 = \cdots = x_n$. Since the exponential function is strictly increasing, exponentiation of the two sides of the inequality yields the desired result:

$$\frac{1}{n}(x_1 + x_2 + \cdots + x_n) \ge \sqrt[n]{x_1 x_2 \cdots x_n},$$

with strict inequality unless $x_1 = x_2 = \cdots = x_n$.

More generally, one can prove in this way that the same inequality holds for the weighted means; that is,

$$\lambda_1 x_1 + \lambda_2 x_2 + \cdots + \lambda_n x_n \geq x_1^{\lambda_1} x_2^{\lambda_2} \cdots x_n^{\lambda_n},$$

where the "weights" $\lambda_1, \lambda_2, \ldots, \lambda_n$ are positive with sum one.

Example 2. Let p and q be positive numbers such that

$$\frac{1}{p} + \frac{1}{q} = 1.$$

Prove that for all positive numbers a and b,

$$\frac{1}{p} a^p + \frac{1}{q} b^q \geq ab.$$

This follows easily from Example 1. The weighted arithmetic mean

$$\frac{1}{p} a^p + \frac{1}{q} b^q$$

is less than or equal to the weighted geometric mean

$$(a^p)^{1/p} (b^q)^{1/q} = ab.$$

Equality holds iff $a^p = b^q$.

Example 3. *Hölder's Inequality.* Let p and q be as above. (Sometimes such a pair of positive numbers are said to be *conjugate*.) If $\mathbf{a} = (a_1, \ldots, a_n)$ and $\mathbf{b} = (b_1, \ldots, b_n)$ are any two n-tuples of positive numbers, then Hölder's inequality states that

$$\sum_{k=1}^{n} a_k b_k \leq \left(\sum_{k=1}^{n} a_k^p \right)^{1/p} \left(\sum_{k=1}^{n} b_k^q \right)^{1/q}.$$

To prove Hölder's inequality, we first set

$$\|\mathbf{a}\|_p = \left(\sum_{k=1}^{n} a_k^p \right)^{1/p} \quad \text{and} \quad \|\mathbf{b}\|_q = \left(\sum_{k=1}^{n} b_k^q \right)^{1/q}.$$

By the previous example,

$$\frac{a_k}{\|\mathbf{a}\|_p} \frac{b_k}{\|\mathbf{b}\|_q} \leq \frac{1}{p} \left(\frac{a_k}{\|\mathbf{a}\|_p} \right)^p + \frac{1}{q} \left(\frac{b_k}{\|\mathbf{b}\|_q} \right)^q$$

$(k = 1, \ldots, n)$. We now sum over both sides, and obtain

$$\frac{\sum_{k=1}^{n} a_k b_k}{\|\mathbf{a}\|_p \|\mathbf{b}\|_q} \leq \frac{1}{p} \frac{\sum_{k=1}^{n} a_k^p}{(\|\mathbf{a}\|_p)_p} + \frac{1}{q} \frac{\sum_{k=1}^{n} b_k^p}{(\|\mathbf{b}\|_q)_q}$$

$$= \frac{1}{p} \cdot 1 + \frac{1}{q} \cdot 1 = 1.$$

Hence

$$\sum_{k=1}^{n} a_k b_k \leq \|\mathbf{a}\|_p \|\mathbf{b}\|_q.$$

It is easy to check that the inequality also holds if the a_k's and b_k's are merely nonnegative (where we make the convention that $0^p = 0^q = 0$). More generally, if $\mathbf{a} = (a_1, \ldots, a_n)$ is any n-tuple of real numbers, we define its *p-norm* to be

$$\|\mathbf{a}\|_p = \left(\sum_{k=1}^n |a_k|^p \right)^{1/p}.$$

If $\mathbf{b} = (b_1, \ldots, b_n)$ is a second such n-tuple, then we have

$$\left| \sum_{k=1}^n a_k b_k \right| \leq \sum_{k=1}^n |a_k b_k| \leq \|\mathbf{a}\|_p \|\mathbf{b}\|_q.$$

In particular, when $p = q = 2$, we have the *Cauchy-Schwarz inequality*.

Example 4. If $p > 1$, prove that

$$n^{p-1} \sum_{k=1}^n |a_k|^p \geq \left(\sum_{k=1}^n |a_k| \right)^p$$

for every n-tuple $\mathbf{a} = (a_1, \ldots, a_n)$ of real numbers.

By the Hölder inequality,

$$\sum_{k=1}^n |a_k| \leq \|\mathbf{a}\|_p \|(1, 1, \ldots, 1)\|_q = \|\mathbf{a}\|_p\, n^{1/q}.$$

Thus

$$\left(\sum_{k=1}^n |a_k| \right)^p \leq n^{p/q} (\|\mathbf{a}\|_p)^p = n^{p-1} \sum_{k=1}^n |a_k|^p.$$

One can obtain the same result via Jensen's inequality by observing that the function $f(x) = x^p$ is convex on $(0, +\infty)$ [since $f''(x) = p(p-1)x^{p-2} > 0$].

Example 5. The *Minkowski* inequality is a companion of the Hölder inequality. It states that

$$\left(\sum_{k=1}^n |a_k + b_k|^p \right)^{1/p} \leq \|\mathbf{a}\|_p + \|\mathbf{b}\|_p,$$

where $p > 1$, $\mathbf{a} = (a_1, \ldots, a_n)$, and $\mathbf{b} = (b_1, \ldots, b_n)$.

To prove Minkowski's inequality, we first observe that

$$|a_k + b_k|^p = |a_k + b_k|\,|a_k + b_k|^{p-1} \leq |a_k|\,|a_k + b_k|^{p-1} + |b_k|\,|a_k + b_k|^{p-1}$$

$(k = 1, \ldots, n)$. Thus

$$\sum|a_k + b_k|^p \leq \sum|a_k|\,|a_k + b_k|^{p-1} + \sum|b_k|\,|a_k + b_k|^{p-1}$$
$$\leq \|\mathbf{a}\|_p (\sum|a_k + b_k|^{pq-q})^{1/q} + \|\mathbf{b}\|_p (\sum|a_k + b_k|^{pq-q})^{1/q}.$$

Here we have applied Hölder's inequality twice. Noting that $pq - q = p$ and $1 - 1/q = 1/p$, we see that the desired inequality follows upon division by

$$(\sum|a_k + b_k|^{pq-q})^{1/q}.$$

Minkowski's inequality can be written more compactly if we introduce the operation of *vector addition* for n-tuples. If

$$\mathbf{a} = (a_1, \ldots, a_n) \quad \text{and} \quad \mathbf{b} = (b_1, \ldots, b_n),$$

we define

$$\mathbf{a} + \mathbf{b} = (a_1 + b_1, a_2 + b_2, \ldots, a_n + b_n).$$

Then Minkowski's inequality can be written

$$\|\mathbf{a} + \mathbf{b}\|_p \leq \|\mathbf{a}\|_p + \|\mathbf{b}\|_p.$$

Example 6. If $\alpha > 1$, $x > 0$, $y > 0$, and $x + y = 1$, show that

$$\left(x + \frac{1}{x}\right)^\alpha + \left(y + \frac{1}{y}\right)^\alpha \geq \frac{5^\alpha}{2^{\alpha-1}}.$$

Let $f(t) = (t + 1/t)^\alpha$. Then

$$f''(t) = \alpha(\alpha - 1)\left(t + \frac{1}{t}\right)^{\alpha-2}\left(1 - \frac{1}{t^2}\right)^2 + \alpha\left(t + \frac{1}{t}\right)^{\alpha-1}\frac{1}{t^3} > 0$$

if $t > 0$. Hence, f is convex on the interval $(0, +\infty)$. Consequently,

$$\frac{5^\alpha}{2^{\alpha-1}} = 2f(\tfrac{1}{2}) = 2f\left(\frac{x+y}{2}\right)$$

$$\leq f(x) + f(y) = \left(x + \frac{1}{x}\right)^\alpha + \left(y + \frac{1}{y}\right)^\alpha,$$

by Jensen's inequality.

We shall conclude this section with a general discussion of mean values. In particular, we shall generalize the inequality between the arithmetic and geometric means.

Let f be any strictly monotone and continuous mapping of the set $\mathbf{R}^+$ of positive real numbers into $\mathbf{R}$, and let n be any positive integer. We shall associate with the function f a mapping M_f of $(\mathbf{R}^+)^n$, the set of all n-tuples of positive numbers, into $\mathbf{R}^+$. This mapping, called the *mean associated with f*, is defined as follows:

$$M_f(\mathbf{a}) = f^{-1}\left(\frac{1}{n}\sum_{k=1}^{n} f(a_k)\right)$$

for all $\mathbf{a} = (a_1, \ldots, a_n) \in (\mathbf{R}^+)^n$. In particular, by letting $f(x)$ be x, $\ln x$, $1/x$, we get, respectively, the arithmetic, geometric, and harmonic means of the numbers $a_1, \ldots, a_n$. Also, if we let $f(x) = x^p$ ($p \neq 0$), we get the *p-mean*

$$M_p(\mathbf{a}) = \left(\frac{1}{n}\sum_{k=1}^{n} a_k^p\right)^{1/p}.$$

Note that if $p > 1$, $M_p(\mathbf{a}) = n^{-1/p}\|\mathbf{a}\|_p$. We also define

$$M_{+\infty}(\mathbf{a}) = \max\,\{a_k \mid 1 \leq k \leq n\},$$
$$M_{-\infty}(\mathbf{a}) = \min\,\{a_k \mid 1 \leq k \leq n\}.$$

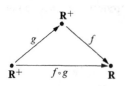

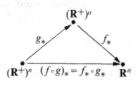

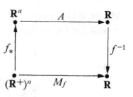

FIG. 9–37   **FIG. 9–38**

We shall build up to the main results through a series of observations.

1) If we have a mapping $f: \mathbf{R}^+ \to \mathbf{R}$, than there exists an induced mapping $f_*: (\mathbf{R}^+)^n \to \mathbf{R}^n$ defined as follows:

$$f_*(\mathbf{a}) = \big(f(a_1), f(a_2), \ldots, f(a_n)\big)$$

for all $\mathbf{a} = (a_1, \ldots, a_n) \in (\mathbf{R}^+)^n$. If we have a second mapping $g: \mathbf{R}^+ \to \mathbf{R}$, then

$$(f \circ g)_* = f_* \circ g_*$$

(see Fig. 9–37). Moreover, f_* is invertible iff f is, in which case

$$(f_*)^{-1} = (f^{-1})_*.$$

2) We define a mapping $A : \mathbf{R}^n \to \mathbf{R}$ as follows:

$$A(\mathbf{a}) = \frac{1}{n}(a_1 + \cdots + a_n)$$

for all $\mathbf{a} = (a_1, \ldots, a_n) \in \mathbf{R}^n$. In other words, A assigns to each n-tuple of real numbers their arithmetic mean. Observe that A maps $(\mathbf{R}^+)^n$ into $\mathbf{R}^+$.

3) We denote by $\mathfrak{M}$ the family of all mappings $f: \mathbf{R}^+ \to \mathbf{R}$ which are continuous and strictly monotone on $\mathbf{R}^+$. If $f \in \mathfrak{M}$, we observe that the mean associated with f can be written as

$$M_f = f^{-1} \circ A \circ f_*$$

(see Fig. 9–38). We shall now prove that M_f maps $(\mathbf{R}^+)^n$ into $\mathbf{R}^+$, as previously advertised.

For the sake of definiteness, let us assume that f is strictly increasing. If $\mathbf{a} = (a_1, \ldots, a_n) \in (\mathbf{R}^+)^n$, we choose indices μ and ν such that

$$a_\mu = M_{-\infty}(\mathbf{a}) \quad \text{and} \quad a_\nu = M_{+\infty}(\mathbf{a}).$$

Thus for all indices k, we have $a_\mu \leq a_k \leq a_\nu$, which implies in turn that

$$f(a_\mu) \leq f(a_k) \leq f(a_\nu).$$

We multiply by $1/n$ and sum. The result is

$$f(a_\mu) \leq (A \circ f_*)(\mathbf{a}) \leq f(a_\nu).$$

By the intermediate-value theorem, the number $(A \circ f_*)(\mathbf{a})$ belongs to $\mathfrak{D}_{f^{-1}} = \mathfrak{R}_f$. Since f^{-1} is strictly increasing,

$$M_{-\infty}(\mathbf{a}) = f^{-1}(f(a_\mu)) \leq f^{-1}(A(f_*(\mathbf{a}))) = M_f(\mathbf{a}) \leq f^{-1}(f(a_\nu)) = M_{+\infty}(\mathbf{a}).$$

In summary, if $f \in \mathfrak{M}$ and $\mathbf{a} \in (\mathbf{R}^+)^n$, then $\mathbf{a}$ belongs to the domain of

$$M_f = f^{-1} \circ A \circ f_*,$$

and

$$M_{-\infty}(\mathbf{a}) \leq M_f(\mathbf{a}) \leq M_{+\infty}(\mathbf{a}).$$

Since $M_f(\mathbf{a}) \geq M_{-\infty}(\mathbf{a}) > 0$, it follows that M_f maps $(\mathbf{R}^+)^n$ into $\mathbf{R}^+$. [Actually, M_f maps $(\mathbf{R}^+)^n$ onto $\mathbf{R}^+$. Why?]

What modifications of the above argument are needed if f is strictly decreasing?

4) If A is a set, we denote by i_A the identity function on A. That is, the domain of i_A is A, and for each $x \in A$, $i_A(x) = x$. Observe that

$$i_A \circ f = f \text{ iff } \mathfrak{R}_f \subset A, \qquad f \circ i_A = f \text{ iff } \mathfrak{D}_f \subset A.$$

Also observe that if f is an invertible function, then $f^{-1} \circ f$ and $f \circ f^{-1}$ are the identity functions on $\mathfrak{D}_f$ and $\mathfrak{R}_f$, respectively.

5) If f and g are real-valued functions, we shall write $f \leq g$ iff f and g have the same domain A and for all $x \in A$, $f(x) \leq g(x)$. If $f \leq g$, then it is easy to see that $f \circ h \leq g \circ h$, where h is any other function. If h is a nondecreasing function defined on some subset of $\mathbf{R}$, then we also have $h \circ f \leq h \circ g$. Also, note that if $\mathfrak{D}_f = I$ is an interval and if f is convex on I, then Jensen's inequality may be written as

$$f \circ A \leq A \circ f_*.$$

It is now relatively easy to state and prove a theorem of Hardy.

Theorem 1. Hardy's Theorem. *Let f and g be functions in $\mathfrak{M}$ such that*

1) f is strictly increasing,

2) $f \circ g^{-1}$ is convex.

Then $M_f \geq M_g$. If $f \circ g^{-1}$ is strictly convex and $\mathbf{a} = (a_1, \ldots, a_n) \in (\mathbf{R}^+)^n$, then $M_f(\mathbf{a}) > M_g(\mathbf{a})$ unless $a_1 = a_2 = \cdots = a_n$.

Proof. By (5) and (1), we have

$$(f \circ g^{-1}) \circ A \leq A \circ (f \circ g^{-1})_* = A \circ f_* \circ (g_*)^{-1}.$$

Using (4) and (5), we can justify each of the following inequalities:

$$(f \circ g^{-1}) \circ A \circ g_* \leq A \circ f_* \circ (g_*)^{-1} \circ g_* = A \circ f_*,$$
$$M_g = g^{-1} \circ A \circ g_* = f^{-1} \circ (f \circ g^{-1}) \circ A \circ g_* \leq f^{-1} \circ A \circ f = M_f.$$

The rest follows from the conditions for equality in Jensen's inequality.

We can apply Hardy's theorem to generalize Example 1. We first define $M_0(\mathbf{a})$ to be the geometric mean of $\mathbf{a}$. [Thus $M_0 = M_f$, where $f(x) = \ln x$.]

Corollary 2. *If $p' > p$ and $\mathbf{a} = (a_1, \ldots, a_n) \in (\mathbf{R}^+)^n$, then*

$$M_{p'}(\mathbf{a}) \geq M_p(\mathbf{a}),$$

with strict inequality unless $a_1 = a_2 = \cdots = a_n$.

Proof. If $\mathbf{a} = (a_1, \ldots, a_n)$ and $\mathbf{b} = (1/a_1, \ldots, 1/a_n)$, then one can check that

$$M_{-p}(\mathbf{a}) = \frac{1}{M_p(\mathbf{b})}.$$

Because of this identity, it suffices to consider two special cases.

Case I. $p > 0$. Then $M_{p'} = M_f$ and $M_p = M_g$, where $f(x) = x^{p'}$ and $g(x) = x^p$. Since $f'(x) = p'x^{p'-1} > 0$ for $x > 0$, f is strictly increasing on $\mathbf{R}^+$. Also,

$$(f \circ g^{-1})(x) = x^\alpha,$$

where $\alpha = p'/p > 1$. Since

$$(f \circ g^{-1})''(x) = \alpha(\alpha - 1)x^{\alpha-2} > 0,$$

$f \circ g^{-1}$ is strictly convex. Thus the hypotheses in Hardy's theorem are satisfied, completing the proof in this case.

Case II. $p = 0$. Here $M_{p'} = M_f$ and $M_p = M_g$, where f is as before and $g(x) = \ln x$. Then

$$(f \circ g^{-1})(x) = (e^x)^{p'} = e^{p'x}, \qquad (f \circ g^{-1})''(x) = (p')^2 e^{p'x} > 0.$$

Again we see that the hypotheses of Hardy's theorem are satisfied.

If we fix $\mathbf{a} = (a_1, \ldots, a_n) \in (\mathbf{R}^+)^n$, then we have seen that the function

$$h(p) = M_p(\mathbf{a})$$

is nondecreasing. Moreover, unless $a_1 = \cdots = a_n$, the function is strictly increasing. The function h is also continuous on $\mathbf{R}$. In fact, at all points other than zero the function is clearly differentiable. Continuity at zero may be established as follows:

$$\lim_{p \to 0} \ln M_p(\mathbf{a}) = \lim_{p \to 0} \frac{1}{p} \ln \frac{1}{n} \sum a_k^p$$

$$= \frac{d}{dp} \left(\ln \frac{1}{n} \sum a_k^p \right) \Big|_{p=0}$$

$$= \frac{(1/n) \sum (\ln a_k) a_k^p}{(1/n) \sum a_k^p} \Big|_{p=0}$$

$$= \frac{1}{n} \sum \ln a_k$$

and

$$\lim_{p \to 0} M_p(\mathbf{a}) = \lim_{p \to 0} \exp [\ln M_p(\mathbf{a})]$$

$$= \exp \left(\frac{1}{n} \sum \ln a_k \right) = M_0(\mathbf{a}).$$

We also have continuity at $\pm \infty$ in the sense that

$$\lim_{p \to +\infty} M_p(\mathbf{a}) = M_{+\infty}(\mathbf{a}) \qquad \text{and} \qquad \lim_{p \to -\infty} M_p(\mathbf{a}) = M_{-\infty}(\mathbf{a}).$$

To prove continuity at $+\infty$, we use the notation of (3):

$$M_p(\mathbf{a}) = \left(\frac{1}{n} \sum_{k=1}^{n} a_k^p\right)^{1/p} \geq \left(\frac{1}{n} a_\nu^p\right)^{1/p}$$

$$= \frac{1}{n^{1/p}} a_\nu = \frac{1}{n^{1/p}} M_{+\infty}(\mathbf{a}).$$

Thus

$$\frac{1}{n^{1/p}} M_{+\infty}(\mathbf{a}) \leq M_p(\mathbf{a}) \leq M_{+\infty}(\mathbf{a}).$$

Since $\lim_{p \to +\infty} n^{1/p} = 1$, the result follows from the pinching theorem.

The results which we have established can be generalized considerably. First, we can replace the averaging operator A by a weighted arithmetic mean, say

$$A_\lambda(a) = \lambda_1 a_1 + \lambda_2 a_2 + \cdots + \lambda_n a_n,$$

where $\lambda = (\lambda_1, \ldots, \lambda_n)$ is a fixed n-tuple of weights (that is, the λ_k's are positive and have sum one). If $f \in \mathfrak{M}$, we define weighted f-means as follows:

$$M_{f,\lambda} = f^{-1} \circ A_\lambda \circ f_*.$$

Hardy's theorem holds for such weighted means, as do the results for p-means. The most powerful and useful generalizations result from replacing the operator A by the integral with respect to an arbitrary probability measure. Similar remarks hold for the Hölder and Minkowski inequalities.

Exercises

1. Prove the inequality in Example 2 by applying Young's inequality to the function $f(x) = x^{p-1}$. (See Exercise 10, Section 9-1.)

2. When does strict inequality hold in Hölder's inequality? in Minkowski's inequality?

3. Let p and q be positive numbers such that $1/p + 1/q = 1$. If f and g are continuous on the interval $[a, b]$, prove the following version of the Hölder inequality:

$$\left| \int_a^b f(x)\, g(x)\, dx \right| \leq \int_a^b |f(x)\, g(x)|\, dx$$

$$\leq \left(\int_a^b |f(x)|^p\, dx \right)^{1/p} \left(\int_a^b |g(x)|^q\, dx \right)^{1/q}.$$

4. State and prove a version of Minkowski's inequality for integrals.

5. The purpose of this exercise is to prove a generalization of the Hölder inequality. Let $p_1, \ldots, p_r$ be positive numbers such that $1/p_1 + 1/p_2 + \cdots + 1/p_r = 1$.

   a) If $x_1, \ldots, x_r$ are positive, show that

$$\frac{1}{p_1} x_1^{p_1} + \cdots + \frac{1}{p_r} x_r^{p_r} \geq x_1 x_2 \cdots x_r.$$

b) Let $a_1, \ldots, a_r$ be elements of $(\mathbf{R}^+)^n$, say, $a_k = (a_{k1}, \ldots, a_{kn})$, $k = 1, \ldots, r$. Prove that

$$\sum_{j=1}^{n} \left(\prod_{k=1}^{r} a_{kj} \right) \le \|a_1\|_{p_1} \|a_2\|_{p_2} \cdots \|a_r\|_{p_r}.$$

6. Suppose that f and g are functions in $\mathfrak{M}$ such that $f(x) \equiv ag(x) + b$, where a and b are constants. Show that $M_f = M_g$. Observe, in particular, that if p is any real number and $f(x) = \int_1^x t^{p-1}\, dt$, then M_f is the p-mean.

7. If $a = (a_1, \ldots, a_n) \in \mathbf{R}^n$, we define

$$\|a\|_\infty = \max \{|a_k| \mid 1 \le k \le n\}.$$

Show that $\|a + b\|_\infty \le \|a\|_\infty + \|b\|_\infty$. Also show that

$$\left| \sum_{k=1}^{n} a_k b_k \right| \le \sum_{k=1}^{n} |a_k b_k| \le \|a\|_1 \|b\|_\infty,$$

where $\|a\|_1 = \sum_{k=1}^{n} |a_k|$.

8. Show that the elements of $\mathbf{R}^n$ form an abelian group under vector addition, i.e., they satisfy the addition axioms for a field. In $\mathbf{R}^2$, show that vector addition has the following geometric interpretation. If a and b are elements of $\mathbf{R}^2$, and we draw arrows from the origin to these two points and construct a parallelogram with these two arrows as sides, then $a + b$ is the fourth vertex of the parallelogram (see Fig. 9-39). Interpret $-b$ and $a - b = a + (-b)$ geometrically. Observe that $\|a\|_2$ is the distance from the origin to a. Interpret geometrically the following special case of Minkowski's inequality:

$$\|a + b\|_2 \le \|a\|_2 + \|b\|_2.$$

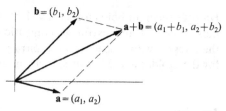

FIG. 9-39

9. (Continuation of Exercise 8.) Observe that the distance between two points $a = (a_1, a_2)$ and $b = (b_1, b_2)$ in $\mathbf{R}^2$ is given by

$$d(a, b) = \|a - b\|_2.$$

It is customary to define the distance between two points a and b in $\mathbf{R}^n$ in exactly the same way. Using Minkowski's inequality, prove that

$$d(a, b) + d(b, c) \ge d(a, c)$$

holds for any three points a, b, c in $\mathbf{R}^n$. Explain why it is appropriate to call this result the *triangle inequality*. Does the triangle inequality also hold if we define the distance between a and b to be $\|a - b\|_p$, where p is any fixed number greater than one?

10. a) Show that the functions $x \ln x$ and x^x are strictly convex on $(0, +\infty)$.

b) If x, y, a, b are positive, show that

$$x \ln \frac{x}{a} + y \ln \frac{y}{b} \ge (x + y) \ln \frac{x + y}{a + b},$$

with strict inequality unless $x/a = y/b$.

c) Find the minimum value of

$$x_1^{x_1} x_2^{x_2} \cdots x_2^{x_2}$$

subject to the condition that $x_1 + x_2 + \cdots x_n = S$, where S is a positive constant.

11. Let $\mathbf{a} = (a_1, \ldots, a_n) \in (\mathbf{R}^+)^n$, where not all the a_k's are equal, and let

$$\varphi(x) = x \ln M_x(\mathbf{a}).$$

a) Show that

$$\varphi''(x) = \frac{(\sum a_k^x)[\sum a_k^x(\ln a_k)^2] - (\sum a_k^x \ln a_k)^2}{(\sum a_k^x)^2}.$$

By applying the Cauchy-Schwarz inequality, deduce that $\varphi''(x) > 0$. Hence φ is convex.

b) Show that

$$x^2 \frac{d}{dx}\left(\frac{\varphi(x)}{x}\right) = x\varphi'(x) - \varphi(x)$$

and

$$\frac{d}{dx}[x\varphi'(x) - \varphi(x)] = x\varphi''(x).$$

Deduce from these identities and part (a) that $\varphi(x)/x$ and thus

$$M_x(\mathbf{a}) = \exp\left(\frac{\varphi(x)}{x}\right)$$

is a strictly increasing function for $x > 0$.

c) Let $p_1, \ldots, p_r$ be positive numbers, and let $\lambda_1, \ldots, \lambda_r$ be positive with sum one. We define

$$S_p(\mathbf{a}) = \sum_{k=1}^{n} a_k^p = n[M_p(\mathbf{a})]^p.$$

Using (a) and Jensen's inequality, prove that

$$S_{\lambda_1 p_1 + \cdots + \lambda_r p_r}(\mathbf{a}) \leq [S_{p_1}(\mathbf{a})]^{\lambda_1} [S_{p_2}(\mathbf{a})]^{\lambda_2} \cdots [S_{p_n}(\mathbf{a})]^{\lambda_n}.$$

The next set of exercises involves rather easy applications of the inequalities established in the text and in the previous exercises. Prove each of the assertions made.

12. $\frac{5}{4} < \int_0^1 (1 + x)^{2/3}(1 + x^3)^{1/3}\, dx < (\frac{3}{2})^{2/3}(\frac{5}{4})^{1/3} < \frac{3}{2}$.

13. $\dfrac{n(n + 1)}{2} < \displaystyle\sum_{k=0}^{n} \sqrt{k(k + 1)} < \dfrac{(n + 1)\sqrt{n(n + 2)}}{2}$.

14. If x, y, z are positive and $x^4 + y^4 + z^4 = 27$, then $x + y + z \leq 3\sqrt{3}$.

15. $|a \cos \theta + b \sin \theta \cos \phi + c \sin \theta \sin \phi| \leq (a^2 + b^2 + c^2)^{1/2}$.

16. $|a \cos^{2/3} \theta + b \sin^{2/3} \theta| \leq (a^{3/2} + b^{3/2})^{2/3}$ $(a, b > 0)$.

17. $(x^2 + y^2)^2 \leq (x + y)(x^3 + y^3) \leq 2(x^4 + y^4)$ $(x, y > 0)$.

18. $(a^4 + b^4 + c^4)^3 \leq (a^2 + b^2 + c^2)(a^3 + b^3 + c^3)(a^7 + b^7 + c^7)$
$$\leq 9(a^{12} + b^{12} + c^{12}) \quad (a, b, c > 0).$$

19. $\sqrt[3]{a^3 + b^3} + \sqrt[3]{c^3 + d^3} \geq \sqrt[3]{(a + c)^3 + (b + d)^3}$ $(a, b, c, d > 0)$.

20. If x, y, z are positive, then

$$\frac{1}{x} + \frac{1}{y} + \frac{1}{z} \geq \frac{9}{x + y + z}.$$

21. If x, y, z are positive, then

$$(\sqrt{x} + \sqrt{y} + \sqrt{z})^2 \geq 9\sqrt[3]{xyz}.$$

22. $\lim_{n \to \infty} \sqrt[n]{a^n + a^{-n}} = \max(a, 1/a)$.

23. $(abc + xyz)^6 \leq (a + x)^6(b^2 + x^2)^3(c^3 + z^3)^2$, where all the quantities are positive.

9-8 An analytic treatment of the trigonometric functions

Our work with the trigonometric functions has involved no small degree of logical circularity. We shall now set matters straight by giving a purely analytic treatment of these functions.

Although several different paths may be chosen, we begin with the definition of the arc tangent.

Definition 1. For each real number x, we define

$$\operatorname{Tan}^{-1} x = \int_0^x \frac{dt}{1 + t^2}.$$

By the fundamental theorem, $d(\operatorname{Tan}^{-1} x)/dx = 1/(1 + x^2)$. Since the latter expression is everywhere positive, the function $\operatorname{Tan}^{-1} x$ is strictly increasing. The fact that $\operatorname{Tan}^{-1} 0 = 0$ while the derivative of Tan^{-1} is even implies that Tan^{-1} is an odd function. Moreover, Tan^{-1} possesses continuous derivatives of all orders. [The derivatives are easily seen to be rational functions having powers of $(1 + x^2)$ as denominators.]

Proposition 2. For all $x > 0$,

$$\operatorname{Tan}^{-1} x + \operatorname{Tan}^{-1}(1/x) = 2 \operatorname{Tan}^{-1} 1.$$

Proof

$$\frac{d}{dx}\left(\operatorname{Tan}^{-1}\frac{1}{x}\right) = \frac{1}{1 + (1/x)^2}\left(-\frac{1}{x^2}\right) = -\frac{1}{1 + x^2}.$$

It follows immediately that $\operatorname{Tan}^{-1} x + \operatorname{Tan}^{-1}(1/x)$ is constant for $x > 0$, and setting $x = 1$, we get the result.

Definition 3

$$\frac{\pi}{4} = \operatorname{Tan}^{-1} 1.$$

The identity above may then be written as

$$\operatorname{Tan}^{-1} x + \operatorname{Tan}^{-1}\frac{1}{x} = \frac{\pi}{2} \quad (x > 0).$$

From this it follows that

$$\lim_{x \to +\infty} \text{Tan}^{-1} x = \lim_{x \to +\infty} \left(\frac{\pi}{2} - \text{Tan}^{-1} \frac{1}{x} \right) = \frac{\pi}{2}.$$

We next define the tangent function on the interval $(-\pi/2, \pi/2)$ to be the inverse of the function Tan^{-1}. From our general theorems on inverse functions, it follows that the tangent function is differentiable, strictly increasing, and odd. Also

$$\lim_{x \to \pi/2-} \tan x = +\infty.$$

Definition 4. We define the functions sin and cos on the interval $(-\pi, \pi)$ as follows:

$$\cos x = \frac{1 - u^2}{1 + u^2}, \qquad \sin x = \frac{2u}{1 + u^2},$$

where $u = \tan (x/2)$.

Proposition 5. *If* $-\pi < x < \pi$, *then*

$$\frac{d}{dx} (\sin x) = \cos x, \qquad \frac{d}{dx} (\cos x) = -\sin x.$$

Proof. As in the definition, we set $u = \tan (x/2)$. Then $x = 2 \, \text{Tan}^{-1} u$, so that

$$1 = \frac{2}{1 + u^2} \frac{du}{dx} \qquad \text{and} \qquad \frac{du}{dx} = \tfrac{1}{2} (1 + u^2).$$

Thus

$$\frac{d}{dx} (\cos x) = \frac{d}{du} \left(\frac{1 - u^2}{1 + u^2} \right) \frac{du}{dx} = \frac{4u}{(1 + u^2)^2} \cdot \tfrac{1}{2} (1 + u^2)$$

$$= \frac{2u}{1 + u^2} = \sin x.$$

The derivation of the other differentiation formula is similar.

As $x \to \pi-$, $u = \tan (x/2) \to +\infty$, and thus

$$\lim_{x \to \pi-} \cos x = \lim_{u \to +\infty} \frac{1 - u^2}{1 + u^2} = -1$$

and

$$\lim_{x \to \pi-} \sin x = \lim_{u \to +\infty} \frac{2u}{1 + u^2} = 0.$$

Similarly,

$$\lim_{x \to -\pi+} \cos x = -1 \qquad \text{and} \qquad \lim_{x \to -\pi+} \sin x = 0.$$

If we now extend the definitions of sin and cos to the entire line by the requirements that

$$\sin \pi = 0, \qquad \sin (x + 2\pi) = \sin x,$$

$$\cos \pi = 1, \qquad \cos (x + 2\pi) = \cos x,$$

then it is easily checked that the resulting functions are everywhere continuous. (Consider, for example, continuity of the cosine function at π. We have

$$\lim_{x \to \pi-} \cos x = -1 = \cos \pi$$

and

$$\lim_{x \to \pi+} \cos x = \lim_{t \to -\pi+} \cos (t + 2\pi)$$
$$= \lim_{t \to -\pi+} \cos t = -1 = \cos \pi,$$

as desired.) It follows immediately that the functions have everywhere continuous derivatives and that the formulas

$$\frac{d}{dx} (\sin x) = \cos x \quad \text{and} \quad \frac{d}{dx} (\cos x) = -\sin x$$

hold on the entire line. [Recall the following consequence of the mean-value theorem: if f is continuous at x_0 and $\lim_{x \to x_0} f'(x)$ exists, then f has a continuous derivative at x_0.]

Proposition 6. Pythagorean Identity. *For all $x \in \mathbf{R}$,*

$$\sin^2 x + \cos^2 x = 1.$$

Proof

$$\frac{d}{dx} (\sin^2 x + \cos^2 x) = 2 \sin x \cos x + 2 \cos x (-\sin x) = 0.$$

Thus $\sin^2 x + \cos^2 x$ is identically constant, and setting $x = \pi$, we see that the constant is one.

We observe that if $y = \sin x$ or $y = \cos x$, then

$$\frac{d^2y}{dx^2} + y = 0. \tag{*}$$

More generally, this differential equation is satisfied by every linear combination of $\sin x$ and $\cos x$; that is, if $y = A \cos x + B \sin x$, where A and B are constants, then (*) holds. We shall now prove an important converse.

Theorem 7. *Suppose that f is a function such that*

$$f''(x) + f(x) = 0$$

for all $x \in \mathbf{R}$. Then there exist constants A and B such that

$$f(x) = A \cos x + B \sin x.$$

Proof. Let

$$g(x) = f(x) - f(0) \cos x - f'(0) \sin x.$$

Then one can verify that

$$g''(x) + g(x) = 0, \qquad g(0) = g'(0) = 0.$$

Hence

$$\frac{d}{dx}\{[g'(x)]^2 + [g(x)]^2\} = 2g'(x)[g''(x) + g(x)] = 0.$$

The expression $[g'(x)]^2 + [g(x)]^2$ must be constant and (setting $x = 0$) that constant must be zero. Both $g'(x)$ and $g(x)$ must be identically zero and, in particular,

$$f(x) = f(0)\cos x + f'(0)\sin x.$$

Corollary 8. Addition Formulas. *For all real numbers a and x, we have*

$$\cos(a + x) = \cos a \cos x - \sin a \sin x,$$
$$\sin(a + x) = \sin a \cos x + \cos a \sin x.$$

Proof. If we let $f(x) = \sin(a + x)$, then $f''(x) = -\sin(a + x) = -f(x)$. Thus by the theorem,

$$\sin(a + x) = A\cos x + B\sin x,$$

where A and B are constants. Differentiating, we get

$$\cos(a + x) = -A\sin x + B\cos x.$$

Finally, we set $x = 0$ and obtain

$$\sin a = A, \qquad \cos a = B.$$

Corollary 9. *The cosine function is even and the sine function is odd.*

Proof. Let $f(x) = \sin(-x)$. Then $f'(x) = -\cos(-x)$ and $f''(x) = -\sin(-x) = f(x)$. Thus, by the proof of the theorem,

$$f(x) = f(0)\cos x + f'(0)\sin x.$$

But $f(0) = 0$ and $f'(0) = -1$. (Why?) Thus $\sin(-x) = f(x) = -\sin x$. By differentiation, $-\cos(-x) = -\cos x$.

If $x = \pi/2$, then $u = \tan(x/2) = 1$, so that

$$\cos\frac{\pi}{2} = \frac{1 - 1^2}{1 + 1^2} = 0 \qquad \text{and} \qquad \sin\frac{\pi}{2} = \frac{2 \cdot 1}{1 + 1^2} = 1.$$

The two preceding corollaries yield the formulas

$$\sin\left(\frac{\pi}{2} - x\right) = \cos x, \qquad \sin(x + \pi) = -\sin x, \qquad \cos(x + \pi) = -\cos x, \quad \text{etc.}$$

Most of the important trigonometric formulas follow in the usual way (double-angle and half-angle formulas, formulas for $\sin a \sin b$, etc.). There is, however, one more identity to be established before we can regard our treatment of analytic trigonometry as complete, namely,

$$\tan x = \frac{\sin x}{\cos x} \quad \left(-\frac{\pi}{2} < x < \frac{\pi}{2} \right).$$

A proof may be given as follows. First, we have the identity

$$2 \operatorname{Tan}^{-1} u = \operatorname{Tan}^{-1} \frac{2u}{1 - u^2} \quad (|u| < 1).$$

[Both sides are zero when $u = 0$, and both sides have the same derivative, namely, $2/(1 + u^2)$.] If we let $x = 2 \operatorname{Tan}^{-1} u$, then $-\pi/2 < x < \pi/2$,

$$u = \tan \frac{x}{2} \quad \text{and} \quad \tan x = \frac{2u}{1 - u^2} = \frac{\sin x}{\cos x}.$$

We define the tangent function outside the interval $(-\pi/2, \pi/2)$ by the identity

$$\tan x = \frac{\sin x}{\cos x}.$$

We also define $\sec x = 1/(\cos x)$, $\csc x$, ...

To this point our development has been purely analytic. We must now show how the trigonometric functions are related to geometry.

Proposition 10. *For each real number θ the point*

$$P_\theta = (\cos \theta, \sin \theta)$$

lies on the unit circle $x^2 + y^2 = 1$. Conversely, each point on the unit circle is of the form P_θ. Moreover, $P_\theta = P_{\theta'}$ iff θ and θ' differ by an integral multiple of 2π.

Proof. The fact that P_θ lies on the unit circle follows from the Pythagorean identity, $\sin^2 \theta + \cos^2 \theta = 1$. Because the sine and cosine functions have period 2π, $P_\theta = P_{\theta'}$ if θ and θ' differ by an integral multiple of 2π.

If (x, y) is any point on the unit circle other than $(-1, 0)$, we set

$$\theta = 2 \operatorname{Tan}^{-1} \frac{y}{x + 1}.$$

Then $-\pi < \theta < \pi$,

$$u = \tan \frac{\theta}{2} = \frac{y}{x + 1},$$

and thus

$$\cos \theta = \frac{1 - u^2}{1 + u^2} = \frac{1 - \dfrac{y^2}{(x + 1)^2}}{1 + \dfrac{y^2}{(x + 1)^2}} = \frac{x^2 + 2x + 1 - y^2}{x^2 + 2x + 1 + y^2} = \frac{2(x^2 + x)}{2(x + 1)} = x;$$

similarly,

$$\sin \theta = y.$$

(Note that we used the fact that $x^2 + y^2 = 1$.) The point $(-1, 0)$ is, of course, P_π.

	$-\pi < \theta < -\pi/2$	$-\pi/2 < \theta < 0$	$0 < \theta < \pi/2$	$\pi/2 < \theta < \pi$
$\cos \theta$	$-$	$+$	$+$	$-$
$\sin \theta$	$-$	$-$	$+$	$+$

FIG. 9-40

We have shown that each point (x, y) on the unit circle is of the form P_θ, where $-\pi < \theta \le \pi$. To complete the proof, it suffices to show that this number θ is unique. We note first the signs of $\cos \theta$ and $\sin \theta$ (see Fig. 9-40). [If $0 < \theta < \pi/2$, then $0 < u = \tan (\theta/2) < 1$, so that

$$\cos \theta = \frac{1 - u^2}{1 + u^2} > 0 \quad \text{and} \quad \sin \theta = \frac{2u}{1 + u^2} > 0.$$

The remaining entries in the table follow from the identities $\sin (x + \pi/2) = \cos x$ and $\cos (x + \pi/2) = -\sin x$.] If $y \ge 0$, then θ must be nonnegative. Since the cosine function is strictly decreasing on the interval $[0, \pi]$, there is only one value of θ such that $x = \cos \theta$. Similarly, we get uniqueness if $y < 0$.

If $P = (x, y)$ is any point in the plane other than the origin, we set $r = \sqrt{x^2 + y^2}$. Then $(x/r, y/r)$ lies on the unit circle, and by the theorem there exists a number θ unique to within integral multiples of 2π such that $(x/r, y/r) = P_\theta$, or equivalently,

$$x = r \cos \theta \quad \text{and} \quad y = r \sin \theta.$$

Any such θ is called the *polar angle* (or *argument*) of P. If $-\pi < \theta \le \pi$, then θ (which is now unique) is called the *principle value* of the polar angle. The polar angle is clearly constant along the ray emanating from the origin and passing through P (see Fig. 9-41).

We defined π to be $4 \int_0^1 (1 + x^2)^{-1} dx$. We shall now give its more usual characterization: the ratio of circumference to diameter. Consider a circle C of radius r. If (a, b) is its center, then, by the preceding discussion, it is easily shown that C has the parametrization

$$x = a + r \cos \theta, \quad y = b + r \sin \theta \quad (-\pi \le \theta \le \pi).$$

As θ increases from $-\pi$ to 0, the point moves over the lower semicircle from left to

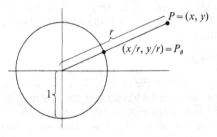

FIG. 9-41

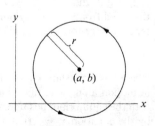

FIG. 9-42

right, and as θ increases from 0 to π, the point moves over the upper semicircle from right to left, as indicated in Fig. 9–42. Thus as θ varies over the interval $[-\pi, \pi]$, the points of C are traced out exactly once in a counterclockwise direction. The circumference of C is therefore

$$\int_{-\pi}^{\pi} \sqrt{(dx/d\theta)^2 + (dy/d\theta)^2}\, d\theta = \int_{-\pi}^{\pi} \sqrt{(-r \sin \theta)^2 + (r \cos \theta)^2}\, d\theta$$

$$= \int_{-\pi}^{\pi} r\, d\theta = 2\pi r.$$

Exercises

1. If $f(x) = \cos x$ and $g(x) = \sin x$, then

$$f(x + y) = f(x)f(y) - g(x)g(y), \qquad g(x + y) = f(x)g(y) + g(x)f(y)$$

for all $x, y \in \mathbf{R}$.

a) Show that if f and g are differentiable functions satisfying the above identities, then either f and g are identically zero or else $f(x) = \cos x$ and $g(x) = \sin x$.

*b) Prove the same statement under the weakened assumption that f and g are continuous.

The remaining exercises are concerned with *simple harmonic motion*.

2. a) Prove the following generalization of a theorem in the text. *If $x(t)$ satisfies the differential equation*

$$x''(t) + \omega^2 x(t) = 0$$

for all $t \in \mathbf{R}$ (where ω is a constant), then there exist constants a and b such that

$$x(t) = a \cos \omega t + b \sin \omega t.$$

b) Observe that the constants are given by

$$a = x(0) \quad \text{and} \quad b = x'(0)/\omega.$$

c) Show that $x(t)$ can also be written in the form

$$x(t) = A \cos (\omega t - \phi)$$

where $[A; \phi]$ are polar coordinates of the point (a, b). Show that the constants A, ϕ, and ω are as advertised on the graph in Fig. 9–43.

3. A spring is said to be *Hookean* if it obeys *Hooke's law*. This law states that if a spring is displaced from its equilibrium position by the amount x, then the tension in the spring tending to restore it to equilibrium position is given by

$$F = kx,$$

where k is a constant known as the *spring constant* of the system. Imagine such a spring in motion along a straight line. We introduce coordinates so that the origin coincides with the equilibrium position. Suppose, then, that the motion of the system is described by the equation $x = f(t)$, $0 \le t \le a$, where t is time and x is displacement. Assume also that $f(0) = 0$, that is, that the spring is initially in its equilibrium position. Show

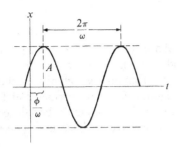

FIG. 9-43

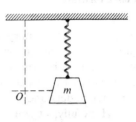

FIG. 9-44

that the work done in overcoming the tension in the spring is

$$\tfrac{1}{2}k[f(a)]^2,$$

and thus depends only on the terminal displacement. For this reason we say that $\tfrac{1}{2}kx^2$ is the *potential energy* of the spring when displaced from its equilibrium position by the amount x. (This potential energy is the amount of work which the spring is capable of doing by virtue of its position.)

4. Imagine that a Hookean spring is suspended vertically and that a weight of mass m is attached to the free end of the spring. We introduce coordinates along the line of the spring so that the origin coincides with the equilibrium position of the weight (see Fig. 9-44). Imagine that the system is in motion, and denote by $x(t)$ the displacement at time t. Neglecting air resistance, apply Newton's second law of motion to deduce that

$$m\frac{d^2x}{dt^2} = -kx(t).$$

(Why the minus sign?) Rewrite the equation in the form

$$x''(t) + \omega^2 x(t) = 0,$$

where $\omega = \sqrt{k/m}$. Apply Exercise 2 to show that the motion of the system is completely determined by its initial displacement $x_0 = x(0)$ and velocity $v_0 = x'(0)$, namely,

$$x(t) = x_0 \cos \omega t + \frac{v_0}{\omega} \sin \omega t.$$

By part (c) of the same exercise, $x(t)$ can be written in the form

$$x(t) = A \cos (\omega t - \phi).$$

The constant A, which represents the maximum displacement from equilibrium, is called the *amplitude* of the vibration. $T = 2\pi/\omega$, the time required for one complete vibration, is called the *period* of vibration. [Note that T may be characterized as the smallest positive number such that $x(t + T) = x(t)$ for all $t \in \mathbf{R}$.] Furthermore, $f = 1/T = \omega/2\pi$, the number of complete vibrations per unit time, is called the *frequency* of the vibration. The constant ϕ is called the *phase lag*. If the mass m is doubled, what effect will this have on the motion of the system?

5. Show that $[x'(t)]^2 + \omega^2[x(t)]^2$, or equivalently, $\tfrac{1}{2}m[x'(t)]^2 + \tfrac{1}{2}k[x(t)]^2$, is a constant. [You may have shown this previously in working Exercise 2(a).] This may be interpreted as saying that the sum of the kinetic and potential energies of the system is constant.

9-9 Euler's formula

The study of the elementary transcendental functions can be unified enormously through the use of complex variable methods. We haven't the time to develop this requisite machinery, but this will not prevent us from taking a glimpse at the remarkable relationships that exist between the trigonometric and exponential functions. Let the reader understand at the onset that what follows should be labeled wishful thinking. Hopefully, at some other time he may have the pleasure of seeing these relationships established properly.

To follow the discussion, the reader must have some acquaintance with complex numbers. Actually, it suffices to know that the set C of complex numbers is a field such that

1) C contains as a subfield the field R of real numbers;

2) C contains an element denoted by i such that $i^2 = -1$;

3) every element of C is of the form $a + bi$, where a and b are real numbers.

It follows that complex numbers can be added, multiplied, and divided as follows:

$$(a + bi) + (c + di) = (a + c) + (b + d)i,$$
$$(a + bi)(c + di) = (ac - bd) + (ad + bc)i,$$
$$\frac{a + bi}{c + di} = \frac{(a + bi)(c - di)}{(c + di)(c - di)} = \frac{ac + bd}{c^2 + d^2} + \frac{bc - ad}{c^2 + d^2}i \qquad (c^2 + d^2 \neq 0).$$

It also follows that the representation of an element α of C in the form $a + bi$ ($a, b \in R$) is unique. [Suppose that

$$a + bi = a' + b'i,$$

where $a, b, a', b' \in R$. Then

$$a - a' = (b' - b)i.$$

If $b' - b$ were not zero, then we would have $i = (a - a')/(b' - b)$, which is impossible, since $(a - a')/(b' - b)$ is a real number. Thus $b' - b = 0$, which implies that

$$a - a' = (b' - b)i = 0.$$

Hence $a = a'$ and $b = b'$.] The real numbers a and b are called the *real* and *imaginary parts* of α, respectively.

In the text and exercises we have established that the power series expansions

$$e^x = 1 + x + \frac{x^2}{2!} + \frac{x^3}{3!} + \cdots,$$

$$\sin x = x - \frac{x^3}{3!} + \frac{x^5}{5!} - \cdots,$$

$$\cos x = 1 - \frac{x^2}{2!} + \frac{x^4}{4!} - \cdots$$

are valid for all real numbers x. We shall assume that they are also meaningful for complex values of x. Let's see what happens if x is pure imaginary, say, $x = i\theta$ ($\theta \in \mathbf{R}$):

$$e^{i\theta} = 1 + i\theta + \frac{(i\theta)^2}{2!} + \frac{(i\theta)^3}{3!} + \frac{(i\theta)^4}{4!} + \cdots$$

$$= 1 + i\theta - \frac{\theta^2}{2!} - i\frac{\theta^3}{3!} + \frac{\theta^4}{4!} + i\frac{\theta^5}{5!} - \cdots$$

$$= \left(1 - \frac{\theta^2}{2!} + \frac{\theta^4}{4!} - \cdots\right) + i\left(\theta - \frac{\theta^3}{3!} + \frac{\theta^5}{5!} - \cdots\right).$$

Thus we have derived a celebrated formula of Euler:

$$e^{i\theta} = \cos\theta + i\sin\theta.$$

This remarkably simple formula provides the key relationship between the exponential and trigonometric functions. We explore some of its consequences in the sequel.

1. *Polar Representation of the Complex Numbers.* We identify the complex number $z = x + iy$ with the point (x, y) in $\mathbf{R}^2$. Because of this identification, the set of complex numbers $\mathbf{C}$ is also called the *complex plane.* If $z \neq 0$, let $[r; \theta]$ ($r > 0$) be polar coordinates of the point (x, y). Then $r = |z|$ is called the *absolute value* (or *modulus*) of z, and $\theta = \arg z$ is called the *argument of z.* By virtue of Euler's formula, we have

$$z = x + iy = r\cos\theta + ir\sin\theta = r(\cos\theta + i\sin\theta) = re^{i\theta} = |z|e^{i\arg z};$$

the latter expression is called the polar representation of z.

The polar representation is useful in interpreting multiplication. If z and z' are any two complex numbers, then

$$zz' = |z|e^{i\arg z}\,|z'|e^{i\arg z'} = |z|\,|z'|e^{i(\arg z + \arg z')}.$$

Thus the *modulus of a product is the product of the moduli, and the argument of a product is the sum of the arguments of the factors.* (We have assumed, of course, that the exponent laws hold for complex exponents—another instance of wishful thinking.)

2. *De Moivre's Formula.* De Moivre's formula states that

$$(\cos x + i\sin x)^n = \sin nx + i\cos nx.$$

It follows easily from Euler's formula:

$$(\cos x + i\sin x)^n = (e^{ix})^n = e^{inx} = \cos nx + i\sin nx.$$

3. $e^{i\pi} = -1$. This remarkable formula relates four of the most important numbers in mathematics: e, π, the imaginary unit, and the negative unit. It follows from Euler's formula by substitution of $\theta = \pi$.

4. *"Proof" of the Addition Formulas for Sines and Cosines.* We have

$$\cos(x+y) + i\sin(x+y) = e^{i(x+y)} = e^{ix}e^{iy}$$
$$= (\cos x + i\sin x)(\cos y + i\sin y)$$
$$= (\cos x \cos y - \sin x \sin y)$$
$$+ i(\sin x \cos y + \cos x \sin y).$$

Equating real and imaginary parts yields the addition formulas

$$\cos(x+y) = \cos x \cos y - \sin x \sin y,$$
$$\sin(x+y) = \sin x \cos y + \cos x \sin y.$$

5. *Relationships between the Trigonometric and Hyperbolic Functions*

$$e^{-\theta} = \cos(-\theta) + i\sin(-\theta)$$
$$= \cos\theta - i\sin\theta.$$

By treating the equations

$$e^{i\theta} = \cos\theta + i\sin\theta, \qquad e^{-i\theta} = \cos\theta - i\sin\theta$$

as simultaneous equations for $\sin\theta$ and $\cos\theta$, we get

$$\cos\theta = \frac{e^{i\theta} + e^{-i\theta}}{2}, \qquad \sin\theta = \frac{e^{i\theta} - e^{-i\theta}}{2i}.$$

The analogy between these equations and the equations

$$\cosh x = \frac{e^x + e^{-x}}{2}, \qquad \sinh x = \frac{e^x - e^{-x}}{2}$$

is clear. Furthermore, we have

$$\cosh i\theta = \cos\theta, \qquad \sinh i\theta = i\sin\theta.$$

6. *Construction of a Regular Pentagon.* Let n be a positive integer. We define n complex numbers $\omega_0, \omega_1, \ldots, \omega_{n-1}$ as follows:

$$\omega_k = e^{2k\pi i/n} = \cos\frac{2k\pi}{n} + i\sin\frac{2k\pi}{n}$$

($k = 0, 1, \ldots, n$). Also note that $\omega_k = \omega_1^k$. These numbers clearly form the vertices of a regular n-gon inscribed in the unit circle, as shown in Fig. 9–45. Moreover,

$$\omega_k^n = e^{2k\pi i} = \cos 2k\pi + i\sin 2k\pi = 1.$$

Thus $\omega_0, \omega_1, \ldots, \omega_{n-1}$ are the n nth roots of unity.

We now set $n = 5$. Since $\omega_0, \ldots, \omega_4$ are the roots of the equation $z^5 - 1 = 0$, we have

$$z^5 - 1 = (z - \omega_0)(z - \omega_1)(z - \omega_2)(z - \omega_3)(z - \omega_4).$$

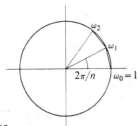

FIG. 9-45   FIG. 9-46

Also $\omega_1\omega_4 = e^{2\pi i/5}e^{8\pi i/5} = e^{2\pi i} = 1$ and $\omega_1 + \omega_4 = 2a$, where $a = \cos 2\pi/5$. Thus

$$(z - \omega_1)(z - \omega_4) = z^2 - 2az + 1.$$

Similarly,

$$(z - \omega_2)(z - \omega_3) = z^2 - 2bz + 1,$$

where $b = \cos 4\pi/5$. Hence

$$(z^2 - 2az + 1)(z^2 - 2bz + 1) = \frac{z^5 - 1}{z - 1} = z^4 + z^3 + z^2 + z + 1.$$

We expand the left-hand side and then equate coefficients of like powers in the resulting equation. This gives

$$-2(a + b) = 1,$$
$$2 + 4ab = 1.$$

Solving these simultaneous equations, we get

$$a = \frac{-1 + \sqrt{5}}{4}, \qquad b = \frac{-1 - \sqrt{5}}{4}.$$

One can easily construct line segments having lengths a and b by using a compass and a straightedge, and this leads to a construction for a regular pentagon (see Fig. 9-46).

7. *Some Sum Formulas.* We shall derive formulas for

$$C = 1 + \cos x + \cos 2x + \cdots + \cos nx$$

and

$$S = \sin x + \sin 2x + \cdots + \sin nx.$$

We have

$$C + iS = 1 + (\cos x + i\sin x) + (\cos 2x + i\sin 2x) + \cdots + (\cos nx + i\sin nx)$$
$$= 1 + \zeta + \zeta^2 + \cdots + \zeta^n,$$

where $\zeta = e^{ix}$. By the well-known formula for geometric progressions,

$$C + iS = \frac{\zeta^{n+1} - 1}{\zeta - 1} = \frac{\zeta^{n+1/2} - \zeta^{-1/2}}{\zeta^{1/2} - \zeta^{-1/2}}.$$

But
$$\zeta^{1/2} - \zeta^{-1/2} = e^{(x/2)i} - e^{-(x/2)i} = 2i \sin \frac{x}{2},$$
$$\zeta^{n+1/2} = \cos (n + \tfrac{1}{2})x + i \sin (n + \tfrac{1}{2})x,$$
$$\zeta^{-1/2} = \cos \frac{x}{2} - i \sin \frac{x}{2}.$$

Thus
$$C + iS = - \frac{\cos (n + \tfrac{1}{2})x - \cos (x/2)}{2 \sin (x/2)} i + \frac{\sin (n + \tfrac{1}{2})x + \sin (x/2)}{2 \sin (x/2)}.$$

Equating real and imaginary parts, we get
$$C = \frac{\sin (n + \tfrac{1}{2})x + \sin (x/2)}{2 \sin (x/2)} \quad \text{and} \quad S = - \frac{\cos (n + \tfrac{1}{2})x - \cos (x/2)}{2 \sin (x/2)}.$$

From the first expression, we get an identity important in the study of Fourier series:
$$\tfrac{1}{2} + \cos x + \cos 2x + \cdots + \cos nx = \frac{\sin (n + \tfrac{1}{2})x}{2 \sin (x/2)}.$$

In closing, it must be stressed that the mathematics in this section does *not* meet current standards. It is offered rather as an example of how mathematics has been done in the past (in the time of Euler, specifically). If it tantalizes the reader into the further study of function theory, then its purpose is served.

Exercises

1. Prove by induction that De Moivre's formula,
$$(\cos \theta + i \sin \theta)^n = \cos n\theta + i \sin n\theta,$$
holds for $n \in \mathbf{Z}^+$. Do not use Euler's formula.

2. Using De Moivre's formula, express $\cos 3\theta$, $\cos 4\theta$, $\sin 3\theta$, and $\cos 3\theta$ in terms of $\sin \theta$ and $\cos \theta$.

3. Describe explicitly a compass and straightedge construction for a regular pentagon.

4. Calculate the three cube roots of one. Devise from them a compass and straightedge construction for an equilateral triangle.

***5. Devise a compass and straightedge construction for a regular seventeen-sided polygon.

6. Prove that every complex number z has exactly n distinct nth roots which form the vertices of a regular n-sided polygon inscribed in a circle of radius $\sqrt[n]{|z|}$.

7. a) Observe that if z and z' are two complex numbers, then $|z - z'|$ is the distance between them in the complex plane.

   *b) Let $A_0, A_1, \ldots, A_{n-1}$ be consecutive vertices of a regular n-sided polygon inscribed in a circle of radius r with center O. Let P be a point outside the circle on the line through O and A_0. Prove that
$$\prod_{k=0}^{n-1} d(P, A_k) = [d(O, P)]^n - r^n.$$

8. Show that

$$\sin^{2n} x = \frac{(-1)^n}{2^{2n}}\left[e^{2nix} - \binom{2n}{1}e^{(2n-2)ix} + \binom{2n}{2}e^{(2n-4)ix} - \cdots + e^{-2nix}\right].$$

[Use the formula $\sin x = (e^{ix} - e^{-ix})/2i$.] Deduce that

$$\int_0^x \sin^{2n} t\, dt = \frac{(-1)^n}{2^{2n}}\left[\frac{1}{n}\sin 2nx - \binom{2n}{1}\frac{1}{n-1}\sin(2n-2)x + \cdots \right.$$
$$\left. + (-1)^{n-1}\binom{2n}{n-1}\sin 2x + (-1)^n\binom{2n}{n}x\right].$$

In particular, we have

$$\int_0^{\pi/2}\sin^{2n} t\, dt = \binom{2n}{n}\frac{1}{2^{2n}}\frac{\pi}{2} = \frac{1\cdot 3\cdot 5\cdots(2n-1)}{2\cdot 4\cdot 6\cdots 2n}\frac{\pi}{2}.$$

9. Derive the identity

$$\cos x + \cos 3x + \cdots + \cos(2n-1)x = \frac{\sin 2nx}{2\sin x}.$$

10. Prove that

$$\frac{d^n}{dx^n}(e^{ax}\cos bx) = r^n e^{ax}\cos(bx + n\theta),$$

where $[r; \theta]$ are polar coordinates of (a, b) and $r \geq 0$.

11. In our study of simple harmonic motion in the previous set of exercises, we neglected air resistance. We shall now attempt to take it into account. Let us suppose that our spring-mass system is immersed in some fluid (such as air) and that the resistance offered by the fluid to motion is directly proportional to velocity. Argue that the equation of motion of the system can now be written as

$$m\frac{d^2x}{dt^2} = -kx - \rho\frac{dx}{dt}, \tag{*}$$

where ρ is a positive constant.

a) Show that

$$\tfrac{1}{2}m[x'(t)]^2 + \tfrac{1}{2}k[x(t)]^2 + \int_0^t \rho[x'(t)]^2\, dt$$

is a constant. Observe that the term $\int_0^t \rho[x'(t)]^2\, dt$ represents the work done in overcoming the resistance of the fluid, and deduce from this that the total energy of the system (sum of kinetic and potential energy) at time t is equal to the total energy at time zero minus the energy expended in overcoming the resistance offered by the fluid. In particular, show that if $x(t)$ satisfies the equation of motion (*) and if $x(0) = x'(0) = 0$, then $x(t)$ is identically zero.

b) We rewrite the equation of motion as follows:

$$x''(t) + 2\lambda x'(t) + \omega^2 x(t) = 0, \tag{**}$$

where $\lambda = \rho/2m$ and $\omega = \sqrt{k/m}$. Show that if $x_1(t)$ and $x_2(t)$ satisfy this differential equation, then so does every function of the form $c_1x_1(t) + c_2x_2(t)$, where c_1 and c_2 are constants. (This so-called *principle of superposition* holds for the large important class of *linear homogeneous differential equations*.)

c) Show that $x(t) = e^{rt}$ satisfies (∗∗) iff the constant r is a root of the *auxiliary equation*

$$r^2 + 2\lambda r + \omega^2 = 0.$$

There are three cases to be considered.

Case i. Damped Harmonic Motion. $\omega > \lambda$. In this case the roots of the auxiliary equations are $-\lambda + i\mu$ and $-\lambda - i\mu$, where $\mu = \sqrt{\omega^2 - \lambda^2}$.

Case ii. Critical Damping. $\omega = \lambda$. In this case $-\lambda$ is the root of the auxiliary equation with multiplicity two.

Case iii. Overdamping. $\omega < \lambda$. In this case the roots of the auxiliary equation, $-\lambda + \sqrt{\lambda^2 - \omega^2}$ and $-\lambda - \sqrt{\lambda^2 - \omega^2}$, are both negative and unequal.

d) For case i, deduce from Euler's formula, (b), and (c) that the functions

$$e^{-\lambda t} \cos \mu t = \tfrac{1}{2} e^{(-\lambda + i\mu)t} + \tfrac{1}{2} e^{(-\lambda - i\mu)t}$$

and $e^{-\lambda t} \sin \mu t$ satisfy (∗∗). Also verify this directly. It follows that every function of the form

$$x(t) = a e^{-\lambda t} \cos \mu t + b e^{-\lambda t} \sin \mu t$$

satisfies (∗∗). Prove, conversely, that every (real-valued) solution of (∗∗) is of this form, with

$$a = x(0) \quad \text{and} \quad b = \frac{x'(0) + \lambda a}{\mu}.$$

Thus the motion of the damped system is completely determined by the initial displacement and velocity. [Let a and b be defined by the equations above. Observe that $X(t) = x(t) - a e^{-\lambda t} \cos \mu t - b e^{-\lambda t} \sin \mu t$ satisfies (∗∗) and that

$$X(0) = X'(0) = 0.$$

Deduce from (a) that $X(t)$ is identically zero.] Sketch the graph of $X(t)$ after first writing it in the form $A e^{-\lambda t} \cos (\mu t - \phi)$.

e) In the case of critical damping, show that $x(t)$ must be of the form $(a + bt)e^{-\lambda t}$, where a and b are constants. Discuss the behavior of the solution and sketch its graph for several choices of a and b.

f) In the case of overdamping, show that $x(t)$ is of the form

$$a e^{(-\lambda + \sqrt{\lambda^2 - \omega^2})t} + b e^{(-\lambda - \sqrt{\lambda^2 - \omega^2})t}.$$

Sketch the graph.

Miscellaneous exercises

1. Show that $xy \le x \ln x + e^{y-1}$ for all $x > 0$ and $y \in \mathbf{R}$. When does equality hold? (Use Young's inequality.)

2. Establish the following inequalities:

a) $2x \operatorname{Tan}^{-1} x \ge \ln (1 + x^2)$,

b) $1 + x \ln (x + \sqrt{1 + x^2}) \ge \sqrt{1 + x^2}$,

c) $\ln (1 + x) > \dfrac{\operatorname{Tan}^{-1} x}{1 + x}$ $(x > 0)$,

d) $\cosh x \ge 1 + \dfrac{x^2}{2}$.

3. Prove that the graph of $x^y = y^x$ consists of the straight line $y = x$ and the curve

$$x = (1 + 1/t)^t, \qquad y = (1 + 1/t)^{t+1}.$$

Sketch the latter curve.

4. If $x, y > 0$, show that $e^{xy/(x+y)} < (1 + x/y)^y$.

5. The curve defined parametrically by the equations

$$x = a[\ln \tan (t/2) + \cos t], \qquad y = a \sin t \qquad (0 < t < \pi)$$

is called a *tractrix*.

a) Show that the tractrix is also defined by the equation

$$x = \pm \left(\sqrt{a^2 - y^2} + a \ln \frac{y}{a + \sqrt{a^2 - y^2}} \right).$$

b) Show that there exists a function f such that the tractrix is the graph of $y = f(x)$. At what points is f differentiable?

c) Prove that each tangent line to the tractrix is such that the distance between the point of contact and the point of intersection with the x-axis is equal to a.

d) Sketch the curve.

e) Show that the length of the arc joining the points $(0, a)$ and (x, y) on the tractrix is $a \ln (y/a)$.

6. a) Let $f(x) = (1 + 1/x)^{x+p}$. Show that f is decreasing on $(0, +\infty)$ iff $p \geq \frac{1}{2}$. Show that f is increasing on $(0, +\infty)$ iff $p \leq 0$. What is the situation if $0 < p < \frac{1}{2}$? Show that in all cases $\lim_{x \to +\infty} f(x) = e$.

b) Discuss in a similar fashion the functions

$$(1 - 1/x)^{x-p}, \qquad (1 + 1/x)^x(1 + p/x), \qquad \text{and} \qquad (1 + p/x)^{x+1}.$$

7. Show that

$$D^n(\mathrm{Tan}^{-1} x) = \frac{(-1)^{n-1}(n - 1)!}{(1 + x^2)^{n/2}} \sin \left(n \, \mathrm{Tan}^{-1} \frac{1}{x} \right).$$

8. Sketch the curve $x = te^t, y = te^{-t}$.

9. Two rods each of mass m and length l lie on the same straight line at a distance l apart. Find the force of attraction between them.

10. Air initially containing a percent of CO_2 is filtered through a cylindrical vessel containing a layer of absorbent material with thickness H. Air leaving the cylinder has b percent of CO_2. Assuming that the amount of absorption of CO_2 by a thin layer is directly proportional to the CO_2 concentration and to the thickness of the layer, find the thickness of absorbent material required to produce air having c percent of CO_2.

10 ☐ TECHNIQUES OF INTEGRATION

This chapter is devoted to the technique of finding antiderivatives. The relationship between differentiation and integration is paradoxical in at least one respect. It is easier for a function to have an antiderivative than a derivative, and, generally speaking, the process of integration is a "smoothing" one, in the sense that an antiderivative will have more pleasant properties than the original function. Yet the problem of computing antiderivatives is much more difficult than the problem of calculating derivatives. For instance, the function $1/(x^4 + 1)$ is easily seen to have the derivative

$$\frac{-4x^3}{(x^4 + 1)^2}.$$

The problem of integrating $1/(x^4 + 1)$, however, proved troublesome even to Leibnitz. The integral is, in fact,

$$\int \frac{dx}{x^4 + 1} = \frac{1}{4\sqrt{2}} \ln \left| \frac{x^2 + \sqrt{2}\,x + 1}{x^2 - \sqrt{2}\,x + 1} \right|$$

$$+ \frac{1}{2\sqrt{2}} \operatorname{Tan}^{-1} (\sqrt{2}\,x + 1)$$

$$+ \frac{1}{2\sqrt{2}} \operatorname{Tan}^{-1} (\sqrt{2}\,x - 1).$$

Moreover, the integral of an elementary function need not be elementary. [For our purposes, an elementary function may be defined as any function which can be obtained from the functions e^x, $\sin x$, $|x|$, the identity function, and the constants by the usual algebraic operations (addition, multiplication, etc.), composition, restriction, and inversion.] The integral $\int e^{-x^2/2}\,dx$, which is of importance in probability and statistics, is an example.

10-1 Reduction to standard formulas

To start things rolling, we shall collect a few integral formulas. The student will probably find it worthwhile to commit them to memory.

$$\int x^n \, dx = \frac{x^{n+1}}{n+1} + C \qquad (n \neq -1)$$

$$\int \frac{dx}{x} = \ln|x| + C$$

$$\int \frac{dx}{\sqrt{a^2 - x^2}} = \text{Sin}^{-1} \frac{x}{a} + C$$

$$\int \frac{dx}{a^2 + x^2} = \frac{1}{a} \text{Tan}^{-1} \frac{x}{a} + C$$

$$\int \frac{dx}{\sqrt{a^2 + x^2}} = \sinh^{-1} \frac{x}{a} + C = \ln(x + \sqrt{a^2 + x^2}) + C'$$

$$\int \frac{dx}{\sqrt{x^2 - a^2}} = \begin{cases} \text{Cosh}^{-1} \frac{x}{a} + C & \text{if } x > a, \\ -\text{Cosh}^{-1} \frac{(-x)}{a} + C & \text{if } x < -a \end{cases}$$

$$= \ln|x + \sqrt{x^2 - a^2}| + C'$$

$$\int \frac{dx}{a^2 - x^2} = \begin{cases} \frac{1}{a} \tanh^{-1} \frac{x}{a} + C & \text{if } |x| < |a|, \\ \frac{1}{a} \tanh^{-1} \frac{a}{x} + C & \text{if } |x| > |a| \end{cases}$$

$$= \frac{1}{2a} \ln \left| \frac{a + x}{a - x} \right| + C$$

The validity of each formula can be quickly checked by differentiation of the right-hand sides of the equations. A word or two must be said, however, concerning the proper interpretation of these formulas. Consider the second one, for example. Interpreted literally, it says that any antiderivative of $1/x$ is of the form $\ln|x| + C$, where C is a constant. This is untrue in the sense that

$$f(x) = \begin{cases} \ln x + 100 & \text{if } x > 0, \\ \ln(-x) - \pi & \text{if } x < 0 \end{cases}$$

is an antiderivative of $1/x$ which is not of the form stated. The difficulty stems from the fact that the domain of $1/x$ is not an interval, but rather the union of two disjoint intervals. To be perfectly correct, one should consider separately the restrictions of $1/x$ to these intervals and write, for instance,

$$\int \frac{dx}{x} = \ln x + C \quad (x > 0) \quad \text{and} \quad \int \frac{dx}{x} = \ln(-x) + C \quad (x < 0).$$

It has become customary in tables of integrals, however, to write

$$\int \frac{dx}{x} = \ln |x| + C,$$

and for this reason we shall accept it (with some uneasiness) as a shorthand for the two correct equations. Similar remarks hold for the first formula when $n < -1$ and for those integrals involving $x^2 - a^2$.

We shall now show by examples how other integrals can be reduced to these formulas by substitutions and completions of squares.

Example 1. Find $\int \tan \theta \, d\theta$.

We note that $\tan \theta = \sin \theta / \cos \theta$ and make the substitution $u = \cos \theta$. Since

$$du = \sin \theta \, d\theta,$$

we get

$$\int \tan \theta \, d\theta = -\int du/u = -\ln |u| + C = \ln |u^{-1}| + C = \ln |\sec \theta| + C.$$

The domain of the tangent function is the union of the intervals $(-\pi/2, \pi/2)$, $(\pi/2, 3\pi/2)$, $(-3\pi/2, -\pi/2)$, ..., and we should interpret the above equation as holding on these intervals separately.

The preceding integral is a special case of

$$\int \sin^m \theta \cos^n \theta \, d\theta,$$

where m and n are integers. Integrals of this form arise frequently. If one of the integers is odd, we can evaluate the integral by making a substitution and by using perhaps the Pythagorean identity $\sin^2 \theta + \cos^2 \theta = 1$. Otherwise, half-angle formulas may be required.

Example 2. Find $\int \sec \theta \, d\theta$.

We note that

$$\sec \theta = \frac{1}{\cos \theta} = \frac{\cos \theta}{\cos^2 \theta} = \frac{\cos \theta}{1 - \sin^2 \theta}.$$

If we let $u = \sin \theta$, then we obtain

$$\int \sec \theta \, d\theta = \int \frac{\cos \theta \, d\theta}{1 - \sin^2 \theta} = \int \frac{du}{1 - u^2}$$

$$= \tfrac{1}{2} \ln \left| \frac{1 + u}{1 - u} \right| + C$$

$$= \tfrac{1}{2} \ln \left| \frac{1 + \sin \theta}{1 - \sin \theta} \right| + C.$$

Example 3. Find $\int dx/\sqrt{x^2 + 2x + 5}$.

Here we complete the square:

$$x^2 + 2x + 5 = (x + 1)^2 + 4.$$

If we then set $u = x + 1$, we get

$$\int \frac{dx}{\sqrt{x^2 + 2x + 5}} = \int \frac{du}{\sqrt{4 + u^2}}$$
$$= \sinh^{-1} \frac{u}{2} + C$$
$$= \sinh^{-1} \frac{x + 1}{2} + C,$$

or, equivalently,

$$\int \frac{dx}{\sqrt{x^2 + 2x + 5}} = \ln (u + \sqrt{4 + u^2}) + C'$$
$$= \ln (x + 1 + \sqrt{x^2 + 2x + 5}) + C'.$$

Example 4. Find $\int (x + 1)/\sqrt{5 + 4x - x^2}\, dx$.

Completing the square, we get

$$5 + 4x - x^2 = 9 - (x - 2)^2.$$

We set $u = x - 2$. Then

$$\int \frac{x + 1}{\sqrt{5 + 4x - x^2}} dx = \int \frac{u + 3}{\sqrt{9 - u^2}} du$$
$$= \int \frac{u\, du}{\sqrt{9 - u^2}} + 3 \int \frac{du}{\sqrt{9 - u^2}}$$
$$= -\sqrt{9 - u^2} + 3 \operatorname{Sin}^{-1} \tfrac{1}{3}u + C$$
$$= -\sqrt{5 + 4x - x^2} + 3 \operatorname{Sin}^{-1} \tfrac{1}{3}(x - 2) + C.$$

Exercises

Evaluate the following integrals.

1. $\displaystyle \int x\sqrt{2x + 3}\, dx$

2. $\displaystyle \int \frac{dx}{x\sqrt{1 - (\ln x)^2}}$

3. $\displaystyle \int xe^{x^2}\, dx$

4. $\displaystyle \int \frac{dx}{\sqrt{1 + e^{-2x}}}$

5. $\displaystyle \int \frac{\operatorname{Tan}^{-1} x}{1 + x^2}\, dx$

6. $\displaystyle \int \frac{dx}{x \ln x}$

7. $\displaystyle \int \frac{x + 1}{\sqrt{1 - x^2}}\, dx$

8. $\displaystyle \int \frac{dx}{\sqrt{1 - 25x^2}}$

9. $\displaystyle \int \sin^5 \theta\, d\theta$

10. $\displaystyle \int e^{\sin^2 x} \sin x\, dx$

11. $\displaystyle \int \sin^3 \theta \cos^4 \theta\, d\theta$

12. $\displaystyle \int \cos^4 \theta \sin 2\theta\, d\theta$

13. $\displaystyle \int \operatorname{ctn} \theta\, d\theta$

14. $\displaystyle \int \frac{\sin x\, dx}{2 - \sin^2 x}$

15. $\displaystyle \int \csc \theta\, d\theta$

16. $\displaystyle \int \frac{d\theta}{1 - \cos \theta}$

17. $\displaystyle \int \cos^4 \theta\, d\theta$

18. $\displaystyle \int \frac{e^x\, dx}{\sqrt{e^{2x} + e^x + 1}}$

19. $\int \dfrac{dx}{\sqrt{x^2 - 4x + 13}}$

20. $\int \dfrac{dx}{2 - 3x^2}$

21. $\int \dfrac{x\,dx}{x^2 - x + 1}$

22. $\int \dfrac{dx}{\sqrt{4x - 3 - x^2}}$

23. $\int \dfrac{dx}{\sqrt{2x - x^2}}$

24. $\int \dfrac{dx}{x^2 - 5x + 6}$

25. $\int \dfrac{x}{3x^2 - 2x + 1}\,dx$

26. $\int \dfrac{dx}{x(\ln x + \ln^2 x)}$

27. $\int \dfrac{dx}{4x^2 + 4x + 5}$

*28. Let $X = ax^2 + bx + c$. Find in all possible cases $\int dx/X$ and $\int dx/\sqrt{X}$.

29. Use the identities

$$\sin \alpha \cos \beta = \tfrac{1}{2}[\sin (\alpha + \beta) + \sin (\alpha - \beta)],$$
$$\sin \alpha \sin \beta = -\tfrac{1}{2}[\cos (\alpha + \beta) - \cos (\alpha - \beta)],$$
$$\cos \alpha \cos \beta = \tfrac{1}{2}[\cos (\alpha + \beta) + \cos (\alpha - \beta)]$$

to calculate the following integrals:

$$\int \sin 5x \cos 3x \, dx, \qquad \int \sin 2x \sin x \, dx, \qquad \int \cos (x + 1) \cos (x - 1) \, dx.$$

Also, prove the orthogonality relations:

$$\int_{-\pi}^{\pi} \sin mx \sin nx \, dx = \begin{cases} 0 & \text{if } m \neq n, \\ \pi & \text{if } m = n, \end{cases}$$

$$\int_{-\pi}^{\pi} \sin mx \cos nx \, dx = 0,$$

$$\int_{-\pi}^{\pi} \cos mx \cos nx \, dx = \begin{cases} 0 & \text{if } m \neq n, \\ \pi & \text{if } m = n. \end{cases}$$

30. By making the substitution $u = \sec x + \tan x$, show that

$$\int \sec x \, dx = \ln |\sec x + \tan x| + C.$$

Reconcile this with the result in Example 2.

31. Find

a) $\displaystyle\int \dfrac{2\,dx}{e^x + e^{-x}} = \int \operatorname{sech} x \, dx,$

b) $\displaystyle\int \dfrac{e^x - e^{-x}}{e^x + e^{-x}} \, dx = \int \tanh x \, dx,$

c) $\displaystyle\int \dfrac{dx}{e^x + 1}.$

10-2 Integration by parts

One complicating feature of the technique of integration is the absence of a rule for integrating the product of two functions. The integration by parts formula comes closest to filling this gap. This formula is based upon the rule for differentiating the product of two functions:

$$(fg)' = fg' + gf'.$$

Taking antiderivatives of each side, we get

$$f(x)g(x) = \int f(x)g'(x)\,dx + \int g(x)f'(x)\,dx$$

or

$$\int f(x)g'(x)\,dx = f(x)g(x) - \int g(x)f'(x)\,dx.$$

This rule is memorable when written in the formalism of differentials. Let

$$u = f(x) \quad \text{and} \quad v = g(x).$$

Then

$$du = f'(x)\,dx \quad \text{and} \quad dv = g'(x)\,dx,$$

so that we get, upon substitution,

$$\int u\,dv = uv - \int v\,du,$$

the integration by parts formula in its usual notation. The reader is advised to commit this formula to memory before reading further.

Example 1. Find $\int xe^x\,dx$.

In applying the integration by parts formula, one tries to choose u and v in such a way that the second integral ($\int v\,du$) is simpler than the first ($\int u\,dv$). In this case the factor e^x does not become more complicated under integration, while the factor x simplifies under differentiation. This suggests that we let

$$u = x \quad \text{and} \quad dv = e^x\,dx.$$

Then

$$du = dx \quad \text{and} \quad v = e^x,$$

and upon substitution into the integration by parts formula, we get

$$\int xe^x\,dx = \int u\,dv = uv - \int v\,du = xe^x - \int e^x\,dx = (x-1)e^x + C.$$

Example 2. Find $\int x^2 \sin x\,dx$.

The factor x^2 simplifies under differentiation, while the factor $\sin x$ does not become more complicated under integration. We therefore set

$$u = x^2 \quad \text{and} \quad dv = \sin x\,dx.$$

Then

$$du = 2x\,dx \quad \text{and} \quad v = -\cos x.$$

Thus

$$\int x^2 \sin x\,dx = -x^2 \cos x + 2\int x \cos x\,dx.$$

The integral on the right-hand side of the equation can be evaluated by a second integration by parts. Let $u = x$ and $dv = \cos x\,dx$. Then $du = dx$, $v = \sin x$, and

$$\int x \cos x\,dx = x \sin x - \int \sin x\,dx = x \sin x + \cos x + C.$$

Hence

$$\int x^2 \sin x\,dx = (2 - x^2)\cos x + 2x \sin x + C.$$

Example 3. Find $\int \ln x \, dx$.

In this case one hasn't much choice. Let

$$u = \ln x \quad \text{and} \quad dv = dx.$$

Then $du = (1/x)\, dx$ and $v = x$. Thus

$$\int \ln x \, dx = x \ln x - \int x(1/x)\, dx = x \ln x - x + C.$$

Example 4. Find $I = \int x^2 \, dx/\sqrt{x^2 - 1}$.

Let

$$u = x \quad \text{and} \quad dv = \frac{x\, dx}{\sqrt{x^2 - 1}}.$$

Then $du = dx$, $v = \sqrt{x^2 - 1}$, and

$$I = \int \frac{x^2\, dx}{\sqrt{x^2 - 1}} = x\sqrt{x^2 - 1} - \int \sqrt{x^2 - 1}\, dx.$$

The second integral can be evaluated by an algebraic device. Note that

$$\int \sqrt{x^2 - 1}\, dx = \int \frac{x^2 - 1}{\sqrt{x^2 - 1}}\, dx = \int \frac{x^2\, dx}{\sqrt{x^2 - 1}} - \int \frac{dx}{\sqrt{x^2 - 1}} = I - \text{Cosh}^{-1} x.$$

Thus we get

$$I = x\sqrt{x^2 - 1} + \text{Cosh}^{-1} x - I,$$

whence

$$I = \tfrac{1}{2}(x\sqrt{x^2 - 1} + \text{Cosh}^{-1} x) + C = \tfrac{1}{2}[x\sqrt{x^2 - 1} + \ln (x + \sqrt{x^2 - 1})] + C.$$

Example 5. Find $I = \int e^x \cos x \, dx$ and $J = \int e^x \sin x \, dx$.

Consider I first. If we let

$$u = e^x \quad \text{and} \quad dv = \cos x \, dx,$$

then $du = e^x\, dx$ and $v = \sin x$, so that

$$I = e^x \sin x - \int e^x \sin x \, dx = e^x \sin x - J.$$

Consider next J. If we let

$$u = e^x \quad \text{and} \quad dv = \sin x \, dx,$$

then $du = e^x\, dx$, $v = -\cos x$, and

$$J = \int e^x \sin x \, dx = -e^x \cos x + \int e^x \cos x \, dx = e^x \cos x + I.$$

It might at first seem that we have not gone anywhere. (We can calculate I if we can calculate J, and we can calculate J if we can calculate I.) Actually, we have obtained two simultaneous equations for I and J:

$$I + J = e^x \sin x, \quad I - J = e^x \cos x.$$

Solving, we get

$$I = \tfrac{1}{2}e^x(\sin x + \cos x) + C, \quad J = \tfrac{1}{2}e^x(\sin x - \cos x) + C.$$

Also, Euler's formula provides an interesting formal technique for finding I and J. Hence we get

$$I + iJ = \int e^x(\cos x + i \sin x)\,dx$$
$$= \int e^x e^{ix}\,dx = \int e^{(i+1)x}\,dx$$
$$= \frac{1}{1+i} e^{(1+i)x} = e^x \frac{\cos x + i \sin x}{1+i}$$
$$= e^x \frac{\cos x + i \sin x}{1+i} \frac{1-i}{1-i}$$
$$= \frac{e^x}{2}[(\cos x + \sin x) + i(\sin x - \cos x)].$$

Equating real and imaginary parts, we get

$$I = \frac{e^x}{2}(\cos x + \sin x), \qquad J = \frac{e^x}{2}(\sin x - \cos x).$$

Example 6. Find a *reduction* (recursion) *formula* for $I_n = \int dx/(x^2 + a^2)^n$.
Observe that

$$I_{n-1} = \int \frac{dx}{(x^2 + a^2)^{n-1}} = \int \frac{(x^2 + a^2)}{(x^2 + a^2)^n}\,dx$$
$$= \int \frac{x^2\,dx}{(x^2 + a^2)^n} + a^2 I_n.$$

We now use integration by parts on $\int x^2\,dx/(x^2 + a^2)^n$. Let

$$u = x \quad \text{and} \quad dv = \frac{x\,dx}{(x^2 + a^2)^n}.$$

Then

$$du = dx, \qquad v = -\frac{1}{2n-2}(x^2 + a^2)^{-n+1},$$

and

$$\int \frac{x^2\,dx}{(x^2 + a^2)^n} = -\frac{x}{(2n-2)(x^2+a^2)^{n-1}} + \frac{1}{2n-2}\int \frac{dx}{(x^2+a^2)^{n-1}}$$
$$= -\frac{x}{(2n-2)(x^2+a^2)^{n-1}} + \frac{I_{n-1}}{2n-2}.$$

Thus we have

$$I_{n-1} = a^2 I_n + \frac{I_{n-1}}{2n-2} - \frac{x}{(2n-2)(x^2+a^2)^{n-1}}$$

or

$$I_n = \frac{x}{(2n-2)a^2(x^2+a^2)^{n-1}} + \frac{2n-3}{(2n-2)a^2}I_{n-1},$$

which is the desired reduction formula. This formula is valid except, of course, when $n = 1$. It is mainly of interest when n is a positive integer greater than one. In that case, I_n can be found by successive applications of this formula. [Note that $I_1 = (1/a)\,\mathrm{Tan}^{-1}(x/a)$.]

We give a couple of examples:

$$\int \frac{dx}{(x^2 + 4)^3} = \frac{x}{16(x^2 + 4)^2} + \frac{3}{16} \int \frac{dx}{(x^2 + 4)^2}$$

$$= \frac{x}{16(x^2 + 4)^2} + \frac{3}{16} \left[\frac{x}{8(x^2 + 4)} + \frac{1}{8} \int \frac{dx}{x^2 + 4} \right]$$

$$= \frac{x}{16(x^2 + 4)^2} + \frac{3x}{128(x^2 + 4)} + \frac{3}{256} \, \text{Tan}^{-1} \frac{x}{2} + C,$$

and

$$\int (x^2 + 1)^{-3/2} \, dx = \frac{x}{\sqrt{x^2 + 1}} + C.$$

Example 7. Obtain a reduction formula for $I_{m,n} = \int \sin^m x \cos^n x \, dx$.
   If we let

$$u = \sin^m x \cos^{n-1} x, \qquad dv = \cos x \, dx,$$

we get

$$du = [m \sin^{m-1} x \cos^n x - (n - 1) \sin^{m+1} x \cos^{n-2} x] \, dx, \qquad v = \sin x,$$

and

$$I_{m,n} = \sin^{m+1} x \cos^{n-1} x - m I_{m,n} + (n - 1) I_{m+2,n-2}.$$

But

$$I_{m+2,n-2} = \int \sin^{m+2} x \cos^{n-2} x \, dx$$

$$= \int \sin^m x (1 - \cos^2 x) \cos^{n-2} x \, dx$$

$$= I_{m,n-2} - I_{m,n}.$$

Thus

$$I_{m,n} = \sin^{m+1} x \cos^{n-1} x - m I_{m,n} + (n - 1) I_{m,n-2} - (n - 1) I_{m,n}.$$

Solving for $I_{m,n}$, we get

$$I_{m,n} = \frac{1}{m + n} [\sin^{m+1} x \cos^{n-1} x + (n - 1) I_{m,n-2}].$$

A special case of this formula ($m = 0$) is the following:

$$\int \cos^n x \, dx = \frac{1}{n} \sin x \cos^{n-1} x + \frac{n - 1}{n} \int \cos^{n-2} x \, dx.$$

One can also derive similarly the reduction formulas

$$I_{m,n} = \frac{1}{m + n} [-\sin^{n-1} x \cos^{n+1} x + (m - 1) I_{m-2,n}]$$

and

$$\int \sin^m x \, dx = -\frac{1}{m} \sin^{m-1} x \cos x + \frac{m - 1}{m} \int \sin^{m-2} x \, dx.$$

Using these reduction formulas, we can evaluate $I_{m,n}$ for all positive integers m and n.

Example 8. The integral sine function

$$\mathrm{Si}\,(x) = \int_0^x \frac{\sin t}{t}\,dt$$

cannot be expressed in terms of the elementary functions. However, its integral can be expressed in terms of elementary functions and the integral sine function itself. Let

$$u = \mathrm{Si}\,(x) \qquad \text{and} \qquad dv = dx.$$

Then $du = [(\sin x)/x]\,dx$, $v = x$, and

$$\int \mathrm{Si}\,(x)\,dx = x\,\mathrm{Si}\,(x) - \int x\,\frac{\sin x}{x}\,dx = x\,\mathrm{Si}\,(x) + \cos x + C.$$

Exercises

Evaluate the following integrals.

1. $\displaystyle\int x^2 e^x$

2. $\displaystyle\int x \ln x\,dx$

3. $\displaystyle\int \sqrt{x}\,\ln x\,dx$

4. $\displaystyle\int \ln^2 x\,dx$

5. $\displaystyle\int \mathrm{Sin}^{-1} x\,dx$

6. $\displaystyle\int \mathrm{Tan}^{-1} x\,dx$

7. $\displaystyle\int x\,\mathrm{Tan}^{-1} x\,dx$

8. $\displaystyle\int x^3 e^{-x^2}\,dx$

9. $\displaystyle\int x^3 \sin x\,dx$

10. $\displaystyle\int x^4 \sqrt{1 - x^2}\,dx$

11. $\displaystyle\int x \tan^2 x\,dx$

12. $\displaystyle\int \frac{\ln \ln x}{x}\,dx$

13. $\displaystyle\int \frac{x \cos x}{\sin^2 x}\,dx$

14. $\displaystyle\int x \cosh x\,dx$

15. $\displaystyle\int \ln\,(x^2 + 1)\,dx$

16. $\displaystyle\int \frac{x^2\,\mathrm{Tan}^{-1} x}{1 + x^2}\,dx$

17. $\displaystyle\int x 3^x\,dx$

18. $\displaystyle\int \mathrm{Tan}^{-1} \sqrt{x}\,dx$

19. $\displaystyle\int \frac{\mathrm{Sin}^{-1} x}{\sqrt{x + 1}}\,dx$

20. $\displaystyle\int x^2 \ln\,(1 + x)\,dx$

21. $\displaystyle\int x^2 a^x\,dx$

22. $\displaystyle\int \frac{x^2\,dx}{(1 + x^2)^2}$

23. $\displaystyle\int (\mathrm{Sin}^{-1} x)^2\,dx$

24. $\displaystyle\int \sqrt{a^2 + x^2}\,dx$

25. $\displaystyle\int x^2 e^x \sin x\,dx$

26. Using the methods of Example 5, find

$$I = \int e^{ax} \cos bx\,dx \qquad \text{and} \qquad J = \int e^{ax} \sin bx\,dx.$$

27. Using the methods of Example 5, find

$$I = \int \cos\,(\ln x)\,dx \qquad \text{and} \qquad J = \int \sin\,(\ln x)\,dx.$$

28. Show that

$$\int e^x P(x)\, dx = e^x[P(x) - P'(x) + P''(x) - \cdots],$$

where P is any polynomial.

29. Using the reduction formula of Example 6, find

a) $\displaystyle\int \frac{dx}{(x^2 + 9)^2}$,

b) $\displaystyle\int \frac{dx}{(x^2 + 4)^{5/2}}$.

30. Prove the following reduction formulas:

a) $\displaystyle\int \sec^n x\, dx = \frac{1}{n-1}\left[\sec^{n-2} x \tan x + (n-2)\int \sec^{n-2} x\, dx\right]$,

b) $\displaystyle\int \sin^m x \cos^n x\, dx$

$$= \frac{1}{m+n}\left[-\sin^{m-1} x \cos^{n+1} x + (m-1)\int \sin^{m-2} x \cos^n x\, dx\right].$$

Use these formulas plus those given in Example 7 to find

c) $\displaystyle\int \sin^2 x\, dx$,

d) $\displaystyle\int \sin^4 x \cos^2 x\, dx$,

e) $\displaystyle\int \sec^3 x\, dx$.

31. Derive the formula

$$\int x^m (\ln x)^n\, dx = \frac{x^{m+1}(\ln x)^n}{m+1} - \frac{n}{m+1}\int x^m (\ln x)^{n-1}\, dx,$$

where $m, n \neq -1$. Discuss also the case in which m or n is -1.

32. a) Show that for all odd positive integers n, the integral $\int x^n e^{-x^2}\, dx$ can be expressed in terms of elementary functions.

b) If n in the above is even, show that the integral can be expressed in terms of elementary functions and the integral $\int e^{-x^2}\, dx$. (The latter integral plays an important role in probability and statistics, and tables for it are readily available.)

*33. a) Prove that

$$\int_a^x \left[\int_a^u g(t)\, dt\right] du = \int_a^x g(u)(x - u)\, du.$$

b) Prove more generally that the nth iterated integral of g is given by

$$\frac{1}{(n-1)!}\int_a^x g(u)(x - u)^{n-1}\, du.$$

c) Given $g = f^{(n)}$, show that the nth iterated integral of g is

$$f(x) - \sum_{k=0}^{n-1} \frac{f^{(k)}(a)}{k!}(x - a)^k.$$

Deduce Taylor's theorem with integral remainder:

$$f(x) = \sum_{k=0}^{n-1} \frac{f^{(k)}(a)}{k!}(x - a)^k + \int_a^x \frac{f^{(n)}(u)}{(n-1)!}(x - u)^{n-1}\, du.$$

34. Suppose that f is a continuous invertible function and that $\int f(x)\,dx = F(x)$. Show that

$$\int f^{-1}(x)\,dx = xf^{-1}(x) - F(f^{-1}(x)).$$

Use this formula to find $\int \ln x\,dx$, $\int \operatorname{Sin}^{-1} x\,dx$, and $\int \sinh^{-1} x\,dx$.

10-3 Rational functions

In this section we shall show how the integral of any rational function can be expressed in terms of elementary functions. The technique which is commonly used for this purpose is that of *partial fraction decomposition*. Suppose, for instance, that we are required to integrate

$$R(x) = \frac{x^4 - 2x^3 - 3x^2 + 15x - 8}{x^3 - 3x + 2}.$$

One can check that

$$R(x) = x - 2 + \frac{2}{x-1} + \frac{1}{(x-1)^2} - \frac{2}{x+2}.$$

The right-hand side of this equation is called the partial fraction decomposition of $R(x)$; it can be used to calculate with ease the integral of $R(x)$. From the above equation, we get

$$\int R(x)\,dx = \frac{x^2}{2} - 2x + 2\ln|x-1| - \frac{1}{x-1} - 2\ln|x+2| + C.$$

Listed below are additional examples of partial fraction decomposition:

$$\frac{x^2 + 6x - 1}{(x-1)(x-3)^2} = \frac{3}{2}\frac{1}{x-1} - \frac{1}{2(x-3)} + \frac{13}{(x-3)^2},$$

$$\frac{5}{(x-1)(x^2+4)} = \frac{1}{x-1} - \frac{x+1}{x^2+4},$$

$$\frac{-x^3 + 6x^2 + x + 2}{(x^2-1)(x^2+1)} = \frac{1}{x-1} - \frac{1}{x+1} - \frac{2}{x^2+1} - \frac{x-2}{(x^2+1)^2}.$$

In general, every rational function can be expressed as a sum of
a) a polynomial;
b) functions of the form $A/(x-a)^n$;
c) functions of the form

$$\frac{Ax + B}{(x^2 + 2bx + c)^2},$$

where $x^2 + 2bx + c$ is an *irreducible quadratic*. By the latter we mean that $x^2 + 2bx + c$ cannot be written as a product of two linear functions with real coefficients. This is equivalent to saying that the equation

$$x^2 + 2bx + c = 0$$

does not have real roots, or (by the quadratic formula) that $b^2 < c$.

Functions of types (a), (b), and (c) can be integrated by methods which we have already discussed. We have, for example,

$$\int \frac{A}{(x-a)^n}\, dx = \begin{cases} A \ln |x-a| & \text{if } n = 1, \\ -\dfrac{A}{(n-1)(x-a)^{n-1}} & \text{if } n \neq 1. \end{cases}$$

To integrate a function of the form

$$\frac{Ax+B}{(x^2 + 2bx + c)^n},$$

we first complete the square on the irreducible quadratic:

$$x^2 + 2bx + c = (x+b)^2 + (c - b^2).$$

If we let $u = x + b$ and $a^2 = c - b^2$, we get

$$\int \frac{Ax+B}{(x^2+2bx+c)^n} = \int \frac{Au+B'}{(u^2+a^2)^n}\, du = A \int \frac{u\,du}{(u^2+a^2)^n} + B' \int \frac{du}{(u^2+a^2)^n},$$

where $B' = B - bA$. The first integral is easily calculated:

$$\int \frac{u\,du}{(u^2+a^2)^n} = \begin{cases} -\dfrac{1}{2(n-1)(u^2+a^2)^{n-1}} & \text{if } n \neq 1, \\ \ln(u^2+a^2) & \text{if } n = 1. \end{cases}$$

For the second integral, we can apply the reduction formula given in Example 6 of the preceding section.

We shall now present a recipe for decomposing a rational function into partial fractions. Its justification will be taken up later.

General rules for decomposing a rational function $R(x)$ into partial fractions

1) Write $R(x)$ as a quotient of two polynomials if it is not already in this standard form. Then examine the degrees of the numerator and denominator. If the degree of the denominator is greater than the degree of the numerator [in which case $R(x)$ is said to be a *proper* rational function], nothing further need be done at this stage. If, however, $R(x)$ is improper (that is, the degree of the denominator is less than that of the numerator), then express $R(x)$ as a polynomial plus a proper rational function by performing long division. In other words, express $R(x)$ in the form

$$R(x) = P_\infty(x) + \frac{P(x)}{Q(x)},$$

where P_∞, P, and Q are polynomials and the degree of P is less than that of Q.

2) Factor the denominator $Q(x)$ into linear terms and irreducible quadratics. (By a corollary of the fundamental theorem of algebra, such a factorization is always possible.) Hence, we may write

$$Q(x) = c(x - a_1)^{m_1} \cdots (x - a_r)^{m_r} X_1^{n_1} \cdots X_s^{n_s},$$

where $a_1, \ldots, a_r$ are distinct real numbers and $X_1, \ldots, X_s$ are distinct irreducible quadratics with leading coefficients one. (Thus, X_k is of the form $x^2 + 2b_k x + c_k$, where $b_k^2 < c_k$.)

3) Write the "proper part" of $R(x)$, namely, $P(x)/Q(x)$, as a sum of partial fractions with undetermined coefficients.

a) If $(x - a)^n$ is the highest power of $(x - a)$ which divides $Q(x)$, then there correspond n terms in the partial fraction decomposition of $P(x)/Q(x)$:

$$\frac{A_1}{x - a} + \frac{A_2}{(x - a)^2} + \cdots + \frac{A_n}{(x - a)^n}.$$

b) If X^n is the highest power of the irreducible quadratic X which divides $Q(x)$, then there correspond n terms in the decomposition, namely,

$$\frac{A_1 x + B_1}{X} + \frac{A_2 x + B_2}{X^2} + \cdots + \frac{A_n x + B_n}{X^n}.$$

4) Clear of fractions in the above.

5) Solve for the undetermined coefficients.

Example 1. Let us consider the rational function given at the beginning of this section, namely,

$$R(x) = \frac{x^4 - 2x^3 - 3x^2 + 15x - 8}{x^3 - 3x + 2}.$$

1) Since the function is improper, we first perform long division using detached coefficients.

```
                              1 - 2
1 + 0 - 3 + 2 | 1 - 2 - 3 + 15 - 8
                1 + 0 - 3 +  2
               ─────────────────
                - 2 + 0 + 13 - 8
                - 2 + 0 +  6 - 4
               ─────────────────
                        7 - 4
```

Thus

$$R(x) = x - 2 + \frac{7x - 4}{x^3 - 3x + 2}.$$

2) Next we factor the denominator:

$$x^3 - 3x + 2 = (x - 1)^2(x + 2).$$

3) By rule three above, the proper part of $R(x)$ has a decomposition of the form

$$\frac{7x - 4}{x^3 - 3x + 2} = \frac{A}{x - 1} + \frac{B}{(x - 1)^2} + \frac{C}{x + 2}.$$

4) We clear of fractions in the above:

$$7x - 4 = A(x - 1)(x + 2) + B(x + 2) + C(x - 1)^2.$$

5) We shall solve for the undetermined coefficients A, B, and C by two methods.

Method 1. This method is always applicable at this stage although it is rarely the most efficient method. First we collect like powers of x. We have

$$7x - 4 = A(x^2 + x - 2) + B(x + 2) + C(x^2 - 2x + 1)$$
$$= (A + C)x^2 + (A + B - 2C)x + (-2A + 2B + C).$$

Since this equation must hold for all real numbers x, the coefficients of like powers of x must be equal. [More generally, if P and Q are polynomials of degree n or less, and if

$$P(x) = Q(x)$$

for at least $n + 1$ different values for x, then $P(x) = Q(x)$ for all real numbers x, and the coefficients of like powers in P and Q must be equal.] This leads to three simultaneous linear equations for A, B, and C, namely,

$$A + C = 0,$$
$$A + B - 2C = 7,$$
$$-2A + 2B + C = -4.$$

Solving, we get $A = 2$, $B = 1$, and $C = -2$. Thus

$$R(x) = x - 2 + \frac{2}{x - 1} + \frac{1}{(x - 1)^2} - \frac{2}{x + 2}.$$

Method 2. The undetermined coefficients can also be determined by substitutions and divisions. Since the equation

$$7x - 4 = A(x - 1)(x + 2) + B(x + 2) + C(x - 1)^2 \qquad (*)$$

must hold for all real numbers x, it must hold if $x = 1$. This substitution gives $3 = 3B$. Hence $B = 1$. Similarly, the substitution $x = -2$ gives $-18 = 9C$, so that $C = -2$. The remaining coefficient A can be determined in a variety of ways. One of these is to substitute the values for B and C into $(*)$. This gives, upon rearrangement of terms,

$$2x^2 + 2x - 4 = A(x^2 + x - 2),$$

from which we see that $A = 2$.

Example 2. By rule (3), the rational function

$$\frac{x^2 - 3x + 7}{(x - 1)^3(x + 3)^2(x^2 + 1)^2}$$

has a partial fraction decomposition of the form

$$\frac{A}{x - 1} + \frac{B}{(x - 1)^2} + \frac{C}{(x - 1)^3} + \frac{D}{x + 3} + \frac{E}{(x + 3)^2} + \frac{Fx + G}{x^2 + 1} + \frac{Hx + I}{(x^2 + 1)^2}.$$

By clearing of fractions and equating coefficients of like powers, we could obtain a system of nine simultaneous linear equations for the undetermined coefficients $A, \ldots, I$.

Example 3. Consider again the integral $\int dx/(a^2 - x^2)$.
   Since
$$a^2 - x^2 = (a + x)(a - x),$$
the integrand has a partial fraction decomposition of the form
$$\frac{1}{a^2 - x^2} = \frac{A}{a + x} + \frac{B}{a - x}.$$

Clearing of fractions, we get
$$1 = A(a - x) + B(a + x).$$

Setting x equal to $+a$ and $-a$, we get
$$A = B = 1/2a.$$

Thus
$$\frac{1}{a^2 - x^2} = \frac{1}{2a}\left(\frac{1}{a + x} + \frac{1}{a - x}\right).$$

Therefore,
$$\int \frac{dx}{a^2 - x^2} = \frac{1}{2a}(\ln|a + x| - \ln|a - x|)$$
$$= \frac{1}{2a}\ln\left|\frac{a + x}{a - x}\right| + C.$$

Example 4. Find
$$\int \frac{-x^3 + 6x^2 + x + 2}{(x^2 - 1)(x^2 + 1)^2}\,dx.$$

Since the integrand is a proper fraction, we can immediately write a partial fraction decomposition with undetermined coefficients:
$$\frac{-x^3 + 6x^2 + x + 2}{(x^2 - 1)(x^2 + 1)^2} = \frac{A}{x - 1} + \frac{B}{x + 1} + \frac{Cx + D}{x^2 + 1} + \frac{Ex + F}{(x^2 + 1)^2}.$$

We clear of fractions, thus obtaining
$$-x^3 + 6x^2 + x + 2 = A(x + 1)(x^2 + 1)^2 + B(x - 1)(x^2 + 1)^2$$
$$+ (Cx + D)(x^2 - 1)(x^2 + 1) + (Ex + F)(x^2 - 1).$$

Setting x equal to $+1$ and -1, we get, respectively,
$$8 = 8A, \qquad A = 1$$
and
$$8 = -8B, \qquad B = -1.$$

We now substitute these values for A and B. After a little algebra, we get
$$-2x^4 - x^3 + 2x^2 - x = (Cx + D)(x^2 - 1)(x^2 + 1) + (Ex + F)(x^2 - 1).$$

Dividing by $x^2 - 1$, we get
$$-2x^2 - x = (Cx + D)(x^2 + 1) + (Ex + F).$$

Clearly, $Cx + D$ and $Ex + F$ must be the quotient and remainder, respectively, of $-2x^2 - x$ divided by $x^2 + 1$.

$$\begin{array}{r} -2 \\ x^2 + 1 \overline{\smash{\big)} -2x^2 - x} \\ \underline{-2x^2 \qquad - 2} \\ - x + 2 \end{array}$$

Thus $C = 0$, $D = -2$, $E = -1$, and $F = 2$. Hence

$$R(x) = \frac{-x^3 + 6x^2 + x + 2}{(x^2 - 1)(x^2 + 1)^2} = \frac{1}{x - 1} - \frac{1}{x + 1} - \frac{2}{x^2 + 1} - \frac{x - 2}{(x^2 + 1)^2}.$$

By the reduction formula from the preceding section,

$$\int \frac{dx}{(x^2 + 1)^2} = \frac{x}{2(x^2 + 1)} + \frac{1}{2} \int \frac{dx}{x^2 + 1} = \frac{x}{2(x^2 + 1)} + \tfrac{1}{2} \operatorname{Tan}^{-1} x.$$

Hence

$$\int R(x)\,dx = \ln \left| \frac{x - 1}{x + 1} \right| - \operatorname{Tan}^{-1} x - \tfrac{1}{2} \ln (x^2 + 1) + \frac{x}{x^2 + 1}.$$

Example 5. Find $\int dx/(x^4 + 1)$.

The denominator factors into two irreducible quadratics:

$$x^4 + 1 = (x^2 + \sqrt{2}\,x + 1)(x^2 - \sqrt{2}\,x + 1).$$

Thus we may write

$$\frac{1}{x^4 + 1} = \frac{Ax + B}{x^2 + \sqrt{2}\,x + 1} + \frac{Cx + D}{x^2 - \sqrt{2}\,x + 1}.$$

Clearing of fractions and equating coefficients of like powers yield the equations

$$A + C = 0,$$
$$-\sqrt{2}\,A + B + \sqrt{2}\,C + D = 0,$$
$$A - \sqrt{2}\,B + C + \sqrt{2}\,D = 0,$$
$$B + D = 1.$$

Solving, we get $A = 1/2\sqrt{2}$, $B = -\tfrac{1}{2}$, $C = -1/2\sqrt{2}$, $D = \tfrac{1}{2}$. Thus

$$\frac{1}{x^4 + 1} = \frac{x - \sqrt{2}}{2\sqrt{2}(x^2 + \sqrt{2}\,x + 1)} - \frac{x - \sqrt{2}}{2\sqrt{2}(x^2 - \sqrt{2}\,x + 1)}.$$

After completing the squares on the quadratics, making substitutions, and dealing with radicals, we get finally

$$\int \frac{dx}{x^4 + 1} = \frac{1}{4\sqrt{2}} \ln \frac{x^2 + \sqrt{2}\,x + 1}{x^2 - \sqrt{2}\,x + 1} + \frac{1}{2\sqrt{2}} \operatorname{Tan}^{-1} (\sqrt{2}\,x + 1)$$

$$+ \frac{1}{2\sqrt{2}} \operatorname{Tan}^{-1} (\sqrt{2}\,x - 1) + C.$$

Exercises

1. Calculate each of the following integrals.

a) $\displaystyle\int \frac{1+x^2}{1+x}\,dx$

b) $\displaystyle\int \frac{x^3\,dx}{x^2+x+1}$

c) $\displaystyle\int \frac{x\,dx}{x^2-1}$

d) $\displaystyle\int \frac{x^2+x+1}{(x-1)(x-2)(x-3)}\,dx$

e) $\displaystyle\int \frac{x^3}{x^2-2x-3}\,dx$

f) $\displaystyle\int \frac{x^2+x+1}{(2x+1)(x^2+1)}\,dx$

g) $\displaystyle\int \frac{x^2-x-8}{(2x-3)(x^2+2x+2)}\,dx$

h) $\displaystyle\int \frac{x\,dx}{(x-1)^2}$

i) $\displaystyle\int \frac{dx}{x^3+1}$

j) $\displaystyle\int \frac{2x^2+1}{(x-2)^3}\,dx$

k) $\displaystyle\int \frac{x^4+4x-16}{(2-x)^2(4+x^2)}\,dx$

l) $\displaystyle\int \frac{x^3-3x^2+2x-3}{(x^2+1)^2}\,dx$

m) $\displaystyle\int \frac{x^4+x^3+18x^2+10x+81}{(x^2+9)^3}\,dx$

n) $\displaystyle\int \frac{dx}{(x+1)^2(x-2)^3}$

o) $\displaystyle\int \frac{dx}{x^2(x^2+1)^2}$

p) $\displaystyle\int \frac{x+2}{(x^2-1)(x^2+1)^2}\,dx$

2. a) Suppose that P is a polynomial of degree less than n. Show that $P(x)/(x-a)^n$ has the partial fraction decomposition

$$\frac{P(x)}{(x-a)^n} = \frac{A_1}{x-a} + \frac{A_2}{(x-a)^2} + \cdots + \frac{A_n}{(x-a)^n},$$

where the coefficients $A_n, A_{n-1}, \ldots, A_1$ are the remainders obtained upon successive divisions of $P(x)$ by $x-a$. Use this result to calculate the integral in Exercise 1(j).

b) Obtain a similar result for irreducible quadratics. Compute again the integral in Exercise 1(l).

3. a) Explain why a polynomial of degree n can have at most n roots ($n \geq 1$).

b) Suppose that P and Q are polynomials of degree $n-1$ or less. Suppose that there exist n distinct numbers $a_1, \ldots, a_n$ such that $P(a_k) = Q(a_k)$, for $k = 1, \ldots, n$. Prove that P and Q must have the same coefficients [and hence $P(x) = Q(x)$ for every number x].

c) Let $a_1, \ldots, a_n$ be distinct real numbers, and let P be a polynomial of degree $n-1$ at most. Prove that

$$\frac{P(x)}{(x-a_1)(x-a_2)\cdots(x-a_n)} = \sum_{k=1}^{n} \frac{C_k}{x-a_k},$$

where

$$C_k = \frac{P(a_k)}{\prod_{j=1,\,j\neq k}^{n} (a_k - a_j)}.$$

This establishes the alogorithm for finding partial fraction decompositions for the case in which the denominator splits into distinct linear factors with real coefficients.

10-4 Some standard substitutions

An integrand can often be transformed into a rational function by a suitable substitution. This is true, for instance, for integrands involving the trigonometric functions. More specifically, consider an integral of the form

$$\int R(\cos\theta, \sin\theta)\, d\theta,$$

where $R(x, y)$ is a rational function in x and y. [That is, $R(x, y)$ is a quotient of two
polynomials in x and y. For example, $R(x, y)$ might be $(x^5 + x^2y^3 + y^4)/(x^2 + y^2)$.]
If we are only concerned with values of θ in the interval $(-\pi/2, \pi/2)$, then the integral
can be rationalized by the substitution

$$t = \tan(\theta/2).$$

For then we have

$$\cos\theta = \frac{1 - t^2}{1 + t^2},$$

$$\sin\theta = \frac{2t}{1 + t^2},$$

$$d\theta = \frac{2\, dt}{1 + t^2}.$$

Consequently,

$$\int R(\cos\theta, \sin\theta)\, d\theta = \int \frac{2}{1 + t^2} R\left(\frac{1 - t^2}{1 + t^2}, \frac{2t}{1 + t^2}\right) dt.$$

The second integral can be calculated by the methods of the preceding section, and
the resulting antiderivative can then be expressed in terms of $\sin\theta$ and $\cos\theta$, since

$$t = \frac{\sin\theta}{1 + \cos\theta} = \frac{1 - \cos\theta}{\sin\theta}.$$

Although the relevant identities have been derived previously, we shall derive
them once again. Let $P = (\cos\theta, \sin\theta)$, where $-\pi/2 < \theta < \pi/2$ (see Fig. 10–1).
Then $t = \tan(\theta/2)$ is the slope of the line joining $(-1, 0)$ and P, and thus we have
$t = \sin\theta/(1 + \cos\theta)$. Now

$$1 + t^2 = 1 + \frac{\sin^2\theta}{(1 + \cos\theta)^2}$$

$$= \frac{1 + 2\cos\theta + \cos^2\theta + \sin^2\theta}{(1 + \cos\theta)^2}$$

$$= \frac{2 + 2\cos\theta}{(1 + \cos\theta)^2} = \frac{2}{1 + \cos\theta}.$$

Solving the latter equation for $\cos\theta$, we get

$$\cos\theta = \frac{1 - t^2}{1 + t^2}.$$

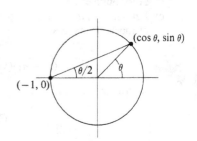

FIG. 10–1

Then

$$\sin \theta = (1 + \cos \theta)t = \frac{2t}{1 + t^2} \cdot$$

Finally, taking differentials of both sides of the equation

$$1 + t^2 = \frac{2}{1 + \cos \theta},$$

we get

$$2t \, dt = \frac{2 \sin \theta \, d\theta}{(1 + \cos \theta)^2} = t(1 + t^2) \, d\theta,$$

and so

$$d\theta = \frac{2 \, dt}{1 + t^2} \cdot$$

Example 1. Find $I = \int dx/(a + b \cos x)$.

Using the substitution, we get

$$I = \int \frac{dt}{a(1 + t^2) + b(1 - t^2)} \cdot$$

We distinguish four cases.

Case I. $a = b$. In this case,

$$I = \frac{1}{a} \int \frac{dx}{1 + \cos x} = \frac{1}{a} \int dt = \frac{1}{a} \tan \frac{x}{2} \cdot$$

Case II. $a = -b$. In this case, we get

$$I = \frac{1}{a} \int \frac{dx}{1 - \cos x} = \frac{1}{a} \int \frac{dt}{t^2} = -\frac{1}{a} \operatorname{ctn} \frac{x}{2} + C.$$

Case III. $|a| > |b|$. Then $(a - b)(a + b) = a^2 - b^2 > 0$. This means that $(a - b)$ and $(a + b)$ are of the same sign, and thus $(a + b)/(a - b) > 0$. Now

$$
\begin{aligned}
I &= \frac{2}{a - b} \int \frac{dt}{(a + b)/(a - b) + t^2} \\
&= \frac{2}{a - b} \sqrt{\frac{a - b}{a + b}} \operatorname{Tan}^{-1} \left(\sqrt{\frac{a - b}{a + b}} \, t \right) + C \\
&= \frac{2}{\sqrt{a^2 - b^2}} \operatorname{Tan}^{-1} \left(\sqrt{\frac{a - b}{a + b}} \tan \frac{x}{2} \right) + C.
\end{aligned}
$$

Case IV. $|a| < |b|$. Then $(b + a)/(b - a) > 0$ and

$$
\begin{aligned}
I &= \frac{2}{b - a} \int \frac{dt}{(b + a)/(b - a) - t^2} \\
&= \frac{1}{b - a} \sqrt{\frac{b - a}{b + a}} \ln \left| \frac{\sqrt{(b - a)/(b + a)} + t}{\sqrt{(b - a)/(b + a)} - t} \right| + C \\
&= \frac{1}{\sqrt{b^2 - a^2}} \ln \left| \frac{\sqrt{|b + a|} + \sqrt{|b - a|} \tan (x/2)}{\sqrt{|b + a|} - \sqrt{|b - a|} \tan (x/2)} \right| + C.
\end{aligned}
$$

Example 2. Consider once again $I = \int \sec x \, dx$.

By the substitution above, the integral becomes

$$I = \int \frac{1 + t^2}{1 - t^2} \frac{2 \, dt}{1 + t^2} = \int \frac{2 \, dt}{1 - t^2}$$

$$= \ln \left| \frac{1 + t}{1 - t} \right|$$

$$= \ln \left| \frac{\tan (\pi/4) + \tan (x/2)}{1 - \tan (\pi/4) \tan (x/2)} \right|$$

$$= \ln \left| \tan \left(\frac{\pi}{4} + \frac{x}{2} \right) \right| + C.$$

Although the substitution $t = \tan (x/2)$ can always be used to integrate a rational function of $\sin x$ and $\cos x$, it is not always the most efficient technique. For example, integrals of the type

$$\int \cos^m x \sin^m x \, dx$$

are better handled by the methods of the preceding sections.

Example 3. Find $I = \int dx/(a^2 \sin^2 x + b^2 \cos^2 x)$.

Although the substitution $t = \tan (x/2)$ can be used, the substitution $t = \tan x$ is better in this case. Substituting, we get

$$I = \frac{1}{a^2} \int \frac{1}{\tan^2 x + b^2/a^2} \frac{dx}{\cos^2 x}$$

$$= \frac{1}{b^2} \int \frac{1}{t^2 + (b/a)^2} \, dt$$

$$= \frac{1}{a^2} \frac{a}{b} \operatorname{Tan}^{-1} \left(\frac{a}{b} t \right)$$

$$= \frac{1}{ab} \operatorname{Tan}^{-1} \left(\frac{a}{b} \tan x \right).$$

(We are assuming that $a > 0$ and $b > 0$.)

Rational functions of $\sinh x$ and $\cosh x$ can be handled in an analogous way. We shall leave to the reader the verification of the following relationships:

$$t = \tanh \frac{x}{2} = \frac{\sinh x}{\cosh x + 1}, \qquad \cosh x = \frac{1 + t^2}{1 - t^2},$$

$$\sinh x = \frac{2t}{1 - t^2}, \qquad dx = \frac{2 \, dt}{1 - t^2}.$$

Note, however, that any rational function of $\sinh x$ and $\cosh x$ is a rational function of e^x, and that any such function can be integrated by use of the substitution $u = e^x$. Thus

$$\int R(e^x) \, dx = \int R(u) \frac{du}{u}.$$

Example 4. Find $I = \int \mathrm{sech}\, x \, dx = \int dx/\cosh x$.

By making the substitution $t = \tanh (x/2)$, we get

$$I = \int \frac{1 - t^2}{1 + t^2} \frac{2\, dt}{1 - t^2}$$

$$= 2\, \mathrm{Tan}^{-1}\, t + C = 2\, \mathrm{Tan}^{-1} [\tanh (x/2)] + C.$$

If we make the substitution $u = e^x$, we get

$$I = \int \frac{2\, dx}{e^x + e^{-x}} = \int \frac{2e^x\, dx}{e^{2x} + 1} = \int \frac{2\, du}{u^2 + 1}$$

$$= 2\, \mathrm{Tan}^{-1}\, u + C = 2\, \mathrm{Tan}^{-1}\, e^x + C.$$

This means, of course, that $\mathrm{Tan}^{-1} [\tanh (x/2)]$ and $\mathrm{Tan}^{-1}\, e^x$ must differ by a constant. This can be verified directly in the following manner:

$$\mathrm{Tan}^{-1} \left(\tanh \frac{x}{2} \right) = \mathrm{Tan}^{-1} \frac{e^{x/2} - e^{-x/2}}{e^{x/2} + e^{-x/2}} = \mathrm{Tan}^{-1} \frac{e^x + (-1)}{1 - e^x(-1)}$$

$$= \mathrm{Tan}^{-1}\, e^x + \mathrm{Tan}^{-1}\, (-1) = \mathrm{Tan}^{-1}\, e^x - \pi/4.$$

Another large class of integrals which can be expressed in terms of the elementary functions are integrals of the form

$$\int R(x, \sqrt{Q(x)})\, dx,$$

where $R(x, y)$ is a rational function of two variables and $Q(x)$ is a quadratic polynomial. By completing the square on $Q(x)$ and making a linear substitution, we can obtain one of the following three forms for the above integral:

$$\int S(u, \sqrt{a^2 - u^2})\, du, \qquad \int S(u, \sqrt{a^2 + u^2})\, du, \qquad \int S(u, \sqrt{u^2 - a^2})\, du.$$

Here S denotes a rational function of two variables. We can obtain the form $\int R^*(\cos \theta, \sin \theta)\, d\theta$ for these integrals by using the trigonometric substitutions

$$u = a\, \mathrm{Sin}\, \theta, \qquad u = a\, \mathrm{Tan}\, \theta, \qquad u = a\, \sec \theta,$$

respectively.

Example 5. Find $I = \int dx/(1 + \sqrt{x^2 + 2x + 2})$.

Since $x^2 + 2x + 2 = (x + 1)^2 + 1$, the substitution $u = x + 1$ yields

$$I = \int \frac{du}{1 + \sqrt{1 + u^2}}.$$

We let $u = \mathrm{Tan}\, \theta$. Then

$$I = \int \frac{\sec^2 \theta\, d\theta}{1 + \sec \theta}.$$

Now

$$\frac{\sec^2 \theta}{1 + \sec \theta} = \frac{1}{\cos \theta(1 + \cos \theta)} = \frac{1}{\cos \theta} - \frac{1}{1 + \cos \theta} = \sec \theta - \frac{1}{1 + \cos \theta}.$$

By previous examples,

$$I = \ln (\sec \theta + \tan \theta) - \tan (\theta/2) + C$$

$$= \ln (\sec \theta + \tan \theta) - \frac{1 - \cos \theta}{\sin \theta} + C.$$

(Note that $\sec \theta + \tan \theta$ is positive if $-\pi/2 < \theta < \pi/2$.) We can express I in terms of u by referring to the triangle in Fig. 10-2. Thus

$$I = \ln (\sqrt{1 + u^2} + u) - \frac{\sqrt{1 + u^2}}{u} + \frac{1}{u} + C$$

$$= \frac{1 - \sqrt{x^2 + 2x + 2}}{x + 1} + \ln (x + 1 + \sqrt{x^2 + 2x + 2}) + C.$$

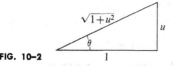

FIG. 10-2

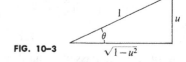

FIG. 10-3

Example 6. Find $I = \int du/(u^2\sqrt{1 - u^2})$.

We set $u = \sin \theta$. Since we shall only be considering values of θ in the interval $(-\pi/2, \pi/2)$, and since $\cos \theta$ is positive in this interval, we can replace $\sqrt{1 - u^2}$ with $\cos \theta$ (see Fig. 10-3). Also, $du = \cos \theta \, d\theta$. Hence

$$I = \int \frac{\cos \theta \, d\theta}{\sin^2 \theta \cos \theta} = \int \csc^2 \theta \, d\theta = -\operatorname{ctn} \theta + C = -\frac{\sqrt{1 - u^2}}{u} + C.$$

In the last two examples we have illustrated the use of the trigonometric substitutions $u = a \sin \theta$ and $u = a \tan \theta$. The third substitution $u = a \sec \theta$ is somewhat more delicate. Before discussing it, let us consider the rationale behind the calculations in these last two examples. (After all, we have been rather free in our use of differentials, etc.) In calculating an integral of the form $\int f(u) \, du$, we have replaced u by an expression of the form $g(\theta)$, and we have subsequently replaced the integral $\int f(u) \, du$ by $\int f(g(\theta))g'(\theta) \, d\theta$. We shall indicate this schematically as follows:

$$\int f(u) \, du$$

$$\Big\downarrow u \to g(\theta)$$

$$\int f(g(\theta))g'(\theta) \, d\theta$$

In each instance g was a differentiable and invertible function. We then calculated $\int f(g(\theta))g'(\theta) \, d\theta$ to get, say, $G(\theta) + C$. Finally, we replaced θ with $g^{-1}(u)$ and concluded that

$$\int f(u) \, du = G(g^{-1}(u)) + C.$$

Hence the whole scheme of the calculation may be indicated schematically as follows:

$$\int f(u)\, du \quad = \quad G(g^{-1}(u)) + C$$

$$\downarrow u \to g(\theta) \qquad\qquad \uparrow \theta \to g^{-1}(u)$$

$$\int f(g(\theta))g'(\theta)\, d\theta \quad = \quad G(\theta) + C$$

The justification of this procedure requires that we be able to prove the following:

$$G' = (f \circ g)g' \Rightarrow (G \circ g^{-1})' = f.$$

As one might expect, the key to the proof is the chain rule. To simplify notation, we set $h = g^{-1}$. Then by the chain rule,

$$(G \circ g^{-1})' = (G \circ h)' = (G' \circ h)h'.$$

Now

$$G' \circ h = [(f \circ g)g'] \circ h = (f \circ g \circ h)(g' \circ h)$$
$$= f(g' \circ h)$$

and

$$h' = \frac{1}{g' \circ h}.$$

Thus $(G \circ g^{-1})' = f$, as asserted.

Consider now an integral of the form

$$\int S(u, \sqrt{u^2 - a^2})\, du,$$

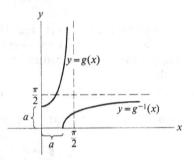

FIG. 10-4

where S is a rational function of two variables. Since the domain of the integrand is a subset of $(-\infty, -a] \cup [a, +\infty)$, we shall treat separately cases in which $u \geq a$ and $u \leq -a$. When $u \geq a$, we let g be the restriction of the function $a \sec \theta$ to the interval $[0, \pi/2)$. Then g is a strictly increasing (and hence invertible) function mapping $[0, \pi/2]$ onto the interval $[a, +\infty)$ (see Fig. 10-4). The inverse function g^{-1} is differentiable except at a. When we make the substitution $u = g(\theta)$, we can then replace $\sqrt{u^2 - a^2}$ with $a \tan \theta$, since the tangent function is nonnegative in the interval $[0, \pi/2)$. The original integral then becomes

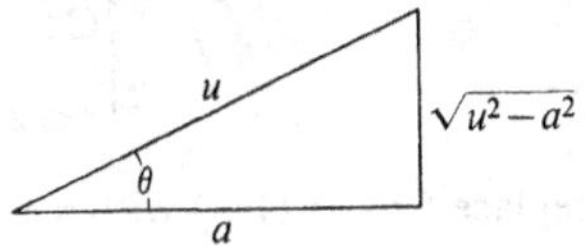

$$a \int S(a \sec \theta, a \tan \theta) \sec \theta \tan \theta\, d\theta,$$

FIG. 10-5

where $0 \leq \theta < \pi/2$. After computing this integral, we can carry out the inverse substitution $\theta \to g^{-1}(u)$ with the aid of the reference triangle given in Fig. 10-5. [So, for example, $\sin \theta$ can be replaced with $\sqrt{u^2 - a^2}/u$ and θ can be replaced with $\text{Cos}^{-1}(a/u)$, etc.] When $u \leq -a$, we can take g to be the restriction of $a \sec \theta$ to

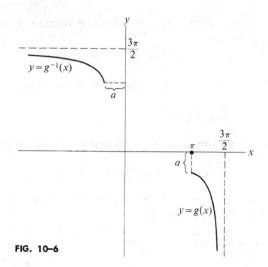

FIG. 10-6

the interval $[\pi, 3\pi/2)$ (see Fig. 10-6). Since the tangent function is positive on this interval, the substitution $u \to g(\theta)$ yields the same result as in the first case. In carrying out the inverse substitution $\theta \to g^{-1}(u)$, we make just one change: we replace θ with

$$g^{-1}(u) = 2\pi - \text{Cos}^{-1}(a/u) = 3\pi/2 + \text{Sin}^{-1}(a/u)$$

rather than $\text{Cos}^{-1}(a/u)$. Otherwise, we have, as in the first case,

$$\sin\theta \to \frac{\sqrt{u^2 - a^2}}{u}, \qquad \text{ctn}\,\theta \to \frac{a}{\sqrt{u^2 - a^2}},$$

etc., so the same reference triangle may be used.

Example 7. Find $I = \int dx/(u\sqrt{u^2 - 1})$.

Using the substitutions just discussed, we transform the integral into

$$\int \frac{\sec\theta\tan\theta\,d\theta}{\sec\theta\tan\theta} = \theta + C.$$

Thus

$$I = \begin{cases} \text{Cos}^{-1}(1/u) + C & \text{if } u > 1, \\ 3\pi/2 + \text{Sin}^{-1}(1/u) + C & \text{if } u < -1. \end{cases}$$

We replace $\text{Cos}^{-1}(1/u)$ with $\pi/2 - \text{Sin}^{-1}(1/u)$ and absorb $3\pi/2$ into C. Thus

$$I = \begin{cases} -\text{Sin}^{-1}(1/u) + C & \text{if } u > 1, \\ \text{Sin}^{-1}(1/u) + C & \text{if } u < -1. \end{cases}$$

A more conventional way to write this is

$$\int \frac{du}{u\sqrt{u^2 - 1}} = -\text{Sin}^{-1}\frac{1}{|u|} + C.$$

Example 8. Find $I = \int \sqrt{x^2 - 2x - 1}\, dx$.

The substitution $u = x - 1$ transforms I into $\int \sqrt{u^2 - 2}\, du$. We let $u = \sqrt{2} \sec \theta$ (with the understanding that we restrict the secant function to the intervals $[0, \pi/2)$ and $[\pi, 3\pi/2)$ according as $u \geq \sqrt{2}$ or $u \leq -\sqrt{2}$.) We get

$$I = \int (\sqrt{2} \tan \theta)(\sqrt{2} \sec \theta \tan \theta)\, d\theta = 2 \int (\sec^3 \theta - \sec \theta)\, d\theta.$$

We can calculate $\int \sec^3 \theta\, d\theta$ by using the reduction formula

$$\int \sec^n \theta\, d\theta = \frac{1}{n-1} \sin \theta \sec^{n-1} \theta + \frac{n-2}{n-1} \int \sec^{n-1} \theta\, d\theta,$$

established in a previous exercise. Thus

$$\int \sec^3 \theta\, d\theta = \tfrac{1}{2} \sin \theta \sec^2 \theta + \tfrac{1}{2} \int \sec \theta\, d\theta.$$

Hence

$$I = \sin \theta \sec^2 \theta - \int \sec \theta\, d\theta = \sin \theta \sec^2 \theta - \ln |\sec \theta + \tan \theta| + C$$

$$= \frac{\sqrt{u^2 - 2}}{u} \frac{u^2}{2} - \ln \left| \frac{u}{\sqrt{2}} + \frac{\sqrt{u^2 - 2}}{\sqrt{2}} \right| + C$$

$$= \tfrac{1}{2}(x-1)\sqrt{x^2 - 2x - 1} - \ln |x - 1 + \sqrt{x^2 - 2x - 2}| + C.$$

Exercises

1. Find $\int dx/(a + b \sin x)$ for all possible values of a and b.
2. Find $\int dx/(a^2 \sin^2 x - b^2 \cos^2 x)$.
3. Find $\int dx/(1 + \cos^2 x)$.
4. a) Show that $a \sin x + b \cos x = r \sin (x + \theta)$, where $[r; \theta]$ are the polar coordinates of the point (a, b).
   b) Deduce that
   $$\int \frac{dx}{a \sin x + b \sin x} = \frac{1}{r} \ln \left| \tan \frac{x + \theta}{2} \right| + C.$$
   c) Find $\int dx/(\sin x - \cos x)$.
5. Compute the following integrals by any convenient method.

a) $\displaystyle \int \frac{dx}{1 + 2 \tan x}$

b) $\displaystyle \int \frac{\sin x - \cos x}{\sin x + \cos x}\, dx$

c) $\displaystyle \int \frac{\sqrt{\tan x}\, dx}{\sin x \cos x}$

d) $\displaystyle \int \frac{\cos 2x}{\sin^4 x}\, dx$

e) $\displaystyle \int \sqrt{\frac{\sin^3 x}{\cos^7 x}}\, dx$

f) $\displaystyle \int \tan^5 x\, dx$

g) $\displaystyle \int \frac{2 - \sin x}{2 + \cos x}\, dx$

h) $\displaystyle \int \frac{\sin^2 x\, dx}{1 - \tan x}$

i) $\displaystyle \int \frac{dx}{1 + \sin^2 x}$

j) $\displaystyle \int \frac{dx}{(\sin x + \cos x)^2}$

k) $\displaystyle \int \frac{dx}{\sin^2 x + \tan^2 x}$

l) $\displaystyle \int \frac{\cos x\, dx}{(1 - \cos x)^2}$

m) $\displaystyle \int \sqrt{1 + \sin x}\, dx$

n) $\displaystyle \int \sqrt{\tan x}\, dx$

o) $\displaystyle \int \frac{dx}{\sqrt{1 - \sin^4 x}}$

6. Compute each of the following by any convenient method.

a) $\displaystyle\int \frac{dx}{\cosh^2 x}$

b) $\displaystyle\int \sinh^2 x\, dx$

c) $\displaystyle\int \frac{dx}{\sinh x}$

d) $\displaystyle\int \sinh^2 x \cosh^3 x\, dx$

e) $\displaystyle\int \frac{dx}{(1 - \cosh x)^2}$

f) $\displaystyle\int \frac{x\, dx}{\cosh^2 x}$

g) $\displaystyle\int \frac{x + \sinh x}{\cosh x - \sinh x}\, dx$

h) $\displaystyle\int \sqrt{\tanh x}\, dx$

7. The *Gudermannian* function may be defined to be $\mathrm{Sin}^{-1} \circ \tanh$; it is denoted by gd. Thus,

$$\mathrm{gd}\, x = \mathrm{Sin}^{-1}(\tanh x).$$

a) Show that the domain of gd is the entire line and that the function is strictly increasing. Also, verify the identities

$$\cos(\mathrm{gd}\, x) = \mathrm{sech}\, x, \qquad \tan(\mathrm{gd}\, x) = \sinh x, \qquad \sec(\mathrm{gd}\, x) = \cosh x.$$

b) Show that

$$\tan\left(\frac{\pi}{4} + \frac{\theta}{2}\right) = \sec\theta + \tan\theta,$$

and then show that

$$\mathrm{gd}^{-1}\theta = \ln\tan\left(\frac{\pi}{4} + \frac{\theta}{2}\right).$$

8. Compute the following integrals.

a) $\displaystyle\int \frac{x^2\, dx}{(a^2 - x^2)^{3/2}}$

b) $\displaystyle\int \frac{x\, dx}{(1 - x^2)\sqrt{1 + x^2}}$

c) $\displaystyle\int \frac{dx}{x\sqrt{2 + x - x^2}}$

d) $\displaystyle\int \frac{\sqrt{2x + x^2}}{x^2}\, dx$

e) $\displaystyle\int \sqrt{a^2 - u^2}\, du$

f) $\displaystyle\int \frac{dx}{x^2(x + \sqrt{1 + x^2})}$

g) $\displaystyle\int \frac{dx}{1 + \sqrt{x^2 + 2x + 2}}$

h) $\displaystyle\int \frac{dx}{x\sqrt{x^2 + 4x - 4}}$

i) $\displaystyle\frac{x^2\, dx}{\sqrt{1 - 2x - x^2}}$

j) $\displaystyle\int \frac{\sqrt{1 + x^2}}{2 + x^2}\, dx$

9. Can the hyperbolic functions be used to compute integrals of the form $\int R(x, \sqrt{Q(x)})\, dx$? Discuss in full and illustrate.

10. Discuss the use of the substitution $u = 1/x$ in evaluating integrals of the form $\int R(u, \sqrt{u^2 - a^2})\, du$, where R is a rational function of two variables.

11. Consider an integral of the form

$$\int R(x, (ax + b)^\alpha)\, dx,$$

where R is a rational function of two variables, a and b are constants, and α is a positive rational number, say, $\alpha = m/n$ $(m, n \in \mathbf{Z}^+)$. Show that the substitution

$$u^n = ax + b$$

transforms the integral into the integral of a rational function. In this way, find the following.

a) $\int x\sqrt{a + x}\,dx$ b) $\int \dfrac{dx}{\sqrt{x}\,(x - 1)}$ *c) $\int \dfrac{dx}{x^3\sqrt{1 - x}}$

12. Consider integrals of the form

$$\int R(x, (ax + b)^{\alpha_1}, \ldots, (ax + b)^{\alpha_n}),$$

where R is a rational function of $n + 1$ variables and $\alpha_1, \ldots, \alpha_n$ are positive rational numbers. Show that the integral can be rationalized by the substitution $u^r = ax + b$, where r is the least common denominator of $\alpha_1, \ldots, \alpha_n$. Calculate the following integrals.

a) $\int \dfrac{dx}{x^{2/3} - x^{1/2}}$ b) $\int \dfrac{x^2 + \sqrt{1 + x}}{\sqrt[3]{1 + x}}\,dx$ c) $\int \dfrac{x\,dx}{\sqrt{x + 1} + \sqrt[3]{x + 1}}$

13. Consider integrals of the form

$$\int R\left[x, \left(\dfrac{ax + b}{cx + d}\right)^\alpha\right],$$

where R is a rational function, a, b, c, and d are constants, and α is a positive rational number, say, $\alpha = (m/n)$ $(m, n \in \mathbf{Z}^+)$. Show that the integral can be rationalized by the substitution

$$u^n = \dfrac{ax + b}{cx + d}.$$

Find

a) $\int \sqrt{\dfrac{1 - x}{1 + x}}\dfrac{dx}{x}$,   b) $\int \sqrt[3]{\dfrac{1 - x}{1 + x}}\dfrac{dx}{x}$,   c) $\int \sqrt{\dfrac{1 + x}{x}}\,dx$.

10-5 Wallis' product and Stirling's formula

Our aim in this section is to present two beautiful and remarkable results. The first of these, Wallis' formula, gives a simple infinite product expansion for π, namely,

$$\frac{\pi}{2} = \frac{2}{1}\cdot\frac{2}{3}\cdot\frac{4}{3}\cdot\frac{4}{5}\cdot\frac{6}{5}\cdots$$

The second, Stirling's formula, gives a widely used asymptotic estimate for $n!$, namely,

$$n! \sim \sqrt{2\pi n}\,n^n e^{-n}.$$

The right-hand side, in contrast to the left-hand side, can be easily calculated for large values of n with the aid of logarithms.

We begin the derivation of Wallis' formula with the recursion formula

$$\int \sin^n x \, dx = -\frac{1}{n} \sin^{n-1} x \cos x + \frac{n-1}{n} \int \sin^{n-2} x \, dx.$$

It follows that if $n > 1$, then

$$I_n = \int_0^{\pi/2} \sin^n x \, dx = \frac{n-1}{n} I_{n-2}. \qquad (*)$$

This second recursion formula and the observation that

$$I_0 = \int_0^{\pi/2} dx = \pi/2 \qquad \text{and} \qquad I_1 = \int_0^{\pi/2} \sin x \, dx = 1$$

together imply that

$$I_{2n} = \frac{2n-1}{2n} \cdot \frac{2n-3}{2n-2} \cdots \frac{1}{2} \cdot \frac{\pi}{2},$$

whereas

$$I_{2n+1} = \frac{2n}{2n+1} \cdot \frac{2n-2}{2n-1} \cdots \frac{2}{3}.$$

Throughout the interval $[0, \pi/2]$, we note that $0 \le \sin x \le 1$, so

$$0 \le \sin^{n-1} x \le \sin^n x.$$

By the monotone property of definite integrals it follows that $I_1, I_2, \ldots$ is a non-increasing sequence of positive numbers. Hence

$$1 \le \frac{I_{2n}}{I_{2n+1}} \le \frac{I_{2n-1}}{I_{2n+1}} = 1 + \frac{1}{2n}.$$

By the pinching theorem,

$$\lim_{n \to \infty} \frac{I_{2n}}{I_{2n+1}} = 1.$$

We note, however, that I_{2n}/I_{2n+1} can be written as $(1/P_n)(\pi/2)$, where

$$P_n = \frac{2}{1} \cdot \frac{2}{3} \cdot \frac{4}{3} \cdot \frac{4}{5} \cdots \frac{2n}{2n-1} \cdot \frac{2n}{2n+1}.$$

It immediately follows that

$$\lim_{n \to \infty} P_n = \frac{\pi}{2},$$

which is what we shall adopt as the meaning of Wallis' formula,

$$\frac{\pi}{2} = \frac{2}{1} \cdot \frac{2}{3} \cdot \frac{4}{3} \cdot \frac{4}{5} \cdots$$

As a corollary of Wallis' formula, we have the following:

$$\lim_{n \to \infty} \frac{(n!)^2 2^{2n}}{(2n!)\sqrt{n}} = \sqrt{\pi}.$$

To see why this is so, multiply and divide the right-hand side of the equation

$$P_n = \frac{2}{1} \cdot \frac{2}{3} \cdot \frac{4}{3} \cdot \frac{4}{5} \cdots \frac{2n}{2n-1} \cdot \frac{2n}{2n+1}$$

by $2 \cdot 2 \cdot 4 \cdot 4 \cdots (2n) \cdot (2n)$. This yields

$$P_n = \frac{(n!)^4 2^{4n}}{[(2n)!]^2(2n+1)}.$$

Thus

$$\lim_{n \to \infty} \frac{(n!)^2 2^{2n}}{(2n)!\sqrt{n}} = \lim_{n \to \infty} \sqrt{P_n} \cdot \sqrt{2 + \frac{1}{n}} = \sqrt{\frac{\pi}{2}} \cdot \sqrt{2} = \sqrt{\pi}.$$

The derivation of Stirling's formula also involves some nonstandard applications of integration. We consider the area A_n of the region bounded by the graph $y = \ln x$, the x-axis, and the line $x = n$, where $n \in \mathbf{Z}^+$ (see Fig. 10–7). We have

$$A_n = \int_1^n \ln x \, dx = (x \ln x - x)\Big|_1^n = n \ln n - n + 1.$$

We can obtain a lower estimate T_n of A_n by adding the areas of $n-1$ trapezoids; the kth trapezoid is indicated in Fig. 10–8. Thus

$$T_n = \sum_{k=1}^{n-1} \tfrac{1}{2}[\ln k + \ln (k+1)]$$
$$= \ln 1 + \ln 2 + \cdots + \ln n - \tfrac{1}{2} \ln n$$
$$= \ln n! - \tfrac{1}{2} \ln n.$$

The fact that $T_n < A_n$ stems from the fact that $\ln x$ is concave and that its graph therefore lies above each chord joining any two of its points. It would seem intuitively plausible that as $n \to \infty$, T_n and A_n have the same order of magnitude. By exponentiating, we would find that $n!$ has the same order of magnitude as $n^{n+1/2}e^{-n}$, which is one of the implications of Stirling's formula.

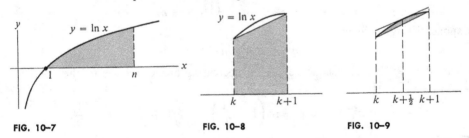

FIG. 10–7   FIG. 10–8   FIG. 10–9

We now consider $\epsilon_n = A_n - T_n$, the error which is made when T_n is used as an approximation to A_n. We will show that ϵ_n is an increasing sequence which is bounded above. We observe that $\epsilon_{k+1} - \epsilon_k$ is simply the area of the region bounded by the graph $y = \ln x$ and the chord joining the points $(k, \ln k)$ and $(k+1, \ln (k+1))$. This region is shown in Fig. 10–9. Thus $\epsilon_{k+1} - \epsilon_k > 0$, thereby showing that $\{\epsilon_n\}$

is an increasing sequence. Since $\ln x$ is concave, the graph $y = \ln x$ lies beneath each of its tangent lines. Consequently, $\epsilon_{k+1} - \epsilon_k$ is less than the area of the quadrilateral determined by the chord previously mentioned, the lines $x = k$, $x = k + 1$, and the tangent line to $y = \ln x$ at the point $(k + \frac{1}{2}, \ln(k + \frac{1}{2}))$. Hence

$$\epsilon_{k+1} - \epsilon_k < \ln(k + \tfrac{1}{2}) - \tfrac{1}{2}[\ln k + \ln(k + 1)]$$
$$= \tfrac{1}{2}\ln\left(1 + \frac{1}{2k}\right) - \tfrac{1}{2}\ln\left(1 + \frac{1}{2(k + \frac{1}{2})}\right)$$
$$< \tfrac{1}{2}\ln\left(1 + \frac{1}{2k}\right) - \tfrac{1}{2}\ln\left(1 + \frac{1}{2(k + 1)}\right).$$

We now add these inequalities, setting $k = 1, 2, \ldots, n - 1$. Both sides telescope to yield

$$\epsilon_n < \tfrac{1}{2}\ln\tfrac{3}{2} - \tfrac{1}{2}\ln(1 + 1/2n) < \tfrac{1}{2}\ln\tfrac{3}{2}.$$

This proves that the sequence $\{\epsilon_n\}$ is bounded.

It follows that the sequence $\{\epsilon_n\}$ has a limit, call it ϵ. Moreover,

$$\epsilon_m - \epsilon_n = \sum_{k=n}^{m-1} (\epsilon_{k+1} - \epsilon_k)$$
$$< \tfrac{1}{2}\ln\left(1 + \frac{1}{2n}\right) - \tfrac{1}{2}\ln\left(1 + \frac{1}{2(m + 1)}\right)$$
$$< \tfrac{1}{2}\ln\left(1 + \frac{1}{2n}\right),$$

provided $m > n$, so that

$$\epsilon - \epsilon_n = \lim_{m \to \infty} (\epsilon_m - \epsilon_n) \leq \tfrac{1}{2}\ln\left(1 + \frac{1}{2n}\right).$$

We are now ready for the kill:

$$\ln n! = T_n + \tfrac{1}{2}\ln n = A_n - \epsilon_n + \tfrac{1}{2}\ln n$$
$$= 1 - \epsilon_n + (n + \tfrac{1}{2})\ln n - n.$$

Exponentiating, we have

$$n! = a_n n^n \sqrt{n}\, e^{-n},$$

where $a_n = e^{1 - \epsilon_n}$. The sequence $\{a_n\}$ is decreasing and converges to the *nonzero* limit $a = e^{1 - \epsilon}$. Hence

$$1 < \frac{a_n}{a} = e^{\epsilon - \epsilon_n} < \exp\left[\tfrac{1}{2}\ln\left(1 + \frac{1}{2n}\right)\right] = \sqrt{1 + \frac{1}{2n}} < 1 + \frac{1}{4n}.$$

Therefore,

$$a < a_n < a\left(1 + \frac{1}{4n}\right)$$

and, consequently,

$$a\left(\frac{n}{e}\right)^n \sqrt{n} < n! < a\left(\frac{n}{e}\right)^n \sqrt{n}\left(1 + \frac{1}{4n}\right).$$

All that remains to be done is the evaluation of the constant a. By the corollary to Wallis' formula,

$$\sqrt{\pi} = \lim_{n \to \infty} \frac{(n!)^2 2^{2n}}{(2n)! \sqrt{n}}.$$

Replacing $n!$ by $a_n(n/e)^n\sqrt{n}$ and $(2n)!$ by $a_{2n}(2n/e)^{2n}\sqrt{2n}$, we get

$$\sqrt{\pi} = \lim_{n \to \infty} \frac{a_n^2}{a_n\sqrt{2}} = \frac{a^2}{a\sqrt{2}},$$

so that $a = \sqrt{2\pi}$. (Note that our knowledge that $a \neq 0$ is essential to the argument.)

Theorem. Stirling's Formula. *We have*

$$\sqrt{2\pi n}\left(\frac{n}{e}\right)^n < n! < \sqrt{2\pi n}\left(\frac{n}{e}\right)^n\left(1 + \frac{1}{4n}\right)$$

for every $n \in \mathbf{Z}^+$. In particular,

$$n! \sim \sqrt{2\pi n}\left(\frac{n}{e}\right)^n.$$

Exercises

1. Show that when $\sqrt{2\pi n}\, n^n e^{-n}$ is used as an approximation to $n!$, then the relative error is less than $1/4n$.

2. Deduce from Stirling's formula the earlier result

$$\lim_{n \to \infty} \frac{\sqrt[n]{n!}}{n} = \frac{1}{e}.$$

3. For a fixed $p \in \mathbf{Z}^+$, show that $(n + p)!/n! \sim n^p$.

4. Find $\lim_{n \to \infty} (n!)^{1/(n \ln n)}$.

5. Find an asymptotic estimate for $\binom{2n}{n}$.

Miscellaneous exercises

Calculate the integrals in Exercises 1 through 27.

1. $\int \frac{e^{\sqrt{x}}}{\sqrt{x}}\, dx$

2. $\int x \cos(x^2)\, dx$

3. $\int \frac{\sqrt{x}\, dx}{1 + \sqrt{x}}$

4. $\int \frac{dx}{e^x(5 - e^{-x})}$

5. $\int x \sin x \cos x\, dx$

6. $\int \mathrm{Cos}^{-1}\sqrt{\frac{x}{x+1}}\, dx$

7. $\int \frac{dx}{e^x + 1}$

8. $\int \frac{2x + 5}{\sqrt{9x^2 + 6x + 2}}\, dx$

9. $\int x^3 e^{5x}\, dx$

10. $\int e^{e^x + x}\, dx$

11. $\int \frac{1 - \tan x}{1 + \tan x}\, dx$

12. $\int \frac{\sqrt{1 + \cos x}}{\sin x}\, dx$

13. $\displaystyle\int \frac{dx}{x\sqrt[4]{x^2+4}}$ 14. $\displaystyle\int \frac{dx}{(ax+b)\sqrt{x}}$ 15. $\displaystyle\int \frac{\text{ctn } x}{\ln \sin x}\,dx$

16. $\displaystyle\int x^3 e^{x^2}\,dx$ 17. $\displaystyle\int \ln(x+\sqrt{1+x^2})\,dx$ 18. $\displaystyle\int \frac{\sin x\,dx}{1+\sin x}$

19. $\displaystyle\int \frac{\ln(x+1)-\ln x}{x(x+1)}\,dx$ 20. $\displaystyle\int \frac{dx}{x^4+x^6}$ 21. $\displaystyle\int \frac{\sqrt[3]{x}\,dx}{x(\sqrt{x}+\sqrt[3]{x})}$

22. $\displaystyle\int \frac{\sqrt{(x^2-a^2)^5}}{x}\,dx$ 23. $\displaystyle\int xe^{-\sqrt{x}}\,dx$ 24. $\displaystyle\int \frac{x^3+x^2}{x-1}\,dx$

25. $\displaystyle\int \ln(a^2+x^2)\,dx$ 26. $\displaystyle\int \frac{3x^2-1}{2x\sqrt{x}}\,\text{Tan}^{-1}x\,dx$ 27. $\displaystyle\int e^x \frac{1+\sin x}{1+\cos x}\,dx$

28. The lemniscate $(x^2+y^2)^2 = a^2(x^2-y^2)$ is revolved about the x-axis. Find the volume of the region bounded by the surface of revolution so generated.

29. Find the area of the region bounded by the curve $\theta = r\,\text{Tan }r$ and the rays $\theta = 0$, $\theta = \pi/\sqrt{3}$.

30. Find the area of the region bounded by the closed curve $x^4 + y^4 = x^2 + y^2$. (Use polar coordinates.)

31. Find the length of the arc of the parabola $y = x^2/2p$ joining the origin and the point $(x, x^2/2p)$.

32. Find the length of the arc of the logarithmic curve $y = \ln x$ between the points whose x-coordinates are $\sqrt{3}$ and $\sqrt{8}$.

*33. We define

$$\varphi(x) = -\int_0^x \ln \cos t\,dt.$$

(In Russia, at least, this integral is known as *Lobachevsky's integral*.)
a) Prove that $\varphi(x) = 2\varphi(\pi/4 + x/2) - 2\varphi(\pi/4 - x/2) - x\ln 2$.
b) Show that $\varphi(\pi/2) = (\pi/2)\ln 2$.

34. Find the length of the spiral of Archimedes, $r = a\theta$, from the origin to the end of the first turn.

35. Find the surface area which is generated by revolving one arch of the curve $y = \sin x$ about the x-axis.

36. Find the center of gravity of the lamina in the right half-plane bounded by the curves $y = \text{Tan}^{-1}x$, $y = -\sin(\pi/2)x$, and the line $x = 1$.

If as $x \to a$, $f(x)$ and $g(x)$ tend to zero, then almost anything can happen to the quotient $f(x)/g(x)$. A few examples should illustrate this point adequately.

$$\lim_{x \to 0+} \frac{\sin ax}{x} = a, \qquad \lim_{x \to 0+} \frac{e^{-1/x}}{x} = 0, \qquad \lim_{x \to 0+} \frac{x}{e^{-1/x}} = +\infty.$$

This state of affairs is summarized by saying that $0/0$ is an *indeterminate form*. The behavior of the quotient $f(x)/g(x)$ is also indeterminate if as $x \to a$, $f(x) \to \pm \infty$ and $g(x) \to \pm \infty$. Other indeterminate forms are $0 \cdot \infty$, $\infty - \infty$, 0^0, ∞^0, 0^∞, and 1^∞. While we have successfully "evaluated" a good many such indeterminate forms, we have had no general techniques for handling these problems. In this section such a general technique is presented, namely, l'Hôpital's rule.

Suppose that $f(a) = g(a) = 0$ and that we are required to calculate

$$\lim_{x \to a} \frac{f(x)}{g(x)}.$$

Suppose also that f and g have continuous derivatives throughout some neighborhood V of a and that $g'(a) \neq 0$. For each $x \in V$, there exists, according to the mean-value theorem, numbers ξ and ξ' between a and x such that

$$f(x) = f(x) - f(a) = (x - a)f'(\xi),$$
$$g(x) = g(x) - g(a) = (x - a)g'(\xi'),$$

and thus

$$\frac{f(x)}{g(x)} = \frac{f'(\xi)}{g'(\xi')}.$$

Since f' and g' are continuous, taking limits, we get

$$\lim_{x \to a} \frac{f(x)}{g(x)} = \frac{f'(a)}{g'(a)}.$$

Consider, for example, the following limit:

$$\lim_{x \to 0} \frac{e^x - 1}{\sin x}.$$

The hypotheses are clearly satisfied, and we may therefore write

$$\lim_{x \to 0} \frac{e^x - 1}{\sin x} = \frac{\dfrac{d}{dx}(e^x - 1)\Big|_{x=0}}{\dfrac{d}{dx}(\sin x)\Big|_{x=0}} = \frac{e^0}{\cos 0} = 1.$$

We have established a special case of l'Hôpital's rule. This rule states that if the limit

$$\lim_{x \to a} \frac{f(x)}{g(x)}$$

is an indeterminate form of type $0/0$ or ∞/∞, then one can consider the limit of the derivatives,

$$\lim_{x \to a} \frac{f'(x)}{g'(x)}.$$

If the latter exists, then the former also exists, and the two are equal. Thus

$$\lim_{x \to a} \frac{f(x)}{g(x)} = \lim_{x \to a} \frac{f'(x)}{g'(x)}.$$

To prove l'Hôpital's rule, we need a generalization of the law of the mean.

Theorem 1. Generalized Mean-Value Theorem. *Let f and g be functions such that*

1) *f and g are continuous throughout the interval* $[a, b]$;

2) *f′ and g′ exist throughout the open interval* (a, b);

3) *g′ does not vanish throughout the open interval* (a, b).

Then there exists a number ξ in the interval (a, b) such that

$$\frac{f(b) - f(a)}{g(b) - g(a)} = \frac{f'(\xi)}{g'(\xi)}.$$

Before beginning the proof, there are a couple of things worth commenting on. Observe, first of all, that if $g(x) = x$, then the theorem reduces to the ordinary mean-value theorem. Observe, secondly, that the ordinary mean-value theorem implies that there exist two numbers ξ and ξ' (possibly different) in the interval (a, b) such that

$$\frac{f(b) - f(a)}{g(b) - g(a)} = \frac{(b - a)f'(\xi)}{(b - a)g'(\xi')} = \frac{f'(\xi)}{g'(\xi')}.$$

The generalized mean-value theorem, however, makes the stronger assertion that ξ and ξ' can be chosen so as to be equal.

We offer two proofs of the theorem.

Proof 1. Since g' is nonzero throughout the open interval (a, b), it follows that $g(a) \ne g(b)$. [For if $g(a)$ and $g(b)$ were equal, then by the law of the mean, there would exist a number ξ in the interval (a, b) such that

$$0 = g(b) - g(a) = (b - a)g'(\xi),$$

and thus we would have $g'(\xi) = 0$, a contradiction.] Consider now the function

$$\varphi(x) = \begin{vmatrix} f(x) & g(x) & 1 \\ f(a) & g(a) & 1 \\ f(b) & g(b) & 1 \end{vmatrix}$$
$$= [g(a) - g(b)]f(x) - [f(a) - f(b)]g(x) + f(a)g(b) - f(b)g(a).$$

The function φ will be continuous throughout the closed interval $[a, b]$ and differentiable throughout the open interval (a, b). Also, it is easily checked that $\varphi(a) = \varphi(b) = 0$. (If we set x equal to a or b, then two rows of the above determinant will be the same and the determinant will therefore be equal to zero.) The function φ therefore satisfies the hypotheses of Rolle's theorem. Hence there exists a number ξ in the interval (a, b) such that

$$0 = \varphi'(\xi) = [g(a) - g(b)]f'(\xi) - [f(a) - f(b)]g'(\xi),$$

from which we get

$$\frac{f(b) - f(a)}{g(b) - g(a)} = \frac{f'(\xi)}{g'(\xi)}.$$

Note that we can dispense with hypothesis (3) if we are willing to write the conclusion in the form

$$[g(b) - g(a)]f'(\xi) = [f(b) - f(a)]'g(\xi).$$

Proof 2. Since g' does not vanish on the interval (a, b), it follows from Darboux's theorem that g' is either strictly positive or strictly negative throughout the interval. In either case, g will be strictly monotone and hence invertible. Let I be the closed interval determined by $g(a)$ and $g(b)$, and let $\varphi = f \circ g^{-1}$. From our general theorems on inverse functions, it follows that φ is continuous on I and differentiable at all interior points of I (see Fig. 11-1). Hence the ordinary mean-value theorem is applicable, and we may conclude that there exists a number η in the interior of I such that

$$\varphi\big(g(b)\big) - \varphi\big(g(a)\big) = \varphi'(\eta)[g(b) - g(a)].$$

There exists a unique ξ in an interval (a, b) such that $\eta = g(\xi)$. But

$$\varphi\big(g(b)\big) = f(b), \qquad \varphi\big(g(a)\big) = f(a),$$

and

$$\varphi'(\eta) = f'\big(g^{-1}(\eta)\big) \frac{1}{g'\big(g^{-1}(\eta)\big)} = \frac{f'(\xi)}{g'(\xi)}.$$

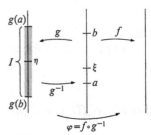

FIG. 11-1

Here we have applied the chain rule and the rule for differentiating inverse functions. Thus we get

$$\frac{f(b) - f(a)}{g(b) - g(a)} = \frac{f'(\xi)}{g'(\xi)}.$$

Theorem 2. L'Hôpital's Rule for Indeterminate Forms of Type 0/0. *Suppose that as* $x \to a+$, $f(x)$ *and* $g(x)$ *tend to zero. Suppose further that*

$$\lim_{x \to a+} \frac{f'(x)}{g'(x)} = L,$$

where L *is either a real number,* $+\infty$, *or* $-\infty$. *Then*

$$\lim_{x \to a+} \frac{f(x)}{g(x)} = L.$$

Moreover, the same holds for left-hand limits, two-sided limits, and limits as $x \to +\infty$ *or* $x \to -\infty$.

Proof. Since

$$\lim_{x \to a+} \frac{f'(x)}{g'(x)} = L,$$

there exists a deleted right-hand neighborhood $(a, b]$, where $b > a$, throughout which f' and g' exist and g' is never zero. We define (or, if necessary, redefine)

$$f(a) = g(a) = 0.$$

Then f and g satisfy the hypotheses of the generalized mean-value theorem. Hence for each x in the interval (a, b), there exists a ξ such that $a < \xi < x < b$ and

$$\frac{f(x)}{g(x)} = \frac{f(x) - f(a)}{g(x) - g(a)} = \frac{f'(\xi)}{g'(\xi)}.$$

Since

$$\lim_{t \to a+} \frac{f'(t)}{g'(t)} = L,$$

it follows from the above equation that

$$\lim_{x \to a+} \frac{f(x)}{g(x)} = L$$

also.

The proof for left-hand limits is similar. Suppose now that

$$\lim_{x \to +\infty} f(x) = \lim_{x \to +\infty} g(x) = 0 \quad \text{and} \quad \lim_{x \to +\infty} \frac{f'(x)}{g'(x)} = L.$$

Let $F(t) = f(1/t)$ and $G(t) = g(1/t)$. Then

$$\lim_{t \to 0+} \frac{F'(t)}{G'(t)} = \lim_{t \to 0+} \frac{f'(1/t)(-1/t^2)}{g'(1/t)(-1/t^2)} = \lim_{x \to +\infty} \frac{f'(x)}{g'(x)} = L.$$

Consequently, by the previously proved case of l'Hôpital's rule,

$$\lim_{x \to +\infty} \frac{f(x)}{g(x)} = \lim_{t \to 0+} \frac{F(t)}{G(t)} = L.$$

We now consider several examples, including some limits previously calculated.

Example 1. Show that $\lim_{x \to 0} (\sin x)/x = 1$.

Clearly $\lim_{x \to 0} x = \lim_{x \to 0} \sin x = 0$ and (using the prime notation for derivatives)

$$\lim_{x \to 0} \frac{(\sin x)'}{(x)'} = \lim_{x \to 0} \frac{\cos x}{1} = 1.$$

Hence, by l'Hôpital's rule,

$$\lim_{x \to 0} \frac{\sin x}{x} = 1.$$

Our application of l'Hôpital's rule is circular in this case, since the limit was needed to show that the sine function is differentiable.

Example 2. Show that $\lim_{x \to 0} (\cos x - 1)/x^2 = -\frac{1}{2}$.

We clearly have an indeterminate form of type $0/0$. Since

$$\lim_{x \to 0} \frac{(\cos x - 1)'}{(x^2)'} = \lim_{x \to 0} \frac{-\sin x}{2x} = -\frac{1}{2},$$

we have, by l'Hôpital's rule,

$$\lim_{x \to 0} \frac{\cos x - 1}{x^2} = -\frac{1}{2}.$$

Example 3. Find $\lim_{x \to 0} (e^{\sin x} - e^x)/(\sin x - x)$.

Since we have an indeterminate form of type $0/0$, we consider the limit of the quotient of the derivatives, namely,

$$\lim_{x \to 0} \frac{(\cos x)e^{\sin x} - e^x}{\cos x - 1}.$$

This expression is also of the form $0/0$, so we consider the limit of the quotient of its derivatives:

$$\lim_{x \to 0} \frac{(\cos^2 x - \sin x)e^{\sin x} - e^x}{-\sin x} = 1.$$

By applying l'Hôpital's rule three times in succession, we conclude that each of the preceding limits is also equal to 1. In particular,

$$\lim_{x \to 0} \frac{e^{\sin x} - e^x}{\sin x - x} = 1.$$

We now make three comments regarding the application of this form of l'Hôpital's rule.

1) One should take care to apply l'Hôpital's rule only when dealing with an indeterminate form of type $0/0$. For instance,

$$\lim_{x \to 1} \frac{x^2 + 2}{x^2 - 3} = \frac{1 + 2}{1 - 3} = -\frac{3}{2},$$

whereas l'Hôpital's procedure yields an incorrect result, namely,

$$\lim_{x \to 1} \frac{x^2 + 2}{x^2 - 3} = \lim_{x \to 1} \frac{2x}{2x} = 1.$$

L'Hôpital's rule is not applicable, of course, since we do not have an indeterminate form of type 0/0.

2) L'Hôpital's rule may not simplify the job of calculating a limit. Consider, for instance,

$$\lim_{x \to +\infty} xe^{-x} = \lim_{x \to +\infty} \frac{e^{-x}}{1/x}.$$

L'Hôpital's rule would require that we be able to calculate

$$\lim_{x \to +\infty} \frac{-e^{-x}}{-1/x^2} = \lim_{x \to +\infty} x^2 e^{-x}.$$

We know from our work in Chapter 9 that this limit, and hence the first one also, is zero. In this case, however, l'Hôpital's rule contributes nothing, since the second limit is more complicated than the first and implies the first trivially:

$$\lim_{x \to +\infty} xe^{-x} = \lim_{x \to +\infty} \frac{1}{x} x^2 e^{-x} = 0 \cdot \lim_{x \to +\infty} x^2 e^{-x} = 0 \cdot 0 = 0.$$

3) Not all limits of type 0/0 can be calculated by l'Hôpital's rule. It may happen that $\lim_{x \to a} f(x)/g(x)$ exists while $\lim_{x \to a} f'(x)/g'(x)$ does not. For example,

$$\lim_{x \to 0+} \frac{x^2 \sin (1/x)}{x} = 0,$$

but

$$\lim_{x \to 0+} \frac{[x^2 \sin (1/x)]'}{x'} = \lim_{x \to 0+} \left(2x \sin \frac{1}{x} - \cos \frac{1}{x} \right)$$

does not exist.

We consider next l'Hôpital's rule for indeterminate forms of type ∞/∞. The statement of the rule is virtually the same as before. Suppose that as $x \to a$, $f(x) \to \pm\infty$ and $g(x) \to \pm\infty$. Suppose further that

$$\lim_{x \to a} \frac{f'(x)}{g'(x)} = L.$$

Then the ∞/∞ version of l'Hôpital's rule states that

$$\lim_{x \to a} \frac{f(x)}{g(x)} = L.$$

Consider, once again, $\lim_{x \to +\infty} xe^{-x}$. Instead of regarding this as an indeterminate form of type 0/0, as we did in (2) above, we regard it as one of type ∞/∞. Thus we have

$$\lim_{x \to +\infty} xe^{-x} = \lim_{x \to +\infty} \frac{x}{e^x}.$$

Since

$$\lim_{x\to+\infty} \frac{(x)'}{(e^x)'} = \lim_{x\to+\infty} \frac{1}{e^x} = 0,$$

we conclude from the ∞/∞ version of l'Hôpital's rule that $\lim_{x\to+\infty} xe^{-x} = 0$.

Theorem 3. L'Hôpital's Rule for ∞/∞ Indeterminate Forms. *Suppose that as* $x \to a+, f(x) \to \pm\infty$ *and* $g(x) \to \pm\infty$. *Suppose also that*

$$\lim_{x\to a+} \frac{f'(x)}{g'(x)} = L,$$

where L is either a real number, $+\infty$, or $-\infty$. Then

$$\lim_{x\to a+} \frac{f(x)}{g(x)} = L.$$

Moreover, the same is true for left-hand limits, two-sided limits, and limits as $x \to +\infty$ or $x \to -\infty$.

Proof. We shall consider only the case in which L is a real number, $f(x) \to +\infty$ and $g(x) \to +\infty$ as $x \to a+$. All other cases can either be treated similarly or be reduced to this one.

Suppose that

$$\lim_{x\to a+} \frac{f(x)}{g(x)} \neq L.$$

Then there exists an $\epsilon > 0$ and a sequence of numbers $x_1, x_2, x_3, \ldots$, each greater than a, such that $\lim_{n\to\infty} x_n = a$ and

$$\left| \frac{f(x_n)}{g(x_n)} - L \right| \geq \epsilon \qquad (*)$$

for all positive integers n. Then $f(x_n) \to +\infty$ and $g(x_n) \to +\infty$, and we can therefore find a subsequence of the x_n's, say, $x_{n_1}, x_{n_2}, x_{n_3}, \ldots$, which converges to a so rapidly that

$$\lim_{m\to\infty} \frac{f(x_m)}{f(x_{n_m})} = \lim_{m\to\infty} \frac{g(x_m)}{g(x_{n_m})} = 0.$$

(One could, for instance, choose the indices $n_1, n_2, \ldots$ according to the following inductive scheme. Let n_1 be the least integer such that

$$n \geq n_1 \Rightarrow \begin{cases} f(x_n) \geq |f(x_1)|, \\ g(x_n) \geq |g(x_1)|. \end{cases}$$

Having chosen n_k, let n_{k+1} be the least integer such that $n_{k+1} > n_k$ and

$$n \geq n_{k+1} \Rightarrow \begin{cases} f(x_n) \geq (k+1)|f(x_k)|, \\ g(x_n) \geq (k+1)|g(x_k)|. \end{cases}$$

Then we would have

$$\left| \frac{f(x_m)}{f(x_{n_m})} \right| \leq \frac{1}{m} \qquad \text{and} \qquad \left| \frac{g(x_m)}{g(x_{n_m})} \right| \leq \frac{1}{m}$$

for all positive integers m.) By the generalized mean-value theorem, we may write

$$\frac{f(x_{n_m})\left[1 - \dfrac{f(x_m)}{f(x_{n_m})}\right]}{g(x_{n_m})\left[1 - \dfrac{g(x_m)}{g(x_{n_m})}\right]} = \frac{f(x_{n_m}) - f(x_m)}{g(x_{n_m}) - g(x_m)} = \frac{f'(\xi_m)}{g'(\xi_m)},$$

where ξ_m is some number between x_{n_m} and x_m. As $m \to \infty$, $\xi_n \to a$ and

$$\frac{f'(\xi_m)}{g'(\xi_m)} \to L.$$

Thus, taking limits as $m \to \infty$, we get

$$\lim_{m \to \infty} \frac{f(x_{n_m})}{g(x_{n_m})} = L.$$

But this is impossible because of (∗).

Example 4. If $a > 0$, show that $\lim_{x \to +\infty} (\ln x)/x^a = 0$.
 Since

$$\lim_{x \to +\infty} \frac{(\ln x)'}{(x^a)'} = \lim_{x \to \infty} \frac{x^{-1}}{ax^{a-1}} = \lim_{x \to +\infty} \frac{1}{ax^a} = 0,$$

l'Hôpital's rule gives the desired result.

 We have discussed only the indeterminate forms $0/0$ and ∞/∞. We shall now illustrate by examples how other indeterminate forms can be reduced to $0/0$ or ∞/∞.

Example 5. $0 \cdot \infty$. Find $\lim_{x \to 1-} \sqrt{1 - x} \ln (\ln 1/x)$.
 We have

$$\lim_{x \to 1-} \sqrt{1 - x} \ln \left(\ln \frac{1}{x}\right) = \lim_{x \to 1-} \frac{\ln (-\ln x)}{(1 - x)^{-1/2}} = \lim_{x \to 1-} \frac{1/(x \ln x)}{\frac{1}{2}(1 - x)^{-3/2}},$$

and

$$\lim_{x \to 1-} \frac{(1 - x)^{3/2}}{\ln x} = \lim_{x \to 1-} \frac{-\frac{3}{2}(1 - x)^{1/2}}{1/x} = 0.$$

Thus

$$\lim_{x \to 1-} \sqrt{1 - x} \ln \left(\ln \frac{1}{x}\right) = 0.$$

 In this example we have reduced to $0/0$ and applied l'Hôpital's rule twice.

Example 6. $\infty - \infty$. Find $\lim_{x \to 0+} (1/x - \text{ctn } x)$.
 This is easily reduced to $0/0$ as follows:

$$\lim_{x \to 0+} \left(\frac{1}{x} - \text{ctn } x\right) = \lim_{x \to 0+} \left(\frac{1}{x} - \frac{\cos x}{\sin x}\right) = \lim_{x \to 0+} \frac{\sin x - x \cos x}{x \sin x}.$$

One can show that the latter limit is 0 by applying l'Hôpital's rule two times.

Example 7. 0^0. Find $\lim_{x\to 0+} (\sin x)^x$.

In such cases it is usually advisable to take logarithms. If we let $y = (\sin x)^x$, then $\ln y = x \ln (\sin x)$. Now

$$\lim_{x\to 0+} \ln y = \lim_{x\to 0+} \frac{\ln (\sin x)}{x^{-1}}.$$

Since

$$\lim_{x\to 0+} \frac{[\ln (\sin x)]'}{(x^{-1})'} = \lim_{x\to 0+} \frac{\text{ctn } x}{-x^{-2}}$$

$$= \lim_{x\to 0+} \frac{x}{\sin x} (-x \cos x) = 0,$$

we have, by the ∞ / ∞ version of l'Hôpital's rule, $\lim_{x\to 0+} \ln y = 0$. Therefore,

$$\lim_{x\to 0+} (\sin x)^x = \lim_{x\to 0+} e^{\ln y} = e^0 = 1.$$

Exercises

1. Let γ be a path with the parametrization

$$x = f(t), \qquad y = g(t) \qquad (a \le t \le b),$$

where f and g are continuous on $[a, b]$ and possess first-order derivatives which do not vanish simultaneously throughout the open interval (a, b). We also assume that the initial and terminal points of γ do not coincide. Show that Cauchy's theorem has this geometric interpretation: there exists a point on γ where the tangent line is parallel to the chord joining the endpoints of γ. In other words, the generalized mean-value theorem has the same geometric interpretation as the ordinary mean-value theorem; the difference is that a more general class of curves is now included.

2. Suppose that the inequality $|f'(x)| \ge |g'(x)|$ holds for all x in the interval $[a, b]$. Show that $|\Delta f| \ge |\Delta g|$ where

$$\Delta f = f(x + \Delta x) - f(x), \qquad \Delta g = g(x + \Delta x) - g(x);$$

here x and $x + \Delta x$ are any two points in the interval $[a, b]$.

3. Use the preceding problem to show that the inequality

$$\text{Tan}^{-1} x - \ln (1 + x^2) \ge \pi/4 - \ln 2$$

holds for all x in the interval $[\frac{1}{2}, 1]$.

4. Calculate each of the following limits.

a) $\lim_{x\to 0} \dfrac{\ln \cos x}{\sin x}$

b) $\lim_{x\to 0+} \dfrac{e^{a\sqrt{x}} - 1}{\sqrt{\sin x}}$

c) $\lim_{x\to 0} \dfrac{e^x - e^{-x}}{\sin x \cos x}$

d) $\lim_{x\to 0} \dfrac{a^x - b^x}{\sqrt[x]{1 - x^2}}$

e) $\lim_{x\to 0} \dfrac{e^x - e^{-x} - 2\ln (1 + x)}{x \sin x}$

f) $\lim_{x\to 0} \dfrac{x(1 + a \cos x) - b \sin x}{x^3}$ (all cases)

g) $\lim\limits_{x \to 3\pi} \dfrac{1 + \tan (x/4)}{\cos (x/4)}$

h) $\lim\limits_{x \to -\infty} \dfrac{\ln (1 + 1/x)}{\sin (1/x)}$

i) $\lim\limits_{x \to 0+} \dfrac{1 - \sec x}{x^3}, \quad \lim\limits_{x \to 0-} \dfrac{1 - \sec x}{x^3}$

j) $\lim\limits_{x \to 0+} \dfrac{x \ln x}{\ln (1 + ax)}$

k) $\lim\limits_{x \to 0} \dfrac{x - \mathrm{Tan}^{-1} x}{x^2 \ln (1 + x)}$

l) $\lim\limits_{x \to 0} \left[\dfrac{1}{x} - \dfrac{1}{x^2} \ln (1 + x) \right]$

m) $\lim\limits_{x \to 0} \dfrac{\sin x - \ln (e^x \cos x)}{x \sin x}$

5. a) Find

$$\lim_{h \to 0} \frac{1}{h^3} \sum_{k=0}^{3} (-1)^k \binom{3}{k} f(x + kh).$$

b) Find

$$\lim_{h \to 0} \frac{1}{h^4} \sum_{k=0}^{4} (-1)^k \binom{4}{k} f(x + kh).$$

c) Generalize (a) and (b).

6. Find

$$\lim_{h \to 0} \frac{1}{h^3} \begin{vmatrix} f(x) & g(x) & \varphi(x) \\ f(x + h) & g(x + h) & \varphi(x + h) \\ f(x + 2h) & g(x + 2h) & \varphi(x + 2h) \end{vmatrix}.$$

7. Calculate each of the following limits.

a) $\lim\limits_{x \to 0} \dfrac{\ln x}{\ln \sin x}$

b) $\lim\limits_{x \to 1} \dfrac{\ln (1 - x) + \tan (\pi x/2)}{\mathrm{ctn}\, \pi x}$

c) $\lim\limits_{x \to \pi/2+} \dfrac{\tan x}{\ln (2x - \pi)}$

d) $\lim\limits_{x \to +\infty} \dfrac{e^{e^x}}{e^x}, \quad \lim\limits_{x \to -\infty} \dfrac{e^{e^x}}{e^x}$

e) $\lim\limits_{x \to +\infty} \dfrac{x^3}{x^2 - 3 \cos x}, \quad \lim\limits_{x \to -\infty} \dfrac{x^3}{x^2 - 3 \cos x}$

f) $\lim\limits_{x \to +\infty} \dfrac{\int_0^x e^{t^2} dt}{e^{x^2}}$

g) $\lim\limits_{x \to 0} \dfrac{1}{x} \displaystyle\int_0^x \dfrac{|\sin t|}{t} dt$

h) $\lim\limits_{x \to +\infty} \dfrac{1}{x} \displaystyle\int_0^x \dfrac{|\sin t|}{t} dt$

8. Prove that

$$\lim_{x \to +\infty} \frac{x - \sin x}{x + \sin x}$$

exists but cannot be evaluated by l'Hôpital's rule.

9. Calculate each of the following limits.

a) $\lim\limits_{x \to +\infty} (\pi - 2 \, \mathrm{Tan}^{-1} x) \ln x$

b) $\lim\limits_{x \to 1} \left(\dfrac{x}{x - 1} - \dfrac{1}{\ln x} \right)$

c) $\lim\limits_{x \to 0+} x \ln \tan x$

d) $\lim\limits_{x \to 0} x \tan (\pi/2 - x)$

e) $\lim_{x \to 0} \left(\dfrac{a}{x} - \mathrm{ctn} \, \dfrac{x}{a} \right)$

f) $\lim_{x \to 0} \left(\dfrac{1}{x} - \dfrac{1}{x^x - 1} \right)$

g) $\lim_{\phi \to a} (a^2 - \phi^2) \tan \dfrac{\pi \phi}{2a}$

h) $\lim_{x \to 1} \dfrac{1}{\cos (\pi x/2) \ln (1 - x)}$

i) $\lim_{x \to +\infty} [\sqrt[3]{(a + x)(b + x)(c + x)} - x]$

j) $\lim_{x \to +\infty} x(e^{1/x} - 1)$

k) $\lim_{x \to \pi/2} (\tan x)^{2x - \pi}$

l) $\lim_{x \to +\infty} x^{\ln (1/x)}$

m) $\lim_{x \to +\infty} (\ln x)^{\ln (1 - x^{-1})}$

n) $\lim_{x \to 0} \left(\dfrac{\tan x}{x} \right)^{1/x^2}$

o) $\lim_{x \to 0} \left(\dfrac{1}{x} \right)^{\tan x}$

p) $\lim_{x \to 0} (e^x + x)^{1/x}$

q) $\lim_{x \to -\infty} e^{ax} e^{be^x}$

r) $\lim_{x \to +\infty} x^a e^{bx} (\ln x)^c$

s) $\lim_{x \to 1-} \sqrt{1 - x} \ln \ln \dfrac{1}{x}$

t) $\lim_{x \to 0+} \sqrt{x} \ln \ln \dfrac{1}{x}$

u) $\lim_{x \to 0} \dfrac{1}{x} \displaystyle\int_0^x (1 + \sin 2t)^{1/t} \, dt$

10. Interpret geometrically the $0/0$ version of l'Hôpital's rule.

11-2 Taylor's theorem

If P is a polynomial of degree n or less, and a is any real number, then we have previously shown that

$$P(x) = \sum_{k=0}^{n} \frac{P^{(k)}(a)}{k!} (x - a)^k.$$

More generally, if f is a function possessing an nth-order derivative at a, then we can consider the *Taylor polynomial of order n about a*, namely,

$$T_n(x) = \sum_{k=0}^{n} \frac{f^{(k)}(a)}{k!} (x - a)^k,$$

and we can ask how well it approximates the function f, or equivalently, how close the *remainder*

$$R_n(x) = f(x) - T_n(x)$$

is to zero.

At the point a, the polynomial T_n agrees with the function f as well as can be expected of a polynomial of degree n or less. Not only is $T_n(a) = f(a)$, but also

$$T_n'(a) = f'(a), \qquad T_n''(a) = f''(a), \qquad \dots, \qquad T_n^{(n)}(a) = f^{(n)}(a).$$

We can show this by calculating successive derivatives of $T_n(x)$ and setting $x = a$.

Thus

$$T_n(x) = f(a) + f'(a)(x - a) + \frac{f''(a)}{2!}(x - a)^2$$
$$+ \frac{f'''(a)}{3!}(x - a)^3 + \cdots + \frac{f^{(n)}(a)}{n!}(x - a)^n,$$

$$T'_n(x) = f'(a) + f''(a)(x - a) + \frac{f'''(a)}{2!}(x - a)^2$$
$$+ \cdots + \frac{f^{(n)}(a)}{(n-1)!}(x - a)^{n-1},$$

$$T''_n(x) = f''(a) + f'''(a)(x - a) + \frac{f^{(4)}(a)}{2!}(x - a)^2$$
$$+ \cdots + \frac{f^{(n)}(a)}{(n-2)!}(x - a)^{n-2},$$

$$\vdots$$

$$T_n^{(n)}(x) = f^{(n)}(a),$$
$$T_n^{(n+1)}(x) = 0.$$

Setting $x = a$, we get $T_n^{(j)}(a) = f^{(j)}(a)$, $j = 0, 1, \ldots, n$. Also observe that $T_n^{(j)}$ is the Taylor polynomial of order $n - j$ for the function $f^{(j)}$, $j = 0, 1, \ldots, n$.

We shall prove next that as $x \to a$, the remainder $R_n(x)$ tends to zero more rapidly than $(x - a)^n$.

Theorem 1. *If $f^{(n)}(a)$ exists, then*

$$\lim_{x \to a} \frac{R_n(x)}{(x - a)^n} = 0.$$

Moreover, if P is any polynomial of degree n or less such that

$$\lim_{x \to a} \frac{f(x) - P(x)}{(x - a)^n} = 0,$$

then $P = T_n$.

Proof. We shall prove the first assertion of the theorem by induction on n. We have

$$\lim_{x \to a} \frac{R_1(x)}{x - a} = \lim_{x \to 0} \left[\frac{f(x) - f(a)}{x - a} - f'(a) \right] = f'(a) - f'(a) = 0.$$

Hence the first statement of the theorem is true when $n = 1$. Assume that the statement is true when $n = k$. Assuming that $f^{(k+1)}(a)$ exists, we set $g = f'$. Now

$$R_{k+1}(x) = f(x) - f(a) - f'(a)(x - a) - \frac{f''(a)}{2!}(x - a)^2$$
$$- \cdots - \frac{f^{(k+1)}(a)}{(k+1)!}(x - a)^{k+1},$$

so that

$$\frac{d}{dx}[R_{k+1}(x)] = f'(x) - f'(a) - f''(a)(x - a) - \cdots - \frac{f^{(k+1)}(a)}{k!}(x - a)^k$$
$$= g(x) - g(a) - g'(a)(x - a) - \cdots - \frac{g^{(k)}(a)}{k!}(x - a)^k.$$

By the inductive hypothesis,

$$\lim_{x \to a} \frac{\frac{d}{dx}[R_{k+1}(x)]}{(x - a)^k} = 0.$$

Since

$$\lim_{x \to a} \frac{\frac{d}{dx}[R_{k+1}(x)]}{\frac{d}{dx}(x - a)^{k+1}} = \lim_{x \to a} \frac{\frac{d}{dx}[R_{k+1}(x)]}{(k + 1)(x - a)^k} = 0,$$

it follows from l'Hôpital's rule that

$$\lim_{x \to a} \frac{R_{k+1}(x)}{(x - a)^{k+1}} = 0,$$

i.e., the first portion of the theorem is true when $n = k + 1$.

We leave the proof of the second part of the theorem to the reader.

The theorem just proved may be interpreted as saying that among all polynomials of degree n or less, the Taylor polynomial T_n provides the best approximation to the function f near the point a. Note, however, that the theorem provides no information on the size of the error $R_n(x)$ for any particular $x \neq a$. The various forms of Taylor's theorem provide expressions for $R_n(x)$.

Theorem 2. Taylor's Theorem with Integral Remainder. *If $f^{(n+1)}$ exists and is continuous on the closed bounded interval having a and x as endpoints, then*

$$R_n(x) = (1/n!) \int_a^x f^{(n+1)}(t)(x - t)^n \, dt.$$

Proof. We shall present a proof which is slightly different from that outlined in Exercise 33, Section 10–2. Our proof will be by induction on n. When $n = 0$, we have

$$\int_a^x f'(t) \, dt = f(x) - f(a) = R_0(x),$$

thereby showing that the theorem is true in this case. Assuming that the theorem is true when $n = k$, we suppose that $f^{(k+2)}$ is continuous on the interval determined by a and x, and we consider the integral

$$\frac{1}{(k + 1)!} \int_a^x f^{(k+2)}(t)(x - t)^{k+1} \, dt.$$

We perform an integration by parts by setting

$$u = \frac{(x - t)^{k+1}}{(k + 1)!}, \qquad dv = f^{(k+2)}(t) \, dt.$$

Then

$$du = -\frac{(x - t)^k}{k!} \, dt, \qquad v = f^{(k+1)}(t) \, dt,$$

so that

$$\frac{1}{(k+1)!} \int_a^x f^{(k+2)}(t)(x-t)^{k+1} \, dt$$

$$= -\frac{(x-t)^{k+1}}{(k+1)!} f^{(k+1)}(t) \Big|_{t=a}^{t=x} + \frac{1}{k!} \int_a^x f^{(k+1)}(t)(x-t)^k \, dt$$

$$= \frac{(x-a)^{k+1}}{(k+1)!} f^{(k+1)}(a) + R_k(x),$$

by the inductive hypothesis. However,

$$f^{(k+1)}(a) \frac{(x-a)^{k+1}}{(k+1)!} + R_k(x) = R_{k+1}(x),$$

thereby completing the induction.

Several other expressions for the remainder can be obtained from the integral expression for $R_n(x)$. Recall the generalized mean-value theorem for integrals. It states that if φ is continuous on $[a, b]$ and g is a Riemann-integrable function which does not change sign on $[a, b]$, then

$$\int_a^b \varphi(x)g(x) \, dx = \varphi(\xi) \int_a^b g(x) \, dx$$

for some ξ in the interval (a, b). First we set

$$\varphi(t) = f^{(n+1)}(t), \qquad g(t) = \frac{(x-t)^n}{n!} .$$

This yields

$$R_n(x) = \int_a^x \varphi(t)g(t) \, dt = \varphi(\xi) \int_a^x g(t) \, dt = \frac{f^{(n+1)}(\xi)}{(n+1)!} (x-a)^{n+1}$$

for some ξ between a and x. The latter expression is called the *Lagrange* form of the remainder. We shall now prove that the Lagrange remainder is valid under slightly weaker hypotheses than those of Taylor's theorem with integral remainder.

Theorem 3. Taylor's Theorem with Lagrange Remainder. *Suppose that*

1) f is continuous throughout the closed interval I determined by a and x;

2) $f^{(n)}(a)$ exists;

3) $f^{(n+1)}$ exists throughout the interior of I.

Then there exists a ξ in the interior of I such that

$$R_n(x) = \frac{f^{(n+1)}(\xi)}{(n+1)!} (x-a)^{n+1}.$$

Proof. To avoid some ugly notation, we shall abbreviate $R_n(t)$ to $R(t)$. For this proof the essential facts about $R(t)$ are the following:

a) $R(a) = R'(a) = \cdots = R^{(n)}(a) = 0$;

b) $R^{(n+1)}(t) = f^{(n+1)}(t)$.

The latter stems from the fact that $R(t) = f(t) - T_n(t)$ and T_n is a polynomial of degree n at most.

We introduce a function φ as follows:

$$\varphi(t) = R(t) - K(t - a)^{n+1},$$

where K is a constant which we choose so that $\varphi(x) = 0$. Thus

$$K = \frac{R(x)}{(x - a)^{n+1}}.$$

(Keep in mind that x is fixed throughout the entire argument.) The function φ has these properties:

    a') $\varphi(x) = \varphi(a) = \varphi'(a) = \cdots = \varphi^{(n)}(a) = 0$;

    b') $\varphi^{(n+1)}(t) = f^{(n+1)}(t) - (n + 1)!K$.

We now apply Rolle's theorem to φ and its derivatives. Since $\varphi(x) = \varphi(a) = 0$, there exists, by Rolle's theorem, a number ξ_1 between a and x such that $\varphi'(\xi_1) = 0$. If $n = 0$, we stop. Otherwise, since $\varphi'(a) = \varphi'(\xi_1) = 0$, there must exist a ξ_2 between a and ξ_1 (and hence between a and x) such that $\varphi''(\xi_2) = 0$ (see Fig. 11–2). By applying Rolle's theorem n times, we conclude that there exists a number ξ between a and x such that $\varphi^{(n+1)}(\xi) = 0$. Thus

$$0 = \varphi^{(n+1)}(\xi) = f^{(n+1)}(\xi) - (n + 1)!K,$$

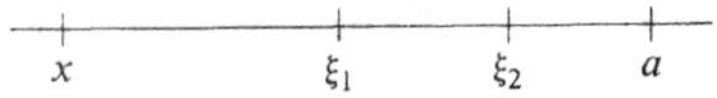

so that

$$R_n(x) = K(x - a)^{n+1} = \frac{f^{(n+1)}(\xi)}{(n + 1)!}(x - a)^{n+1}.$$

FIG. 11–2

Corollary 4. *Let I be an interval containing a and suppose that*

$$|f^{(n+1)}(x)| \le M$$

for all $x \in I$, where M is a constant. Then

$$|R_n(x)| \le \frac{M}{(n + 1)!}|x - a|^{n+1}$$

for all $x \in I$.

Note that the conclusion of the corollary is stronger than the assertion

$$\lim_{x \to a} \frac{R_n(x)}{(x - a)^n} = 0$$

of the first theorem.

We shall take up several applications of Taylor's theorem.

1. Representation of functions by Taylor series

If a function f possesses derivatives of all orders at a point a, then the infinite series

$$f(a) + f'(a)(x - a) + \frac{f''(a)}{2!}(x - a)^2 + \frac{f'''(a)}{3!}(x - a)^3 + \cdots$$

is called the *Taylor series of f about a.* If *a* is the number zero, the series is also known as the *Maclaurin series of f.* For the case in which $f(x) = e^x$, we have $f^{(n)}(0) = e^0 = 1$ for $n = 0, 1, 2, \ldots$, so that the Maclaurin series of e^x is

$$1 + x + \frac{x^2}{2!} + \frac{x^3}{3!} + \frac{x^4}{4!} + \cdots$$

We have previously shown that for each $x \in \mathbf{R}$, the series converges to e^x. For the case in which $f(x) = \mathrm{Tan}^{-1} x$, it is easy to show that the Maclaurin series is

$$x - \frac{x^3}{3} + \frac{x^5}{5} - \frac{x^7}{7} + \cdots;$$

we have shown that the series converges to $\mathrm{Tan}^{-1} x$ for each x in the interval $[-1, +1]$.

We shall be interested in the following general question: *For what values of x does the Taylor series of f about a converge to f(x)?* Sometimes, as in the case of the function e^x, the answer is that the Taylor series converges to the function for every value of x. At the other extreme, the series may only converge to the function at the point a itself. If, for instance,

$$f(x) = \begin{cases} e^{-1/x^2} & \text{if } x \neq 0, \\ 0 & \text{if } x = 0, \end{cases}$$

then every coefficient in the Maclaurin series for f is zero, and hence, while the series converges for each $x \in \mathbf{R}$, it only converges to $f(x)$ when $x = 0$. (See Example 4, Section 9–6.)

Saying that the Taylor series converges to $f(x)$ is equivalent to saying that

$$f(x) = \lim_{n \to \infty} T_n(x) \qquad \text{or that} \qquad \lim_{n \to \infty} R_n(x) = 0.$$

Taylor's theorem can often be used effectively to estimate $R_n(x)$.

Example 1. Prove again that

$$\cos x = 1 - \frac{x^2}{2!} + \frac{x^4}{4!} - \frac{x^6}{6!} + \cdots,$$

$$\sin x = x - \frac{x^3}{3!} + \frac{x^5}{5!} - \frac{x^7}{7!} + \cdots$$

for all $x \in \mathbf{R}$.

One can verify, first of all, that the infinite series are the Maclaurin series for the respective functions. If f is either the sine or the cosine function, then

$$|f^{(n)}(x)| = |f(x + n\pi/2)| \leq 1$$

for all $x \in \mathbf{R}$, and thus by the corollary to Taylor's theorem with Lagrange remainder,

$$|R_n(x)| \leq \frac{|x|^{n+1}}{(n+1)!} \qquad (x \in \mathbf{R}).$$

We have previously shown, however, that

$$\lim_{n \to \infty} \frac{|x|^{n+1}}{(n + 1)!} = 0.$$

Hence, $\lim_{n \to \infty} R_n(x) = 0$, which proves the validity of the series expansions.

Example 2. Find the Maclaurin series for the function $e^x \sin x$, and prove that the series converges to the function on the entire real line.

We set $f(x) = e^x \sin x$. Then

$$f'(x) = (\sin x + \cos x)e^x,$$
$$f''(x) = 2 \cos x e^x,$$
$$f'''(x) = (2 \cos x - 2 \sin x)e^x,$$
$$f''''(x) = -4 \sin x e^x = -4 f^{(0)}(x).$$

From the latter equation, it follows by induction that

$$f^{(n+4)}(x) = -4 f^{(n)}(x)$$

for all $n \in \mathbf{Z}^+$, $x \in \mathbf{R}$. The sequence $f(0), f'(0), f''(0), \ldots$ is as follows:

$$0, 1, 2, 2, 0, -4, -8, -8, 0, 16, 32, \ldots$$

Hence the Maclaurin series for $e^x \sin x$ is

$$x + \frac{2x^2}{2!} + \frac{2x^3}{3!} - \frac{4x^5}{5!} - \frac{8x^6}{6!} - \cdots$$

To show that the sequence converges to $f(x)$, we first seek bounds on $|f^{(n)}(x)|$. Clearly,

$$|f(x)| = |\sin x| e^x \le e^x.$$

Since $\sin x + \cos x = \sqrt{2} \sin (x + \pi/4)$, it follows that

$$|f'(x)| \le \sqrt{2}\, e^x.$$

Similarly,

$$|f''(x)| \le 2e^x \quad \text{and} \quad |f'''(x)| \le 2^{3/2}e^x.$$

Since $f^{(n+4)}(x) = -4 f^{(n)}(x)$, one can show by induction that

$$|f^{(n)}(x)| \le 2^{n/2}e^x.$$

Suppose now that $0 \le x$. Given $n \in \mathbf{Z}^+$, there exists by Taylor's theorem a ξ_n such that

$$R_n(x) = \frac{f^{(n+1)}(\xi_n)}{(n + 1)!} x^{n+1}$$

and $0 \le \xi_n \le x$. Since the exponential function is strictly increasing,

$$|R_n(x)| \le \frac{2^{(n+1)/2}e^x}{(n + 1)!} x^{n+1} = e^x \frac{(\sqrt{2}\, x)^{n+1}}{(n + 1)!}.$$

The latter tends to zero as $n \to \infty$. Thus $\lim_{n \to \infty} R_n(x) = 0$. When $x < 0$, we have

$$|R_n(x)| \le \frac{(\sqrt{2}\,|x|)^{n+1}}{(n+1)!},$$

so that once again we have $\lim_{n \to \infty} R_n(x) = 0$.

We have proved therefore that

$$e^x \sin x = x + \frac{2x^2}{2!} + \frac{2x^3}{3!} - \frac{4x^5}{5!} - \cdots$$

for each $x \in \mathbf{R}$.

We can formally obtain the Maclaurin expansion for $e^x \sin x$ by multiplying together the Maclaurin expansions for e^x and $\sin x$, as though they were ordinary polynomials:

$$1 + x + \frac{x^2}{2!} + \frac{x^3}{3!} + \frac{x^4}{4!} + \cdots$$
$$x \qquad - \frac{x^3}{3!} + \frac{x^5}{5!} - \cdots$$

$$\overline{\qquad\qquad\qquad\qquad\qquad\qquad\qquad}$$

$$x + x^2 + \frac{x^3}{2} + \frac{x^4}{6} + \frac{x^5}{24} + \cdots$$
$$- \frac{x^3}{6} - \frac{x^4}{6} - \frac{x^5}{12} - \cdots$$
$$+ \frac{x^5}{120} + \cdots$$

$$\overline{\qquad\qquad\qquad\qquad\qquad\qquad\qquad}$$

$$x + x^2 + \frac{x^3}{3} \qquad\quad - \frac{x^5}{30} - \cdots$$

(See Exercise 1.)

In the exercises, the reader will be asked to apply Taylor's theorem to (re)establish the validity of the following Maclaurin expansions.

$$\ln(1+x) = x - \frac{x^2}{2} + \frac{x^3}{3} - \frac{x^4}{4} + \cdots \qquad (-1 < x \le 1),$$

$$\mathrm{Tan}^{-1}\, x = x - \frac{x^3}{3} + \frac{x^5}{5} - \cdots \qquad (|x| \le 1),$$

$$(1+x)^\alpha = 1 + \binom{\alpha}{1} x + \binom{\alpha}{2} x^2 + \cdots \qquad (|x| < 1).$$

We shall use these expansions unhesitatingly from now on.

2. Approximation of functions by polynomials

The approximation of functions by polynomials is of immense practical importance. At the present time, computations of any consequential size or complexity are done on digital computers. While these machines are very fast, they are quite limited in what they are built to do: they can perform the basic operations of arithmetic (addition, subtraction, multiplication, and division), but little else. In particular, table reference, while not impossible, is usually not practical. Consequently, if values of the exponential or sine function are needed in the course of a computer calculation, these values must be generated within the computer itself through the exclusive use of the simple arithmetic operations. In most cases this procedure consists of replacing the function by an approximating polynomial.

We shall consider the approximation of functions by their Taylor polynomials.

Example 3. Show that values of the sine function may be computed to four-place accuracy for angles between 41° and 49° by means of the approximation

$$\sin\left(\frac{\pi}{4} + h\right) \approx \frac{\sqrt{2}}{2}\left(1 + h - \frac{h^2}{2}\right),$$

where h is in radians.

Observe first of all that Taylor's theorem with Lagrange remainder may be written as

$$f(a + h) = f(a) + f'(a)h + \frac{f''(a)}{2!}h^2 + \cdots + \frac{f^{(n)}(a)}{n!}h^n + \frac{f^{(n+1)}(a + \theta h)}{(n + 1)!}h^{n+1},$$

where $0 < \theta < 1$. In particular, setting $f(x) = \sin x$, $a = \pi/4$, and $n = 2$, we have

$$\sin\left(\frac{\pi}{4} + h\right) = \sin\frac{\pi}{4} + \left(\cos\frac{\pi}{4}\right)h - \frac{\sin(\pi/4)}{2!}h^2 - \frac{\cos(\pi/4 + \theta h)}{3!}h^3$$

$$= \frac{\sqrt{2}}{2}\left(1 + h - \frac{h^2}{2}\right) - \frac{\cos(\pi/4 + \theta h)}{3!}h^3.$$

For values of h between $-\pi/45$ and $\pi/45$, we have

$$\left|\frac{\cos(\pi/4 + \theta h)}{3!}h^3\right| \leq \frac{\cos 41°}{3!}\left(\frac{\pi}{45}\right)^3 = 0.000043...$$

Hence the accuracy of the approximation is as advertised.

Example 4. Calculate $\cos^2(0.1)$ by using the first three nonzero terms of the Maclaurin expansion for $\cos^2 x$. Find an upper estimate on the error.

By Taylor's theorem applied to the cosine function, we have

$$\cos^2 x = \tfrac{1}{2}(1 + \cos 2x)$$

$$= \frac{1}{2}\left[1 + 1 - \frac{(2x)^2}{2!} + \frac{(2x)^4}{4!} - \frac{\cos\xi}{6!}(2x)^6\right]$$

$$= 1 - x^2 + \frac{x^4}{3} - \frac{2\cos\xi}{45}x^6,$$

where ξ is between 0 and $2x$. Hence

$$\lim_{x\to 0}\frac{\cos^2 x - (1 - x^2 + x^4/3)}{x^4} = 0,$$

which shows that $1 - x^2 + x^4/3$ is the Taylor polynomial of order four for $\cos^2 x$ about zero. (Why?) Setting $x = 0.1$, we have

$$\cos^2(0.1) \approx 1 - 0.01 + \frac{0.001}{3} = 0.9900333...,$$

with an error which is less than

$$\tfrac{2}{45} \times 10^{-6} < 5 \times 10^{-8}.$$

Hence, to seven places, $\cos^2 (0.1) = 0.9900333$.

3. The evaluation of indeterminate forms

The Taylor polynomials can often be used to calculate indeterminate forms of the type $0/0$, where several applications of l'Hôpital's rule would otherwise be needed.

Example 5. Find

$$\lim_{x \to 0} \frac{\cos x - e^{x^2}}{x^2}.$$

We shall first use illegal tactics, and obtain

$$\cos x - e^{x^2} = \left(1 - \frac{x^2}{2!} + \frac{x^4}{4!} - \cdots\right) - \left(1 + x^2 + \frac{x^4}{2!} + \frac{x^6}{3!} + \cdots\right)$$

$$= -\tfrac{3}{2}x^2 - \tfrac{11}{24}x^4 + \cdots$$

Thus

$$\frac{\cos x - e^{x^2}}{x^2} = -\frac{3}{2} - \frac{11}{24}x^2 + \cdots \qquad (x \neq 0).$$

Hence

$$\lim_{x \to 0} \frac{\cos x - e^{x^2}}{x^2} = -\frac{3}{2}.$$

We shall eventually have enough theorems on infinite series to justify the preceding calculation. In the meantime, we can salvage a legitimate argument leading to the same result. By Theorem 1,

$$\cos x = 1 - \frac{x^2}{2} + o(x^2), \qquad e^{x^2} = 1 + x^2 + o(x^2),$$

where in each case $o(x^2)$ is used to denote a term which tends to zero more rapidly than x^2 as $x \to 0$. More generally, we shall write

$$f(x) = g(x) + o((x - a)^n) \quad \text{as} \quad x \to a \qquad \text{iff} \qquad \lim_{x \to a} \frac{f(x) - g(x)}{(x - a)^n} = 0.$$

It follows that

$$\cos x - e^{x^2} = -\tfrac{3}{2}x^2 + o(x^2),$$

so that

$$\lim_{x \to 0} \frac{\cos x - e^{x^2}}{x^2} = \lim_{x \to 0} \left(-\frac{3}{2} + \frac{o(x^2)}{x^2}\right) = -\frac{3}{2}.$$

Example 6. Find

$$\lim_{x \to 0} \frac{e^x - e^{\sin x}}{x - \sin x}.$$

Again we use some illegal tactics. We get

$$e^{\sin x} = 1 + (\sin x) + \frac{(\sin x)^2}{2!} + \cdots$$

$$= 1 + \left(x - \frac{x^3}{3!} + \cdots\right) + \frac{1}{2}\left(x - \frac{x^3}{3!} + \cdots\right)^2 + \frac{1}{6}\left(x - \frac{x^3}{3!} + \cdots\right)^3 + \cdots$$

$$= 1 + \left(x - \frac{x^3}{3!} + \cdots\right) + \frac{1}{2}\left(x^2 - \frac{x^4}{3} + \cdots\right) + \frac{1}{6}(x^3 - \cdots) + \cdots$$

$$= 1 + x + \tfrac{1}{2}x^2 + 0 \cdot x^3 + \cdots$$

By calculating the first three derivatives of $e^{\sin x}$, one can verify that the first four terms of the Maclaurin expansion are indeed $1 + x + \tfrac{1}{2}x^2 + 0 \cdot x^3$. Hence

$$e^{\sin x} = 1 + x + \tfrac{1}{2}x^2 + o(x^3).$$

Also

$$e^x = 1 + x + \tfrac{1}{2}x^2 + \tfrac{1}{6}x^3 + o(x^3).$$

It follows that

$$e^x - e^{\cos x} = \tfrac{1}{6}x^3 + o(x^3).$$

Finally,

$$\lim_{x \to 0} \frac{e^x - e^{\sin x}}{x - \sin x} = \lim_{x \to 0} \frac{\tfrac{1}{6}x^3 + o(x^3)}{x - [x - \tfrac{1}{6}x^3 + o(x^3)]}$$

$$= \lim_{x \to 0} \frac{\tfrac{1}{6} + o(x^3)/x^3}{\tfrac{1}{6} + o(x^3)/x^3} = 1.$$

In this case there is a simpler solution. Set $y = x - \sin x$. Then as $x \to 0$, $y \to 0$ properly. Hence

$$\lim_{x \to 0} \frac{e^x - e^{\sin x}}{x - \sin x} = \lim_{x \to 0} e^x \frac{1 - e^{-(x - \sin x)}}{x - \sin x}$$

$$= e^0 \lim_{y \to 0} \frac{1 - e^{-y}}{y} = \frac{d}{dy}(-e^{-y})\Big|_{y=0} = e^{-y}\Big|_{y=0} = 1.$$

4. Applications to curve sketching

We begin by reproving a result established earlier.

Proposition 5. *If $f'' > 0$ throughout a neighborhood U of a point a, then the tangent line to the curve $y = f(x)$ at the point $(a, f(a))$ lies beneath the curve (at least near the point in question).*

Proof. The tangent line has the equation

$$y = f(a) + f'(a)(x - a).$$

Observe that the right-hand side of the equation is simply the Taylor polynomial for f of order one at a. Saying that the tangent line lies beneath the curve is equivalent to asserting that

$$R_1(x) = f(x) - f(a) - f'(a)(x - a) \geq 0.$$

However, for each $x \in U$,

$$R_1(x) = \frac{f''(\xi)}{2}(x - a)^2$$

for some ξ between a and x. Hence, for all $x \in U$, $R_1(x) \geq 0$ and strict inequality holds unless $x = a$.

We next generalize earlier theorems concerning maxima and minima.

Theorem 6. Higher-Derivative Test for Relative Extrema. *Suppose that*

$$f'(x_0) = f''(x_0) = \cdots = f^{(n-1)}(x_0) = 0 \quad \text{and} \quad f^{(n)}(x_0) \neq 0.$$

Then

1) *f has a strict relative minimum at x_0 if $f^{(n)}(x_0) > 0$ and n is even;*
2) *f has a strict relative maximum at x_0 if $f^{(n)}(x_0) < 0$ and n is even;*
3) *f does not have a relative extremum at x_0 if n is odd.*

Proof. Suppose that $f^{(n)}(x_0) > 0$. Since

$$f^{(n)}(x_0) = \lim_{x \to x_0} \frac{f^{(n-1)}(x) - f^{(n-1)}(x_0)}{x - x_0} = \lim_{x \to x_0} \frac{f^{(n-1)}(x)}{x - x_0},$$

it follows that

$$f^{(n-1)}(x) > 0 \quad \text{if} \quad x \text{ is immediately to the right of } x_0,$$
$$f^{(n-1)}(x) < 0 \quad \text{if} \quad x \text{ is immediately to the left of } x_0.$$

By Taylor's theorem,

$$f(x) = f(x_0) + \frac{f^{(n-1)}(\xi)}{(n-1)!}(x - x_0)^{n-1},$$

where ξ is between x_0 and x.

Suppose n is even. Then for x immediately to the right of x_0,

$$(x - x_0)^{n-1} > 0 \quad \text{and} \quad f^{(n-1)}(\xi) > 0,$$

which implies that $f(x) > f(x_0)$. For x immediately to the left of x_0,

$$(x - x_0)^{n-1} < 0 \quad \text{and} \quad f^{(n-1)}(\xi) < 0,$$

which implies that $f(x) > f(x_0)$. This proves that f has a strict relative minimum at x_0.

If n is odd, then $(x - x_0)^{n-1} > 0$ if $x \neq x_0$. It follows that

$$f(x) > f(x_0) \quad \text{if} \quad x \text{ is immediately to the right of } x_0,$$
$$f(x) < f(x_0) \quad \text{if} \quad x \text{ is immediately to the left of } x_0.$$

Hence f does not have a relative extremum at x_0. (In this instance, we sometimes say that f is *increasing* at x_0.)

The case in which $f^{(n)}(x_0) < 0$ can be reduced to the previous one by replacing f with $-f$.

Example 7. Find the relative extrema of the polynomial $P(x) = 8x^5 - 15x^4 + 10x^2$.
The derivative

$$P'(x) = 40x^4 - 60x^3 + 20x^2 = 20x(x-1)^2(2x+1)$$

is zero at 0, 1, and $-\frac{1}{2}$. Now

$$P''(x) = 20(8x^3 - 9x^2 + 1),$$

so that we have, in particular, $P''(0) = 20$, $P''(1) = 0$, and $P''(-\frac{1}{2}) = -45$. It follows that P has a relative minimum at 0 and a relative maximum at $-\frac{1}{2}$. To determine the behavior at 1, we calculate the third derivative:

$$P'''(x) = 480x^2 - 360x.$$

In particular, $P'''(1) = 120 \neq 0$. Hence the function does not have a relative extremum at 1.

Example 8. Describe the behavior of the function $f(x) = x \operatorname{Tan}^{-1} x - (\sin x)^2$ at zero.
We calculate the first few terms of the Maclaurin series of $f(x)$ by "doing what comes naturally." We get

$$f(x) = x\left(x - \frac{x^3}{3} + \frac{x^5}{5} - \cdots\right) - \left(x - \frac{x^3}{3!} + \frac{x^5}{5!} - \cdots\right)^2$$

$$= \left(x^2 - \frac{x^4}{3} + \frac{x^6}{5} - \cdots\right) - \left(x^2 - \frac{x^4}{3} + \frac{2}{45}x^6 + \cdots\right)$$

$$= \frac{11}{45}x^6 + \cdots$$

Hence $f'(0) = f''(0) = \cdots = f^{(5)}(0) = 0$ and $f^{(6)}(0) = \frac{11}{45} \times (6!) > 0$. (The assertions in the last sentence can be established more directly.) It follows that f has a relative minimum at 0.

Exercises

1. a) Prove the second assertion of Theorem 1. More specifically, if

$$P(x) = a_0 + a_1(x-a) + \cdots + a_n(x-a)^n$$

has the property that

$$\lim_{x \to a} \frac{f(x) - P(x)}{(x-a)^n} = 0,$$

show that the coefficients are given by the recurrence relations

$$a_0 = f(a),$$

$$a_1 = \lim_{x \to a} \frac{f(x) - f(a)}{x - a},$$

$$a_2 = \lim_{x \to a} \frac{f(x) - f(a) - a_1(x-a)}{(x-a)^2},$$

$$\vdots$$

$$a_n = \lim_{x \to a} \frac{f(x) - f(a) - a_1(x-a) - \cdots - a_{n-1}(x-a)^{n-1}}{(x-a)^n}$$

(and hence are unique).

b) Suppose that f and g possess derivatives of order n at a. Let S_n and T_n be the nth order Taylor polynomials for f and g about the point a. Show that $S_n + T_n$ is the Taylor polynomial of order n for $f + g$ about a. Show that one can obtain the nth order Taylor polynomial for fg about a by truncating the polynomial $S_n T_n$ after terms involving $(x - a)^n$.

c) Deduce Leibnitz's rule for calculating the nth derivative of a product from the second part of (b).

*2. Prove or disprove the following. If

$$\lim_{x \to a} \frac{f(x) - P(x)}{(x - a)^n} = 0,$$

where P is a polynomial of degree n or less, then $f^{(n)}(a)$ exists and P is the Taylor polynomial for f of order n about a.

3. Deduce from Taylor's theorem with remainder that if P is a polynomial of degree n or less, then

$$P(x) = P(a) + P'(a)(x - a) + \cdots + \frac{P^{(n)}(a)}{n!} (x - a)^n.$$

4. Derive the *Cauchy* form of the remainder

$$R_n(x) = f^{(n+1)}(\xi) \frac{(x - \xi)^n}{n!} (x - a),$$

where ξ is between a and x. [Apply the generalized mean-value theorem for integrals to the integral expression for $R_n(x)$, by setting $g(t) \equiv 1$.]

5. a) Show that the Maclaurin expansion for $\ln (1 + x)$ is

$$x - \frac{x^2}{2} + \frac{x^3}{3} - \frac{x^4}{4} + \cdots$$

b) Use the Lagrange remainder to show that the series converges to $\ln (1 + x)$ if $0 \le x \le 1$.

c) Use the Cauchy remainder to show that the series converges to $\ln (1 + x)$ if $-1 < x \le 0$.

6. Use the Cauchy form of the remainder to prove that the binomial expansion

$$(1 + x)^\alpha = 1 + \binom{\alpha}{1} x + \binom{\alpha}{2} x^2 + \cdots$$

is valid if $-1 < x < +1$. [For n sufficiently large, show that

$$|R_n(x)| \le \left| \binom{\alpha}{n + 1} x^{n+1} \right|.$$

Show that this tends to zero as $n \to \infty$, by the procedure outlined in Exercise 9(c), Section 9–4.]

7. From the identity

$$\frac{1}{1 - x} = 1 + x + x^2 + \cdots + x^n + \frac{x^{n+1}}{1 + x},$$

deduce via Theorem 1 that $1 + x + x^2 + \cdots$ is the Maclaurin series of $1/(1 - x)$.

Show similarly that

$$x - \frac{x^3}{3} + \frac{x^5}{5} - \frac{x^7}{7} + \cdots$$

is the Maclaurin series for $\text{Tan}^{-1} x$.

8. Use the formula (Exercise 7, p. 447)

$$\frac{d^n}{dx^n} (\text{Tan}^{-1} x) = \frac{(-1)^{n-1}(n-1)!}{(1+x^2)^{n/2}} \sin\left(n \, \text{Tan}^{-1} \frac{1}{x}\right)$$

to show that

$$\text{Tan}^{-1} x = x - \frac{x^3}{3} + \frac{x^5}{5} - \frac{x^7}{7} + \cdots$$

if $-1 \le x \le 1$.

*9. a) Show that for $|x| < 1$,

$$(1 - x^2)^{-1/2} = 1 + \tfrac{1}{2}x^2 + \tfrac{1}{2} \cdot \tfrac{3}{4}x^4 + \cdots + \frac{1 \cdot 3 \cdot 5 \cdots (2n-1)}{2 \cdot 4 \cdot 6 \cdots 2n} x^{2n} + r_n(x),$$

where

$$0 \le r_n(x) \le \frac{1 \cdot 3 \cdot 5 \cdots (2n+1)}{2 \cdot 4 \cdot 6 \cdots 2n} \frac{x^{2n}}{\sqrt{1-x^2}}.$$

b) By integration, deduce that

$$\text{Sin}^{-1} x = x + \frac{1}{2}\frac{x^3}{3} + \frac{1 \cdot 3}{2 \cdot 4}\frac{x^5}{5} + \cdots + \frac{1 \cdot 3 \cdot 5 \cdots (2n-1)}{2 \cdot 4 \cdot 6 \cdots 2n}\frac{x^{2n+1}}{2n+1} + S_n(x),$$

where

$$|S_n(x)| \le \frac{1 \cdot 3 \cdot 5 \cdots (2n-1)}{2 \cdot 4 \cdot 6 \cdots 2n}\frac{|x|^{2n+1}}{\sqrt{1-x^2}} \le \frac{|x|^{2n+1}}{\sqrt{1-x^2}} \qquad (|x| < 1).$$

Deduce from this that

$$\text{Sin}^{-1} x = x + \frac{1}{2}\frac{x^3}{3} + \frac{1 \cdot 3}{2 \cdot 4}\frac{x^5}{5} + \cdots,$$

provided $|x| < 1$.

c) Note that $\pi/6$ can be computed from the above series by setting $x = \tfrac{1}{2}$. Estimate the number of terms which would be required to compute $\pi/6$ to seven decimal places.

10. Generalize Example 2 by replacing $e^x \sin x$ with $e^{ax} \sin bx$.

11. a) Using Exercise 9, Section 9–3, show that the Taylor series for $(\ln x)/x$ about the point 1 is

$$(x - 1) - (1 + \tfrac{1}{2})(x - 1)^2 + (1 + \tfrac{1}{2} + \tfrac{1}{3})(x - 1)^3$$
$$- (1 + \tfrac{1}{2} + \tfrac{1}{3} + \tfrac{1}{4})(x - 1)^4 + \cdots$$

b) Obtain the above series by formally multiplying the series

$$\frac{1}{x} = \frac{1}{1 + (x - 1)} = 1 - (x - 1) + (x - 1)^2 - \cdots$$

and

$$\ln x = \ln[1 + (x - 1)] = (x - 1) - \frac{(x - 1)^2}{2} + \frac{(x - 1)^3}{3} - \cdots$$

*c) Show that the series in (a) converges to $(\ln x)/x$ if $0 < x < 2$. (When $x \geq 1$, use the Lagrange remainder; when $x \leq 1$, use the Cauchy remainder.)

12. a) Use Exercise 8, Section 9-3, to find the Maclaurin series of $(\mathrm{Sin}^{-1} x)^2$.

   b) Also obtain the series by formally squaring the Maclaurin series for $\mathrm{Sin}^{-1} x$ (see Exercise 9).

13. Let $f(x) = e^{m\,\mathrm{Sin}^{-1} x}$.

   a) Show that f satisfies the differential equation

   $$(1 - x^2)f''(x) - xf'(x) - m^2 f(x) = 0.$$

   Deduce from Leibnitz's theorem that

   $$(1 - x^2)f^{(n+2)}(x) - (2n + 1)xf^{(n+1)}(x) - (n^2 + m^2)f^{(n)}(x) = 0.$$

   Hence we obtain the recursion formula

   $$f^{(n+2)}(0) = (n^2 + m^2)f^{(n)}(0).$$

   b) Using (a), show that the Maclaurin series for $f(x)$ is

   $$1 + mx + \frac{m^2}{2!} x^2 + \frac{m(1^2 + m^2)}{3!} x^3 + \frac{m^2(2^2 + m^2)}{4!} x^4 + \frac{m(3^2 + m^2)}{5!} x^5 + \cdots$$

   c) Obtain the series formally by using the Maclaurin expansions for e^z and $\mathrm{Sin}^{-1} x$.

14. a) Show that the Maclaurin series for $\tan x$ begins

   $$x + \frac{x^3}{3} + \tfrac{2}{15}x^5 + \cdots$$

   (You might find it helpful to use Exercise 7, Section 4-6.)

   b) Derive the same result formally by dividing the Maclaurin series for $\sin x$ by the Maclaurin series for $\cos x$.

15. In each case, a function and the first few terms of its Maclaurin series are given. Verify these terms through any convenient method (including the formal and as yet unjustified methods of the text and the previous problems).

   a) $\ln (1 + \sin x)$; $x - \dfrac{x^2}{2} + \dfrac{x^3}{6} - \dfrac{x^4}{12} + \cdots$

   b) $e^x \sin^2 x$; $x^2 + x^3 + \tfrac{1}{6}x^4 + \cdots$

   c) $\ln (x + \sqrt{1 + x^2})$; $x - \dfrac{1}{2} \dfrac{x^3}{3} + \dfrac{1 \cdot 3}{2 \cdot 4} \dfrac{x^5}{5} - \dfrac{1 \cdot 3 \cdot 5}{2 \cdot 4 \cdot 6} \dfrac{x^6}{6} + \cdots$

   d) $(1 + x)^x$; $1 + x^2 - \dfrac{x^3}{2} + \dfrac{5x^4}{6} + \cdots$

   e) $\cosh x$; $1 + \dfrac{x^2}{2!} + \dfrac{x^4}{4!} + \dfrac{x^6}{6!} + \cdots$ f) $\sinh x$; $x + \dfrac{x^3}{3!} + \dfrac{x^5}{5!} + \dfrac{x^7}{7!} + \cdots$

   g) $e^{\cos x}$; $e\left(1 - \dfrac{x^2}{2} + \dfrac{x^4}{6} - \cdots\right)$ h) $-\ln \cos x$; $\dfrac{x^2}{2} + \dfrac{x^4}{12} + \cdots$

   i) $(x - \tan x) \cos x$; $-\dfrac{2x^3}{3!} + \dfrac{4x^5}{5!} - \dfrac{6x^7}{7!} + \cdots$

16. Show that the Maclaurin series for $\cos x \cosh x$ is

$$\sum_{n=0}^{\infty} (-1)^{n+1} \frac{(4x^4)^{n-1}}{(4n - 4)!}.$$

Prove that the series converges to $\cos x \cosh x$ for each real number x.

17. Given that f is an even (odd) function possessing derivatives of all orders at zero, show that the Maclaurin series for f involves only even (odd) powers of x.

*18. Show that Taylor's formula with Lagrange remainder can be written as

$$f(a + h) = \sum_{k=0}^{n} \frac{f^{(k)}(a)}{k!} h^k + \frac{f^{(n+1)}(a + \theta h)}{(n + 1)!} h^{n+1},$$

where $0 < \theta < 1$. Assuming that $f^{(n+2)}$ is continuous and nonzero at a, prove that

$$\lim_{h \to 0} \theta = \frac{1}{n + 2}.$$

19. Prove that if we calculate e^x by the approximate formula

$$e^x \approx 1 + x + x^2/2 + x^3/6,$$

then the error will not exceed 0.01 if $0 \le x \le \frac{1}{2}$.

20. Express the polynomial $P(x) = x^3 - 6x + 2$ in powers of $x - 2$ by
    a) calculating the derivatives of P at 2, and
    b) dividing $P(x)$ successively by $x - 2$.
    Compute $f(2.003)$ to four places.

21. Use the first four nonvanishing terms of the Maclaurin expansion for $\ln (1 + x)$ to compute $\ln 1.5$. Estimate the error.

22. With what accuracy can $\cos 10°$ be computed by use of the first two nonvanishing terms of the Maclaurin expansion for $\cos x$?

23. Compute $\sin 33°$ to four-place accuracy by using sufficiently many terms of the Taylor series for $\sin x$ about the point $\pi/6$. Prove that your answer has the required accuracy. How many terms of the Maclaurin expansion of $\sin x$ would you need to obtain the same accuracy?

24. Using the Maclaurin expansion for e^x, compute $1/e$ accurately to four decimal places. Prove the accuracy of your answer.

*25. Prove that

$$\int_0^1 x^x \, dx = 1 - \frac{1}{2^2} + \frac{1}{3^3} - \frac{1}{4^4} + \cdots$$

26. Estimate the error involved in the approximation

$$e^{\cos x} \approx e\left(1 - \frac{x^2}{2} + \frac{x^4}{6}\right)$$

for $0 \le x \le \frac{1}{2}$. Using this approximation, compute $\int_0^{1/2} e^{\cos x} \, dx$ and estimate the error involved.

27. Calculate each of the following limits.

a) $\lim\limits_{x\to0} \dfrac{e^x \sin x - x - x^2}{x^2 + x \ln (1 - x)}$

b) $\lim\limits_{x\to0} \dfrac{(1 + x)^{1/x} - e}{x^2}$

c) $\lim\limits_{x\to0} \dfrac{\ln (1 + \sin x) - \ln (1 + x)}{x - \tan x}$

d) $\lim\limits_{x\to0} \dfrac{(1 + x)^x - \cosh \sqrt{2}\, x}{e^x \sin^2 x - \mathrm{Tan}^{-1} x}$

e) $\lim\limits_{x\to0} \dfrac{\ln \sec x - \sin^2 x}{x(x - \tan x) \cos x}$

28. Suppose that $f''(x_0) = f'''(x_0) = \cdots = f^{(2n)}(x_0) = 0$ and $f^{(2n+1)}(x_0) \neq 0$. Assume also that $f^{(2n+1)}$ is continuous at x_0. Prove then that the curve $y = f(x)$ crosses its tangent line at the point $(x_0, f(x_0))$. (Use the Cauchy form of the remainder.)

29. State whether the following functions have relative extrema at zero.

a) $x(1 - e^x) \sin x$

b) $x \sin x - \sin^2 x$

c) $x \,\mathrm{Tan}^{-1} x - x^2$

d) $(3 - x)e^{2x} - 4xe^x - x$

30. If $f''(a)$ exists, then it follows from l'Hôpital's rule that

$$\lim_{h\to0} \frac{f(a + h) - 2f(a) + f(a - h)}{h^2} = f''(a).$$

In numerical work one sometimes uses $[f(a + h) - 2f(a) + f(a - h)]/2h^2$ as an approximation to $f''(a)$. Assuming that f possesses a fourth derivative on the interval from $a - h$ to $a + h$, show that the error

$$e(h) = \frac{f(a + h) - 2f(a) + f(a - h)}{h^2} - f''(a)$$

can be written in the form

$$e(h) = \frac{f^{(4)}(\xi_1) + f^{(4)}(\xi_2)}{24} h^2,$$

where ξ_1 lies between a and $a + h$, and ξ_2 lies between a and $a - h$. Using *Darboux's theorem* (Exercise 8, Section 6–2), show that there exists a ξ between $a - h$ and $a + h$ such that

$$e(h) = \frac{f^{(4)}(\xi)}{12} h^2.$$

31. Fill in the details of the following proof of Taylor's theorem. We set

$$F_n(t) = f(x) - f(t) - (x - t)f'(t) - \cdots - \frac{(x - t)^n}{n!} f^{(n)}(t)$$

and

$$\varphi(t) = F_n(t) - \left(\frac{x - t}{x - a}\right)^{n+1} F_n(a).$$

Apply Rolle's theorem to φ.

32. *More about Big $\mathfrak{O}$ and Little o.* We recall some earlier definitions. If x_0 is a fixed real number, we consider three classes of functions $\mathfrak{F}_{x_0}$, $\mathfrak{B}_{x_0}$, and $\mathfrak{N}_{x_0}$. To say that a function f belongs to one of these classes means, respectively, that

1) $\mathfrak{D}_f$ includes some deleted neighborhood of x_0,
2) f is bounded throughout some deleted neighborhood of x_0, and
3) $\lim_{x\to x_0} f(x) = 0$. (See Exercise 6, Section 3–5.)

If $g \in \mathcal{F}_{x_0}$, then we shall write

$$f(x) = \mathfrak{O}(g(x)) \qquad \text{as} \qquad x \to x_0$$

iff there exists a function $h \in \mathcal{B}_{x_0}$ such that $f(x) = g(x)h(x)$. Similarly, we shall write

$$f(x) = \mathfrak{o}(g(x)) \qquad \text{as} \qquad x \to x_0$$

iff there exists a function $h \in \mathfrak{N}_{x_0}$ such that $f(x) = g(x)h(x)$. These definitions are of interest mainly when $g \in \mathfrak{N}_{x_0}$, in which case we interpret "$f(x) = \mathfrak{O}(g(x))$ as $x \to x_0$" to mean that $f(x)$ tends to zero *as rapidly* as $g(x)$ (as $x \to x_0$); similarly, we then interpret "$f(x) = \mathfrak{o}(g(x))$ as $x \to x_0$" to mean that $f(x)$ tends to zero *more rapidly* than $g(x)$.

The $\mathfrak{O}\mathfrak{o}$-notation is used rather freely in some parts of mathematics, but it is usually easy to decipher what is meant. Consider, for instance, the equation

$$\mathfrak{O}(g(x)) + \mathfrak{O}(g(x)) = \mathfrak{O}(g(x)).$$

We will interpret this as a shorthand for the following: If

$$f_1(x) = \mathfrak{O}(g(x)) \qquad \text{as} \qquad x \to x_0$$

and

$$f_2(x) = \mathfrak{O}(g(x)) \qquad \text{as} \qquad x \to x_0,$$

then

$$(f_1 + f_2)(x) = \mathfrak{O}(g(x)) \qquad \text{as} \qquad x \to x_0.$$

Prove this assertion. Interpret and prove similarly the following.

a) $\mathfrak{o}(g(x)) \pm \mathfrak{o}(g(x)) = \mathfrak{o}(g(x))$ b) $\mathfrak{O}(g_1(x))\mathfrak{O}(g_2(x)) = \mathfrak{O}(g_1(x)g_2(x))$
c) $\mathfrak{o}(g_1(x))\mathfrak{o}(g_2(x)) = \mathfrak{o}(g_1(x)g_2(x))$ d) $\mathfrak{O}(\mathfrak{O}(g(x))) = \mathfrak{O}(g(x))$
e) $\mathfrak{O}(\mathfrak{o}(g(x))) = \mathfrak{o}(g(x))$ f) $\mathfrak{o}(\mathfrak{O}(g(x))) = \mathfrak{o}(g(x))$

*33. Suppose that $f(x)$ has the Taylor series

$$a_1 + a_2(x - a) + a_2(x - a)^2 + \cdots,$$

and $g(x) - a$ has the Taylor series

$$c_1(x - b) + c_2(x - b)^2 + c_3(x - b)^3 + \cdots$$

Prove that the Taylor series for $(f \circ g)(x)$ about b can be obtained by formally substituting the second series into the first in place of $x - a$. More precisely, show that if we let

$$S_n(x) = c_1(x - b) + c_2(x - b)^2 + \cdots + c_n(x - b)^n,$$

then we can obtain the first $n + 1$ terms of the Taylor series for $f \circ g$ about b by expanding

$$a_0 + a_1 S_n(x) + a_2(S_n(x))^2 + \cdots + a_n(S_n(x))^n,$$

collecting powers of $x - b$, and dropping out all terms of order higher than $(x - b)^n$. (Use of the $\mathfrak{O}\mathfrak{o}$-notation should be helpful.)

34. Use the second-derivative test to prove that Proposition 5 holds under the weakened hypothesis that $f''(a) > 0$.

11-3 Polynomial interpolation

Interpolation has been aptly called the art of reading between the lines of a function table. Suppose, for example, that we need to calculate $\log_{10} 1.9847$, given the table entries shown in Fig. 11–3. In high school one is taught to reason in the following way: Since 1.9847 is $7/10$ of the way from 1.984 to 1.985, $\log_{10} 1.9847$ will be a number which is approximately $7/10$ of the way between 0.28870 and 0.28892. Thus

$$\log_{10} 1.9847 \approx 0.28870 + \tfrac{7}{10}(0.28892 - 0.28870) = 0.28885.$$

This method is actually only a special kind of interpolation known as *linear interpolation*. In general, suppose we need to calculate $f(x)$ from the table entries given in Fig. 11–4. Using linear interpolation, we take the number $p(x)$ as an approximate value of $f(x)$, where

$$p(x) = y_0 + \frac{(x - x_0)}{(x_1 - x_0)}(y_1 - y_0)$$

or, in a more symmetric form,

$$p(x) = y_0 \frac{(x - x_1)}{(x_0 - x_1)} + y_1 \frac{(x - x_0)}{(x_1 - x_0)}.$$

It is clear that $p(x)$ is a polynomial of degree one or less in x. Also,

$$p(x_0) = y_0 = f(x_0), \qquad p(x_1) = y_1 = f(x_1).$$

Thus in linear interpolation one approximates a function $f(x)$ by a polynomial $p(x)$ of degree one (or less) which agrees with the function for two values of x. Graphically, this amounts to approximating points on a curve by a straight line joining two points of the curve (see Fig. 11–5).

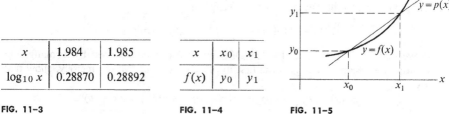

x	1.984	1.985
$\log_{10} x$	0.28870	0.28892

x	x_0	x_1
$f(x)$	y_0	y_1

FIG. 11–3   FIG. 11–4   FIG. 11–5

When the process of linear interpolation is analyzed in this way, some natural questions arise. Need we limit ourselves to using only two functional values and polynomials $p(x)$ of degree one? By using a greater number of table entries and higher degree polynomials $p(x)$, might we not improve our accuracy?

We might approximate $f(x)$, for example, by a quadratic polynomial $p(x)$ which agrees with three tabulated values of the function (see Fig. 11–6). The polynomial which we need in this case is

$$p(x) = y_0 \frac{(x - x_1)(x - x_2)}{(x_0 - x_1)(x_0 - x_2)} + y_1 \frac{(x - x_0)(x - x_2)}{(x_1 - x_0)(x_1 - x_2)}$$
$$+ y_2 \frac{(x - x_0)(x - x_1)}{(x_2 - x_0)(x_2 - x_1)}.$$

This is because $p(x)$ is obviously of degree two (or less) in x, and by direct substitution,

$$p(x_0) = y_0 = f(x_0), \qquad p(x_1) = y_1 = f(x_1), \qquad p(x_2) = y_2 = f(x_2).$$

(Don't read further until you have convinced yourself that these assertions are true!) This is called *parabolic* interpolation.

x	x_0	x_1	x_2
$f(x)$	y_0	y_1	y_2

FIG. 11-6

x	x_0	x_1	$\cdots$	x_r
$f(x)$	y_0	y_1	$\cdots$	y_r

FIG. 11-7

In the most general case of polynomial interpolation we approximate a function $f(x)$ with a polynomial $p(x)$ of degree r or less which agrees with $r + 1$ tabulated values of the function (see Fig. 11-7). The polynomial $p(x)$ is completely determined by these conditions (see Exercise 3), and it may be written explicitly in a form which bears the name of Lagrange, namely,

$$p(x) = \sum_{k=0}^{r} y_k \prod_{\substack{j=0 \\ j \neq k}}^{r} \frac{(x - x_j)}{(x_k - x_j)},$$

where

$$\prod_{\substack{j=0 \\ j \neq k}}^{r} \frac{(x - x_j)}{(x_k - x_j)}$$

denotes the product of all terms of the form $(x - x_j)/(x_k - x_j)$, where j is any integer between 0 and r except k. Thus

$$\prod_{\substack{j=0 \\ j \neq k}}^{r} \frac{(x - x_j)}{(x_k - x_j)} = \frac{(x - x_0) \cdots (x - x_{k-1})(x - x_{k+1}) \cdots (x - x_r)}{(x_k - x_0) \cdots (x_k - x_{k-1})(x_k - x_{k+1}) \cdots (x_k - x_r)}.$$

When $r = 1$ or 2, this formula reduces to those obtained earlier for linear and parabolic interpolation.

In the important special case of equally spaced intervals, $x_n = x_0 + nh$, it is convenient to express these polynomials in a form due to Newton. In particular, the formula for parabolic interpolation can be written

$$p(x_1 + k) = y_1 + \frac{k}{2h}\left(y_2 - y_0 + \frac{k}{h}\delta^2(y_1)\right),$$

where

$$\delta^2(y_1) = (y_2 - y_1) - (y_1 - y_0)$$

x	$x_0 = x_1 - h$	x_1	$x_1 + h$
$f(x)$	y_0	y_1	y_2

FIG. 11-8

(see Fig. 11–8). If, as in many tables, second differences $\delta^2 y$ are tabulated, this formula involves only 2 multiplications and 3 additions per value.

Let us try to compute sin 28° from the table given in Fig. 11–9. Using linear interpolation, we get

$$\sin 28° \approx 0.46905;$$

using parabolic interpolation,

$$\sin 28° \approx 0.46950.$$

The correct value is sin 28° = 0.46947. This example seems to indicate that parabolic interpolation is more accurate than linear interpolation. In investigating whether or not this is true, we are led to consider *truncation errors*.

x	25° (= x_0)	30° (= x_1)	35° (= x_2)
$\sin x$	0.42262 (= y_0)	0.50000 (= y_1)	0.57358 (= y_2)

FIG. 11-9

Suppose now that we approximate a function $f(x)$ by a polynomial $p(x)$ which agrees with $f(x)$ for certain values of x. We can then write

$$f(x) = p(x) + e(x).$$

We shall call the function $e(x)$ the truncation error. The next theorem, which is an analog of Taylor's theorem with Lagrange remainder, enables us to estimate in many cases the size of $e(x)$.

Theorem 1. *Let $p(x)$ be a polynomial of degree r or less such that $p(x_i) = f(x_i)$, where $i = 0, 1, \ldots, r$. Let u be a number, and let I be the smallest closed interval containing the numbers $u, x_0, x_1, \ldots, x_r$. Let us suppose that $f(x)$ is continuous on I and that $f^{(r+1)}(x)$ exists throughout the interior of I. Then there exists a number ξ in the interior of I such that*

$$e(u) = f(u) - p(u) = \frac{(u - x_0)(u - x_1) \cdots (u - x_r)}{(r + 1)!} f^{(r+1)}(\xi).$$

The proof is analogous to that given for Taylor's theorem with Lagrange remainder.

Proof. We begin by making two observations concerning the error function

$$e(x) = f(x) - p(x).$$

Since $p(x)$ agrees with $f(x)$ when $x = x_i$ $(i = 0, 1, \ldots, r)$, the error function must vanish at these points, that is,

$$e(x_0) = e(x_1) = \cdots = e(x_r) = 0.$$

We also note that since $p(x)$ is a polynomial of degree r at most, $p^{(r+1)}(x) = f^{(r+1)}(x)$.

We next define a function φ in the following way:

$$\varphi(x) = e(x) - \frac{(x - x_0) \cdots (x - x_r)}{(u - x_0) \cdots (u - x_n)} e(u).$$

[The reader will note that the definition of $\varphi(x)$ is not valid if u is equal to one of the x_i. However, since the theorem is then trivially true, we need not consider this case.] By inspection, we see that $\varphi(x)$ vanishes when $x = x_i$ $(i = 0, 1, \ldots, r)$, and also when $x = u$. In total $\varphi(x)$ vanishes at least $r + 2$ times in the interval I. By Rolle's theorem, $\varphi'(x)$ must vanish at least $r + 1$ times in I. Applying Rolle's theorem to $\varphi'(x)$, we conclude that $\varphi''(x)$ must vanish at least r times in I. More generally, it follows that $\varphi^{(k)}(x)$ must vanish $r + 2 - k$ times in I, and, in particular, $\varphi^{(r+1)}(x)$ must vanish at least once in I, say $\varphi^{(r+1)}(\xi) = 0$.

Hence we have

$$0 = \varphi^{(r+1)}(\xi) = e^{(r+1)}(\xi) - \frac{(r + 1)! e(u)}{(u - x_0) \cdots (u - x_r)},$$

and since $e^{(r+1)}(\xi) = f^{(r+1)}(\xi)$, this yields

$$e(u) = \frac{(u - x_0) \cdots (u - x_r)}{(r + 1)!} f^{(r+1)}(\xi).$$

To illustrate the use of this error formula, let us consider parabolic interpolation with equally spaced intervals. Suppose, that is, that we are using a function table in which the successive entries for the independent variable have a constant difference h. Then using our former notation, we have

$$e(x_1 + k) = k(k^2 - h^2) f'''(\xi)/6.$$

Unless one is working at the end of a function table, one can choose the table entries so that $|k| \leq h/2$. In this case $|k(k^2 - h^2)| < 3h^3/8$, so that

$$|e(x_1 + k)| \leq \frac{h^3 |f'''(\xi)|}{16},$$

where ξ lies between $x_1 - h$ and $x_1 + h$. While we can't compute $f'''(\xi)$—since we don't know what ξ is in general—we can often get useful bounds on it. Suppose, for example, that $f(x) = \sin x$. Then $f'''(x) = -\cos x$, so that $|f'''(\xi)| \leq 1$. If $h = 1°$ or 0.01745 radians, then

$$|e(x)| < \frac{(0.02)^3}{16} < 5 \times 10^{-7}.$$

This shows that using parabolic interpolation, one can obtain six-place accuracy from a table in which $\sin x$ is tabulated at intervals of a degree (rather than a minute, which is the usual case).

Exercises

1. Find the quadratic polynomial $p(x)$ such that $p(-1) = 8$, $p(0) = 2$, and $p(1) = -2$. Do this by two methods.

   a) Use the Lagrange interpolation formula.
   b) Let $p(x) = C_0 + C_1 x + C_2 x^2$, and obtain three simultaneous linear equations for C_0, C_1, and C_2.

2. Write out in full the Lagrange interpolation formula for the case in which $r = 3$.

3. Let the numbers $x_0, x_1, \ldots, x_r$ and $y_0, y_1, \ldots, y_r$ be given, and suppose that the x_k's are distinct. Using the theorem in this section, show that the polynomial $p(x)$ of degree r or less which satisfies the $r + 1$ conditions $p(x_i) = y_i$ $(i = 0, 1, \ldots, r)$ is unique.

4. Show that the formula for linear interpolation in the case of equally spaced intervals can be written

$$p(x_0 + k) = y_0 + (k/h)(y_1 - y_0) \qquad (x_1 = x_0 + h).$$

Show also that the truncation error is

$$e(x_0 + k) = \frac{k(h - k)}{2} f''(\xi).$$

Suppose that $0 < k < h$, so that $x_0 + k$ (and also ξ) lie between the "tabular" values x_0 and $x_1 = x_0 + h$. Show that $|k(h - k)|$ is largest when $k = h/2$ and hence that

$$|e(x_0 + k)| \le (h^2/8)|f''(\xi)|.$$

5. At what intervals must $\log x$ be tabulated for $1 \le x \le 2$ to get four-place accuracy by
   a) linear interpolation?   b) parabolic interpolation?

6. Obtain as a special case of the Lagrange formula with remainder the *Newton forward difference formula:*

$$f(x) = \sum_{k=0}^{n} \frac{\Delta^k f(0)}{k!} x^{(k)} + \frac{f^{(n+1)}(\xi)}{(n+1)!} x^{(n+1)}.$$

(For notations, see pp. 156, 158 and also Exercise 16, Section 4-6.) Note the similarity to Taylor's formula with remainder.

7. The Taylor polynomials can be regarded as limiting cases of interpolation polynomials.

   a) Assume that f is differentiable at a. For sufficiently small h, we let P_h be that polynomial of degree one or less which agrees with f at the points a and $a + h$. Write P_h in the form

$$P_h(x) = a_0(h) + a_1(h)(x - a).$$

   Show that as $h \to 0$, $a_0(h) \to f(a)$ and $a_1(h) \to f'(a)$. Thus, as $h \to 0$, the polynomial P_h approaches the first-order Taylor polynomial for f at a.
   b) Assume that f has a second derivative at a. Let P_h be the polynomial of degree two or less such that P_h agrees with f at the points $a - h$, a, and $a + h$. Show that as $h \to 0$, the polynomial P_h approaches [in the same sense as in (a)] the second-order Taylor polynomial for f at a.

8. Suppose that f possesses a third derivative throughout an interval containing x_0 and x_1, where $x_0 < x_1$.

   a) Show that there exists a unique polynomial P of degree two or less such that

   $$P(x_0) = f(x_0), \qquad P'(x_0) = f'(x_0), \qquad P(x_1) = f(x_1).$$

   (Thus P is a cross between a Taylor polynomial and an interpolation polynomial.)

   b) If $x_0 < x < x_1$, show that there exists a ξ between x_0 and x_1 such that

   $$f(x) - P(x) = \frac{f'''(\xi)}{6} (x - x_0)^2 (x - x_1).$$

   [Set $e(t) = f(t) - P(t)$ and

   $$\varphi(t) = e(t) - \frac{(t - x_0)^2 (t - x_1)}{(x - x_0)^2 (x - x_1)} e(x).$$

   Observe that φ has simple zeros at x_1 and x and a double zero at x_0. Apply Rolle's theorem four times to deduce that there exists a ξ between x_0 and x_1 such that $\varphi'''(\xi) = 0$.]

9. Suppose that f possesses a fourth derivative throughout an interval containing three points, x_0, x_1, and x_2, with $x_0 < x_1 < x_2$. Show that there exists a unique polynomial of degree three or less such that

   $$P(x_0) = f(x_0), \qquad P(x_1) = f(x_1), \qquad P(x_2) = f(x_2), \qquad \text{and} \qquad P'(x_1) = f'(x_1).$$

   Show that $f(x) - P(x)$ can be written in the form

   $$\frac{f^{(4)}(\xi)}{4!} (x - x_0)(x - x_1)^2 (x - x_2).$$

10. Formulate a general theorem which encompasses Taylor's theorem (with Lagrange remainder), the Lagrange formula with remainder, and the previous two problems.

11. Using the error formula for linear interpolation, prove that if a function f possesses a positive second derivative throughout an interval I, then f is strictly convex on I.

12. Suppose that f possesses a bounded derivative of order $n + 1$ throughout the interval $[-1, +1]$. We seek to approximate the function f as well as possible on the interval $[-1, +1]$ by means of a polynomial P of degree n or less, where P is to be obtained by interpolation. The following question then arises. At which $n + 1$ points should we interpolate so as to minimize the error $f(x) - P(x)$? Observe that the error may be written

$$f(x) - P(x) = \frac{f^{(n+1)}(\xi)}{(n + 1)!} L(x),$$

where $L(x) = (x - x_0)(x - x_1) \cdots (x - x_n)$, and $x_0, \ldots, x_n$ are points at which the interpolation is done. In the absence of any further information regarding the function f, it is reasonable to choose $x_0, \ldots, x_n$ so as to minimize

$$\|L\| = \max \{|L(x)| \mid -1 \le x \le 1\}.$$

Conclude that $x_0, \ldots, x_n$ should be the zeros of the Chebyshev polynomial T_{n+1}. What is $\|L\|$ in this case? (See Exercise 11, Section 9-2.)

13. Suppose that f possesses derivatives of all orders on the entire line, and suppose that there exists a constant M such that $|f^{(n)}(x)| \leq M$ for all $x \in \mathbf{R}$ and $n \in \mathbf{Z}^+$. Let $h > 0$ be fixed and for each n, let P_n be the unique polynomial of degree $2n - 1$ or less which agrees with f at the $2n$ points $-(n - 1)h$, $-(n - 2)h$, $\ldots$, h, 0, h, $2h$, $\ldots$, nh. We shall be concerned with the convergence of $P_n(x)$ to $f(x)$ as $n \to \infty$ for $0 \leq x \leq h$. The remainder $|f(x) - P_n(x)|$ is bounded by $M|L(x)|/(2n)!$, where

$$L(x) = [x + (n - 1)h][x + (n - 2)h] \cdots (x - nh).$$

It can be shown that

$$\max \{|L(x)| \mid 0 \leq x \leq h\} = |L(h/2)|.$$

Show that

$$\frac{1}{(2n)!} \left| L\left(\frac{h}{2}\right) \right| = \frac{(2n)!}{2^{4n}(n!)^2} h^{2n}.$$

Use Stirling's formula to show that the latter is asymptotic to

$$\left(\frac{h}{2}\right)^{2n} \frac{1}{\sqrt{\pi n}}.$$

Conclude that convergence is guaranteed only if $0 \leq h \leq 2$.

11-4 Numerical integration

Many definite integrals—$\int_0^1 e^{-x^2}\, dx$ and $\int_{-1}^{1/2} t^{-1} \sin t\, dt$, for example—cannot be calculated in terms of elementary functions via the second fundamental theorem. To calculate such integrals, one must often resort to numerical methods. We consider two of these methods, the *trapezoidal rule* and *Simpson's rule*. In both cases interpolation is used and the overall procedure is the same. If we are given a definite integral $\int_a^b f(x)\, dx$ ($a < b$) to calculate, we first split the interval of integration $[a, b]$ into a number of fairly short subintervals, say, $I_1, \ldots, I_n$. Then

$$\int_a^b f(x)\, dx = \sum_{k=1}^n \int_{I_k} f(x)\, dx.$$

To calculate one of the integrals $\int_{I_k} f(x)\, dx$, we replace the function f by a polynomial P_k obtained by interpolating at certain points of the interval I_k. This leads to the approximation

$$\int_a^b f(x)\, dx \approx \sum_{k=1}^n \int_{I_k} P_k(x)\, dx.$$

In the case of the trapezoidal rule, the intervals $I_1, \ldots, I_n$ are of equal length, namely, $h = (b - a)/n$. The endpoints of the intervals may then be written

$$x_k = a + kh \qquad (k = 0, 1, \ldots, n).$$

We let P_k be the polynomial of degree one or less which agrees with f at the points x_{k-1} and x_k; thus, $hP_k(x_{k-1} + t) = (h - t)f(x_{k-1}) + t f(x_k)$. Since

$$\int_{x_{k-1}}^{x_k} P_k(x) \, dx = h^{-1} \int_0^h hP_k(x_{k-1} + t) \, dt$$

$$= -f(x_{k-1}) \frac{(h - t)^2}{2h}\Big|_0^h + f(x_k) \frac{t^2}{2h}\Big|_0^h$$

$$= (h/2)[f(x_{k-1}) + f(x_k)],$$

we are led to the trapezoidal formula:

$$\int_a^b f(x) \, dx \approx T = \sum_{k=1}^n \frac{h}{2}[f(x_{h-1}) + f(x_k)]$$

$$= \tfrac{1}{2}[f(x_0) + 2f(x_1) + 2f(x_2) + \cdots + 2f(x_{n-1}) + f(x_n)].$$

To apply the trapezoidal rule intelligently, we need to be able to estimate the error involved. We shall assume that f possesses a continuous second derivative throughout the interval $[a, b]$. If $x_{k-1} \leq x \leq x_k$, then by the theorem of the preceding section, there exists a number ξ between x_{k-1} and x_k such that

$$f(x) - P_k(x) = \frac{(x - x_{k-1})(x - x_k)}{2} f''(\xi).$$

Since the expression $(x - x_{k-1})(x - x_k)$ is negative on the interval (x_{k-1}, x_k), we have

$$\frac{(x - x_{k-1})(x - x_k)}{2} M_2 \leq f(x) - P_k(x) \leq \frac{(x - x_{k-1})(x - x_k)}{2} m_2,$$

where M_2 and m_2 denote the maximum and minimum values, respectively, of f'' on the interval $[x_{k-1}, x_k]$. Now

$$\int_{x_{k-1}}^{x_k} \frac{(x - x_{k-1})(x - x_k)}{2} \, dx = \int_0^h \frac{t(h - t)}{2} \, dt = -\frac{h^3}{12}.$$

Hence

$$\frac{h^3}{12} m_2 \leq \int_{x_{k-1}}^{x_k} f(x) \, dx - \int_{x_{k-1}}^{x_k} P_k(x) \, dx \leq \frac{h^3}{12} M_2.$$

By the intermediate-value theorem,

$$\int_{x_{k-1}}^{x_k} f(x) \, dx - \int_{x_{k-1}}^{x_k} P_k(x) \, dx = \frac{h^3}{12} f''(\xi)$$

for some ξ between x_{k-1} and x_k. By summing over the above, one can readily see that the error in trapezoidal approximation cannot exceed

$$\frac{nh^3}{12} M = \frac{(b - a)^3}{12n^2} M,$$

where

$$M = \max \{|f''(x)| \mid a \leq x \leq b\}.$$

If f is positive over the interval, then the trapezoidal rule has a simple geometric interpretation. The integral $\int_a^b f(x)\,dx$ is the "area beneath the graph of f" and T is the sum of the areas of n trapezoids.

We summarize the discussion to this point.

Theorem 1. Trapezoidal Rule. *Suppose that f has a continuous second derivative on the interval $[a, b]$, and set*

$$M^{(2)} = \max \{|f''(x)| \mid a \le x \le b\}.$$

Then $\int_a^b f(x)\,dx$ is given approximately by the trapezoidal formula

$$T = (h/2)[f(x_0) + 2f(x_1) + \cdots + 2f(x_{n-1}) + f(x_n)],$$

where $h = (b - a)/n$ and $x_k = a + kh$ ($k = 0, 1, \ldots, n$), with an error which does not exceed

$$\frac{nh^3 M^{(2)}}{12} = \frac{(b - a)^3}{12n^2} M^{(2)}.$$

Example 1. Plan the computation of $\ln 2 = \int_1^2 (dt/t)$ to five decimal places using the trapezoidal rule.

In this case $M^{(2)}$, the maximum of the absolute value of $2/t^3$ on the interval $1 \le t \le 2$, is 2. The required accuracy is assured if

$$2/12n^2 < 5 \times 10^{-6} \quad \text{or} \quad n > 183.$$

In the case of Simpson's rule, the intervals $I_1, \ldots, I_n$ are also of equal length, say, $2h = (b - a)/n$. If we set

$$x_k = a + kh \quad (k = 0, 1, \ldots, 2n),$$

as before, then the endpoints of the intervals $I_1, \ldots, I_n$ are $x_0, x_2, x_4, \ldots, x_{2n}$. In this case, the polynomial P_k is that polynomial of degree two or less which agrees with f at the three points $x_{2k-2}, x_{2k-1}, x_{2k}$. The reader can then show that

$$\int_{x_{2k-2}}^{x_{2k}} P_k(x)\,dx = (h/3)[f(x_{2k-2}) + 4f(x_{2k-1}) + f(x_{2k})].$$

Thus we are led to Simpson's formula:

$$\int_a^b f(x)\,dx \approx S = \sum_{k=1}^{n} \frac{h}{3}[f(x_{2k-2}) + 4f(x_{2k-1}) + f(x_{2k})]$$

$$= (h/3)[y_0 + y_{2n} + 2(y_2 + y_4 + \cdots + y_{2n-2})$$
$$+ 4(y_1 + y_3 + \cdots + y_{2n-1})],$$

where $y_k = f(x_k)$.

The derivation of an error formula for Simpson's rule is somewhat delicate. It suffices to consider the error involved in the approximate formula

$$\int_{-h}^{h} f(x)\,dx \approx (h/3)[f(-h) + 4f(0) + f(h)].$$

The formula is exact if f is a polynomial of degree two or less. It is also exact if $f(x) = x^3$, since both sides are zero. It follows by linearity that the approximation formula is exact for all polynomials of degree three or less.

Suppose now that f is any function possessing a continuous fourth derivative on the interval $[-h, h]$. We let P be that unique polynomial of degree three or less such that

$$P(-h) = f(-h), \qquad P(0) = f(0), \qquad P(h) = f(h), \qquad \text{and} \qquad P'(0) = f'(0).$$

Then by the remarks of the preceding paragraph,

$$(h/3)[f(-h) + 4f(0) + f(h)] = \int_{-h}^{h} P(x)\,dx.$$

Thus the error in the approximation

$$\int_{-h}^{h} f(x)\,dx \approx (h/3)[f(-h) + 4f(0) + f(h)]$$

is equal to $|\int_{-h}^{h} [f(x) - P(x)]\,dx|$. If $-h \le x \le h$, then by Exercise 9, Section 11-3, we may write

$$f(x) - P(x) = \frac{x^2(x^2 - h^2)}{24} f^{(4)}(\xi)$$

for some $\xi \in (-h, h)$. If we let

$$M^{(4)} = \max\{|f^{(4)}(x)| \,|\, -h \le x \le h\},$$

then

$$\left| \int_{-h}^{h} [f(x) - P(x)]\,dx \right| \le \int_{-h}^{h} |f(x) - P(x)|\,dx$$

$$\le -\int_{-h}^{h} \frac{x^2(x^2 - h^2)}{24} M^{(4)}\,dx = \frac{h^5}{90} M^{(4)}.$$

It follows by linear substitutions and addition that the error in Simpson's approximation cannot exceed

$$\frac{nh^5}{90} M^{(4)} = \frac{(b-a)^5}{2880n^4} M^{(4)},$$

where $M^{(4)} = \max\{|f^{(4)}(x)| \,|\, a \le x \le b\}$.

Theorem 2. Simpson's Rule. *Suppose that f has a continuous fourth derivative on the interval $[a, b]$, and set*

$$M^{(4)} = \max\{|f^{(4)}(x)| \,|\, a \le x \le b\}.$$

Then $\int_a^b f(x)\,dx$ is given approximately by Simpson's formula,

$$S = (h/3)[y_0 + y_{2n} + 2(y_2 + y_4 + \cdots + y_{2n-2}) + 4(y_1 + y_3 + \cdots + y_{2n-1})],$$

where $h = (b-a)/2n$ and $y_k = f(a + kh)$ $(k = 0, 1, \ldots, 2n)$, with an error which does not exceed

$$\frac{nh^5}{90} M^{(4)} = \frac{(b-a)^5}{2880n^4} M^{(4)}.$$

Example 2. Plan the calculation of $\ln 2 = \int_1^2 (dt/t)$ to five decimal places using Simpson's rule.

Here

$$M^{(4)} = \max \{24/x^4 \mid 1 \le x \le 2\} = 24.$$

The required accuracy is assured if

$$\frac{24}{2880n^4} < 5 \times 10^{-6} \qquad \text{or} \qquad n > \frac{10}{\sqrt[4]{6}} = 6+.$$

Hence to calculate $\ln 2$ to five places using Simpson's rule, we need to compute only 15 values of the function $1/x$, as opposed to the 183 values we would need if we used the trapezoidal rule. Using Riemann sums, we find that the required number of values is on the order of 10^5! (See Exercise 8, Section 8–3.)

Exercises

1. Plan the calculation of $\int_0^1 [(\sin t)/t] \, dt$ to five decimal places using
   a) the trapezoidal rule,   b) Simpson's rule.

2. Establish the formula

   $$\int_{x_{2k-2}}^{x_{2k}} P_k(x) \, dx = (h/3)[f(x_{2k-2}) + 4f(x_{2k-1}) + f(x_{2k})]$$

   in the development of Simpson's rule.

3. a) Consider approximations of the form

   $$\int_0^h f(x) \, dx \approx w_1 f(h) + w_2 f(0) + w_3 f(-h),$$

   where the weights w_1, w_2, w_3 are constants. Show that the approximation is exact for the polynomials 1, x, and x^2 iff $w_1 = \frac{5}{12}$, $w_2 = \frac{8}{12}$, and $w_3 = -\frac{1}{12}$. Thus we have *Simpson's five-eight-minus-one rule:*

   $$\int_0^h f(x) \, dx \approx (h/12)[5f(h) + 8f(0) - f(-h)].$$

   Explain why the formula is exact for all polynomials of degree two or less.

   b) Prove that the error involved in this rule can be written in the form

   $$\frac{h^4 f'''(\xi)}{24}$$

   for some ξ in the interval $(-h, h)$.

*4. Establish the formula

   $$\int_0^h f(x) \, dx = \frac{h}{2}[y_0 + y_1] + \frac{h^2}{12}[y_0' - y_1'] + \frac{h^5}{720} f^{(4)}(\xi),$$

   where

   $$y_0 = f(0), \qquad y_1 = f(h), \qquad y_0' = f'(0), \qquad y_1' = f'(h), \qquad \text{and} \qquad 0 < \xi < h.$$

5. Compute π from the formula

$$\frac{\pi}{4} = \int_0^1 \frac{dx}{1 + x^2}$$

by

a) using the trapezoidal formula with $h = 0.1$,
b) using Simpson's formula with $h = 0.1$.

In each case compare the error with the error bounds given in the theorems.

6. We shall consider approximations of the form

$$\int_{-1}^{+1} f(x)\, dx \approx \sum_{k=1}^{r} w_k f(x_k),$$

where $w_1, \ldots, w_r$ are fixed weights and $x_1, \ldots, x_r$ are fixed points in $[-1, +1]$. By interpolating at the points $x_1, \ldots, x_r$, one can choose the weights so that the approximation is exact for all polynomials of degree r or less. Occasionally, however, the formula is exact for polynomials of higher order. For instance, Simpson's rule

$$\int_{-1}^{+1} f(x)\, dx \approx \tfrac{1}{3}[f(-1) + 4f(0) + f(+1)]$$

is exact for all polynomials of degree three or less. The purpose of this exercise is to show that if we take $x_1, \ldots, x_r$ to be the zeros of the Legendre polynomial of order r, then an approximation of the above type can be found which is exact for all polynomials of degree $2r - 1$ or less.

In Exercise 15, Section 4–6, the Legendre polynomial of order n was defined to be

$$P_n(x) = \frac{1}{2^n n!} \frac{d^n}{dx^n} (x^2 - 1)^n.$$

The polynomial $P_n(x)$ is of degree n. Also P_n satisfies the differential equation

$$\frac{d}{dx}[(x^2 - 1)P_n'(x)] - n(n + 1)P_n(x) = 0.$$

[See the same exercise, part (c).]

a) Let $\varphi(x) = (x - a)^n (x - b)^n$. If $1 \le k < n$, show that $\varphi^{(k)}$ has zeros of order $n - k$ at a and b, together with k simple zeros between a and b. (Use Rolle's theorem, etc.) Deduce that $\varphi^{(n)}$ has exactly n simple zeros which lie strictly between a and b. Deduce in particular, that the Legendre polynomial P_n has exactly n simple zeros in the interval $(-1, +1)$.

b) Derive the *orthogonality relation*

$$\int_{-1}^1 P_m(x)P_n(x)\, dx = 0 \quad \text{if} \quad m \ne n.$$

(Use the differential equation above, and perform two integrations by parts.)

c) Given that P is a polynomial of degree n or less, show that P can be uniquely written

$$P(x) = c_0 P_0(x) + \cdots + c_n P_n(x),$$

where $c_0, \ldots, c_n$ are constants. Deduce that $\int_{-1}^1 Q(x)P_r(x)\, dx = 0$ if Q is a polynomial of degree less than r.

d) Let $x_1, \ldots, x_r$ be the zeros of P_r. If f is any function on the interval $[-1, 1]$, we let P_f be the unique polynomial of degree $r - 1$ or less such that $P_f(x_k) = f(x_k)$, $k = 1, \ldots, r$. Show that the approximation

$$\int_{-1}^{1} f(x)\, dx \approx \int_{-1}^{1} P_f(x)\, dx$$

is of the form

$$\int_{-1}^{1} f(x)\, dx \approx \sum_{k=1}^{r} w_k f(x_k),$$

where the weights $w_1, \ldots, w_r$ are independent of f. In particular, if $r = 3$, show that the approximation formula is

$$\int_{-1}^{1} f(x)\, dx = \tfrac{1}{9}[5f(-\alpha) + 8f(0) + 5f(\alpha)],$$

where $\alpha = \sqrt{\tfrac{3}{5}}$.

e) Show that the approximation formula in (d) is exact if f is a polynomial of degree $2r - 1$ or less. [Observe that $f(x) - P_f(x)$ is a polynomial which vanishes at $x_1, \ldots, x_r$ and must therefore be divisible by $(x - x_1)(x - x_2) \cdots (x - x_r)$. Deduce that

$$f(x) - P_f(x) = Q(x)P_r(x)$$

for some polynomial Q. What is the degree of Q? Deduce from (c) that

$$\int_{-1}^{1} f(x)\, dx = \int_{-1}^{1} P_f(x)\, dx,$$

that is, that the approximation is exact for f.]

f) For the case in which $r = 3$, verify (e) directly for each of the polynomials x^4 and x^5.

7. Modify the derivation of the trapezoidal formula by letting P_k be the Taylor polynomial of order one at $(x_{k-1} + x_k)/2 = x_{k-1} + h/2$. Show that this leads to the *tangent formula*

$$\int_{a}^{b} f(x)\, dx \approx h[y_{1/2} + y_{3/2} + \cdots + y_{(2n-1)/2}],$$

where $y_{1/2} = f(x_0 + h/2)$, $y_{3/2} = f(x_1 + h/2)$, $\ldots$ Interpret the tangent approximation geometrically. If f is convex on $[a, b]$, show that the tangent rule underestimates $\int_a^b f(x)\, dx$, while the trapezoidal rule overestimates it. Prove finally that the error involved in the tangent formula approximation does not exceed $M^{(2)}(b - a)h^2/24$, where $M^{(2)} = \max \{|f''(x)| \,|\, a \leq x \leq b\}$.

11-5 Newton's method

An important problem is that of finding numerical solutions of equations. Of the many techniques which have been developed to solve problems of this sort, Newton's method is one of the most effective and widely used. We cannot deal with Newton's method in complete generality. (Like most good things in mathematics, its generalizations have reached stratospheric heights of abstraction.) A sacrifice of generality is in this case, however, no cause for sadness. We can best acquire an appreciation of the rationale behind the method, its simplicity and effectiveness, by examining the special case considered below.

Let f be a real-valued function defined on some subset of the real line. Newton's method is a technique for finding roots of the equation $f(x) = 0$. [A root of the equation $f(x) = 0$ is also said to be a *zero* of the function f.] Occasionally, as in the case of a quadratic polynomial, one can find such roots by using formulas. Similar formulas also exist for polynomials of degree three and four, although they are too complicated to be of much computational use. Formulas for finding the roots of polynomials of degree five or more do not exist.

The rationale behind Newton's method is simple. Suppose that x^* is a root of the equation $f(x) = 0$, and that the graph of the function resembles Fig. 11-10. Although we do not know the value of x^*, we have some rough estimate of x^*, say, x_0. We now use a straight line, specifically the tangent line to the graph of the function at the point $(x_0, f(x_0))$, as an approximation to the graph of f, and we find the point x_1 where this line intersects the x-axis. It would be highly fortuitous if x_1 turned out to be x^*, and yet it may very well happen, as in the case shown in Fig. 11-10, that x_1 will be a closer approximation to x^* than our initial estimate x_0. When this is the case, we might get an even better estimate of x^* be repeating the process with x_1 taking over the role previously played by x_0. This will yield a third estimate of x^*, x_2, and continuing in this fashion, we can generate an entire sequence of real numbers $\{x_n\}$. Hopefully, the sequence will converge to x^*, and the sequence will converge rapidly enough so that after a relatively small number of iterations, we will get an estimate for x^* within the required degree of accuracy.

We shall now get a formula for the elements of the sequence $\{x_n\}$. Using the slope-point formula, we see at once that the equation of the tangent line to the graph of the function at the point $(x_0, f(x_0))$ is

$$y - f(x_0) = f'(x_0)(x - x_0).$$

To find x_1, the point where the tangent line crosses the x-axis, we set y equal to zero in the above equation and solve for x. This yields

$$x_1 = x_0 - \frac{f(x_0)}{f'(x_0)}.$$

To get x_2, we simply replace x_0 by x_1 in the above equation. Thus

$$x_2 = x_1 - \frac{f(x_1)}{f'(x_1)}.$$

More generally,

$$x_{n+1} = x_n - \frac{f(x_n)}{f'(x_n)}. \qquad (*)$$

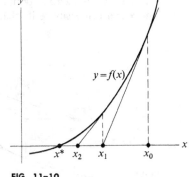

FIG. 11-10

Let us consider some examples.

Example 1. Let $f(x) = x^2 - 2$. Then the two zeros of f are $\pm\sqrt{2}$. The recursion formula (∗) is in this instance

$$x_{n+1} = x_n - \frac{x_n^2 - 2}{2x_n} = \frac{1}{2}\left(x_n + \frac{2}{x_n}\right).$$

We showed previously (Example 12, Section 3–3) that if x_0 is any positive number, then the sequence $\{x_n\}$ obtained from the above equation converges to $\sqrt{2}$. If x_0 is negative, the sequence converges to $-\sqrt{2}$.

Example 2. Find a root x^* of the equation

$$f(x) = x \log_{10} x - 2 = 0$$

to six decimal places.

Since $f(3) = -0.6$ and $f(4) = 0.4$, it follows from the intermediate-value theorem that f has a zero in the interval $(3, 4)$. We let $x_0 = 3.5$. The succeeding values of the sequence $\{\lambda_n\}$ are then computed from the formula

$$x_{n+1} = x_n - \frac{f(x_n)}{f'(x_n)} = \frac{2 - x_n \log_{10} e}{\log_{10} x_n - \log_{10} e}.$$

We get

$$x_1 = 3.598, \qquad x_2 = 3.5972849, \qquad x_3 = 3.5972850.$$

Comparing the last two values of x_k, we can expect that $x^* = 3.597285$ is the correct value of x^* to six places. To check, we compute the values of the function at the points 3.597285 and 3.5972855. The values turn out to be negative and positive, respectively. Hence by the intermediate-value theorem, 3.597285 is the correct value of x^* to six decimal places.

Example 3. There can usually be found cases where any numerical method fails, and fails spectacularly.

Suppose that we were foolish enough to attempt to solve the equation

$$f(x) = \sqrt[3]{x} = 0$$

by Newton's method. The result would be disastrous. No matter how good an initial approximation x_0 we took, the sequence of iterates $\{x_n\}$ which we would get could not converge to the root of the equation (which, of course, is zero in this case). In fact, one can see immediately from the graph of the function (Fig. 11–11) that the sequence would frantically run away from zero in both directions.

The reader can easily verify that in this case $x_n = (-2)^n x_0$.

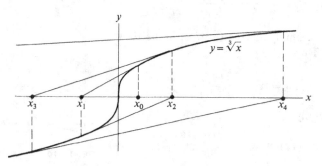

FIG. 11–11

We turn next to theorems which are concerned with when and how well Newton's method works.

In what follows we shall consider a real-valued function f defined on a subset of the real line. We shall define a second function T by the formula

$$T(x) = x - \frac{f(x)}{f'(x)}.$$

Equation (∗) can be written more compactly as $x_{n+1} = T(x_n)$. It follows that

$$x_n = T^n(x_0),$$

where T^n denotes the composition product of T with itself n times. In such a case, we say that $x_n = T^n(x_0)$ is the nth *iterate* of x_0 under T. The questions which we shall consider now are these: Given an x_0, when will its sequence of iterates under T converge to a root of the equation $f(x) = 0$? If $\{T^n(x_0)\}$ does converge to such a root, what can be said about the rate of convergence?

Proposition 1. *If the sequence $\{T^n(x_0)\}$ converges to a number x^*, and if f' is bounded throughout some neighborhood of x^*, then $f(x^*) = 0$.*

Proof. From (∗) we have

$$f(x_n) = f'(x_n)(x_n - x_{n+1}).$$

Since f' is by assumption bounded near x^*, and since $x_n \to x^*$, it follows that the sequence $\{f'(x_n)\}$ is bounded. Also, $(x_n - x_{n+1}) \to x^* - x^* = 0$ as $n \to \infty$. Hence the sequence $\{f(x_n)\}$ is null, since it is product of a bounded sequence and a null sequence. Therefore,

$$f(x^*) = \lim_{n \to \infty} f(x_n) = 0.$$

Proposition 2. *Suppose that $f(x^*) = 0$, $f'(x^*) \neq 0$, and that $f''(x^*)$ exists. Then there exists a neighborhood U of x^* such that*

$$\lim_{n \to \infty} T^n(x) = x^*$$

for each $x \in U$.

Proof. We observe that under the above hypothesis x^* is a *fixed point* of T, that is, $T(x^*) = x^*$. Moreover,

$$T(x) = 1 - \frac{[f'(x)]^2 - f(x)f''(x)}{[f'(x)]^2} = \frac{f(x)f''(x)}{[f'(x)]^2};$$

in particular, $T'(x^*) = 0$. Hence it suffices to prove the following result.

Proposition 3. *Suppose that x^* is a fixed point of a function φ and that $|\varphi'(x^*)| < 1$. Then there exists a neighborhood U of x^* such that*

$$\lim_{n \to \infty} \varphi^n(x) = x^*$$

for all $x \in U$.

Proof. Without loss of generality, we may suppose that $x^* = 0$. (Why?) We select
a number r such that $|\varphi'(0)| = |\varphi'(x^*)| < r < 1$. Since

$$\lim_{x\to 0} \frac{\varphi(x) - \varphi'(0)x}{x} = \varphi'(0) - \varphi'(0) = 0,$$

there exists a $\delta > 0$ such that

$$0 < |x| < \delta \Rightarrow \left|\frac{\varphi(x) - \varphi'(0)x}{x}\right| < r - |\varphi'(0)|.$$

It follows from the triangle inequality that

$$|x| < \delta \Rightarrow |\varphi(x) - \varphi'(0)x| < (r - |\varphi'(0)|)|x|$$
$$\Rightarrow |\varphi(x)| \le |\varphi(x) - \varphi'(0)x| + |\varphi'(0)x| \le r|x|.$$

We let $U = \{x \mid |x| < \delta\}$. Then U is a neighborhood of $x^* = 0$ and

$$x \in U \Rightarrow |\varphi(x)| \le r|x|.$$

Since $r < 1$, the above inequality implies that $\varphi(x) \in U$. It follows by induction that
if $x \in U$, then the sequence of iterates $\{\varphi^n(x)\}$ is entirely contained in U, and for
each $n \in \mathbf{Z}^+$,

$$|\varphi^n(x)| \le r^n|x|.$$

As $n \to \infty$, $r^n \to 0$, and thus

$$\lim_{n\to\infty} \varphi^n(x) = 0 = x^*.$$

   In summary, we have shown (under the hypotheses of Proposition 2) that Newton's
method works provided our initial estimate x_0 is not too far off from x^*. Moreover,
we have seen that the sequence of iterates $\{T^n(x_0)\}$ converges to x^* more rapidly
than r^n tends to zero, where r is *any* number with absolute value less than one. We
shall now deduce from Taylor's theorem that the rate of convergence is more on the
order of r^{2^n} (where, as before, $|r| < 1$).
   Let us assume now that $f(x^*) = 0$, $f'(x^*) \ne 0$, and that f''' exists throughout
some neighborhood of x^*. If x_n is sufficiently close to x^*, we can apply Taylor's
theorem to the function T to get

$$x_{n+1} = T(x_n) = T(x^*) + T'(x^*)(x_n - x^*) + \frac{T''(\xi)}{2}(x_n - x^*)^2$$
$$= x^* + \frac{T''(\xi)}{2}(x_n - x^*)^2,$$

where ξ is between x_n and x^*. If we set $e_n = x_n - x^*$, the error after n iterations of
Newton's method, then

$$\frac{e_{n+1}}{e_n^2} = \frac{T''(\xi)}{2}.$$

If x_0 is sufficiently close to x^*, then as $n \to \infty$, $x_n \to x^*$, and

$$\frac{e_{n+1}}{e_n^2} \to \frac{T''(x^*)}{2} = \frac{f''(x^*)}{2f'(x^*)}.$$

Thus

$$e_{n+1} \sim K e_n^2,$$

where K is the constant $f''(x^*)/[2f'(x^*)]$. Because of this asymptotic relation for the error, Newton's method is said to *converge quadratically*.

Exercises

1. The purpose of this exercise is to give a nonlocal criterion for convergence under Newton's method. The theorem goes as follows: *Suppose that f is twice differentiable on the interval $[a, b]$, and that the following are satisfied:*

   i) $f(a)f(b) < 0$.
   ii) *For all* $x \in [a, b]$, $f'(x) \neq 0$.
   iii) *f'' does not assume both positive and negative values.*
   iv) *If c denotes the endpoint of $[a, b]$ at which $|f'(x)|$ is smaller, then*

   $$\left| \frac{f(c)}{f'(c)} \right| \leq b - a.$$

   Then f has a unique zero x^* in the interval $[a, b]$ and for each $x \in [a, b]$,

   $$\lim_{n \to \infty} T^n(x) = x^*.$$

   a) Use the intermediate-value theorem and Rolle's theorem to prove the first assertion of the theorem.
   b) Show that there are four possibilities to be considered:
      Case I. $f(a) < 0$, $f(b) > 0$, $f''(x) \geq 0$, $c = a$.
      Case II. $f(a) > 0$, $f(b) < 0$, $f''(x) \geq 0$, $c = b$.
      Case III. $f(a) < 0$, $f(b) > 0$, $f''(x) \leq 0$, $c = b$.
      Case IV. $f(a) > 0$, $f(b) < 0$, $f''(x) \leq 0$, $c = a$.

      Prove that one can reduce cases II and IV to cases I and III by replacing f by $-f$. Show that one can reduce case III to case I by replacing $f(x)$ by $f(-x)$.
   c) Prove the theorem for case I. Draw a picture of the situation. Argue geometrically that if $x^* < x_0 \leq b$, then the sequence of iterates $\{T^n(x_0)\}$ is strictly decreasing and bounded below by x^*. Argue that the sequence must therefore have a limit, and that the limit must be x^*. If $a \leq x_0 < x^*$, show that $x^* < T(x_0) \leq b$, so that the succeeding iterates must form a decreasing sequence which converges to x^*.

2. Let f satisfy the conditions of the preceding problem. Let $a \leq x_0 \leq b$, and set

   $$x_n = T^n(x_0) \qquad \text{and} \qquad e_n = x^* - x_n.$$

   a) By Taylor's theorem,

   $$0 = f(x^*) = f(x_n) + f'(x_n)(x^* - x_n) + \frac{f''(\xi)}{2}(x^* - x_n)^2,$$

for some ξ between x_n and x^*. Deduce that

$$e_{n+1} = \frac{f''(\xi)}{2f'(x_n)} e_n^2,$$

and hence that

$$e_{n+1} \sim K e_n^2,$$

where

$$K = \frac{f''(x^*)}{2f'(x^*)}.$$

b) Assume now that f'' is bounded on the interval $[a, b]$. We set

$$M = \max \{|f''(x)| \mid a \leq x \leq b\},$$
$$m = \min \{|f'(x)| \mid a \leq x \leq b\}.$$

Show that if $M \leq 4m$, and if x_n is the value of x^* to k decimal places, then x_{n+1} is the value of x^* to $2k$ decimal places.

3. a) Let $A > 0$. By considering the function $f(x) = 1/x - A$, obtain, via Newton's method, an iterative scheme for computing A^{-1} *without performing any division.*

   *b) Apply the theorem of Exercise 1 to show that convergence is assured if

$$0 < x_0 < 2A^{-1}.$$

c) If $x_0 = 1$, show that the nth iterate is

$$x_n = \frac{1 - (1 - A)^{2^n}}{A}.$$

   Deduce that the sequence converges if $0 < A < 2$. What if $A \geq 2$?

d) Starting with $x_0 = 0.3$, compute e^{-1} to five decimal places by the iterative technique discussed here ($e = 2.7182183...$).

4. Kepler's equation is of the form

$$m = x - E \sin x,$$

where m and E are constants. The equation plays a rather considerable role in dynamical astronomy.

a) Given that $m = 0.8$ and $E = 0.2$, solve the equation to four places by Newton's method.

b) Observe that the solution to Kepler's equation is a fixed point of the function

$$\varphi(x) = m + E \sin x.$$

For the case considered in (a), show that Proposition 3, Section 11–5, is applicable. Show that the sequence defined by the recursion formula

$$x_0 = m, \qquad x_{n+1} = \varphi(x_n)$$

converges to the root of Kepler's equation. Estimate the number of iterations required to obtain four-place accuracy.

5. a) When φ is an invertible function, observe that x^* is a fixed point for f iff it is a fixed point for φ^{-1}. Suppose that x^* is a fixed point of φ, but $|\varphi(x^*)| > 1$. Then, although Proposition 3, Section 11–5, cannot be applied to φ, show that it can be applied to φ^{-1}.

   b) Show that the equation $x = \tan x$ has a unique solution x^* in the interval $(\pi/2, 3\pi/2)$. Let

   $$x_0 = \pi, \qquad x_{n+1} = \pi + \mathrm{Tan}^{-1} x_n.$$

   Show that the sequence converges to x^*. What can you say about the sequence defined by

   $$y_0 = \pi, \qquad y_{n+1} = \tan y_n?$$

   c) Solve the equation to three places using Newton's method.

12 □ PLANE CURVES

We shall resume our earlier study of plane analytic geometry. We begin with a study of the conic sections. Following this are two sections on linear algebra in which we consider certain self-mappings of the plane which are of geometric interest. We then apply the results of these two sections to the classification of plane sets which are defined by equations of the form

$$Ax^2 + 2Bxy + Cy^2 + Dx + Ey + F = 0.$$

Except for some degenerate cases, these sets all turn out to be conic sections. Finally, we give a brief account of the differential geometry of plane curves.

12-1 The conics in central position

Much of elementary plane geometry is concerned with straight lines and circles. A more inclusive family of plane figures which has been studied from very early times is the family of *conic sections*. As the name suggests, a conic section is a plane figure obtained by the intersection of a right circular cone C with a plane π. [For the benefit of readers who may be rusty on their solid geometry, we recall some definitions. Let γ be a circle, let l be the line perpendicular to the plane of γ which passes through the center of γ, and let V be any point on l other than the center of γ (see Fig. 12–1). Then γ and V determine a right circular cone C in the following way. We define a *ruling* of C to be any line which passes through V and a point on γ. The cone C itself is defined to be the union of all its rulings. The line l is called the *axis* of C, and the point V is called the *vertex* of C. If we take the union of those rays which emanate from V and pass through γ, we get a *nappe* of the cone. Clearly, a cone has two nappes; these nappes are symmetric with respect to the plane through the vertex which is perpendicular to the axis.] If the plane π intersects the vertex of the cone, then the resulting figure is either the union of

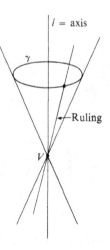

FIG. 12–1

530

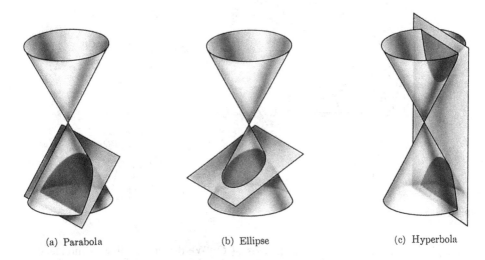

(a) Parabola   (b) Ellipse   (c) Hyperbola

FIG. 12-2

two intersecting lines, a straight line (if π is tangent to the cone), or a singleton set. These are called *degenerate* conics. When π intersects at some place other than the vertex, three different kinds of figures can arise. If π is parallel to a ruling of the cone, the curve of intersection is a *parabola*. If π intersects only one nappe of the cone but is not parallel to a ruling, then we get an *ellipse*. If π intersects both nappes of C, then the intersection is a *hyperbola*; it is a union of two curves called its *branches*. The parabola, ellipse, and hyperbola are pictured in Fig. 12–2.

Although conic sections are plane curves, it would appear that three-dimensional geometry is required for their study. It turns out, however, that the conic sections can be given two-dimensional characterizations. Specifically, if K is any conic section in a plane π other than a circle, then there exists a straight line D, a point F in π, and a positive number e such that K is the set of points P in π, with the property that

$$|FP| = e\,d(P, D), \qquad\qquad (*)$$

where $d(P, D)$ denotes the distance from P to the line D. The line D is called the *directrix* of K, the point F is called its *focus*, and the number e is called its *eccentricity*. In degenerate conics, F lies on D.

We shall now establish the *intrinsic equation* $(*)$ for K. We shall suppose that π intersects the axis of C obliquely at some point other than the vertex. We denote the angle made by the axis of C with any of its rulings by α, and the angle between the axis of C and the plane π by β. We now construct a sphere in the following way. Imagine a very small spherical balloon nestled inside a nappe of C intersected by π [see Fig. 12–3(a)]. We inflate the balloon, keeping it tangent to C, until the moment when the balloon is also tangent to π. We denote the balloon at this stage by S and we let F be its point of tangency with π. The sphere S is tangent to C along a circle

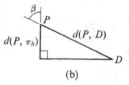

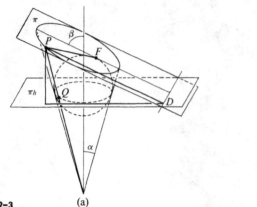

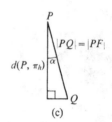

FIG. 12–3   (a)   (c)

whose plane π_h is perpendicular to the axis of C. We denote the line of intersection of π and π_h by D. Consider now an arbitrary point P on $K = \pi \cap C$. We take the ruling of C through P, and denote its intersection with π_h by Q. Now the line segments FP and PQ, which are tangent to S, have the same length. By referring to the cross sections shown in Figs. 12–3(b) and 12–3(c), we see that the distance from P to the plane π_h can be written in two ways, namely, $d(P, D) \cos \beta$ and $|PF| \cos \alpha$. Thus

$$|FP| = e\, d(P, D),$$

where $e = (\cos \beta)/(\cos \alpha)$. If K is a parabola, then $\alpha = \beta$, so that $e = 1$. If K is an ellipse, then $\alpha < \beta < \pi/2$, so that $0 < e < 1$. If K is a hyperbola, then $\beta < \alpha$, so that $e > 1$. To complete the discussion, one must show that if P satisfies (∗), then P lies on K. In addition, one must consider the case in which π intersects at the vertex of C and the case in which π is parallel to the axis of C. We omit these matters.

We shall now discuss separately the cases in which K is a parabola, an ellipse, and a hyperbola.

1. K is a parabola

We obtain the simplest equation for K by introducing coordinate axes as follows. We let the x-axis be parallel to D and midway between F and D. We then choose the y-axis so that F lies on the positive y-axis. Then for some $p > 0$, the coordinates of F are $(0, p)$, and the equation of D is $y = -p$ (see Fig. 12–4). Let $P = (x, y)$ be any point on K. Then since the eccentricity of K is one,

$$\sqrt{x^2 + (y - p)^2} = |FP| = d(P, D) = y + p.$$

Only a small amount of trivial algebra separates us from the standard equation for K, namely,

$$4py = x^2.$$

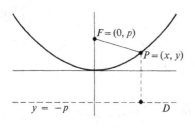

FIG. 12–4

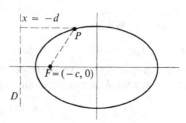

FIG. 12–5

The y-axis is clearly an axis of symmetry for K; it is simply called the *axis* of K. The point of K on this axis, namely, the origin, is called the *vertex* of K.

The most interesting property of a parabola is its *reflection property*. Let P be any point on K. We construct two rays emanating from P, one through F and the other parallel to the axis of K, as indicated in Fig. 12–5. Then the reflection property states that these rays make equal angles with the tangent line to K at P. Consequently, if a headlight reflector is made by revolving a parabola about its axis of symmetry, and if the headlight is located at the focus, then the reflected beams of light will be parallel. The proof of the reflection property is left to the reader. We suggest that he examine Fig. 12–6 and prove first that the quadrilateral $FPQG$ is a rhombus.

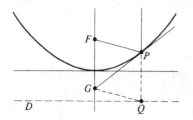

FIG. 12–6

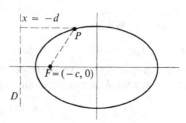

FIG. 12–7

2. *K* is an ellipse

We introduce coordinate axes so that F has the coordinates $(-c, 0)$, and D has the equation $x = -d$ (see Fig. 12–7). Then, if we set $P = (x, y)$, our intrinsic equation for K, namely, $|FP| = e\, d(P, D)$, becomes

$$(x + c)^2 + y^2 = e^2(x + d)^2.$$

The choice of c and d can be adjusted so that $c = e^2 d$, with the result that the first-order terms in the above equation cancel. [Specifically, we can let d be the distance between F and D divided by $(1 - e^2)$.] The resulting equation for K may then be rewritten in the familiar form

$$\left(\frac{x}{a}\right)^2 + \left(\frac{y}{b}\right)^2 = 1,$$

where

$$a = ed \quad \text{and} \quad b = \sqrt{1 - e^2}\, a.$$

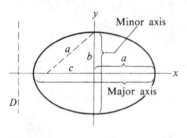

FIG. 12-8

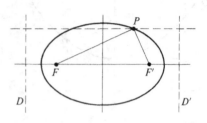

FIG. 12-9

(Note that we must use the fact that $0 < e < 1$.) Also observe that

$$a^2 - b^2 = e^2 a^2 = e^4 d^2 = c^2.$$

The numbers a and b are easily interpreted geometrically. The points $(\pm a, 0)$ lie on K, and the line segment joining them is called the *major axis* of K. It can be shown that among all line segments joining two points of K, the major axis has the greatest length. The points $(0, \pm b)$ also lie on K, and the line segment joining them is called the *minor axis* of K. Among all chords of K which pass through the origin, the minor axis has least length. The equation $a^2 - b^2 = c^2$ shows that the distance from F to either endpoint of the minor axis is a. We observe, moreover, that the original parameters e, c, and d are easily expressed in terms of the new parameters a and b. We have

$$e = \sqrt{1 - (b/a)^2}, \qquad c = \sqrt{a^2 - b^2} = ea, \qquad d = a^2/\sqrt{a^2 - b^2} = a/e.$$

We can quickly derive basic geometric properties for K from its equation. We observe first that K is bounded, since all its points lie inside the rectangle

$$\{(x, y) \mid |x| \le a, |y| \le b\}.$$

It is also clear that K is symmetric about its major and minor axes, and hence also about their point of intersection (the origin, by our choice of coordinates), which is called the *center* of K (see Fig. 12–8). This is a bit surprising; without first looking at some careful drawings one might imagine that K would be smaller at the end closer to the vertex of the cone. It follows, at any rate, that the point F' and the line D', obtained by the reflection of F and D in the y-axis, serve as focus and directrix, respectively, for K. (See Fig. 12–9.) The *string property* of K is an immediate consequence of this observation. If P is any point on K, then

$$|FP| + |F'P| = e\, d(P, D) + e\, d(P, D') = e\, d(D, D') = 2ed = 2a.$$

In other words, the sum of the lengths of the *focal radii* drawn to points on the ellipse is a constant, namely, $2a$. We shall presently show that, conversely, a point P in π satisfying the equation

$$|FP| + |F'P| = 2a$$

must lie on K. The string property provides a simple mechanical construction for K as Fig. 12–10 should suggest.

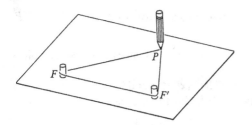

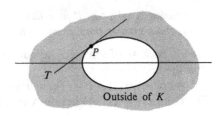

Outside of K

FIG. 12-10   FIG. 12-11

We have previously observed that K has the smooth parametrization

$$x = a \cos t, \qquad y = b \sin t \qquad (0 \leq t \leq 2\pi),$$

and that as t varies over the interval $[0, 2\pi)$, the point described by the equations passes through each point of K exactly once. Thus K is a *simple closed curve*. The famous Jordan curve theorem states that such a curve γ divides the plane into two regions having γ as common boundary, namely, the set of points "inside" γ and the set of points "outside" γ. In the case of K, the set of points inside K is described by the inequality $(x/a)^2 + (y/b)^2 < 1$, while the points outside K are described by the opposite inequality $(x/a)^2 + (y/b)^2 > 1$. Alternatively, the points inside K can be described as the points which lie between the graph of the function

$$f(x) = a^{-1}b\sqrt{a^2 - x^2}$$

and its negative. One can check that the function f has a negative second derivative on the interval $(-a, a)$. Thus f is concave. It follows that if T is the line which is tangent to K at some point P in the upper half-plane, then all points of T except P itself lie above K and hence outside K. This proves part of the next theorem.

Proposition 1. *Suppose that T is a line tangent to K at a point P. Then all points of T except P itself lie outside K (see Fig. 12-11).*

We have established this result for the case in which P lies (strictly) in the upper half-plane. By reflection in the x-axis, the result also holds if P lies (strictly) in the lower half-plane. Finally, one can check that the theorem is true at the endpoints of the major axes, in which case T is vertical.

Another consequence of the fact that f is concave is that the set

$$K^-(f) = \{(x, y) \mid |x| \leq a, y \leq f(x)\}$$

is convex. Similarly,

$$K^+(-f) = \{(x, y) \mid |x| \leq a, y \geq -f(x)\}$$

is convex. Consequently, the union of K with the set of points inside K, which is $K^-(f) \cap K^+(-f)$, is convex.

We also have a generalization of the string property.

Proposition 2. *A point P in π lies inside, outside, or on K according as $|FP| + |F'P|$ is less than, greater than, or equal to 2a.*

Proof. Because of symmetry about the coordinate axes, it suffices to consider points P in the first quadrant.

Suppose first that $P = (x, y)$ lies inside K. Then $|x| < a$ and $0 \leq y < f(x) = y^*$. The point $P^* = (x, y^*)$ lies on K (see Fig. 12–12). Furthermore,

$$|FP| = \sqrt{(x + c)^2 + y^2} < \sqrt{(x + c)^2 + (y^*)^2} = |FP^*|,$$

and, similarly, $|F'P| < |F'P^*|$. Hence

$$|FP| + |F'P| < |FP^*| + |F'P^*| = 2a.$$

Let us next suppose that $P = (x, y)$ lies outside K. If $x \geq a$, then

$$|FP| \geq x + c \geq a + c \quad \text{and} \quad |FP| \geq x - c \geq a - c,$$

with strict inequality in at least one instance. As a result, we have $|FP| + |F'P| > 2a$. If $0 \leq x < a$, we observe that $y > f(x) = y^*$. Then $P^* = (x, y^*)$ lies on K, and by an argument similar to the one in the preceding paragraph,

$$|FP| + |F'P| > |FP^*| + |F'P^*| = 2a.$$

It follows that if $|FP| + |F'P| = 2a$, then P lies on K.

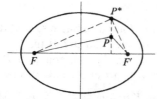

FIG. 12–12

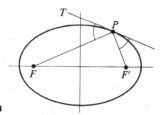

FIG. 12–13

We can now prove the reflection property for the ellipse; this property is the principle underlying the phenomenon of whispering galleries.

Proposition 3. *A tangent line to K makes equal angles with the focal radii drawn to the point of tangency.*

Proof. Let T be a tangent to K, and let P be the point of tangency (see Fig. 12–13). Among all points Q on T, the point P is the one for which $|FQ| + |F'Q|$ is a minimum. The desired result then follows from the solution to Osgood's barnyard problem.

The computation of the arc length of an ellipse leads to an important class of transcendental functions called *elliptic functions*. Consider the arc of the ellipse

$$x = a \cos t, \quad y = b \sin t \quad (0 \leq t \leq \phi),$$

where $a > b > 0$ and $0 \le \phi \le \pi/2$. Its path length is

$$s = \int_0^\phi \sqrt{a^2 \sin^2 t + b^2 \cos^2 t}\, dt$$

$$= a \int_0^\phi \sqrt{1 - e^2 \cos^2 t}\, dt.$$

The substitution $t = \pi/2 - u$ leads to

$$s = a \int_{\pi/2-\phi}^{\pi/2} \sqrt{1 - e^2 \sin^2 u}\, du$$

$$= a[E(e) - E(e, \pi/2 - \phi)],$$

where

$$E(k, \phi) = \int_0^\phi \sqrt{1 - k^2 \sin^2 \phi}\, d\phi \qquad \text{and} \qquad E(k) = E(k, \pi/2).$$

The function $E(k, \phi)$ is called the *Legendre normal form of the incomplete elliptic integral of the second kind;* $E(k)$ is the *complete* elliptic integral of the second kind. Tables listing values of these integrals are readily available.

3. *K* is a hyperbola

We shall limit our discussion to statements of the most pertinent facts. Proofs are omitted.

We introduce coordinate axes exactly as in the previous case. If we let $P = (x, y)$, then the intrinsic equation for K leads to the equation

$$\frac{x^2}{a^2} - \frac{y^2}{b^2} = 1,$$

where we have the following relationships among the parameters a, b, c, d, and e:

$$c = ae, \qquad d = a/e, \qquad c = \sqrt{a^2 + b^2}.$$

The reader can then check that the parameters are as advertised in Fig. 12–14.

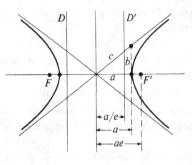

FIG. 12–14

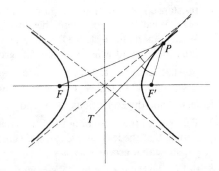

FIG. 12–15

As in the case of the ellipse, the hyperbola K has two foci and two directrices situated symmetrically about the y-axis. Figure 12–14 suggests that the lines

$$\frac{x}{a} - \frac{y}{b} = 0$$

and

$$\frac{x}{a} + \frac{y}{b} = 0$$

are asymptotes. This can be checked (Exercise 9, Section 6–7). The equations are easily remembered: Replace the number 1 by 0 in the equation for K and factor.

The hyperbola K has a string property, namely, K is the set of points P in π for which $|FP| - |F'P|$, the difference in the lengths of the focal radii drawn to P, is in absolute value equal to $2a$. The hyperbola K also has a reflection property: A tangent line to K bisects the angle made by the focal radii drawn to the point of tangency, as shown in Fig. 12–15.

Exercises

1. Show that $y = Ax^2 + Bx + C$ $(A \neq 0)$ is a parabola with axes parallel to the y-axis. Determine the equations for the axis and the directrix, and find the coordinates of the focus.

2. Show that the equation of a parabola with directrix $x + y = 0$ and focus (p, p) is $(x - y)^2 - 4p(x + y) + 4p^2 = 0$.

3. Prove the reflection property of a parabola.

4. A projectile is fired at ground level. Its initial speed is v_0 and its angle of inclination is ϕ (see Fig. 12–16). Argue that if coordinates are suitably chosen, the path of the projectile is described by the parametric equations

$$x = (v_0 \cos \phi)t, \qquad y = (v_0 \sin \phi)t - \tfrac{1}{2}gt^2,$$

where g is the acceleration due to gravity. (Neglect air resistance.) Show that the path is an arc of a parabola. What is the maximum height of the projectile? **FIG. 12–16**
What is its range (distance from the point where it was fired to the point where it strikes the ground)?

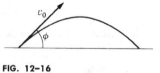

5. In what sense can a circle be regarded as a limiting case of an ellipse? In what sense is it appropriate to regard a circle as a conic with eccentricity zero?

6. Determine the eccentricity, foci, and directrices of the ellipse $4x^2 + 25y^2 = 100$.

7. The path of the earth about the sun is nearly an ellipse, with the sun as one focus. Assuming that the least and greatest distances between the earth and the sun have the ratio $\frac{29}{30}$, show that eccentricity is $\frac{1}{50}$.

8. Fill in the details for the "ice-cream cone proof" of the string property on an ellipse. Using the notation of the text, we suppose that $\alpha < \beta < \pi/2$. Imagine two spherical

scoops of ice cream nestled in the nappe of C intersected by π (see Fig. 12–17). Both spheres S and S' are tangent to C and to π but they lie on different sides of π. We denote the points where S and S' are tangent to π by F and F', respectively. Argue that S and S' intersect C in two circles γ and γ' which lie in parallel planes. Let P be any point on $K = C \cap \pi$. Let Q and Q' be the points where the ruling of C through P meets γ and γ'. Show that

$$|FP| + |F'P| = |QP| + |PQ'| = |QQ'|.$$

Deduce that $|FP| + |F'P|$ is the same for all points P on K.

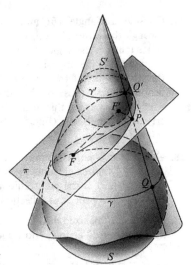

FIG. 12–17

9. a) Prove that the midpoints of parallel chords of an ellipse lie on a line which passes through the center of the ellipse.

   b) Prove conversely that if l is a chord which passes through the center of an ellipse, then the chords of the ellipse bisected by l are parallel. Show, moreover, that the tangent lines constructed at the endpoints of l are also parallel to the chords bisected by l.

10. Let l be a chord of an ellipse which passes through a focus F. Prove that the tangent lines constructed at the endpoints of l intersect at a point on the directrix nearest F.

11. Prove that among all chords of an ellipse, the major axis has greatest length. Also prove that among all chords passing through the center of an ellipse, the minor axis has least length.

12. Derive the equation $(x/a)^2 - (y/b)^2 = 1$ for a hyperbola.

13. Prove the string property for a hyperbola.

14. Describe the sets defined by the inequalities

$$\left(\frac{x}{a}\right)^2 - \left(\frac{y}{b}\right)^2 < 1 \quad \text{and} \quad \left(\frac{x}{a}\right)^2 - \left(\frac{y}{b}\right)^2 > 1.$$

    Which is convex? Characterize these sets by string properties.

15. Show that the hyperbolas $(x/a)^2 - (y/b)^2 = 1$ and $(y/b)^2 - (x/a)^2 = 1$ have the same asymptotes. Sketch a pair of these hyperbolas. Such hyperbolas are said to be *conjugate*. Show that the sum of the squares of the reciprocals of the eccentricities of two conjugate hyperbolas is equal to one.

16. The points where a hyperbola intersects the line through its foci are called the *vertices* of the hyperbola. Prove that each tangent line to a hyperbola intersects the line segment determined by its vertices. Show that the asymptotes of a hyperbola can be regarded as limiting positions of tangent lines to the hyperbola.

17. Consider a triangle ABC and a line l which passes through the vertex A. A theorem of plane geometry states that l is the angle bisector of the angle at A iff l intersects the opposite side BC in such a way that the lengths of the resulting line segments are proportional to those of the adjacent sides of the triangle. Assuming this, prove the reflection property of a hyperbola. (Figure 12–18 will be appropriate.)

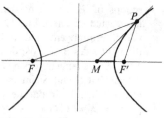

FIG. 12–18

18. a) Given two points F and F' and a line l which intersects the line segment joining F and F', find by calculus the point P on l for which $|FP| - |F'P|$ has minimum absolute value. Is the point unique?

   b) Use (a) to prove the reflection property of a hyperbola.

19. Let F and F' be fixed distinct points in a plane. We let $\mathcal{E}$ denote the family of all ellipses having F and F' as foci, and we let $\mathcal{H}$ denote the family of all hyperbolas having F and F' as foci.

   a) If $E \in \mathcal{E}$ and $H \in \mathcal{H}$, prove that E and H intersect orthogonally. Prove, in other words, that $E \cap H$ is nonempty and that for each $P \in E \cap H$, the tangent lines to E and H at P are perpendicular.

   b) To what extent is the following true: For each point P in the plane, there exists a unique $E \in \mathcal{E}$ and $H \in \mathcal{H}$ such that P belongs to $E \cap H$?

   c) Show that the reflection properties for ellipses and hyperbolas are essentially equivalent [i.e., one follows from the other by virtue of (a) and (b)].

20. If π is a plane parallel to the axis of a right circular cone C, show that $\pi \cap C$ is a hyperbola.

21. Let K be a conic with eccentricity $e > 0$, and let p be the distance from a focus to the corresponding directrix. Let coordinates be introduced so that the origin is a focus and so that the corresponding directrix is parallel to the y-axis and lies in the left-hand plane. Prove that

$$r = \frac{ep}{1 - e \cos \theta} \qquad (r > 0)$$

is the polar equation of K if K is an ellipse or a parabola. What does the equation represent if K is a hyperbola?

22. a) Suppose that C_1 and C_2 are curves having the polar equations $r = f_1(\theta)$, $r = f_2(\theta)$. Prove that if the curves intersect at a point $[r; \theta]$, then the curves intersect orthogonally iff

$$f_1'(\theta)f_2'(\theta) + f_1(\theta)f_2(\theta) = 0.$$

   b) Apply (a) and Exercise 21 to reprove the result in Exercise 19(a).

23. Prove that in any conic the sum of the reciprocals of the lengths of the segments of a focal chord is a constant.

24. Prove that the product of the distances from an arbitrary point on a hyperbola to the asymptotes is a constant.

12-2 R² as a vector space

The object of the next two sections is to study certain self-mappings of $\mathbf{R}^2$. We can do this most easily by first giving $\mathbf{R}^2$ an algebraic structure. We begin by defining an operation of addition.

Definition 1. Given two elements $\alpha = (a_1, a_2)$ and $\beta = (b_1, b_2)$ in $\mathbf{R}^2$, we define their **vector sum** to be

$$\alpha + \beta = (a_1 + b_1, a_2 + b_2).$$

This definition has important interpretations in geometry and physics. First of all, given the points α and β, one can construct the point $\alpha + \beta$ entirely by drawing parallel lines. (The measurement of distances is unnecessary.) We leave the proof of this to the reader; the two relevant figures are shown in Fig. 12–19. Students of physics will immediately recognize in Fig. 12–19(a) a "parallelogram of forces": if the arrows drawn from O to α and to β represent two forces acting on O, then the arrow joining O to $\alpha + \beta$ represents the *resultant force* acting on O.

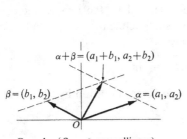
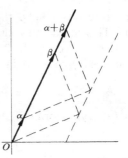

FIG. 12–19   Case 1   (O, α, β noncollinear)   Case 2   (O, α, β collinear)

Proposition 2. Properties of Vector Addition.   *Vector addition has the following properties:*

1) It is associative, that is,

$$\alpha + (\beta + \gamma) = (\alpha + \beta) + \gamma \quad \text{for all} \quad \alpha, \beta, \gamma \text{ in } \mathbf{R}^2.$$

2) There exists an element in $\mathbf{R}^2$ denoted by 0 such that

$$0 + \alpha = \alpha + 0 = \alpha \quad \text{for all} \quad \alpha \in \mathbf{R}^2.$$

3) For each $\alpha \in \mathbf{R}^2$, there exists an element in $\mathbf{R}^2$ denoted by $-\alpha$ such that

$$\alpha + (-\alpha) = (-\alpha) + \alpha = 0.$$

4) Addition is commutative, that is,

$$\alpha + \beta = \beta + \alpha \quad \text{for all} \quad \alpha, \beta \in \mathbf{R}^2.$$

Proof. Properties (1) and (4) follow from the corresponding properties of the real numbers. The *null vector* 0 of property (2) is simply the origin (0, 0). If

$$\alpha = (a_1, a_2),$$

then the vector $-\alpha$ of property (3) is simply $(-a_1, -a_2)$. Referring to Fig. 12–20, note that α and $-\alpha$ are situated symmetrically with respect to the origin. If we draw arrows from 0 to α and $-\alpha$, then the arrows have the same length, lie on the same line, but point in opposite directions. (As the reader may have already deduced from context, the elements of $\mathbf{R}^2$ are often called vectors in discussions where vector addition appears.)

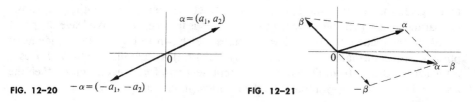

FIG. 12-20   FIG. 12-21

Given two vectors α and β, their *difference* $\alpha - \beta$ may be defined to be the vector $\alpha + (-\beta)$. This vector may also be described as the unique vector which when added to β gives α (see Fig. 12–21). We shall define the *norm* of an element $\alpha \in \mathbf{R}^2$ to be its distance from the origin, namely,

$$\|\alpha\| = (a_1^2 + a_2^2)^{1/2},$$

where $\alpha = (a_1, a_2)$. Norms are related to vector addition by the *triangle inequality:*

$$\|\alpha + \beta\| \le \|\alpha\| + \|\beta\|.$$

The geometric content of this inequality is that in any nondegenerate triangle the length of each side is less than the sum of the lengths of the other two sides. The inequality is a special case of the Minkowski inequality proved in Chapter 9. Also note that $\|-\alpha\| = \|\alpha\|$. Thus the norm on $\mathbf{R}^2$ functions like an absolute value. Furthermore, it should be noted that $\|\alpha - \beta\|$ is equal to the distance between α and β.

The final ingredient of our algebraic structure will be multiplication of vectors by real numbers.

Definition 3. Given a vector $\alpha = (a_1, a_2)$ and a real number c, we define

$$c\alpha = (ca_1, ca_2).$$

Then $c\alpha$ is called a **scalar multiple** of α.

Note then that the product of a vector with a *scalar* (a term often used in such contexts to designate a real number) is again a vector.

Proposition 4. Properties of Multiplication by Scalars. *Let α and β be vectors in* **R**^{2}, *and let c and d be real numbers. Then we have*

1) $c(\alpha + \beta) = c\alpha + c\beta$,
2) $(c + d)\alpha = c\alpha + d\alpha$,
3) $c(d\alpha) = (cd)\alpha$,
4) $\|c\alpha\| = |c|\, \|\alpha\|$,
5) *the points 0, α, and $c\alpha$ are collinear; if $c > 0$, they lie on the same side of 0; if $c < 0$, they lie on opposite sides of 0.*

Proof. The derivations of the first four assertions are perfectly straightforward. The last two properties are of interest since they lead to a geometric interpretation of multiplication by scalars:

$$\|c\alpha\| = \|(ca_1, ca_2)\| = \sqrt{(ca_1)^2 + (ca_2)^2} = |c|\sqrt{a_1^2 + a_2^2} = |c|\, \|\alpha\|.$$

Suppose $c > 0$. If $0 < c < 1$, then $0 < 1 - c < 1$ and

$$d(0, c\alpha) + d(c\alpha, \alpha) = \|c\alpha\| + \|\alpha - c\alpha\| = c\|\alpha\| + (1 - c)\|\alpha\| = \|\alpha\| = d(0, \alpha).$$

Here we use $d(\alpha, \beta)$ to denote the distance from α to β. Thus 0, $c\alpha$, and α must be collinear, with $c\alpha$ lying between 0 and α. In particular, $c\alpha$ and α lie on the same side of 0 (see Fig. 12–22). If $c = 1$, then $c\alpha = \alpha$. If $c > 1$, then

$$d(0, \alpha) + d(\alpha, c\alpha) = \|\alpha\| + (c - 1)\|\alpha\| = \|c\alpha\| = d(0, c\alpha).$$

Thus 0, $c\alpha$, and α must be collinear, with α lying between 0 and $c\alpha$ (see Fig. 12–23).

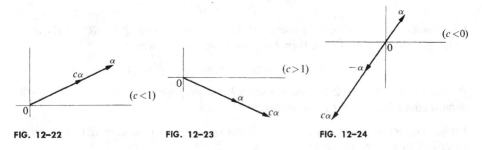

FIG. 12–22   **FIG. 12–23**   **FIG. 12–24**

If $c < 0$, then since $c\alpha = (-c)(-\alpha)$, it follows by the preceding case that 0, $-\alpha$, and $c\alpha$ are collinear, with $-\alpha$ and $c\alpha$ lying on the same side of 0. Hence 0, α, and $c\alpha$ are collinear, with α and $c\alpha$ lying on opposite sides of 0 (see Fig. 12–24).

Thus we obtain the following geometric interpretation of $c\alpha$. If we think of the vectors in **R**^{2} as arrows emanating from 0, then $c\alpha$ is an arrow whose length is $|c|$ times the length of α; if $c > 0$, $c\alpha$ points in the same direction as α, whereas if $c < 0$, then $c\alpha$ points in the direction opposite to that of α.

The reader will find it instructive to give a geometric interpretation to the distributive law $c(\alpha + \beta) = c\alpha + c\beta$ in terms of similar triangles.

Part of the usefulness of vectors, especially in higher dimensions, comes about from the fact that a single vector equation may be used in place of an entire system of equations. Consider, for instance, the parametric equations of a straight line. The straight line which passes through the points $\zeta_1 = (x_1, y_1)$ and $\zeta_2 = (x_2, y_2)$ is described by the two scalar equations

$$x = (1 - t)x_1 + tx_2, \qquad y = (1 - t)y_1 + ty_2.$$

If we set $\zeta = (x, y)$, then these two equations are equivalent to the single vector equation

$$\zeta = (1 - t)\zeta_1 + t\zeta_2.$$

In the next section we shall be concerned with certain mappings of the plane $\mathbf{R}^2$ into itself. Of special interest are those mappings which preserve the vector operations which we have just introduced.

Definition 5. A mapping $T: \mathbf{R}^2 \to \mathbf{R}^2$ is called a **linear transformation** iff

1) $T(\alpha + \beta) = T(\alpha) + T(\beta)$,

2) $T(c\alpha) = cT(\alpha)$

for all $\alpha, \beta \in \mathbf{R}^2$ and $c \in \mathbf{R}$.

It is a fairly easy matter to describe explicitly the linear mappings of $\mathbf{R}^2$ into itself. We shall show that each such mapping can be uniquely associated with a 2×2 *matrix* of real numbers. By the latter we shall mean a square array of constants,

$$A = \begin{pmatrix} a_{11} & a_{12} \\ a_{21} & a_{22} \end{pmatrix}.$$

Given such an array of real numbers, we shall define a mapping $T_A: \mathbf{R}^2 \to \mathbf{R}^2$ in the following way: If $\zeta = (x, y)$, then $T_A(\zeta) = \zeta' = (x', y')$, where

$$x' = a_{11}x + a_{12}y, \qquad y' = a_{21}x + a_{22}y.$$

Proposition 6. *If A is a 2×2 matrix of real numbers, then the mapping $T_A: \mathbf{R}^2 \to \mathbf{R}^2$ defined above is a linear transformation.*

Proof. We set $\alpha^1 = (a_{11}, a_{21})$ and $\alpha^2 = (a_{12}, a_{22})$ and observe that T_A can be described by the vector equation

$$T_A(\xi) = x_1\alpha^1 + x_2\alpha^2,$$

where $\xi = (x_1, x_2)$.

We check first that T_A preserves vector addition. Let

$$\xi = (x_1, x_2) \qquad \text{and} \qquad \eta = (y_1, y_2).$$

Then $\xi + \eta = (x_1 + y_1, x_2 + y_2)$ and

$$\begin{aligned} T_A(\xi + \eta) &= (x_1 + y_1)\alpha^1 + (x_2 + y_2)\alpha^2 \\ &= (x_1\alpha^1 + x_2\alpha^2) + (y_1\alpha^1 + y_2\alpha^2) = T_A(\xi) + T_A(\eta). \end{aligned}$$

It is also easily checked that T preserves multiplication by scalars. If $c \in \mathbf{R}$, then $c\xi = (cx_1, cx_2)$, and so

$$T_A(c\xi) = (cx_1)\alpha^1 + (cx_2)\alpha^2 = c(x_1\alpha^1 + x_2\alpha^2) = cT_A(\xi).$$

Thus T_A is linear.

We observe that $T_A(\epsilon_1) = \alpha^1$ and $T_A(\epsilon_2) = \alpha^2$, where $\epsilon_1 = (1, 0)$ and $\epsilon_2 = (0, 1)$. We shall now show that T_A is the unique linear transformation of $\mathbf{R}^2$ into itself which carries ϵ_1 and ϵ_2 into α^1 and α^2, respectively. Suppose that T is such a linear transformation. If $\xi = (x_1, x_2)$, then

$$\begin{aligned} T(\xi) &= T(x_1\epsilon_1 + x_2\epsilon_2) = T(x_1\epsilon_1) + T(x_2\epsilon_2) \\ &= x_1 T(\epsilon_1) + x_2 T(\epsilon_2) = x_1\alpha^1 + x_2\alpha^2 = T_A(\xi). \end{aligned}$$

Thus $T = T_A$, as asserted.

Proposition 7. *Every linear transformation* $T: \mathbf{R}^2 \to \mathbf{R}^2$ *is of the form* T_A *for some* 2×2 *matrix of real numbers. Every such linear transformation is determined by its behavior on the vectors* ϵ_1 *and* ϵ_2, *and the action of the transformation on* ϵ_1 *and* ϵ_2 *can be prescribed arbitrarily.*

The correspondence between linear transformation and matrices thus obtained is easily seen to be one-to-one, where we define

$$\begin{pmatrix} a_{11} & a_{12} \\ a_{21} & a_{22} \end{pmatrix} = \begin{pmatrix} b_{11} & b_{12} \\ b_{21} & b_{22} \end{pmatrix}$$

to mean that $a_{11} = b_{11}$, $a_{12} = b_{12}$, $a_{21} = b_{21}$, and $a_{22} = b_{22}$.

We conclude this section by considering compositions and inverses of linear mappings.

Proposition 8. *If* $S, T: \mathbf{R}^2 \to \mathbf{R}^2$ *are linear transformations, then* $S \circ T$ *is also a linear transformation.*

Proof. If $\xi, \eta \in \mathbf{R}^2$, we have by the linearity of S and T,

$$\begin{aligned} (S \circ T)(\xi + \eta) &= S(T(\xi + \eta)) = S(T(\xi) + T(\eta)) \\ &= S(T(\xi)) + S(T(\eta)) = (S \circ T)(\xi) + (S \circ T)(\eta). \end{aligned}$$

Also, if $c \in \mathbf{R}$,

$$\begin{aligned} (S \circ T)(c\xi) &= S(T(c\xi)) = S(cT(\xi)) \\ &= cS(T(\xi)) = c(S \circ T)(\xi). \end{aligned}$$

This proves that $S \circ T$ is a linear transformation.

The composition of linear mappings leads to a corresponding operation for matrices. Given two matrices

$$A = \begin{pmatrix} a_{11} & a_{12} \\ a_{21} & a_{22} \end{pmatrix}, \qquad B = \begin{pmatrix} b_{11} & b_{12} \\ b_{21} & b_{22} \end{pmatrix},$$

we define their *matrix product AB* to be the unique 2×2 matrix for which

$$T_{AB} = T_A \circ T_B.$$

To compute the elements of AB, we consider the action of $T_A \circ T_B$ on ϵ_1 and ϵ_2. The components of $(T_A \circ T_B)(\epsilon_1)$ and $(T_A \circ T_B)(\epsilon_2)$ will yield the first and second column, respectively, of AB. Now

$$\epsilon_1 \xrightarrow{T_B} (b_{11}, b_{21}) \xrightarrow{T_A} (a_{11}b_{11} + a_{12}b_{21}, a_{21}b_{11} + a_{22}b_{21})$$

and

$$\epsilon_2 \xrightarrow{T_B} (b_{12}, b_{22}) \xrightarrow{T_A} (a_{11}b_{12} + a_{12}b_{22}, a_{21}b_{12} + a_{22}b_{22}).$$

Thus

$$AB = \begin{pmatrix} a_{11}b_{11} + a_{12}b_{21} & a_{11}b_{12} + a_{12}b_{22} \\ a_{21}b_{11} + a_{22}b_{21} & a_{21}b_{12} + a_{22}b_{22} \end{pmatrix}.$$

This expression for AB need not be memorized. Observe that the element of AB in the jth row and kth column may be obtained by extracting the jth row from A and the kth column from B and forming a sum of products of these elements. This is illustrated below for the case in which $j = 1$ and $k = 2$.

$$AB = \begin{pmatrix} - & a_{11}b_{12} + a_{12}b_{22} \\ - & - \end{pmatrix} = \left(\boxed{\begin{matrix} a_{11} & a_{12} \end{matrix}} \begin{matrix} \\ a_{21} & a_{22} \end{matrix} \right) \begin{pmatrix} b_{11} & \boxed{b_{12}} \\ b_{21} & \boxed{b_{22}} \end{pmatrix}$$

Before going further, we suggest that the reader check the following result:

$$\begin{pmatrix} 2 & -1 \\ 0 & 3 \end{pmatrix} \begin{pmatrix} -1 & 3 \\ 4 & 1 \end{pmatrix} = \begin{pmatrix} -6 & 5 \\ 12 & 3 \end{pmatrix}.$$

We observe that the identity mapping $I: \mathbf{R}^2 \to \mathbf{R}^2$, defined by $I(\xi) = \xi$ for all $\xi \in \mathbf{R}^2$, is linear. The corresponding matrix, which we shall call the *identity matrix* and shall also denote by I, is

$$I = \begin{pmatrix} 1 & 0 \\ 0 & 1 \end{pmatrix}.$$

The identity matrix is characterized by the fact that $AI = IA = A$ for all 2×2 matrices A.

We shall associate with a given matrix

$$A = \begin{pmatrix} a_{11} & a_{12} \\ a_{21} & a_{22} \end{pmatrix}$$

a certain real number called its *determinant*, namely, the number

$$\det A = \begin{vmatrix} a_{11} & a_{12} \\ a_{21} & a_{22} \end{vmatrix} = a_{11}a_{22} - a_{12}a_{21}.$$

In the section following we shall give a geometric interpretation of this number. In the meantime we observe that whether or not the number is zero determines whether or not the linear transformation corresponding to A is invertible.

Proposition 9. *Let $T: \mathbf{R}^2 \to \mathbf{R}^2$ be a linear transformation, and let A be the corresponding 2×2 matrix. The following statements are logically equivalent:*

1) $T(\xi) = 0$ iff $\xi = 0$;

2) T is injective (one-to-one);

3) T is surjective (onto);

4) $\det A \neq 0$;

5) the points $0, T(\epsilon_1)$, and $T(\epsilon_2)$ are noncollinear.

We shall prove $(1) \Rightarrow (2)$, $(2) \Rightarrow (1)$, $(1) \Rightarrow (5)$, $(5) \Rightarrow (4)$, $(4) \Rightarrow (3)$, and $(3) \Rightarrow (1)$.

Proof of $(1) \Rightarrow (2)$. Suppose that $T(\xi) = T(\xi')$. Then $0 = T(\xi) - T(\xi') = T(\xi - \xi')$. Since we are assuming that (1) holds, it follows that $\xi - \xi' = 0$, or that $\xi = \xi'$. Hence T is one-to-one.

Proof of $(2) \Rightarrow (1)$. Suppose that $T(\xi) = 0$. Since we also know that $T(0) = 0$, the fact that T is one-to-one implies that $\xi = 0$.

Proof of $(1) \Rightarrow (5)$. We prove the contrapositive. Let us assume that $0, T(\epsilon_1)$, and $T(\epsilon_2)$ are collinear. If either $T(\epsilon_1)$ or $T(\epsilon_2)$ is zero, then (1) clearly fails. If both $T(\epsilon_1)$ and $T(\epsilon_2)$ are nonzero, then each is a scalar multiple of the other. Hence we may write $T(\epsilon_1) = \lambda T(\epsilon_2)$. Then $T(\epsilon_1 - \lambda \epsilon_2) = T(\epsilon_1) - \lambda T(\epsilon_2) = 0$. But

$$\epsilon_1 - \lambda \epsilon_2 = (1, -\lambda) \neq 0,$$

which shows that (1) does not hold.

Proof of $(5) \Rightarrow (4)$. Again we prove the contrapositive. Suppose that

$$\det A = a_{11}a_{22} - a_{12}a_{21} = 0.$$

If $T(\epsilon_1) = (a_{11}, a_{21}) = 0$, then $0, T(\epsilon_1)$, and $T(\epsilon_2)$ are collinear. If $T(\epsilon_1) \neq 0$, then either $a_{11} \neq 0$ or $a_{21} \neq 0$. In the first instance we have

$$T(\epsilon_2) = (a_{12}, a_{22}) = \lambda T(\epsilon_1),$$

where $\lambda = a_{12}/a_{11}$; in the second instance we have

$$T(\epsilon_2) = \lambda T(\epsilon_1),$$

where $\lambda = a_{22}/a_{21}$. In either case $0, T(\epsilon_1), T(\epsilon_2)$ are collinear.

Proof of $(4) \Rightarrow (3)$. Given $\eta = (y_1, y_2)$, we must show that there exists a $\xi = (x_1, x_2)$ such that $T(\xi) = \eta$. This vector equation is equivalent to the two scalar equations

$$a_{11}x_1 + a_{12}x_2 = y_1, \qquad a_{21}x_1 + a_{22}x_2 = y_2.$$

Since, by assumption, det $A \neq 0$, the equations have the solutions

$$x_1 = \frac{\begin{vmatrix} y_1 & a_{12} \\ y_2 & a_{22} \end{vmatrix}}{\det A} = \frac{a_{22}y_1 - a_{12}y_2}{\det A},$$

$$x_2 = \frac{\begin{vmatrix} a_{11} & y_1 \\ a_{21} & y_2 \end{vmatrix}}{\det A} = \frac{-a_{21}y_1 + a_{11}y_2}{\det A}.$$

Proof of (3) $\Rightarrow$ *(1)*. Assuming that T is surjective, we may choose vectors ξ_1 and ξ_2 such that $T(\xi_1) = \epsilon_1$ and $T(\xi_2) = \epsilon_2$. We let S be the unique linear transformation carrying ϵ_1 and ϵ_2 onto ξ_1 and ξ_2, respectively. Then $T \circ S$ fixes ϵ_1 and ϵ_2 and hence must be the identity transformation. It follows that S satisfies condition (1):

$$S(\xi) = 0 \Rightarrow \xi = I(\xi) = T(S(\xi)) = T(0) = 0.$$

But by what we have proved thus far, S must also be surjective. Suppose then that $T(\eta) = 0$. We can write $\eta = S(\xi)$. Hence

$$0 = T(\eta) = T(S(\xi)) = \xi \quad \text{and} \quad \eta = S(\xi) = 0.$$

Definition 10. A linear mapping $T: \mathbf{R}^2 \to \mathbf{R}^2$ is said to be **nonsingular** iff the conditions of the above theorem are satisfied.

A nonsingular linear transformation T is invertible. The inverse mapping T^{-1} is also linear. In fact, if

$$A = \begin{pmatrix} a_{11} & a_{12} \\ a_{21} & a_{22} \end{pmatrix}$$

is the matrix corresponding to T, then our proof of (4) $\Rightarrow$ (3) shows that T^{-1} is a linear transformation whose matrix is

$$A^{-1} = \begin{pmatrix} a_{22}\Delta^{-1} & -a_{12}\Delta^{-1} \\ -a_{21}\Delta^{-1} & a_{11}\Delta^{-1} \end{pmatrix},$$

where $\Delta = \det A$. The transformation T^{-1} is nonsingular, and the determinant of the corresponding matrix is

$$\det A^{-1} = (a_{22}a_{11} - a_{12}a_{21})\Delta^{-2} = \Delta\Delta^{-2} = \Delta^{-1} = \frac{1}{\det A},$$

the reciprocal of the determinant of the matrix for T.

The nonsingular linear transformations are closed under composition. If S and T are nonsingular linear mappings, then $S \circ T$ is linear (because it is the product of linear maps) and injective (because it is the product of injective maps). Hence $S \circ T$ is nonsingular. Moreover, since

$$(S \circ T) \circ (T^{-1} \circ S^{-1}) = S \circ (T \circ T^{-1}) \circ S^{-1} = S \circ I \circ S^{-1} = S \circ S^{-1} = I,$$

we have

$$(S \circ T)^{-1} = T^{-1} \circ S^{-1};$$

that is, the inverse of a product is the product of the inverses in the reverse order. The fact that $S \circ T$ is nonsingular also follows from the *multiplication rule for determinants:*

$$\det AB = (\det A)(\det B)$$

for any two 2×2 matrices A and B. The multiplication rule can be verified by a straightforward but ugly computation.

Exercises

1. Draw in a coordinate plane the vectors $\alpha = (-2, 3)$ and $\beta = (1, 2)$. Do the same for the vectors $-\alpha$, 3β, $(-2)\alpha$, $\alpha + \beta$, $3\beta - \alpha$, and $\beta - 2\alpha$. Can every vector in $\mathbf{R}^2$ be written in the form $c_1\alpha + c_2\beta$, where c_1 and c_2 are scalars?

2. Prove in detail that vector addition can be performed through the construction of parallel lines.

3. a) Show that $\alpha, \beta, \gamma, \delta$ are consecutive vertices of a parallelogram iff $\beta - \alpha = \gamma - \delta$.
   b) Using (a), show that a quadrilateral is a parallelogram iff the diagonals bisect each other.

4. Interpret geometrically the vector $(1/\|\alpha\|)\alpha$, where $\alpha \neq 0$.

5. Using vector methods, prove that the medians of a triangle are concurrent. Express the point of intersection (the center of gravity) in terms of the vertices.

6. Derive from the triangle equality the *reverse triangle inequality:*

$$\big|\, \|\alpha\| - \|\beta\| \,\big| \leq \|\alpha - \beta\|.$$

Interpret geometrically.

7. When is $\|\alpha + \beta\| = \|\alpha\| + \|\beta\|$?

8. Verify properties (1), (2), (3), and (4) in Proposition 4.

9. Let

$$A = \begin{pmatrix} 2 & -1 \\ 3 & 7 \end{pmatrix}.$$

Find $T_A(\epsilon_1)$ and $T_A(\epsilon_2)$. Also find $T_A(\xi)$, where $\xi = (3, -1)$. Find a vector η such that $T_A(\eta) = (8, -5)$.

10. Let $\alpha = (1, -1)$ and $\beta = (2, 5)$. Show that every vector $\xi \in \mathbf{R}^2$ can be uniquely written

$$\xi = c_1\alpha + c_2\beta,$$

where c_1 and c_2 are scalars. If we define $T(\xi) = (c_1, c_2)$, show that the mapping so obtained is linear. What is the matrix corresponding to T?

11. If $T: \mathbf{R}^2 \to \mathbf{R}^2$ is linear, show that the range of T is either the entire plane, a straight line, or the singleton set consisting of the zero vector. What can you say regarding the *kernel* or *null space* of T, namely, the set

$$\ker T = \mathfrak{N}_T = \{\xi \in \mathbf{R}^2 \mid T(\xi) = 0\} ?$$

12. If

$$A = \begin{pmatrix} 2 & 3 \\ -1 & 7 \end{pmatrix} \quad \text{and} \quad B = \begin{pmatrix} 1 & 3 \\ 4 & -1 \end{pmatrix},$$

compute both AB and BA. Find $T_A T_B(\alpha)$, where $\alpha = (-1, 2)$.

13. Compute $\det A$, $\det B$, $\det AB$, and $\det BA$, where A and B are the matrices of the preceding exercise.

14. Prove directly that (4) $\Rightarrow$ (1) in Proposition 9.

15. Let α, β and α', β' be two pairs of vectors, where 0, α, and β are noncollinear.

    a) Prove that there exists a unique linear transformation T carrying α, β into α', β', respectively.

    b) Prove that T is nonsingular iff 0, α', and β' are noncollinear. (Simple proofs are possible.)

16. Let A and B be as in Exercise 12. Compute A^{-1}, B^{-1}, $(AB)^{-1}$, and $B^{-1}A^{-1}$.

17. If $\det A \neq 0$, we define A^{-1} to be the matrix such that

$$T_{A^{-1}} = (T_A)^{-1}.$$

From this identity and the identity $T_A \circ T_B = T_{AB}$, establish the following matrix identities:

$$A(BC) = (AB)C, \qquad AA^{-1} = A^{-1}A = I, \qquad (AB)^{-1} = B^{-1}A^{-1},$$

where in the second two cases A and B have nonzero determinants.

18. Prove the product rule for determinants. Deduce from it that $\det A^{-1} = (\det A)^{-1}$, wherever $\det A \neq 0$.

19. If

$$A = \begin{pmatrix} a_{11} & a_{12} \\ a_{21} & a_{22} \end{pmatrix}$$

has nonzero determinant, we associate with it a function f_A defined as follows:

$$f_A(x) = \frac{a_{11}x + a_{12}}{a_{21}x + a_{22}}.$$

Show that $f_A \circ f_B = f_{AB}$. Is the mapping $A \to f_A$ one-to-one?

20. Let α and β be two vectors such that 0, α, and β are noncollinear.

    a) Prove that any vector $\xi \in \mathbf{R}^2$ can be uniquely written as $\xi = c_1\alpha + c_2\beta$, where c_1 and c_2 are scalars. [Let T be the unique linear transformation carrying ϵ_1 and ϵ_2 onto α and β, respectively. State why T must be nonsingular. Show that

$$(c_1, c_2) = T^{-1}(\xi)$$

does the job.]

    b) Explain how the coefficients c_1 and c_2 can be found through the construction of parallel lines. (Figure 12-25 is relevant!)

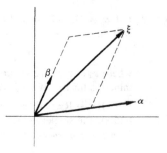

FIG. 12-25

12-3 Affine mappings of the plane

Let S be any set. By a *transformation group on S* we shall mean a family $\mathcal{G}$ of functions such that

1) each $\varphi \in \mathcal{G}$ is a one-to-one mapping of S onto itself;
2) $\mathcal{G}$ is closed under composition of mappings (that is, if φ and ψ belong to $\mathcal{G}$, so does $\varphi \circ \psi$);
3) $\mathcal{G}$ is closed under inversion (that is, $\varphi \in \mathcal{G} \Rightarrow \varphi^{-1} \in \mathcal{G}$).

In this section we shall study certain transformation groups of $\mathbf{R}^2$ which arise in the study of geometry. One of these groups is the *general linear group $GL(\mathbf{R}^2)$* consisting of all nonsingular linear transformations of $\mathbf{R}^2$ onto itself. [The fact that $GL(\mathbf{R}^2)$ is a transformation group follows, of course, from theorems in the preceding section.] Observe that each element T of $GL(\mathbf{R}^2)$ carries lines into lines. For if ζ lies on the line determined by ζ_1 and ζ_2, then $\zeta = (1 - t)\zeta_1 + t\zeta_2$ for some $t \in \mathbf{R}$, and so

$$T(\zeta) = T\big((1 - t)\zeta_1\big) + T(t\zeta_2) = (1 - t)T(\zeta_1) + tT(\zeta_2),$$

that is, $T(\zeta)$ lies on the line determined by $T(\zeta_1)$ and $T(\zeta_2)$. It is clear, moreover, that T maps line segments onto line segments and preserves betweenness. [In other words, if ζ_1, ζ_2, and ζ_3 are collinear and ζ_2 lies between ζ_1 and ζ_3, then $T(\zeta_1)$, $T(\zeta_2)$, and $T(\zeta_3)$ are collinear with $T(\zeta_2)$ lying between $T(\zeta_1)$ and $T(\zeta_3)$.]

Definition 1. By a **nonsingular affine mapping** of $\mathbf{R}^2$ we shall mean a one-to-one mapping of $\mathbf{R}^2$ onto itself which carries lines onto lines.

By our previous remarks, every nonsingular linear map is affine. Another important class of affine mappings is the class of *translations*. If $\alpha \in \mathbf{R}^2$, then we shall denote by T_α the mapping on $\mathbf{R}^2$ defined as follows:

$$T_\alpha(\xi) = \alpha + \xi.$$

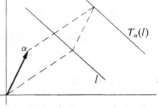

We call T_α the *translation* by α. The identity

$$T_\alpha \circ T_\beta = T_\beta \circ T_\alpha = T_{\alpha+\beta}$$

$(\alpha, \beta \in \mathbf{R}^2)$ follows directly. In particular, $T_\alpha \circ T_{-\alpha} = T_{-\alpha} \circ T_\alpha = T_0 = I$, from which it follows that T_α is one-to-one and onto with

FIG. 12-26

$$(T_\alpha)^{-1} = T_{-\alpha}.$$

The two displayed identities make it clear that the translation operators form a transformation group. Each translation operator is affine. In fact, any given line will be carried onto *parallel* lines by the translation operators (see Fig. 12–26). In addition, we have

$$\|T_\alpha(\xi) - T_\alpha(\eta)\| = \|(\alpha + \xi) - (\eta + \alpha)\| = \|\xi - \eta\|,$$

that is, distances are preserved under translation. One can check, moreover, that translations carry line segments onto line segments and preserve betweenness.

We shall now undertake to describe all the affine mappings of the plane.

Proposition 2. *The nonsingular affine mappings of* $\mathbf{R}^2$ *form a transformation group* $AF(\mathbf{R}^2)$.

Proof. Let $S, T \in AF(\mathbf{R}^2)$. Then $S \circ T$ is a one-to-one mapping of $\mathbf{R}^2$ onto itself. (Why?) If l is any line, then T carries l onto a second line l', and S carries l' onto a line l''. Hence $S \circ T$ must carry l onto l''. The mapping $S \circ T$ is therefore affine. Inverses may be handled similarly.

Corollary 3. *The composition of a nonsingular linear transformation and a translation is a nonsingular affine mapping.*

We shall presently prove the converse of this corollary, namely, every non-singular affine mapping can be expressed as the composition product of a nonsingular linear transformation and a translation. We build up to the result with a series of lemmas.

Lemma 4. *If* α, β, *and* γ *are any three noncollinear points in* $\mathbf{R}^2$, *there exists a linear transformation* S *such that the affine mapping* $T = T_\alpha \circ S = S \circ T_{S^{-1}(\alpha)}$ *carries* 0, ϵ_1, ϵ_2 *into* α, β, γ, *respectively.*

Proof. Since α, β, γ are noncollinear, the same will be true of $0 = T_{-\alpha}(\alpha)$, $\beta - \alpha = T_{-\alpha}(\beta)$, and $\gamma - \alpha = T_{-\alpha}(\gamma)$. Let S be the unique nonsingular linear transformation such that $S(\epsilon_1) = \beta - \alpha$ and $S(\epsilon_2) = \gamma - \alpha$. Then

$$(T_\alpha \circ S)(0) = T_\alpha(0) = \alpha, \qquad (T_\alpha \circ S)(\epsilon_1) = T_\alpha(\beta - \alpha) = \beta,$$

and

$$(T_\alpha \circ S)(\epsilon_2) = T_\alpha(\gamma - \alpha) = \gamma.$$

We also observe that for each $\xi \in \mathbf{R}^2$,

$$(S \circ T_{S^{-1}(\alpha)})(\xi) = S(\xi + S^{-1}(\alpha)) = S(\xi) + S(S^{-1}(\alpha))$$
$$= S(\xi) + \alpha = (T_\alpha \circ S)(\xi).$$

Thus $T_\alpha \circ S = S \circ T_{S^{-1}(\alpha)}$.

Lemma 5. *Every nonsingular affine transformation carries parallel lines into parallel lines.*

Proof. Let $T \in AF(\mathbf{R}^2)$, and let l_1 and l_2 be any two parallel lines. Then T carries l_1 and l_2 onto two lines l_1' and l_2'. These lines cannot intersect, for if they did, then T^{-1} would map a point of their intersection onto a point common to both l_1 and l_2, which is impossible. Thus l_1' and l_2' must be parallel.

Note that we have not as yet proved that elements of $AF(\mathbf{R}^2)$ carry line segments onto line segments.

Lemma 6. *If $T \in AF(\mathbf{R}^2)$ and $T(0) = 0$, then T preserves addition, that is,*

$$T(\alpha + \beta) = T(\alpha) + T(\beta)$$

for all α and β in $\mathbf{R}^2$.

Proof. The assertion follows quickly from the fact that T preserves parallelism and that one can construct $\alpha + \beta$ from α and β by drawing parallel lines (see Fig. 12–27).

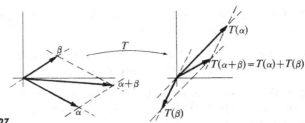

FIG. 12–27

Lemma 7. *If $\varphi \colon \mathbf{R} \to \mathbf{R}$ preserves both addition and multiplication* [*i.e.,*

$$\varphi(x + y) = \varphi(x) + \varphi(y) \qquad and \qquad \varphi(xy) = \varphi(x)\varphi(y)$$

for all $x, y \in \mathbf{R}$], and if φ is not identically zero, then φ is the identity map [$\varphi(x) = x$ *for all $x \in \mathbf{R}$*].

A proof is outlined in Exercise 10, Section 5–3.

Lemma 8. *If $T \in AF(\mathbf{R}^2)$ fixes 0, ϵ_1, and ϵ_2, then T is the identity mapping.*

Proof. Since T fixes both 0 and $\epsilon_1 = (1, 0)$, it follows that T must map the x-axis onto itself. Hence there exists a function $\varphi \colon \mathbf{R} \to \mathbf{R}$ such that $T(x\epsilon_1) = \varphi(x)\epsilon_1$ for all $x \in \mathbf{R}$. Similarly, there exists a function $\psi \colon \mathbf{R} \to \mathbf{R}$ *such that* $T(y\epsilon_2) = \psi(y)\epsilon_2$ *for all $y \in \mathbf{R}$.* Moreover,

$$\varphi(0) = \psi(0) = 0 \qquad and \qquad \varphi(1) = \psi(1) = 1.$$

We also know that T preserves addition. Hence

$$T(\epsilon_1 + \epsilon_2) = T(\epsilon_1) + T(\epsilon_2) = \epsilon_1 + \epsilon_2,$$

that is, T fixes $\epsilon_1 + \epsilon_2 = (1, 1)$. It follows that T maps the line $y = x$ onto itself. Thus, for all $x \in \mathbf{R}$,

$$T(x\epsilon_1 + x\epsilon_2) = T(x\epsilon_1) + T(x\epsilon_2) = \varphi(x)\epsilon_1 + \psi(x)\epsilon_2 = \big(\varphi(x), \psi(x)\big)$$

lies on the line $y = x$, which implies that $\varphi(x) = \psi(x)$.

To summarize, then, we have

$$T(x\epsilon_1 + y\epsilon_2) = \varphi(x)\epsilon_1 + \varphi(y)\epsilon_2$$

for all $x, y \in \mathbf{R}$. Moreover, since T preserves addition, it follows that φ preserves addition. If k is any real number, then T maps the line $y = kx$ onto a line which also passes through the origin, say, $y = k'x$. Thus, for all $x \in \mathbf{R}$,

$$\varphi(x)\epsilon_1 + k'\varphi(x)\epsilon_2 = T(x\epsilon_1 + kx\epsilon_2) = \varphi(x)\epsilon_1 + \varphi(kx)\epsilon_2,$$

which implies that $k'\varphi(x) = \varphi(kx)$. Setting $x = 1$, we get $k' = \varphi(k)$. Therefore, we have

$$\varphi(kx) = \varphi(k)\varphi(x)$$

for all $k, x \in \mathbf{R}$. Since φ preserves addition and multiplication and is not identically zero, $\varphi(x) = x$ for all $x \in \mathbf{R}$. It follows that $T = I$.

Theorem 9. *Every element in $AF(\mathbf{R}^2)$ may be written as the product of a nonsingular linear transformation and a translation (in either order).*

Proof. Let $T \in AF(\mathbf{R}^2)$. We set $\alpha = T(0)$, $\beta = T(\epsilon_1)$, and $\gamma = T(\epsilon_2)$. Then α, β, and γ are noncollinear. (Why?) Hence there exists a nonsingular linear transformation S such that $T_\alpha \circ S$ maps 0, ϵ_1, and ϵ_2 onto α, β, and γ, respectively. The mapping $T^{-1} \circ (T_\alpha \circ S)$ is a nonsingular affine transformation which fixes 0, ϵ_1, and ϵ_2. Hence $T^{-1} \circ (T_\alpha \circ S) = I$, and multiplication by T on the left yields

$$T = T_\alpha \circ S = S \circ T_{S^{-1}(\alpha)}.$$

Corollary 10. *The nonsingular affine mappings of $\mathbf{R}^2$ preserve betweenness and carry line segments onto line segments.*

We have already commented that both nonsingular linear transformations and translations have the stated properties, and hence so do products of these mappings.

We can also strengthen the statement of one of the lemmas.

Corollary 11. *If T is a nonsingular affine mapping of $\mathbf{R}^2$ which fixes the origin, then T is a nonsingular linear transformation.*

Proof. We may write $T = T_\alpha \circ S$, where S is a nonsingular linear transformation. Then $0 = T(0) = T_\alpha(S(0)) = T_\alpha(0) = \alpha$. Consequently, $T = T_0 \circ S = I \circ S = S$.

Proposition 12. *If α, β, γ and α', β', γ' are any two triples of noncollinear points, then there exists a unique $T \in AF(\mathbf{R}^2)$ such that $\alpha' = T(\alpha)$, $\beta' = T(\beta)$, $\gamma' = T(\gamma)$.*

Proof. We can choose (by Lemma 4) mappings $T_1, T_2 \in AF(\mathbf{R}^2)$ such that

$$T_1(0) = \alpha, \qquad T_1(\epsilon_1) = \beta, \qquad T_1(\epsilon_2) = \gamma,$$
$$T_2(0) = \alpha', \qquad T_2(\epsilon_1) = \beta', \qquad T_2(\epsilon_2) = \gamma'.$$

Then $T = T_2 \circ T_1^{-1}$ is the required mapping.

Suppose T^* is a second such mapping. Then $T_2^{-1} \circ T^* \circ T_1$ is an element of $AF(\mathbf{R}^2)$ which fixes 0, ϵ_1, and ϵ_2. Hence $T_2^{-1} \circ T^* \circ T_1 = I$. We multiply on the left by T_2 and on the right by T_1^{-1}. We get $T^* = T_2 \circ T_1^{-1} = T$.

We have given, then, a complete description of the nonsingular affinities of the plane. *Plane affine geometry* may be described as that branch of geometry which deals with those properties of figures which remain invariant under the application

of affine mappings. These properties clearly include incidence, parallelism, and betweenness. Another affine invariant is convexity. (See the exercises.) On the other
hand, such notions as distance and angular measure do not belong to affine geometry,
since affine mappings can distort distances and angles.

There are several important subgroups of $AF(\mathbf{R}^2)$. The one most pertinent to
Euclidean geometry is the group of *rigid motions* or *isometries*.

Definition 13. By an **isometry (congruence)** of $\mathbf{R}^2$ we shall mean a one-to-one map-
ping T of $\mathbf{R}^2$ onto itself which preserves distances, that is, $\|T(\alpha) - T(\beta)\| = \|\alpha - \beta\|$
for all α and β in $\mathbf{R}^2$.

The fact that every isometry is affine stems from the fact that line segments can
be characterized in terms of distance; specifically, the line segment joining α and β
consists of all points γ such that

$$\|\alpha - \gamma\| + \|\gamma - \beta\| = \|\alpha - \beta\|.$$

Thus an isometry must map line segments onto line segments and hence lines onto
lines. It is also clear that products and inverses of isometries are isometries.

Proposition 14. *The isometries of $\mathbf{R}^2$ form a subgroup of $AF(\mathbf{R}^2)$.*

We now consider examples of isometries. We have already observed that translations are isometries. The linear transformation V corresponding to the matrix

$$\begin{pmatrix} 1 & 0 \\ 0 & -1 \end{pmatrix},$$

namely, reflection in the x-axis, is also an isometry. Another extremely important
class of examples is the group of *rotations*. For each $\theta \in \mathbf{R}$, we let R_θ be the linear
transformation corresponding to the matrix

$$\begin{pmatrix} \cos\theta & -\sin\theta \\ \sin\theta & \cos\theta \end{pmatrix}.$$

We shall endeavor to show that R_θ is an isometry, and that it corresponds to our intuitive notion of a rotation about the origin through an angle of θ radians.

To obtain the basic properties of rotations, we begin with the observation that

$$\begin{pmatrix} \cos\theta & -\sin\theta \\ \sin\theta & \cos\theta \end{pmatrix}\begin{pmatrix} \cos\phi & -\sin\phi \\ \sin\phi & \cos\phi \end{pmatrix}$$
$$= \begin{pmatrix} \cos\theta\cos\phi - \sin\theta\sin\phi & -\cos\theta\sin\phi - \sin\theta\cos\phi \\ \sin\theta\cos\phi + \cos\theta\sin\phi & -\sin\theta\sin\phi + \cos\theta\cos\phi \end{pmatrix}$$
$$= \begin{pmatrix} \cos(\theta+\phi) & -\sin(\theta+\phi) \\ \sin(\theta+\phi) & \cos(\theta+\phi) \end{pmatrix},$$

from which we get the basic identity

$$R_\theta \circ R_\phi = R_{\theta+\phi} = R_\phi \circ R_\theta.$$

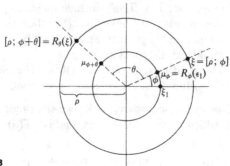

FIG. 12-28

In particular, $R_\theta \circ R_{-\theta} = R_{-\theta} \circ R_\theta = R_0 = I$, from which we see that R_θ is non-singular and that

$$(R_\theta)^{-1} = R_{-\theta}.$$

The last two displayed identities show that *the rotations of $\mathbf{R}^2$ form a commutative subgroup $SO(\mathbf{R}^2)$ of the general linear group $GL(\mathbf{R}^2)$.* The subgroup $SO(\mathbf{R}^2)$ is called the *special orthogonal group* of the plane. Let us consider next the effect of rotations upon polar coordinates. We note that

$$R_\theta(\epsilon_1) = (\cos\theta, \sin\theta) = \mu_\theta,$$

the point on the unit circle with polar angle θ. Consequently, if $[\rho; \phi]$ are polar coordinates of a point $\xi \in \mathbf{R}^2$, then $\xi = \rho\mu_\phi$ and

$$R_\theta(\xi) = \rho R_\theta(\mu_\phi) = \rho R_\theta\big(R_\phi(\epsilon_1)\big) = \rho R_{\theta+\phi}(\epsilon_1) = \rho\mu_{\theta+\phi}.$$

Thus, $[\rho; \phi + \theta]$ are polar coordinates of $R_\theta(\xi)$ (see Fig. 12–28). Hence R_θ may be described as a rotation about the origin through θ radians. Observe, in particular, that $\|R_\theta(\xi)\| = \|\rho\mu_{\theta+\phi}\| = \rho = \|\xi\|$, that is, R_θ preserves norms. This property, plus linearity, immediately implies that R_θ is an isometry:

$$\|R_\theta(\alpha) - R_\theta(\beta)\| = \|R_\theta(\alpha - \beta)\| = \|\alpha - \beta\|.$$

We shall now describe all isometries.

Theorem 15. *Every isometry of $\mathbf{R}^2$ can be written as a product of isometries of the following types: rotations, translations, and reflection in the x-axis.*

Proof. Let T be any isometry of $\mathbf{R}^2$. We set $\alpha = T(0)$. Then $T_{-\alpha} \circ T$ is an isometry which fixes 0. Therefore, $(T_{-\alpha} \circ T)(\epsilon_1)$ must be some point on the unit circle, say, μ_θ. Then $R_{-\theta} \circ T_{-\alpha} \circ T = T^*$ is an isometry which fixes both 0 and ϵ_1. It follows that $T^*(\epsilon_2)$ is a point in the plane whose distances from $0 = T^*(0)$ and $\epsilon_1 = T^*(\epsilon_1)$ are 1 and $\sqrt{2}$, respectively. There are only two possibilities: either $T^*(\epsilon_2) = \epsilon_2$ or $T^*(\epsilon_2) = -\epsilon_2$. Hence either T^* or $V \circ T^*$ is an isometry (and, in particular, an affinity) which fixes 0, ϵ_1, and ϵ_2. (As before, V denotes reflection in the x-axis.)

Thus either $T^* = I$ or $V \circ T^* = I$. Equivalently, $T = T_\alpha \circ R_\theta$ or $T = T_\alpha \circ R_\theta \circ V$. (Note that $V^{-1} = V$.)

Plane Euclidean geometry may be described as that branch of geometry which deals with those properties of figures which are invariant under the group of isometries of $\mathbf{R}^2$. Thus, in addition to dealing with the affine invariants, Euclidean geometry concerns distances, angles, areas, etc. The fact that areas are preserved under isometries is not entirely obvious. We shall presently give a proof of this. In the meantime, let us observe that arc lengths are preserved under rigid motions.

Let γ be a smooth path defined by the parametric equations

$$x = f(t), \qquad y = g(t) \qquad (a \le t \le b).$$

We can replace these two equations by a single vector equation,

$$\zeta = \varphi(t) = \big(f(t), g(t)\big) = f(t)\epsilon_1 + g(t)\epsilon_2.$$

We shall define the derivative of this vector-valued function by the equation

$$\frac{d\zeta}{dt} = \varphi'(t) = \big(f'(t), g'(t)\big) = f'(t)\epsilon_1 + g'(t)\epsilon_2.$$

Then the arc length of γ can be written as

$$l(\gamma) = \int_a^b \|\varphi'(t)\| \, dt.$$

Suppose now that T is any rigid motion of $\mathbf{R}^2$. Then T carries γ onto a path $T(\gamma)$ whose parametric equation is

$$\zeta = T\big(\varphi(t)\big) \qquad (a \le t \le b).$$

Let us write $T = T_\alpha \circ S$, where S is either a rotation or else a rotation composed with reflection in the x-axis, in either case, a nonsingular linear transformation which preserves norms [that is, $\|S(\xi)\| = \|\xi\|$ for all $\xi \in \mathbf{R}^2$]. Then

$$T\big(\varphi(t)\big) = S\big(\varphi(t)\big) + \alpha = f(t)S(\epsilon_1) + g(t)S(\epsilon_2) + \alpha$$

and

$$\frac{d}{dt}[T(\varphi(t))] = f'(t)S(\epsilon_1) + g'(t)S(\epsilon_2)$$

$$= S\big(f'(t)\epsilon_1 + g'(t)\epsilon_2\big) = S\big(\varphi'(t)\big).$$

(General rules for differentiating vector-valued functions will be developed in the exercises.) Thus

$$l(T(\gamma)) = \int_a^b \left\| \frac{d}{dt}[T(\varphi(t))] \right\| dt$$

$$= \int_a^b \|S(\varphi'(t))\| \, dt = \int_a^b \|\varphi'(t)\| \, dt$$

$$= l(\gamma).$$

This proves that T preserves arc lengths.

One might hope to be able to describe the behavior of arc lengths under an arbitrary affine mapping. Unfortunately, such a description is quite out of the question. Consider the linear transformation corresponding to the *diagonal matrix*

$$\begin{pmatrix} a & 0 \\ 0 & b \end{pmatrix} \quad (a, b > 0, a \neq b).$$

Its effect is that of altering the scale on the x- and y-axes. This affine mapping carries the unit circle $x^2 + y^2 = 1$ onto the ellipse

$$\frac{x^2}{a^2} + \frac{y^2}{b^2} = 1,$$

and there is no simple relationship between the arc lengths of these two curves.

By contrast, we shall be able to describe quite completely the behavior of areas under a general affine mapping. Essential to this purpose is the decomposition of a nonsingular linear transformation into *shears* parallel to the coordinate axes and a transformation given by a diagonal matrix. If $k \in \mathbf{R}$, then we denote by S_k the linear transformation having the matrix

$$\begin{pmatrix} 1 & k \\ 0 & 1 \end{pmatrix}.$$

We call S_k the *shear in the x-direction by k*. Since

$$\begin{pmatrix} 1 & k \\ 0 & 1 \end{pmatrix}\begin{pmatrix} 1 & k' \\ 0 & 1 \end{pmatrix} = \begin{pmatrix} 1 & k + k' \\ 0 & 1 \end{pmatrix},$$

we have the identities

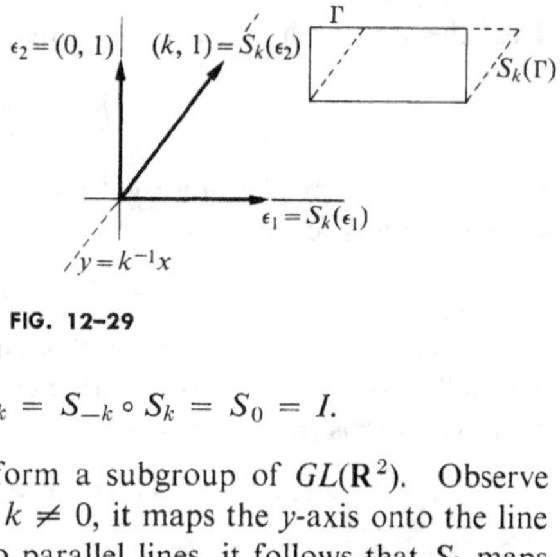

FIG. 12–29

$$S_{k+k'} = S_k \circ S_{k'}, \qquad S_k \circ S_{-k} = S_{-k} \circ S_k = S_0 = I.$$

It follows that the mappings $\{S_k \mid k \in \mathbf{R}\}$ form a subgroup of $GL(\mathbf{R}^2)$. Observe that S_k fixes each point on the x-axis, and if $k \neq 0$, it maps the y-axis onto the line $y = k^{-1}x$. Since S_k maps parallel lines into parallel lines, it follows that S_k maps each rectangle with sides parallel to the coordinate axes onto a parallelogram whose base is parallel to the x-axis; moreover, the base and height of the image parallelogram are equal to those of the original rectangle (see Fig. 12–29). Similar remarks hold for shears in the y-direction; these transformations are given by matrices of the form

$$\begin{pmatrix} 1 & 0 \\ k & 1 \end{pmatrix}.$$

Proposition 16. *Every nonsingular linear transformation T can be expressed as the product of a linear transformation having a diagonal matrix and shears in the directions of the coordinate axes.*

Proof. Let

$$\begin{pmatrix} a & b \\ c & d \end{pmatrix}$$

be the matrix for T. Then $\Delta = ad - bc \neq 0$.

Case I. $a \neq 0$. We have

$$\begin{pmatrix} a & b \\ c & d \end{pmatrix} = \begin{pmatrix} 1 & 0 \\ a^{-1}c & 1 \end{pmatrix} \begin{pmatrix} 1 & ab\Delta^{-1} \\ 0 & 1 \end{pmatrix} \begin{pmatrix} a & 0 \\ 0 & a^{-1}\Delta \end{pmatrix}.$$

Case II. $a = 0$. Then b and c are both nonzero, and we have

$$\begin{pmatrix} 0 & b \\ c & d \end{pmatrix} = \begin{pmatrix} 1 & -b^{-1}(b+d) \\ 0 & 1 \end{pmatrix} \begin{pmatrix} 1 & 0 \\ 1 & 1 \end{pmatrix} \begin{pmatrix} 1 & -1 \\ 0 & 1 \end{pmatrix} \begin{pmatrix} c & 0 \\ 0 & -b \end{pmatrix}.$$

We are now prepared to show how areas behave under affine mappings. If T is an affine mapping, we can write $T = T_\alpha \circ S$, where $\alpha \in \mathbf{R}^2$ and $S \in GL(\mathbf{R}^2)$. Moreover, this particular decomposition of T is unique. We call S the *linear part* of T, and we define det S to be the *determinant* of T.

*Theorem 17. *Let* $T \in AF(\mathbf{R}^2)$, *and let* B *be a Jordan-measurable subset of* $\mathbf{R}^2$. *Then* $T(B)$, *the image of* B *under* T, *is Jordan-measurable and*

$$m(T(B)) = |\det T| m(B).$$

Less precisely, T alters areas by the constant factor $|\det T|$.

Proof. Let us first observe that it suffices to prove the theorem for the case in which B is a bounded rectangle with sides parallel to the coordinate axes. For it will then immediately follow (from the fact that the Jordan-measurable sets form a ring and that m is finitely additive) that the theorem holds for all blocks. We go from blocks to the general case as follows. If B is Jordan-measurable, then corresponding to any $\epsilon > 0$, we can find blocks A and C such that

$$A \subset B \subset C \quad \text{and} \quad m(C) - m(A) < \epsilon/|\det T|.$$

Then $T(A)$ and $T(C)$ are Jordan-measurable, $T(A) \subset T(B) \subset T(C)$, and

$$m(T(C)) - m(T(A)) = |\det T|(m(C) - m(A)) < \epsilon.$$

Thus $T(B)$ is Jordan-measurable. If we define

$$\mu(A) = \frac{1}{|\det T|} m(T(A))$$

for each Jordan-measurable set A, then μ is an area content, and hence (by Proposition 8, Section 8–5), $\mu(A) = m(A)$ for each Jordan-measurable A.

Secondly, we observe that if the theorem holds for S and T in $AF(\mathbf{R}^2)$, then it also holds for $S \circ T$. (Here one needs the rule for determinants of matrix products.)

Consequently, it suffices to prove the theorem for the special cases in which T is a translation, a shear in the direction of one of the coordinate axes, or a transformation with a diagonal matrix.

Suppose now that T is a shear in the x-direction and that B is a rectangle with sides parallel to the coordinate axes. Let B have width w and height h. Then $T(B)$ is a parallelogram with the same width and height (see Fig. 12–30). The parallelogram $T(B)$ is Jordan-measurable, since it is bounded by the graphs of two continuous (and hence Riemann-integrable) functions. By the Fubini theorem for Jordan-measurable sets, one can obtain the area (Jordan content) of $T(B)$ by integrating the lengths of horizontal cross sections. This yields the familiar result $m\big(T(B)\big) = wh = m(B)$. Since $\det T = 1$, the proof is complete in this case.

Suppose next that T has the matrix

$$\begin{pmatrix} \lambda & 0 \\ 0 & \mu \end{pmatrix},$$

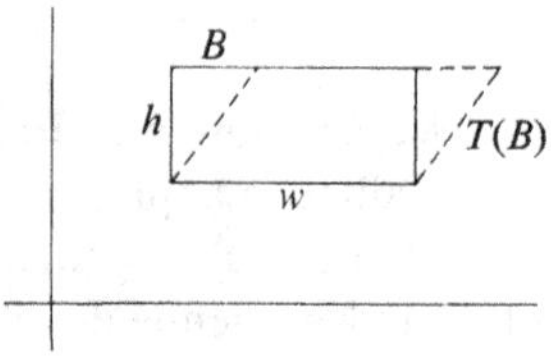

and that B is the same as in the preceding case. Then $T(B)$ is a rectangle with sides parallel to the coordinate axes; its width is $|\lambda|w$, and its height is $|\mu|h$. Hence

FIG. 12–30

$$m\big(T(B)\big) = |\lambda|w|\mu|h = |\lambda\mu|wh = |\det T|m(B).$$

The proof for the case of translations is simplest of all.

Corollary 18. *Jordan content is invariant under rigid motions of the plane.*

Proof. The linear part of an isometry is either a matrix of the form

$$\begin{pmatrix} \cos\theta & -\sin\theta \\ \sin\theta & \cos\theta \end{pmatrix}$$

or else the product of such a matrix and the matrix

$$\begin{pmatrix} 1 & 0 \\ 0 & -1 \end{pmatrix}.$$

Since the determinants of these matrices have absolute value one, the result is immediate.

Observe that we can apply our theorem to obtain the area of an ellipse from that of a circle. If T is the linear transformation defined by the matrix

$$\begin{pmatrix} a & 0 \\ 0 & b \end{pmatrix} \qquad (a, b > 0),$$

then T carries the unit circle $x^2 + y^2 = 1$ onto the ellipse $(x/a)^2 + (y/b)^2 = 1$. According to the theorem, the area of the region bounded by the ellipse is equal to $|\det T| = ab$ times the area of the region bounded by the circle. Hence the area is πab.

Given a nonsingular affine mapping T of $\mathbf{R}^2$, we now have a geometric interpretation for $|\det T|$: it is the constant factor by which T alters areas. What then can be said of the sign of T? It can be shown that T is *orientation-preserving* if $\det T > 0$, and *orientation-reversing* if $\det T < 0$. For instance, if γ is a simple closed path whose points are traversed in a counterclockwise direction, then the points on the image path $T(\gamma)$ will be traversed in a counterclockwise direction if $\det T > 0$, and in a clockwise direction if $\det T < 0$. We shall not attempt to make these motions rigorous. The nonsingular linear transformations with positive determinants (i.e., the orientation-preserving mappings) form a subgroup $SL(\mathbf{R}^2)$ of $GL(\mathbf{R}^2)$ known as the *special linear group*. It is also easy to show that the only orientation-preserving rigid motions (which we shall call *proper isometries*) are compositions of rotations and translations; intuitively, these motions of the plane can be realized physically without ever leaving the plane. By contrast, the rigid motion consisting of reflection in the x-axis, which is orientation-reversing, can be realized by a 180° rotation of the plane about the x-axis, but it cannot be realized by physical motion confined in the plane.

Exercises

1. Determine the affine mapping T which carries the points $(0, 1)$, $(-1, 2)$, and $(1, 1)$ onto the points $(3, -3)$, $(1, -6)$, and $(6, -2)$, respectively. Write T as the product of a nonsingular linear transformation and a translation in both possible orders.

2. Given that $S, T \in AF(\mathbf{R}^2)$, show that the linear part of their product $S \circ T$ is the product of the linear parts of S and T.

3. a) Show that the proper isometries form a transformation group. Show that every proper isometry may be written as a product of rotations and translations. What is the determinant of a proper isometry?

   b) Show that $T_\alpha \circ R_\theta \circ T_{-\alpha}$ is a rotation about the point α through an angle of θ radians.

   c) Prove that every proper isometry is either a translation or a rotation about some point in $\mathbf{R}^2$. [First write the isometry as $R_\theta \circ T_\beta$. If $\theta = 0$, there is nothing to prove. If $\theta \neq 0$ (or any multiple of 2π), show that $R_\theta \circ T_\beta$ has a unique fixed point α, that is, $(R_\theta \circ T_\beta)(\alpha) = \alpha$. Show then that $R_\theta \circ T_\beta = T_\alpha \circ R_\theta \circ T_{-\alpha}$.]

4. a) Show that the linear transformation $R_\phi \circ V \circ R_{-\phi}$ represents a reflection of the plane in the straight line through the origin with angle of inclination ϕ. What is the matrix of the transformation if $\phi = \pi/4$?

   b) Show, more generally, that $T_\alpha \circ R_\phi \circ V \circ R_{-\phi} \circ T_{-\alpha}$ represents a reflection in a straight line. Do all reflections arise in this way?

   c) Show that every translation may be written as the product of reflections in two parallel lines and, conversely, that the product of any two reflections in parallel lines is a translation.

   d) Show that every rotation R_θ can be written as the product of reflections in two lines which pass through the origin. More generally, what can you say about the product of reflections in any two intersecting lines?

   e) Show that the product of an even number of reflections is a proper isometry and, conversely, that every proper isometry is the product of an even number of reflections.

   f) What can you say about the product of an odd number of reflections?

5. If $T: \mathbf{R}^2 \to \mathbf{R}^2$, show that the following are logically equivalent:

   a) $T((1 - t)\xi + t\eta) = (1 - t)T(\xi) + tT(\eta)$ for all $\xi, \eta \in \mathbf{R}^2$ and $t \in \mathbf{R}$.

   b) T preserves centroids, that is,

   $$T(t_1\xi_1 + t_2\xi_2 + \cdots + t_n\xi_n) = t_1T(\xi_1) + t_2T(\xi_2) + \cdots + t_nT(\xi_n)$$

   for all $\xi_1, \ldots, \xi_n$ in $\mathbf{R}^2$ and all $t_1, \ldots, t_n$ in $\mathbf{R}$ such that $t_1 + t_2 + \cdots + t_n = 1$.

   c) T is a composition product of a linear mapping (not necessarily nonsingular) and a translation.

   A mapping with these properties is said to be *affine*.

6. a) Show that an affine mapping of $\mathbf{R}^2$ into itself carries convex sets onto convex sets.

   b) Show that an affine mapping of $\mathbf{R}^2$ either maps $\mathbf{R}^2$ onto itself, maps $\mathbf{R}^2$ onto a straight line, or else collapses the entire space onto a single point.

7. a) Let ζ_0, ζ_1, and ζ_2 be any three noncollinear points in $\mathbf{R}^2$. Prove that every point ζ in $\mathbf{R}^2$ can be uniquely written as

   $$\zeta = t_0\zeta_0 + t_1\zeta_1 + t_2\zeta_2,$$

   where $t_0 + t_1 + t_2 = 1$. The coefficients t_0, t_1, t_2 are then called the *baracentric coordinates* of ζ with respect to the points $\zeta_0, \zeta_1, \zeta_2$. [Let T be the nonsingular affine mapping which carries $0, \epsilon_1, \epsilon_2$ into $\zeta_0, \zeta_1, \zeta_2$, respectively. If $\zeta \in \mathbf{R}^2$, let $(t_1, t_2) = T^{-1}(\zeta)$ and show that $1 - t_1 - t_2, t_1, t_2$ are the required baracentric coordinates.]

   b) Show that a nonsingular affine mapping preserves baracentric coordinates. (First, of course, you must decide exactly what this statement means.)

   c) Given that $\zeta_0, \zeta_1, \zeta_2$ are noncollinear points in $\mathbf{R}^2$, describe the set of points in $\mathbf{R}^2$ whose baracentric coordinates are all nonnegative. What is the point whose baracentric coordinates are $\frac{1}{3}, \frac{1}{3}, \frac{1}{3}$?

   d) Let ABC and $A'B'C'$ be nondegenerate triangles in $\mathbf{R}^2$, and let T be the unique nonsingular affine mapping which carries the vertices A, B, C onto the vertices A', B', C', respectively. Show that T carries the center of gravity of one triangle onto the center of gravity of the other.

8. The *inner product* of two vectors $\alpha = (a_1, a_2)$ and $\beta = (b_1, b_2)$ is defined to be

   $$\langle \alpha, \beta \rangle = a_1b_1 + a_2b_2.$$

The term *scalar product* is also used, since the inner product of two vectors is always a scalar. Prove that the inner product has the following algebraic properties:

1) *Bilinearity.* We have the identities

$$\langle c_1\alpha_1 + c_2\alpha_2, \beta \rangle = c_1\langle \alpha_1, \beta \rangle + c_2\langle \alpha_2, \beta \rangle,$$
$$\langle \alpha, c_1\beta_1 + c_2\beta_2 \rangle = c_1\langle \alpha, \beta_1 \rangle + c_2\langle \alpha, \beta_2 \rangle$$

   for all vectors $\alpha_1, \alpha_2, \alpha, \beta_1, \beta_2, \beta$ and all scalars c_1, c_2.

2) *Symmetry.* $\langle \alpha, \beta \rangle = \langle \beta, \alpha \rangle$ for all vectors α and β.

3) *Positivity.* $\langle \alpha, \alpha \rangle \geq 0$, with strict inequality unless $\alpha = 0$.

4) *Cauchy-Schwarz Inequality.* For all vectors α and β,

$$|\langle \alpha, \beta \rangle| \leq \|\alpha\| \, \|\beta\|,$$

   with strict inequality unless one vector is a scalar multiple of the other.

9. a) Two vectors α and β are *orthogonal* iff $\langle \alpha, \beta \rangle = 0$. Prove that α and β are orthogonal iff one of the vectors is the zero vector or both vectors are nonzero and perpendicular (when regarded as arrows, say).

   b) Using the properties of inner products listed above, show that α and β are orthogonal iff $\|\alpha + \beta\|^2 = \|\alpha\|^2 + \|\beta\|^2$. Explain why this result is sometimes called the Pythagorean theorem.

   c) Prove the *parallelogram law:*

$$\|\alpha + \beta\|^2 + \|\alpha - \beta\|^2 = 2(\|\alpha\|^2 + \|\beta\|^2).$$

   Interpret geometrically.

10. Let $T: \mathbf{R}^2 \to \mathbf{R}^2$ be linear. Prove that the following conditions are logically equivalent:

   a) T is an isometry.

   b) T preserves norms [that is, $\|T(\zeta)\| = \|\zeta\|$ for all $\zeta \in \mathbf{R}^2$].

   c) T preserves inner products [that is, $\langle T(\xi), T(\eta) \rangle = \langle \xi, \eta \rangle$ for all $\xi, \eta \in \mathbf{R}^2$].

   d) $T(\epsilon_1)$ and $T(\epsilon_2)$ are orthogonal vectors of unit length.

11. Verify that $SL(\mathbf{R}^2)$ is a group and that $SO(\mathbf{R}^2)$ is a subgroup.

12. Among all the concepts in elementary plane geometry, the notion of angles and their measurement is one of the most intuitively simple and yet one of the most difficult to formulate precisely. We are in as good a position now as we shall ever be to do this.

First let us review what has already been done. We now adopt a completely "arithmetic" attitude toward plane geometry. By a "point" we shall mean a member of $\mathbf{R}^2$. By the line through two distinct points ζ_1 and ζ_2 in $\mathbf{R}^2$ we mean the set

$$\{(1 - t)\zeta_1 + t\zeta_2 \mid t \in \mathbf{R}\},$$

etc. In Chapter 9 we gave a completely arithmetic development of the trigonometric functions, a treatment in which the notions of angles and triangles were completely extraneous. We then linked the trigonometric functions to geometry by proving that each point on the unit circle $\{\xi \in \mathbf{R}^2 \mid \|\xi\| = 1\}$ can be written in the form

$$\mu_\theta = (\cos \theta, \sin \theta),$$

where the number θ is unique to within an integral multiple of 2π. It follows that if $\xi \in \mathbf{R}^2$ is any nonzero vector, then, since $\xi/\|\xi\|$ is a vector on the unit circle, we may write $\xi/\|\xi\| = \mu_\theta$ or

$$\xi = \|\xi\|\mu_\theta.$$

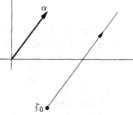

We then defined θ to be a *polar angle* of the point ξ. Moreover, we showed in this section that if θ is a polar angle of ξ, then $\theta + \phi$ is a polar angle of $R_\phi(\xi)$.

FIG. 12-31

   a) By a *ray* emanating from a point $\zeta_0 \in \mathbf{R}^2$ we shall mean any set $\mathfrak{r}$ of the form $\{\zeta_0 + t\alpha \mid t \geq 0\}$, where α is a fixed nonzero vector. If ζ_0 and ζ_1 are distinct points in $\mathbf{R}^2$ show that there is a unique ray which emanates from ζ_0 and passes through ζ_1 (see Fig. 12–31). Also show that if $\mathfrak{r}$ is a ray emanating from the origin, then the polar angle is constant along $\mathfrak{r}$ (i.e., if θ and ϕ are polar angles for two points on $\mathfrak{r}$ distinct from 0, then θ and ϕ differ by an integral multiple of 2π). Hence we can speak of the polar angle of such a ray.

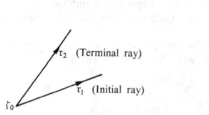

FIG. 12–32

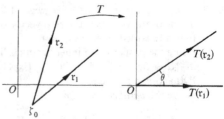

FIG. 12–33

b) If $\mathfrak{r}$ is any ray, prove that there is a unique proper isometry of $\mathbf{R}^2$ which maps $\mathfrak{r}$ onto the positive x-axis.

c) By a *directed angle* we shall mean an ordered pair of rays $(\mathfrak{r}_1, \mathfrak{r}_2)$ emanating from a common point $\mathfrak{s}_0$. Then $\mathfrak{s}_0$ is called the *vertex* of the directed angle; $\mathfrak{r}_1$ is called the *initial ray*, and $\mathfrak{r}_2$ is called the *terminal ray* (see Fig. 12–32). We define the radian measure of such an angle as follows. Let T be the unique proper isometry which carries $\mathfrak{r}_1$ onto the x-axis. Then T carries $\mathfrak{r}_2$ onto a ray emanating from the origin (see Fig. 12–33). If θ is a polar angle of the latter ray, then we say that θ is a *radian measure* of the directed angle $(\mathfrak{r}_1, \mathfrak{r}_2)$, and we write $\theta = \angle(\mathfrak{r}_1, \mathfrak{r}_2)$. Thus the radian measure of a directed angle is unique to within an integral multiple of 2π. Prove that radian measure of directed angles is preserved under proper isometries.

d) Prove the following addition rule for directed angles:

$$\angle(\mathfrak{r}_1, \mathfrak{r}_2) + \angle(\mathfrak{r}_2, \mathfrak{r}_3) = \angle(\mathfrak{r}_1, \mathfrak{r}_3).$$

Deduce that $\angle(\mathfrak{r}_2, \mathfrak{r}_1) = -\angle(\mathfrak{r}_1, \mathfrak{r}_2)$. (Don't rest until you have proofs which are airtight and elegant.)

e) If α and β are nonzero vectors, then by the directed angle (α, β) we shall mean the directed angle $(\mathfrak{r}_\alpha, \mathfrak{r}_\beta)$, where $\mathfrak{r}_\alpha = \{t\alpha \mid t \geq 0\}$ and $\mathfrak{r}_\beta = \{t\beta \mid t \geq 0\}$, and we shall write $\angle(\alpha, \beta) = \angle(\mathfrak{r}_\alpha, \mathfrak{r}_\beta)$. (See Fig. 12–34.) Prove that for all vectors α and β in $\mathbf{R}^2$,

$$\langle \alpha, \beta \rangle = \|\alpha\| \, \|\beta\| \cos \angle(\alpha, \beta).$$

[Treat first the case in which one of the vectors is zero. Then consider the case in which both vectors are nonzero. By using (a) and Exercise 12(c), show that it suffices to consider the case in which α lies on the positive x-axis. Observe that

$$\alpha = (\|\alpha\|, 0) \quad \text{and} \quad \beta = \|\beta\| \mu_{\angle(\alpha,\beta)} = (\|\beta\| \cos \angle(\alpha, \beta), \|\beta\| \sin \angle(\alpha, \beta)).$$

Then compute $\langle \alpha, \beta \rangle$.]

f) Formulate a precise version of the law of cosines and then prove it using (e). [By translating to the origin, show that there is no loss of generality in supposing that

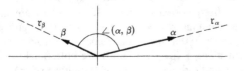

FIG. 12–34

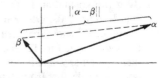

FIG. 12–35

one of the vertices of the triangle is the origin. Using the diagram and notations of Fig. 12–35, observe that the square of the length of the line segment joining α and β is $\|\alpha - \beta\|^2 = (\alpha - \beta, \alpha - \beta) = \cdots]$

g) Given a straight line $y = mx + b$, define its angle of inclination appropriately, and show that its tangent is equal to m.

h) How is the radian measure of a directed angle affected by reflection in the x-axis? by any improper isometry?

One can carry this program further. In particular, the reader might find it profitable to think about the following matters: the formula for the area of a circular sector, the usual interpretation of the trigonometric functions in terms of right triangles, the law of sines, etc.

13. Suppose that T is a mapping of $\mathbf{R}^2$ onto itself and that there exists a constant $k > 0$ such that

$$\|T(\xi) - T(\eta)\| = k\|\xi - \eta\|.$$

In other words, T multiplies all distances by the constant factor k. Then we say that T is a *similarity* transformation.

a) Show that every similarity transformation is one-to-one. Also show that if T and k are as above, then the mapping S defined by

$$S(\xi) = (1/k)(T \circ T_{-T^{-1}(0)})(\xi)$$

is a linear isometry. Infer that T is an affine mapping.

b) Show that the family of all similarity transformations forms a transformation group. Show that every similarity transformation can be written as the product of translations, rotations, reflection in the x-axis, and linear transformations whose matrices are of the form

$$kI = \begin{pmatrix} k & 0 \\ 0 & k \end{pmatrix} \quad (k > 0).$$

c) Show that the *proper similarities* (i.e., similarities with positive determinants) form a subgroup of the group of all similarities. Prove that a proper similarity preserves the measures of directed angles.

d) Describe the effect of a similarity transformation on areas.

e) We say that two sets A and A' in $\mathbf{R}^2$ are (*properly*) *similar* iff there exists a (proper) similarity $T: \mathbf{R}^2 \to \mathbf{R}^2$ which carries the set A onto the set A'. Discuss the following theorem from plane geometry: Two triangles are similar iff their corresponding angles are equal.

f) Prove that two conic sections are similar iff they have the same eccentricity. Is the statement true if we substitute "properly similar" for "similar"?

14. Prove that the area of the triangle whose vertices are (x_0, y_0), (x_1, y_1), (x_2, y_2) is equal to the absolute value of

$$\frac{1}{2} \begin{vmatrix} x_0 & y_0 & 1 \\ x_1 & y_1 & 1 \\ x_2 & y_2 & 1 \end{vmatrix}.$$

(Let T be the affine mapping which carries 0, ϵ_1, ϵ_2 onto the three given points. Determine the linear part of T, and find its determinant. Then use the theorem concerning the behavior of areas under affine mappings.)

15. Let φ and ψ be functions from $\mathbf{R}$ into $\mathbf{R}^2$. Let f be a function from $\mathbf{R}$ to $\mathbf{R}$, and let $T: \mathbf{R}^2 \to \mathbf{R}^2$ be linear. State precise versions of the following differentiation rules, and then prove them.

a) $\dfrac{d}{dt}(\varphi + \psi) = \dfrac{d\varphi}{dt} + \dfrac{d\psi}{dt}$

b) $\dfrac{d}{dt}(T \circ \varphi) = T\left(\dfrac{d\varphi}{dt}\right)$

c) $\dfrac{d}{dt}(f\varphi) = \dfrac{df}{dt}\varphi + f\dfrac{d\varphi}{dt}$

d) $\dfrac{d}{dt}\langle \varphi, \psi \rangle = \left\langle \dfrac{d\varphi}{dt}, \psi \right\rangle + \left\langle \varphi, \dfrac{d\psi}{dt} \right\rangle$

16. Show that an affine mapping preserves tangency. More specifically, show that if l is a tangent line to a curve γ, then $T(l)$ is tangent to $T(\gamma)$, where T is any nonsingular affine mapping.

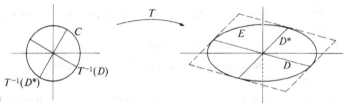

FIG. 12–36

17. Let $a > b > 0$, and let T be the linear transformation with the matrix

$$\begin{pmatrix} a & 0 \\ 0 & b \end{pmatrix}.$$

We have observed that T maps the circle C, $x^2 + y^2 = 1$, onto the ellipse E, $(x/a)^2 + (y/b)^2 = 1$.

a) Let D and D^* be two chords of E whose preimages $T^{-1}(D)$ and $T^{-1}(D^*)$ are perpendicular diameters of C (see Fig. 12–36). We say that D and D^* are *conjugate diameters* of the ellipse. Observe that D and D^* pass through the center of the ellipse. Show that D bisects every chord of E which is parallel to D^*, and, conversely, that D^* bisects every chord of E which is parallel to D. [Observe that the preimages $T^{-1}(D)$ and $T^{-1}(D^*)$ have these properties. Then use the fact that affine maps preserve parallelism, carry midpoints into midpoints, etc.]

b) Prove similarly that the tangent lines at the endpoints of D are parallel to D^*.

c) Prove that the area of the parallelogram formed by the tangent lines at the endpoints of a pair of conjugate diameters is $4ab$.

d) Prove that the legs of the angle subtended by a diameter at any point of an ellipse give the directions of a pair of conjugate diameters (see Fig. 12–37).

e) Prove that the product of the lengths of the line segments defined on a tangent to an ellipse by two conjugate diameters (see Fig. 12–38) is equal to one-fourth the square of the length of the diameter of the ellipse parallel to the tangent.

$|AB| \cdot |BC| = l^2$

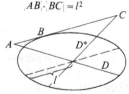

FIG. 12–37   **FIG. 12–38**

12-4 The general second-degree equation

Consider a second-degree polynomial in two variables:

$$P(x, y) = Ax^2 + 2Bxy + Cy^2 + 2Dx + 2Ey + F.$$

One objective is to describe in all possible cases the set of points (x, y) in $\mathbf{R}^2$ such that $P(x, y) = 0$. We shall see that this set is either an ellipse, a parabola, or a hyperbola, except in some degenerate cases. The technique which we shall use is that of *transformation of coordinates*.

Imagine two analytic geometers situated at two points of a plane. Each establishes his own system of Cartesian coordinates. One geometer, say Mr. Z, labels his coordinate axes x and y; the other geometer, whom we shall call Mr. W, labels his coordinate axes u and v (see Fig. 12–39). To facilitate the communication of their results, they agree to use the same unit of length, and they also agree to use right-handed systems. (That is, they agree to use systems such that counterclockwise rotations about the origins through $\pi/2$ radians carry the positive x- and u-axes onto the positive y- and v-axes, respectively.) For the geometers to interpret each other's work, it is essential for them to know how to convert the coordinates of a point with respect to one system into the coordinates of the point with respect to the other system. This correspondence between coordinates can be described by a mapping of $\mathbf{R}^2$ onto itself: For each $\zeta = (x, y) \in \mathbf{R}^2$, we let $P(\zeta)$ be the point whose coordinates are (x, y) in Mr. Z's system, and we let $T(\zeta) = (u, v)$ be the coordinates of $P(\zeta)$ in Mr. W's system. The mapping $T: \mathbf{R}^2 \to \mathbf{R}^2$ is one-to-one and onto; it is also distance-preserving (since Mr. Z and Mr. W use the same unit of length), and it also preserves orientation (since both men use right-handed systems). Hence T must be the composition product of a translation and a rotation. It therefore suffices to discuss separately the cases in which T is a translation or a rotation.

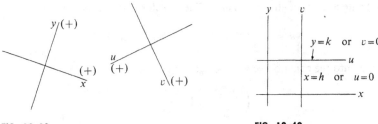

FIG. 12–39   FIG. 12–40

If T is a translation, and if $\alpha = (h, k)$ are the coordinates in Mr. Z's system of the origin in Mr. W's system, as in Fig. 12–40, then $T = T_{-\alpha}$. Consequently, the conversion of coordinates is described by the single vector equation $(u, v) = T_{-\alpha}(x, y)$ or the two scalar equations

$$u = x - h, \qquad v = y - k.$$

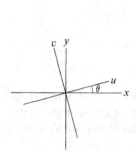

FIG. 12-41   FIG. 12-42

If T is a rotation, and if θ is the polar angle of the positive u-axis in Mr. Z's system, then $T = R_{-\theta}$. Consequently, the equations of transformation are

$$u = (\cos \theta)x + (\sin \theta)y, \qquad v = -(\sin \theta)x + (\cos \theta)y,$$

or equivalently [since $(x, y) = R_\theta(u, v)$],

$$x = (\cos \theta)u - (\sin \theta)v, \qquad y = (\sin \theta)u + (\cos \theta)v$$

(see Fig. 12-41).

What then is the point of changing coordinates? The answer is that the equation describing a set of points may be simpler in one system than in the other. We illustrate with an example.

Example 1. Show that $xy - 2y + 3x - 7 = 0$ is an equilateral hyperbola. Find its foci. If we make the substitutions

$$u = x - 2, \qquad v = y + 3$$

(which amounts, of course, to a translation of coordinate axes), then the equation becomes $uv = 1$. The curve is symmetric about the line $u = v$ and, consequently, if $uv = 1$ is a hyperbola, then $u = v$ must be one of its axes of symmetry. This suggests that we rotate our coordinate axes through $\pi/4$ radians. The transformation equations are

$$u = \frac{1}{\sqrt{2}}(X - Y), \qquad v = \frac{1}{\sqrt{2}}(X + Y).$$

Substituting into the equation $uv = 1$, we get

$$\left(\frac{X}{\sqrt{2}}\right)^2 - \left(\frac{Y}{\sqrt{2}}\right)^2 = 1,$$

which is the equation of an equilateral hyperbola. The foci of the hyperbola have the coordinates $X = \pm 2$, $Y = 0$, as shown in Fig. 12-42. The coordinates relative to the given system can be calculated from the transformation equations:

$$x = u + 2 = \frac{1}{\sqrt{2}}(X - Y) + 2, \qquad y = v - 3 = \frac{1}{\sqrt{2}}(X + Y) - 3.$$

The coordinates of the foci are seen to be $(2 + \sqrt{2}, -3 + \sqrt{2})$ and $(2 - \sqrt{2}, -3 - \sqrt{2})$.

We now tackle the general second-order equation,

$$Ax^2 + 2Bxy + Cy^2 + 2Dx + 2Ey + F = 0.$$

We shall assume that A, B, and C are not all zero. The term $2Bxy$ is the most troublesome. We try to get rid of it by rotating coordinate axes. The substitution

$$x = (\cos \theta)u - (\sin \theta)v, \qquad y = (\sin \theta)u + (\cos \theta)v$$

yields the equation

$$A'u^2 + 2B'uv + C'v^2 + 2D'u + 2E'v + F' = 0,$$

where

$$A' = A \cos^2 \theta + 2B \sin \theta \cos \theta + C \sin^2 \theta,$$
$$B' = (C - A) \sin \theta \cos \theta + B(\cos^2 \theta - \sin^2 \theta),$$
$$C' = A \sin^2 \theta - 2B \sin \theta \cos \theta + C \cos^2 \theta;$$

the remaining coefficients are of no interest at the moment. A small amount of trigonometry yields the equations

$$A' + C' = A + C, \qquad A'C' - (B')^2 = AC - B^2.$$

The second equation is especially important. The expression $AC - B^2$ is called the *discriminant* of the original quadratic polynomial. Thus we see that the *discriminant of a quadratic polynomial in two variables is invariant under rotations.* The equation for B' may be written as

$$B' = \tfrac{1}{2}(C - A) \sin 2\theta + B \cos 2\theta.$$

From this equation, it is clear that θ can be chosen so that $B' = 0$. If $A = C$, we can let $\theta = \pi/2$; if $A \neq C$, we can let

$$\theta = \tfrac{1}{2} \operatorname{Tan}^{-1} \left(\frac{2B}{A - C} \right).$$

Observe further that when $B' = 0$, we have

$$(\lambda - A')(\lambda - C') \equiv \lambda^2 - (A + C)\lambda + (AC - B^2).$$

Thus A' and C' are the roots of the quadratic equation

$$\lambda^2 - (A + C)\lambda + (AC - B^2) = 0.$$

Proposition 1. *The polynomial*

$$Ax^2 + 2Bxy + Cy^2 + 2Dx + 2Ey + F$$

can be reduced to the form

$$A'u^2 + C'v^2 + 2D'u + 2E'v + F'$$

by a rotation of coordinates. The coefficients A' and C' are the roots of the quadratic equation

$$\lambda^2 - (A + C)\lambda + (AC - B^2) = \begin{vmatrix} A - \lambda & B \\ B & C - \lambda \end{vmatrix} = 0.$$

Once the xy-term is gone, completion of squares and translations of coordinate axes lead to further simplifications.

Proposition 2. *The set of points satisfying the equation* $P(x, y) = 0$ *is either an ellipse, a parabola, a hyperbola, the union of two lines, a single line, a singleton set, or the null set.*

Proof. Because of the previous theorem, we may assume that

$$P(x, y) = Ax^2 + Cy^2 + 2Dx + 2Ey + F,$$

where $AC \neq 0$.

Case I. Suppose that neither A nor C is zero. Then the translation of coordinate axes

$$x = u - D/A, \qquad y = u - E/C$$

yields the equation

$$Au^2 + Cv^2 = F',$$

where F' is a constant. If A and C have opposite signs, then the set is either a hyperbola or the union of two intersecting lines. If A and C have the same sign, then the set is either an ellipse, a singleton set, or the null set.

Case II. Suppose that $C = 0$. The linear term in x can be eliminated by translation, and the equation can be reduced to the form

$$Au^2 + E'v + F' = 0.$$

If $E' \neq 0$, we have a parabola. If $E = 0$, we have either two parallel lines, a single line, or the empty set.

Case III. Suppose that $A \neq 0$. This case reduces to case II by the interchange of x and y.

We now consider some examples.

Example 2. Identify the curve $13x^2 - 8xy + 7y^2 = 30$.
    We rotate axes so as to rid ourselves of the xy-term. According to our main theorem, the coefficients of the reduced form are roots of the equation

$$\begin{vmatrix} 13 - \lambda & -4 \\ -4 & 7 - \lambda \end{vmatrix} = \lambda^2 - 20\lambda + 75 = 0.$$

The roots are 15 and 5. Thus the reduced equation is either

$$5u^2 + 15v^2 = 30 \qquad \text{or} \qquad 15u^2 + 5v^2 = 30,$$

depending on the particular choice of the angle of rotation. In either case we see that the curve is an ellipse. The major axis has length $2\sqrt{6}$; the minor axis has length $2\sqrt{2}$.

Note that in the previous example it was unnecessary to determine the angle of rotation and the equations for the transformation of variables. If, however, we wish to determine the foci or directrices of the ellipse, then the transformation equations are required. We can find these equations by first computing

$$\theta = \tfrac{1}{2}\operatorname{Tan}^{-1}\left(\frac{2B}{A-C}\right)$$

and then substituting this value of θ into the rotation matrix

$$R_\theta = \begin{pmatrix} \cos\theta & -\sin\theta \\ \sin\theta & \cos\theta \end{pmatrix}.$$

Actually, we can determine R_θ without first calculating θ. It suffices, of course, to determine the vector $\mu_\theta = (\cos\theta, \sin\theta)$. A proof of the following result is outlined in the exercises. *If the quadratic form $Ax^2 + 2Bxy + Cy^2$ is transformed to $A'u^2 + C'v^2$ by the change of variables $(x, y) = R_\theta(u, v)$, then the vector μ_θ satisfies the equation $T(\mu_\theta) = A'\mu_\theta$, where T is the linear transformation having the matrix*

$$\begin{pmatrix} A & B \\ B & C \end{pmatrix}.$$

[One can also show that the second column of R_θ, namely, $\nu_\theta = (-\sin\theta, \cos\theta)$ satisfies the equation $T(\nu_\theta) = C'\nu_\theta$.]

Example 3. Find the foci of the ellipse given in Example 2.

Let us take the transformed equation to be $5u^2 + 15v^2 = 30$. To find the rotation matrix, we first solve the vector equation $T(\zeta) = 5\zeta$, where T has the matrix

$$\begin{pmatrix} 13 & -4 \\ -4 & 7 \end{pmatrix}.$$

If we set $\zeta = (x, y)$, we see that the vector equation $T(\zeta) = 5\zeta$ is equivalent to the simultaneous scalar equations

$$13x - 4y = 5x, \qquad -4x + 7y = 5y,$$

which in turn are equivalent to the single scalar equation $y = 2x$. Thus ζ satisfies the equation $T(\zeta) = 5\zeta$ iff ζ is a scalar multiple of the vector $(1, 2)$. Among such vectors there are two of unit length: $\pm(1/\sqrt{5}, 2/\sqrt{5})$. We may set μ_θ equal to either of these unit vectors; we choose the one with the plus sign. It follows that the rotation matrix is

$$\begin{pmatrix} 1/\sqrt{5} & -2/\sqrt{5} \\ 2/\sqrt{5} & 1/\sqrt{5} \end{pmatrix},$$

and the transformation equations are

$$x = \frac{1}{\sqrt{5}}(u - 2v), \qquad y = \frac{1}{\sqrt{5}}(2u + v).$$

The coordinates of the foci in the uv-system are $u = \pm\sqrt{a^2 - b^2} = \pm\sqrt{6 - 2} = \pm 2$ and $v = 0$ (see Fig. 12–43). Hence the coordinates of the foci in the given coordinate system are

$$x = \pm\frac{2}{\sqrt{5}} \quad \text{and} \quad y = \pm\frac{4}{\sqrt{5}}.$$

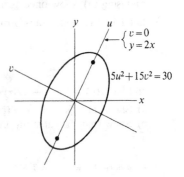

FIG. 12-43

Exercises

1. Investigate thoroughly each of the following curves. (Find the eccentricity, foci, and directrices of the nondegenerate conics. Find the asymptotes of any hyperbola. Sketch the curves, and label the new and old coordinate axes.)

   a) $x^2 + 5y^2 + 6x - 20y - 71 = 0$
   b) $x^2 - 6xy + 9y^2 + 8x - 4y + 5 = 0$
   c) $2x^2 - xy - 6y^2 + 5x + 11y - 3 = 0$
   d) $x^2 - xy = 1$
   e) $y = x + (1/x)$
   f) $2x^2 + xy - 6y^2 - 7y - 2 = 0$
   g) $5x^2 + 2xy + 10y^2 + 2x + 6y + 1 = 0$
   h) $x^2 - 6xy + 9y^2 - x + 3y - 2 = 0$

2. If $AC - B^2 \neq 0$, show that the linear terms of the quadratic form

$$Ax^2 + 2Bxy + Cy^2 + 2Dx + 2Ey + F$$

   can be eliminated by a translation of coordinate axes. What if $AC - B^2 = 0$?

3. In the discussion of the reduction of quadratic forms, show that

$$D' = D \cos \theta + E \sin \theta, \qquad E' = -D \sin \theta + E \cos \theta.$$

4. We set

$$A_\theta = A \cos^2 \theta + 2B \sin \theta \cos \theta + C \sin^2 \theta,$$
$$B_\theta = (C - A) \sin \theta \cos \theta + B(\cos^2 \theta - \sin^2 \theta),$$
$$C_\theta = A \sin^2 \theta - 2B \sin \theta \cos \theta + C \cos^2 \theta.$$

Thus $A_\theta u^2 + 2B_\theta uv + C_\theta v^2$ is the transformation of the quadratic form

$$Ax^2 + 2Bxy + Cy^2$$

under the rotation $(x, y) = R_\theta(u, v)$.

a) Show that the following differential equations are satisfied:

$$\frac{d}{d\theta} (A_\theta) = 2B_\theta, \qquad \frac{d}{d\theta} (C_\theta) = -2B_\theta, \qquad \frac{d}{d\theta} (B_\theta) = (C_\theta - A_\theta).$$

b) Using (a), show once again that $A_\theta C_\theta - B_\theta^2$ and $A_\theta + C_\theta$ are independent of θ.

5. Let T be the linear transformation corresponding to the matrix

$$\begin{pmatrix} A & B \\ B & C \end{pmatrix},$$

and let $Q(\zeta) = Ax^2 + 2Bxy + Cy^2$, where $\zeta = (x, y)$.

a) Show that $Q(\zeta) = \langle \zeta, T(\zeta) \rangle$.

b) Show that the substitution $\zeta = R_\theta(\zeta')$ transforms $Q(\zeta)$ into

$$Q'(\zeta') = \langle \zeta', T'(\zeta') \rangle,$$

   where $T' = R_{-\theta} \circ T \circ R_\theta$.

c) Suppose that T' has the diagonal matrix

$$\begin{pmatrix} A' & 0 \\ 0 & C' \end{pmatrix}.$$

Show that $R_\theta \circ T = T \circ R_\theta$, and then apply both sides to the vector $\epsilon_1 = (1, 0)$ to deduce that

$$T(\mu_\theta) = A'\mu_\theta.$$

What information do you obtain by applying both sides to ϵ_2?

In general, if there exists a nonzero vector ξ and a scalar λ such that $T(\xi) = \lambda\xi$, we say that ξ is an *eigenvector* for the linear mapping T and that λ is the corresponding *eigenvalue*. Thus we have proved that the nonzero coefficients of the reduced quadratic form are the eigenvalues of T and that the columns of the rotation matrix are the corresponding eigenvectors.

*6. The most satisfactory scheme for classifying the set

$$\{(x, y) \mid A^2x^2 + 2Bxy + Cy^2 + 2Dx + 2Ey + F = 0\}$$

is the following. Set

$$\Delta_1 = \begin{vmatrix} A & B \\ B & C \end{vmatrix}, \qquad \Delta_2 = \begin{vmatrix} A & B & D \\ B & C & E \\ D & E & F \end{vmatrix}.$$

Then the nature of the set can be shown by the following table.

	$\Delta_1 > 0$	$\Delta_1 < 0$	$\Delta_1 = 0$
$\Delta_2 \neq 0$	Ellipse	Hyperbola	Parabola
$\Delta_2 = 0$	Point or $\varnothing$	Intersecting lines	Parallel lines

Prove that the scheme is valid.

12-5 A little more about vectors

In preparation for the concluding section, we turn our attention once again to vectors and vector-valued functions. Very little new material is involved: we shall adopt new notations, reinterpret old results, and collect for handy reference material previously relegated to the exercises.

Up to this point we have used lower-case Greek letters for vectors and lower-case italic letters for scalars. We shall now adopt the more commonly accepted convention of designating vectors by boldface letters; we shall continue to designate scalars by italic letters as is customary. For example, we shall use $c\mathbf{a}$ to designate the product of the scalar c and the vector $\mathbf{a}$. We small denote by $\mathbf{0}$ the null vector, and by $\mathbf{i}$ and $\mathbf{j}$ the vectors previously labeled ϵ_1 and ϵ_2. Thus

$$\mathbf{i} = (1, 0) \qquad \text{and} \qquad \mathbf{j} = (0, 1)$$

are the vectors of unit length which point in the directions
of the positive x- and y-axes, as shown in Fig. 12–44. We
shall continue to let $\|\mathbf{a}\|$ denote the length of the vector $\mathbf{a}$.
We remind the reader of the triangle inequality:

$$\|\mathbf{a} + \mathbf{b}\| \le \|\mathbf{a}\| + \|\mathbf{b}\|,$$

and one of its easily derived consequences, the reverse
triangle inequality:

$$\big|\, \|\mathbf{a}\| - \|\mathbf{b}\| \,\big| \le \|\mathbf{a} - \mathbf{b}\|.$$

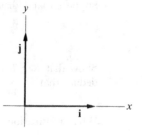

FIG. 12–44

Also we have $\|c\mathbf{a}\| = |c|\,\|\mathbf{a}\|$.

We shall also adopt a different notation for inner products. Given two vectors
$\mathbf{a} = (a_1, a_2) = a_1\mathbf{i} + a_2\mathbf{j}$ and $\mathbf{b} = b_1\mathbf{i} + b_2\mathbf{j}$, we shall use $\mathbf{a} \cdot \mathbf{b}$ rather than $\langle \mathbf{a}, \mathbf{b}\rangle$
to denote their inner product:

$$\mathbf{a} \cdot \mathbf{b} = a_1 b_1 + a_2 b_2.$$

(This product is also called the *dot product*.) The properties of inner products which
one should keep in mind are the following:

1) *Distributivity.* $(\mathbf{a} + \mathbf{b}) \cdot \mathbf{c} = \mathbf{a} \cdot \mathbf{c} + \mathbf{b} \cdot \mathbf{c}; \quad \mathbf{a} \cdot (\mathbf{b} + \mathbf{c}) = \mathbf{a} \cdot \mathbf{b} + \mathbf{a} \cdot \mathbf{c}.$

2) *Symmetry.* $\mathbf{a} \cdot \mathbf{b} = \mathbf{b} \cdot \mathbf{a}.$

3) *Homogeneity.* $(c\mathbf{a}) \cdot \mathbf{b} = c(\mathbf{a} \cdot \mathbf{b}) = \mathbf{a} \cdot (c\mathbf{b}).$

The geometric interpretation of the inner product is equally important:

$$\mathbf{a} \cdot \mathbf{b} = \|\mathbf{a}\|\,\|\mathbf{b}\| \cos \angle(\mathbf{a}, \mathbf{b}),$$

where $\angle(\mathbf{a}, \mathbf{b})$ denotes the angle between $\mathbf{a}$ and $\mathbf{b}$. In particular, we have

$$\mathbf{a} \cdot \mathbf{a} = \|\mathbf{a}\|^2.$$

Also, it follows that two nonzero vectors $\mathbf{a}$ and $\mathbf{b}$ are perpendicular iff $\mathbf{a} \cdot \mathbf{b} = 0$.

We consider next vector-valued functions of a real variable and limits of such
functions.

Definition 1

$$\lim_{t \to t_0} \mathbf{f}(t) = \mathbf{L} \quad \text{iff} \quad \lim_{t \to t_0} \|\mathbf{f}(t) - \mathbf{L}\| = 0.$$

The components of a limit are the limits of the components.

Proposition 2. *Let* $\mathbf{f}(t) = f_1(t)\mathbf{i} + f_2(t)\mathbf{j}$ *and* $\mathbf{L} = L_1\mathbf{i} + L_2\mathbf{j}$. *Then*

$$\lim_{t \to t_0} \mathbf{f}(t) = \mathbf{L} \Leftrightarrow \begin{cases} \displaystyle\lim_{t \to t_0} f_1(t) = L_1, \\[1mm] \displaystyle\lim_{t \to t_0} f_2(t) = L_2. \end{cases}$$

Proof. The implication $\Leftarrow$ follows from the inequalities

$$|f_j(t) - L_j| \leq \|\mathbf{f}(t) - \mathbf{L}\| \qquad (j = 1, 2).$$

The converse follows from an application of the triangle inequality:

$$\begin{aligned}
\|\mathbf{f}(t) - \mathbf{L}\| &= \|[f_1(t) - L_1]\mathbf{i} + [f_2(t) - L_2]\mathbf{j}\| \\
&\leq \|[f_1(t) - L_1]\mathbf{i}\| + \|[f_2(t) - L_2]\mathbf{j}\| \\
&= |f_1(t) - L_1| + |f_2(t) - L_2|.
\end{aligned}$$

Another characterization is possible for the case of nonzero limits.

Proposition 3. *Let* $\mathbf{L} \neq \mathbf{0}$. *Then*

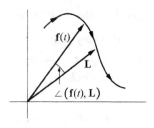

$$\lim_{t \to t_0} \mathbf{f}(t) = \mathbf{L} \Leftrightarrow \begin{cases} \lim_{t \to t_0} \|\mathbf{f}(t)\| = \|\mathbf{L}\|, \\ \lim_{t \to t_0} \cos \angle (\mathbf{f}(t), \mathbf{L}) = 1. \end{cases}$$

We shall interpret the last two expressions as saying that $\mathbf{f}(t)$ approaches $\mathbf{L}$ both in length and direction, as indicated in Fig. 12–45.

FIG. 12-45

Proof. Suppose $\lim_{t \to t_0} \mathbf{f}(t) = \mathbf{L}$. From the reverse triangle inequality

$$\big|\, \|\mathbf{f}(t)\| - \|\mathbf{L}\| \,\big| \leq \|\mathbf{f}(t) - \mathbf{L}\|,$$

it immediately follows that $\|\mathbf{f}(t)\| \to \|\mathbf{L}\|$ as $t \to t_0$. The fact that $\cos \angle (\mathbf{f}(t), \mathbf{L})$ tends to one then follows from the identity

$$\begin{aligned}
\|\mathbf{f}(t) - \mathbf{L}\|^2 &= [\mathbf{f}(t) - \mathbf{L}] \cdot [\mathbf{f}(t) - \mathbf{L}] = \|\mathbf{f}(t)\|^2 + \|\mathbf{L}\|^2 - 2\mathbf{f}(t) \cdot \mathbf{L} \\
&= \|\mathbf{f}(t)\|^2 + \|\mathbf{L}\|^2 - 2\|\mathbf{f}(t)\| \, \|\mathbf{L}\| \cos \angle (\mathbf{f}(t), \mathbf{L}).
\end{aligned}$$

Moreover, the converse is also an immediate consequence of the same identity.

The analogs of the standard theorems concerning limits of sums and products hold for vector-valued functions. For instance, given that $\lim_{t \to t_0} \mathbf{f}(t) = \mathbf{L}$ and $\lim_{t \to t_0} \varphi(t) = c$, it follows that $\lim_{t \to t_0} \varphi(t)\mathbf{f}(t) = c\mathbf{L}$. Using the previous theorems, we can easily give proofs. We have

$$\begin{aligned}
0 \leq \;&\|\varphi(t)\mathbf{f}(t) - c\mathbf{L}\| \\
&= \|[\varphi(t) - c]\mathbf{f}(t) + c[\mathbf{f}(t) - \mathbf{L}]\| \\
&\leq \|[\varphi(t) - c]\mathbf{f}(t)\| + \|c[\mathbf{f}(t) - \mathbf{L}]\| \\
&= |\varphi(t) - c| \, \|\mathbf{f}(t)\| + |c| \, \|\mathbf{f}(t) - \mathbf{L}\|.
\end{aligned}$$

As $t \to t_0$, the last expression tends to $0 \cdot \|\mathbf{L}\| + |c| \cdot 0 = 0$. Hence

$$\|\varphi(t)\mathbf{f}(t) - c\mathbf{L}\| \to 0$$

or, equivalently, $\varphi(t)\mathbf{f}(t) \to c\mathbf{L}$. Alternatively, we can prove the same result by considering coordinates. Let $\mathbf{f}(t) = f_1(t)\mathbf{i} + f_2(t)\mathbf{j}$ and $\mathbf{L} = L_1\mathbf{i} + L_2\mathbf{j}$. Then

$$\varphi(t)\mathbf{f}(t) = \varphi(t)f_1(t)\mathbf{i} + \varphi(t)f_2(t)\mathbf{j} \to cL_1\mathbf{i} + cL_2\mathbf{j} = c\mathbf{L}.$$

Other examples are given in the exercises.

The derivative of a vector-valued function may be defined as a limit of difference quotients, as in the case of real-valued functions.

Definition 4. The derivative of $\mathbf{f}(t)$ is

$$\mathbf{f}'(t_0) = \lim_{h \to 0} \frac{1}{h}[\mathbf{f}(t_0 + h) - \mathbf{f}(t_0)],$$

provided the limit exists.

It is easy to show that this notion of a derivative coincides with our previous one.

Proposition 5. *Let* $\mathbf{f}(t) = f_1(t)\mathbf{i} + f_2(t)\mathbf{j}$. *Then* $\mathbf{f}'(t_0)$ *exists iff* $f_1'(t_0)$ *and* $f_2'(t_0)$ *exist, in which case*

$$\mathbf{f}'(t_0) = f_1'(t_0)\mathbf{i} + f_2'(t_0)\mathbf{j}.$$

We shall now give a dynamic interpretation of these derivatives. Let us assume that the functions f_1 and f_2 have continuous derivatives which do not vanish simultaneously on the interval $[a, b]$. We may then regard $\mathbf{f}(t) = f_1(t)\mathbf{i} + f_2(t)\mathbf{j}$ as specifying the position of a particle at time t. Then

$$s(t) = \int_a^t \sqrt{[f_1'(u)]^2 + [f_2'(u)]^2} \, du = \int_a^t \|\mathbf{f}'(u)\| \, du$$

represents the path length traced out by the particle during the time interval $[a, t]$. Thus $\|\mathbf{f}'(t)\| = ds/dt$ represents the instantaneous rate of change of the path length—intuitively, and also by definition, the *speed* of the particle at time t. What about the direction of $\mathbf{f}'(t)$? It is the limit of the direction of the vector $h^{-1}[\mathbf{f}(t + h) - \mathbf{f}(t)]$ as $h \to 0$. The vector in question, however, is parallel to the secant line joining $\mathbf{f}(t)$ and $\mathbf{f}(t + h)$. Thus $\mathbf{f}'(t)$ ought to be parallel to the tangent line to the path of the particle at the point $\mathbf{f}(t)$; we *define* the tangent line so that this is the case (see Fig. 12–46). We have argued, then, that the vector $\mathbf{f}'(t)$ specifies both the instantaneous speed and the direction of the particle at time t; this vector is defined to be the *velocity* of the particle at time t, and we frequently denote it by $\mathbf{V}$ or $\mathbf{V}(t)$. The vector

$$\mathbf{A} = \frac{d\mathbf{V}}{dt} = \frac{d^2\mathbf{f}}{dt^2}$$

is called the *acceleration* of the particle. Its importance stems mainly from Newton's

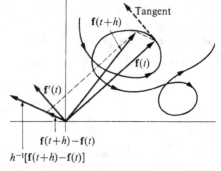

FIG. 12–46

second law, which states that a force $\mathbf{F}$ acting on a particle of mass m produces an acceleration $\mathbf{A}$ such that

$$\mathbf{F} = m\mathbf{A}.$$

(Here $\|\mathbf{F}\|$ is the strength of the force, and the direction of $\mathbf{F}$ is the direction in which the force acts.)

Most of the rules for differentiating real-valued functions have valid analogs for vector-valued functions. In particular, we have

$$\frac{d}{dt}(\mathbf{f} + \mathbf{g}) = \frac{d\mathbf{f}}{dt} + \frac{d\mathbf{g}}{dt}, \qquad \frac{d}{dt}(\varphi\mathbf{f}) = \varphi\frac{d\mathbf{f}}{dt} + \frac{d\varphi}{dt}\mathbf{f},$$

$$\frac{d}{dt}(\mathbf{f} \cdot \mathbf{g}) = \mathbf{f} \cdot \frac{d\mathbf{g}}{dt} + \frac{d\mathbf{f}}{dt} \cdot \mathbf{g}, \qquad \frac{d}{dt}(\mathbf{f} \circ \varphi) = (\mathbf{f}' \circ \varphi)\frac{d\varphi}{dt},$$

where in each instance the equation is valid at those points where the expression on the right-hand side is well-defined. In addition, if $T: \mathbf{R}^2 \to \mathbf{R}^2$ is linear, then

$$\frac{d}{dt}(T \circ \mathbf{f}) = T \circ \mathbf{f}',$$

that is, linear mappings behave much like constant factors.

We shall also have occasion to differentiate functions whose values are linear transformations. We define the derivative of a matrix-valued function

$$A(t) = \begin{pmatrix} a_{11}(t) & a_{12}(t) \\ a_{21}(t) & a_{22}(t) \end{pmatrix}$$

by

$$A'(t) = \begin{pmatrix} a'_{11}(t) & a'_{12}(t) \\ a'_{21}(t) & a'_{22}(t) \end{pmatrix}.$$

We then define the derivative of the corresponding linear transformation to be the linear transformation of the derived matrix:

$$\frac{d}{dt}[T_{A(t)}] = T_{A'(t)}.$$

For example,

$$\frac{d}{d\theta}\begin{pmatrix} \cos\theta & -\sin\theta \\ \sin\theta & \cos\theta \end{pmatrix} = \begin{pmatrix} -\sin\theta & -\cos\theta \\ \cos\theta & -\sin\theta \end{pmatrix}$$

$$= \begin{pmatrix} \cos(\theta + \pi/2) & -\sin(\theta + \pi/2) \\ \sin(\theta + \pi/2) & \cos(\theta + \pi/2) \end{pmatrix},$$

so that

$$\frac{d}{d\theta}(R_\theta) = R_{\theta + \pi/2}.$$

We shall require the following differentiation formula:

$$\frac{d}{dt}[T_{A(t)}(\mathbf{f}(t))] = T_{A'(t)}(\mathbf{f}(t)) + T_{A(t)}(\mathbf{f}'(t)).$$

This is a consequence of the usual rule for differentiating products applied to the components of

$$T_{A(t)}\big(\mathbf{f}(t)\big) = [a_{11}(t)f_1(t) + a_{12}(t)f_2(t)]\mathbf{i} + [a_{21}(t)f_1(t) + a_{22}(t)f_2(t)]\mathbf{j}.$$

We shall now use the machinery we have developed to study motion along a circle. Suppose that $\mathbf{f}(t)$ is the position at time t of a particle which moves along the unit circle $x^2 + y^2 = 1$. We consider the motion over the time interval $a \le t \le b$, and we assume that $\mathbf{f}$ has a continuous derivative throughout the interval. Then we have

$$\mathbf{f}(t) \cdot \mathbf{f}(t) = [f_1(t)]^2 + [f_2(t)]^2 \equiv 1.$$

Differentiating, we get

$$2\mathbf{f}(t) \cdot \mathbf{f}'(t) = 2f_1(t)f_1'(t) + 2f_2(t)f_2'(t) \equiv 0.$$

To get additional information, we consider polar angles. For each $t \in [a, b]$, we can choose a value $\theta(t)$ of the polar angle of the point $\mathbf{f}(t)$. At each point of the path, the choice can be made in infinitely many different ways. It should be intuitively clear, however, that once a value of $\theta(a)$ has been chosen, the other values of $\theta(t)$ can be chosen in only one way so that the resulting function θ is continuous. We shall now prove that this is so.

We define

$$\theta(t) = \theta(a) + \int_a^t [f_1(u)f_2'(u) - f_1'(u)f_2(u)]\, du,$$

where $\theta(a)$ is any value of the polar angle of the point $\mathbf{f}(a)$. The function θ is differentiable on the interval $[a, b]$. We shall prove that for each $t \in [a, b]$, the number $\theta(t)$ is a polar angle of the point $\mathbf{f}(t)$, or, equivalently, that $R_{-\theta(t)}\big(\mathbf{f}(t)\big) \equiv \mathbf{i}$. Now

$$\frac{d}{dt}[R_{-\theta(t)}\big(\mathbf{f}(t)\big)] = R_{-\theta(t)}\big(\mathbf{f}'(t)\big) - \frac{d\theta}{dt} R_{-\theta(t)+\pi/2}\big(\mathbf{f}(t)\big)$$

$$= R_{-\theta(t)}\left[\mathbf{f}'(t) - \frac{d\theta}{dt} R_{\pi/2}\big(\mathbf{f}(t)\big)\right].$$

Also,

$$\mathbf{f}'(t) = f_1'(t)\mathbf{i} + f_2'(t)\mathbf{j}, \qquad R_{\pi/2}\big(\mathbf{f}(t)\big) = -f_2(t)\mathbf{i} + f_1(t)\mathbf{j}$$

and

$$\frac{d\theta}{dt} = f_1(t)f_2'(t) - f_1'(t)f_2(t).$$

From these equations and the identities in the preceding paragraph, we get

$$\mathbf{f}'(t) - \frac{d\theta}{dt} R_{\pi/2}\big(\mathbf{f}(t)\big) = 0.$$

Thus $(d/dt)[R_{-\theta(t)}\big(\mathbf{f}(t)\big)] \equiv 0$. It follows that

$$R_{-\theta(t)}\big(\mathbf{f}(t)\big) \equiv R_{-\theta(a)}\big(\mathbf{f}(a)\big) = \mathbf{i}.$$

Proposition 6. *Suppose that* **f** *possesses a continuous derivative on the interval* $[a, b]$ *and that* $\|\mathbf{f}(t)\| \equiv 1$. *Then there exists a real-valued function* θ *of class* C^1 *on* $[a, b]$ *such that for each* $t \in [a, b]$, *the number* $\theta(t)$ *is a polar angle of the point* $\mathbf{f}(t)$. *Furthermore,*

$$\mathbf{f}'(t) = \frac{d\theta}{dt} R_{\pi/2}(\mathbf{f}(t)).$$

Exercises

1. Given that **a** and **b** are nonzero orthogonal vectors, prove that every vector **r** can be uniquely written $\mathbf{r} = c_1\mathbf{a} + c_2\mathbf{b}$, where c_1 and c_2 are scalars. Show, moreover, that c_1 and c_2 may be computed as follows:

$$c_1 = (\mathbf{r} \cdot \mathbf{a})/(\mathbf{a} \cdot \mathbf{a}) \quad \text{and} \quad c_2 = (\mathbf{r} \cdot \mathbf{b})/(\mathbf{b} \cdot \mathbf{b}).$$

2. a) Establish the formula

$$\frac{d}{dt}(T_{A[\varphi(t)]}) = \varphi'(t)T_{A'[\varphi(t)]}.$$

On the right-hand side, we have a scalar times a linear transformation. Such a thing is defined as follows: $(cT)(\mathbf{a}) = cT(\mathbf{a})$.
   b) Using (a), prove that

$$\frac{d}{dt}\left[R_{-\theta(t)}(\mathbf{f}(t))\right] = R_{-\theta(t)}(\mathbf{f}'(t)) - \frac{d\theta}{dt}R_{-\theta(t)+\pi/2}(\mathbf{f}(t)).$$

3. Prove that if $\mathbf{f}'$ is zero throughout some interval I, then **f** is constant on I.

4. Explain why in Proposition 6 it does not suffice to let θ be the principal value of the polar angle.

5. If A and B are differentiable matrix-valued functions, show that

$$(AB)' = A'B + AB'.$$

6. The *exterior product* of two vectors in $\mathbf{R}^2$ is defined as

$$\mathbf{a} \times \mathbf{b} = R_{\pi/2}(\mathbf{a}) \cdot \mathbf{b}.$$

   a) Show that the exterior product is linear in each of its variables, i.e., that

$$(\mathbf{a}_1 + \lambda\mathbf{a}_2) \times \mathbf{b} = \mathbf{a}_1 \times \mathbf{b} + \lambda\mathbf{a}_2 \times \mathbf{b}$$

and

$$\mathbf{a} \times (\mathbf{b}_1 + \lambda\mathbf{b}_2) = \mathbf{a} \times \mathbf{b}_1 + \lambda\mathbf{a} \times \mathbf{b}_2.$$

   b) Show that the product is anticommutative: $\mathbf{a} \times \mathbf{b} = -\mathbf{b} \times \mathbf{a}$.
   c) Show that

$$(a_1\mathbf{i} + a_2\mathbf{j}) \times (b_1\mathbf{i} + b_2\mathbf{j}) = \begin{vmatrix} a_1 & b_1 \\ a_2 & b_2 \end{vmatrix}.$$

   d) Show that $\mathbf{a} \times \mathbf{b} = \|\mathbf{a}\| \, \|\mathbf{b}\| \sin \angle(\mathbf{a}, \mathbf{b})$.
   e) Show that $R_\theta(\mathbf{a}) \times R_\theta(\mathbf{b}) = \mathbf{a} \times \mathbf{b}$.
   f) Derive a formula for $(d/dt)[\mathbf{a}(t) \times \mathbf{b}(t)]$.

7. State precise (minimal) hypotheses under which the formula

$$\frac{d}{dt}(A^{-1}) = -A^{-1}\frac{dA}{dt}A^{-1}$$

is true, and then establish the formula for matrix-valued functions.

8. The integral of a vector-valued function can be defined as follows:

$$\int_a^b [f_1(t)\mathbf{i} + f_2(t)\mathbf{j}]\,dt = \left(\int_a^b f_1(t)\,dt\right)\mathbf{i} + \left(\int_a^b f_2(t)\,dt\right)\mathbf{j}.$$

a) What properties of ordinary definite integrals continue to hold for vector-valued functions?

b) If $T: \mathbf{R}^2 \to \mathbf{R}^2$ is linear, prove that

$$T\left(\int_a^b \mathbf{f}(t)\,dt\right) = \int_a^b (T \circ \mathbf{f})(t)\,dt.$$

c) Show that

$$\mathbf{a} \cdot \int_a^b \mathbf{f}(t)\,dt = \int_a^b \mathbf{a} \cdot \mathbf{f}(t)\,dt,$$

where $\mathbf{a}$ is a constant vector.

d) Show that

$$\left\|\int_a^b \mathbf{f}(t)\,dt\right\| \le \int_a^b \|\mathbf{f}(t)\|\,dt.$$

[Use (c) and the Cauchy-Schwarz inequality.]

e) Use (d) to show that "the shortest distance between two points is a straight line."

9. If A is a differentiable matrix-valued function, and if for each t, $[A(t)]^{-1}$ exists, then the *Cartan matrix* of A is defined to be

$$C(A) = \frac{dA}{dt}A^{-1}.$$

In a sense, $C(A)$ is the logarithmic derivative of A.

a) Show that $C(AB) = C(A) + AC(B)A^{-1}$.

b) Prove that $C(A) = C(B)$ iff there exists a constant nonsingular matrix M such that $B = AM$.

c) Prove that the Cartan matrix of R_t is the constant matrix $R_{\pi/2}$. (Here we are identifying matrices and the linear transformations which they define.)

12–6 Curvature of plane curves

We have seen that the sign of the second derivative of a function determines the direction of bending of its graph. We shall presently see that the absolute value of a second derivative is related to the rate of bending of the graph as measured by its *curvature*. We shall also prove that the curvature of a plane curve as a function of path length determines the curve up to proper isometry. This means that curvature may be regarded as the key notion in the study of plane curves.

Consider a particle moving through the plane. Let $\mathbf{f}(t) = f_1(t)\mathbf{i} + f_2(t)\mathbf{j}$ specify its position at time t. We shall assume that f_1 and f_2 have continuous second-order derivatives and that f_1 and f_2 vanish simultaneously only finitely often. We observe the particle over the time interval $[a, b]$ and set

$$s(t) = \int_a^t \|\mathbf{f}'(t)\|\, dt$$

for all $t \in [a, b]$; thus s describes the distance traveled by the particle along its path of motion γ during the time interval $[a, t]$. We now consider the velocity and acceleration of the particle. The velocity may be written

$$\mathbf{V}(t) = \frac{ds}{dt}\mathbf{T}(t),$$

where $\mathbf{T}$ is a vector of unit length which is parallel to the tangent line to γ at the point $\mathbf{f}(t)$. The vector $\mathbf{T}(t)$ is called the *forward unit tangent vector*. We then define

$$\mathbf{N}(t) = R_{\pi/2}\big(\mathbf{T}(t)\big);$$

it is called the *unit normal*. As t varies over the interval $[a, b]$, the vector $\mathbf{T}(t)$ traces out a path on the unit circle, called the *tangent image* of γ. We let θ be a continuous function which assigns to each $t \in [a, b]$ a polar angle of $\mathbf{T}(t)$. Whenever convenient, we shall regard $\mathbf{T}(t)$ and $\mathbf{N}(t)$ as arrows attached (via the translation $T_{\mathbf{f}(t)}$) to the point $\mathbf{f}(t)$ rather than to the origin (see Fig. 12–47). The pair of unit vectors $\mathbf{T}(t)$, $\mathbf{N}(t)$ is then often called the *moving frame* for the path. Note that these vectors are the image of the unit vectors $\mathbf{i}$ and $\mathbf{j}$ under the proper isometry $T_{\mathbf{f}(t)} \circ R_{\theta(t)}$. We can think of $\theta(t)$ as the angle between the moving frame and the fixed frame $\mathbf{i}, \mathbf{j}$.

We shall first express the acceleration of the particle in terms of our moving frame. Now

$$\mathbf{A}(t) = \frac{d}{dt}\mathbf{V}(t) = \frac{d}{dt}\left(\frac{ds}{dt}\mathbf{T}(t)\right)$$

$$= \frac{d^2 s}{dt^2}\mathbf{T}(t) + \frac{ds}{dt}\frac{d}{dt}\mathbf{T}(t)$$

$$= \frac{d^2 s}{dt^2}\mathbf{T}(t) + \frac{ds}{dt}\frac{d\theta}{dt}R_{\pi/2}\big(\mathbf{T}(t)\big)$$

$$= \frac{d^2 s}{dt^2}\mathbf{T}(t) + \frac{ds}{dt}\frac{d\theta}{dt}\mathbf{N}(t).$$

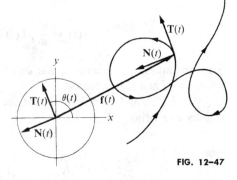

FIG. 12-47

The coefficients of $\mathbf{T}(t)$ and $\mathbf{N}(t)$ are called the *tangential* and *normal components*, respectively, of the acceleration. The equation above tells us that the tangential component of $\mathbf{A}$ is the derivative of the speed, and the normal component of $\mathbf{A}$ is the product of the speed and the rate of turning of the moving frame. We now write

$$\frac{d\theta}{dt} = \frac{d\theta}{ds}\frac{ds}{dt},$$

where $d\theta/ds$ is a shorthand for $d(\theta \circ s^{-1})/dt$. The quantity $d\theta/ds$ may be thought of as the rate of turning of the moving frame with respect to path length. By contrast to the quantity $d\theta/dt$, it depends only on the geometry of the curve and not on the speed of the particle. Further insight can be gained by showing that

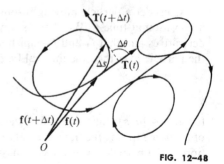

FIG. 12-48

$$\frac{d\theta}{ds} = \lim_{\Delta t \to 0} \frac{\Delta\theta}{\Delta s},$$

where $\Delta\theta = \theta(t + \Delta t) - \theta(t)$ and $\Delta s = s(t + \Delta t) - s(t)$, and by observing that the quantities $\Delta\theta$ and Δs are properly labeled on Fig. 12-48.

Definition 1. If $\mathbf{f}'(t) \neq \mathbf{0}$, then

$$k(t) = \frac{d\theta}{dt} = (\theta \circ s^{-1})'(t)$$

is called the **curvature** of the path γ at the point $\mathbf{f}(t)$.

From our previous discussion, it is clear that the curvature may be regarded as the rate of bending of a curve. As further evidence of this, we shall presently show that circles possess constant curvature. Namely, the curvature of a circle of radius R is equal to $\pm 1/R$; the sign depends on whether the points are traced out in a clockwise or counterclockwise direction. For this reason, the quantity $\rho(t) = 1/|k(t)|$ is called the *radius of curvature*.

Returning to our earlier discussion, we see that the acceleration of the particle can be written

$$\mathbf{A}(t) = \frac{d^2s}{dt^2}\mathbf{T}(t) + \left(\frac{ds}{dt}\right)^2 k(t)\mathbf{N}(t).$$

Thus the *normal component of the acceleration is equal to the product of the square of the speed and the curvature.*

We shall now find a formula for the curvature. Let us first alter our notation slightly and write

$$\mathbf{f}(t) = x(t)\mathbf{i} + y(t)\mathbf{j}.$$

Then

$$\frac{ds}{dt}\mathbf{T}(t) = \mathbf{f}'(t) = x'\mathbf{i} + y'\mathbf{j},$$

so that

$$\mathbf{T}(t) = \frac{1}{ds/dt}(x'\mathbf{i} + y'\mathbf{j}) \quad \text{and} \quad \mathbf{N} = \frac{1}{ds/dt}R_{\pi/2}(\mathbf{T}) = \frac{-y'\mathbf{i} + x'\mathbf{j}}{ds/dt}.$$

Also,

$$\mathbf{A} = x''\mathbf{i} + y''\mathbf{j}.$$

Consequently,

$$\left(\frac{ds}{dt}\right)^2 k(t) = \mathbf{A} \cdot \mathbf{N} = (x''\mathbf{i} + y''\mathbf{j}) \cdot \frac{(-y'\mathbf{i} + x'\mathbf{j})}{ds/dt} = \frac{x'y'' - x''y'}{ds/dt}.$$

Thus

$$k(t) = \frac{x'y'' - x''y'}{(ds/dt)^3} = \frac{x'y'' - x''y'}{[(x')^2 + (y')^2]^{3/2}},$$

the desired formula.

Example 1. Determine the curvature of the positively oriented ellipse,

$$\frac{x^2}{a^2} + \frac{y^2}{b^2} = 1.$$

The ellipse is positively oriented by the parametrization

$$x = a \cos t, \qquad y = b \sin t;$$

that is, its points are traced out in a counterclockwise direction. The curvature is then given by

$$k(t) = \frac{ab}{(a^2 \sin^2 t + b^2 \cos^2 t)^{3/2}} = \frac{ab}{(a^2 + b^2 - x^2 - y^2)^{3/2}}.$$

If $a \neq b$, the minimum curvature occurs at the extremities of the minor axis and the maximum curvature occurs at the extremities of the major axis. If $a = b$, the curvature is equal to $1/a$ at all points.

We point out a special case of our curvature formula. If γ is the graph of a function f (the points being traced out from left to right), then the curvature of γ at the point $(x, f(x))$ is given by

$$k(x) = \frac{f''(x)}{\{1 + [f'(x)]^2\}^{3/2}}.$$

We consider next the sign of the curvature. If $k(t) > 0$ throughout an interval, then θ is an increasing function throughout the interval; this means that the moving frame rotates in a counterclockwise direction. If $k(t) < 0$, the frame rotates clockwise. We gain further insight by making a change of coordinates. We introduce a second set of coordinate axes labeled u and v via the transformation

$$(x, y) = (T_{f(t_0)} \circ R_{\theta(t_0)})(u, v)$$

(see Fig. 12–49). Then immediately prior to and immediately following t_0, the path of the particle is described by an equation of the form $v = G(u)$, where G is of class C^2 and $G(0) = G'(0) = 0$. (Why?) Consequently, $G''(0)$ is equal to the curvature of the path at the instant t_0. If the

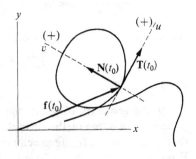

FIG. 12–49

curvature is positive for t near t_0, then G'' must be positive throughout a neighborhood of zero, and thus the path must lie above the u-axis (at least locally), since the u-axis is tangent to the path. In the case of negative curvature, the path lies beneath the u-axis.

If $k(t_0) \neq 0$, then there exists a unique oriented circle C such that

1) C passes through the point $\mathbf{f}(t_0)$ and has $\mathbf{T}(t_0)$, $\mathbf{N}(t_0)$ as its moving frame at the point, and

2) the curvature of C is equal to $k(t_0)$.

This circle is called the *osculating circle* (or circle of curvature) at the point. The circle has $1/|k(t_0)|$ as radius and the point

$$\mathbf{E}(t_0) = \mathbf{f}(t_0) + \frac{1}{k(t_0)} \mathbf{N}(t_0)$$

as center. The circle lies on the same side of the tangent line at $\mathbf{f}(t_0)$ as the path of the particle. The path defined by the vector-valued function $\mathbf{E}$ is called the *evolute* of the original path, and is denoted by $\mathcal{E}(\gamma)$.

Example 2. Find the evolute of the ellipse $x^2/a^2 + y^2/b^2 = 1$.

We set

$$\mathbf{f}(t) = a \cos t \mathbf{i} + b \sin t \mathbf{j}.$$

Then

$$\frac{1}{k(t)} = \frac{(a^2 \sin^2 t + b^2 \cos^2 t)^{3/2}}{ab},$$

by the previous example. One can also easily check that

$$\mathbf{N}(t) = \frac{R_{\pi/2}(\mathbf{f}'(t))}{\|\mathbf{f}'(t)\|} = -\frac{b \cos t \mathbf{i} + a \sin t \mathbf{j}}{\sqrt{a^2 \sin^2 t + b^2 \cos^2 t}}.$$

Thus

$$\mathbf{E}(t) = \mathbf{f}(t) + \frac{1}{k(t)} \mathbf{N}(t)$$

$$= \frac{a^2 - b^2}{a} \cos^3 t \mathbf{i} - \frac{a^2 - b^2}{b} \sin^3 t \mathbf{j}.$$

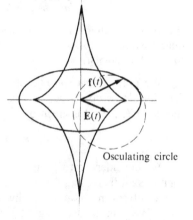

Osculating circle

FIG. 12-50

An implicit equation for the evolute is the following:

$$(ax)^{2/3} + (by)^{2/3} = (a^2 - b^2)^{2/3}.$$

(see Fig. 12–50).

In the geometry of paths, the use of path lengths as parameters leads to a simplification of some of the formulas. Such a parametrization is always possible; all that is involved is the replacement of the given vector-valued function $\mathbf{f}$ by $\mathbf{f} \circ s^{-1}$. From a dynamic point of view, this simply means that we travel along the curve with a constant speed of one. At points along the path where the given function $\mathbf{f}$ has zero derivative, the new function $\mathbf{f} \circ s^{-1}$ may fail to be differentiable. Except for such a finite number of singular points, however, the new function is of class C^2.

Suppose then that $\mathbf{f}(s)$ represents a point on a path γ such that the path length from the point to the initial point $\mathbf{f}(0)$ is s. If we also think of s as a time variable, then the speed is identically one, and the velocity and acceleration are given by

$$\mathbf{f}'(s) = \mathbf{T}(s) \quad \text{and} \quad \mathbf{f}''(s) = k(s)\mathbf{N}(s).$$

Thus we have the first of the *Frenet formulas:*

$$\frac{d\mathbf{T}}{ds} = k(s)\mathbf{N}(s), \quad \frac{d\mathbf{N}}{ds} = -k(s)\mathbf{T}(s).$$

The second of these formulas follows easily from the first. We have

$$\begin{aligned} \frac{d\mathbf{N}}{ds} &= \frac{d}{ds}[R_{\pi/2}(\mathbf{T})] = R_{\pi/2}\left(\frac{d\mathbf{T}}{ds}\right) \\ &= R_{\pi/2}(k(s)\mathbf{N}) = k(s)R_{\pi/2}(R_{\pi/2}(\mathbf{T})) \\ &= k(s)R_{\pi}(\mathbf{T}) = -k(s)\mathbf{T}. \end{aligned}$$

We now prove an assertion made at the very beginning of this section.

Theorem 2. *Let γ be a path without singular points defined by functions of class C^2. Then γ is determined to within a proper isometry by its curvature as a function of path length.*

Proof. Let $\mathbf{f}$ be the parametrization of γ by path length. We must show that if $\mathbf{f}^*$ and γ^* are similarly related, and if $k^*(s) \equiv k(s)$, then there exists a proper isometry which carries γ onto γ^*. We set $\varphi(s) = \theta^*(s) - \theta(s)$, where, as before, the functions θ^* and θ make a continuous selection of polar angles for the tangent vectors $\mathbf{T}^*$ and $\mathbf{T}$. We have

$$\mathbf{T}^*(s) = R_{\varphi(s)}[\mathbf{T}(s)].$$

Hence

$$\begin{aligned} k(s)\mathbf{N}^*(s) = k^*(s)\mathbf{N}^*(s) &= \frac{d\mathbf{T}^*}{ds} \\ &= R_{\varphi(s)}\left(\frac{d\mathbf{T}}{ds}\right) + R_{\varphi(s)+\pi/2}[\mathbf{T}(s)]\varphi'(s) \\ &= k(s)R_{\varphi(s)}[\mathbf{N}(s)] + R_{\varphi(s)+\pi/2}[\mathbf{T}(s)]\varphi'(s) \\ &= k(s)\mathbf{N}^*(s) + R_{\varphi(s)+\pi/2}[\mathbf{T}(s)]\varphi'(s). \end{aligned}$$

Thus $R_{\varphi(s)+\pi/2}[\mathbf{T}(s)]\varphi'(s) \equiv 0$. However, since $R_{\varphi(s)+\pi/2}[\mathbf{T}(s)] = \mathbf{N}^*(s)$ is a unit vector, we must have $\varphi'(s) \equiv 0$. This implies that φ is constant, say $\varphi(s) \equiv \varphi_0$. Now

$$(\mathbf{f}^*)'(s) = \mathbf{T}^*(s) = R_{\varphi_0}[\mathbf{T}(s)] = (R_{\varphi_0} \circ \mathbf{f})'(s).$$

This implies that $\mathbf{f}^*$ and $R_{\varphi_0} \circ \mathbf{f}$ differ by a constant vector $\mathbf{a}$. Thus

$$\mathbf{f}^*(s) = (R_{\varphi_0} \circ \mathbf{f})(s) + \mathbf{a} = (T_\mathbf{a} R_{\varphi_0} \circ \mathbf{f})(s).$$

This proves that the proper isometry $T_\mathbf{a} \circ R_{\varphi_0}$ carries γ onto γ^*.

One consequence of the theorem is that straight lines and circles are the only curves with constant curvature.

We have defined the evolute of a path γ to be the path $\mathcal{E}(\gamma)$ traced out by the center of the circle of curvature. We shall now explore some of the relationships between these two curves. If the function $\mathbf{f}$ parametrizes γ by path length, then the function $\mathbf{E}$ defined by

$$\mathbf{E}(s) = \mathbf{f}(s) + \frac{1}{k(s)} \mathbf{N}(s) = \mathbf{f}(s) + \rho(s)\mathbf{N}(s)$$

gives a parametrization of $\mathcal{E}(\gamma)$, although the parameter s does not represent the path length of $\mathcal{E}(\gamma)$. We define the function s_e by

$$s_e(s) = \int_0^s \|\mathbf{E}'(t)\| \, dt.$$

Thus s_e represents the path length measured along $\mathcal{E}(\gamma)$. Now

$$\begin{aligned}
\frac{d\mathbf{E}}{ds} &= \mathbf{T}(s) + \frac{1}{k(s)} \frac{d\mathbf{N}}{ds} + \rho'(s)\mathbf{N}(s) \\
&= \mathbf{T}(s) + \frac{1}{k(s)} (-k(s)\mathbf{T}(s)) + \rho'(s)\mathbf{N}(s) \\
&= \rho'(s)\mathbf{N}(s).
\end{aligned}$$

From this equation we can conclude that the tangent line to $\mathcal{E}(\gamma)$ at the point $\mathbf{E}(s)$ is parallel to the normal line to γ at $\mathbf{f}(s)$. However, $\mathbf{E}(s)$ also lies on this normal line to γ. Consequently, *if P is any point on γ, and if Q is the corresponding point on $\mathcal{E}(\gamma)$, then the normal to γ at P coincides with the tangent line to $\mathcal{E}(\gamma)$ at Q*, as shown in Fig. 12–51. (Detailed hypotheses under which this statement is true will appear later.) Another consequence of the equation above is that $ds_e/ds = |\rho'(s)|$. If ρ' is

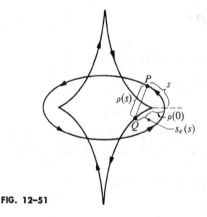

FIG. 12–51

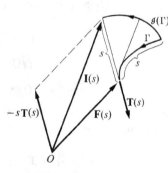

FIG. 12–52

of constant sign, then either $s_e(s) = \rho(s) - \rho(0)$ or $s_e(s) = \rho(0) - \rho(s)$. In either case,

$$s_e(s) = |\rho(s) - \rho(0)|.$$

Thus, if γ has monotone curvature, then as P moves along γ, the change in the distance $d(P, Q)$ is equal to the path length traced out by Q along $\mathcal{E}(\gamma)$.

This leads us to what is sometimes called the string property of an evolute. This property is most easily formulated in terms of *involutes*. Suppose that the function $\mathbf{F}$ parametrizes a path Γ by arc length. Imagine that an inextensible string of length $l(\Gamma)$ is attached at the terminal point of Γ and wound tightly around the curve. Beginning at the initial point of Γ, we then unwind the string while keeping it taut. The path $\mathcal{I}(\Gamma)$ traced out by the endpoint of the string is called the involute of the curve Γ. Reference to Fig. 12–52 should help convince the reader that this involute is parametrized by the function

$$\mathbf{I}(s) = \mathbf{F}(s) - s\mathbf{T}(s).$$

More generally, by an involute of Γ we mean any path which has the parametrization

$$\mathbf{I}(s) = \mathbf{F}(s) - (s + c)\mathbf{T}(s),$$

where c is a constant. If $c > 0$, we can obtain the involute by adding an extra bit of string of length c to the free end and following the new endpoint.

The string property of an evolute can now be formulated.

Proposition 3. *Let $\mathbf{f}$ be a function of class C^3 which parametrizes a path γ by path length. Suppose further that the curvature is never zero and that the radius of curvature is an increasing function of the path length. Then γ is an involute of its evolute.*

Proof. Since ρ is an increasing function, $ds_e/ds = +\rho'(x)$, so

$$s_e(s) = \rho(s) - \rho(0).$$

We let $\mathbf{E}^* = \mathbf{E} \circ s_e^{-1}$ be the parametrization of $\mathcal{E}(\gamma)$ by path length. Then the forward unit tangent to $\mathcal{E}(\gamma)$ at the point $\mathbf{E}^*(\sigma)$ is $\mathbf{T}_e(\sigma) = \mathbf{N}\big(s_e^{-1}(\sigma)\big)$. Thus, upon substituting $s = s_e^{-1}(\sigma)$ into the equation

$$\mathbf{E}(s) = \mathbf{f}(s) + \rho(s)\mathbf{N}(s) = \mathbf{f}(s) + [s_e(s) + \rho(0)]\mathbf{N}(s),$$

we get

$$(\mathbf{f} \circ s_e^{-1})(\sigma) = \mathbf{E}^*(\sigma) - [\sigma + \rho(0)]\mathbf{T}_e(\sigma).$$

The left-hand side is a reparametrization of γ; the right-hand side defines an involute of $\mathcal{E}(\gamma)$. Hence the theorem follows.

If the radius of curvature is a decreasing function of path length, then the theorem applies to the path $-\gamma$ parametrized by the function $\mathbf{g}(s) = \mathbf{f}\big(l(\gamma) - s\big)$.

The fact that a curve is an involute of its evolute is analogous to the fact that a function is an antiderivative of its derivative. (These loosely stated "facts" have about the same degree of truth.) One might expect that a curve is also the evolute of its involutes. A proof is outlined in the exercises.

Exercises

1. The position of a moving particle at time t is $t^2\mathbf{i} + t^3\mathbf{j}$. Does the path of the particle have any singular points? Compute the velocity, acceleration, speed, and moving frame. Also calculate the tangential and normal components of the acceleration. What is the curvature?

2. Check that the curvature of a straight line is zero.

3. The position of a projectile at time t is

$$v_0 t \cos \alpha \mathbf{i} + (v_0 t \sin \alpha - \tfrac{1}{2} g t^2)\mathbf{j},$$

where g is the acceleration due to gravity. Compute the speed, velocity, and acceleration of the particle. Also compute the tangential and normal components of the acceleration.

4. If a particle moves along a circle with constant angular velocity, show that the tangential component of the acceleration is zero.

5. The position of a particle at time t is

$$a[(\cos t + t \sin t)\mathbf{i} + (\sin t - t \cos t)\mathbf{j}].$$

Show that the tangential component of the acceleration is constant and that the component of the normal acceleration is at.

6. a) From the formula for curvature given in the text, deduce that the curvature is independent of parametrization.
   b) Also deduce from the formula that the curvature is invariant under proper isometries.

7. What is the effect of an improper isometry on curvature?

8. Prove the assertions made in the text regarding the circle of curvature.

9. Find the circle of curvature for the folium of Descartes,

$$x = \frac{3t}{1 + t^3}, \qquad y = \frac{3t^2}{1 + t^3},$$

when $t = 1$. Sketch carefully both the curve and the circle.

10. We have proved that a plane curve is characterized to within a rigid motion (proper isometry) by its curvature. Prove that if k is a real-valued function which is continuous and nonzero on some interval $[0, S]$, then there exists a path γ for which k represents the curvature as a function of path length. [Define

$$\theta(s) = \int_0^s k(\sigma)\, d\sigma, \qquad x(s) = \int_0^s \cos \theta(\sigma)\, d\sigma, \qquad y(s) = \int_0^s \sin \theta(\sigma)\, d\sigma$$

for all $s \in [0, S]$. Show that the curve parametrized by $x(s)\mathbf{i} + y(s)\mathbf{j}$ has the required property.]

11. An equation expressing either the curvature or the radius of curvature as a function of arc length is called the *natural equation* of the curve. (It is "natural" in the sense that it does not depend on the choice of coordinate system.) For instance $k \equiv 0$ is the natural equation of a straight line.

a) Show that the natural equation of the involute of a circle

$$a[(\cos t + t \sin t)\mathbf{i} + (\sin t - t \cos t)\mathbf{j}]$$

is

$$\rho^2 = 2as.$$

b) The equation $r = a \exp(\theta \, \text{ctn} \, \alpha)$ is the polar equation of a *logarithmic spiral*. Show that the natural equation is

$$\rho = s \, \text{ctn} \, \alpha + a \csc \alpha.$$

c) Show that the natural equation of the catenary $y = a \cosh(x/a)$ is $\rho = a + s^2/a$.

d) Find the natural equation of the cycloid $a[(t - \sin t)\mathbf{i} + (1 - \cos t)\mathbf{j}]$.

12. Show that the evolute of the tractrix

$$x = -a[\ln \tan(t/2) + \cos t], \qquad y = a \sin t$$

is a catenary.

13. Show that the evolute of a logarithmic spiral is a logarithmic spiral which differs from the original one only by a rotation. Is there any logarithmic spiral which is its own evolute?

14. Show that the evolute of the cycloid $a[(t - \sin t)\mathbf{i} + (1 - \cos t)\mathbf{j}]$ is the curve

$$a[(t + \sin t)\mathbf{i} - (1 - \cos t)\mathbf{j}].$$

Show that this curve is also a cycloid. Graph both cycloids.

15. Show that the radius of curvature of a parabola is twice the length of that segment of the normal lying between its points of intersection and the directrix.

16. Show that the radius of curvature of the lemniscate $r^2 = a^2 \cos 2\theta$ is inversely proportional to the distance from the point to the origin.

17. Determine the involutes of the four-cusp hypocycloid $x = a \cos^3 t$, $y = a \sin^3 t$.

18. Suppose that the curvature of a curve γ attains either a strict maximum or a strict minimum at a point P. Show that the corresponding point of the evolute is a singular point.

19. Suppose that γ is parametrized by arc length by a function $\mathbf{f}$ of class C^2, and suppose that γ has an everywhere positive curvature. Prove that the involute of γ defined by $\mathbf{I}(s) = \mathbf{f}(s) - s\mathbf{T}(s)$ is a curve of class C^3 whose evolute is γ. [Let

$$s_i(s) = \int_0^s \|\mathbf{I}'(\sigma)\| \, d\sigma \qquad \text{and} \qquad \mathbf{I}^* = \mathbf{I} \circ s_i^{-1}.$$

Show first that $ds_i/ds = sk(s)$. Show that the forward unit tangent to the involute at the point $\mathbf{I}^*(\sigma)$ is $\mathbf{T}_i(\sigma) = -\mathbf{N}(s_i^{-1}(\sigma))$. Deduce that $\mathbf{N}_i(\sigma) = \mathbf{T}(s_i^{-1}(\sigma))$. Show that

$$k_i(\sigma)\mathbf{N}_i(\sigma) = \mathbf{T}_i'(\sigma) = \frac{1}{s_i^{-1}(\sigma)} \, \mathbf{N}_i(\sigma),$$

where k_i denotes the curvature of the involute. Substitute $s = s_i^{-1}(\sigma)$ into the equation $\mathbf{I}(s) = \mathbf{f}(s) - s\mathbf{T}(s)$.]

*20. Suppose that a path γ has monotone curvature. Prove that if P_1 and P_2 are any two points of γ, then one of the osculating circles lies inside the other.

13 □ INFINITE SERIES

In this chapter we shall deal with the general theory of infinite series. The basic question underlying our work is this: To what extent can infinite series be treated like finite sums? When can we add or multiply them, insert or remove parentheses, change the order of the terms? Is the derivative of a sum still the sum of the derivatives when infinitely many summands are involved? Some of these questions are easily answered; others are more subtle, and their answers are sometimes surprising. We begin with several of the easy questions.

13-1 A humble beginning

Let us review some definitions and establish some notation. Given an infinite series

$$\sum_{k=1}^{\infty} a_k = a_1 + a_2 + a_3 + \cdots,$$

we shall generally denote by S_n the nth partial sum:

$$S_n = \sum_{k=1}^{n} a_k.$$

If there exists a real number S such that

$$S = \lim_{n \to \infty} S_n,$$

then we say that the series converges and that S is its sum; otherwise, the series is said to diverge.

We begin with a necessary condition for convergence.

Proposition 1. Necessary Condition for Convergence. *If the infinite series $\sum_{k=1}^{\infty} a_k$ converges, then $\lim_{n \to \infty} a_n = 0$.*

Proof. Using the notation above, we have

$$\lim_{n \to \infty} a_n = \lim_{n \to \infty} (S_n - S_{n-1}) = S - S = 0.$$

It follows from this theorem that the infinite series

$$\sum_{k=1}^{\infty} \left(1 + \frac{1}{k}\right)^k$$

diverges, since

$$\lim_{n\to\infty} \left(1 + \frac{1}{n}\right)^n = e \neq 0.$$

It must be stressed, however, that the convergence of the nth term to zero is a necessary *but not sufficient* condition for the convergence of a series. A classical counterexample is provided by the *harmonic series*

$$\sum_{k=1}^{\infty} \frac{1}{k}.$$

The nth term clearly tends to zero, but, as we shall now show, the sequence of partial sums $\{S_n\}$ is unbounded. Clearly,

$$S_{2^n} = 1 + \frac{1}{2} + \left(\frac{1}{3} + \frac{1}{4}\right) + \left(\frac{1}{5} + \frac{1}{6} + \frac{1}{7} + \frac{1}{8}\right) + \cdots + \left(\frac{1}{2^{n-1} + 1} + \cdots + \frac{1}{2^n}\right)$$

$$> 1 + \frac{1}{2} + \left(\frac{1}{4} + \frac{1}{4}\right) + \left(\frac{1}{8} + \frac{1}{8} + \frac{1}{8} + \frac{1}{8}\right) + \cdots + \underbrace{\left(\frac{1}{2^n} + \frac{1}{2^n} + \cdots + \frac{1}{2^n}\right)}_{2^{n-1} \text{ terms}}$$

$$= 1 + \underbrace{\frac{1}{2} + \frac{1}{2} + \cdots + \frac{1}{2}}_{n \text{ terms}} = 1 + \frac{n}{2}.$$

Since $1 + n/2 \to +\infty$ as $n \to \infty$, it follows that the sequence $\{S_n\}$ is unbounded.

Convergent series can be multiplied by constants and added in the obvious way.

Proposition 2. Linearity Property of Infinite Series. *If the infinite series $\sum_{k=1}^{\infty} a_k$ and $\sum_{k=1}^{\infty} b_k$ converge, then so does any infinite series of the form $\sum_{k=1}^{\infty} (\alpha a_k + \beta b_k)$, where α and β are constants. Moreover,*

$$\sum_{k=1}^{\infty} (\alpha a_k + \beta b_k) = \alpha \sum_{k=1}^{\infty} a_k + \beta \sum_{k=1}^{\infty} b_k.$$

Proof. We denote the nth partial sums of $\sum_{k=1}^{\infty} a_k$ and $\sum_{k=1}^{\infty} b_k$ by S_n and T_n, respectively, and we denote the sums of these series by S and T. Then $\alpha S_n + \beta T_n$ is the nth partial sum of the series $\sum_{k=1}^{\infty} (\alpha a_k + \beta b_k)$. Since

$$\lim_{n\to\infty} (\alpha S_n + \beta T_n) = \alpha S + \beta T,$$

the assertions of the theorem follow immediately.

Since the series expansions

$$e^x = 1 + x + \frac{x^2}{2!} + \frac{x^3}{3!} + \cdots, \qquad e^{-x} = 1 - x + \frac{x^2}{2!} - \frac{x^3}{3!} + \cdots$$

hold for all $x \in \mathbf{R}$, it follows from the linearity property just established that the expansions

$$\cosh x = \frac{e^x + e^{-x}}{2} = 1 + \frac{x^2}{2!} + \frac{x^4}{4!} + \cdots$$

and

$$\sinh x = \frac{e^x - e^{-x}}{2} = x + \frac{x^3}{3!} + \frac{x^5}{5!} + \cdots$$

are also valid for all $x \in \mathbf{R}$.

The proofs of the next result and its corollaries are left to the reader.

Proposition 3. Positivity. *If the series $\sum_{k=1}^{\infty} a_k$ converges, and if $a_n \geq 0$ for each $n \in \mathbf{Z}^+$, then $\sum_{k=1}^{\infty} a_k \geq 0$, with strict inequality unless each term of the series is zero.*

Corollary 4. Monotone Property. *If the infinite series $\sum_{k=1}^{\infty} a_k$ and $\sum_{k=1}^{\infty} b_k$ both converge, and if $a_n \leq b_n$ for each $n \in \mathbf{Z}^+$, then*

$$\sum_{k=1}^{\infty} a_k \leq \sum_{k=1}^{\infty} b_k,$$

with strict inequality unless $a_n = b_n$ for all $n \in \mathbf{Z}^+$.

Corollary 5. Triangle Inequality. *If each of the series $\sum_{k=1}^{\infty} a_k$ and $\sum_{k=1}^{\infty} |a_k|$ converges, then*

$$\left| \sum_{k=1}^{\infty} a_k \right| \leq \sum_{k=1}^{\infty} |a_k|.$$

Moreover, strict inequality holds if the series $\sum a_k$ has two terms of opposite sign.

We shall later prove that the convergence of the series $\sum a_k$ is implied by the convergence of the series $\sum |a_k|$.

If simple parentheses are inserted into an infinite series, then the partial sums of the resulting series will be a subsequence of the sequence of partial sums of the original series. For instance, if S_n is the nth partial sum of the series $\sum a_k$, then the nth partial sum of the series

$$(a_1 + a_2) + (a_3 + a_4) + \cdots$$

is S_{2n}, and the nth partial sum of the series

$$a_1 + a_2 + (a_3 + a_4) + (a_5 + a_6 + a_7 + a_8) + \cdots$$

is S_{2^n}. It follows that *if parentheses are inserted into a convergent series, then the resulting series will converge and will have the same sum as the original series.* Care must be taken, however, in removing parentheses. Thus the series

$$(1 - 1) + (1 - 1) + \cdots$$

converges (with sum zero), whereas the series $1 - 1 + 1 - 1 + \cdots$ diverges.

Exercises

1. Explain why the convergence of an infinite series is unaffected if only a finite number of terms are tampered with (added, deleted, changed, etc.).

2. If the infinite series $\sum_{k=1}^{\infty} a_k$ converges, prove that the *tail* of the series, namely, $\sum_{k=n+1}^{\infty} a_k$, tends to zero as $n \to \infty$.

3. Show that an infinite number of zeros can be inserted into an infinite series without affecting its convergence or (in the case of a convergent series) its sum.

4. Prove Proposition 3 and its corollaries.

5. Suppose that
   a) $\sum_{k=1}^{\infty} (a_k + b_k)$ converges, and
   b) $\lim_{n \to \infty} b_n = 0$.

   Prove that the series $a_1 + b_1 + a_2 + b_2 + \cdots$ converges. Show by a counterexample that hypothesis (b) is essential.

6. Suppose that the series $\sum_{k=1}^{\infty} a_k$ and $\sum_{k=1}^{\infty} b_k$ converge. By the linearity property, the series $\sum_{k=1}^{\infty} (a_k + b_k)$ also converges. What can you say about the series

$$a_1 + b_1 + a_2 + b_2 + \cdots$$

13-2 Series with nonnegative terms

A question of general interest is the following. Suppose that we are given an infinite series. How can we tell whether or not the series converges? In this section our concern is with tests of convergence for series whose terms are entirely nonnegative. The convergence of infinite series containing an infinite number of both positive and negative terms (a more delicate matter) will be taken up in the succeeding section.

Suppose that we are given a series $\sum a_k$ in which all terms are nonnegative. Then each of the partial sums is also nonnegative, and the sequence $\{S_n\}$ of partial sums is nondecreasing. Consequently, there are only two possibilities:

1) The sequence of partial sums is unbounded, in which case

$$\lim_{n \to \infty} S_n = +\infty,$$

   and the series diverges.

2) The sequence of partial sums is bounded, in which case the series converges, and its sum can be characterized as the least upper bound of the sequence of partial sums.

In the first case we write

$$\sum_{k=1}^{\infty} a_k = +\infty.$$

Proposition 1. Comparison Test. *Suppose that $\sum_{k=1}^{\infty} a_k$ and $\sum_{k=1}^{\infty} b_k$ are infinite series such that $0 \le a_n \le b_n$ for all $n \in \mathbf{Z}^+$ (or, more generally, for all sufficiently large $n \in \mathbf{Z}^+$). If the series $\sum_{k=1}^{\infty} b_k$ converges, then the series $\sum_{k=1}^{\infty} a_k$ converges also.*

Proof. Let S_n and T_n be the nth partial sums of the series $\sum a_k$ and $\sum b_k$, respectively. If $\sum b_k$ converges, let its sum be T. Then for each $n \in \mathbf{Z}^+$,

$$S_n \leq T_n \leq T.$$

Hence $\{S_n\}$ is a bounded sequence, and the series $\sum a_k$ must therefore converge.

We point out the contrapositive of the theorem. *If $\sum a_k$ diverges, then $\sum b_k$ does also.*

Example 1. Show that the series $\sum_{k=1}^{\infty} 1/k2^k$ converges.
For each $n \in \mathbf{Z}^+$,

$$\frac{1}{n2^n} \leq \left(\frac{1}{2}\right)^n.$$

Since the geometric series $\sum_{k=1}^{\infty} (\frac{1}{2})^k$ converges (with sum one), the result follows. Moreover, by monotonicity,

$$0 < \sum_{k=1}^{\infty} \frac{1}{k2^k} < \sum_{k=1}^{\infty} \left(\frac{1}{2}\right)^k = 1.$$

Example 2. Show that the series $\sum_{k=1}^{\infty} 1/\sqrt{k}$ diverges.
Since

$$\frac{1}{\sqrt{n}} > \frac{1}{n},$$

the series diverges, by comparison with the harmonic series.

Proposition 2. *Asymptotic Test. Suppose that $\{a_n\}$ and $\{b_n\}$ are sequences of non-negative terms such that $a_n \sim b_n$ or, more generally, that a_n/b_n tends to some nonzero number as $n \to \infty$. Then the infinite series $\sum a_k$ and $\sum b_k$ converge or diverge together.*

Proof. Because of the linearity property, it suffices to consider the case in which $a_n \sim b_n$. Since $\lim_{n \to \infty} (a_n/b_n) = 1$,

$$\frac{1}{2} < \frac{a_n}{b_n} < \frac{3}{2}$$

for all n sufficiently large. This implies the separate inequalities

$$a_n < \tfrac{3}{2}b_n \quad \text{and} \quad b_n < 2a_n,$$

so that the result follows via the comparison test.

Example 3. Discuss the convergence of the infinite series $\sum_{k=1}^{\infty} \sin (1/k)$.
Since $\lim_{x \to 0} (\sin x)/x = 1$, we have

$$\sin \frac{1}{n} \sim \frac{1}{n}.$$

Since the harmonic series diverges, so does the given series.

Example 4. Discuss the convergence of the series $\sum_{k=1}^{\infty} 1/k^2$.
Now

$$\frac{1}{n^2} \sim \frac{1}{n(n+1)}$$

and the series $\sum_{k=1}^{\infty} 1/[k(k+1)]$ converges with sum one. [The nth partial sum is $n/(n+1)$.]
It follows that the series

$$\sum_{k=1}^{\infty} \frac{1}{k^2}$$

converges. (Its sum, incidentally, turns out to be $\pi^2/6$.)

So far we have seen that the infinite series

$$\sum_{k=1}^{\infty} \frac{1}{k^p}$$

converges if $p = 2$ and diverges if $p = 1$. By the comparison test, the series must
diverge if $p \leq 1$ and converge if $p \geq 2$. What happens then if $1 < p < 2$? This
question is easily settled by the next test.

Proposition 3. Integral Test. *Let f be a nonnegative and nonincreasing function on
the interval $[1, +\infty)$, and let*

$$R = \lim_{x \to +\infty} \int_1^x f(t)\, dt.$$

*Then the infinite series $\sum_{k=1}^{\infty} f(k)$ converges or diverges according as $R < +\infty$ or
$R = +\infty$. In the first event we have*

$$R \leq \sum_{k=1}^{\infty} f(k) \leq R + f(1).$$

Proof. First we remark that for each $x \geq 1$, $\int_1^x f(t)\, dt$ exists as a Riemann integral,
since the integrand is monotone. Moreover, $\int_1^x f(t)\, dt$ is an increasing function of x.
Hence

$$R = \lim_{x \to +\infty} \int_1^x f(t)\, dt$$

exists, although R may be $+\infty$.
For each integer $n \geq 2$,

$$\int_1^n f(x)\, dx = \sum_{k=1}^{n-1} \int_k^{k+1} f(x)\, dx.$$

Since f is nonincreasing, $f(k) \geq f(x) \geq f(k+1)$ if $k \leq x \leq k+1$, and thus

$$f(k) = \int_k^{k+1} f(k)\, dx \geq \int_k^{k+1} f(x)\, dx \geq f(k+1).$$

Hence

$$S_n - f(1) = \sum_{k=1}^{n-1} f(k+1) \leq \int_1^n f(x)\, dx \leq \sum_{k=1}^{n-1} f(k) = S_{n-1}.$$

From these inequalities, it is immediately clear that the sequence of partial sums $\{S_n\}$ is bounded or unbounded (and the corresponding series convergent or divergent) according as $R = \lim_{n\to\infty} \int_1^n f(x)\, dx$ is less than or equal to $+\infty$. By rearranging the inequalities, we have

$$\int_1^n f(x)\, dx \le S_{n-1} \quad \text{and} \quad S_n \le f(1) + \int_1^n f(x)\, dx.$$

The inequality

$$R \le \sum_{k=1}^{\infty} f(k) \le f(1) + R$$

follows, when we take limits as $n \to \infty$.

Example 5. Discuss the convergence of the series $\sum_{k=1}^{\infty} 1/k^p$ where $1 < p < 2$.

If we set $f(t) = 1/t^p$, then the hypotheses of the theorem are satisfied. Now

$$R_p = \lim_{x\to+\infty} \int_1^x f(t)\, dt = \lim_{x\to+\infty} \left(-\frac{1}{p-1} \right)\left(\frac{1}{x^{p-1}} - 1 \right) = \frac{1}{p-1}.$$

Hence for $1 < p < 2$, the series converges and

$$\frac{1}{p-1} \le \sum_{k=1}^{\infty} \frac{1}{k^p} \le 1 + \frac{1}{p-1} = \frac{p}{p-1}.$$

Example 6. Show that the series $\sum_{k=2}^{\infty} 1/(k \ln k)$ diverges.

The function $x \ln x$ is increasing for $x > e^{-1}$ and positive for $x > 1$. Hence the hypotheses of the theorem are satisfied if

$$f(x) = \frac{1}{(x+1)\ln(x+1)}.$$

Since

$$\lim_{x\to+\infty} \int_1^x \frac{dt}{(t+1)\ln(t+1)} = \lim_{x\to+\infty} [\ln \ln (x+1) - \ln \ln 2] = +\infty,$$

it follows that the infinite series

$$\sum_{k=1}^{\infty} \frac{1}{(k+1)\ln(k+1)} = \sum_{k=2}^{\infty} \frac{1}{k \ln k}$$

diverges.

Another important set of tests are the *ratio* and *root tests*. To state them in full generality, we shall need to refine the notion of a limit of a sequence.

Let $a_1, a_2, a_3, \ldots$ be any sequence of real numbers. We may associate with this sequence a second sequence of extended real numbers in the following way. For each $n \in \mathbf{Z}^+$, we let M_n be the least upper bound of all terms of the original sequence beyond a_n. Thus

$$M_n = \text{lub } \{a_k \mid k \ge n+1\}$$

($+\infty$ is a permissible value for M_n). The resulting sequence $M_1, M_2, \ldots$ is easily seen to be nondecreasing, and therefore has a limit ($+\infty$ and $-\infty$ are possible values).

This limit is called the *limit superior* of the original sequence and is denoted by either

$$\overline{\lim_{n \to \infty}} \, a_n \quad \text{or} \quad \lim_{n \to \infty} \sup a_n.$$

If the sequence is not bounded above, then the limit superior is $+\infty$; the limit superior is $-\infty$ iff the limit of the sequence is $-\infty$; otherwise, the limit superior is a real number.

Example 7. Find the limits superior for each of the following sequences:

$$1, -1, 1, -1, \ldots,$$
$$1, -1, \tfrac{1}{2}, -2, \tfrac{1}{3}, -3, \ldots,$$
$$1, -1, 2, -2, 3, -3, \ldots,$$
$$-1, -2, -3, -4, \ldots$$

The derived sequences of least upper bounds are

$$1, 1, 1, 1, \ldots,$$
$$1, \tfrac{1}{2}, \tfrac{1}{2}, \tfrac{1}{3}, \tfrac{1}{3}, \ldots,$$
$$+\infty, +\infty, +\infty, \ldots,$$
$$-1, -2, -3, -4, \ldots,$$

respectively. Hence the limits superior are 1, 0, $+\infty$, and $-\infty$.

For our purposes, the definition of the limit superior is not as important as its epsilonic characterization.

Proposition 4. *Suppose that*

$$-\infty < \overline{\lim_{n \to \infty}} \, a_n < +\infty.$$

Then the number $L = \overline{\lim}_{n \to \infty} a_n$ *may be characterized in the following way: For each* $\epsilon > 0$, *the inequality*

$$a_n > L + \epsilon$$

holds for a finite number of positive integers n at most; the inequality

$$a_n > L - \epsilon$$

holds for infinitely many positive integers n.

We suggest that the reader mentally check this characterization for the first two sequences of the preceding example and then prove the theorem. Limits inferior can be defined and characterized similarly. Also, we have

$$\underline{\lim_{n \to \infty}} \, a_n = - \overline{\lim_{n \to \infty}} \, (-a_n).$$

Proposition 5. *Ratio Test. Let $\{a_n\}$ be a sequence of positive numbers; let*

$$r = \underline{\lim_{n \to \infty}} \frac{a_{n+1}}{a_n} \quad \text{and} \quad R = \overline{\lim_{n \to \infty}} \frac{a_{n+1}}{a_n}.$$

If $R < 1$, then the series $\sum a_k$ converges; if $r > 1$, the series diverges.

Proof. Suppose that $R < 1$. We choose a number t such that $R < t < 1$. Then the inequality

$$\frac{a_{n+1}}{a_n} > t$$

holds for a finite number of n at most. Equivalently, there exists an integer N such that

$$n \geq N \Rightarrow \frac{a_{n+1}}{a_n} \leq t.$$

Then

$$a_{N+1} \leq a_N t, \qquad a_{N+2} \leq a_{N+1} t \leq a_N t^2, \qquad a_{N+3} \leq a_{N+2} t \leq a_N t^3,$$

and by induction,

$$a_{N+k} \leq a_N t^k.$$

Thus

$$n \geq N \Rightarrow a_n \leq M t^n, \qquad \text{where} \qquad M = a_N t^{-N}.$$

Since the geometric series $\sum t^k$ converges, the series $\sum a_k$ converges, by comparison.

Suppose next that $r > 1$. Then for all n sufficiently large,

$$\frac{a_{n+1}}{a_n} > 1.$$

It follows that the sequence $\{a_n\}$ is eventually increasing. Since the terms are all positive, a_n cannot tend to zero, and hence the series cannot converge.

Example 8. Discuss the convergence of the series $\sum k^3 e^{-k}$.

Since

$$\lim_{n \to \infty} \frac{(n+1)^3 e^{-(n+1)}}{n^3 e^{-n}} = e^{-1} < 1,$$

it follows from the ratio test that the series converges.

Note that the ratio test says nothing about what happens if $r \leq 1 \leq R$. Its silence is well-justified. Consider the infinite series

$$\sum_{k=1}^{\infty} \frac{1}{k^p} \qquad (p > 0).$$

For all values of p, $r = R = 1$. Yet the series converges if $p > 1$ and diverges if $p \leq 1$.

The root test is a slightly stronger companion of the ratio test. Like the root test, its proof involves comparison with the geometric series.

Proposition 6. Root Test. *Let $\{a_n\}$ be a sequence of nonnegative numbers, and let*

$$r = \overline{\lim_{n \to \infty}} \sqrt[n]{a_n}.$$

If $r < 1$, then the series $\sum a_k$ converges; if $r > 1$, then the series diverges.

The proof is omitted.

Example 9. Show that the series $\sum_{k=2}^{\infty} 1/(\ln k)^k$ converges.

Since we have

$$\varlimsup_{n \to \infty} \sqrt[n]{\frac{1}{(\ln n)^n}} = \lim_{n \to \infty} \frac{1}{\ln n} = 0,$$

the series converges by the root test.

We consider now several miscellaneous examples.

Example 10. $\sum_{k=1}^{\infty} \sqrt{k(k+1)}/2^k$.

We note first that

$$\frac{\sqrt{n(n+1)}}{2^n} \sim \frac{n}{2^n}.$$

The series $\sum k/2^k$ is easily seen to converge by the ratio test, since

$$\lim_{n \to \infty} \frac{(n+1)/2^{n+1}}{n/2^n} = \frac{1}{2} < 1.$$

Thus the given series converges.

Example 11. $\sum_{k=1}^{\infty} 1/\sqrt[k]{k!}$.

By Stirling's formula,

$$n! < \sqrt{2\pi n}\,(n/e)^n(1 + 1/4n).$$

Thus

$$\frac{1}{\sqrt[n]{n!}} > \frac{e/n}{(2\pi n)^{1/2n}(1 + 1/4n)^{1/n}}.$$

Since the denominator of the right-hand side tends to one as $n \to \infty$, the right-hand side is asymptotic to e/n. Since the harmonic series diverges, we conclude that the original series diverges.

We can reach the same conclusion by using a less sophisticated result, namely,

$$\lim_{n \to \infty} \frac{\sqrt[n]{n!}}{n} = \frac{1}{e}.$$

(See Example 9, Section 9–6.)

Example 12. $\sum_{k=1}^{\infty} a^n \sin^2 n\alpha$ for $0 < a < 1$.

In this case it easily follows from the root test that the series converges, since

$$\varlimsup_{n \to \infty} \sqrt[n]{a^n \sin^2 n\alpha} \leq a < 1.$$

In this instance, however, the ratio test is (for some α's at least) inconclusive.

Most of our concern in this section has been with tests for convergence. We close with a theorem of a different sort.

Proposition 7. *If $\sum a_k$ is a convergent series of nonnegative terms, then any infinite series obtained from it by the rearrangement of terms must also converge and have the same sum.*

Proof. By a rearrangement of the series $\sum a_k$, we shall mean any series of the form $\sum_{k=1}^{\infty} a_{\varphi(k)}$, where φ is a one-to-one mapping of $\mathbf{Z}^+$ onto itself. We shall prove that such a series converges to $S = \sum_{k=1}^{\infty} a_k$.

Let $\epsilon > 0$ be given. We can choose M such that

$$S - \epsilon < \sum_{k=1}^{M} a_k.$$

We set $N = \max \{\varphi^{-1}(k) \mid k = 1, 2, \ldots, M\}$. Suppose then that $n \geq N$. Certainly the indices $\varphi(1), \varphi(2), \ldots, \varphi(n)$ must include the integers $1, 2, \ldots, M$. Consequently,

$$A - \epsilon < \sum_{k=1}^{M} a_k \leq \sum_{k=1}^{n} a_{\varphi(k)} \leq \sum_{k=1}^{M'} a_k \leq S,$$

where $M' = \max \{\varphi(k) \mid k = 1, 2, \ldots, n\}$. From this we may conclude that

$$\lim_{n \to \infty} \sum_{k=1}^{n} a_{\varphi(k)} = S.$$

Exercises

1. Let $\{a_n\}$ and $\{b_n\}$ be sequences of positive numbers such that the series $\sum b_k$ converges. Suppose also that

$$\lim_{n \to \infty} \frac{a_n}{b_n} = 0.$$

   Prove that the series $\sum a_k$ converges.

2. Test each of the following series for convergence.

   a) $\displaystyle\sum_{k=1}^{\infty} \frac{|\sin k|}{k^2}$

   b) $\displaystyle\sum_{k=1}^{\infty} \frac{2 + (-1)^k}{3^k}$

   c) $\displaystyle\sum_{k=1}^{\infty} \sin^2 \frac{1}{k}$

   d) $\displaystyle\sum_{k=1}^{\infty} \frac{1}{5 + 79k}$

   e) $\displaystyle\sum_{k=2}^{\infty} \frac{1}{k(\ln k)^p}$

   f) $\displaystyle\sum_{k=3}^{\infty} \frac{1}{k \ln k (\ln \ln k)^p}$

   g) $\displaystyle\sum_{k=1}^{\infty} k e^{-k^2}$

   h) $\displaystyle\sum_{k=1}^{\infty} \frac{1}{(2k - 1)2^k}$

   i) $\displaystyle\sum_{k=1}^{\infty} \frac{(k!)^2}{(2k)!}$

   j) $\displaystyle\sum_{k=1}^{\infty} \frac{k!}{3^k}$

   k) $\displaystyle\sum_{k=1}^{\infty} \frac{k^{k+1/k}}{(k + 1/k)^k}$

   l) $\displaystyle\sum_{k=1}^{\infty} \frac{(k!)^2}{2^{k^2}}$

   m) $\displaystyle\sum_{k=2}^{\infty} \frac{1}{(\ln k)^{\ln k}}$

   n) $\displaystyle\sum_{k=1}^{\infty} \left(\frac{\pi}{2} - \mathrm{Tan}^{-1} k\right)$

o) $\displaystyle\sum_{k=1}^{\infty} \frac{k^3[\sqrt{2} + (-1)^k]^k}{3^k}$

p) $\displaystyle\sum_{k=1}^{\infty} \frac{(1 + 10/k)^{k^2}}{k!}$

q) $\displaystyle\sum_{k=1}^{\infty} \frac{1}{k} \ln\left(1 + \frac{1}{k}\right)$

r) $\displaystyle\sum_{k=1}^{\infty} (\ln k)^p$

s) $\displaystyle\sum_{k=1}^{\infty} \frac{\ln(k+1) - \ln k}{\mathrm{Tan}^{-1}(2/k)}$

t) $\displaystyle\sum_{k=3}^{\infty} \frac{\ln(1 + k^{-1})}{|a|^{\ln \ln k}}$ $(a \neq 0)$

u) $\displaystyle\sum_{k=1}^{\infty} (\sqrt[k]{k} - 1)^k$

v) $\displaystyle\sum_{k=1}^{\infty} p^k k^p$ $(p > 0)$

w) $\displaystyle\sum_{k=2}^{\infty} k^p \left(\frac{1}{\sqrt{k-1}} - \frac{1}{\sqrt{k}}\right)$

x) $\dfrac{2}{1} + \dfrac{2 \cdot 5}{1 \cdot 5} + \cdots + \dfrac{2 \cdot 5 \cdots (3n-1)}{1 \cdot 5 \cdots (4n-3)} + \cdots$

y) $\displaystyle\sum_{k=1}^{\infty} \frac{1}{k} (\sqrt{k^2 + k + 1} - \sqrt{k^2 - k - 1})$

z) $\displaystyle\sum_{k=1}^{\infty} k^2 e^{-\sqrt{k}}$

3. Prove Proposition 4.

4. Prove the root test.

5. Generally speaking, the task of finding the sum of an infinite series selected at random is almost impossible. In some instances, however, the terms telescope to permit explicit calculation of the nth partial sum. For instance, the sum of the infinite series $\sum_{k=1}^{\infty} 1/[k(k+1)]$ is one, since

$$\sum_{k=1}^{n} \frac{1}{k(k+1)} = \sum_{k=1}^{n} \left(\frac{1}{k} - \frac{1}{k+1}\right) = 1 - \frac{1}{n+1}.$$

This problem deals with such series.

a) Show that if α is any number other than $0, -1, -2, -3, \ldots$, then

$$\sum_{k=0}^{\infty} \frac{1}{(\alpha + k)(\alpha + k + 1)} = \frac{1}{\alpha}.$$

b) Show that

$$\sum_{k=0}^{\infty} \frac{1}{(\alpha + k)(\alpha + k + 1)(\alpha + k + 2)} = \frac{1}{2\alpha(\alpha + 1)}.$$

c) Show more generally that

$$\sum_{k=0}^{\infty} \frac{1}{(\alpha + k)(\alpha + k + 1) \cdots (\alpha + k + r)} = \frac{1}{r} \frac{1}{\alpha(\alpha + 1) \cdots (\alpha + r - 1)}.$$

d) Deduce from (b), with $\alpha = \frac{1}{3}$, that

$$\frac{1}{1 \cdot 4 \cdot 7} + \frac{1}{4 \cdot 7 \cdot 10} + \frac{1}{7 \cdot 10 \cdot 13} + \cdots = \frac{1}{24}.$$

e) Show that

$$\sum_{k=0}^{\infty} \frac{1}{\binom{r+k+1}{r+1}} = \frac{r+1}{r}.$$

f) Find the sum of

$$\frac{1}{1\cdot 3} + \frac{1}{2\cdot 4} + \frac{1}{3\cdot 5} + \cdots$$

*g) Show that

$$\frac{x}{1+x} + \frac{2x^2}{1+x^2} + \frac{4x^4}{1+x^4} + \frac{8x^8}{1+x^8} + \cdots = \frac{x}{1-x} \quad \text{if } |x| < 1.$$

[Show that the nth partial sum is

$$x\mathcal{L}\left(\frac{x^{2^{n+1}}-1}{x-1}\right),$$

where $\mathcal{L}(f(x)) = f'(x)/f(x)$, the logarithmic derivative of f.]

h) Show that

$$\sum_{n=1}^{\infty} \text{Tan}^{-1}\frac{2}{k^2} = \frac{3\pi}{4}.$$

(Show that $\text{Tan}^{-1}(2/n^2) = \text{Tan}^{-1}[1/(n-1)] - \text{Tan}^{-1}[1/(n+1)]$.)

i) Show that

$$\sum_{k=1}^{\infty} \text{Tan}^{-1}\frac{1}{k^2+k+1} = \frac{\pi}{4}.$$

j) Show that

$$\frac{1}{y} + \frac{x}{y(y+1)} + \frac{x(x+1)}{y(y+1)(y+2)} + \cdots = \frac{1}{y-x} \quad \text{if } y > x > 0.$$

k) Show that

$$1 + \frac{a}{b} + \frac{a(a+1)}{b(b+1)} + \frac{a(a+1)(a+2)}{b(b+1)(b+2)} + \cdots = \frac{b-1}{b-a-1} \quad \text{if } b > a+1 > 1.$$

6. Let $\{a_n\}$ be any sequence, and set

$$u = \varliminf_{n\to\infty} a_n \quad \text{and} \quad v = \varlimsup_{n\to\infty} a_n.$$

a) Prove that $\lim_{n\to\infty} a_n = L$ iff $u = v = L$.

b) Prove that there exists a subsequence of $\{a_n\}$ whose limit is u. The same holds, of course, for v.

c) Deduce from (b) that if $\{a_n\}$ is bounded, then $\{a_n\}$ contains a convergent subsequence.

7. Let $\{a_n\}$ and $\{b_n\}$ be any two sequences of real numbers.

a) If $a_n \le b_n$ for each $n \in \mathbf{Z}^+$, show that

$$\varliminf_{n\to\infty} a_n \le \varliminf_{n\to\infty} b_n \quad \text{and} \quad \varlimsup_{n\to\infty} a_n \le \varlimsup_{n\to\infty} b_n.$$

b) Show that

$$\overline{\lim_{n\to\infty}} (a_n + b_n) \le \overline{\lim_{n\to\infty}} a_n + \overline{\lim_{n\to\infty}} b_n$$

and

$$\underline{\lim_{n\to\infty}} (a_n + b_n) \ge \underline{\lim_{n\to\infty}} a_n + \underline{\lim_{n\to\infty}} b_n.$$

c) If $a_n, b_n > 0$ for all $n \in \mathbf{Z}^+$, show that

$$\overline{\lim_{n\to\infty}} a_n b_n \le (\overline{\lim_{n\to\infty}} a_n)(\overline{\lim_{n\to\infty}} b_n).$$

d) Find examples in which strict inequality holds in (b) and (c).

8. a) It has been asserted that the root test is stronger than the ratio test. Prove this by showing that

$$\underline{\lim_{n\to\infty}} \frac{a_{n+1}}{a_n} \le \underline{\lim_{n\to\infty}} \sqrt[n]{a_n} \le \overline{\lim_{n\to\infty}} \sqrt[n]{a_n} \le \overline{\lim_{n\to\infty}} \frac{a_{n+1}}{a_n}$$

whenever $\{a_n\}$ is a sequence of positive numbers.

b) Let $\{a_n\}$ be the sequence $2^{-2}, 2^{-1}, 2^{-4}, 2^{-3}, \ldots$ Show that the outer two inequalities in (a) are strict. Observe that in this case the root test succeeds while the ratio test fails.

c) Set $a_n = n^n/n!$ and deduce from (a) that

$$\lim_{n\to\infty} \frac{n}{\sqrt[n]{n!}} = e.$$

9. Let f satisfy the hypothesis of the integral test. Let

$$S_n = \sum_{k=1}^{n} f(k),$$

$$T_n = \int_{1}^{n} f(t) \, dt,$$

and

$$\Delta_n = S_n - T_n.$$

a) Prove that

$$0 < f(n + 1) \le \Delta_{n+1} \le \Delta_n \le f(1),$$

and deduce that $\Delta = \lim_{n\to\infty} \Delta_n$ exists.

b) Show further that for each $n \in \mathbf{Z}^+$,

$$\int_{1}^{n} f(t) \, dt + \Delta \le \sum_{k=1}^{n} f(k) \le f(n) + \int_{1}^{n} f(t) \, dt + \Delta.$$

c) Deduce that

$$\gamma = \lim_{n\to\infty} \left(\sum_{k=1}^{n} \frac{1}{k} - \ln n \right)$$

exists. γ is known as *Euler's constant*. Deduce also that

$$\sum_{k=1}^{n} \frac{1}{k} = \ln n + \gamma + \mathcal{O}\left(\frac{1}{n}\right).$$

10. Let $a_1, a_2, \ldots$ be a nonincreasing sequence of nonnegative numbers. Prove that the series $\sum_{k=1}^{\infty} a_k$ converges iff the series $\sum_{k=1}^{\infty} 2^k a_{2^k}$ converges. This is known as *Cauchy's condensation test*. (Show that

$$\frac{1}{2} \sum_{k=1}^{n} 2^k a_{2^k} \leq \sum_{k=1}^{2^n} a_k$$

and

$$\sum_{k=1}^{2^n-1} a_k \leq \sum_{k=1}^{n} 2^k a_{2^k}.$$

Note that the essential idea was previously used in showing the divergence of the harmonic series.)

11. Use Cauchy's condensation test to determine the convergence of the series $\sum_{k=1}^{\infty} k^{-s}$ and $\sum_{k=1}^{\infty} k^{-1}(\ln k)^{-s}$.

12. Let $\{a_n\}$ and $\{b_n\}$ be sequences of positive numbers, and suppose that for all sufficiently large n,

$$\frac{b_{n+1}}{b_n} \leq \frac{a_{n+1}}{a_n}.$$

a) Prove that if the series $\sum a_k$ converges, then the series $\sum b_k$ also converges.
b) Prove that if the series $\sum b_k$ diverges, then so does the series $\sum a_k$.

13. Let $\{a_n\}$ be a sequence of positive numbers, and suppose that there exists a number a such that

$$\frac{a_{n+1}}{a_n} = 1 - \frac{a}{n} + \mathfrak{O}\left(\frac{1}{n^2}\right).$$

Observe that $\lim_{n \to \infty} (a_{n+1}/a_n) = 1$, so that the ratio test does not apply. Prove that the series $\sum a_k$ converges if $a > 1$ and diverges if $a < 1$. This test is a variant of *Gauss'* test and *Raabe's test*. (Compare the given series with the infinite series $\sum k^{-s}$, using the preceding problem.)

14. Using Gauss' test, prove that the *hypergeometric series*

$$1 + \frac{\alpha \cdot \beta}{1 \cdot \gamma} + \frac{\alpha(\alpha+1)\beta(\beta+1)}{1 \cdot 2 \cdot \gamma(\gamma+1)} + \frac{\alpha(\alpha+1)(\alpha+2)\beta(\beta+1)(\beta+2)}{1 \cdot 2 \cdot 3 \cdot \gamma(\gamma+1)(\gamma+2)} + \cdots$$

converges if $\gamma > \alpha + \beta$ and diverges if $\gamma < \alpha + \beta$. In particular, the series

$$1 + \frac{\alpha}{1} + \frac{\alpha(\alpha+1)}{1 \cdot 2} + \cdots$$

converges when $\alpha < 0$ and diverges when $\alpha > 0$.

15. Given a sequence $\{s_n\}$, we define a second sequence $\{\sigma_n\}$ by taking averages:

$$\sigma_n = (s_1 + s_2 + \cdots + s_n)/n.$$

Show that

$$\underline{\lim}\, s_n \leq \underline{\lim}\, \sigma_n \leq \overline{\lim}\, \sigma_n \leq \overline{\lim}\, s_n.$$

13-3 Absolute versus conditional convergence

Let us assume for the moment that we can rearrange at will the terms of any convergent series without destroying its convergence or altering its sum. We know that

$$\ln 2 = 1 - \tfrac{1}{2} + \tfrac{1}{3} - \tfrac{1}{4} + \cdots$$

It follows from our assumption that

$$
\begin{aligned}
\ln 2 &= 1 - \tfrac{1}{2} - \tfrac{1}{4} + \tfrac{1}{3} - \tfrac{1}{6} - \tfrac{1}{8} + \tfrac{1}{5} - \cdots \\
&= (1 - \tfrac{1}{2}) - \tfrac{1}{4} + (\tfrac{1}{3} - \tfrac{1}{6}) - \tfrac{1}{8} + \cdots \\
&= \tfrac{1}{2} - \tfrac{1}{4} + \tfrac{1}{6} - \tfrac{1}{8} + \tfrac{1}{10} - \cdots \\
&= \tfrac{1}{2}(1 - \tfrac{1}{2} + \tfrac{1}{3} - \tfrac{1}{4} + \cdots) \\
&= \tfrac{1}{2}\ln 2.
\end{aligned}
$$

This implies that $\ln 2 = 0$, which is absurd. It must follow that in rearranging the terms of the original series, we either destroyed its convergence or altered its sum. In general, both phenomena can occur; in this case it happens to be the latter. Thus

$$\tfrac{1}{2}\ln 2 = 1 - \tfrac{1}{2} - \tfrac{1}{4} + \tfrac{1}{3} - \tfrac{1}{6} - \tfrac{1}{8} + \tfrac{1}{5} - \cdots$$

A proof will be outlined in the exercises.

The pathology encountered above is related to the fact that although the series

$$1 - \tfrac{1}{2} + \tfrac{1}{3} - \tfrac{1}{4} + \cdots$$

converges, the series consisting of the absolute values, namely, the harmonic series, diverges. In exploring such matters, we begin by proving a theorem stated previously.

Theorem 1. *If the infinite series $\sum_{k=1}^{\infty} |a_k|$ converges, then the infinite series $\sum_{k=1}^{\infty} a_k$ also converges. Moreover,*

$$\sum_{k=1}^{\infty} a_k = \sum_{k=1}^{\infty} P_k + \sum_{k=1}^{\infty} N_k,$$

where

$$P_n = \max(a_n, 0), \qquad N_n = \min(a_n, 0)$$

for each $n \in \mathbf{Z}^+$.

Since zero terms can be added or deleted at will, the second part of the theorem says that one can find the sum of the series $\sum a_k$ by first adding separately the positive and the negative terms and then adding these two resulting sums. Note that this is not true of the series

$$\ln 2 = 1 - \tfrac{1}{2} + \tfrac{1}{3} - \tfrac{1}{4} + \tfrac{1}{5} - \cdots,$$

since both $1 + \tfrac{1}{3} + \tfrac{1}{5} + \cdots$ and $-\tfrac{1}{2} - \tfrac{1}{4} - \tfrac{1}{6} - \cdots$ are divergent.

Proof. For each n, we have $a_n = P_n + N_n$, $0 \le P_n \le |a_n|$, and $0 \le -N_n \le |a_n|$. By the comparison test, both $\sum P_k$ and $\sum(-N_k)$ converge; the rest follows immediately from the linearity property of infinite series.

Definition 2. An infinite series $\sum_{k=1}^{\infty} a_k$ is said to **converge absolutely** iff the series $\sum_{k=1}^{\infty} |a_k|$ converges.

The previous theorem tells us that an absolutely convergent series does indeed converge.

Example 1. The infinite series

$$1 - \frac{1}{2^2} + \frac{1}{3^2} - \frac{1}{4^2} + \cdots$$

converges absolutely. The series

$$1 - \tfrac{1}{2} + \tfrac{1}{3} - \tfrac{1}{4} + \cdots$$

converges, but not absolutely.

We have shown that if we have a convergent series of nonnegative terms, then one can rearrange its terms without destroying its convergence or altering its sum. It follows that any rearrangement of an absolutely convergent series is absolutely convergent. Furthermore, such a rearrangement cannot alter the sum of the positive terms and the sum of the negative terms. Therefore, we have the following corollary.

Corollary 3. *One can rearrange the terms of an absolutely convergent series without destroying its (absolute) convergence or altering its sum.*

The situation is quite the reverse for a "conditionally" convergent series.

Definition 4. An infinite series is said to be **conditionally convergent** if it is convergent, but not absolutely convergent.

We have observed that the conditionally convergent series

$$\ln 2 = 1 - \tfrac{1}{2} + \tfrac{1}{3} - \tfrac{1}{4} + \cdots$$

can be rearranged so as to converge to $\tfrac{1}{2} \ln 2$. Similarly, one can show that

$$1 + \tfrac{1}{3} - \tfrac{1}{2} + \tfrac{1}{5} + \tfrac{1}{7} - \tfrac{1}{4} + \cdots = \tfrac{3}{2} \ln 2$$

and

$$1 - \tfrac{1}{2} - \tfrac{1}{4} - \tfrac{1}{6} - \tfrac{1}{8} + \tfrac{1}{3} - \tfrac{1}{10} - \tfrac{1}{12} - \tfrac{1}{14} - \tfrac{1}{16} + \tfrac{1}{5} - \cdots = 0.$$

We shall now prove that by rearranging terms, we can make this series converge to anything we please!

Theorem 5. Riemann. *A conditionally convergent series can be made either to converge to any prescribed real number or to diverge by rearrangement.*

Proof. Suppose that $\sum_{k=1}^{\infty} a_k$ is conditionally convergent. Let P_n and N_n be defined as before. Then each of the series $\sum P_k$ and $\sum N_k$ must diverge. [If both series converged, then $\sum|a_k| = \sum(P_k - N_k)$ would have to converge, contradicting the assumption that $\sum a_k$ converges *conditionally*. If one of the series converged and the other diverged, then $\sum a_k$ would diverge.] We delete from the sequence $\{P_n\}$ all zero terms (of which there will be infinitely many) and relabel the remaining terms (of which there will also be infinitely many) so that the resulting sequence is $P_1, P_2, \ldots$ We do the same to the sequence $\{N_n\}$. We may also assume without loss of generality that the original series has no zero terms. We place the two sequences $\{P_n\}$ and $\{N_n\}$ into two infinitely long filing drawers labeled $\mathcal{P}$ and $\mathfrak{N}$, respectively.

Let A be any real number. We shall now produce a rearrangement of the original series whose sum is A. First we pull out of $\mathcal{P}$ just enough terms $P_1, \ldots, P_{m_1}$ so that

$$S_1 = \sum_{k=1}^{m_1} P_k > A.$$

(Thus, if $m_1 > 1$, $\sum_{k=1}^{m_1-1} P_k \leq A$. Furthermore, such an m_1 exists, since $\sum P_k = +\infty$.) Next we pull out of drawer $\mathfrak{N}$ just enough terms $N_1, \ldots, N_{n_1}$ so that

$$T_1 = S_1 + \sum_{k=1}^{n_1} N_k < A.$$

Next we return to drawer $\mathcal{P}$ and pull out just enough of the remaining terms P_{m_1+1}, $\ldots, P_{m_2}$ so that

$$S_2 = T_1 + \sum_{k=m_1+1}^{m_2} P_k > A.$$

(This is possible, since $\sum_{k=m_1+1}^{\infty} P_k = +\infty$.) We then return to drawer $\mathfrak{N}$, and so on. (See Fig. 13-1.) Continuing in this way, we finally obtain the following rearrangement of the original series:

$$P_1 + \cdots + P_{m_1} + N_1 + \cdots + N_{n_1} + P_{m_1+1} + \cdots$$

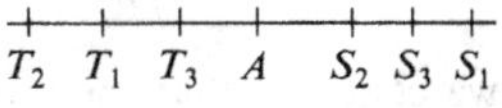

FIG. 13-1

We will prove that this series converges to A.

If S is any partial sum of the rearranged series, then for some k, either

$$T_k \leq S \leq S_k \quad \text{or} \quad T_k \leq S \leq S_{k+1}$$

(where we set $T_0 = 0$). Hence it suffices to show that the sequences $\{S_n\}$ and $\{T_n\}$ both converge to A. Now by the way S_k was formed,

$$S_k - P_{m_k} \leq A < S_k \quad \text{or} \quad A < S_k \leq A + P_{m_k}.$$

Since the series $\sum a_k$ converges, the sequence $\{a_n\}$ converges to zero, and hence so does the subsequence $\{P_{m_k}\}$. By the pinching theorem, $\lim_{n \to \infty} S_n = A$. Similarly, $T_n \to A$.

The job of producing a rearrangement which diverges to $+\infty$ is simpler. We form

$$P_1 + \cdots + P_{m_1} + N_1 + P_{m_1+1} + \cdots + P_{m_2} + N_2 + P_{m_2+1} + \cdots,$$

where m_n is chosen inductively so that

$$P_1 + \cdots + P_{m_1} + N_1 + \cdots + P_{m_n} \geq n.$$

We next consider tests for absolute and conditional convergence.

To test for absolute convergence, we can apply all the machinery of the preceding section. Let us consider the ratio test in particular. If we wish to test the series $\sum a_k$ for absolute convergence, and if none of the terms are zero, we can compute

$$r = \varliminf_{n \to \infty} \left| \frac{a_{n+1}}{a_n} \right| \quad \text{and} \quad R = \varlimsup_{n \to \infty} \left| \frac{a_{n+1}}{a_n} \right|.$$

If $R < 1$, we can conclude from the ratio test that $\sum |a_k|$ converges, i.e., that $\sum a_k$ converges absolutely. If $r > 1$, we can similarly conclude that the series does not converge *absolutely*. If we reexamine the proof of the ratio test, we see that the reason for the divergence of the series $\sum |a_k|$ is the fact that the nth term $|a_n|$ does not tend to zero. Since the same must be true of the term a_n, we conclude that if $r > 1$, then the series $\sum a_k$ does not converge *at all*. A similar observation holds for the root test.

Example 2. Discuss the convergence of the infinite series

$$x - \frac{x^2}{2} + \frac{x^3}{3} - \frac{x^4}{4} + \cdots$$

We have previously shown that the series converges to $\ln (1 + x)$ if $-1 < x \leq +1$. We have ventured no opinion, however, concerning the behavior of the series for other values of x.

We apply the ratio test. Since

$$\lim_{n \to \infty} \left| - \frac{x^{n+1}/(n + 1)}{x^n/n} \right| = |x|,$$

we conclude that the series converges absolutely if $|x| < 1$ and diverges if $|x| > 1$. The ratio test gives no information if $x = \pm 1$.

Although we know that the series

$$1 - \tfrac{1}{2} + \tfrac{1}{3} - \tfrac{1}{4} + \cdots$$

converges, we have as yet no general convergence test which tells us so. One of the best tests for series which are not absolutely convergent is due to Dirichlet. A special case, Leibnitz's test for alternating series, is also quite useful.

Lemma 6. *Abel's Partial Summation Formula. If $1 < m < n$, then*

$$\sum_{k=m}^{n} a_k b_k = a_{n+1} B_{n+1} - a_m B_m - \sum_{k=m}^{n} B_{k+1}(a_{k+1} - a_k),$$

where

$$B_k = \sum_{j=1}^{k-1} b_j.$$

Proof. This formula is analogous to the integration by parts formula

$$\int_a^b f(x)g'(x)\,dx = f(x)g(x)\Big|_a^b - \int_a^b g(x)f'(x)\,dx,$$

and can be derived by analogous methods. Note that $b_k = \Delta B_k$, where Δ is the difference operator: $\Delta c_k = c_{k+1} - c_k$. Consequently, the formula may be rewritten as follows:

$$\sum_{k=m}^n a_k\,\Delta B_k = a_{n+1}B_{n+1} - a_m B_m - \sum_{k=m}^n B_{k+1}\,\Delta a_k,$$

thereby making the analogy to the integration formula more apparent.

To derive the formula, we calculate Δ applied to a product:

$$\begin{aligned}
\Delta(a_k B_k) &= a_{k+1}B_{k+1} - a_k B_k \\
&= (a_{k+1} - a_k)B_{k+1} + a_k(B_{k+1} - B_k) \\
&= B_{k+1}\,\Delta a_k + a_k\,\Delta B_k.
\end{aligned}$$

Hence

$$\sum_{k=m}^n B_{k+1}\,\Delta a_k + \sum_{k=m}^n a_k\,\Delta B_k = \sum_{k=m}^n \Delta(a_k B_k) = a_{n+1}B_{n+1} - a_m B_m,$$

due to the telescoping of the sum $\sum \Delta(a_k B_k)$.

Theorem 7. *Dirichlet's Test. Suppose that $\sum b_k$ is a series whose sequence of partial sums is bounded, and suppose that $\{a_n\}$ is a nonincreasing sequence converging to zero. Then the series $\sum a_k b_k$ converges.*

Proof. The overall strategy is to show via Abel's partial summation formula that the partial sums of the series $\sum a_k b_k$ form a Cauchy sequence.

By assumption, the sequence $\{B_n\}$ is bounded, say $|B_n| \le M$, where, as in the lemma,

$$B_n = \sum_{j=1}^{n-1} b_j.$$

Also $a_{k+1} - a_k \ge 0$. It follows that

$$\left| \sum_{k=m}^n B_k(a_{k+1} - a_k) \right| \ge M \sum_{k=m}^n (a_{k+1} - a_k) = M(a_{n+1} - a_m).$$

The latter tends to zero as m and n tend to ∞.

By Abel's partial summation formula, we therefore have

$$\sum_{k=m}^n a_k b_k = a_{n+1}B_{n+1} - a_m B_m + R_{m,n},$$

where $\lim_{m,n\to\infty} R_{m,n} = 0$. Each of the terms $a_{n+1}B_{n+1}$ and $a_m B_m$, being the product of a bounded sequence and a null sequence, tends to zero as m and n tend to ∞.

Thus

$$\lim_{m,n\to\infty} (S_n - S_{m-1}) = \lim_{m,n\to\infty} \sum_{k=m}^{n} a_k b_k = 0,$$

where S_n is the nth partial sum of the series $\sum a_k b_k$. Therefore the sequence $\{S_n\}$ is Cauchy, and hence convergent.

Let us observe that the boundedness of the sequence $\{B_n\}$ is assured if the series $\sum b_k$ converges. If, in addition, we merely assume that the sequence $\{a_n\}$ has a limit α, then we have, as before,

$$\lim_{m,n\to\infty} (a_{n+1}B_{n+1} - a_m B_m) = \alpha B - \alpha B = 0,$$

where $B = \sum_{k=1}^{\infty} b_k$. Finally, if we add the assumption that $\{a_n\}$ is monotone, then

$$R_{m,n} \leq M|a_{n+1} - a_m|,$$

from which it again follows that $R_{m,n} \to 0$ as $m, n \to \infty$. Thus we have a second convergence test.

Theorem 8. Abel's Test. *If the series $\sum b_n$ converges, and if $\{a_n\}$ is a monotone convergent sequence, then the series $\sum a_k b_k$ converges.*

Before giving examples, we deduce as a corollary to Dirichlet's theorem a test for the convergence of a series whose terms alternate in sign. Observe that the partial sums of the infinite series $\sum (-1)^{k+1}$ form a bounded sequence, namely,

$$1, 0, 1, 0, 1, \ldots$$

Corollary 9. Leibnitz's Alternating Sequence Test. *If $\{a_n\}$ is a sequence which converges monotonically to zero, then the alternating series $\sum_{k=1}^{\infty} (-1)^{k+1} a_k$ converges.*

A more direct proof is outlined in the exercises. An important upshot of the proof is the fact that *the partial sum $\sum_{k=1}^{n} (-1)^{k+1} a_k$ differs from the sum of the series by an amount which is in absolute value less than $|a_{n+1}|$, the absolute value of the first neglected term.*

Example 3. Prove that the series

$$\sum_{k=1}^{\infty} (-1)^{k+1} \frac{(1 + 1/k)^k}{\sqrt{k}}$$

converges.

The sequence $\{1/\sqrt{n}\}$ converges monotonically to zero. By Leibnitz's test, the series

$$\sum_{k=1}^{\infty} (-1)^{k+1} \frac{1}{\sqrt{k}}$$

therefore converges. Since the sequence $\{(1 + 1/n)^n\}$ converges monotonically to e, it follows from Abel's test that the given series converges.

Example 4. If the sequence $\{a_n\}$ tends monotonically to zero, show that the infinite series

$$\sum_{k=1}^{\infty} a_k \sin kx$$

converges for every real number x.

If x is an integral multiple of π, then the series converges, since each term is zero. For other values of x, it suffices to show that the partial sums of the series $\sum_{k=1}^{\infty} \sin kx$ are bounded. The formula

$$\sum_{k=1}^{n} \sin kx = \frac{\sin \dfrac{nx}{2} \sin \dfrac{(n+1)x}{2}}{\sin (x/2)}$$

makes it apparent that $|\csc (x/2)|$ is a bound of the partial sums.

Example 5. The *Bessel function* J_0 may be defined by the series expansion

$$J_0(x) = 1 - \frac{x^2}{2^2} + \frac{x^4}{2^2 \cdot 4^2} - \frac{x^6}{2^2 \cdot 4^2 \cdot 6^2} + \cdots$$

Justify this definition, and compute $J_0(1)$ to four decimal places.

First we show that the series converges absolutely for all $x \in \mathbf{R}$. The absolute value of the nth term can be written

$$\frac{1}{(n!)^2} \left(\frac{x}{2}\right)^{2n};$$

hence for $x \neq 0$, the ratio of the $(n+1)$th term to the nth is in absolute value

$$\left[\frac{x}{2(n+1)}\right]^2.$$

Since this ratio tends to zero as $n \to \infty$, it follows from the ratio test that the series converges absolutely for each real number $x \neq 0$. If $x = 0$, all terms except the first are zero, and hence the series converges absolutely.

We shall compute $J_0(1)$ by truncating the series

$$1 - \frac{1}{2^2} + \frac{1}{2^2 \cdot 4^2} - \cdots$$

If we truncate at the $(n-1)$th term, then, by the comment following Leibnitz's test, the error will be less than the absolute value of the nth term. We therefore require that

$$\frac{1}{2^{2n}(n!)^2} < 5 \times 10^{-5},$$

or equivalently that

$$2^n n! > \sqrt{2} \times 10^2 = 141.4 \ldots$$

The smallest such $n \in \mathbf{Z}^+$ is 4. Hence to four decimal places,

$$J_0(1) = 1 - \frac{1}{2^2} + \frac{1}{2^2 \cdot 4^2} = 0.7656.$$

Exercises

1. If $\sum a_k$ and $\sum b_k$ are absolutely convergent, show that the infinite series $\sum (a_k + b_k)$ is also.

2. If $\sum a_k$ is absolutely convergent, show that the same is true of the series $\sum a_k^2$ and $\sum a_k/(1 + a_k)$.

3. Let S_n and T_n denote the nth partial sums of the series

$$1 - \tfrac{1}{2} + \tfrac{1}{3} - \tfrac{1}{4} + \cdots$$

and the rearranged series

$$1 - \tfrac{1}{2} - \tfrac{1}{4} + \tfrac{1}{3} - \tfrac{1}{6} - \tfrac{1}{8} + \tfrac{1}{5} - \cdots$$

Show that $T_{3n} = \tfrac{1}{2} S_{2n}$. Deduce that $\lim_{n \to \infty} T_{3n} = \tfrac{1}{2} \ln 2$. Show that

$$\lim T_{3n} = \lim T_{3n+1} = \lim T_{3n+2}.$$

Conclude finally that the series

$$1 - \tfrac{1}{2} - \tfrac{1}{4} + \tfrac{1}{3} - \tfrac{1}{6} - \tfrac{1}{8} + \tfrac{1}{5} - \cdots$$

converges and that its sum is $\tfrac{1}{2} \ln 2$.

4. Prove that

$$1 + \tfrac{1}{3} - \tfrac{1}{2} + \tfrac{1}{5} + \tfrac{1}{7} - \tfrac{1}{4} + \cdots = \tfrac{3}{2} \ln 2.$$

5. Prove that

$$1 - \tfrac{1}{2} - \tfrac{1}{4} - \tfrac{1}{6} - \tfrac{1}{8} + \tfrac{1}{3} - \tfrac{1}{10} - \tfrac{1}{12} - \tfrac{1}{14} - \tfrac{1}{16} + \cdots = 0.$$

6. Point out the fallacy in the following "proof" that $\ln 2 = 0$.

$$\ln 2 = 1 - \tfrac{1}{2} + \tfrac{1}{3} - \tfrac{1}{4} + \tfrac{1}{5} - \tfrac{1}{6} + \cdots$$
$$= 1 + \tfrac{1}{2} + \tfrac{1}{3} + \tfrac{1}{4} + \tfrac{1}{5} + \tfrac{1}{6} + \cdots - 2(\tfrac{1}{2} + \tfrac{1}{4} + \tfrac{1}{6} + \cdots)$$
$$= (1 + \tfrac{1}{2} + \tfrac{1}{3} + \cdots) - (1 + \tfrac{1}{2} + \tfrac{1}{3} + \cdots)$$
$$= 0.$$

7. If $p > 1$, we let

$$\zeta(p) = \sum_{k=1}^{\infty} \frac{1}{k^p}.$$

(The function ζ is known as the *Riemann zeta function*; it plays an important role in analytic number theory.) Using the methods of Exercise 6, show that

$$1 - \frac{1}{2^p} + \frac{1}{3^p} - \frac{1}{4^p} + \cdots = \left(1 - \frac{1}{2^{p-1}}\right) \zeta(p).$$

8. By inserting parentheses into the series

$$1 - \tfrac{1}{2} + \tfrac{1}{3} - \tfrac{1}{4} + \cdots,$$

obtain the following:

$$\ln 2 = \frac{1}{1 \cdot 2} + \frac{1}{3 \cdot 4} + \frac{1}{5 \cdot 6} + \cdots, \qquad \ln 2 = 1 - \frac{1}{2 \cdot 3} - \frac{1}{4 \cdot 5} - \cdots$$

Observe that both series are absolutely convergent. Deduce that

$$\tfrac{7}{12} < \ln 2 < \tfrac{10}{12}.$$

By adding the two series and inserting parentheses, show that

$$\ln 2 = \frac{1}{2} + \frac{1}{1 \cdot 2 \cdot 3} + \frac{1}{3 \cdot 4 \cdot 5} + \cdots$$

and

$$\ln 2 = \frac{3}{4} - \frac{1}{2 \cdot 3 \cdot 4} + \frac{1}{4 \cdot 5 \cdot 6} - \cdots$$

Observe that these series converge more rapidly than the original series.

9. Prove that every conditionally convergent series can be transformed into an absolutely convergent series by the insertion of parentheses. (Use the fact that every sequence of real numbers contains a monotone subsequence.)

10. Let $\{a_n\}$ be a nondecreasing sequence of nonnegative terms, and let S_n be the nth partial sum of the infinite series $\sum(-1)^{k-1} a_k$.

   a) Show that $S_1, S_3, S_5, \ldots$ is a nonincreasing sequence bounded below by each term of the sequence $S_2, S_4, S_6, \ldots$. Set $\overline{S} = \lim_{n \to \infty} S_{2n+1}$. Show similarly that $S_2, S_4, S_6, \ldots$ is a nondecreasing sequence bounded above by each number of the sequence $S_1, S_3, S_5, \ldots$. Set $\underline{S} = \lim_{n \to \infty} S_{2n}$.

   b) Show that $\overline{S} - \underline{S} = \lim_{n \to \infty} a_n$. Deduce Leibnitz's alternating series test. Also prove the italicized statement following the statement of Leibnitz's test.

11. Let $\{a_n\}$ be a sequence tending monotonically to zero. Show that the infinite series $\sum a_k \cos kx$ converges, provided x is not an integral multiple of π. What is the situation if x is an integral multiple of π?

12. Discuss the convergence of each of the following series. State when the series converges absolutely and when it converges conditionally.

   a) $\displaystyle\sum_{k=2}^{\infty} \frac{(-1)^k}{k(\ln k)^p}$

   b) $x - \dfrac{x^3}{3} + \dfrac{x^5}{5} - \dfrac{x^7}{7} + \cdots$

   c) $\sin x + \tfrac{1}{2}\sin 2x + \tfrac{1}{3}\sin 3x + \cdots$

   d) $\cos x + \tfrac{1}{2}\cos 2x + \tfrac{1}{3}\cos 3x + \cdots$

   e) $\sin x + \dfrac{1}{2}\left(1 + \dfrac{1}{2}\right)\sin 2x$
   $$+ \frac{1}{3}\left(1 + \frac{1}{2!} + \frac{1}{3!}\right)\sin 3x + \frac{1}{4}\left(1 + \frac{1}{2!} + \frac{1}{3!} + \frac{1}{4!}\right)\sin 4x + \cdots$$

   f) $\displaystyle\sum_{k=1}^{\infty} (-1)^k \sin \frac{x}{k}$

   g) $\displaystyle\sum_{k=1}^{\infty} \left(1 + \frac{1}{2} + \cdots + \frac{1}{k}\right)\frac{\sin kx}{k}$

   h) $\displaystyle\sum_{k=1}^{\infty} (-1)^k (k+a)^{-s} \quad (a > 0)$

   i) $\displaystyle\sum_{k=1}^{\infty} (-1)^k \frac{\alpha(\alpha+1)\cdots(\alpha+k+1)}{\beta(\beta+1)\cdots(\beta+k+1)}$

   j) $\displaystyle\sum_{k=1}^{\infty} (-1)^k k \left(1 - \cos\frac{a}{k}\right)$

   k) $\displaystyle\sum_{k=1}^{\infty} \frac{(-1)^k}{k - \ln k}$

   l) $\displaystyle\sum_{k=1}^{\infty} \frac{(-1)^k}{k}(\pi - \mathrm{Tan}^{-1} k)$

13. If $\sum a_k$ is a conditionally convergent series, and if S is any real number, prove that there exists a sequence $\epsilon_1, \epsilon_2, \ldots$ such that $|\epsilon_n| = 1$ for every $n \in \mathbf{Z}^+$, and

$$S = \sum \epsilon_k a_k.$$

(Use the same procedure as in the proof of the Riemann rearrangement theorem.)

*14. Let $\{a_n\}$ be a nonincreasing sequence of nonnegative numbers such that $\sum_{k=1}^{\infty} a_k < +\infty$. Let S denote the set of all numbers which are sums of subseries of $\sum_{k=1}^{\infty} a_k$. Prove that S is an interval iff for each $n \in \mathbf{Z}^+$,

$$a_n \leq \sum_{k=n+1}^{\infty} a_k.$$

For what values of r does the geometric series $\sum_{k=1}^{\infty} r^k$ satisfy the above condition?

15. a) If $\sum_{k=1}^{\infty} a_k^2$ and $\sum_{k=1}^{\infty} b_k^2$ converge, prove that $\sum_{k=1}^{\infty} a_k b_k$ converges absolutely and that

$$\left| \sum_{k=1}^{\infty} a_k b_k \right| \leq \sum_{k=1}^{\infty} |a_k b_k| \leq \left(\sum_{k=1}^{\infty} a_k^2 \right)^{1/2} \left(\sum_{k=1}^{\infty} b_k^2 \right)^{1/2}.$$

[*Hint:* By the Cauchy-Schwarz inequality,

$$\left| \sum_{k=1}^{n} a_k b_k \right| \leq \sum_{k=1}^{n} |a_k b_k|$$

$$\leq \left(\sum_{k=1}^{n} a_k^2 \right)^{1/2} \left(\sum_{k=1}^{n} b_k^2 \right)^{1/2}$$

$$\leq \left(\sum_{k=1}^{\infty} a_k^2 \right)^{1/2} \left(\sum_{k=1}^{\infty} b_k^2 \right)^{1/2}.$$

Deduce first that $\sum_{k=1}^{\infty} a_k b_k$ converges absolutely. Then let $n \to \infty$, to prove the rest.]

b) The result in (a) can be generalized. If $p \geq 1$, we say that a sequence $\alpha = \{a_n\}$ is *p-summable* iff $\sum_{k=1}^{\infty} |a_k|^p < +\infty$, in which case we define the *p-norm* of α to be

$$\|\alpha\|_p = \left(\sum_{k=1}^{\infty} |a_k|^p \right)^{1/p}.$$

If α is p-summable, and if $q > p$, prove that α is also q-summable.

c) Let p and q be positive numbers related by the equation $p^{-1} + q^{-1} = 1$. If $\alpha = \{a_n\}$ is p-summable and $\beta = \{\beta_n\}$ is q-summable, prove that $\{a_n b_n\}$ is 1-summable (that is, $\sum_{k=1}^{\infty} a_k b_k$ is absolutely convergent), and

$$\left| \sum_{k=1}^{\infty} a_k b_k \right| \leq \sum_{k=1}^{\infty} |a_k b_k| \leq \|\alpha\|_p \|\beta\|_q.$$

(Use Hölder's inequality, Section 9–7.)

d) If $\alpha = \{a_n\}$ and $\beta = \{b_n\}$ are both p-summable, prove that $\alpha + \beta = \{a_n + b_n\}$ is also p-summable and that

$$\|\alpha + \beta\|_p \leq \|\alpha\|_p + \|\beta\|_p.$$

   e) If $\sum_{k=1}^{\infty} |a_k|^{3/2} < +\infty$, show that $\sum_{k=1}^{\infty} |a_k| k^{-2/5} < +\infty$.

   f) If $\sum_{k=1}^{\infty} a_k^2 < +\infty$, show that $n^{-1} \sum_{k=1}^{\infty} \sqrt{k}\, a_k$ is bounded.

16. Let p and q be fixed positive integers with $p \geq q$. We set

$$x_n = \sum_{k=qn+1}^{pn} \frac{1}{k}, \qquad S_n = \sum_{k=1}^{n} \frac{(-1)^{k+1}}{k}.$$

   a) Show that $x_n \to \ln (p/q)$ as $n \to \infty$. [Use Exercise 9(c), Section 13–2.] Deduce that the harmonic series $\sum k^{-1}$ diverges.

   b) When $q = 1$ and $p = 2$, then $S_{2n} = x_n$. [See Exercise 30(b), Section 1–9.] Deduce that

$$\sum_{k=1}^{\infty} \frac{(-1)^{k+1}}{k} = \ln 2.$$

   c) Rearrange the series $\sum (-1)^{k+1}/k$ by writing alternately p positive terms followed by q negative terms, and prove that this rearrangement has sum

$$\ln 2 + \tfrac{1}{2} \ln (p/q).$$

Deduce the results given in Exercises 3, 4, and 5.

13–4 Double series

The product of two finite sums $\sum_{j=1}^{\infty} a_j$ and $\sum_{k=1}^{\infty} b_k$ can be found by the addition of the $m \times n$ products

$$
\begin{array}{cccc}
a_1 b_1 & a_1 b_2 & \dots & a_1 b_n \\
a_2 b_1 & a_2 b_2 & \dots & a_2 b_n \\
\vdots & \vdots & & \vdots \\
a_m b_1 & a_m b_2 & \dots & a_m b_n
\end{array}
$$

in any order. The proper generalization of this result to the case of two infinite series $\sum_{k=1}^{\infty} a_k$ and $\sum_{k=1}^{\infty} b_k$ is not obvious. In this case the products of the terms form a doubly semi-infinite array

$$
\begin{array}{cccc}
a_1 b_1 & a_1 b_2 & a_1 b_3 & \dots \\
a_2 b_1 & a_2 b_2 & a_2 b_3 & \dots \\
a_3 b_1 & a_3 b_2 & a_3 b_3 & \dots \\
\vdots & \vdots & \vdots &
\end{array}
$$

and it is not clear how we should attempt to form their sum. More generally, we shall consider ways of summing a doubly indexed set of numbers $c_{j,k}$, where the ordered pair (j, k) ranges over the set $\mathbf{Z}^+ \times \mathbf{Z}^+$. There are many possibilities. We can, for instance, arrange the elements of $\mathbf{Z}^+ \times \mathbf{Z}^+$ in a sequence and form an ordinary infinite series with the corresponding terms. For example, by threading our way through $\mathbf{Z}^+ \times \mathbf{Z}^+$, as shown,

$$
\begin{array}{cccc}
(1,1) & (1,2) \longrightarrow (1,3) & (1,4) & \cdots \\
\downarrow & \uparrow \qquad \downarrow & \uparrow & \\
(2,1) \longrightarrow (2,2) & (2,3) & (2,4) & \cdots \\
\downarrow & & & \uparrow \\
(3,1) \longleftarrow (3,2) \longleftarrow (3,3) & (3,4) & \cdots \\
\downarrow & & & \uparrow \\
(4,1) \longrightarrow (4,2) \longrightarrow (4,3) \longrightarrow (4,4) & \cdots \\
\vdots \qquad \vdots \qquad \vdots \qquad \vdots
\end{array}
$$

we get the series

$$c_{1,1} + c_{2,1} + c_{2,2} + c_{1,2} + c_{1,3} + c_{2,3} + \cdots$$

Another procedure is to sum first the columns of the array

$$
\begin{array}{cccc}
c_{1,1} & c_{1,2} & c_{1,3} & \cdots \\
c_{2,1} & c_{2,2} & c_{2,3} & \cdots \\
c_{3,1} & c_{3,2} & c_{3,3} & \cdots \\
\vdots & \vdots & \vdots &
\end{array}
$$

and then to add these sums. This leads to the *iterated series*

$$\sum_{k=1}^{\infty} \left(\sum_{j=1}^{\infty} c_{j,k} \right).$$

By summing first the rows, we are led to a second iterated series

$$\sum_{j=1}^{\infty} \left(\sum_{k=1}^{\infty} c_{j,k} \right).$$

Still another possibility is to sum first along the diagonals. This leads to

$$\sum_{k=1}^{\infty} \left(\sum_{j=1}^{n-1} c_{j,n-j} \right) = \sum_{k=1}^{\infty} \left(\sum_{i+j=n} c_{i,j} \right).$$

On the basis of our experience with sums of conditionally convergent series, we might expect these various procedures for summing a double series to lead to different results. This is indeed the case. Consider, for instance, the array

$$
\begin{array}{ccccc}
0 & 1 & 0 & 0 & 0 \cdots \\
-1 & 0 & 1 & 0 & 0 \cdots \\
0 & -1 & 0 & 1 & 0 \cdots \\
0 & 0 & -1 & 0 & 1 \cdots \\
\vdots & \vdots & \vdots & \vdots & \vdots
\end{array}
$$

It has sum 1 by rows, −1 by columns, and 0 by diagonals.

We begin by developing a general theory of infinite sums—one encompassing both double infinite series and ordinary ones—in which the order of terms plays absolutely no role.

Let Λ be any set. We shall consider mappings $\varphi: \Lambda \to \mathbf{R}$, but we shall use φ_λ rather than $\varphi(\lambda)$ to denote the image of λ under φ. Given such a function φ, our problem will be to decide on a suitable meaning for the sum of the terms φ_λ as the index λ ranges over the index set Λ. When Λ is finite, there is no problem; when Λ is infinite, then we shall use "generalized partial sums." Let $\mathfrak{F}$ denote the family of all finite subsets of Λ. If $F \in \mathfrak{F}$, we shall use S_F to denote the finite sum

$$\sum_{\lambda \in F} \varphi_\lambda = \sum_F \varphi.$$

We shall then define $\sum_{\lambda \in \Lambda} \varphi_\lambda = \sum \varphi$ to be a certain kind of limit of these generalized partial sums.

Definition 1. A function $\varphi: \Lambda \to \mathbf{R}$ is said to be **summable over** Λ iff there exists a number S such that

$$S = \lim \left\{ \sum_F \varphi \mid F \in \mathfrak{F} \right\}$$

in the following sense: $(\forall \epsilon > 0)(\exists F_0 \in \mathfrak{F})(\forall F \in \mathfrak{F})(F_0 \subset F \Rightarrow |\sum_F \varphi - S| < \epsilon)$. The number S (which is easily shown to be unique) is called the **sum of** φ **over** Λ, and we write

$$S = \sum_{\lambda \in \Lambda} \varphi_\lambda = \sum_\Lambda \varphi.$$

Let us first consider the case in which $\Lambda = \mathbf{Z}^+$.

Proposition 2. *A sequence φ is summable over $\mathbf{Z}^+$ iff the infinite series $\sum_{k=1}^{\infty} \varphi_k$ is absolutely convergent, in which case the sum of φ over $\mathbf{Z}^+$ is the usual sum of the infinite series.*

Proof. Let us first suppose that φ is summable over $\mathbf{Z}^+$ with sum S. Given $\epsilon > 0$, we can then find a finite set $F_0 \subset \mathbf{Z}^+$ such that

$$\left| \sum_F \varphi - S \right| < \epsilon$$

whenever F is a finite subset of $\mathbf{Z}^+$ which includes F_0. We set

$$N = \max \{ k \mid k \in F_0 \}.$$

If $n \geq N$, then $F_0 \subset \{1, 2, \dots, n\}$, so

$$\left| \sum_{k=1}^{n} \varphi_k - S \right| < \epsilon.$$

This proves that the infinite series $\sum_{k=1}^{\infty} \varphi_k$ converges with sum S. Since our notion of summability does not involve the notion of order, any rearrangement of the series $\sum_{k=1}^{\infty} \varphi_k$ will also converge to S. Because of the Riemann theorem concerning the rearrangement of conditionally convergent series, it follows that the infinite series must converge absolutely.

Suppose now that the infinite series converges absolutely, and let S denote its sum. Given any $\epsilon > 0$, we can choose an $N \in \mathbf{Z}^+$ such that

$$\sum_{k=N+1}^{\infty} |a_k| < \epsilon.$$

(Why?) Set $F_0 = \{1, 2, \ldots, N\}$. Suppose that F is any finite subset of $\mathbf{Z}^+$ including F_0. Set $n = \max \{k \mid k \in F\}$. Then $\{1, 2, \ldots, N\} \subset F \subset \{1, 2, \ldots, n\}$, so that

$$\left| \sum_F \varphi - S \right| \leq \left| \sum_F \varphi - \sum_{k=1}^{n} \varphi_k \right| + \left| \sum_{k=1}^{n} \varphi_k - S \right|$$

$$\leq \sum_{k=N+1}^{n} |\varphi_k| + \left| \sum_{k=n+1}^{\infty} \varphi_k \right|$$

$$\leq \sum_{k=N+1}^{n} |\varphi_k| + \sum_{k=n+1}^{\infty} |\varphi_k| = \sum_{k=N+1}^{\infty} |\varphi_k| < \epsilon.$$

This proves that φ is summable over $\mathbf{Z}^+$ and that S is its sum over $\mathbf{Z}^+$.

We turn now to general theorems on summability.

Proposition 3. *Suppose that φ and ψ are summable over Λ.*

1) Linearity. If a and b are constants, then $a\varphi + b\psi$ is summable over Λ, and

$$\sum_\Lambda (a\varphi + b\psi) = a \sum_\Lambda \varphi + b \sum_\Lambda \psi.$$

2) Positivity. If $\varphi \geq 0$ on Λ, then

$$\sum_\Lambda \varphi \geq 0,$$

with strict inequality unless $\varphi \equiv 0$.

3) Monotonicity. If $\varphi \leq \psi$ on Λ, then

$$\sum_\Lambda \varphi \leq \sum_\Lambda \psi,$$

with strict inequality unless $\varphi \equiv \psi$.

A proof using *nets* is outlined in the exercises.

Lemma 4. *Let $\varphi \colon \Lambda \to \mathbf{R}$ be nonnegative. The following are equivalent:*

1) φ is summable over Λ.

2) The set of generalized partial sums of φ is bounded.

3) There are at most a countable number of indices λ for which $\varphi_\lambda \neq 0$, and if these indices each appear in the sequence $\lambda_1, \lambda_2, \lambda_3, \ldots$, where the λ_k's are distinct, then the infinite series $\sum_{k=1}^{\infty} \varphi_{\lambda_k}$ converges. Moreover, if φ is summable, then its sum over Λ may be characterized as both the least upper bound of the set of generalized partial sums and the sum of the infinite series $\sum_{k=1}^{\infty} \varphi_{\lambda_k}$.

The arguments required in the proof essentially duplicate earlier arguments for infinite series consisting of nonnegative terms. Only one novel feature is involved. Consider the implication (2) $\Rightarrow$ (3). Let us suppose that M is an upper bound of the set of generalized partial sums. Then for each $n \in \mathbf{Z}^+$, the set

$$\Lambda_n = \{\lambda \in \Lambda \mid \varphi_\lambda > 1/n\}$$

is finite; in fact, it cannot contain more than Mn points. (Why?) It follows that the set

$$\{\lambda \in \Lambda \mid \varphi_\lambda \neq 0\} = \bigcup_{n=1}^{\infty} \Lambda_n$$

is countable. (See Section 8–6.)

Theorem 5. *Let $\varphi: \Lambda \to \mathbf{R}$. The following are equivalent:*

1) φ is summable over Λ.

2) φ^+ and φ^- are summable over Λ.

3) $|\varphi|$ is summable over Λ.

4) The set of generalized partial sums of φ is bounded.

5) There are at most a countable number of indices λ for which $\varphi_\lambda \neq 0$, and if these indices each appear in the sequence $\lambda_1, \lambda_2, \ldots$, where the λ_k's are distinct, then the infinite series $\sum_{k=1}^{\infty} \varphi_{\lambda_k}$ converges absolutely. Moreover, if φ is summable over Λ, then its sum over Λ is just the sum of the infinite series $\sum_{k=1}^{\infty} \varphi_{\lambda_k}$.

Proof. Because of linearity, (2) implies both (1) and (3). Using the lemma and the triangle inequality, we can readily see that (3) $\Rightarrow$ (4). From the lemma it is also clear that (3) is equivalent to (5).

We set

$$\Lambda^+ = \{\lambda \in \Lambda \mid \varphi_\lambda > 0\}, \qquad \Lambda^- = \{\lambda \in \Lambda \mid \varphi_\lambda < 0\}.$$

Then Λ^+ and Λ^- are disjoint sets, and for each $F \in \mathfrak{F}$, we have

$$\sum_F \varphi = \sum_{F \cap \Lambda^+} \varphi + \sum_{F \cap \Lambda^-} \varphi,$$

$$\sum_F \varphi^+ = \sum_{F \cap \Lambda^+} \varphi^+ = \sum_{F \cap \Lambda^+} \varphi,$$

and

$$\sum_F \varphi^- = \sum_{F \cap \Lambda^-} \varphi^- = -\sum_{F \cap \Lambda^-} \varphi.$$

It follows that if the set $\{\sum_F \varphi \mid F \in \mathfrak{F}\}$ is bounded, then the same will be true of the sets $\{\sum_F \varphi^+ \mid F \in \mathfrak{F}\}$ and $\{\sum_F \varphi^- \mid F \in \mathfrak{F}\}$. Because of the lemma, we therefore have (4) $\Rightarrow$ (2).

To complete the proof of the equivalence of (1), $\ldots$, (5), it suffices to show that (1) $\Rightarrow$ (4') and $\sim$(2) $\Rightarrow$ $\sim$(4'), where (4') is a formally weaker condition than (4), namely,

4') There exists an $F_0 \in \mathfrak{F}$ such that $\{\sum_F \varphi \mid F \in \mathfrak{F} \text{ and } F_0 \subset F\}$ is bounded.

That $(1) \Rightarrow (4')$ follows immediately from the definition of summability with, say, ϵ set equal to one. (Compare Proposition 2, Section 3–5.) Assume now that (2) does not hold; for definiteness, assume that φ^+ is not summable. Let $F_0 \in \mathfrak{F}$ be arbitrary. If F is any finite subset of Λ^+, then

$$\sum_{F_0 \cup F} \varphi = \sum_{F_0 \cup F} \varphi^+ - \sum_{F_0} \varphi^- \geq \sum_F \varphi^+ - \sum_{F_0} \varphi^-.$$

Since φ^+ is not summable, the set $\{\sum_F \varphi^+ \mid F \in \mathfrak{F}\} = \{\sum_F \varphi^+ \mid F \in \mathfrak{F} \text{ and } F \subset \Lambda^+\}$ is unbounded. It is clear from the inequality above that

$$\left\{ \sum_F \varphi \mid F \in \mathfrak{F} \text{ and } F_0 \subset F \right\} \supset \left\{ \sum_{F_0 \cup F} \varphi \mid F \in \mathfrak{F} \text{ and } F \subset \Lambda^+ \right\}$$

is unbounded. Hence $\sim(2) \Rightarrow \sim(4')$.

The remaining assertions follow easily from the linearity property, together with the lemma applied to φ^+ and φ^-.

The equivalence of (1) and (4) is a bit surprising. Observe, for instance, that for an infinite series, the sequence of partial sums can be bounded without the series being convergent.

Corollary 6. *If $|\varphi| \leq \psi$ on Λ, and if ψ is summable over Λ, then φ is summable over Λ.*

Corollary 7. *If φ is summable over Λ, and if $\Lambda_0 \subset \Lambda$, then φ is summable over Λ_0 and*

$$\sum_{\Lambda_0} \varphi = \sum_{\Lambda} \varphi \cdot \varphi_{\Lambda_0},$$

where φ_{Λ_0} denotes the characteristic function of Λ_0.

Proof. $|\varphi \cdot \varphi_{\Lambda_0}| \leq |\varphi|$. Hence $\varphi \cdot \varphi_{\Lambda_0}$ is summable over Λ. The rest follows from the observation that the set of generalized partial sums of φ over Λ_0 is equal to the set of generalized partial sums of $\varphi \cdot \varphi_{\Lambda_0}$ over Λ.

Suppose that φ is summable over Λ, and that A and B are disjoint subsets of Λ. Then $\varphi_{A \cup B} = \varphi_A + \varphi_B$, so that

$$\sum_{A \cup B} \varphi = \sum \varphi \cdot \varphi_{A \cup B} = \sum \varphi \cdot (\varphi_A + \varphi_B)$$

$$= \sum \varphi \cdot \varphi_A + \sum \varphi \cdot \varphi_B = \sum_A \varphi + \sum_B \varphi.$$

This additivity property can be extended by induction to the case of any finite number of disjoint subsets $A_1, \ldots, A_n$ of Λ. The *raison d'être* for summability comes from the fact that this additivity property also holds for any countable number of disjoint sets.

Theorem 8. Countable Additivity. *Suppose that $\Lambda_1, \Lambda_2, \ldots$ are disjoint sets whose union is Λ. Then φ is summable over Λ iff*

1) for each $n \in \mathbf{Z}^+$, φ is summable over Λ_n, and

2) the infinite series $\sum_{k=1}^{\infty} (\sum_{\Lambda_k} |\varphi|)$ converges.

Moreover, if φ is summable over Λ, then

$$\sum_\Lambda \varphi = \sum_{k=1}^\infty \left(\sum_{\Lambda_k} \varphi \right).$$

A somewhat more elegant version of this theorem is formulated in the exercises. The present version is, however, a little better suited to the applications which we shall consider.

Proof. It suffices to consider the case where φ is nonnegative. (Why?) In the proof we shall use a very simple observation. If $A \subset B \subset \Lambda$, then $\varphi \cdot \varphi_A \leq \varphi \cdot \varphi_B$, so that

$$\sum_A \varphi = \sum_\Lambda \varphi \cdot \varphi_A \leq \sum_\Lambda \varphi \cdot \varphi_B = \sum_B \varphi.$$

Suppose that φ is summable over Λ. By Corollary 7, φ is summable over each of the sets $\Lambda_1, \Lambda_2, \ldots$ Set $\Lambda_{(n)} = \Lambda_1 \cup \Lambda_2 \cup \cdots \cup \Lambda_n$. Then

$$\sum_{k=1}^n \left(\sum_{\Lambda_k} \varphi \right) = \sum_{\Lambda_{(n)}} \varphi \leq \sum_\Lambda \varphi.$$

It follows that the infinite series $\sum_{k=1}^\infty \left(\sum_{\Lambda_k} \varphi \right)$ converges and that

$$\sum_{k=1}^\infty \left(\sum_{\Lambda_k} \varphi \right) \leq \sum_\Lambda \varphi.$$

The reverse inequality is easily established. Let $\epsilon > 0$ be arbitrary. Since $\sum_\Lambda \varphi$ is the least upper bound of its set of generalized partial sums, there exists a finite set $F \subset \Lambda$ such that

$$\sum_F \varphi \geq \sum_\Lambda \varphi - \epsilon.$$

Now choose N so that $F \subset \Lambda_{(N)}$. If $n \geq N$, then

$$\sum_{k=1}^n \left(\sum_{\Lambda_k} \varphi \right) = \sum_{\Lambda_{(n)}} \varphi \geq \sum_{\Lambda_{(N)}} \varphi$$
$$\geq \sum_F \varphi \geq \sum_\Lambda \varphi - \epsilon.$$

Hence

$$\sum_{k=1}^\infty \left(\sum_{\Lambda_k} \varphi \right) \geq \sum_\Lambda \varphi - \epsilon,$$

and since ϵ was arbitrary, $\sum_{k=1}^\infty \left(\sum_{\Lambda_k} \varphi \right) \geq \sum_\Lambda \varphi$. Thus

$$\sum_{k=1}^\infty \left(\sum_{\Lambda_k} \varphi \right) = \sum_\Lambda \varphi.$$

Suppose now that conditions (1) and (2) hold. It is easily shown that each of the functions $\varphi \cdot \varphi_{\Lambda_1}$, $\varphi \cdot \varphi_{\Lambda_2}$, ... is summable over Λ. It follows that for each $n \in \mathbf{Z}^+$, $\varphi \cdot \varphi_{\Lambda_{(n)}} = \varphi \cdot \varphi_{\Lambda_1} + \cdots + \varphi \cdot \varphi_{\Lambda_n}$ is summable over Λ and

$$\sum_{\Lambda_{(n)}} \varphi = \sum_{\Lambda} \varphi \cdot \varphi_{\Lambda_{(n)}} = \sum_{k=1}^{n} \varphi \cdot \varphi_{\Lambda_k} = \sum_{k=1}^{n} \left(\sum_{\Lambda_k} \varphi \right).$$

If F is any finite subset of Λ, then there exists an n such that $F \subset \Lambda_{(n)}$, and consequently

$$\sum_{F} \varphi \le \sum_{\Lambda_{(n)}} \varphi = \sum_{k=1}^{n} \left(\sum_{\Lambda_k} \varphi \right) \le \sum_{k=1}^{\infty} \left(\sum_{\Lambda_k} \varphi \right).$$

Thus the set of generalized partial sums of φ is bounded, and φ is therefore summable over Λ.

We shall now apply the general theory of summability to double series.

Theorem 9. Fubini Theorem for Double Series. *Let* $\varphi: \mathbf{Z}^+ \times \mathbf{Z}^+ \to \mathbf{R}$, *and let* ψ *be a one-to-one mapping of* $\mathbf{Z}^+$ *onto* $\mathbf{Z}^+ \times \mathbf{Z}^+$. *The following are logically equivalent:*

1) φ *is summable over* $\mathbf{Z}^+ \times \mathbf{Z}^+$.
2) The infinite series $\sum_{k=1}^{\infty} \varphi_{\psi(k)}$ *converges absolutely.*
3) The iterated series $\sum_{j=1}^{\infty} \left(\sum_{k=1}^{\infty} |\varphi_{j,k}| \right)$ *converges.*
4) The iterated series $\sum_{k=1}^{\infty} \left(\sum_{j=1}^{\infty} |\varphi_{j,k}| \right)$ *converges.*
5) The infinite series $\sum_{k=2}^{\infty} \left(\sum_{i+j=k} |\varphi_{i,j}| \right)$ *converges.*
Moreover, if φ *is summable over* $\mathbf{Z}^+ \times \mathbf{Z}^+$, *then*

$$\sum_{\mathbf{Z}^+ \times \mathbf{Z}^+} \varphi = \sum_{k=1}^{\infty} \varphi_{\psi(k)} = \sum_{j=1}^{\infty} \left(\sum_{k=1}^{\infty} \varphi_{j,k} \right)$$
$$= \sum_{k=1}^{\infty} \left(\sum_{j=1}^{\infty} \varphi_{j,k} \right) = \sum_{k=2}^{\infty} \left(\sum_{i+j=k} \varphi_{i,j} \right).$$

Proof. This theorem is a trivial consequence of the theorem on countable additivity. All that is needed is the suitable selection of the sets $\Lambda_1, \Lambda_2, \ldots$. To prove the equivalence of (1) and (5), we let

$$\Lambda_1 = \varnothing, \quad \Lambda_2 = \{(1,1)\}, \quad \Lambda_3 = \{(1,2),(2,1)\}, \quad \Lambda_4 = \{(1,3),(2,2),(3,1)\},$$

i.e., the diagonals of $\mathbf{Z}^+ \times \mathbf{Z}^+$. To prove the equivalence of (1) and (3), we let

$$\Lambda_k = \{(k,j) \mid j \in \mathbf{Z}^+\},$$

the kth row of $\mathbf{Z}^+ \times \mathbf{Z}^+$.

Definition 10. If φ is summable over $\mathbf{Z}^+ \times \mathbf{Z}^+$, we say that the double series $\sum_{j,k=1}^{\infty} \varphi_{j,k}$ **converges absolutely**; $S = \sum_{\mathbf{Z}^+ \times \mathbf{Z}^+} \varphi$ is called the **sum of the double series**, and we write

$$\sum_{j,k=1}^{\infty} \varphi_{j,k} = S.$$

Our previous theorem tells us that all procedures for adding the terms of an absolutely convergent double series yield the same sum.

Corollary 11. *Multiplication Rule for Absolutely Convergent Series. Suppose that* $\sum_{k=0}^{\infty} a_k$ *and* $\sum_{k=0}^{\infty} b_k$ *are absolutely convergent series with sums A and B. Then the double series* $\sum_{j,k=0}^{\infty} a_j b_k$ *converges and has sum AB. In particular,*

$$\sum_{k=0}^{\infty} \left(\sum_{i+j=k} a_i b_j \right) = AB.$$

Proof. Since

$$\sum_{j=0}^{\infty} \left(\sum_{k=0}^{\infty} |a_j b_k| \right) = \sum_{j=0}^{\infty} \left(|a_j| \sum_{k=0}^{\infty} |b_k| \right) = \sum_{j=0}^{\infty} |a_j| \sum_{k=0}^{\infty} |b_k| < +\infty,$$

the double series $\sum_{j,k=0}^{\infty} a_j b_k$ converges absolutely. Its sum is

$$\sum_{j=0}^{\infty} \left(\sum_{k=0}^{\infty} a_j b_k \right) = \sum_{j=0}^{\infty} \left(a_j \sum_{k=0}^{\infty} b_k \right) = \left(\sum_{j=0}^{\infty} a_j \right) \left(\sum_{b=0}^{\infty} b_k \right) = AB.$$

The infinite series $\sum_{k=0}^{\infty} \left(\sum_{i+j=k} a_i b_j \right)$ is called the *Cauchy product* of the infinite series $\sum_{k=0}^{\infty} a_k$ and $\sum_{k=0}^{\infty} b_k$. It is of interest primarily because the Cauchy product of two power series $\sum_{k=0}^{\infty} a_k (z - c)^k$ and $\sum_{k=0}^{\infty} b_k (z - c)^k$ is that power series which one can obtain formally by multiplying the two given series in the manner of polynomials. Thus

$$a_0 + a_1(z - c) + a_2(z - c)^2 + \cdots$$
$$b_0 + b_1(z - c) + b_2(z - c)^2 + \cdots$$

$$a_0 b_0 + a_1 b_0(z - c) + a_2 b_0(z - c)^2 + \cdots$$
$$+ a_0 b_1(z - c) + a_1 b_1(z - c)^2 + \cdots$$
$$+ a_0 b_2(z - c)^2 + \cdots$$
$$\cdots$$

$$a_0 b_0 + (a_0 b_1 + a_1 b_0)(z - c) + (a_0 b_2 + a_1 b_1 + a_2 b_0)(z - c)^2 + \cdots$$
$$= \sum_{k=0}^{\infty} \left(\sum_{i+j=k} a_i b_j \right) (z - c)^k.$$

According to the multiplication rule, it follows that if the two power series

$$\sum_{k=0}^{\infty} a_k (z - c)^k \quad \text{and} \quad \sum_{k=0}^{\infty} b_k (z - c)^k$$

converge absolutely for some value of z to, say, $f(z)$ and $g(z)$, then for the same value of z, the product series $\sum_{k=0}^{\infty} \left(\sum_{i+j=k} a_i b_j \right)(z - c)^k$ also converges absolutely, and its sum is $f(z) g(z)$.

Example 1. Show that the series expansion

$$(\mathrm{Sin}^{-1} x)^2 = \frac{1}{2} \sum_{k=1}^{\infty} \frac{(k-1)!!}{(2k)!} (2x)^{2k}$$

is valid iff $|x| \le 1$. [Here $n!! = (n!)^2$.]

We first show that the infinite series is the Maclaurin series for $(\mathrm{Sin}^{-1} x)^2$. If we let $y = (\mathrm{Sin}^{-1} x)^2$, then by Exercise 8, Section 9–3, the derivatives at zero satisfy the recursion formula

$$y^{(n+2)}(0) = n^2 y^{(n)}(0) \qquad (n \ge 1).$$

Moreover, one can verify directly that $y(0) = y'(0) = 0$ and $y''(0) = 2$. Consequently, all odd-order derivatives vanish at zero. [This is also clear from the fact that $(\mathrm{Sin}^{-1} x)$ is an even function.] Applying the recursion formula, we get

$$y^{(4)}(0) = 2^2 \cdot 2,$$
$$y^{(6)}(0) = 4^2 \cdot 2^2 \cdot 2,$$
$$\vdots$$
$$y^{(2k)}(0) = (2k-2)^2 (2k-4)^2 \cdots 2^2 \cdot 2$$
$$= 2^{2k-1}(k-1)!!.$$

Hence the Maclaurin expansion for $(\mathrm{Sin}^{-1} x)^2$ is

$$\sum_{k=1}^{\infty} \frac{2^{2k-1}(k-1)!!}{(2k)!} x^{2k} = \frac{1}{2} \sum_{k=1}^{\infty} \frac{(k-1)!!}{(2k)!} (2x)^{2k}.$$

Next we investigate convergence of the series. The ratio of the $(n+1)$th term to the nth is

$$\frac{(2nx)^2}{2n(2n-1)},$$

which, as $n \to \infty$, tends to x^2. By the ratio test, the series diverges if $|x| > 1$ and converges absolutely if $|x| < 1$.

We can show finally that for $|x| \le 1$, the series converges (absolutely) to $(\mathrm{Sin}^{-1} x)^2$. In a previous exercise, we showed that if $\sum_{k=0}^{\infty} a_k(x-a)^k$ and $\sum_{k=0}^{\infty} b_k(x-a)^k$ are Taylor series for the functions f and g, respectively, then their Cauchy product is the Taylor series for fg about the point a. Hence, by the multiplication rule, it suffices to prove that the Maclaurin series for $\mathrm{Sin}^{-1} x$, namely,

$$x + \frac{1}{2}\frac{x^3}{3} + \frac{1}{2}\cdot\frac{3}{4}\frac{x^5}{5} + \cdots,$$

converges absolutely to $\mathrm{Sin}^{-1} x$ if $|x| \le 1$. Since the terms are all of the same sign, convergence implies absolute convergence. This result was established for $|x| < 1$ in Exercise 9, Section 11–2. The behavior at $\ne 1$ will be treated later in this chapter.

Example 2. Prove that

$$\zeta(2) = \sum_{k=1}^{\infty} \frac{1}{k^2} = \frac{\pi^2}{6}.$$

By summing over the identity

$$-\frac{(j-1)!}{k(k+1)\cdots(k+j)} = \frac{1}{k}\frac{j!}{(k+1)\cdots(k+j)} - \frac{(j-1)!}{k\cdots(k+j-1)},$$

where $j = 1, 2, \ldots, k-1$, we get

$$\frac{1}{k^2} = \sum_{j=1}^{k-1}\frac{(j-1)!}{k(k+1)\cdots(k+j)} + \frac{(k-1)!}{k^2(k+1)\cdots(2k-1)}.$$

Thus

$$1 = 1,$$

$$\frac{1}{2^2} = \frac{0!}{2\cdot3} + \frac{1!}{2^2\cdot3},$$

$$\frac{1}{3^2} = \frac{0!}{3\cdot4} + \frac{1!}{3\cdot4\cdot5} + \frac{2!}{3^2\cdot4\cdot5},$$

$$\frac{1}{4^2} = \frac{0!}{4\cdot5} + \frac{1!}{4\cdot5\cdot6} + \frac{2!}{4\cdot5\cdot6\cdot7} + \frac{3!}{4^2\cdot5\cdot6\cdot7},$$

$$\vdots$$

By adding zeros to this triangular array, we get a double series. Since it is absolutely convergent by rows, it is absolutely convergent, and it can be summed by columns. By Exercise 5, Section 13-3, the sum of the terms in the nth column is

$$\frac{(n-1)!}{n^2(n+1)\cdots(2n-1)} + \frac{(n-1)!}{n(n+1)\cdots(2n)} = 3\frac{(n-1)!}{n(n+1)\cdots(2n)} = 3\frac{(n-1)!!}{(2n)!}.$$

Therefore, since the row and column sums are the same, we have

$$\sum_{k=1}^{\infty}\frac{1}{k^2} = 3\sum_{n=1}^{\infty}\frac{(n-1)!!}{(2n)!}.$$

However, by the preceding example,

$$(\text{Sin}^{-1}\tfrac{1}{2})^2 = \frac{1}{2}\sum_{n=1}^{\infty}\frac{(n-1)!!}{(2n)!}.$$

Thus

$$\sum_{k=1}^{\infty}\frac{1}{k^2} = 6(\text{Sin}^{-1}\tfrac{1}{2})^2 = \frac{\pi^2}{6}.$$

Example 3. Using series methods, prove the basic identity

$$e^x \cdot e^y = e^{x+y}.$$

Since the series

$$e^x = 1 + x + \frac{x^2}{2!} + \frac{x^3}{3!} + \cdots,$$

$$e^y = 1 + y + \frac{y^2}{2!} + \frac{y^3}{3!} + \cdots$$

converge absolutely (by, say, the ratio test), their Cauchy product converges absolutely to $e^x e^y$. Thus

$$e^x e^y = \sum_{n=0}^{\infty} \left(\sum_{j+k=n} \frac{x^j}{j!} \frac{y^k}{k!} \right)$$

$$= \sum_{n=0}^{\infty} \frac{1}{n!} \left[\sum_{j+k=n} \binom{n}{j} x^j y^k \right]$$

$$= \sum_{n=0}^{\infty} \frac{1}{n!} (x + y)^n = e^{x+y}.$$

In working with infinite series, we have somewhat arbitrarily limited our attention to series whose terms are real numbers. Many of our theorems, however, are also valid for series with complex terms. We define e^z for complex values of z by the series

$$e^z = 1 + z + \frac{z^2}{2!} + \frac{z^3}{3!} + \cdots$$

One can then prove by the above method that the identity

$$e^{z+w} = e^z e^w$$

holds for all *complex* numbers z and w.

Example 4. The *Dirichlet product* of two infinite series $\sum_{k=1}^{\infty} a_k$ and $\sum_{k=1}^{\infty} b_k$ is defined to be the infinite series $\sum_{k=1}^{\infty} c_k$, where

$$c_k = \sum_{j \cdot k = n} a_j b_k.$$

If $\sum a_k$ and $\sum b_k$ converge absolutely, prove that their Dirichlet product does also and that

$$\sum_{k=1}^{\infty} c_k = \left(\sum_{k=1}^{\infty} a_k \right) \left(\sum_{k=1}^{\infty} b_k \right).$$

We must return to the theorem on countable additivity. We know that the double series $\sum_{j,k=1}^{\infty} a_j b_k$ is absolutely convergent, or, equivalently, that φ is summable over $\mathbf{Z}^+ \times \mathbf{Z}^+$, where $\varphi_{j,k} = a_j b_k$. If we let

$$\Lambda_n = \{(j, k) \mid j, k \in \mathbf{Z}^+ \text{ and } jk = n\},$$

then the sets $\Lambda_1, \Lambda_2, \ldots$ are disjoint, and their union is $\mathbf{Z}^+ \times \mathbf{Z}^+$. Consequently,

$$\sum_{k=1}^{\infty} \left(\sum_{\Lambda_k} |\varphi| \right) < +\infty.$$

However, $c_k = \sum_{\Lambda_k} \varphi$ and $|c_k| \le \sum_{\Lambda_k} |\varphi|$. It follows that $\sum_{k=1}^{\infty} c_k$ is absolutely convergent. Moreover,

$$\left(\sum_{k=1}^{\infty} a_k \right) \left(\sum_{k=1}^{\infty} b_k \right) = \sum_{\mathbf{Z}^+ \times \mathbf{Z}^+} \varphi = \sum_{k=1}^{\infty} \left(\sum_{\Lambda_k} \varphi \right) = \sum_{k=1}^{\infty} c_k.$$

A *Dirichlet series* is any series of the form

$$\sum_{k=1}^{\infty} a_k k^{-s}$$

where s is variable. In particular, the Riemann zeta function has the Dirichlet series expansion

$$\zeta(s) = \sum_{k=1}^{\infty} k^{-s} \qquad (s > 1).$$

Dirichlet products are related to Dirichlet series in the same way that Cauchy products are related to power series: if the coefficients a_n, b_n, and c_n are related as above, then the Dirichlet product of the two Dirichlet series $\sum_{k=1}^{\infty} a_k k^{-s}$ and $\sum_{k=1}^{\infty} b_k k^{-s}$ is the Dirichlet series $\sum_{k=1}^{\infty} c_k k^{-s}$.

Example 5. If $n \in \mathbf{Z}^+$, we denote the number of distinct factors of n (including the numbers 1 and n) by $d(n)$. For· instance, $d(18) = 6$, since 1, 2, 3, 6, 9, and 18 are the distinct factors of 18. Prove that if $s > 1$, then

$$\zeta(s)^2 = \sum_{k=1}^{\infty} d(k) k^{-s}.$$

We know that

$$\zeta(s) = \sum_{k=1}^{\infty} k^{-s},$$

because the convergence is absolute. By the preceding example,

$$\zeta(s)^2 = \sum_{n=1}^{\infty} \left(\sum_{jk=n} 1 \cdot 1 \right) n^{-s} = \sum_{n=1}^{\infty} d(n) n^{-s}.$$

Example 6. Show that

$$\frac{x}{1+x} + \frac{2x^2}{1+x^2} + \frac{4x^4}{1+x^4} + \frac{8x^8}{1+x^8} + \cdots = \frac{x}{1-x} \quad \text{if } |x| < 1.$$

We expand each term in its Maclaurin series:

$$\frac{x}{1+x} = x - x^2 + x^3 - x^4 + x^5 - x^6 + x^7 - x^8 + \cdots,$$

$$\frac{2x^2}{1+x^2} = 2x^2 \qquad - 2x^4 \qquad + 2x^6 \qquad - 2x^8 + \cdots,$$

$$\frac{4x^4}{1+x^4} = \qquad\qquad 4x^4 \qquad\qquad\qquad - 4x^8 + \cdots,$$

$$\frac{8x^8}{1+x^8} = \qquad\qquad\qquad\qquad\qquad 8x^8 + \cdots,$$

$$\vdots$$

Summing by columns, we get

$$x + x^2 + x^3 + x^4 + \cdots = \frac{x}{1 - x}.$$

To justify this calculation, it suffices to show that if we replace each term of the array by its absolute value, then the resulting double series is convergent by rows. The sum of the absolute values of the terms in the nth row is

$$\frac{2^n |x|^{2^n}}{1 - |x|^{2^n}}.$$

If $|x| < 1$, then the infinite series

$$\sum_{k=1}^{\infty} \frac{2^k |x|^{2^k}}{1 - |x|^{2^k}}$$

is easily seen to converge by the ratio test, since the ratio of the $(n + 1)$th term to the nth term, namely,

$$\frac{2^n |x|^{2^n} (1 - |x|^{2^{n+1}})}{1 - |x|^{2^n}},$$

tends to zero as $n \to \infty$.

Exercises

1. We have encountered many different kinds of limits: limits of sequences, one-sided continuous limits, two-sided limits, etc. In the theory of the Riemann integral, we encountered another type of limit: $\lim_{\|\pi\| \to 0} I_{\pi,\sigma}(f)$ (see p. 325); in the present section, we encounter yet another: $\sum_\Lambda \varphi = \lim \{\sum_F \varphi \mid F \in \mathcal{F}\}$. There is a single notion of limit which generalizes all of these.

   Recall the definition of the limit of a sequence: $\lim_{n \to \infty} a_n = L$ iff for every neighborhood V of L there exists an $N \in \mathbf{Z}^+$ such that for all $n \in \mathbf{Z}^+$,

   $$n > N \Rightarrow a_n \in V.$$

   We shall generalize this definition by replacing the set $\mathbf{Z}^+$ of positive integers by a *directed set*. By the latter we shall mean a set D, together with a relation "$\prec$," satisfying the following two conditions:

   D1. Transitivity. *For all α, β, γ in D, $(\alpha \prec \beta$ and $\beta \prec \gamma) \Rightarrow \alpha \prec \gamma$.*

   D2. *For any two elements α and β in D, there exists an element $\gamma \in D$ such that $\alpha \prec \gamma$ and $\beta \prec \gamma$.*

   If D is such a directed set then we define the limit as follows: $L = \lim_{\alpha \in D} a_\alpha$ iff for each neighborhood V of L there exists an $\alpha_0 \in D$ such that for all $\alpha \in D$,

   $$\alpha_0 \prec \alpha \Rightarrow a_\alpha \in V.$$

   (Here L may be an extended real number or a symbol of the form $a+$ or $a-$, where $a \in \mathbf{R}$.) The indexed set of numbers $\{a_\lambda\}$, which can be thought of as a generalized sequence, is called a *net*. We consider several examples.

a) Let $D = \mathbf{Z}^+$, and let "$m \prec n$" mean "$m < n$." Observe that D is a directed set and that $\lim_D a_n = \lim_{n \to \infty} a_n$.

b) Let $D = \mathbf{R}$ and let "$x \prec y$" mean "$x > y$." Show that D is a directed set and that $\lim_{x \in D} f(x) = \lim_{x \to -\infty} f(x)$.

c) Let $x_0 \in \mathbf{R}$ be fixed. Let $D = (-\infty, x_0)$, and let "$x \prec y$" mean "$x < y$." Show that D is a directed set and that $\lim_{x \in D} f(x) = \lim_{x \to x_0-} f(x)$.

d) Let $x_0 \in \mathbf{R}$ be fixed. Let $D = \mathbf{R} \backslash \{x_0\} = (-\infty, x_0) \cup (x_0, +\infty)$, and let "$x \prec y$" mean "$|y - x_0| < |x - x_0|$." Show that D is a directed set and that

$$\lim_{x \in D} f(x) = \lim_{x \to x_0} f(x).$$

*e) Let a and b be fixed numbers with $a < b$. Let D be the set of all ordered pairs (π, σ), where π is a partition of $[a, b]$ and σ is a selector for π. Let

$$\text{"}(\pi, \sigma) \prec (\pi', \sigma')\text{"}$$

mean "$\|\pi'\| < \|\pi\|$." Show that D is a directed set and that

$$\lim_{(\pi,\sigma) \in D} I_{\pi,\sigma}(f) = \lim_{\|\pi\| \to 0} I_{\pi,\sigma}(f).$$

According to Theorem 5, Section 8–2, f is Riemann-integrable on $[a, b]$ iff this limit exists, in which case the limit is equal to $\int_a^b f$.

**f) Let D be as in (e), but let "$(\pi, \sigma) \prec (\pi', \sigma')$" mean "$\pi'$ is a refinement of π." Prove that f is Riemann-integrable on $[a, b]$ iff $\lim_{(\pi,\sigma) \in D} I_{\pi,\sigma}(f)$ exists, in which case the limit is equal to $\int_a^b f$.

g) Let Λ be any set. Let D be the family of all finite subsets of Λ, and let "$F_0 \prec F$" mean "$F_0 \subset F$." Observe that φ is summable over Λ iff $\lim_{F \in D} \sum_F \varphi$ exists, in which case the limit is $\sum_\Lambda \varphi$.

2. Let D be a directed set, and let $\{a_\alpha\}$ and $\{b_\alpha\}$ be nets directed by D. Suppose that

$$\lim_{\alpha \in D} a_\alpha = A \qquad \text{and} \qquad \lim_{\alpha \in D} b_\alpha = B,$$

where A and B are real numbers.

a) Prove that $\lim_{\alpha \in D} (a_\alpha + b_\alpha) = A + B$.

b) Prove that $\lim_{\alpha \in D} a_\alpha b_\alpha = AB$.

c) Suppose that there exists an $\alpha_0 \in D$ such that $a_\alpha \le b_\alpha$ whenever $\alpha_0 \prec \alpha$. Prove that $A \le B$.

d) Prove that there exists an $\alpha_0 \in D$ such that the set $\{a_\alpha \mid \alpha_0 \prec \alpha\}$ is bounded. Show by example, however, that the entire net $\{a_\alpha\}$ need not be bounded.

3. a) Use Exercise 2 to prove Proposition 3.

b) Use Exercise 2 to prove the analog of Proposition 3 for Riemann integrals.

*4. Formulate and prove an analog of the substitution rule for nets.

5. Give a detailed proof of Lemma 4.

*6. Suppose that Λ is a disjoint union of a family of subsets $\{\Lambda_\alpha \mid \alpha \in A\}$. (In other words, for each $p \in \Lambda$, there is a unique $\alpha \in A$ such that $p \in \Lambda_\alpha$.) Prove that φ is summable over Λ iff

a) φ is summable over Λ_α for each $\alpha \in A$, and

b) the mapping $\sigma: A \to \mathbf{R}$ defined by $\sigma_\alpha = \sum_{A_\alpha} |\varphi|$ is summable over A. Show, moreover, that if φ is summable over Λ, then

$$\sum_\Lambda \varphi = \sum_{\alpha \in \Lambda} \left(\sum_{\Lambda_\alpha} \varphi \right).$$

This generalization of Theorem 8 is sometimes called the *generalized associative law*.

7. Given that $\sum_{k=0}^\infty a_k$ is absolutely convergent, show that

$$\sum_{k=0}^\infty a_k = \sum_{k=0}^9 \left(\sum_{j=0}^\infty a_{10j+k} \right).$$

Observe that this does not follow from the rearrangement theorem of Section 13–3.

8. The limit of a *double sequence* $\{a_{m,n}\}$ is defined as follows: $L = \lim_{m,n \to \infty} a_{m,n}$ iff for every neighborhood V of L there exists an $N \in \mathbf{Z}^+$ such that for all m and n in $\mathbf{Z}^+$,

$$\left. \begin{array}{c} m \geq N \\ n \geq N \end{array} \right\} \Rightarrow a_{m,n} \in V.$$

Show that such a limit can be interpreted as the limit of a net.

9. It is customary to say that a double series $\sum_{j,k=1}^\infty a_{j,k}$ is convergent iff there exists a real number S such that

$$S = \lim_{m,n \to \infty} \sum_{j=1}^m \sum_{k=1}^n a_{jk},$$

in which case S is called the *sum* of the double series.

a) Prove that the double series $\sum_{j,k=1}^\infty a_{jk}$ is absolutely convergent iff the series

$$\sum_{j,k=1}^\infty |a_{jk}|$$

is convergent in the sense just defined.

b) Prove that an absolutely convergent double series converges and that its sum in the sense just defined is equal to its sum by rows, columns, etc.

c) Prove that if the columns and rows of a double series $\sum a_{ij}$ converge, and if in addition the double series itself converges, then each of the iterated series

$$\sum_{j=1}^\infty \left(\sum_{k=1}^\infty a_{jk} \right) \quad \text{and} \quad \sum_{k=1}^\infty \left(\sum_{j=1}^\infty a_{jk} \right)$$

converges, and in each case the sum is equal to the sum of the double series. (This is known as *Pringsheim's theorem*.)

d) Show that the double series

$$(a_0 + b_0) + (a_1 - b_0) + a_2 + a_3 + a_4 + \cdots,$$
$$(-a_0 + b_1) + (-a_1 - b_1) + a_2 - a_3 - a_4 + \cdots,$$
$$b_2 \qquad\qquad - b_2 + 0 + 0 + 0 + \cdots,$$
$$b_3 \qquad\qquad - b_3 + 0 + 0 + 0 + \cdots,$$
$$\vdots$$

always converges with sum zero. Observe, however, that the sum by rows exists iff the series $\sum a_n$ converges. What about the sum by columns? Show that partial sums $\sum_{j=1}^{m} \sum_{k=1}^{n}$ need not be bounded in this case.

e) Show that the double series

$$2 + 0 - 1 + 0 + 0 + 0 + \cdots,$$
$$0 + 2 + 0 - 1 + 0 + 0 + \cdots,$$
$$-1 + 0 + 2 + 0 - 1 + 0 + \cdots,$$
$$0 - 1 + 0 + 2 + 0 - 1 + \cdots,$$
$$0 + 0 - 1 + 0 + 2 + 0 + \cdots,$$
$$0 + 0 + 0 - 1 + 0 + 2 + \cdots,$$
$$\vdots$$

has sum two by columns, rows, and diagonals, but nevertheless diverges.

10. a) Suppose that $\sum_{k=0}^{\infty} a_k$ converges absolutely and that the sequence $\beta_0, \beta_1, \ldots$ tends to zero. Set $r_n = a_0 \beta_n + a_1 \beta_{n-1} + \cdots + a_n \beta_0$. Prove that r_n converges to zero. (Let M be a bound for the terms of the sequence $\beta_0, \beta_1, \ldots$ Let $\epsilon > 0$ be arbitrary. Choose N such that $\sum_{k=N+1}^{\infty} |a_k| \leq \epsilon/N$. If $n > N$, observe that

$$|r_n| \leq |a_0| \, |\beta_n| + \cdots + |a_n| \, |\beta_{n-N}| + \epsilon,$$

and deduce that $\overline{\lim}_{n \to \infty} |r_n| \leq \epsilon$.)

b) Suppose that $\sum_{k=0}^{\infty} a_k$ and $\sum_{k=0}^{\infty} b_k$ are convergent with sums A and B, respectively. Prove that if at least one of the series is absolutely convergent, then the Cauchy product of the two series $\sum_{k=0}^{\infty} c_k$ converges with sum AB. (This is known as *Merten's theorem.*) [Assume that $\sum a_k$ is absolutely convergent. Set

$$A_n = \sum_{k=0}^{n} a_k, \qquad B_n = \sum_{k=0}^{n} b_k, \qquad C_n = \sum_{k=0}^{n} c_k.$$

Show that

$$c_n = A_n B + a_0 \beta_n + a_1 \beta_{n-1} + \cdots + a_n \beta_0,$$

where $\beta_n = B - B_n$. Then apply (a).]

11. Prove that the Cauchy product of the series $\sum_{k=0}^{\infty} (-1)^{k+1}/\sqrt{k+1}$ with itself diverges.

12. "Prove" by series methods that

$$\sin (x + y) = \sin x \cos y + \cos x \sin y$$

for all real numbers x and y.

13. Obtain once again the series expansion for $e^x \sin x$. Use the multiplication theorem.

14. For each $n \in \mathbf{Z}^+$, we let $\sigma(n)$ denote the sum of the integers which divide n. For instance, $\sigma(10) = 18$, since 1, 2, 5, and 10 are the divisors of 10. Prove that

$$\zeta(s)\zeta(s - 1) = \sum_{k=1}^{\infty} \sigma(k) k^{-s} \quad \text{if} \quad s > 2.$$

*15. a) Show that

$$\frac{1}{x(x+1)\cdots(x+n)} = \frac{1}{n!}\frac{1}{x} - \frac{0}{1!(n-1)!}\frac{1}{x+1}$$
$$+ \frac{1}{2!(n-2)!}\frac{1}{x+2} - \cdots \pm \frac{1}{n!}\frac{1}{x+n}.$$

b) Prove *Prym's identity:*

$$\frac{1}{x} + \frac{1}{x(x+1)} + \frac{1}{x(x+1)(x+2)} + \cdots$$
$$= e\left[\frac{1}{x} - \frac{1}{1!}\frac{1}{x+1} + \frac{1}{2!}\frac{1}{x+2} - \cdots\right].$$

For what values of x is the identity valid?

*16. a) Prove that

$$\frac{1}{t^2} = \frac{0!}{t(t+1)} + \frac{1!}{t(t+1)(t+2)} + \frac{2!}{t(t+1)(t+2)(t+3)} + \cdots$$

b) Prove the following identity of Stirling:

$$\frac{1}{t^2} + \frac{1}{(t+1)^2} + \frac{1}{(t+2)^2} + \cdots = \frac{0!}{t} + \frac{1!}{2t(t+1)} + \frac{2!}{3t(t+1)(t+2)} + \cdots$$

For what values of t is it valid?

*17. If $|x| < 1$, show that

$$\frac{x}{(1-x)^2} + \frac{x^2}{(1-x^2)^2} + \frac{x^3}{(1-x^3)^2} + \cdots = \sum_{k=1}^{\infty} \sigma(k)x^k,$$

where, as in Exercise 14, $\sigma(n)$ denotes the sum of the divisors of n.

13–5 Pointwise versus uniform convergence

Suppose that a function f has an infinite series expansion:

$$f(x) = \sum_{k=1}^{\infty} u_k(x).$$

The following questions are then of interest. Is term-by-term differentiation possible? In other words, is

$$f'(x) = \sum_{k=1}^{\infty} u'_k(x)?$$

When will it be true that

$$\int_a^b f(x)\, dx = \sum_{k=1}^{\infty} \int_a^b u_k(x)\, dx?$$

Assuming that the individual terms $u_1(x)$, $u_2(x)$, ... are continuous, is f necessarily continuous? The same questions can be posed if, more generally, f is the limit of a sequence of functions. Here, as in our discussion of double series, the issue is the interchangeability of certain limit operations.

Consider the sequence of functions $f_1, f_2, \ldots$, where $f_n(x) = x^n$. For each $x \in [0, 1]$, the sequence $\{f_n(x)\}$ has a limit:

$$\lim_{n \to \infty} f_n(x) = f(x) = \begin{cases} 0 & \text{if } 0 \le x < 1, \\ 1 & \text{if } x = 1. \end{cases}$$

In this case we say that the sequence $\{f_n\}$ converges *pointwise* (that is, at each point) to f on the interval $[0, 1]$. In this instance each term of the sequence is continuous on the interval $[0, 1]$, whereas the limit function is not (see Fig. 13–2).

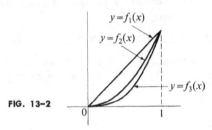

FIG. 13-2

Definition 1. A sequence of real-valued functions $f_1, f_2, \ldots$ is said to **converge pointwise to a function f on a set S** iff

$$(\forall x \in S)\big(\lim_{n \to \infty} f_n(x) = f(x)\big)$$

or, in a more expanded form,

$$(\forall x \in S)(\forall \epsilon > 0)(\exists N \in \mathbf{Z}^+)(\forall n \in \mathbf{Z}^+)(n > N \Rightarrow |f_n(x) - f(x)| < \epsilon).$$

An infinite series $\sum_{k=1}^{\infty} u_k(x)$ is said to **converge pointwise to $f(x)$ on a set S** iff the sequence of partial sums converges pointwise to f on S.

Our example shows that a pointwise limit of continuous functions need not be continuous. There is, however, a stronger kind of convergence under which continuity is preserved; its definition differs from that of pointwise convergence only in the order of the quantifiers.

Definition 2. A sequence of real-valued functions $f_1, f_2, \ldots$ is said to **converge uniformly to a function f on a set S** iff

$$(\forall \epsilon > 0)(\exists N \in \mathbf{Z}^+)(\forall n \in \mathbf{Z}^+)(\forall x \in S)(n > N \Rightarrow |f_n(x) - f(x)| < \epsilon).$$

By the uniform convergence of a series of functions we mean uniform convergence of its sequence of partial sums.

The difference between these two notions of convergence is simply that in the case of pointwise convergence, the N can vary so as to accommodate both ϵ and x, whereas in the case of uniform convergence, it must be possible to select N independently of x. (Compare with continuity versus uniform continuity.)

Example 1. Prove that the sequence of functions $\{x^n\}$ converges uniformly to zero on each interval of the form $[0, b]$, where $0 < b < 1$, but that it does not converge uniformly on either of the intervals $[0, 1]$ or $[0, 1)$.

Let $0 < b < 1$, and let $\epsilon > 0$ be given. Since $\lim_{n \to \infty} b^n = 0$, then there exists an $N \in \mathbf{Z}^+$ such that $|b^n| < \epsilon$ wherever $n > N$. If $x \in [0, b]$ and $n > N$, then

$$|x^n - 0| = x^n \leq b^n < \epsilon.$$

This proves that $\{x^n\}$ converges uniformly to zero on the interval $[0, b]$.

Let us suppose that the sequence converges uniformly on $[0, 1)$ to some function g. Since g must then, in particular, be the pointwise limit of this sequence, g must be identically zero on $[0, 1)$. Hence it suffices to show that the sequence does not converge uniformly to zero on the interval $[0, 1)$, or equivalently, that

$$(\exists \epsilon > 0)(\forall N \in \mathbf{Z}^+)(\exists n \in \mathbf{Z}^+)(\exists x)(n > N, 0 \leq x < 1, \text{ and } |x^n| \geq \epsilon).$$

We set $\epsilon = \frac{1}{2}$. Given $N \in \mathbf{Z}^+$, we can let $n = N + 1$, and we can let x be any number in the interval $(2^{-(1/n)}, 1)$. Then $n > N$, $0 \leq x < 1$, and $|x^n| = x^n > \frac{1}{2} = \epsilon$.

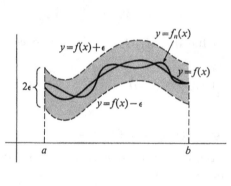

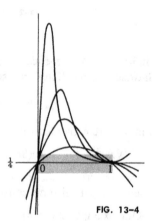

FIG. 13–3   FIG. 13–4

Uniform convergence is easily interpreted geometrically. Suppose that the functions $f, f_1, f_2, \ldots$ have as domain the interval $[a, b]$. To say that the sequence f_1, $f_2, \ldots$ converges uniformly to f on $[a, b]$ means the following: If we surround the graph of f by any band of constant width, say 2ϵ, in the manner indicated in Fig. 13–3, then for sufficiently large n, the graph of f_n will lie entirely inside this band. Glancing back at the graphs of the functions x^n, it is clear that the sequence does not converge uniformly on the interval $[0, 1]$. A more striking example is provided by the sequence of functions $f_1, f_2, \ldots$ with domain $[0, 1]$, where $f_n(x) = n^2 x(1 - x)^n$. In this case the sequence converges pointwise to zero. Yet if we surround the x-axis with a band of width $\frac{1}{4}$, then absolutely none of the graphs will be wholly within this band, as shown in Fig. 13–4.

Suppose that we wish to calculate the values of some function $f(x)$ to ten-place accuracy for all values of x in some interval $[a, b]$. If we have a sequence of functions $f_1, f_2, \ldots$ converging uniformly to f on $[a, b]$, then we can select an n such that $f_n(x)$ agrees with $f(x)$ to ten decimal places for each $x \in [a, b]$. The function f_n can then be used to replace f, for the purpose of the calculation. Should f_n be a simpler function than f (such as a polynomial of fairly low degree), the replacement is advantageous. In this context the appropriate measure of the deviation of f_n from f is given by the number $\|f - f_n\|$, where the symbol $\|\ \ \|$ is defined in the following way.

Definition 3. If g is a bounded real-valued function on an interval $[a, b]$, then the **uniform norm of g on the interval** is defined to be

$$\|g\| = \text{lub } \{|g(x)| \mid a \leq x \leq b\}.$$

The following is an easy exercise.

Proposition 4. *A sequence of functions $f_1, f_2, \ldots$ converges uniformly to a function f on an interval $[a, b]$ iff*

$$\lim_{n \to \infty} \|f_n - f\| = 0.$$

Example 2. Prove once again that the sequence $\{x^n\}$ does not converge uniformly on the interval $[0, 1]$.

We set $f_n(x) = x^n$. Since $\{f_n\}$ converges pointwise to the function

$$f(x) = \begin{cases} 0 & \text{if } 0 \leq x < 1, \\ 1 & \text{if } x = 1 \end{cases}$$

on the interval $[0, 1]$, it suffices to show that $\{f_n\}$ does not converge uniformly to f on the interval $[0, 1]$. Now

$$f_n(x) - f(x) = \begin{cases} x^n & \text{if } 0 \leq x < 1, \\ 0 & \text{if } x = 1. \end{cases}$$

Since $\|f_n - f\| = \text{lub } \{|f_n(x) - f(x)| \mid 0 \leq x \leq 1\} = 1$ for each $n \in \mathbf{Z}^+$,

$$\lim_{n \to \infty} \|f_n - f\| = 1 \neq 0.$$

Hence f_n does not converge uniformly to f on the interval $[0, 1]$.

Example 3. Let $f_n(x) = n^2 x (1 - x)^n$. Prove that $\{f_n\}$ does not converge uniformly on $[0, 1]$.

The sequence converges pointwise to zero on the interval $[0, 1]$. Hence it suffices to show that $\{f_n\}$ does not converge uniformly to zero on $[0, 1]$. For each $n \in \mathbf{Z}^+$, $\|f_n - 0\|$ is just the maximum value of f_n on the interval $[0, 1]$, and the latter is easily seen to occur at the point $1/(n + 1)$. Thus

$$\|f_n\| = f_n\left(\frac{1}{n + 1}\right) = n\left(1 - \frac{1}{n + 1}\right)^{n+1} \sim e^{-1} n.$$

Since $\|f_n\|$ tends to $+\infty$ rather than zero as $n \to \infty$, uniform convergence of $\{f_n\}$ to zero is clearly out of the question.

Example 4. If f is continuous on the interval $[a, b]$, prove that f is the uniform limit of a sequence of step functions.

Let $n \in \mathbf{Z}^+$. Since f is uniformly continuous on the interval, there exists a $\delta > 0$ such that

$$|f(x) - f(x')| < 1/n$$

whenever x and x' are points in $[a, b]$ such that $|x - x'| < \delta$, We now choose points

$$a = x_0 < x_1 < x_2 < \cdots < x_r = b$$

such that $x_k - x_{k-1} < \delta$ $(k = 1, \ldots, r)$. We let f_n be the step function

$$f_n = \sum_{k=1}^{r-1} f(x_{k-1}) \varphi_{[x_{k-1}, x_k)} + f(x_{r-1}) \varphi_{[x_{r-1}, x_r]}.$$

It can then be verified that $\|f_n - f\| \le 1/n$. [Suppose $a \le x \le b$. Then for some k,

$$x_{k-1} \le x \le x_k$$

or else $x = b$. In the first case,

$$|f_n(x) - f(x)| = |f(x_{k-1}) - f(x)| < 1/n,$$

since $|x_{k-1} - x| < \delta$; in the second case,

$$|f_n(b) - f(b)| = |f(x_{r-1}) - f(x_r)| < 1/n.$$

Hence $\|f_n - f\| = \text{lub } \{|f_n(x) - f(x)| \mid a \le x \le b\} \le 1/n.]$ Clearly, $\|f_n - f\| \to 0$, which proves that $\{f_n\}$ converges uniformly to f.

We now proceed to the development of the theory of uniform convergence. We observe first that the uniform norms behave much like absolute values.

Proposition 5. Properties of Uniform Norms. *Let f and g be bounded real-valued functions on the interval $[a, b]$, and let $c \in \mathbf{R}$. Then we have the following properties.*

1) Triangle inequality. $\|f + g\| \le \|f\| + \|g\|$.

2) Homogeneity. $\|cf\| = |c| \|f\|$.

3) Positivity. $\|f\| \ge 0$, *with strict inequality unless $f \equiv 0$.*

Proof. For each $x \in [a, b]$,

$$|f(x) + g(x)| \le |f(x)| + |g(x)| \le \|f\| + \|g\|.$$

Hence

$$\|f + g\| = \text{lub } \{|f(x) + g(x)| \mid a \le x \le b\} \le \|f\| + \|g\|.$$

The rest is trivial.

Proposition 6. Cauchy Criterion. *A sequence of functions $\{f_n\}$ converges uniformly on $[a, b]$ iff*

$$\lim_{m, n \to \infty} \|f_m - f_n\| = 0.$$

Proof. Suppose $f_n \to f$ uniformly on $[a, b]$. Then $\lim_{n \to \infty} \|f_n - f\| = 0$. Since

$$\|f_m - f_n\| = \|(f_m - f) + (f - f_n)\| \leq \|f_m - f\| + \|f_n - f\|,$$

it follows immediately that $\lim_{m,n \to \infty} \|f_m - f_n\| = 0$.

Suppose now that $\lim_{m,n \to \infty} \|f_m - f_n\| = 0$. For each $x \in [a, b]$,

$$|f_m(x) - f_n(x)| \leq \|f_m - f_n\|,$$

from which it is clear that $\{f_n(x)\}$ is Cauchy. We set

$$f(x) = \lim_{n \to \infty} f_n(x) \qquad (a \leq x \leq b).$$

We shall now prove that f_n converges uniformly to f on $[a, b]$. Let $\epsilon > 0$ be given. We choose N such that

$$\left.\begin{array}{c} m \\ n \end{array}\right\} > N \Rightarrow \|f_m - f_n\| < \epsilon.$$

Let $n > N$ and $x \in [a, b]$. Then

$$|f_n(x) - f(x)| = \lim_{m \to \infty} |f_n(x) - f_m(x)|.$$

Since $|f_n(x) - f_m(x)| \leq \|f_m - f_n\| < \epsilon$, we have

$$|f_n(x) - f(x)| \leq \epsilon.$$

Since this holds uniformly for $x \in [a, b]$, we have shown that $f_n \to f$ uniformly on $[a, b]$.

Theorem 7. *Suppose that $f_n \to f$ uniformly throughout some neighborhood V of $x_0 \in \mathbf{R}$, and suppose that each function f_n is continuous at x_0. Then f is continuous at x_0.*

Proof. Let $\epsilon > 0$ be given. We can then choose $n \in \mathbf{Z}^+$ such that

$$|f_n(x) - f(x)| < \epsilon/3$$

for all $x \in V$. Since f_n is continuous at x_0, there exists a neighborhood $U \subset V$ of x_0 such that

$$x \in U \Rightarrow |f_n(x) - f_n(x_0)| < \epsilon/3.$$

Consequently, if $x \in U$, we get, via the triangle equality,

$$|f(x) - f(x_0)| \leq |f(x) - f_n(x)| + |f_n(x) - f_n(x_0)|$$
$$+ |f_n(x_0) - f(x_0)| < \epsilon/3 + \epsilon/3 + \epsilon/3 = \epsilon.$$

Therefore, f is continuous at x_0.

The same proof may be given for a modified form of the theorem in which x_0 is replaced by x_0+ or x_0- and the term "continuous" is replaced by "right-continuous" or "left-continuous," respectively. In particular, we have the following corollary.

Corollary 8. *If $f_n \to f$ uniformly on $[a, b]$, and if each of the functions $f_1, f_2, \ldots$ is continuous on $[a, b]$, then f is continuous on $[a, b]$.*

A series version of this corollary easily follows.

Corollary 8'. *If we have*

$$f(x) = \sum_{k=1}^{\infty} u_k(x),$$

where the convergence is uniform on $[a, b]$, and if each of the terms $u_1, u_2, \ldots$ is continuous on $[a, b]$, then f is continuous on $[a, b]$.

Proof. Apply the previous corollary with $f_n = \sum_{k=1}^{n} u_k$.

It is an easy matter to show that under the hypotheses of these corollaries we have

$$\int_a^b f(x)\, dx = \lim_{n \to \infty} \int_a^b f_n(x)\, dx$$

and

$$\int_a^b f(x)\, dx = \sum_{k=1}^{\infty} \int_a^b u_k(x)\, dx.$$

Let us prove the first assertion. We have

$$\left| \int_a^b f_n(x)\, dx - \int_a^b f(x)\, dx \right| = \left| \int_a^b [f_n(x) - f(x)]\, dx \right|$$

$$\le \int_a^b |f_n(x) - f(x)|\, dx \le \int_a^b \|f_n - f\|\, dx$$

$$= \|f_n - f\|(b - a).$$

Since the last member in this string of inequalities tends to zero as $n \to \infty$, so does the first. Thus

$$\lim_{n \to \infty} \int_a^b f_n(x)\, dx = \int_a^b f(x)\, dx.$$

The second assertion follows from the first:

$$\int_a^b f(x)\, dx = \lim_{n \to \infty} \int_a^b \left(\sum_{k=1}^{n} u_k(x) \right) dx$$

$$= \lim_{n \to \infty} \sum_{k=1}^{n} \left(\int_a^b u_k(x)\, dx \right) = \sum_{k=1}^{\infty} \int_a^b u_k(x)\, dx.$$

In these proofs we have assumed, of course, that continuous functions are integrable (in either the Newton or the Riemann sense). We shall next prove a theorem which has as corollaries not only the two results just established, but also the fact that continuous functions are Newton-integrable.

Theorem 9. *Suppose that $f_n \to f$ uniformly on $[a, b]$. Suppose that each of the functions f_n is Newton-integrable on $[a, b]$ and that the function f is continuous at all points of*

$[a, b]$, *with at most a finite number of exceptions. Then f is Newton-integrable on $[a, b]$ and*

$$\int_a^b f(x)\,dx = \lim_{n \to \infty} \int_a^b f_n(x)\,dx.$$

Proof. We set

$$F_n(x) = \int_a^x f_n(x)\,dx \qquad (n \in \mathbf{Z}^+, a \le x \le b).$$

Then F_n is an antiderivative of f_n, so, in particular, it is continuous on $[a, b]$. Since

$$
\begin{aligned}
|F_m(x) - F_n(x)| &= \left| \int_a^x [f_m(t) - f_n(t)]\,dt \right| \\
&\le \int_a^x |f_m(t) - f_n(t)|\,dt \\
&\le \int_a^x \|f_m - f_n\|\,dt = (x - a)\|f_m - f_n\| \\
&\le (b - a)\|f_m - f_n\|
\end{aligned}
$$

for all $x \in [a, b]$, we have

$$\|F_m - F_n\| \le (b - a)\|f_m - f_n\|.$$

By the Cauchy criterion, the sequence $\{F_n\}$ converges uniformly on $[a, b]$ to some function F, which, by the first corollary to the preceding theorem, must be continuous on $[a, b]$. We will now show that F is an antiderivative of f. It suffices to show that if $a < x_0 < b$ and f is continuous at x_0, then $F'(x_0) = f(x_0)$. To this end we consider

$$e(h) = \frac{F(x_0 + h) - F(x_0)}{h} - f(x_0).$$

Now

$$
\begin{aligned}
e(h) &= \lim_{n \to \infty} \frac{F_n(x_0 + h) - F_n(x_0)}{h} - \frac{1}{h} \int_{x_0}^{x_0+h} f(x_0)\,dt \\
&= \lim_{n \to \infty} \frac{1}{h} \int_{x_0}^{x_0+h} [f_n(t) - f(x_0)]\,dt.
\end{aligned}
$$

Also

$$
\begin{aligned}
|f_n(t) - f(x_0)| &\le |f_n(t) - f(t)| + |f(t) - f(x_0)| \\
&\le \|f_n - f\| + |f(t) - f(x_0)|.
\end{aligned}
$$

Hence, if we let $\omega(h)$ denote the least upper bound of $|f(t) - f(x_0)|$ for values of t between x_0 and $x_0 + h$, we have

$$\left| \frac{1}{h} \int_{x_0}^{x_0+h} [f_n(t) - f(x_0)]\,dt \right| \le \|f_n - f\| + \omega(h).$$

Thus

$$|e(h)| \le \lim_{n \to \infty} [\|f_n - f\| + \omega(h)] = \omega(h).$$

Since f is continuous at x_0, it follows that $\lim_{h\to 0} \omega(h) = 0$, so that $\lim_{h\to 0} e(h) = 0$, or equivalently,

$$F'(x_0) = \lim_{h\to 0} \frac{F(x_0 + h) - F(x_0)}{h} = f(x_0).$$

We have established that f is Newton-integrable on $[a, b]$. Finally,

$$\int_a^b f(x)\, dx = F(b) = \lim_{n\to\infty} F_n(b) = \lim_{n\to\infty} \int_a^b f_n(x)\, dx.$$

This theorem has several important corollaries. First we have the fact that continuous functions possess antiderivatives.

Corollary 10. *If f is continuous on $[a, b]$, then f possesses an antiderivative on $[a, b]$.*

Proof. This follows from the theorem if we let $\{f_n\}$ be any sequence of step functions converging uniformly to f.

Corollary 11. *If $\{f_n\}$ is a sequence of continuous functions on $[a, b]$, and if $f_n \to f$ uniformly on $[a, b]$, then*

$$\int_a^b f(x)\, dx = \lim_{n\to\infty} \int_a^b f_n(x)\, dx.$$

In particular, if

$$f(x) = \sum_{k=1}^{\infty} u_k(x),$$

where the convergence of the series is uniform on $[a, b]$ and each term is continuous on $[a, b]$, then

$$\int_a^b f(x)\, dx = \sum_{k=1}^{\infty} \int_a^b u_k(x)\, dx.$$

We also get a theorem concerning term-by-term differentiation.

Corollary 12. *Suppose that $\{\varphi_n\}$ is a sequence of functions satisfying the following conditions:*

1) *Each function φ_n possesses a continuous derivative throughout an open interval J.*

2) *There exists a point $x_0 \in J$ such that the sequence $\{\varphi_n(x_0)\}$ converges.*

3) *The sequence of derivatives $\{\varphi_n'\}$ converges to a function ψ uniformly on every closed bounded subinterval of J.*

Then the sequence φ_n converges to a function φ uniformly on every closed bounded subinterval of J, the function φ is differentiable, and

$$\varphi'(x) = \psi(x) = \lim_{n\to\infty} \varphi_n'(x).$$

Proof. If we let

$$F_n(x) = \int_{x_0}^{x} \varphi_n'(t) \, dt = \varphi_n(x) - \varphi_n(x_0),$$

then it follows, as in the proof of the theorem, that F_n converges uniformly to some function F on every closed bounded subinterval of J. Moreover, $F'(x) = \psi(x)$ for each $x \in J$. If we set $\varphi(x) = F(x) + A$, where $A = \lim_{n \to \infty} \varphi_n(x_0)$, then it is easy to check that φ has the required properties.

The series version may be formulated as follows.

Corollary 12'. *Suppose that*

1) *for each* $k \in \mathbf{Z}^+$, u_k *possesses a continuous derivative throughout an open interval J;*

2) *for some* $x_0 \in J$, *the series* $\sum_{k=1}^{\infty} u_k(x_0)$ *converges;*

3) *the derived series* $\sum_{k=1}^{\infty} u_k'(x)$ *converges uniformly to a function g on every closed bounded subinterval of J.*

Then the infinite series $\sum_{k=1}^{\infty} u_k$ *converges uniformly to a function f on every closed bounded subinterval of J, the function f is differentiable, and*

$$f'(x) = \sum_{k=1}^{\infty} u_k'(x)$$

for each $x \in J$.

Before giving applications of these theorems, we present one of the most useful tests for uniform convergence.

Theorem 13. Weierstrass M-test. *Let* $u_1, u_2, \ldots$ *be a sequence of functions on a set S, and suppose that there exists a sequence of constants $M_1, M_2, \ldots$ such that*

1) $|u_k(x)| \le M_k$ *for all* $k \in \mathbf{Z}^+$ *and* $x \in S$, *and*

2) $\sum_{k=1}^{\infty} M_k < +\infty$.

Then the series $\sum_{k=1}^{\infty} u_k$ *converges uniformly (and absolutely) on S.*

Proof. Let

$$S_n(x) = \sum_{k=1}^{n} u_k(x) \qquad \text{and} \qquad T_n = \sum_{k=1}^{n} M_k.$$

Then if $n > m$, we have

$$|S_n(x) - S_m(x)| = \left| \sum_{k=m+1}^{n} u_k(x) \right| \le \sum_{k=m+1}^{n} M_k = |T_n - T_m|$$

for all $x \in S$. Hence

$$\|S_n - S_m\| \le |T_n - T_m|.$$

Since $\{T_n\}$ is a Cauchy sequence,

$$\lim_{m,n \to \infty} \|S_n - S_m\| = 0.$$

Thus $\{S_n\}$ converges uniformly on S.

Example 5. Prove that for each $x \in \mathbf{R}$,

$$\int_0^x \frac{\sin t}{t}\, dt = x - \frac{1}{3!}\frac{x^3}{3} + \frac{1}{5!}\frac{x^5}{5} - \cdots$$

Since all the terms appearing are odd, we may assume that $x > 0$. We know of course that

$$\sin t = t - \frac{t^3}{3!} + \frac{t^5}{5!} - \cdots,$$

so that

$$\frac{\sin t}{t} = 1 - \frac{t^2}{3!} + \frac{t^4}{5!} - \cdots,$$

where we understand the left-hand side to be one when $t = 0$. If $0 \leq t \leq x$, then each term of the series is in absolute value less than the corresponding term of the series

$$1 + \frac{x^2}{3!} + \frac{x^4}{5!} + \cdots$$

The latter series converges to $(\sinh x)/x$. Hence by the Weierstrass M-test, the series

$$1 - \frac{t^2}{3!} + \frac{t^4}{5!} - \cdots$$

converges uniformly to $(\sin t)/t$ on the interval $[0, x]$. Term-by-term integration is therefore justified:

$$\int_0^x \frac{\sin t}{t}\, dt = \int_0^x dt - \int_0^x \frac{t^2}{3!}\, dt + \int_0^x \frac{t^4}{5!}\, dt - \cdots$$

$$= x - \frac{1}{3!}\frac{x^3}{3} + \frac{1}{5!}\frac{x^5}{5} - \cdots$$

Example 6. Prove once again that

$$e^x = 1 + x + \frac{x^2}{2!} + \frac{x^3}{3!} + \cdots$$

We observe first that the series converges uniformly on every closed bounded interval $[a, b]$. If $a \leq x \leq b$, then each term of the series is in absolute value less than the corresponding term of the series

$$1 + M + \frac{M^2}{2!} + \frac{M^3}{3!} + \cdots,$$

where $M = \max(|a|, |b|)$. Since the latter series is seen to converge by the ratio test, it follows by the Weierstrass theorem that the given series converges uniformly on $[a, b]$.

Let $E(x) = 1 + x + x^2/2! + \cdots$ The series obtained formally by term-by-term differentiation is the given series itself. The hypotheses of our theorem concerning the differentiation of series are satisfied, and we may therefore conclude that $E'(x) = E(x)$ for all $x \in \mathbf{R}$. Also, $E(0) = 1$. Since the exponential function e^x is the only function having these two properties, we conclude that

$$e^x = E(x) = 1 + x + \frac{x^2}{2!} + \frac{x^3}{3!} + \cdots$$

Example 7. Discuss regularity properties of the Riemann zeta function

$$\zeta(s) = \sum_{k=1}^{\infty} k^{-s} \qquad (s > 1).$$

If we take the rth derivative of each term of the series, we get the series

$$\sum_{k=1}^{\infty} (-\ln k)^r k^{-s}.$$

We shall now prove that the series converges uniformly on every interval of the form $[a, +\infty)$, where $1 < a$. If $s \in [a, +\infty)$, then

$$|(-\ln k)^r k^{-s}| \leq (\ln k)^r k^{-a}.$$

Since the series $\sum_{k=1}^{\infty} (\ln k)^r k^{-a}$ converges (by comparison with the series $\sum_{k=1}^{\infty} k^{-a'}$, where $1 < a' < a$), we may conclude via the Weierstrass M-test that the rth derived series,

$$\sum_{k=1}^{\infty} (-\ln k)^r k^{-s},$$

does indeed converge uniformly on the interval $[a, +\infty)$.

Setting $r = 1$, we are assured by our theorem on term-by-term differentiation that ζ is differentiable and that

$$\zeta'(s) = \sum_{k=1}^{\infty} (-\ln k)k^{-s} \qquad (s > 1).$$

Setting $r = 2$, we next conclude that

$$\zeta''(s) = \sum_{k=1}^{\infty} (-\ln k)^2 k^{-s} \qquad (s > 1).$$

By induction on r, we get, more generally,

$$\zeta^{(r)}(s) = \sum_{k=1}^{\infty} (-\ln k)^r k^{-s} \qquad (s > 1).$$

This proves that the zeta function possesses continuous derivatives of all orders throughout the interval $(0, +\infty)$.

Example 8. Here we shall apply the Weierstrass M-test to produce an example of a continuous but nowhere differentiable function.

We begin by defining $\{x\}$ to be the distance from x to the nearest integer. The function so obtained is a continuous function with period one having a sawtooth graph, as shown in Fig. 13–5. We then define

$$f(x) = \sum_{k=0}^{\infty} 10^{-k}\{10^k x\}.$$

Since each term is in absolute value less than the corresponding term of the convergent geometric series $\sum_{k=0}^{\infty} 10^{-k}$, the Weierstrass M-test tells us that the series converges uniformly

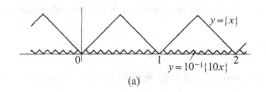

$$y = \{x\}$$

$$y = 10^{-1}\{10x\}$$

(a)

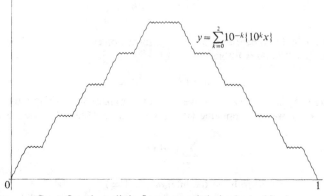

$$y = \sum_{k=0}^{2} 10^{-k}\{10^{k}x\}$$

"Great fleas have little fleas upon their backs to bite 'em
And little fleas have lesser fleas and so *ad infinitum*."

A. De Morgan

FIG. 13–5
(b)

on the entire real line. Since each term is continuous and has period one, the function f also has these two properties.

We shall now prove that for each $x \in \mathbf{R}$, $f'(x)$ does not exist. In each instance we shall produce a sequence $\{h_n\}$ converging to zero such that the sequence of difference quotients $\{[f(x + h_n) - f(x)]/h_n\}$ does not have a limit. Since f has period one, we may assume that $0 \leq x < 1$. Let $x = 0.a_1a_2a_3 \ldots$ be a decimal expansion of x. Observe that

$$\{10^n x\} = \{0.a_{n+1}a_{n+2} \ldots\}$$

and thus

$$\{10^n x\} = \begin{cases} 0.a_{n+1}a_{n+2} \ldots & \text{if} \quad 0.a_{n+1}a_{n+2} \ldots \leq \tfrac{1}{2}, \\ 1 - 0.a_{n+1}a_{n+2} \ldots & \text{otherwise.} \end{cases}$$

We define

$$h_n = \begin{cases} -10^{-n} & \text{if} \quad a_n \text{ is 4 or 9,} \\ 10^{-n} & \text{otherwise.} \end{cases}$$

Then

$$r_n = \frac{f(x + h_n) - f(x)}{h_n} = 10^n \sum_{k=0}^{\infty} \pm \frac{\{10^k(x + h_n)\} - \{10^k x\}}{10^k}.$$

If $k \geq n$, then $10^k h_n$ is an integer, so $\{10^k(x + h_n)\} - \{10^k x\} = 0$. If $k < n$, then because of the way in which h_n was defined, the numbers $\{10^k(x + h_n)\}$ and $\{10^k x\}$ lie on the same side of the number $\tfrac{1}{2}$; hence their difference is $\pm 10^k h_n = \pm 10^{k-n}$. It follows that in our series for r_n, the first n terms are ± 1 and the remaining terms are zero. It follows that r_n is an integer which is even or odd according as n is even or odd. Since $\{r_n\}$ is a sequence of integers which are alternately even and odd, it can have no limit, and the proof is complete.

Geometrically, we have formed the function f by stacking little peaks on top of each other. The result is that the graph of f is infinitely crinkly, but has no breaks.

Example 9. Show that the series

$$x - \frac{x^3}{3} + \frac{x^5}{5} - \cdots$$

converges uniformly but not absolutely in the interval $[-1, 1]$.

We know that the series converges pointwise to $\text{Tan}^{-1} x$ throughout the interval. If $s_n(x)$ is the nth partial sum of the series, then our argument on p. 381 shows that

$$\|\text{Tan}^{-1} - s_n\| = \text{lub } \{|\text{Tan}^{-1} x - s_n(x)| \mid |x| \leq 1\} \leq \frac{1}{2n + 1}.$$

Since the latter term tends to zero as $n \to \infty$, we conclude that the series converges uniformly to $\text{Tan}^{-1} x$ on the interval $[-1, +1]$. The series does not, however, converge absolutely at the endpoints of the interval.

The above example shows that uniform convergence does not imply absolute convergence. Consequently, the Weierstrass M-test, while extremely useful, cannot always be used to prove the uniform convergence of a series. There are tests for establishing conditional uniform convergence analogous to Dirichlet's and Abel's tests which are occasionally useful. These statements require a couple of preliminary definitions.

Definition 14. A sequence of functions $f_1, f_2, \ldots$ is **uniformly bounded** on a set S iff there exists an M such that

$$|f_n(x)| \leq M$$

for all $x \in S$ and $n \in \mathbf{Z}^+$.

Definition 15. A sequence of real-valued functions $f_1, f_2, \ldots$ is **nonincreasing** on S iff $f_1(x) \geq f_2(x) \geq f_3(x) \geq \cdots$ for all $x \in S$.

Proposition 16. Dirichlet's Test for Uniform Convergence. *Suppose that the partial sums of the series $\sum u_k(x)$ are uniformly bounded on a set S and that the sequence $v_1(x), v_2(x), \ldots$ is nondecreasing and tends uniformly to zero on S. Then $\sum u_k(x) v_k(x)$ converges uniformly on S.*

Proposition 17. Abel's Test for Uniform Convergence. *Suppose $\sum_{k=1}^{\infty} u_k(x)$ converges uniformly on S. Suppose also that $v_1(x), v_2(x), \ldots$ is a sequence of functions which are uniformly bounded, nonincreasing, and nonnegative on S. Then*

$$\sum_{k=1}^{\infty} u_k(x) v_k(x)$$

converges uniformly on S.

The proofs involve no essentially new ideas, so they are omitted. In the next section we shall use the following corollary to Abel's test.

Corollary 18. *Suppose*

    1) $\sum_{k=1}^{\infty} a_k$ *converges,*

    2) $M \geq v_1(x) \geq v_2(x) \geq \cdots \geq 0$ *for all* $x \in S.$

Then $\sum_{k=1}^{\infty} a_k v_k(x)$ *converges uniformly on* $S.$

Example 10. Show that the series

$$\sin x + \frac{\sin 3x}{3} + \frac{\sin 5x}{5} + \cdots$$

converges uniformly on every interval of the form $[\epsilon, 2\pi - \epsilon]$, where $\delta < \epsilon < \pi$.
Since

$$\sin x + \sin 3x + \cdots + \sin (2n - 1) = \frac{\sin^2 nx}{\sin x},$$

we have

$$\left| \sum_{k=1}^{n} \sin kx \right| \leq \frac{1}{|\sin x|}.$$

From the above inequality, it is clear that the partial sums of the series $\sin x + \sin 3x + \cdots$ will be uniformly bounded on any interval of the sort considered. Since the sequence $1, \frac{1}{3}, \frac{1}{5}, \ldots$ tends monotonically to zero, Dirichlet's test applies.

Exercises

1. In Example 3 we observed that the sequence of functions defined by $f_n(x) = n^2 x(1 - x)^n$ converges pointwise to zero on the interval $[0, 1]$. Prove that

$$\lim_{n \to \infty} \int_0^1 f_n(x)\, dx \neq \int_0^1 [\lim_{n \to \infty} f_n(x)]\, dx.$$

2. Given $f_n(x) = x^{2n}/(1 + x^{2n})$, prove that f_n converges pointwise, but not uniformly, on the entire line.

3. Let $f_n(x) = (\sin nx)/\sqrt{n}$. Show that f_n converges uniformly to zero on the entire line. Prove, however, that $\lim_{n \to \infty} f_n'(x)$ does not exist for any x. (Why does this not violate Corollary 12?)

4. Show that as $n \to \infty$, $(1 + x/n)^n$ converges uniformly to e^x on every closed bounded interval. Is the convergence uniform on the entire line?

5. Let $f_n(x) = x/(1 + nx^2)$. Show that f_n converges uniformly to a function f and that $f'(x) = \lim_{n \to \infty} f_n'(x)$ except if $x = 0$.

6. Prove Proposition 4.

7. Suppose that $f_n \to f$ uniformly on a set S and that each of the functions is bounded on S. Prove that f is bounded on S.

8. a) Prove that if $f_n \to f$ and $g_n \to g$ uniformly on a set S, then $(f_n + g_n) \to (f + g)$ uniformly on S.

   b) Prove that $f_n g_n \to fg$ uniformly on S, provided that the functions $f_1, f_2, \ldots, g_1, g_2, \ldots$ are bounded on S.

   c) Show by the construction of a counterexample that the assumption of boundedness in part (b) is essential.

*9. Suppose that $\{f_n\}$ is a sequence of functions which are Riemann-integrable on the interval $[a, b]$, and suppose that $f_n \to f$ uniformly on $[a, b]$. Prove that f is Riemann-integrable on $[a, b]$ and that $\int_a^b f(x)\, dx = \lim_{n \to \infty} \int_a^b f_n(x)\, dx$. (Use the necessary and sufficient conditions for Riemann integrability given in Section 8-6.)

10. A function f is said to be *regulated* on $[a, b]$ iff there exists a sequence of step functions which converges uniformly to f on $[a, b]$.

   a) Using previous theorems and exercises, prove that the class of regulated functions on $[a, b]$ is closed under addition, multiplication, the lattice operations, and the formation of uniform limits. Observe also that every piecewise continuous function is regulated.

   b) Observe that every regulated function on $[a, b]$ is Riemann-integrable on the interval. Let I denote the Riemann integral restricted to the class of regulated functions on $[a, b]$. Observe that I is a definite integral on $[a, b]$ (in the sense of Chapter 8). We shall call I the *Bourbaki integral* on $[a, b]$.

   c) Show that the function

$$f(x) = \begin{cases} \sin (1/x) & \text{if } x \neq 0, \\ 0 & \text{if } x = 0 \end{cases}$$

   is Riemann-integrable on $[0, 1]$ but not regulated on the interval.

11. The purpose of this exercise is to develop the theory of the Bourbaki integral from scratch (that is, without using the Riemann theory). Let I_0 be the unique definite integral on $[a, b]$ whose domain is the class of all step functions on the interval. If f is regulated on $[a, b]$, define the Bourbaki integral of f as follows. Let $\{f_n\}$ be any sequence of step functions converging uniformly to f on $[a, b]$. Show that the sequence of numbers $\{I_0(f_n)\}$ is Cauchy. Define $I(f) = \lim_{n \to \infty} I_0(f_n)$ and observe that this limit is independent of a particular choice of the step functions $\{f_n\}$. (Suppose that $\{g_n\}$ is a second sequence of step functions converging uniformly to f, and then consider the interlaced sequence $f_1, g_1, f_2, g_2, \ldots$) Prove directly that the function I so defined is a definite integral on $[a, b]$.

12. a) If a and b are positive, prove that

$$\frac{1}{a} - \frac{1}{a + b} + \frac{1}{a + 2b} - \frac{1}{a + 3b} + \cdots = \int_0^1 \frac{t^{a-1}}{1 + t^b}\, dt.$$

   (This result is due to Gauss.) Prove this by generalizing the argument used in first establishing the power series expansion for $\text{Tan}^{-1} x$.

   b) Show that $1 - \frac{1}{4} + \frac{1}{7} - \frac{1}{10} + \cdots = \frac{1}{3}(\pi/\sqrt{3} + \ln 2)$.

   c) Show that $\frac{1}{2} - \frac{1}{5} + \frac{1}{8} - \frac{1}{11} + \cdots = \frac{1}{3}(\pi/\sqrt{3} - \ln 2)$.

   d) Show that $1 - \frac{1}{5} + \frac{1}{9} - \frac{1}{13} + \cdots = (1/4\sqrt{2})[\pi + 2\ln(\sqrt{2} + 1)]$.

*13. Show that

$$\frac{1}{2} - \frac{1 \cdot 3}{2 \cdot 4} + \frac{1 \cdot 3 \cdot 5}{2 \cdot 4 \cdot 6} - \cdots = \frac{2 - \sqrt{2}}{2}.$$

[Use the formula

$$\int_0^{\pi/2} \cos^{2n} x \, dx = \frac{\pi}{2} \frac{1 \cdot 3 \cdot 5 \cdots (2n - 1)}{2 \cdot 4 \cdot 6 \cdots (2n)},$$

integrate term by term, etc.]

14. Prove that the following series converge uniformly on the entire line.

a) $\displaystyle\sum_{k=1}^{\infty} \frac{\sin kx}{k!}$

    b) $\displaystyle\sum_{k=1}^{\infty} \frac{1}{n^2[1 + (nx)^2]}$

c) $\displaystyle\sum_{k=1}^{\infty} \frac{\cos kx}{3^k}$

    d) $\displaystyle\sum_{k=1}^{\infty} \frac{e^{-k^2 x^2}}{k(k + 1)}$

15. If $\omega > 0$, we define

$$f(x) = \sum_{k=0}^{\infty} \frac{1}{1 + (x + k\omega)^2} + \sum_{k=1}^{\infty} \frac{1}{1 + (x - k\omega)^2}.$$

Show that f is defined for every real number x. Show also that f is a continuous function with period ω.

16. For positive x, we define

$$f(x) = \sum_{k=1}^{\infty} ke^{-kx}.$$

Justify the definition, and show that f is continuous on $(0, +\infty)$. Find $\int_{\ln 2}^{\ln 3} f(x) \, dx$.

17. Show that while the series

$$\sum_{k=1}^{\infty} \frac{\sin 2k\pi x}{k^2}$$

converges uniformly on the entire real line, the series cannot be differentiated term by term on any open interval.

*18. Show that the series

$$\sum_{k=1}^{\infty} \frac{\ln (1 + kx)}{kx^k}$$

converges uniformly on every interval of the form $[a, +\infty)$, where $a > 1$.

19. Let S be a subset of $\mathbf{R}$, and let $f_1, f_2, \ldots$ be a sequence of functions which are uniformly continuous on S. If $f_n \to f$ uniformly on S, prove that f is uniformly continuous on S.

20. Prove Dirichlet's test for uniform convergence.

21. Prove Abel's test for uniform convergence.

22. Formulate and prove an analog of the alternating series test for uniform convergence.

23. Prove or disprove the following. If a series $\sum u_k(x)$ converges pointwise on S, and if the nth term tends uniformly to zero on S, then the series converges uniformly on S.

24. Let

$$f_n(x) = \begin{cases} 2n^3x & \text{if } 0 \leq x \leq 1/n, \\ n^2 - 2n^3(x - 1/2n) & \text{if } 1/2n \leq x \leq 1/n, \\ 0 & \text{if } 1/n < x \leq 1. \end{cases}$$

Prove that $\lim_{n \to \infty} \int_0^1 f_n(x)\, dx = +\infty$, whereas $\int_0^1 \lim_{n \to \infty} f_n(x)\, dx = 0$.

25. Prove that each of the following series converges uniformly on the intervals indicated.

a) $x - \dfrac{x^2}{2} + \dfrac{x^3}{3} - \cdots$; [0, 1] b) $\displaystyle\sum_{k=1}^{\infty} (-1)^k \dfrac{x^2 + k}{k^2}$; every bounded interval

c) $\displaystyle\sum_{k=2}^{\infty} \dfrac{(-1)^k}{(\ln k)k^x}$; $[0, +\infty)$ d) $\displaystyle\sum_{k=1}^{\infty} \dfrac{\cos kx}{\sqrt{k}}$; $[\epsilon, 2\pi - \epsilon]$, where $0 < \epsilon < \pi$

e) $\displaystyle\sum_{k=3}^{\infty} \dfrac{(-1)^k x^k}{(\ln \ln k)(1 + x^k)}$; [0, 1] f) $\displaystyle\sum_{k=1}^{\infty} \dfrac{x}{k(1 + kx^2)}$; $(-\infty, +\infty)$

26. *Dini's Theorem.* Suppose that $f, f_1, f_2, \ldots$ are continuous functions on $[a, b]$ and that the sequence $f_1, f_2, \ldots$ tends pointwise and monotonically to f on $[a, b]$. Prove that $f_n \to f$ uniformly on $[a, b]$. [Show that it suffices to consider the case in which $f = 0$ and $f_1(x) \geq f_2(x) \geq \cdots \geq 0$ for all $x \in [a, b]$. Choose $x_n \in [a, b]$ such that

$$f_n(x_n) = \|f_n\| = \max \{f_n(x) \mid a \leq x \leq b\}.$$

Argue that $\{x_n\}$ has a convergent subsequence, say $x_{n_m} \to x_0$. Given $\epsilon > 0$, show that there exists an N such that $f_N(x_0) < \epsilon/2$ and a $\delta > 0$ such that

$$|x - x_0| < \delta \Rightarrow f_N(x) < \epsilon.$$

Show further that there exists an M such that $|x_{n_M} - x_0| < \delta$ and that

$$\|f_n\| \leq \|f_{n_M}\| = f_{n_M}(x_{n_M}) < \epsilon$$

whenever $n > n_M$.]

27. Suppose that

a) $\sum_{k=1}^{\infty} u_k(x)$ converges pointwise on $[a, b]$ to $f(x)$;
b) the functions $u_1, u_2, \ldots$ are nonnegative and continuous on $[a, b]$;
c) f is continuous on $[a, b]$.

Deduce from Dini's theorem that the series converges uniformly on $[a, b]$.

28. Rework Exercise 4 using Dini's theorem.

*29. Construct a sequence of everywhere continuous functions which converges pointwise to the ruler function.

*30. Show that

$$\lim_{m \to \infty} \lim_{n \to \infty} (\cos m!\pi x)^{2n} = \begin{cases} 0 & \text{if } x \text{ is irrational,} \\ 1 & \text{if } x \text{ is rational.} \end{cases}$$

(A function which is the pointwise limit of everywhere continuous functions is said to be of *Baire class one.* Pointwise limits of functions of Baire class one are said to be of *Baire class two.* It can be shown that the points of discontinuity of a function of Baire class one must form a set which is "small" in a certain sense. This problem shows, however, that a function of Baire class two can be everywhere discontinuous.)

13-6 Power series

We shall now apply the general theory developed in preceding sections to the study of power series. Given a power series

$$\sum_{k=0}^{\infty} a_k(x - x_0)^k,$$

we shall consider questions such as these. What can be said regarding the convergence of the series? If we let $f(x)$ be the sum of the series, what properties will the function f have? Assuming that the function f has derivatives of all orders, how does the given series compare with the Taylor series for f about x_0? To what extent can power series be treated like oversized polynomials? When, for instance, can we multiply power series together or substitute one power series into another?

We begin with the matter of convergence.

Theorem 1. Convergence Properties of Power Series. *Given the power series*

$$\sum_{k=0}^{\infty} a_k(x - x_0)^k,$$

we set

$$\frac{1}{r} = \varlimsup_{n \to \infty} \sqrt[n]{|a_n|}.$$

The set of all x for which the series converges is an interval I. Specifically, if $r = +\infty$, then I is the entire line $(-\infty, +\infty)$; if $r = 0$, then I is the singleton set $\{a\}$; and if $0 < r < +\infty$, then I is a bounded interval with endpoints $x_0 - r$ and $x_0 + r$. At interior points of I the series converges absolutely. Moreover, convergence is uniform on every closed bounded interval contained in the interior of I.

Proof. Since

$$\varlimsup_{n \to \infty} \sqrt[n]{|a_n(x - x_0)^n|} = \frac{1}{r} |x - x_0|,$$

it follows from the ratio test that the series converges absolutely if $|x - x_0| < r$ and diverges if $|x - x_0| > r$. All the assertions of the theorem follow from this, except the statement concerning uniform convergence.

Suppose, then, that $[a, b]$ is contained in the interior of I. Then an interior point x_1 of I may be chosen such that

$$|x - x_0| \le |x_1 - x_0| < r$$

for all $x \in [a, b]$; indeed we can take x_1 to be either a or b depending on which of these is further from x_0. Then

$$|a_n(x - x_0)^n| \le |a_n(x_1 - x_0)^n|$$

for all $x \in [a, b]$. Since the power series converges absolutely at x_1, the Weierstrass M-test tells us that the series must converge uniformly on $[a, b]$.

The interval I is called the *interval of convergence* of the power series. The number r is called the *radius of convergence*. If we set

$$f(x) = \sum_{k=0}^{\infty} a_k(x - x_0)^k$$

for all $x \in I$, then it follows from our theorems on uniform convergence that the function f must be continuous at all interior points of I. It is natural to inquire next about differentiability of f. The series obtained by term-by-term differentiation is again a power series, namely,

$$\sum_{k=0}^{\infty} ka_k(x - x_0)^{k-1}.$$

Its radius of convergence is the same as the radius of convergence of the series

$$\sum_{k=0}^{\infty} ka_k(x - x_0)^k,$$

namely, the reciprocal of

$$\overline{\lim_{n \to \infty}} \sqrt[n]{n|a_n|} = \overline{\lim_{n \to \infty}} \sqrt[n]{|a_n|} = \frac{1}{r},$$

since $\lim_{n \to \infty} \sqrt[n]{n} = \lim_{n \to \infty} e^{(\ln n)/n} = e^0 = 1$. In other words, the radius of convergence of the derived series is the same as the radius of convergence of the original series. By applying both our convergence theorem on power series and our theorem concerning term-by-term differentiation, we conclude that f must be differentiable at interior points of I and that

$$f'(x) = \sum_{k=0}^{\infty} ka_k(x - x_0)^{k-1} = \sum_{k=0}^{\infty} (k + 1)a_{k+1}(x - x_0)^k$$

for all interior points x in I.

The argument can then be repeated. The upshot is that f possesses continuous derivatives of all orders throughout the interior of I and that power series expansions for the derivatives of f can be obtained through term-by-term differentiation of the original series. Thus

$$f(x) = a_0 + a_1(x - x_0) + a_2(x - x_0)^2 + \cdots,$$
$$f'(x) = a_1 + 2a_2(x - x_0) + 3a_3(x - x_0)^2 + \cdots,$$
$$f''(x) = 2a_2 + 6a_3(x - x_0) + 12a_4(x - x_0)^2 + \cdots,$$
$$f'''(x) = 6a_3 + 24a_4(x - x_0) + 60a_5(x - x_0)^2 + \cdots,$$
$$f^{(4)}(x) = 24a_4 + 120a_5(x - x_0) + \cdots,$$
$$\vdots$$

where the expansions are valid at all interior points of I. Setting $x = x_0$, we get

$$a_0 = f(x_0), \qquad a_1 = f'(x_0), \qquad a_2 = f''(x_0)/2!,$$
$$a_3 = f'''(x_0)/3!, \qquad a_4 = f^{(4)}(x_0)/4!, \qquad \dots$$

More generally, $a_n = f^{(n)}(x_0)/n!$. Therefore,

$$f(x) = \sum_{k=0}^{\infty} a_k(x - x_0)^k = \sum_{k=0}^{\infty} \frac{f^{(k)}(x_0)}{k!} (x - x_0)^k,$$

or in other words, the given series turns out to be the Taylor series for f about x_0. Let us summarize this discussion.

Theorem 2. Differentiability of Power Series. *If we set*

$$f(x) = \sum_{k=0}^{\infty} a_k(x - x_0)^k$$

for each x in the interval of convergence I, then f has continuous derivatives of all orders throughout the interior of I; power series expansions for these derivatives can be obtained by term-by-term differentiation of the given series. The radii of convergence of all these derived series are equal to the radius of convergence of the given series. Moreover, the given series is just the Taylor series for f about x_0.

Corollary 3. Uniqueness of Power Series Expansions. *A given function can have at most one power series expansion near any given point. More precisely, if the equation*

$$f(x) = \sum_{k=0}^{\infty} a_k(x - x_0)^k = \sum_{k=0}^{\infty} b_k(x - x_0)^k$$

holds for all x in some neighborhood of x_0, then $a_n = b_n$ ($n = 0, 1, 2, \dots$). In particular, if

$$\sum_{k=0}^{\infty} c_k(x - x_0)^k = 0$$

for all x in some neighborhood of x_0, then all the coefficients c_k are zero.

Proof. The first assertion follows from the fact $a_n = b_n = f^{(n)}(x_0)/n!$.

Because of uniform convergence on closed bounded intervals within I, term-by-term integration is possible on such intervals.

Theorem 4. Integrability of Power Series. *If $[a, b]$ is a closed bounded subinterval of the interior of I, then*

$$\int_a^b f(x)\, dx = \sum_{k=0}^{\infty} \int_a^b a_k(x - x_0)^k\, dx$$

$$= \sum_{k=0}^{\infty} \frac{a_k}{k + 1} [(b - x_0)^{k+1} - (a - x_0)^{k+1}].$$

So far we have said nothing about behavior at the endpoints of I, $x_0 - r$ and $x_0 + r$, for the case in which $0 < r < +\infty$. These endpoints may or may not belong to I, as shown by previous examples. [Consider, for instance, the Maclaurin series for $1/(1 - x)$, $\ln (1 + x)$, $\ln (1 - x)$, and $\mathrm{Tan}^{-1} x$.] If I does contain one of its endpoints, then it is natural to inquire whether the function

$$f(x) = \sum_{k=0}^{\infty} a_k(x - x_0)^k$$

is continuous at the endpoint. (So far we'only know that f is continuous at the interior points of I.) The answer is affirmative, as can be deduced from our next theorem.

Theorem 5. Abel's Theorem. *Suppose that the series $\sum_{k=0}^{\infty} a_k$ converges. Then*

$$\sum_{k=0}^{\infty} a_k = \lim_{x \to 1^-} \sum_{k=0}^{\infty} a_k x^k.$$

Proof. An equivalent assertion is, of course, that the function

$$f(x) = \sum_{k=0}^{\infty} a_k x^k$$

is left-continuous at 1. Hence it suffices to prove that the power series converges uniformly on the interval $[0, 1]$. Observe, however, that $1 \geq x \geq x^2 \geq x^3 \geq \cdots \geq 0$ for all $x \in [0, 1]$. The fact that $\sum_{k=0}^{\infty} a_k x^k$ converges uniformly on $[0, 1]$ then follows from Corollary 18, Section 13-5.

Example 1. Discuss once again the binomial series $\sum_{k=0}^{\infty} \binom{\alpha}{k} x^k$.

We consider only the case where α is not one of the integers $0, 1, 2, \ldots$

To determine the radius of convergence of the series, we use the ratio test. The ratio of the $(n + 1)$th term to the nth term is $(\alpha - n + 1)x/n$; its absolute value tends to $|x|$ as $n \to \infty$. Therefore, the series converges if $|x| < 1$ and diverges if $|x| > 1$. The radius of convergence is therefore one. We set

$$f(x) = \sum_{k=0}^{\infty} \binom{\alpha}{k} x^k.$$

We shall next prove that $f(x) = (1 + x)^\alpha$ whenever $|x| < 1$. First we show that for such x we have

$$(1 + x)f'(x) = \alpha f(x).$$

Since power series may be differentiated term by term,

$$f'(x) = \sum_{k=0}^{\infty} k \binom{\alpha}{k} x^{k-1}$$

$$= \sum_{k=0}^{\infty} (k + 1) \binom{\alpha}{k + 1} x^k$$

$$= \alpha \sum_{k=0}^{\infty} \binom{\alpha - 1}{k} x^k \qquad (|x| < 1).$$

Thus we have

$$(1 + x)f'(x) = \alpha \sum_{k=0}^{\infty} \binom{\alpha - 1}{k} x^k + \alpha \sum_{k=0}^{\infty} \binom{\alpha - 1}{k} x^{k+1}$$

$$= \alpha + \alpha \sum_{k=1}^{\infty} \binom{\alpha - 1}{k} x^k + \alpha \sum_{k=1}^{\infty} \binom{\alpha - 1}{k - 1} x^k$$

$$= \alpha + \alpha \sum_{k=1}^{\infty} \left[\binom{\alpha - 1}{k} x^k + \binom{\alpha - 1}{k - 1} \right] x^k$$

$$= \alpha + \alpha \sum_{k=1}^{\infty} \binom{\alpha}{k} x^k = \alpha \sum_{k=0}^{\infty} \binom{\alpha}{k} x^k$$

$$= \alpha f(x).$$

To prove now that $f(x) = (1 + x)^\alpha$, we set $\varphi(x) = (1 + x)^{-\alpha} f(x)$. Then

$$\varphi'(x) = \frac{(1 + x)f'(x) - \alpha f(x)}{(1 + x)^{\alpha+1}} = 0$$

for all $x \in (-1, +1)$. The function φ must be constant on $(-1, +1)$, and by setting $x = 0$, we see that the constant must be one. Therefore, we have

$$(1 + x)^\alpha = f(x) = \sum_{k=0}^{\infty} \binom{\alpha}{k} x^k \qquad (|x| < 1).$$

Finally we discuss the behavior of the series at the endpoints ± 1. Suppose that $\alpha > 0$. Then at the endpoints, the ratios of the $(n + 1)$th term to the nth term is in absolute value equal to $1 - (\alpha + 1)/n$ for sufficiently large n. The series therefore converges absolutely, by the variation of *Raabe's test* given in Exercise 13, Section 13–2. Since the function $(1 + x)^\alpha$ is continuous at ± 1, we have from Abel's theorem

$$(1 + x)^\alpha = \sum_{k=0}^{\infty} \binom{\alpha}{k} x^k \qquad (|x| \leq 1, \alpha > 0).$$

Suppose next that $-1 \leq \alpha < 0$. Then the binomial coefficients alternate in sign. If $x = +1$, then the series converges conditionally, by Leibnitz's theorem. If $x = -1$, the series diverges, by Raabe's test. Hence we have from Abel's theorem

$$(1 + x)^\alpha = \sum_{k=0}^{\infty} \binom{\alpha}{k} x^k \qquad (-1 < x \leq 1, -1 \leq \alpha < 0).$$

If $\alpha < -1$, then at both endpoints, each of the terms is in absolute value greater than or equal to one. Hence the series cannot converge. Thus

$$(1 + x)^\alpha = \sum_{k=0}^{\infty} \binom{\alpha}{k} x^k \qquad (|x| < 1, \alpha < -1).$$

Example 2. Show again that

$$\ln (1 + x) = x - \frac{x^2}{2} + \frac{x^3}{3} - \frac{x^4}{4} + \cdots,$$

where $-1 < x \leq 1$.

We know that $1/(1 + x) = 1 - x + x^2 - x^3 + \cdots$ if $-1 < x < 1$. Consequently, if $|x| < 1$,

$$\ln(1 + x) = \int_0^x \frac{dt}{1 + t}$$

$$= \int_0^x dt - \int_0^x t\,dt + \int_0^x t^2\,dt - \cdots$$

$$= x - \frac{x^2}{2} + \frac{x^3}{3} - \frac{x^4}{4} + \cdots$$

The latter series converges when $x = 1$, by Leibnitz's test. Since $\ln(1 + x)$ is continuous at one, it follows from Abel's theorem that the series expansion is also valid when $x = 1$. For all other values of x, the series diverges.

Example 3. Obtain a power series expansion for $\operatorname{Sin}^{-1} x$.

By Example 1,

$$\frac{1}{\sqrt{1 + x}} = (1 + x)^{-1/2} = 1 - \tfrac{1}{2}x + \tfrac{1}{2}\cdot\tfrac{3}{4}x^2 - \tfrac{1}{2}\cdot\tfrac{3}{4}\cdot\tfrac{5}{6}t^6 + \cdots$$

if $-1 < x \le 1$. Hence

$$\frac{1}{\sqrt{1 - t^2}} = 1 + \tfrac{1}{2}t^2 + \tfrac{1}{2}\cdot\tfrac{3}{4}t^4 + \tfrac{1}{2}\cdot\tfrac{3}{4}\cdot\tfrac{5}{6}t^6 + \cdots$$

if $-1 < t < +1$. If $|x| < 1$, we have

$$\operatorname{Sin}^{-1} x = \int_0^x \frac{dt}{\sqrt{1 - t^2}} = \int_0^x dt + \int_0^x \tfrac{1}{2}t^2\,dt + \cdots$$

$$= x + \frac{1}{2}\frac{x^3}{3} + \frac{1}{2}\cdot\frac{3}{4}\frac{x^5}{5} + \frac{1}{2}\cdot\frac{3}{4}\cdot\frac{5}{6}\frac{x^7}{7} + \cdots$$

The series expansion is also valid if $x = \pm 1$. This may be seen as follows. Let $S_n(x)$ denote the nth partial sum of the series. If $0 < x < 1$, then all terms of the series are positive, and we therefore have

$$S_n(x) < \operatorname{Sin}^{-1} x < \operatorname{Sin}^{-1} 1 = \pi/2$$

for all $n \in \mathbf{Z}^+$. Thus for any fixed $n \in \mathbf{Z}^+$,

$$S_n(1) = \lim_{x\to 1-} S_n(x) \le \pi/2.$$

When $x = 1$, we therefore have a series of positive terms whose partial sums are bounded above by $\pi/2$. The series therefore converges, and the fact that its limit is $\operatorname{Sin}^{-1} 1$ is once again a consequence of Abel's theorem. Finally, the series expansion is valid if $x = -1$, since all terms are odd.

In summary,

$$\operatorname{Sin}^{-1} x = x + \frac{1}{2}\frac{x^3}{3} + \frac{1}{2}\cdot\frac{3}{4}\frac{x^5}{5} + \cdots$$

iff $|x| \le 1$.

Example 4. Obtain an infinite series expansion for the *complete elliptic integral*

$$K = \int_0^{\pi/2} \frac{d\phi}{\sqrt{1 - k^2 \sin^2 \phi}} \qquad (k^2 < 1).$$

By the preceding example,

$$\frac{1}{\sqrt{1 - k^2 \sin^2 \phi}} = 1 + \tfrac{1}{2}k^2 \sin^2 \phi + \tfrac{1}{2} \cdot \tfrac{3}{4}k^4 \sin^4 \phi + \tfrac{1}{2} \cdot \tfrac{3}{4} \cdot \tfrac{5}{6}k^6 \sin^6 \phi + \cdots$$

For all values of ϕ, the terms of this series are in absolute value less than or equal to those of the series

$$1 + \tfrac{1}{2}k^2 + \tfrac{1}{2} \cdot \tfrac{3}{4}k^4 + \cdots$$

The latter converges since $k^2 < 1$, and thus the first series converges uniformly (in ϕ) on the entire line. Consequently, term-by-term integration is permissible. Since

$$\int_0^{\pi/2} \sin^{2n} \phi \, d\phi = \frac{1}{2} \cdot \frac{3}{4} \cdots \frac{2n-1}{2n} \frac{\pi}{2}$$

(see p. 476), we get

$$K = \frac{\pi}{2} \left[1 + \left(\frac{1}{2}\right)^2 k^2 + \left(\frac{1 \cdot 3}{2 \cdot 4}\right)^2 k^4 + \left(\frac{1 \cdot 3 \cdot 5}{2 \cdot 4 \cdot 6}\right)^2 k^6 + \cdots \right].$$

Example 5. Suppose that $\sum_{k=0}^{\infty} c_k$ is the Cauchy product of the infinite series $\sum_{k=0}^{\infty} a_k$ and $\sum_{k=0}^{\infty} b_k$. If the three series converge with sums C, A, and B, respectively, prove that $C = AB$. (This result is due to Abel.)

Since the series $\sum a_k$ and $\sum b_k$ both converge, the power series

$$\sum_{k=0}^{\infty} a_k x^k \qquad \text{and} \qquad \sum_{k=0}^{\infty} b_k x^k$$

must have radii of convergence greater than or equal to one. If $|x| < 1$, both power series converge absolutely and, consequently, the series $\sum_{k=0}^{\infty} c_k x^k$, which is their Cauchy product, does also. Moreover,

$$\sum_{k=0}^{\infty} c_k x^k = \left(\sum_{k=0}^{\infty} a_k x^k\right)\left(\sum_{k=0}^{\infty} b_k x^k\right).$$

The proof is now easily completed by Abel's theorem:

$$C = \lim_{x \to 1-} \sum_{k=0}^{\infty} c_k x^k = \lim_{x \to 1-} \left(\sum_{k=0}^{\infty} a_k x^k\right)\left(\sum_{k=0}^{\infty} b_k x^k\right)$$

$$= \left(\lim_{x \to 1-} \sum_{k=0}^{\infty} a_k x^k\right)\left(\lim_{x \to 1-} \sum_{k=0}^{\infty} b_k x^k\right)$$

$$= AB.$$

Example 6. Obtain a power series expansion for $(\text{Sin}^{-1} x)/\sqrt{1 - x^2}$.

Example 3 closes the gap in Example 1 of Section 13-4. Consequently, we have the expansion

$$(\text{Sin}^{-1} x)^2 = \frac{1}{2} \sum_{k=1}^{\infty} \frac{(k-1)!!}{(2k)!} (2x)^{2k},$$

where $|x| \leq 1$. We differentiate term by term and multiply by 2, obtaining

$$\frac{\text{Sin}^{-1} x}{\sqrt{1 - x^2}} = x + \frac{2}{3} x^3 + \frac{2 \cdot 4}{3 \cdot 5} x^5 + \cdots$$

The expansion is certainly valid if $|x| < 1$. (Why?) Moreover, the series diverges if $|x| \geq 1$. If this weren't the case, then the series would have to converge with either $x = +1$ or $x = -1$, and Abel's theorem would therefore imply that the function $(\text{Sin}^{-1} x)/\sqrt{1 - x^2}$ must have a limit as either $x \to +1-$ or $x \to -1+$.

Example 7. Solve the differential equation

$$xy'' + y' + xy = 0$$

subject to the initial conditions $y(0) = 1$, $y'(0) = 0$. This differential equation is known as the *Bessel differential equation of order zero*.

We shall use the so-called *method of undetermined coefficients*. We seek a solution which has a Maclaurin expansion. Accordingly, we let

$$y = a_0 + a_1 x + a_2 x^2 + a_3 x^3 + \cdots$$

Differentiating term by term, we obtain

$$y' = a_1 + 2a_2 x + 3a_3 x^2 + \cdots, \qquad y'' = 2a_2 + 6a_3 x + 12a_4 x^2 + \cdots,$$

and we substitute into the differential equation. After collecting like powers of x, we have

$$a_1 + (4a_2 + a_0)x + (9a_3 + a_1)x^2 + \cdots + [(k+1)^2 a_{k+1} + a_{k-1}]x^k + \cdots = 0.$$

For this equation to hold for all values of x in some neighborhood of 0, all the coefficients on the left-hand side must be zero. Thus we see that $a_1 = 0$ and

$$a_{k+1} = -\frac{1}{(k+1)^2} a_{k-1},$$

for $k = 1, 2, \ldots$ In order that $y(0) = 1$, we must have $a_0 = 1$. Then

$$a_2 = -\frac{1}{2^2}, \qquad a_4 = \frac{1}{2^2 \cdot 4^2}, \qquad a_6 = -\frac{1}{2^2 \cdot 4^2 \cdot 6^2}, \qquad \cdots$$

and, more generally,

$$a_{2n} = (-1)^n \frac{1}{2^2} \cdot \frac{1}{4^2} \cdots \frac{1}{(2n)^2} = (-1)^n \frac{1}{4^n n!!}.$$

Moreover, $a_1 = a_3 = a_5 = \cdots = 0$. This suggests that

$$y = 1 - \frac{x^2}{2^2} + \frac{x^4}{2^2 \cdot 4^2} - \frac{x^6}{2^2 \cdot 4^2 \cdot 6^2} + \cdots$$

may be the solution to our problem. To check, we first observe (via the ratio test) that the series converges on the entire real line. Consequently, the series can be differentiated term by term as often as we please, thereby justifying our previous calculations.

It can be shown that the function which we have obtained is the only solution to the problem. (Our calculations show only that it is the unique solution possessing a Maclaurin expansion.) This function is denoted by $J_0(x)$. We shall let the reader show that the function

$$J_n(x) = \frac{x^n}{2^n n!}\left[1 - \frac{x^2}{2^2(n+1)} + \frac{x^4}{2^4 \cdot 2!(n+1)(n+2)}\right.$$
$$\left. - \frac{x^6}{2^6 \cdot 3!(n+1)(n+2)(n+3)} + \cdots\right]$$

is a solution of the nth-*order Bessel equation*

$$x^2 y'' + xy' + (x^2 - n^2)y = 0.$$

Exercises

1. Find the interval of convergence of each of the following power series.

   a) $\displaystyle\sum_{k=0}^{\infty} k!(x-3)^k$ b) $\displaystyle\sum_{k=0}^{\infty} \frac{x^k}{k(k+1)}$ c) $\displaystyle\sum_{k=1}^{\infty} \frac{[k(x+2)]^k}{k!}$

   d) $\displaystyle\sum_{k=1}^{\infty} [k(x-1)]^k$ e) $\displaystyle\sum_{k=1}^{\infty} x^{2^k}$ f) $\displaystyle\sum_{k=1}^{\infty} \left(\frac{k+1}{k}\right)^{k^2}(x+1)^k$

   g) $\displaystyle\sum_{k=2}^{\infty} \frac{\ln k}{k} x^k$

2. The purpose of this exercise is to reestablish the convergence properties of power series without using the root test. Let $\sum a_k(x - x_0)^k$ be given, and let I be the set of all x for which the series converges.

   a) Prove that if $x_1 \in I$ and if $0 < \rho < |x_1 - x_0|$, then the interval $[x_0 - \rho, x_0 + \rho]$ is a subset of I and the series converges absolutely and uniformly on this interval. (Argue that there exists an M such that $|a^n(x_1 - x_0)^n| \leq M$ for all $n \in \mathbf{Z}^+$. Apply the Weierstrass M-test by comparing the given series with a geometric series.)

   b) Set

   $$r = \text{lub } \{|x - x_0| \mid x \in I\}.$$

   Deduce from (a) that I is an interval with endpoints $x_0 - r$ and $x_0 + r$ and the series converges uniformly and absolutely on every closed bounded subinterval contained in $(x_0 - r, x_0 + r)$.

3. The purpose of this exercise is to establish convergence properties of a *generalized Dirichlet series* by methods analogous to those of the preceding exercise. By a generalized Dirichlet series we mean any series of the form

   $$\sum_{k=1}^{\infty} a_k e^{-\lambda_k s}, \qquad (*)$$

   where $0 \leq \lambda_1 < \lambda_2 < \cdots$

a) Prove that there exists a number γ_a, known as the *abscissa of absolute convergence*, such that the series converges absolutely if $s > \gamma_a$ and does not converge absolutely if $s < \gamma_a$.

b) Suppose that for some $\epsilon > 0$, $\sum_{k=1}^{\infty} e^{-\epsilon \lambda_k} < +\infty$. Prove that the Dirichlet series (*) diverges if $s < \gamma_a - \epsilon$.

c) Prove that there exists a number $\gamma_c \leq \gamma_a$, called the *abscissa of convergence*, such that (*) converges if $s > \gamma_c$ and diverges if $s < \gamma_c$. Prove, moreover, that if the series converges at the point s', then it converges uniformly on the interval $[s', +\infty)$.

d) State and prove an analog of Abel's theorem for generalized Dirichlet series.

e) Using (b), show that for an ordinary Dirichlet series ($\lambda_k = \ln k$), $\gamma_c \leq \gamma_a \leq \gamma_c + 1$.

f) Prove that a generalized Dirichlet series can be differentiated term by term throughout the interior of its interval of convergence.

4. Prove by the methods of Example 2 that

$$\text{Tan}^{-1} x = x - \frac{x^3}{3} + \frac{x^5}{5} - \cdots \qquad (|x| \leq 1).$$

5. Derive the expansion

$$\log \frac{1+x}{1-x} = 2\left(x + \frac{x^3}{3} + \frac{x^5}{5} + \cdots\right).$$

For what values of x is it valid?

6. Show that

$$\int_0^x t^{-1} \sin t \, dt = \sum_{k=0}^{\infty} \frac{(-1)^k x^{2k+1}}{(2k+1)(2k+1)!}$$

for all $x \in \mathbf{R}$.

7. Show that

$$\int_0^{\pi/2} \frac{\cos t}{1 - x \sin^2 t} \, dt = \sum_{k=0}^{\infty} \frac{x^k}{2k+1} \quad \text{if } |x| < 1.$$

8. Show that if $k^2 < 1$,

$$\int_0^{\pi/2} \sqrt{1 - k^2 \sin^2 \phi} \, d\phi = \frac{\pi}{2}\left[1 - \left(\frac{1}{2}\right)^2 \frac{k^2}{1} - \left(\frac{1 \cdot 3}{2 \cdot 4}\right)^2 \frac{k^4}{3} - \cdots\right].$$

9. We have defined the nth-order Bessel function $J_n(x)$ by the series expansion

$$J_n(x) = \sum_{k=0}^{\infty} \frac{(-1)^k}{k!(k+n)!}\left(\frac{x}{2}\right)^{2k+n}.$$

a) Prove that the radius of convergence is $+\infty$.

b) Prove that

$$\frac{d}{dx}[x^n J_n(x)] = x^n J_{n-1}(x) \quad \text{and} \quad \frac{d}{dx}[x^{-n} J_n(x)] = -x^{-n} J_{n+1}(x).$$

c) Deduce from (b) that

$$x^2 J_n''(x) + x J_n'(x) + (x^2 - n^2) J_n(x) = 0.$$

d) Show that

$$\int_0^{\pi/2} \cos (x \sin \phi)(\cos \phi)^{2n} \, d\phi = \sum_{k=0}^{\infty} \frac{(-1)^k x^{2k}}{(2k)!} \int_0^{\pi/2} (\sin \phi)^{2k}(\cos \phi)^{2n} \, d\phi,$$

and then use the integration formula

$$\int_0^{\pi/2} (\sin \phi)^{2k}(\cos \phi)^{2n} \, d\phi = \frac{\pi(2k)!(2n)!}{2^{2k+2n+1}n!k!(n+k)!}$$

to show that

$$J_n(x) = \frac{2^{n+1}n!}{\pi(2n)!} \left(\frac{x}{2}\right)^n \int_0^{\pi/2} \cos (x \sin \phi)(\cos \phi)^{2n} \, d\phi.$$

10. Find a power series solution of the initial-value problem

$$xy'' + y' - y = 0, \qquad y(0) = 1, \qquad y'(0) = 1.$$

11. Obtain a Maclaurin expansion for $\ln (x + \sqrt{1 + x^2})$. [Integrate the Maclaurin series for $(1 + x^2)^{-1/2}$.]

13-7 Real analytic functions

We shall now study the algebra of power series and the functions they represent.

Definition 1. A function f is said to be **(real) analytic** on an open interval I iff for each $x_0 \in I$ there exists a power series about x_0, $\sum_{k=0}^{\infty} a_k(x - x_0)^k$, and a neighborhood U of x_0 such that

$$f(x) = \sum_{k=0}^{\infty} a_k(x - x_0)^k$$

holds for all $x \in U$.

The power series in the definition must, of course, be the Taylor series of f about x_0. An analytic function may therefore be regarded as a C^{∞}-function (one possessing continuous derivatives of all orders) which can be represented locally by its Taylor series.

Example 1. The function $1/x$ is analytic on each of the open intervals $(-\infty, 0)$ and $(0, +\infty)$. Suppose $x_0 \in (0, +\infty)$. If $|x - x_0| < x_0$, then $|(x - x_0)/x_0| < 1$, so that

$$\frac{1}{x} = \frac{1}{x_0} \frac{1}{1 + (x - x_0)/x_0} = \frac{1}{x_0} \left[1 - \frac{x - x_0}{x} + \left(\frac{x - x_0}{x_0}\right)^2 - \cdots \right].$$

This proves that $1/x$ is analytic on the interval $(0, +\infty)$.

Our first theorem will provide many more examples of analytic functions.

Theorem 2. *Let $\sum_{k=0}^{\infty} a_k(x - x_0)^k$ be a power series whose radius of convergence r is positive. If we let*

$$f(x) = \sum_{k=0}^{\infty} a_k(x - x_0)^k$$

for all x in the interval of convergence, then f is analytic throughout the interior I of the interval of convergence. Specifically, if $x_1 \in I$, then f can be expanded in a power series about x_1 whose radius of convergence is at least $r - |x_1 - x_0|$.

Proof. Let us suppose that $|x - x_1| < r - |x_1 - x_0|$. Then

$$|x - x_0| \leq |x - x_1| + |x_1 - x_0| < r,$$

so

$$f(x) = \sum_{k=0}^{\infty} a_k(x - x_0)^k = \sum_{k=0}^{\infty} a_k[(x - x_1) + (x_1 - x_0)]^k$$

$$= \sum_{k=0}^{\infty} a_k \left[\sum_{i+j=k} \binom{k}{j} (x - x_1)^i (x_1 - x_0)^j \right].$$

The latter is the sum by diagonals of the double series

$$\sum_{j,k=0}^{\infty} a_{j+k} \binom{j+k}{j} (x_1 - x_0)^j (x - x_1)^k.$$

We shall prove that this double series converges absolutely. Now

$$\sum_{k=0}^{\infty} |a_k| \left[\sum_{i+j=k} \binom{k}{j} |x - x_1|^i |x_1 - x_0|^j \right] = \sum_{k=0}^{\infty} |a_k| (|x - x_1| + |x_1 - x_0|)^k.$$

Since $|x - x_1| + |x_1 - x_0| < r$, and since r is the radius of convergence of the power series $\sum_{k=0}^{\infty} |a_k| x^k$, it follows that

$$\sum_{k=0}^{\infty} |a_k| \left[\sum_{i+j=k} \binom{k}{j} |x - x_1|^i |x_1 - x_0|^j \right] < +\infty.$$

By the Fubini theorem, the double series converges and may be summed by columns. Thus

$$f(x) = \sum_{k=0}^{\infty} \left[\sum_{j=0}^{\infty} \binom{j+k}{j} a_{j+k}(x_1 - x_0)^j \right] (x - x_1)^k,$$

thereby completing the proof.

It follows from the theorem that the functions e^x, $\sin x$, $\cos x$, $\sinh x$, $\cosh x$, $J_0(x)$, and all polynomials are analytic on the entire line. It also follows that the functions $\text{Sin}^{-1} x$, $\text{Tan}^{-1} x$, and $\ln (1 + x)$ are analytic on the interval $(-1, +1)$. We shall presently prove stronger assertions about the last two functions, namely, that $\text{Tan}^{-1} x$ is analytic on the entire line and that $\ln (1 + x)$ is analytic on the interval $(-1, +\infty)$.

We shall now consider closure properties of analytic functions.

Proposition 3. *If f and g are analytic on an interval I, then the same is true of the functions* $f + g$ *and* fg.

Proof. Let us prove that fg is analytic on I. Let $x_0 \in I$. Then f and g admit power series expansions about x_0:

$$f(x) = \sum_{k=0}^{\infty} a_k(x - x_0)^k, \qquad |x - x_0| < r_a,$$

$$g(x) = \sum_{k=0}^{\infty} b_k(x - x_0)^k, \qquad |x - x_0| < r_b.$$

If $|x - x_0| < r = \min (r_a, r_b)$, then both series converge absolutely. Hence their Cauchy product does also, and we have

$$f(x)\, g(x) = \sum_{k=0}^{\infty} \left(\sum_{i+j=k} a_i b_j \right) (x - x_0)^k.$$

This proves that fg is analytic on I.

Proposition 4. *If f is analytic in I, then so are its derivatives and antiderivatives.*

Proof. Suppose that F is an antiderivative of f in I. Let $x_0 \in I$. Then f has a power series expansion

$$f(x) = \sum_{k=0}^{\infty} a_k(x - x_0)^k, \qquad |x - x_0| < r.$$

If $|x - x_0| < r$, then

$$F(x) = F(x_0) + \int_{x_0}^{x} f(t)\, dt$$

$$= F(x_0) + \sum_{k=0}^{\infty} a_k \int_{x_0}^{x} (t - x_0)^k\, dt$$

$$= F(x_0) + \sum_{k=1}^{\infty} \frac{a_{k-1}}{k} (x - x_0)^k.$$

This proves that F is analytic on I. The fact that the derivatives of f are analytic on I follows even more easily from previous theorems.

Example 2. It follows from Example 1 and the last theorem that $\ln x$ is analytic on the interval $(0, +\infty)$, or by translation, that $\ln (1 + x)$ is analytic on the interval $(-1, +\infty)$.

We consider next composition products of analytic functions and the somewhat delicate matter of substituting one power series into another.

Proposition 5. *Suppose that we have*

$$f(x) = \sum_{k=0}^{\infty} a_k x^k, \qquad |x| < R,$$

$$g(x) = \sum_{k=0}^{\infty} b_k x^k, \qquad |x| < r.$$

Then if $|x| < r$ and if $\sum_{k=0}^{\infty} |b_k x^k| < R$, we have

$$f(g(x)) = \sum_{j=0}^{\infty} a_j \left(\sum_{k=0}^{\infty} b_k^{(j)} x^k \right) = \sum_{k=0}^{\infty} \left(\sum_{j=0}^{\infty} a_j b_k^{(j)} \right) x^k,$$

where

$$[g(x)]^j = \sum_{k=0}^{\infty} b_k^{(j)} x^k.$$

Proof. Because of the Fubini theorem, it suffices to prove that

$$\sum_{j=0}^{\infty} |a_j| \left(\sum_{k=0}^{\infty} |b_k^{(j)} x^k| \right) < +\infty.$$

Now

$$b_k^{(j)} = \sum_{i_1 + \cdots + i_j = k} b_{i_1} b_{i_2} \cdots b_{i_j},$$

so that

$$|b_k^{(j)}| \le \sum_{i_1 + \cdots + i_j = k} |b_{i_1}| \, |b_{i_2}| \cdots |b_{i_i}| = B_k^{(j)};$$

the latter expression is the coefficient of x^k in power series expansion of $(\sum_{n=0}^{\infty} |b_n| x^n)^j$. Thus

$$\sum_{k=0}^{\infty} |b_k^{(j)} x^k| \le \sum_{k=0}^{\infty} B_k^{(j)} |x^k| = \left(\sum_{k=0}^{\infty} |b_k x^k| \right)^j < R^j.$$

If we set $y = \sum_{k=0}^{\infty} |b_k x^k|$, then $|y| = y < R$. Since a power series converges absolutely throughout the interior of its interval of convergence, $\sum_{j=0}^{\infty} |a_j| y^j < +\infty$, and so by comparison,

$$\sum_{j=0}^{\infty} |a_j| \left(\sum_{k=0}^{\infty} |b_k^{(j)} x^k| \right) < +\infty.$$

Observe that if $R = +\infty$ (in which case f is said to be an *entire* function), then the power series for $f(g(x))$ converges to the function whenever $|x| < r$. If both f and g are entire functions, then $f \circ g$ is entire.

Example 3. Show that $e^{\sin x}$ is an entire function, and find the first six terms of its Maclaurin expansion.

The function is the composition product of two entire functions—the exponential function and the sine function—and is therefore an entire function also. Thus

$$
\begin{aligned}
e^{\sin x} &= 1 + \sin x + \frac{(\sin x)^2}{2!} + \frac{(\sin x)^3}{3!} + \cdots \\
&= 1 + \left(x - \frac{x^3}{3!} + \cdots \right) + \frac{1}{2!}\left(x - \frac{x^3}{3!} + \cdots \right)^2 + \frac{1}{3!}\left(x - \frac{x^3}{3!} + \cdots \right)^3 + \cdots \\
&= 1 + \left(x + 0 - \frac{x^3}{6} + 0 + \frac{x^5}{120} + \cdots \right) + \left(0 + \frac{x^2}{2} + 0 - \frac{x^4}{6} + 0 + \cdots \right) \\
&\quad + \left(0 + 0 + \frac{x^3}{6} + 0 - \frac{x^5}{12} + \cdots \right) + \left(0 + 0 + 0 + \frac{x^4}{24} + 0 + \cdots \right) \\
&\quad + \left(0 + 0 + 0 + 0 + \frac{x^5}{120} + \cdots \right) + \cdots \\
&= 1 + x + \frac{x^2}{2} - \frac{x^4}{8} - \frac{3x^5}{20} + \cdots
\end{aligned}
$$

Proposition 6. *Suppose that we have*

$$ I \xrightarrow{g} J \xrightarrow{f} R, $$

where f and g are analytic on J and I, respectively. Then $f \circ g$ is analytic on I.

Proof. Let $x_0 \in I$, and set $y_0 = g(x_0)$. Then for suitable r and R, we have

$$ g(x) = y_0 + \sum_{k=1}^{\infty} b_k(x - x_0)^k, \qquad |x - x_0| < r, $$

$$ f(y) = \sum_{k=0}^{\infty} a_k(y - y_0)^k, \qquad |y - y_0| < R. $$

If $|x - x_0| < R$, we may define

$$ G(x) = \sum_{k=1}^{\infty} |b_k(x - x_0)^k|. $$

(Why is this definition justified?) The function G is continuous at x_0. (Why?) We may therefore choose Δ such that $0 < \Delta < r$ and

$$ |x - x_0| < \Delta \Rightarrow \sum_{k=1}^{\infty} |b_k(x - x_0)^k| = G(x) < R. $$

By our previous theorem it follows that $f(g(x))$ has a power series expansion about the point x_0 which is valid whenever $|x - x_0| < \Delta$.

Corollary 7. *If f and g are analytic on an interval I, and if g is nonvanishing on I, then f/g is analytic on I.*

Proof. The function g must map I into either the interval $(0, +\infty)$ or the interval $(-\infty, 0)$. The function $h(x) = 1/x$ is analytic on both these intervals, so $1/g = h \circ g$ is analytic on I. Consequently, f/g is analytic on I, since it is the product of two analytic functions.

One consequence of this corollary is the fact that *a rational function is analytic throughout each open interval contained in its domain.* For instance, the function

$$\frac{3x^4 - 2x + 1}{(x^2 - 1)(x^2 + 1)}$$

is analytic on the intervals $(-\infty, -1)$, $(-1, +1)$, and $(1, +\infty)$.

Example 4. Show that $\mathrm{Tan}^{-1} x$ is analytic on the entire line.

The rational function $1/(1 + x^2)$ is analytic on the entire line. Since $\mathrm{Tan}^{-1} x$ is an antiderivative of $1/(1 + x^2)$, it too is analytic on **R**.

Example 5. Show that $(1 + x)^\alpha$ is analytic on the interval $(-1, +\infty)$ for each real number α.

The function is the composition product of the exponential function and the function $\alpha \ln (1 + x)$.

The next three examples involve the *Bernoulli numbers.*

Example 6. Observe that

$$e^x - 1 = x\left(1 + \frac{x}{2!} + \frac{x^2}{3!} + \frac{x^3}{4!} + \cdots\right).$$

The series in parentheses converges for all x and consequently defines an entire function which we shall denote by $(e^x - 1)/x$. The function is never zero, and its reciprocal is therefore analytic on the entire line. Discuss the Maclaurin expansion of this reciprocal.

We shall write

$$\frac{x}{e^x - 1} = B_0 + \frac{B_1}{1!} x + \frac{B_2}{2!} x^2 + \frac{B_3}{3!} x^3 + \cdots$$

The numbers $B_0, B_1, B_2, \ldots$, which we shall presently determine, are called the Bernoulli numbers. From the general theory, we know that such numbers exist and are unique and that the expansion is valid throughout some neighborhood U of zero. Hence, for all $x \in U$,

$$\left(x + \frac{x}{2!} + \frac{x^2}{3!} + \cdots\right)\left(B_0 + \frac{B_1}{1!} x + \frac{B_2}{2!} x^2 + \cdots\right) = x.$$

We expand the left-hand side into a power series about zero, and we equate the coefficients of like powers of x from the two sides of the equation. This yields a series of equations:

$$B_0 = 1,$$

$$\frac{1}{2!} \frac{B_0}{0!} + \frac{1}{1!} \frac{B_1}{1!} = 0,$$

$$\frac{1}{3!} \frac{B_0}{0!} + \frac{1}{2!} \frac{B_1}{1!} + \frac{1}{1!} \frac{B_2}{2!} = 0,$$

and, more generally, if $n \geq 2$,

$$\frac{1}{n!} \frac{B_0}{0!} + \frac{1}{(n-1)!} \frac{B_1}{1!} + \cdots + \frac{1}{1!} \frac{B_{n-1}}{(n-1)!} = 0.$$

If we multiply the last equation by $n!$, we obtain a somewhat more memorable form:

$$\binom{n}{0} B_0 + \binom{n}{1} B_1 + \cdots + \binom{n}{n-1} B_{n-1} = 0.$$

Thus we have

$$B_0 = 1,$$
$$2B_1 + B_0 = 0,$$
$$3B_2 + 3B_1 + B_0 = 0,$$
$$4B_3 + 6B_2 + 4B_1 + B_0 = 0,$$
$$5B_4 + 10B_3 + 10B_2 + 5B_1 + B_0 = 0,$$
$$\vdots$$

from which we get $B_0 = 1$, $B_1 = -\frac{1}{2}$, $B_2 = \frac{1}{6}$, $B_3 = 0$, $B_4 = -\frac{1}{30}$, $B_5 = 0$, $B_6 = \frac{1}{42}$, $B_8 = -\frac{1}{30}$, $B_{10} = \frac{5}{66}$, $B_{12} = -\frac{691}{2730}$, $B_{14} = \frac{7}{6}, \ldots$ As we shall show in the next example, $B_3 = B_5 = B_7 = \cdots = 0$. It should also be clear that the Bernoulli numbers are all rational.

Example 7. Obtain a Maclaurin expansion for $x \operatorname{ctnh} x$.
  Let $x = y/2$. Then

$$x \operatorname{ctnh} x = \frac{y}{2} \frac{e^{y/2} + e^{-y/2}}{e^{y/2} - e^{-y/2}}$$

$$= \frac{y}{2} \frac{e^y + 1}{e^y - 1} = \frac{y}{2} \left(\frac{2}{e^y - 1} + 1 \right)$$

$$= \frac{y}{e^y - 1} + \frac{y}{2}$$

$$= 1 + \frac{B_2 y^2}{2!} + \frac{B_3 y^3}{3!} + \cdots$$

$$= 1 + \frac{B_2}{2!} (2x)^2 + \frac{B_3}{3!} (2x)^3 + \cdots$$

Since the function $x \operatorname{ctnh} x$ is even, $B_3 = B_5 = B_7 = \cdots = 0$. Hence

$$x \operatorname{ctnh} x = 1 + \frac{B_2}{2!} (2x)^2 + \frac{B_4}{4!} (2x)^4 + \cdots$$

Example 8. Obtain a Maclaurin expansion for $x \operatorname{ctn} x$.
  By the previous example,

$$x \operatorname{ctnh} x = x \frac{1 + \dfrac{x^2}{2!} + \dfrac{x^4}{4!} + \cdots}{x + \dfrac{x^3}{3!} + \dfrac{x^5}{5!} + \cdots} = 1 + \frac{B_2}{2!} (2x)^2 + \frac{B_4}{4!} (2x)^4 + \cdots$$

By alternating signs throughout the above equation, we get

$$x \operatorname{ctn} x = x \frac{1 - \dfrac{x^2}{2!} + \dfrac{x^4}{4!} - \cdots}{x - \dfrac{x^3}{3!} + \dfrac{x^5}{5!} - \cdots} = 1 - \frac{B_2}{2!}(2x)^2 + \frac{B_4}{4!}(2x)^4 - \cdots$$

The justification of this procedure will be pursued in the exercises.

Derivations for the following expansions will also be outlined in the exercises:

$$\tan x = \sum_{k=1}^{\infty} (-1)^{k-1} \frac{2^{2k}(2^{2k} - 1)B_{2k}}{(2k)!} x^{2k-1}$$

and

$$\frac{x}{\sin x} = \sum_{k=0}^{\infty} (-1)^{k-1} \frac{(2^{2k} - 2)B_{2k}}{(2k)!} x^{2k}.$$

These expansions involving the Bernoulli numbers are unsatisfactory in that we are not in a position to determine their radii of convergence. This much can be said, however: At any interior point of the interval of convergence, the series converges to the value of the function at that point. This is a consequence of the *identity principle*. A variant of the identity principle is the fact that the zeros of an analytic function are *isolated*.

Proposition 8. The Isolation of Zeros of Analytic Functions. *Suppose that f is analytic on an open interval I and that $f(x_0) = 0$ for some $x_0 \in I$. Then either f is identically zero throughout I or else there exists a deleted neighborhood of x_0 throughout which f is nonzero.*

Proof. We know that there exists a neighborhood U of x_0 throughout which f has a series expansion about x_0:

$$f(x) = \sum_{k=0}^{\infty} a_k(x - x_0)^k.$$

Since $f(x_0) = 0$, it follows that $a_0 = 0$. We consider two cases.

Case I. Not all the coefficients a_k are zero. If a_m is the first nonvanishing coefficient, then $m \geq 1$ and

$$f(x) = (x - x_0)^m[a_m + a_{m+1}(x - x_0) + \cdots]$$

for all $x \in U$. If we set

$$g(x) = a_m + a_{m+1}(x - x_0) + \cdots \qquad (x \in U),$$

then g is analytic throughout U and $g(x_0) \neq 0$. Since g is continuous at x_0, in particular, there exists a neighborhood $V \subset U$ of x_0 such that $g(x) \neq 0$ for all

$x \in V$. Thus, if x belongs to the deleted neighborhood $V \setminus \{x_0\}$, then

$$f(x) = (x - x_0)^m g(x) \neq 0.$$

Case II. All the coefficients a_k are zero. Then f is identically zero throughout U. We shall now prove that f is identically zero throughout I. Consider the set

$$S = \{x \in I \mid x > x_0 \text{ and } f(x) \neq 0\}.$$

We wish to prove, of course, that S is empty. Assume the contrary. Then S has a greatest lower bound $x_1 > x_0$. The function f is zero on the interval $[x_0, x_1)$, and by continuity, it is also zero at x_1. We know that f has a power series expansion about x_1 which is valid throughout a neighborhood V of x_1. By case I the coefficients must all be zero. This implies, however, that f is identically zero immediately to the right of x_1, contradicting the fact that $x_1 = \text{glb } S$. Thus $S = \emptyset$. Similarly, the set $\{x \in I \mid x < x_0 \text{ and } f(x) \neq 0\}$ is empty.

Corollary 9. Identity Principle for Analytic Functions. *Suppose that f and g are analytic on an open interval I. Then the following are logically equivalent:*

1) $f(x) = g(x)$ for all $x \in I$.

2) $f(x) = g(x)$ for all x in some open subinterval of I (no matter how small).

3) There exists a convergent sequence $x_1, x_2, \ldots$ of distinct points of I such that $\lim_{n \to \infty} x_n$ also belongs to I and $f(x_k) = g(x_k)$, for $k = 1, 2, \ldots$

4) There exists an $x_0 \in I$ such that $f^{(k)}(x_0) = g^{(k)}(x_0)$, for $k = 1, 2, \ldots$

Proof. Clearly, (1) $\Rightarrow$ (2), (2) $\Rightarrow$ (3), (2) $\Rightarrow$ (4), (4) $\Rightarrow$ (2). The nontrivial implication (3) $\Rightarrow$ (1) can be established by application of the theorem to $h = f - g$.

Example 9. Show that the interval of convergence I of the series

$$1 - \frac{B_2}{2!}(2x)^2 + \frac{B_4}{4!}(2x)^4 - \cdots$$

is a subinterval of $(-\pi, \pi)$, and that for each $x \in I$,

$$x \operatorname{ctn} x = 1 - \frac{B_2}{2!}(2x)^2 + \frac{B_4}{4!}(2x)^4 - \cdots$$

For each $x \in I$, we define $g(x)$ to be the sum of the infinite series. Then g is analytic throughout the interior of I. The function $x \operatorname{ctn} x = (x \sin x)/(\cos x)$ is analytic throughout the interval $(-\pi, \pi)$. We also know that $x \operatorname{ctn} x = g(x)$ for all x in *some* neighborhood of zero. If $(-\pi, \pi)$ is contained in I, then by the identity principle, $x \operatorname{ctn} x = g(x)$ for all $x \in (-\pi, \pi)$. The inclusion cannot be proper, however, since Abel's theorem would then imply the existence of either $\lim_{x \to -\pi+} x \operatorname{ctn} x$ or $\lim_{x \to \pi-} x \operatorname{ctn} x$. Since I is symmetric about zero, we must have $I \subset (-\pi, \pi)$. The fact that

$$x \operatorname{ctn} x = g(x) = 1 - \frac{B_2}{2!}(2x)^2 + \frac{B_4}{4!}(2x)^4 - \cdots$$

for all $x \in I$ then follows from the identity principle.

We shall conclude our discussion of real analytic functions by proving that this class of functions is closed under inversion.

Theorem 10. *Suppose that f is analytic on the interval I and that f′ does not vanish on I. Then f restricted to I is invertible, and its inverse is analytic.*

Proof. From the intermediate-value theorem, we know that f' must either be positive throughout I or negative throughout I. In either case f restricted to I is invertible and the image J of I under f is an open interval. We shall prove that $g = (f \mid I)^{-1}$ is analytic on J.

Let $y_0 \in J$. Without loss of generality, we may suppose that $y_0 = 0 = g(y_0)$ and that $f'(0) = 1$. Then $g'(0) = 1$, and we know that f has a series expansion

$$f(x) = x + a_2 x^2 + a_3 x^3 + a_4 x^4 + \cdots$$

valid throughout some neighborhood of zero. We must prove that $g(y)$ may similarly be written

$$g(y) = y + b_2 y^2 + b_3 y^3 + \cdots,$$

where the equation is valid throughout some neighborhood of zero. Let us assume momentarily that this is true. By our theorem on substitution of power series, the equation

$$x = g\big(f(x)\big) = x + (b_2 + a_2)x^2 + (b_3 + 2b_2 a_2 + a_3)x^3 \\ + (b_4 + b_2^2 a_2 + 2b_3 a_2 + 3b_2 a_3 + a_4)x^4 + \cdots$$

holds for all x in some neighborhood of zero. Hence the coefficients of $x^2, x^3, \ldots$ must all be zero. This leads to the recursion formula

$$b_2 = -a_2,$$
$$b_3 = -2b_2 a_2 - a_3,$$
$$b_4 = -b_2^2 a_2 - 2b_3 a_2 - 3b_2 a_3 - a_4,$$
$$\vdots$$

It now suffices to prove that if the coefficients $b_2, b_3, \ldots$ satisfy the equations above, then the series

$$y + b_2 y^2 + b_3 y^3 + \cdots,$$

which we shall call the *formal inverse* of the original series $x + a_2 x^2 + a_3 x^3 + \cdots$, has a positive radius of convergence. (Why?)

Suppose that $A_2, A_3, \ldots$ is any sequence of numbers such that $|a_k| \leq A_k$, for $k = 2, 3, \ldots$ If

$$y + B_2 y + B_3 y^2 + \cdots$$

is the formal inverse of the series

$$x - A_2 x^2 - A_3 x^3 - \cdots,$$

then it is immediately apparent from the equations above that $|b_k| \leq B_k$, for $k = 2$, $3, \ldots$ Hence if we can choose the sequence $A_2, A_3, \ldots$ in such a way that the series

$$y + B_1 y^2 + B_2 y^2 + \cdots$$

has a positive radius of convergence, then the proof will be complete.

Choose $\rho > 0$ so that the series $\rho + a_2 \rho^2 + \cdots$ converges. Then the terms of the series are bounded, say

$$|a_k \rho|^k \leq M \qquad (k = 2, 3, \ldots),$$

where M is a constant. We set $A_k = \rho^{-k} M$. Then, clearly, $|a_k| \leq A_k$, for $k = 2, 3, \ldots$ We let

$$F(x) = x - A_2 x^2 - A_3 x^3 - \cdots$$

$$= x - M \frac{x^2}{\rho^2} \left(1 + \frac{x}{\rho} + \frac{x^2}{\rho^2} + \cdots \right)$$

$$= x - \frac{Mx^2}{\rho(\rho - x)}.$$

The graph of

$$y = F(x) = M + \left(1 + \frac{M}{\rho} \right) x + \frac{M\rho}{x - \rho}$$ **FIG. 13-6**

is easily sketched (see Fig. 13-6). In particular, one can show that F is strictly increasing on the interval $(-\infty, x_1)$, where $x_1 = \rho[1 - \sqrt{M/(M + \rho)}]$. If G is the inverse of F restricted to $(-\infty, x_1)$, then a simple application of the quadratic formula shows that

$$G(y) = \frac{\rho}{2(M + \rho)} [\rho + y - \sqrt{y^2 - 2(2M + \rho)y + \rho^2}]$$

$$= \frac{\rho^2}{2(M + \rho)} \left[1 + \frac{y}{\rho} - \left(1 - \frac{y}{y_1} \right)^{1/2} \left(1 - \frac{y}{y_2} \right)^{1/2} \right],$$

where $y_1 = 2M + \rho - 2\sqrt{M(M + \rho)}$ and $y_2 = 2M + \rho + 2\sqrt{M(M + \rho)}$. Using the expansion

$$(1 - z)^{1/2} = 1 - z/2 + \cdots \qquad (|z| < 1),$$

we see that $G(y)$ can be expanded in a power series about zero which is valid if $|y| < y_1$. But this series must be

$$y + B_2 y^2 + B_3 y^3 + \cdots$$

Hence the proof is complete.

Algorithms for determining the formal inverse of a given power series due to Lagrange and Jacobi are given in the exercises.

We have gone about as far as we can without revealing the unfortunate truth that the real line is an unnatural setting for the study of power series and analytic functions. It is only when one can get off the tightrope of the real line and wander freely in the realm of the complex plane that the theory makes complete sense. Consider the function $f(x) = 1/(1 + x^2)$. It is analytic on the entire line, yet its Maclaurin expansion is valid only in the interval $(-1, +1)$. Our theory provides no explanation of why the radius of convergence should be *a priori* one rather than 23 or $+\infty$. The theory of analytic functions of a *complex* variable, by contrast, offers a completely satisfactory explanation. If we are given a power series

$$\sum_{k=0}^{\infty} a_k(z - z_0)^k,$$

where the a_k's, z, and z_0 may be complex numbers, then one can show that the set of z for which the series converges is either the singleton set $\{z_0\}$, the entire complex plane, or a disk centered at z_0. Furthermore, in the latter case the disk of convergence extends out to the "first singularity" of the function defined by the series. The function $1/(1 + z^2)$ has no singularities on the real line, but it has two in the complex plane, namely, $+i$ and $-i$. The disk of convergence of the Maclaurin series for the function extends out to these points and hence has radius one (see Fig. 13-7). Similarly, using complex variable methods, one can easily show that the set of real numbers x for which

$$x \operatorname{ctn} x = 1 - \frac{B_2}{2!}(2x)^2 + \frac{B_4}{4!}(4x)^4 - \cdots$$

is precisely the interval $(-\pi, \pi)$.

FIG. 13-7

Exercises

1. Determine the intervals of analyticity for each of the following functions. Justify your answers.

   a) $\sqrt{x}$

   b) x^x

   c) $f(x) = \begin{cases} e^{-1/x^2} & \text{if } x \neq 0, \\ 0 & \text{if } x = 0 \end{cases}$

   d) $\tan x$

   e) $|x|$

   f) $\operatorname{Sin}^{-1} x$

   g) $\sqrt{1 - x^2}$

2. Verify that each of the following expansions is valid throughout some neighborhood of zero. Determine as best you can the intervals throughout which the expansions are valid.

   a) $\frac{1}{2}(\operatorname{Tan}^{-1} x)^2 = \sum_{k=1}^{\infty} (-1)^{k-1} b_k \frac{x^{2k}}{2k}$, where $b_k = 1 + \frac{1}{3} + \frac{1}{5} + \cdots + \frac{1}{2k - 1}$

   b) $\frac{1}{2}(\operatorname{Tan}^{-1} x) \ln(1 + x^2) = \sum_{k=1}^{\infty} (-1)^{k-1} h_{2k} \frac{x^{2k+1}}{2k + 1}$, where $h_k = 1 + \frac{1}{2} + \cdots + \frac{1}{n}$

   c) $\frac{1}{2}\left(\ln \frac{1}{1 - x}\right)^2 = \sum_{k=2}^{\infty} \frac{h_{k-1} x^k}{k}$

3. Verify the following.

   a) $\dfrac{x}{\ln\,[1/(1\,-\,x)]} = 1 - \dfrac{x}{2} - \dfrac{x^2}{12} - \dfrac{x^3}{24} - \cdots$

   b) $(1 - x)e^{x+x^2/2+\cdots+x^m/m} = 1 - \dfrac{x^{m+1}}{m+1} + \cdots$ $(m \geq 1)$

   c) $\tan\,(\sin x) - \sin\,(\tan x) = \frac{1}{30}x^7 + \frac{29}{756}x^9 + \cdots$

   d) $(1/e)(1 + x)^{1/x} = 1 - x/2 + \frac{11}{24}x^2 - \frac{7}{16}x^3 + \cdots$

4. Show that the power series expansions for $1/f$ can be obtained from those of f through long division.

5. Suppose that throughout some neighborhood of zero we have

$$\frac{1 + a_2x^2 + a_4x^4 + \cdots}{1 + b_2x^2 + b_4x^4 + \cdots} = 1 + c_2x^2 + c_4x^4 + \cdots$$

   Prove that we have

$$\frac{1 - a_2x^2 + a_4x^4 - \cdots}{1 - b_2x^2 + b_4x^4 - \cdots} = 1 - c_2x^2 + c_4x^4 - \cdots$$

   throughout the same neighborhood. Note that this justifies Example 8.

6. Prove the identity

$$\tan x = \operatorname{ctn} x - 2 \operatorname{ctn} 2x.$$

   Use this identity and Example 8 to show that

$$\tan x = \sum_{k=1}^{\infty} (-1)^{k-1} \frac{2^{2k}(2^{2k} - 1)B_{2k}}{(2k)!} x^{2k-1}.$$

7. Prove the identity

$$\frac{1}{\sin x} = \operatorname{ctn} x + \tan \frac{x}{2}.$$

   Show that

$$\frac{x}{\sin x} = \sum_{k=0}^{\infty} (-1)^{k-1} \frac{(2^{2k} - 2)B_{2k}}{(2k)!} x^{2k}.$$

*8. Prove that the Riemann zeta function is real analytic on the interval $(1, +\infty)$.

9. The purpose of this problem is to generalize somewhat our discussion of the Bernoulli numbers.

   a) For each t, we let

$$f_t(x) = \begin{cases} \dfrac{xe^{tx}}{e^x - 1} & \text{if } x \neq 0, \\ 1 & \text{if } x = 0. \end{cases}$$

   Observe that f_t is analytic on the entire line and hence has an expansion

$$f_t(x) = \sum_{k=0}^{\infty} B_k(t) \frac{x^k}{k!}$$

which is valid throughout some neighborhood of zero. From the identity

$$f_t(x) = e^{tx}f_0(x),$$

conclude that

$$B_n(t) = \sum_{k=0}^{n} \binom{n}{k} B_k t^{n-k},$$

where $B_k = B_k(0)$ is the kth Bernoulli number. The polynomial $B_n(t)$ is called the nth *Bernoulli polynomial*. Prove the following.

b) $B_0 = 1, B_1 = -\frac{1}{2}, \sum_{k=0}^{n-1} \binom{n}{k} B_k = 0 \qquad (n = 2, 3, \ldots)$

c) $B_n'(t) = nB_{n-1}(t) \qquad (n \in \mathbf{Z}^+)$

d) $B_n(t+1) - B_n(t) = nt^{n-1} \qquad (n \in \mathbf{Z}^+)$

e) $B_n(1 - t) = (-1)^n B_n(t) \qquad (n \in \mathbf{Z}^+)$

f) $B_{2n+1} = 0 \qquad (n \in \mathbf{Z}^+)$

g) $\displaystyle\sum_{k=1}^{m} k^n = \frac{B_{n+1}(m) - B_{n+1}}{n+1} \qquad (n = 2, 3, \ldots)$

10. Suppose that the functions f and h are both analytic throughout some neighborhood of zero, and suppose that $f(0) \neq 0$. Observe that the function $g(x) = xf(x)$ is analytic and invertible when restricted to a suitably small neighborhood of zero.

a) Show that $h \circ g^{-1}$ is analytic near zero and that

$$\frac{d^n}{dx^n} (h \circ g^{-1}) \bigg|_{x=0} = \frac{d^{n-1}}{dx^{n-1}} \left\{ \frac{h'(x)}{[f(x)]^n} \right\} \bigg|_{x=0}$$

(*Lagrange's formula*).

b) If $g(x) = x(1 + x)^m$, use (a) to show that g^{-1} has the Maclaurin expansion

$$g^{-1}(x) = x - \frac{2m}{2!} x^2 + \frac{3m(3m + 1)}{3!} x^3 - \frac{4m(4m + 1)(4m + 2)}{4!} x^4 + \cdots$$

c) If $g(x) = xe^{bx}$ and $h(x) = e^{ax}$, use (a) to verify the expansion

$$(h \circ g^{-1})(x) = 1 + ax + \frac{a(a - 2b)}{2!} x^2 + \frac{a(a - 3b)^2}{3!} x^3 + \cdots$$

Prove that the expansion is valid if $|xb| < e^{-1}$.

d) From part (c), deduce Eisenstein's solution of the transcendental equation $\ln x = \alpha x$, namely,

$$x = 1 + \alpha + 3\frac{\alpha^2}{2!} + 4^2\frac{\alpha^3}{3!} + 5^3\frac{\alpha^4}{4!} + \cdots,$$

where $|\alpha| < e^{-1}$.

*e) We define a function f by the equation

$$f(x) = x^{x^{x^{\cdot^{\cdot^{\cdot}}}}}$$

Give an appropriate meaning to the right-hand side of the equation. Determine the domain and range of f and find f^{-1}. Are f and f^{-1} analytic functions?

13-8 Fourier series

A *trigonometric series* is of the form

$$\frac{a_0}{2} + \sum_{k=1}^{\infty} (a_k \cos kx + b_k \sin kx),$$

where the a_k's and b_k's are constants. A great deal of attention has been lavished upon such series; the literature devoted to them is extensive and deep. We offer here only a modest introduction to the subject.

To get a feeling for these series, let us first assume that the series consisting of only the coefficients converges absolutely, i.e., that

$$\frac{|a_0|}{2} + \sum_{k=1}^{\infty} (|a_k| + |b_k|) < +\infty.$$

By the Weierstrass M-test, it then follows that the series converges absolutely and uniformly on the entire line. Let $f(x)$ be the sum of the series. Since each term of the series is continuous and periodic with period 2π, the function f also has these two properties. Let us think of "x" as a time variable, and let us regard the functions as specifying the displacements of moving particles. The individual terms of the series represent *pure oscillations*; they describe the simplest type of periodic movement, namely, simple harmonic motion. The series itself may then be interpreted as representing the *superposition* of simple vibrations, and the equation

$$f(x) = \frac{a_0}{2} + \sum_{k=1}^{\infty} (a_k \cos kx + b_k \sin kx) \qquad (*)$$

may be thought of as describing a resolution of the (possibly very complicated) periodic motion described by f into pure oscillations.

A relation between the coefficients and the function f is easily established. Term-by-term integration yields

$$\int_{-\pi}^{\pi} f(x)\,dx = \frac{a_0}{2} \int_{-\pi}^{\pi} dx + \sum_{k=1}^{\infty} \left(a_k \int_{-\pi}^{\pi} \cos kx\,dx + b_k \int_{-\pi}^{\pi} \sin kx\,dx \right) = a_0 \pi.$$

(The term-by-term integration is justified, of course, because of the uniform convergence.) Next we multiply $(*)$ by $\cos nx$; since the resulting equation holds uniformly in x, term-by-term integration is again possible. We get

$$\int_{-\pi}^{\pi} f(x) \cos nx\,dx = \frac{a_0}{2} \int_{-\pi}^{\pi} \cos nx\,dx$$

$$+ \sum_{k=1}^{\infty} \left(a_k \int_{-\pi}^{\pi} \cos kx \cos nx\,dx + b_k \int_{-\pi}^{\pi} \sin kx \cos nx\,dx \right).$$

The right-hand side simplifies to a single term as a result of the *orthogonality relations*:

$$\int_{-\pi}^{\pi} \cos mx \cos nx \, dx = \begin{cases} 0 & \text{if} \quad m \neq n, \\ \pi & \text{if} \quad m = n, \end{cases}$$

$$\int_{-\pi}^{\pi} \sin mx \sin nx \, dx = \begin{cases} 0 & \text{if} \quad m \neq n, \\ \pi & \text{if} \quad m = n, \end{cases}$$

$$\int_{-\pi}^{\pi} \cos mx \sin nx \, dx = 0.$$

Thus

$$\int_{-\pi}^{\pi} f(x) \cos nx \, dx = a_n \pi.$$

Similarly,

$$\int_{-\pi}^{\pi} f(x) \sin nx \, dx = b_n \pi.$$

In summary, then,

$$a_k = \frac{1}{\pi} \int_{-\pi}^{\pi} f(x) \cos kx \, dx \qquad (k = 0, 1, 2, \ldots),$$

$$b_k = \frac{1}{\pi} \int_{-\pi}^{\pi} f(x) \sin kx \, dx \qquad (k = 1, 2, \ldots).$$

This leads to a definition.

Definition 1. If f is Riemann-integrable on the interval $[-\pi, \pi]$, and if the numbers $a_0, a_1, a_2, \ldots, b_1, b_2, \ldots$ are defined by the above equations, then these numbers are called the **Fourier coefficients** of f, and the trigonometric series

$$\frac{a_0}{2} + \sum_{k=1}^{\infty} (a_k \cos kx + b_k \sin kx)$$

is called the **Fourier series** of f. We write

$$f \sim \frac{a_0}{2} + \sum_{k=1}^{\infty} (a_k \cos kx + b_k \sin kx).$$

Our preceding discussion may be summarized as follows:

Proposition 2. *If the trigonometric series*

$$\frac{a_0}{2} + \sum_{k=1}^{\infty} (a_k \cos kx + b_k \sin kx)$$

converges uniformly on **R**, *and if* $f(x)$ *is the sum of the series, then the given series is the Fourier series of* f.

The main question which we shall consider is this: If f is Riemann-integrable on $[-\pi, \pi]$, when and in what sense does the Fourier series of f converge to f? If one prefers, the question can be reformulated in the language of physics: When and in what sense can an arbitrary periodic motion be synthesized by the superposition of pure oscillations?

We list three functions together with their Fourier series. The examples were not randomly selected; later we shall see that they are of genuine interest. The computations have been omitted, but the reader is urged to supply them before plunging into the more abstract discussion which follows.

Example 1. Let

$$f(x) = \begin{cases} \pi/4 & \text{if} \quad 0 \le x \le \pi, \\ -\pi/4 & \text{if} \quad -\pi \le x < 0. \end{cases}$$

Then

$$f \sim \sin x + \frac{\sin 3x}{3} + \frac{\sin 5x}{5} + \cdots$$

Example 2. Let $f(x) = |x|$ $(-\pi \le x \le \pi)$. Then

$$f \sim \frac{\pi}{2} - \frac{4}{\pi}\left(\cos x + \frac{\cos 3x}{3^2} + \frac{\cos 5x}{5^2} + \cdots\right).$$

Example 3. If $f(x) = \cos cx$ $(-\pi \le x \le \pi)$, where c is not an integer, then

$$f \sim \frac{2c}{\pi}\sin c\pi \left(\frac{1}{2c^2} + \sum_{k=1}^{\infty} \frac{(-1)^k \cos kx}{c^2 - k^2}\right).$$

1. General properties of orthogonal expansions

Some of the ideas and techniques involved in the study of Fourier series apply to a more general class of series. Accordingly, we shall do some of our spade work in a more general setting.

We shall work over an arbitrary closed bounded interval $[a, b]$. We let $\Re$ denote the class of all Riemann-integrable functions on $[a, b]$. If f and g belong to $\Re$, we define their *inner product* to be

$$\langle f, g \rangle = \int_a^b f(x)\, g(x)\, dx.$$

[The more usual notation is (f, g).] Inner products have the following properties.

1) *Bilinearity.* The inner product is linear in each of its variables separately; that is,

$$\langle f_1 + f_2, g \rangle = \langle f_1, g \rangle + \langle f_2, g \rangle,$$
$$\langle f, g_1 + g_2 \rangle = \langle f, g_1 \rangle + \langle f, g_2 \rangle,$$

and

$$\langle cf, g \rangle = c\langle f, g \rangle = \langle f, cg \rangle,$$

where c is a constant.

2) *Symmetry.* $\langle f, g \rangle = \langle g, f \rangle$.

3) *Positivity.* For every $f \in \mathfrak{R}$, $\langle f, f \rangle \geq 0$.

It follows by induction on (1) that

$$\left\langle \sum_{j=1}^{n} a_k f_k, \sum_{k=1}^{n} b_k g_k \right\rangle = \sum_{j,k=1}^{m,n} a_j b_k \langle f_j, g_k \rangle,$$

where $a_1, \ldots, a_m, b_1, \ldots, b_n$ are constants and $f_1, \ldots, f_m, g_1, \ldots, g_n$ belong to $\mathfrak{R}$. For this reason, the term "product" is used.

If $f \in \mathfrak{R}$, we shall also define its L_2-*norm* to be

$$\|f\|_2 = \sqrt{\langle f, f \rangle} = \left\{ \int_a^b [f(x)]^2 \, dx^{1/2} \right\}.$$

We shall regard $\|f\|_2$ as one possible measure of the size of f. If $f(t)$ represents the intensity of an alternating current at time t, then $\|f\|_2$ is proportional to the so-called *effective current* during the time interval from a to b. The L_2-norm has two important properties.

1) *Cauchy-Schwarz Inequality.* For all f and g in $\mathfrak{R}$,

$$|\langle f, g \rangle| \leq \|f\|_2 \|g\|_2.$$

2) *Minkowski Inequality.* For all f and g in $\mathfrak{R}$,

$$\|f + g\|_2 \leq \|f\|_2 + \|g\|_2.$$

(See Exercises 3 and 4, Section 9–7.) We have the more trivial equation

$$\|cf\|_2 = |c| \, \|f\|_2,$$

where c is a constant.

We shall be concerned with series of the form

$$\sum_{k=1}^{\infty} c_k \varphi_k(x),$$

where the φ_k's are functions defined on the interval $[a, b]$ and have the following property:

$$\langle \varphi_j, \varphi_k \rangle = \delta_{jk} = \begin{cases} 0 & \text{if } j \neq k, \\ 1 & \text{if } j = k. \end{cases}$$

Such a sequence of functions $\varphi_1, \varphi_2, \ldots$ is said to be *orthonormal* on $[a, b]$. We point out two important examples of orthonormal families of functions.

Example 4. Let $[a, b] = [-\pi, \pi]$. Then the functions

$$\frac{1}{\sqrt{2\pi}}, \quad \frac{\cos x}{\sqrt{\pi}}, \quad \frac{\sin x}{\sqrt{\pi}}, \quad \frac{\cos 2x}{\sqrt{\pi}}, \quad \frac{\sin 2x}{\sqrt{\pi}}, \quad \cdots$$

are orthonormal.

Example 5. Let $[a, b] = [-1, +1]$, and let

$$\varphi_n(x) = \sqrt{\frac{2}{n+1}} P_n(x),$$

where P_n is the Legendre polynomial of order n. Then $\varphi_1, \varphi_2, \ldots$ is an orthonormal sequence of functions. (The proof is partially outlined in Exercise 6, Section 11–4.)

Suppose now that $\varphi_1, \varphi_2, \ldots$ is orthonormal on $[a, b]$ and that the infinite series $\sum_{k=1}^{\infty} c_k \varphi_k(x)$ converges uniformly on the interval $[a, b]$. We let $f(x)$ be the sum of the series. For each $n \in \mathbf{Z}^+$, we then have

$$f(x)\varphi_n(x) = \sum_{k=1}^{\infty} c_k \varphi_k(x)\varphi_n(x),$$

convergence being uniform. By integrating term by term, we get

$$\langle f, \varphi_n \rangle = \sum_{k=0}^{\infty} c_k \langle \varphi_k, \varphi_n \rangle = c_n.$$

Definition 1′. Let $\varphi_1, \varphi_2, \ldots$ be orthonormal on $[a, b]$. If $f \in \mathfrak{R}$, then

$$c_n = \langle f, \varphi_n \rangle = \int_a^b f(x)\varphi_n(x)\, dx$$

is called the **nth Fourier coefficient** of f relative to the orthonormal sequence $\{\varphi_n\}$. The series $\sum_{k=1}^{\infty} c_k \varphi_k(x)$ is called the **Fourier series** of f, and we write

$$f \sim \sum_{k=1}^{\infty} c_k \varphi_k.$$

In particular, if the functions $\varphi_1, \varphi_2, \ldots$ are those of Example 4, then the Fourier series of this definition is the trigonometric Fourier series previously defined. Furthermore, we have, as before, the following result.

Proposition 2′. *If*

$$f(x) = \sum_{k=1}^{\infty} c_n \varphi_n(x)$$

uniformly on $[a, b]$, then $\sum_{k=1}^{\infty} c_k \varphi_k$ is the Fourier series of f.

If $f \in \mathfrak{R}$, it is natural to ask how well it can be approximated by the partial sums of its Fourier series. Our first theorem deals with the size of the remainders as measured by the L_2-norm.

Proposition 3. *Best Approximations in the Sense of Least Squares. Let $\varphi_1, \varphi_2, \ldots$ be orthonormal on $[a, b]$. Let $f \in \mathfrak{R}$, and let $\sum_{k=1}^{\infty} c_k \varphi_k$ be its Fourier series. Then among all functions of the form $g = \sum_{k=1}^{n} a_k \varphi_k$, where the a_k's are constants, the nth partial sum of the Fourier series of f, $s_n = \sum_{k=1}^{n} c_k \varphi_k$, is the unique function for which $\|f - g\|_2$ is a minimum.*

Proof

$$0 \leq \|f - g\|_2^2 = \langle f - g, f - g \rangle = \langle f, f \rangle - 2\langle f, g \rangle + \langle g, g \rangle$$

$$= \langle f, f \rangle - 2 \sum_{k=1}^{n} a_k \langle f, \varphi_k \rangle + \sum_{j,k=1}^{n} a_j a_k \langle \varphi_j, \varphi_k \rangle$$

$$= \langle f, f \rangle - 2 \sum_{k=1}^{n} a_k c_k + \sum_{k=1}^{n} a_k^2$$

$$= \langle f, f \rangle - \sum_{k=1}^{n} c_k^2 + \sum_{k=1}^{n} (a_k - c_k)^2.$$

The latter expression is clearly a minmum when $a_k = c_k$ ($k = 1, \ldots, n$), in which case

$$\|f - g\|_2^2 = \|f - s_n\|_2^2 = \langle f, f \rangle - \sum_{k=1}^{n} c_k^2 = \|f\|_2^2 - \sum_{k=1}^{n} c_k^2.$$

In summary, we have

$$0 \leq \|f\|_2^2 - \sum_{k=1}^{n} c_k^2 = \|f - s_n\|_2^2 \leq \|f - g\|_2^2;$$

the second inequality is strict unless $a_k = c_k$ ($k = 1, \ldots, n$). This proves the assertion.

An expression for the L_2-norm of the remainder in a Fourier series is a by-product of the above calculation.

Proposition 4. *Under the hypotheses of the previous theorem,*

$$\left\| f - \sum_{k=1}^{n} c_k \varphi_k \right\|_2^2 = \|f\|_2^2 - \sum_{k=1}^{n} c_k^2.$$

There are two immediate consequences of the above equation. First, since

$$\sum_{k=1}^{n} c_k^2 \leq \|f\|_2^2$$

for each $n \in \mathbf{Z}^+$, the series $\sum_{k=1}^{\infty} c_k^2$ converges and

$$\sum_{k=1}^{\infty} c_k^2 \leq \|f\|_2^2.$$

Second, the inequality $\sum_{k=1}^{\infty} c_k^2 \leq \|f\|^2$ is strict unless

$$\lim_{n \to \infty} \left\| f - \sum_{k=1}^{n} c_k \varphi_k \right\|_2 = 0,$$

in which case the Fourier series $\sum_{k=1}^{\infty} c_k \varphi_k$ is said to converge to f *in the mean.*

Definition 5. A sequence $f_1, f_2, \ldots$ of functions in $\mathcal{R}$ is said to **converge in the mean** to a function f in $\mathcal{R}$ iff

$$\lim_{n\to\infty} \|f - f_n\|_2 = 0.$$

Convergence in the mean is a much weaker type of convergence than uniform convergence.

Proposition 6. *If $f_n \to f$ uniformly on $[a, b]$, then $\{f_n\}$ converges in the mean to f.*

Proof

$$\|f - f_n\|_2 = \left\{\int_a^b [f(x) - f_n(x)]^2 \, dx\right\}^{1/2}$$

$$\leq \left(\int_a^b \|f - f_n\|^2 \, dx\right)^{1/2} = \sqrt{b - a} \, \|f - f_n\|,$$

where $\| \ \|$ indicates the uniform norm:

$$\|f\| = \text{lub } \{|f(x)| \mid a \leq x \leq b\}.$$

Since $f_n \to f$ uniformly on $[a, b]$, it follows that $\|f - f_n\| \to 0$ and hence

$$\|f - f_n\|_2 \to 0$$

also.

It will be shown in the exercises that convergence in the mean does not even imply pointwise convergence.

Proposition 7. *Bessel's Inequality. Let $\varphi_1, \varphi_2, \ldots$ be orthonormal on $[a, b]$. Let $f \in \mathcal{R}$, and let $\sum_{k=1}^{\infty} c_k\varphi_k$ be the Fourier series of f relative to $\{\varphi_n\}$. Then*

$$\sum_{k=1}^{\infty} c_k^2 \leq \|f\|_2^2,$$

with equality iff the Fourier series converges to f in the mean.

The theorem has several important corollaries.

Corollary 8. *If f is Riemann-integrable on $[-\pi, \pi]$, and if*

$$f \sim \frac{a_0}{2} + \sum_{k=1}^{\infty} (a_k \cos kx + b_k \sin kx),$$

then

$$\frac{a_0^2}{2} + \sum_{k=1}^{\infty} (a_k^2 + b_k^2) \leq \frac{1}{\pi} \int_{-\pi}^{\pi} [f(x)]^2 \, dx.$$

We shall later prove that equality actually holds.

Corollary 9. *Riemann-Lebesgue Lemma. Under the hypotheses of the theorem,* $\lim_{n\to\infty} c_n = 0$. *In particular, if f is Riemann-integrable on the interval $[-\pi, \pi]$, then*

$$\lim_{n\to\infty} \int_{-\pi}^{\pi} f(x) \cos nx \, dx = 0 \quad and \quad \lim_{n\to\infty} \int_{-\pi}^{\pi} f(x) \sin nx \, dx = 0.$$

Corollary 10. *If under the hypothesis of the theorem the Fourier series of f converges to f uniformly on* [a, b], *then*

$$\sum_{k=1}^{\infty} c_k^2 = \|f\|_2^2.$$

2. Pointwise convergence of Fourier series

We return now to the study of trigonometric Fourier series. A representation for the nth partial sum as an integral is essential to the study of pointwise convergence of these series.

Suppose

$$f \sim \frac{a_0}{2} + \sum_{k=1}^{\infty} (a_k \cos kx + b_k \sin kx).$$

Now

$$s_n(x) = \frac{a_0}{2} + \sum_{k=1}^{n} (a_k \cos kx + b_k \sin kx)$$

$$= \frac{1}{2\pi} \int_{-\pi}^{\pi} f(t) \, dt + \frac{1}{\pi} \sum_{k=1}^{n} \int_{-\pi}^{\pi} f(t)(\cos kt \cos kx + \sin kt \sin kx) \, dt$$

$$= \frac{1}{2\pi} \int_{-\pi}^{\pi} f(t) \, dt + \frac{1}{\pi} \sum_{k=1}^{n} \int_{-\pi}^{\pi} f(t) \cos k(x - t) \, dt$$

$$= \frac{1}{\pi} \int_{-\pi}^{\pi} f(t) D_n(x - t),$$

where

$$D_n(x) = \tfrac{1}{2} + \cos x + \cos 2x + \cdots + \cos nx$$

$$= \frac{\sin (n + \tfrac{1}{2})x}{2 \sin (x/2)}.$$

If necessary, we shall redefine the function at the point $+\pi$ so that $f(\pi) = f(-\pi)$; this will not alter either the integrability of f or the Fourier coefficients of f. We then (re)define the function f outside the interval $[-\pi, \pi]$ so that it has period 2π. Having done this, we may derive other expressions for $s_n(x)$. We make the substitution $u = t - x$ and observe that D_n (sometimes called the *Dirichlet kernel*) is an even function with period 2π. Thus we have

$$s_n(x) = \frac{1}{\pi} \int_{-\pi}^{\pi} f(t) D_n(x - t) \, dt$$

$$= \frac{1}{\pi} \int_{-\pi-x}^{\pi-x} f(x + u) D_n(-u) \, dt$$

$$= \frac{1}{\pi} \int_{-\pi}^{\pi} f(x + u) D_n(u) \, du.$$

The substitution $u = x - t$ leads to

$$s_n(x) = \frac{1}{\pi} \int_{x-\pi}^{x+\pi} f(x-u)D_n(u)\,du$$

$$= \frac{1}{\pi} \int_{-\pi}^{\pi} f(x-u)D_n(u)\,du.$$

We can obtain another expression for $s_n(x)$ by averaging the previous two, namely,

$$s_n(x) = \frac{1}{\pi} \int_{-\pi}^{\pi} \frac{f(x+u) + f(x-u)}{2} D_n(u)\,du.$$

Proposition 11. Dirichlet Formula. *Suppose f has period 2π and is Riemann-integrable on $[-\pi, \pi]$. Then the nth partial sum of its Fourier series may be written*

$$s_n(x) = \frac{1}{\pi} \int_{-\pi}^{\pi} f(t)D_n(x-t)\,dt = \frac{1}{\pi} \int_{-\pi}^{\pi} \frac{f(x+t) + f(x-t)}{2} D_n(t)\,dt,$$

where

$$D_n(x) = \tfrac{1}{2} + \cos x + \cos 2x + \cdots + \cos nx = \frac{\sin (n + \tfrac{1}{2})t}{2 \sin (t/2)}.$$

Whether or not the Fourier series converges to f at a particular point x_0 depends only on the behavior of f near the point. (This is in marked contrast to the situation with power series.) Let us suppose that the following exist (and are finite):

$$f(x_0+) = \lim_{x \to x_0} f(x), \qquad f(x_0-) = \lim_{x \to x_0} f(x),$$

$$f'_+(x_0) = \lim_{x \to x_0} \frac{f(x) - f(x_0+)}{x - x_0},$$

and

$$f'_-(x_0) = \lim_{x \to x_0-} \frac{f(x) - f(x_0-)}{x - x_0}.$$

In effect, we are assuming that f possesses one-sided limits and one-sided derivatives at x_0. We then say that f satisfies the *Dirichlet condition* at x_0. We shall now prove that the Dirichlet condition is a sufficient condition for pointwise convergence of a Fourier series.

Theorem 12. *Suppose that f has period 2π and is Riemann-integrable on the interval $[-\pi, \pi]$. If f satisfies the Dirichlet condition at a point x_0, then the Fourier series of f converges at x_0 to $[f(x_0+) + f(x_0-)]/2$.*

Proof. Since

$$D_n(t) = \tfrac{1}{2} + \cos t + \cos 2t + \cdots + \cos nt,$$

we get, upon integrating both sides,

$$\frac{1}{\pi} \int_{-\pi}^{\pi} D_n(t)\,dt = 1.$$

Thus

$$s_n(x_0) - \frac{f(x_0+) + f(x_0-)}{2} = \frac{1}{\pi} \int_{-\pi}^{\pi} \frac{f(x_0 + t) + f(x_0 - t)}{2} D_n(t)\, dt$$

$$- \frac{1}{\pi} \int_{-\pi}^{\pi} \frac{f(x_0+) + f(x_0-)}{2} D_n(t)\, dt$$

$$= \frac{1}{\pi} \int_{-\pi}^{\pi} g(t) \sin (n + \tfrac{1}{2})t\, dt,$$

where

$$g(t) = \frac{1}{2} \frac{f(x_0 + t) + f(x_0 - t) - f(x_0+) - f(x_0-)}{t} \frac{t}{2 \sin (t/2)}.$$

We observe that the function g is Riemann-integrable on the interval $[-\pi, \pi]$. Since f is Riemann-integrable, the only thing that could conceivably prevent g from being Riemann-integrable is the unboundedness of g near zero. However, since f satisfies the Dirichlet condition at x_0, it is very easy to show that

$$\lim_{t \to 0} g(t) = \tfrac{1}{2}[f'_+(x_0) + f'_-(x_0)],$$

so, in particular, g is bounded near zero.

Thus we have

$$s_n(x_0) - \frac{f(x_0+) + f(x_0-)}{2} = \frac{1}{\pi} \int_{-\pi}^{\pi} g(t) \sin (n + \tfrac{1}{2})t\, dt$$

$$= \frac{1}{\pi} \int_{-\pi}^{\pi} g(t) \sin \frac{t}{2} \cos nt\, dt$$

$$+ \frac{1}{\pi} \int_{-\pi}^{\pi} g(t) \cos \frac{t}{2} \sin nt\, dt.$$

The last two terms, however, tend to zero as $n \to \infty$, by the Riemann-Lebesgue lemma. Hence the theorem is proved.

Corollary 13. *Suppose that*

1) f has period 2π and is Riemann-integrable on $[-\pi, \pi]$;

2) f satisfies the Dirichlet condition at x_0;

3) f is continuous at x_0.

Then the Fourier series for f converges at x_0 to $f(x_0)$.

Example 6. Let f be the function of Example 1. Then f satisfies the Dirichlet condition at each of its points. Hence for values of x in $[-\pi, \pi]$,

$$\sin x + \frac{\sin 3x}{3} + \frac{\sin 5x}{5} + \cdots = \begin{cases} \pi/4 & \text{if } 0 < x < \pi, \\ 0 & \text{if } x = -\pi, 0, \pi, \\ -\pi/4 & \text{if } -\pi < x < 0. \end{cases}$$

Setting $x = \pi/2$, we have, in particular,

$$1 - \frac{1}{3} + \frac{1}{5} - \frac{1}{7} + \cdots = \frac{\pi}{4}.$$

Example 7. Let f be the function of Example 2. We conclude from the convergence theorem that

$$\frac{\pi}{2} - \frac{4}{\pi}\left(\cos x + \frac{\cos 3x}{3^2} + \frac{\cos 5x}{5^2} + \cdots\right) = |x| \quad \text{if} \quad -\pi \le x \le \pi.$$

Setting $x = 0$, we get

$$1 + \frac{1}{3^2} + \frac{1}{5^2} + \cdots = \frac{\pi^2}{8}.$$

From the series we can once again see that

$$\zeta(2) = \sum_{k=1}^{\infty} \frac{1}{k^2} = \frac{\pi^2}{6},$$

for

$$\frac{1}{2^2} + \frac{1}{4^2} + \frac{1}{6^2} + \cdots = \tfrac{1}{4}\zeta(2),$$

so by addition,

$$\zeta(2) = \frac{\zeta(2)}{4} + \frac{\pi^2}{8},$$

from which the desired result follows immediately.

Example 8. Applying the convergence theorem to the function in Example 3, we have

$$\cos cx = \frac{2c}{\pi}\sin c\pi\left(\frac{1}{2c^2} + \sum_{k=1}^{\infty}\frac{(-1)^k\cos kx}{c^2 - k^2}\right) \quad \text{if} \quad -\pi \le x \le \pi.$$

We set $x = \pi$ and switch notation slightly to get

$$\operatorname{ctn}\pi t = \frac{2t}{\pi}\left(\frac{1}{2t^2} + \frac{1}{t^2 - 1^2} + \frac{1}{t^2 - 2^2} + \cdots\right) \quad (t \ne 0, \pm 1, \pm 2, \ldots).$$

This "expansion of $\operatorname{ctn}\pi t$ into partial fractions" has many remarkable consequences. Among them is an infinite product expansion for the sine function, namely,

$$\frac{\sin \pi x}{\pi x} = \left(1 - \frac{x^2}{1^2}\right)\left(1 - \frac{x^2}{2^2}\right)\left(1 - \frac{x^2}{3^2}\right)\cdots,$$

from which, in particular, Wallis' product

$$\frac{\pi}{2} = \frac{2}{1}\cdot\frac{2}{3}\cdot\frac{4}{3}\cdot\frac{4}{5}\cdots$$

follows (with $x = \frac{1}{2}$). We can also obtain a formula for the values of the Riemann zeta function at the positive even integers, namely,

$$\zeta(2n) = (-1)^{n-1} \frac{B_{2n}(2\pi)^{2n}}{2(2n)!},$$

where $B_1, B_2, B_3, \ldots$ is the sequence of Bernoulli numbers. The derivations of these and other similar results will be outlined in the exercises.

3. Fejér's theorem

The pointwise convergence of a Fourier series is a delicate matter; in some respects it is not the "natural" type of convergence to consider. We shall prove, however, that the Fourier series of a continuous function converges to the function in the L_2-sense. We shall also show that if we take averages of the partial sums, these averages converge uniformly to the function.

Let f be Riemann-integrable on $[-\pi, \pi]$, and let

$$f \sim \frac{a_0}{2} + \sum_{k=1}^{\infty} (a_k \cos kx + b_k \sin kx).$$

We denote by $s_0(x), s_1(x), s_2(x), \ldots$ the partial sums of the Fourier series, and for each $n \in \mathbf{Z}^+$, we define

$$\sigma_n(x) = (1/n)[s_0(x) + s_1(x) + \cdots + s_{n-1}(x)].$$

Then $\sigma_n(x)$ is called the nth *Cesàro mean* of the Fourier series. We shall now derive an integral formula for σ_n:

$$\sigma_n(x) = \frac{1}{n} \sum_{k=0}^{n-1} s_k(x)$$

$$= \frac{1}{n} \sum_{k=0}^{n-1} \frac{1}{\pi} \int_{-\pi}^{\pi} f(t) D_k(x - t)\, dt$$

$$= \frac{1}{\pi} \int_{-\pi}^{\pi} f(t) K_n(x - t),$$

where

$$K_n(t) = \frac{1}{n} \sum_{k=0}^{n-1} D_k(t)$$

is called the nth *Fejér kernel*. The identity

$$K_n(t) = \frac{1}{2n}\left[\frac{\sin(nx/2)}{\sin(x/2)}\right]^2$$

is left as a challenging computational problem for the reader.

Lemma 14. Properties of the Fejér Kernel

    1) K_n is a continuous nonnegative even function with period 2π.

    2) $(1/\pi) \int_{-\pi}^{\pi} K_n(t)\, dt = 1$.

    3) If $0 < \delta < \pi$, then

$$\lim_{n \to \infty} \operatorname{lub} \{K_n(x) \mid \delta \le |x| \le \pi\} = 0.$$

Proof. Property (2) is a evident from the definition of K_n and the properties of the Dirichlet kernels:

$$\frac{1}{\pi} \int_{-\pi}^{\pi} K_n(t)\, dt = \frac{1}{n} \sum_{k=0}^{n-1} \frac{1}{\pi} \int_{-\pi}^{\pi} D_k(t)\, dt = \frac{1}{n} \cdot n = 1.$$

Property (3) requires a couple of simple inequalities. If $0 < \delta \le |x| \le \pi$, then

$$[\sin (x/2)]^2 \ge [\sin (\delta/2)]^2,$$

so that

$$K_n(x) = \frac{1}{2n} \left[\frac{\sin (nx/2)}{\sin (x/2)} \right]^2 \le \frac{1}{2n[\sin (\delta/2)]^2}.$$

Theorem 15. Fejér's Theorem. *If f is continuous and has period 2π, then the sequence $\sigma_1, \sigma_2, \ldots$ of the Cesàro means of the Fourier series of f converges uniformly to f on **R**.*

Proof. We have

$$\sigma_n(x) = (1/\pi) \int_{-\pi}^{\pi} f(t) K_n(x - t)\, dt = (1/\pi) \int_{-\pi}^{\pi} f(x + t) K_n(t)\, dt$$

and

$$\sigma_n(x) - f(x) = (1/\pi) \int_{-\pi}^{\pi} [f(x + t) - f(x)] K_n(t)\, dt.$$

If $0 < \delta < \pi$, we can split up the last integral into three:

$$\int_{-\pi}^{\pi} = \int_{-\pi}^{-\delta} + \int_{-\delta}^{\delta} + \int_{\delta}^{\pi}.$$

We estimate the absolute values of each of the three integrals and apply the triangle inequality to get

$$|\sigma_n(x) - f(x)| \le \operatorname{lub} \{|f(x + t) - f(x)| \mid |t| < \delta\}$$
$$+ 4\|f\| \operatorname{lub} \{K_n(t) \mid \delta \le |t| \le \pi\},$$

where $\|f\|$ is the uniform norm of f on $[-\pi, \pi]$.

    Let $\epsilon > 0$ be given. Since f is uniformly continuous, there exists a $\delta > 0$ such that

$$|y - z| < \delta \Rightarrow |f(y) - f(z)| < \epsilon/2.$$

By property (3) of the lemma, we can choose N such that

$$\text{lub } \{K_n(t) \mid \delta \leq |t| \leq \pi\} < \frac{\epsilon}{8\|f\|}$$

whenever $n \geq N$. (Why may we assume that $f \neq 0$?) If $n \geq N$ and x is any real number, it follows from the preceding paragraph that

$$|\sigma_n(x) - f(x)| < \frac{\epsilon}{2} + 4\|f\| \frac{\epsilon}{8\|f\|} = \epsilon.$$

Thus $\sigma_n \to f$ uniformly, as asserted.

Corollary 16. *If f is continuous and periodic with period 2π, then the Fourier series of f converges to f in the mean, so*

$$\frac{a_0^2}{2} + \sum_{k=1}^{\infty} (a_k^2 + b_k^2) = \frac{1}{\pi} \int_{-\pi}^{\pi} [f(x)]^2 \, dx.$$

The last equation is known as *Parseval's identity.* The corollary is actually true under the much weaker assumption that f is Riemann-integrable on $[-\pi, \pi]$. A proof for the case in which f is piecewise continuous is outlined in the exercises.

Proof. Since $\sigma_n(x)$ is a linear combination of the functions $1, \cos x, \sin x, \ldots,$ $\cos (n - 1)x, \sin (n - 1)x$, the best approximation of the theorem, in the least-squares sense, tells us that

$$\|f - s_{n-1}\|_2 \leq \|f - \sigma_n\|_2 \leq \sqrt{b - a} \, \|f - \sigma_n\|.$$

By Fejér's theorem, $\|f - \sigma_n\| \to 0$ and hence $\|f - s_{n-1}\|_2 \to 0$ also.

Corollary 17. Weierstrass Approximation Theorem. *If f is continuous on $[a, b]$, then there exists a sequence of polynomials which converges uniformly to f on $[a, b]$.*

Proof. An equivalent assertion is that for each $\epsilon > 0$ there exists a polynomial P such that $|f(x) - P(x)| < \epsilon$ for all $x \in [a, b]$. Without loss of generality, we may assume that $[a, b] = [0, 1]$. (We can make a linear change of variable to accomplish this reduction.) We define f outside $[0, 1]$, so that it is continuous and periodic with period 2π. By Fejér's theorem, we can choose a Cesàro mean σ_n of the Fourier series of f such that

$$|\sigma_n(x) - f(x)| < \delta/2$$

for all $x \in \mathbf{R}$. The function σ_n is entire, and the partial sums of its Maclaurin expansion converge uniformly to f on $[0, 1]$. We can therefore choose a partial sum P such that

$$|P(x) - \sigma_n(x)| < \epsilon/2$$

for all $x \in [0, 1]$. Hence

$$|P(x) - f(x)| \leq |P(x) - \sigma_n(x)| + |\sigma_n(x) - f(x)| < \epsilon/2 + \epsilon/2 = \epsilon$$

for all $x \in [0, 1]$.

Corollary 18. *A continuous function with period 2π is completely determined by its Fourier coefficients.*

Proof. Suppose that f and g are continuous functions with period 2π which have the same Fourier coefficients. Then $h = f - g$ is also continuous and has period 2π, and its Fourier coefficients are all zero. It follows from Parseval's identity that $\int_{-\pi}^{\pi} [h(x)]^2 \, dx = 0$. Since h is continuous, this can occur only if h is identically zero, or equivalently, $f = g$.

Corollary 19. *Suppose that f and g are both continuous and periodic with period 2π. If*

$$f \sim \frac{a_0}{2} + \sum_{k=1}^{\infty} (a_k \cos kx + b_k \sin kx)$$

and

$$g \sim \frac{\alpha_0}{2} + \sum_{k=1}^{\infty} (\alpha_k \cos kx + \beta_k \sin kx),$$

then

$$\frac{a_0 \alpha_0}{2} + \sum_{k=1}^{\infty} (a_k \alpha_k + b_k \beta_k) = \frac{1}{\pi} \int_{-\pi}^{\pi} f(x) \, g(x) \, dx.$$

Proof. Apply Parseval's equation to f, g and $f + g$, and then use the identity

$$2fg = (f + g)^2 - f^2 - g^2.$$

Like Parseval's equation (which it clearly generalizes), the hypothesis can be weakened to Riemann integrability.

We present now several relatively trivial applications.

Example 9. Use Parseval's equation to compute

$$1 + \frac{1}{3^2} + \frac{1}{5^2} + \cdots$$

By Example 1,

$$\sin x + \frac{\sin 3x}{3} + \frac{\sin 5x}{5} + \cdots$$

is the Fourier series for

$$f(x) = \begin{cases} \pi/4 & \text{if} \quad 0 \le x \le \pi, \\ -\pi/4 & \text{if} \quad -\pi \le x < 0. \end{cases}$$

Hence

$$1 + \frac{1}{3^2} + \frac{1}{5^2} + \cdots = \frac{1}{\pi} \int_{-\pi}^{\pi} [f(x)]^2 \, dx$$

$$= \frac{1}{\pi} \left(\frac{\pi}{4}\right)^2 \cdot 2\pi = \frac{\pi^2}{8}.$$

(We are using a stronger form of Parseval's equation than the one proved.)

Example 10. Prove that the trigonometric series

$$\sum_{k=1}^{\infty} \frac{\sin kx}{\sqrt{k}}$$

is not the Fourier series of any Riemann-integrable function.

If the series were the Fourier series of some function, then by Bessel's inequality, the series consisting of the sum of the squares of the coefficients, namely, the harmonic series, would have to converge.

Exercises

1. Work out the calculations for Example 1.
2. Work out the calculations for Example 2.
3. Work out the calculations for Example 3.
4. Let $f_n(x) = \cos^n x$.
   a) Prove that on the interval $[0, \pi]$, the sequence converges in the mean to the function which is identically zero.
   b) Observe that $\{f_n(\pi)\}$ has no limit. Thus f_n does not converge pointwise on $[0, \pi]$.
   c) Prove that f_n converges pointwise but not uniformly on the interval $[0, \pi/2]$.
5. We work over a fixed interval $[a, b]$. Let I be any nontrivial subinterval of $[a, b]$. Let c and d be the endpoints of I, where $c < d$. Define

$$f_n(x) = \begin{cases} 0 & \text{if} \quad a \le x \le c, \\ n(x - c) & \text{if} \quad c \le x \le c + 1/n, \\ 1 & \text{if} \quad c + 1/n \le x \le d - 1/n, \\ -n(x - d) & \text{if} \quad d - 1/n \le x \le d, \\ 0 & \text{if} \quad d \le x \le b, \end{cases}$$

provided $2/n < d - c$. Observe that f_n can be extended to a continuous function on the entire line having period $b - a$. Prove that $f_n \to \varphi_I$ in the mean on the interval $[a, b]$.

6. Given an interval $[a, b]$, we define a family of functions $\mathfrak{F}$ as follows. We say $f \in \mathfrak{F}$ iff f is Riemann-integrable on $[a, b]$ and there exists a sequence of functions $f_1, f_2, f_3, \ldots$ such that

   i) each f_n can be extended to the entire line so as to be continuous and have period $b - a$;
   ii) the sequence $\{f_n\}$ converges to f in the mean.

It can be shown that $\mathfrak{F}$ is the entire set of Riemann-integrable functions. You are asked to prove considerably less.

   a) Show that $\mathfrak{F}$ is closed under addition and multiplication by constants.
   b) Using Exercise 5 and (a), show that $\mathfrak{F}$ contains all step functions on $[a, b]$ and hence all piecewise continuous functions on $[a, b]$.

7. Let $\{\varphi_n\}$ be an orthonormal sequence of functions on an interval $[a, b]$. Let Φ denote the set of all Riemann-integrable functions f such that the Fourier series of f relative to $\{\varphi_n\}$ converges to f in the mean.

a) Prove that $f \in \Phi$ iff for every $\epsilon > 0$ there exists a function φ of the form

$$\varphi = \sum_{k=1}^{n} a_k \varphi_k$$

such that $\|f - \varphi\|_2 < \epsilon$.

b) Prove that Φ is closed under addition and multiplication by scalars.

c) Prove that if $f_1, f_2, \ldots$ is a sequence of functions in Φ and if $\{f_n\}$ converges in the mean to a function $f \in \mathcal{R}$, then $f \in \Phi$. (In other words, Φ is closed under convergence in the mean.)

d) If $f, g \in \Phi$, prove that

$$\langle f, g \rangle = \sum_{k=1}^{\infty} \langle f, \varphi_k \rangle \langle g, \varphi_k \rangle.$$

(Use the argument given in the proof of Corollary 19.)

8. Use the theorem on pointwise convergence of Fourier series to prove each of the following.

a) $\sin x + \dfrac{\sin 2x}{2} + \dfrac{\sin 3x}{3} + \cdots = \begin{cases} (\pi - x)/2 & \text{if } 0 < x < 2\pi, \\ 0 & \text{if } x = 0 \text{ or } 2\pi \end{cases}$

b) $\sin x = \dfrac{\sin 2x}{2} + \dfrac{\sin 3x}{3} - \dfrac{\sin 4x}{4} + \cdots = \begin{cases} x/2 & \text{if } 0 \le x < \pi, \\ 0 & \text{if } x = \pi, \\ \frac{1}{2}x - \pi & \text{if } \pi < x \le 2\pi \end{cases}$

c) $\dfrac{1}{2\alpha} - \dfrac{\alpha}{\alpha^2 + 1^2} \cos x + \dfrac{\alpha}{\alpha^2 + 2^2} \cos 2x - \cdots = \dfrac{\pi}{2} \dfrac{e^{\alpha x} + e^{-\alpha x}}{e^{\alpha \pi} - e^{-\alpha \pi}}$

9. a) Use the Dirichlet test for uniform convergence to show that the series in Exercise 8(a) converges uniformly on every closed subinterval of $(0, 2\pi)$. Given $0 < x < 2\pi$, integrate the series term by term over the interval from π to x and make a change of variable to show that

$$\sum_{k=1}^{\infty} \frac{\cos 2\pi kx}{k^2} = (x^2 - x + \tfrac{1}{6})\pi^2 \quad \text{if } 0 < x < 1.$$

Does the expansion also hold if $x = 0$ or 1? Is the expansion valid for negative values of x?

b) By integrating the series in (a) term by term, show that

$$\sum_{k=1}^{\infty} \frac{\sin 2\pi kx}{k^3} = (\tfrac{2}{3}x^3 - x^2 + \tfrac{1}{3}x)\pi^3.$$

c) Sum the series

$$\sum_{k=1}^{\infty} \frac{\cos 2\pi kx}{k^4}.$$

10. From the expansion

$$\frac{\cos x}{1^2} + \frac{\cos 3x}{3^2} + \frac{\cos 5x}{5^2} + \cdots = \begin{cases} \dfrac{\pi^2}{8} - \dfrac{\pi x}{4} & \text{if } 0 \le x \le \pi, \\[2mm] \dfrac{\pi x}{4} - \dfrac{3\pi^2}{8} & \text{if } \pi \le x \le 2\pi \end{cases}$$

derived in the text, deduce the following.

a) $\displaystyle\sum_{k=1}^{\infty} \frac{\sin (2k-1)x}{(2k-1)^3} = \frac{\pi x}{8}(\pi - x)$

b) $\displaystyle\sum_{k=1}^{\infty} \frac{\cos (2k-1)x}{(2k-1)^4} = \frac{\pi}{48}\left(\frac{\pi}{2} - x\right)(\pi^2 + 2\pi x - 2x^2)$

11. a) From the Fourier expansion for $\cos cx$, deduce that

$$\frac{\pi}{\sin c\pi} = \frac{1}{c} - \frac{2c}{c^2 - 1^2} + \frac{2c}{c^2 - 2^2} - \frac{2c}{c^2 - 3^2} - \cdots$$

b) By combining the expansion in (a) with the partial fraction decomposition for the cotangent function, deduce that

$$\pi \tan \frac{c\pi}{2} = -\frac{4c}{c^2 - 1^2} - \frac{4c}{c^2 - 3^2} - \frac{4c}{c^2 - 5^2} - \cdots$$

c) Replace c by $\frac{1}{2} - c$ in (a) and do some regrouping to show that

$$\frac{\pi}{\cos c\pi} = \frac{4 \cdot 1}{1^2 - 4c^2} - \frac{4 \cdot 3}{3^2 - 4c^2} + \frac{4 \cdot 5}{5^2 - 4c^2} + \cdots$$

12. a) Expand each term on the right-hand side of the equation

$$\pi x \operatorname{ctn} \pi x = 1 - \sum_{k=1}^{\infty} \frac{2x^2}{k^2 - x^2}$$

into a Maclaurin series, and then combine into a single Maclaurin series via the Fubini theorem. Also show that the Maclaurin expansion is valid if $|x| < 1$. Compare with the expansion

$$\pi x \operatorname{ctn} \pi x = 1 + \sum_{k=1}^{\infty} (-1)^k \frac{2^{2k} B_{2k}}{(2k)!} (\pi x)^{2k}$$

to obtain the formula

$$\zeta(2p) = \sum_{k=1}^{\infty} \frac{1}{k^{2p}} = (-1)^{p-1} \frac{B_{2p}(2\pi)^{2p}}{2(2p!)}.$$

b) Deduce from (a) the following:

$$\sum_{k=1}^{\infty} \frac{1}{k^2} = \frac{\pi^2}{6}, \qquad \sum_{k=1}^{\infty} \frac{1}{k^4} = \frac{\pi^4}{90}, \qquad \sum_{k=1}^{\infty} \frac{1}{k^6} = \frac{\pi^6}{945}.$$

c) Also deduce from (a) the formulas

$$1 + \frac{1}{3^{2p}} + \frac{1}{5^{2p}} + \cdots = (-1)^{p-1} \frac{2^{2p} - 1}{2(2p!)} B_{2p} \pi^{2p}$$

and

$$1 - \frac{1}{2^{2p}} + \frac{1}{3^{2p}} - \frac{1}{4^{2p}} + \cdots = (-1)^{p-1} \frac{2^{2p-1} - 1}{(2p)!} B_{2p} \pi^{2p}.$$

d) Conclude from (a) that the Maclaurin series for $\pi x \operatorname{ctn} \pi x$, $\tan x$, and $x/\sin x$ have radii of convergence 1, $\pi/2$, and π, respectively.

e) Observe that the Bernoulli numbers alternate in sign, and $(-1)^p B_{2p}$ is always positive. Observe that $1 < \zeta(2p) < 2$ and hence that

$$\frac{2(2p)!}{(2\pi)^{2p}} < (-1)^p B_{2p} < \frac{4(2p)!}{(2\pi)^{2p}}.$$

Deduce that $|B_{2p+2}/B_{2p}| \to +\infty$ as $p \to \infty$. This shows that the Bernoulli numbers increase quite rapidly.

13. Assuming that f has period 2π and possesses a continuous second derivative, express the Fourier coefficients of $f'(x)$ and $f''(x)$ in terms of those for f.

14. Suppose that
   i) $f^{(p)}$ is continuous on $[-\pi, \pi]$,
   ii) $f^{(k)}(-\pi) = f^{(k)}(\pi)$, for $k = 0, 2, \ldots, p - 1$,
   iii) $|f^{(p)}(x)| \leq M$ if $-\pi \leq x \leq \pi$,
   iv) $f \sim a_0/2 + \sum_{k=1}^{\infty} (a_k \cos kx + b_k \sin kx)$.
   Prove that $|a_k| \leq 2Mk^{-p}$, $|b_k| \leq 2Mk^{-p}$ for all $k \in \mathbb{Z}^+$. (Use integration by parts.)

15. Suppose that f has period 2π and possesses a continuous second derivative. Prove that the Fourier series for f converges uniformly to f.

16. Deduce from the identity

$$\sum_{k=1}^{n} \sin kx = \frac{\cos (x/2) - \cos (n + \frac{1}{2})x}{2 \sin (x/2)}$$

that

$$\sum_{k=1}^{n} \frac{(-1)^k}{k} - \sum_{k=1}^{n} \frac{\cos kx}{k} = \frac{1}{2} \int_{\pi}^{x} \operatorname{ctn} \frac{t}{2} \, dt - \frac{1}{2} \int_{\pi}^{x} \frac{\cos (n + \frac{1}{2})t}{\sin (t/2)} \, dt,$$

and hence that

$$\ln \left(2 \sin \frac{x}{2} \right) = - \sum_{k=1}^{\infty} \frac{\cos kx}{k} \qquad (0 < x < 2\pi).$$

Can the same result be obtained from our theorem on the pointwise convergence of a Fourier series?

17. If s_n denotes the nth partial sum of the series $\sum_{k=1}^{\infty} a_k$, we define

$$\sigma_n = \frac{s_1 + s_2 + \cdots + s_n}{n}.$$

We say the series is *Cesàro-summable*, or $(C, 1)$-*summable*, iff $\lim_{n \to \infty} \sigma_n = S$ exists (and is finite), in which case we write

$$S = \sum_{k=1}^{\infty} a_k \quad (C, 1).$$

a) Using Exercise 15, Section 13–3, prove that if $\sum_{k=1}^{\infty} a_k$ converges, then the series is also $(C, 1)$-summable and its sum in the $(C, 1)$-sense is equal to its sum in the usual sense.

b) Show that

$$1 - 1 + 1 - 1 + \cdots = \tfrac{1}{2} \quad (C, 1).$$

Thus there exist divergent series which are $(C, 1)$-summable.

c) Show that

$$\frac{1}{2} + \sum_{k=1}^{\infty} \cos kx = 0 \quad (C, 1), \qquad x \in \mathbf{Z}.$$

d) Show that

$$\frac{1}{2} \operatorname{ctn} \frac{x}{2} = \sum_{k=1}^{\infty} \sin kx \quad (C, 1), \qquad x \in \mathbf{Z}.$$

e) Show that

$$1 + 0 - 1 + 1 + 0 - 1 + 1 + \cdots = \tfrac{2}{3} \quad (C, 1).$$

Thus the interpolation of zeros into a series may alter its Cesàro sum.

18. Suppose that f has period 2π and is Riemann-integrable on $[-\pi, \pi]$. Prove that if f is continuous at x_0, then $f(x_0)$ is the $(C, 1)$-sum of the Fourier series for f at x_0.

19. If f is piecewise continuous on $[-\pi, \pi]$, prove that f is the limit of its Fourier series, where convergence is in the mean. [Use Fejér's theorem and Exercises 6 and 7.]

20. Let f be a continuous function on $[-1, +1]$. Show that the Fourier series for f with respect to the normalized Legendre polynomials converges to f in the mean. (Use the Weierstrass approximation theorem and the theorem concerning approximations in the least-squares sense.)

21. Find the unique polynomial of degree two or less which best approximates e^x on the interval $[-1, +1]$ in the least-squares sense. [Use the first three normalized Legendre polynomials:

$$\varphi_0(x) = 1/\sqrt{2}, \qquad \varphi_1(x) = \sqrt{\tfrac{3}{2}}\, x, \qquad \varphi_2(x) = \tfrac{1}{2}\sqrt{\tfrac{5}{2}}\, (3x^2 - 1)$$

and Proposition 3.]

22. We have seen that the function

$$f(x) = \begin{cases} e^{-1/x^2} & \text{if } x \neq 0, \\ 0 & \text{if } x = 0 \end{cases}$$

is of class C^∞ on the entire line, that its Maclaurin series converges at all points of $\mathbf{R}$ (trivially), but that it represents f only at zero. The purpose of this exercise is to give an example of a function of class C^∞ on the entire line whose Maclaurin series has zero radius of convergence. Let

$$f(x) = \sum_{k=0}^{\infty} e^{-k} \cos k^2 x.$$

Prove that we may differentiate term by term infinitely often. Show that the Maclaurin series of f has terms only of even degree and that the absolute value of the term of degree $2k$ is

$$\sum_{n=0}^{\infty} \frac{x^{2k} e^{-n} n^{4k}}{(2k)!} > \left(\frac{n^2 x}{2k}\right)^{2k} e^{-n} \qquad (n = 0, 1, 2, \ldots).$$

In particular, we may set $n = 2k$. If $x \neq 0$, then for all integers $k > |e/2x|$, deduce that

$$\left(\frac{n^2 x}{2k}\right)^{2k} e^{-n} = \left(\frac{2kx}{e}\right)^{k2} > 1,$$

and hence that the Maclaurin series diverges at x.

23. Let f be as in Fejér's theorem. Show that

$$\|f - \sigma_n\|_2^2 = \|f - s_n\|_2^2 + \frac{1}{(n+1)^2} \sum_{k=1}^{n} k^2(a_k^2 + b_k^2).$$

Then deduce that

$$\lim_{n \to \infty} \frac{1}{(n+1)^2} \sum_{k=1}^{n} k^2(a_k^2 + b_k^2) = 0.$$

24. The sums of the following series can be obtained from the series expansions presented in this section and in the exercises. Find them.
a) $1 + \frac{1}{5} - \frac{1}{7} - \frac{1}{11} + \frac{1}{13} + \frac{1}{17} - \cdots$
b) $1 - \frac{1}{5} + \frac{1}{7} - \frac{1}{11} + \frac{1}{13} - \cdots$
c) $1 - 1/3^3 + 1/5^3 - 1/7^3 + \cdots$

25. Compute the sum of the series

$$1/1^6 + 1/3^6 + 1/5^6 + \cdots$$

by using 10(a) and Parseval's identity.

13-9 Infinite products

Given an infinite product

$$p_1 p_2 p_3 \cdots = \prod_{k=1}^{\infty} p_k,$$

we shall call $P_n = \prod_{k=1}^{n} p_k$ the nth *partial product*. By analogy with infinite series, one would expect the infinite product to be defined as the limit of the sequence $\{P_n\}$. To get a satisfactory theory, however, special attention must be given to the number zero.

Definition 1. The infinite product above is said to **converge** iff either

1) the sequence of partial products $\{P_n\}$ tends to some nonzero real number P, which is then called the **value** of the product, or

2) the product contains a finite number of zero factors, and when these are deleted, the resulting product converges in the sense of (1); in this case the value of the given product is said to be zero.

Example 1. If $|x| < 1$, show that the product $\prod_{k=1}^{\infty} (1 + x^{2^k})$ converges to $1/(1 - x)$. This follows immediately from the fact that

$$\prod_{k=1}^{n} (1 + x^{2^k}) = \sum_{j=0}^{2^{n+1}-1} x^j.$$

Example 2. The product $\prod_{k=1}^{\infty} 1/k$ *diverges* to zero.

Example 3. Show that the product $\prod_{k=1}^{\infty} (1 - 1/k^2)$ converges to zero.
The first term is zero. Hence it suffices to prove that as $n \to 0$, the product

$$\prod_{k=2}^{n} \left(1 - \frac{1}{k^2}\right)$$

tends to a nonzero real number. But

$$\prod_{k=2}^{n} \left(1 - \frac{1}{k^2}\right) = \prod_{k=2}^{n} \frac{(k-1)(k+1)}{k^2} = \frac{n+1}{2n} \to \frac{1}{2}.$$

As immediate consequences of our definition, we have the following.

1) The convergence of an infinite product is unaffected by the alteration of a finite number of its terms.

2) If an infinite product converges, then its nth factor tends to one as $n \to \infty$ (and thus all factors with possibly a finite number of exceptions are positive).

As a consequence of (2), we usually write infinite products in the form

$$\prod_{k=1}^{\infty} (1 + a_k).$$

Consider an infinite product $\prod_{k=1}^{\infty} (1 + a_k)$ each of whose factors is positive. The infinite series

$$\sum_{k=1}^{\infty} \ln (1 + a_k)$$

is closely related to this product. If P_n denotes the nth partial product, and if S_n denotes the nth partial sum of the series, then $P_n = e^{S_n}$. Consequently, the product converges simultaneously with the series, and in the case of convergence, the value P of the product and the sum S of the series are related by the equation $P = e^S$. In this way the study of infinite products reduces to the study of infinite series. In particular, the value of the product will be independent of the order of the factors iff the infinite series converges absolutely. In this connection we have the following result.

Proposition 2. *Let* $\prod_{k=1}^{\infty} (1 + a_k)$ *be a product whose terms are all positive. Then the product* $\prod_{k=1}^{\infty} (1 + |a_k|)$ *and the series*

$$\sum_{k=1}^{\infty} \ln (1 + |a_k|), \qquad \sum_{k=1}^{\infty} |\ln (1 + a_k)|, \qquad \sum_{k=1}^{\infty} |a_k|$$

are simultaneously convergent.

Proof. Because of our previous observations, it suffices to show that the series

$$\sum |\ln (1 + a_k)| \qquad \text{and} \qquad \sum |a_k|$$

are simultaneously convergent. We know, however, that

$$\lim_{x \to 0} \frac{\ln (1 + x)}{x} = 1.$$

Consequently, if either $\sum |\ln (1 + a_k)|$ or $\sum |a_k|$ converges, then $a_n \to 0$ and, consequently,

$$|\ln (1 + a_n)| \sim |a_n|.$$

This implies the convergence of the other series.

If $\prod_{k=1}^{\infty} (1 + |a_k|)$ converges, we say that the product $\prod_{k=1}^{\infty} (1 + a_k)$ *converges absolutely.*

Uniform convergence of products also has an obvious definition, and one can easily show that the limit of a uniformly convergent product of continuous functions is continuous. We give a theorem concerning the differentiation of infinite products.

Proposition 3. *Let J be an open interval, and let* $f_1, f_2, \ldots$ *be a sequence of functions possessing continuous first-order derivatives on J. Suppose also that the infinite series*

$$\sum_{k=1}^{\infty} |f_k(x)| \qquad \text{and} \qquad \sum_{k=1}^{\infty} |f_k'(x)|$$

converge uniformly on closed bounded subintervals of J. If we define

$$F(x) = \prod_{k=1}^{\infty} [1 + f_k(x)]$$

for each $x \in J$, *then F' exists and is continuous on J, and at points where* $F(x) \neq 0$, *we have*

$$\frac{F'(x)}{F(x)} = \sum_{k=1}^{\infty} \frac{f_k'(x)}{1 + f_k(x)}.$$

Proof. Without loss of generality, we may assume that the series

$$\sum |f_k(x)| \qquad \text{and} \qquad \sum |f'(x)|$$

converge uniformly on *J*. Then $f_n \to 0$ uniformly on *J*, so that for some $N \in \mathbf{Z}^+$, we have $|f_n(x)| < \frac{1}{2}$ for all $n \geq N$ and $x \in J$, and for the same n and x, we have

$$\left|\frac{f_n'(x)}{1 + f_n(x)}\right| \leq 2|f_n'(x)|.$$

It follows that the series

$$\sum_{k=N}^{\infty} \frac{f_k'(x)}{1 + f_k(x)}.$$

converges absolutely and uniformly on J. We set

$$F_N(x) = \prod_{k=N}^{\infty} [1 + f_k(x)].$$

Then $\ln F_N(x) = \sum_{k=N}^{\infty} \ln [1 + f_k(x)]$. It follows from our theorem concerning term-by-term differentiation that $\ln F_N$ and hence $F_N = \exp (\ln F_N)$ are differentiable on J, and that

$$\frac{F_N'(x)}{F_N(x)} = \frac{d}{dx} \ln F_N(x) = \sum_{k=N}^{\infty} \frac{f_k'(x)}{1 + f_k(x)}.$$

The rest follows from the identity

$$F(x) = [1 + f_1(x)] \cdots [1 + f_{N-1}(x)]F_N(x).$$

As an application of the preceding theorem, we shall establish the product expansion for the sine function:

$$\sin \pi x = \pi x \left(1 - \frac{x^2}{1^2}\right)\left(1 - \frac{x^2}{2^2}\right)\left(1 - \frac{x^2}{3^2}\right) \cdots$$

Because of the convergence of the series $\sum k^{-2}$, the product converges absolutely on the entire line and the hypotheses of the theorem are satisfied with $J = (-\infty, \infty)$. We let $F(x)$ denote the value of the product. The function F, like $\sin \pi x$, has simple zeros at the integers; therefore the quotient $F(x)/(\sin x)$ may be defined at the integers so that the resulting function is continuous on the entire line. If x is not an integer, we have

$$\frac{F'(x)}{F(x)} = \frac{1}{x} + \sum_{k=1}^{\infty} \frac{2}{x^2 - k^2} = \pi \operatorname{ctn} \pi x = \frac{(\sin \pi x)'}{\sin \pi x}.$$

Thus

$$\frac{d}{dx}\left(\frac{F(x)}{\sin \pi x}\right) = \frac{F(x)}{\sin \pi x}\left(\frac{F'(x)}{F(x)} - \frac{(\sin \pi x)'}{\sin \pi x}\right) = 0$$

if x is not an integer. It follows that $F(x)/(\sin \pi x)$ is constant on the entire line; call the constant c. Hence

$$c \sin \pi x = \pi x \left(1 - \frac{x^2}{1^2}\right)\left(1 - \frac{x^2}{2^2}\right) \cdots$$

We divide both sides by πx and then take limits as $x \to 0$ (which is possible since the right-hand side will be continuous). We get $c = 1$, and the proof is complete.

Exercises

1. Prove that each of the following products converges and has the value indicated.

a) $\displaystyle\prod_{k=2}^{\infty}\left(1 - \frac{2}{k(k+1)}\right) = \frac{1}{3}$ b) $\displaystyle\prod_{k=2}^{\infty}\left(1 - \frac{2}{n^3+1}\right) = \frac{2}{3}$

c) $\displaystyle\prod_{k=0}^{\infty}[1 + (\tfrac{1}{2})^{2k}] = 2$ d) $\displaystyle\prod_{k=2}^{\infty}\left(1 + \frac{1}{2^k-2}\right) = 2\sum_{k=1}^{\infty} 2^{-k} = 2$

e) $\displaystyle\prod_{k=2}^{\infty}\left(1 + \frac{1}{n^2-1}\right) = 2\sum_{k=1}^{\infty}\frac{1}{k(k+1)} = 2$

2. Show that each of the following products diverges.

a) $\displaystyle\prod_{k=2}^{\infty}\left(1 - \frac{1}{k}\right)$ b) $\displaystyle\prod_{k=1}^{\infty}\frac{x}{k}$ $(x \in \mathbf{R})$ c) $\displaystyle\prod_{k=2}^{\infty}\left(1 + \frac{1}{\ln k}\right)$

3. Prove that for every $x \in \mathbf{R}$,

$$\cos\frac{x}{2}\cos\frac{x}{4}\cos\frac{x}{8}\cdots = \frac{\sin x}{x},$$

where, of course, we understand the right-hand side to be 1 when $x = 0$. [Prove that the product converges to a differentiable function C. Observe that

$$C(2x) = (\cos x)C(x) \qquad \text{and} \qquad C(0) = 1.$$

Set $h(x) = xC(x)/(\sin x)$. Show that $h(2x) = h(x)$, and infer that h is a constant, etc.] By setting $x = \pi/2$, obtain *Vieta's formula:*

$$2/\pi = \sqrt{\tfrac{1}{2}} \cdot \sqrt{\tfrac{1}{2} + \tfrac{1}{2}\sqrt{\tfrac{1}{2}}} \cdot \sqrt{\tfrac{1}{2} + \tfrac{1}{2}\sqrt{\tfrac{1}{2} + \tfrac{1}{2}\sqrt{\tfrac{1}{2}}}} \cdots .$$

4. Deduce Wallis' product from the infinite product expansion for the sine function.

5. a) Let p_n be the nth prime number. (Thus $p_1 = 2, p_3 = 3, p_4 = 5, p_5 = 7, \ldots$) Prove Euler's celebrated product expansion for the Riemann zeta function:

$$\zeta(s) = \sum_{k=1}^{\infty} k^{-s} = \prod_{k=1}^{\infty}\frac{1}{1 - p_k^{-s}}.$$

[Write each factor of the product as a geometric series. Prove that

$$0 \leq \zeta(s) - P_n(s) \leq \sum_{k=n}^{\infty} k^{-s},$$

where $P_n(s)$ is the nth partial product.]

b) Prove that the series $\sum_{k=1}^{\infty} 1/p_k$ diverges. (Suppose this to be false. Show that the Euler product must then converge uniformly on the interval $[1, +\infty)$ and hence that ζ could be defined at 1 so as to be continuous on the same interval. Obtain a contradiction.) This proves, in particular, that p_n tends to $+\infty$ at a slower rate than $n^{1+\epsilon}$ for every $\epsilon > 0$. The famous *prime number theorem* asserts that p_n is asymptotic to $n \ln n$.

6. In this exercise we outline a second proof of Euler's product expansion for $\zeta(s)$. Here we introduce an important arithmetic function μ, known as the *Möbius function*. By definition, $\mu(1) = 1$. If $n \in \mathbf{Z}^+$ and $n > 1$, we define $\mu(n)$ to be zero if n has a squared factor, and we define $\mu(n)$ to be $(-1)^r$ if n is the product of r distinct primes. Thus $\mu(12) = \mu(2^2 \cdot 3) = 0$ and $\mu(1001) = \mu(7 \cdot 11 \cdot 13) = (-1)^3 = -1$.

a) For each $n \in \mathbf{Z}^+$, prove that

$$\sum_{d \mid n} \mu(d) = \begin{cases} 1 & \text{if } n = 1, \\ 0 & \text{if } n > 1. \end{cases}$$

Here "$d \mid n$" means "d divides n." Thus the left-hand side of the above equation indicates the sum of all numbers of the form $\mu(d)$ as d ranges over the divisors of n, the numbers 1 and n included. [If $n > 1$, let $n = P_1^{r_1} P_2^{r_2} \cdots P_s^{r_s}$ be its factorization into distinct primes $P_1, \ldots, P_s$. Argue that

$$\sum_{d \mid n} \mu(d) = 1 + \sum_j \mu(P_j) + \sum_{j<k} \mu(P_j P_k) + \cdots$$

$$= 1 + \binom{s}{1}(-1) + \binom{s}{2}(-1)^2 + \cdots + \binom{s}{s}(-1)^s$$

$$= (1 - 1)^s = 0$$

as desired.]

b) Show that the Dirichlet series $\sum \mu(k) k^{-s}$ converges absolutely for $s > 1$. Using (a) and our results concerning Dirichlet products, prove that

$$\sum_{k=1}^{\infty} \mu(k) k^{-s} = \frac{1}{\zeta(s)} \quad (s > 1).$$

c) Prove that

$$\prod_{k=1}^{n} (1 - p_k^{-s}) = \sum_{k \in S_n} \mu(k) k^{-s},$$

where S_n is the set of all integers k such that $1 \leq k \leq p_1 p_2 \cdots p_n$ and k is a product of powers of $p_1, \ldots, p_n$.

d) Deduce from (b) and (c) that

$$\prod_{k=1}^{\infty} (1 - p_k^{-s}) = \frac{1}{\zeta(s)} \quad (s > 1).$$

Taking reciprocals, we have Euler's expansion for ζ.

The remaining exercises are concerned with the *gamma function*.

7. a) We define

$$K(x) = x \prod_{k=1}^{\infty} (1 + x/k)e^{-x/k}.$$

Show that the product converges absolutely and uniformly on every bounded subinterval of $\mathbf{R}$. Conclude also that K is differentiable on $\mathbf{R}$.

b) For each $x \in \mathbf{R}$, prove that

$$\lim_{n \to \infty} \frac{x(x + 1)(x + 2) \cdots (x + n)}{n!n^x} = e^{\gamma x}K(x),$$

where $\gamma = \lim_{n \to \infty} (\sum_{k=1}^{\infty} k^{-1} - \ln n)$ is *Euler's constant*. (See Exercise 9, Section 13-2.) The *gamma function* Γ is defined as follows. If $x \ne 0, -1, -2, \ldots$, then

$$\Gamma(x) = \lim_{n \to \infty} \frac{n!n^x}{x(x + 1) \cdots (x + n)} = \frac{e^{-2x}}{K(x)} = \frac{e^{-2x}}{x} \prod_{k=1}^{\infty} \frac{e^{x/k}}{1 + x/k}.$$

The first expression is due to Gauss.

c) Show that Γ satisfies the functional relation

$$\Gamma(x + 1) = x\Gamma(x).$$

Show also that $\Gamma(1) = 1$. Then prove by induction that

$$\Gamma(n) = (n - 1)!$$

for each $n \in \mathbf{Z}^+$. Thus the Γ function can be thought of as a continuous extension of the factorial function.

d) Prove that for $x > 0$,

$$\frac{\Gamma'(x)}{\Gamma(x)} = -\gamma + \frac{1}{x} + \sum_{k=1}^{\infty} \left(\frac{1}{k} - \frac{1}{x + k} \right).$$

Prove also that

$$\frac{d^n}{dx^n}[\ln \Gamma(x)] = \sum_{k=0}^{\infty} \frac{(-1)^n(n - 1)!}{(x + k)^n}$$

if $n \ge 2$ and $x \ne 0, -1, -2, \ldots$ Deduce, in particular, that Γ is *logarithmically convex* on $(0, +\infty)$. [That is, $\ln \Gamma(x)$ is convex.] It can be shown (see Bourbaki, *Eléments de Mathématique*, Part I, Book IV, Chapter 7, p. 160, published by Hermann, Paris) that Γ is the only function which is logarithmically convex on $(0, +\infty)$ such that $\Gamma(x + 1) = x\Gamma(x)$.

e) Use the infinite product expansion of the sine function to prove the *complementation relation for* Γ:

$$\frac{1}{\Gamma(x)\Gamma(1 - x)} = \frac{\sin \pi x}{\pi}.$$

Deduce, in particular, that $\Gamma(\frac{1}{2}) = \sqrt{\pi}$.

f) Show that $\Gamma(x)$ is never zero and that its sign is as indicated:

$$
\begin{array}{ccccc}
(+) & (-) & (+) & (-) & (+) \\
\end{array}
$$

$$\Gamma: \quad \begin{array}{ccccc} \rule{3cm}{0.4pt} \\ -3 \quad -2 \quad -1 \quad \quad 0 \end{array}$$

Show that $\lim_{x \to a} |\Gamma(x)| = +\infty$ if $a = 0, -1, -2, \ldots$ From the fact that Γ is logarithmically convex, deduce that $\Gamma''(x)$ and $\Gamma(x)$ have the same sign. On what intervals is Γ convex? concave? Show that Γ has a strict relative minimum in the interval $(1, 2)$. The graph of Γ can be sketched from this information (see Fig. 13-8).

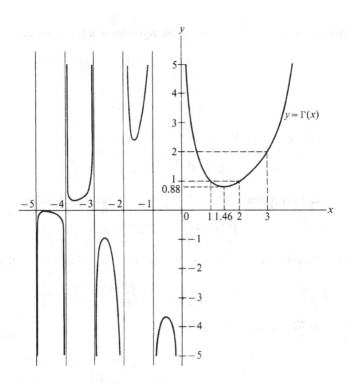

FIG. 13-8. Adapted from N. Bourbaki, *Eléments de Mathématique,* Part I, Book IV, Chapter 7, Figure 1, with permission of Hermann, Paris.

g) Show that

$$\lim_{n \to \infty} \frac{\Gamma(x + n + 1)}{n! n^x} = 1.$$

8. The object of this exercise is to establish a representation for Γ as an *improper* integral, namely,

$$\Gamma(x) = \int_0^{+\infty} t^{x-1} e^{-t} \, dt \qquad (x > 0).$$

a) We interpret the above integral to be

$$\lim_{\epsilon \to 0+} \int_\epsilon^1 t^{x-1} e^{-t} \, dt + \lim_{R \to +\infty} \int_1^R t^{x-1} e^{-t} \, dt.$$

Show that the limits exist for all $x > 0$.

b) For each $n \in \mathbf{Z}^+$, set

$$\Gamma_n(x) = \int_0^n t^{x-1} (1 - t/n)^n \, dt.$$

By a change of variable, show that

$$\Gamma_n(x) = n^x C_n(x),$$

where $C_n(x) = \int_0^1 s^{x-1}(1-s)^n \, ds$. Then prove each of the following:

$$C_{n+1}(x) = C_n(x) - C_n(x+1), \qquad C_n(x) = \frac{n!}{x(x+1)\cdots(x+n)},$$

and

$$\lim_{n\to\infty} \Gamma_n(x) = \Gamma(x).$$

c) Show that for $x > 0$, $\Gamma(x) = \int_0^\infty t^{x-1}e^{-t} \, dt$. [Show that for each $n \in \mathbf{Z}^+$,

$$\Gamma_n(x) \le \int_0^n t^{x-1}e^{-t} \, dt \le \int_0^\infty t^{x-1}e^{-t} \, dt.$$

Deduce that $\Gamma(x) = \lim \Gamma_n(x) \le \int_0^\infty t^{x-1}e^{-t} \, dt$. Show next that if $0 < M < n$, then

$$\Gamma_n(x) > \int_0^M t^{x-1}(1 - t/n)^n \, dt.$$

Using the fact that $(1 - t/n)^n \to e^{-t}$ uniformly on $[0, M]$ as $n \to \infty$, show that

$$\Gamma(x) = \lim_{n\to\infty} \Gamma_n(z) \ge \int_0^M t^{x-1}e^{-t} \, dt.$$

Deduce that $\Gamma(x) \ge \int_0^\infty t^{x-1}e^{-t} \, dt$.]
d) Show that

$$\int_0^\infty e^{-t^2} \, dt = \tfrac{1}{2}\sqrt{\pi}.$$

ANSWERS TO SELECTED EXERCISES

Chapter 1

Section 1-1

7. The first set has no elements. The second two sets are singleton sets (that is, each contains exactly one element) and are equal. The fourth set has two elements.

Section 1-2

3. b) We interpret 2 to be $1 + 1$; -2 is the additive inverse of 2.
   e) $(-a)[-(a^{-1})] = aa^{-1} = 1$; that is, $-(a^{-1})$ is the unique number whose product with $-a$ is 1. Hence $(-a)^{-1} = -(a^{-1})$.
   h) Let $x = (a/b)/c$. Then $cx = a/b = ab^{-1} = b^{-1}a$. Thus $(bc)x = b(cx) = b(b^{-1}a) = (bb^{-1})a = 1 \cdot a = a$. Equivalently, $x = a/(bc)$.

4. If $a + b = a + c$, then

$$b = 0 + b = [(-a) + a] + b = (-a) + (a + b)$$
$$= (-a) + (a + c) = [(-a) + a] + c = 0 + c = c.$$

If $ab = ac$ and if $a \neq 0$, then $b = c$ by an argument similar to the above. Observe, however, that in any field $0 \cdot 0 = 0 \cdot 1 = 0$, but $0 \neq 1$.

7. a) If K is a subfield, then in particular it must be closed under the operations mentioned. Moreover, it must have at least two elements, a zero and a unit (which *a priori* could be different from the zero and unit elements for the entire field F), and at least one of these is nonzero (with respect to F).
      Suppose now that K has at least one nonzero element a and that K is closed under the operations named. It follows that $0 = a + (-a)$ and $1 = aa^{-1}$ must belong to K. Since the identities $0 + x = x + 0 = x$ and $1 \cdot x = x \cdot 1 = x$ hold for all $x \in F$, they hold *a fortiori* for all $x \in K$. Thus Axioms A3 and M3 hold for K. The remaining axioms are then seen to hold for K *a fortiori*.
   b) That the rationals are closed under addition follows from the identity

$$\frac{m}{n} + \frac{m'}{n'} = \frac{mn' + m'n}{nn'}$$

and the assumption that the integers are closed under addition and multiplication.

703

c) The least trivial step consists of showing that the set is closed under the formation of reciprocals. Assume $0 \neq a + b\sqrt{2}$ (a, b rational). Deduce that $0 \neq a - b\sqrt{2}$, and then use the identity

$$\frac{1}{a + b\sqrt{2}} = \frac{a - b\sqrt{2}}{(a + b\sqrt{2})(a - b\sqrt{2})} = \frac{a - b\sqrt{2}}{a^2 - 2b^2}$$

$$= \frac{a}{a^2 - 2b^2} + \left(\frac{-b}{a^2 - 2b^2}\right)\sqrt{2}.$$

d) Show that $(\sqrt[3]{2})^2$ does not belong to the set, thereby showing that the set is not closed under multiplication.

8. If $a \in F$, then $a = 1 \cdot a = 0 \cdot a = 0$.

9. All except M4

14. Set $x = 1.2176176\ldots$ Then $1000x = 1217.6176\ldots$ and $999x = 1216.4$.   Hence $x = 12164/9990$.

15. Set $x = \sqrt[3]{1 + \sqrt{2}}$.   Then $x^3 = 1 + \sqrt{2}$ and $x^6 - 2x^3 + 1 = (x^3 - 1)^2 = 2$. Thus x satisfies the equation $x^6 - 2x^3 - 1 = 0$.

17. a) True. If $x + y$ were rational, then $y = (x + y) - x$ would be rational.
    b) False. Consider $0 \cdot \sqrt{2}$. On the other hand, if it is assumed that $x \neq 0$, then the statement is true.
    c) False.

Section 1-3

1. Since $a/b < c/d$, multiplication by bd (which is positive) yields $ad < cb$. Hence

$$a(b + d) = ab + ad < ab + cb = (a + c)b.$$

We then multiply by $(b + d)^{-1}b^{-1}$ (which is also positive). This yields half the stated result, namely,

$$\frac{a}{b} < \frac{a + c}{b + d}.$$

For the numerical example, write $\frac{1}{2} = \frac{3}{6}$.

2. To show that $A \geq G$, consider $(\sqrt{x} - \sqrt{y})^2$. The inequality $G \leq H$ can be deduced from the inequality $A \geq G$ by the formation of reciprocals. Equality can hold only when $x = y$.

3. Use Example 2 of the text. Define $A = (x + y + z)/3$ and $H^{-1} = (x^{-1} + y^{-1} + z^{-1})/3$.

5. Multiply the desired inequality by x^3 and rearrange terms to get an inequality of the form $P(x) \geq 0$. Factor $P(x)$. Reverse your steps.

8. $Q(x) = a\left(x + \dfrac{b}{a}\right)^2 + \dfrac{ac - b^2}{a} \geq \dfrac{ac - b^2}{a}$ with equality iff $x = -b/a$. If $Q(x) \geq 0$ for all $x \in \mathbf{R}$, then $ac - b^2 \geq 0$.

12. Suppose to the contrary that $\mathbf{Z}_3$ can be ordered (i.e., that there exists a subset P of $\mathbf{Z}_3$ with the required properties). Then we would have $1 > 0$. This would imply that $2 = 1 + 1 > 0$ and hence that $0 = 3 = 2 + 1 > 0$, contradicting the trichotomy law. Essentially the same argument shows that any field with only a finite number of elements cannot be ordered.

15. c) $(-\infty, 0) \cup (1, +\infty)$ e) $\varnothing$ f) $(\frac{5}{2}, 3)$

Section 1–4

2. b) $|x| = |y + (x - y)| \leq |y| + |x - y|$. Thus $|x| - |y| \leq |x - y|$. By interchanging the roles of x and y we have $|y| - |x| \leq |y - x| = |x - y|$. Multiplying by -1 we get $|x| - |y| \geq -|x - y|$. Therefore, we have

$$-|x - y| \leq |x| - |y| \leq |x - y|$$

or, equivalently, $||x| - |y|| \leq |x - y|$.

4. b) $(\frac{5}{2}, \frac{7}{2})$ d) **R**   f) **R**   h) $[1, 5]$ i) $(-\infty, \frac{1}{2})$
   k) $(-\infty, -1) \cup (1, +\infty)$.

Section 1–5

1. It is understood that the universal set is **R**.
   a) None   c) $(\exists x)(\exists y)$ e) $(\exists x)(\forall y)$, $(\forall y)(\exists x)$, $(\forall x)(\exists y)$, $(\exists x)(\exists y)$, the first implying the second, and the first three implying the last.

2. a) All three   c) The first   e) None

3. a) $A = \{0, 1, 2, 3, 5, 6, 7\}$, $B = \{2\}$

Section 1–6

1. All are tautologies except the third and the last.

3. b)

$\sim$	$(P$	$\vee$	$Q)$	$\Leftrightarrow$	$(\sim P)$	$\wedge$	$(\sim Q)$
F	T	T	T	T	F	F	F
F	T	T	F	T	F	F	T
F	F	T	T	T	T	F	F
T	F	F	F	T	T	T	T

d) Although the most straightforward procedure is the construction of a truth table, we shall instead make use of the equivalences previously established:

$$P \Rightarrow (Q \vee R) \Leftrightarrow (\sim P) \vee Q \vee R,$$

$$[(P \wedge \sim Q)] \Rightarrow R \Leftrightarrow \sim(P \wedge \sim Q) \vee R \Leftrightarrow [\sim P \vee \sim(\sim Q)] \vee R \Leftrightarrow (\sim P) \vee Q \vee R.$$

If you choose this technique of proof, carefully justify each of the steps.

4. No

6. The operation is sometimes called the *symmetric difference*. It may be defined as follows:

$$A \triangledown B = \{x \mid x \in A \text{ or } x \in B, \text{ but not both}\}.$$

Equivalently, $A \triangledown B = (A \backslash B) \cup (B \backslash A) = (A \cup B) \backslash (A \cap B)$.

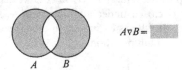

$A \triangledown B =$

A B

7. a) **R**   b) $(-\infty, 1) \cup [3, +\infty)$ d) **R**   e) $[-1, 1]$
g) $(-\infty, 0] \cup [5, +\infty)$

Section 1–7

1. $(\forall \epsilon > 0)(\exists N > 0)(\forall n > N)(\forall x \in S)(|f_n(x) - f(x)| < \epsilon)$.
   Negation: $(\exists \epsilon > 0)(\forall N > 0)(\exists n > N)(\exists x \in S)(|f_n(x) - f(x)| \geq \epsilon)$.

Section 1–8

1. c) Use the identity

$$\frac{1}{k(k+1)} = \frac{1}{k} - \frac{1}{k+1}.$$

e) Assume true for n. Consider a set S having $n + 1$ elements $\{p_1, p_2, \ldots, p_{n+1}\}$, and set $T = \{p_1, \ldots, p_n\}$. The subsets of S can be split into two disjoint classes: those sets which do not contain p_{n+1} and those which do. The first class is simply $S(T)$ which, by assumption, has 2^n elements. The second class consists of all sets of the form $A \cup \{p_{n+1}\}$, where $A \in S(T)$. Hence, it too has 2^n elements. Thus $S(S)$ has $2^n + 2^n = 2^{n+1}$ elements.

f) The crux of the proof is the following:

$$(1 + h)^{n+1} \equiv (1 + h)^n(1 + h) \geq (1 + nh)(1 + h)$$
$$= 1 + (n + 1)h + nh^2 \geq 1 + (n + 1)h.$$

2. Assume true for n. Consider a stack of $n + 1$ disks. By assumption, we can transfer the top n disks to one of the other pegs in $2^n - 1$ moves. We then move the one remaining disk to the third peg. Finally, we move the stack of n disks onto the third peg in $2^n - 1$ moves. The total number of moves is $(2^n - 1) + 1 + (2^n - 1) = 2^{n+1} - 1$.

3. The argument breaks down when $n = 2$.

Section 1–9

3. a) 14   b) 16   c) 32   d) 10

4. $\displaystyle\prod_{k=1}^{n} \left(1 - \frac{1}{(k+1)^2}\right) = \frac{n+2}{2n+2}$

6. Assume true for n. Then

$$(1 - x)\prod_{k=1}^{n+1} (1 + x^{2^{k-1}}) = (1 - x)(1 + x^{2^0})\prod_{k=1}^{n} (1 + x^{2^{k-1}})$$

$$= (1 + x^{2^n})(1 - x^{2^n}) = 1 - (x^{2^n})^2 = 1 - x^{2^{n+1}}.$$

8. Let $P(n)$ be the statement '$(\forall m \in \mathbf{Z}^+)(mn \in \mathbf{Z}^+)$'. Then $P(1)$ is clearly true. Suppose $P(n)$ is true for some positive integer n. If $m \in \mathbf{Z}^+$, then $mn \in \mathbf{Z}^+$ by the inductive hypothesis. Since $\mathbf{Z}^+$ is closed under addition, $m(n + 1) = mn + n \in \mathbf{Z}^+$. Hence '$P(n + 1)$' must be true. It follows by induction that $(\forall n \in \mathbf{Z}^+)(\forall m \in \mathbf{Z}^+)(mn \in \mathbf{Z}^+)$, that is, $\mathbf{Z}^+$ is closed under multiplication.

11. a) The sum $\sum_{k=1}^{n} [(k+1)^2 - k^2]$ telescopes to $(n+1)^2 - 1$. By linearity,

$$\sum_{k=1}^{n} (2k+1) = 2\sum_{k=1}^{n} k + \sum_{k=1}^{n} 1 = n + 2\sum_{k=1}^{n} k.$$

Thus $(n+1)^2 - 1 = n + 2\sum_{k=1}^{n} k$. Solve for $\sum_{k=1}^{n} k$.

b) $(n+1)^3 - 1 = \sum_{k=1}^{n} [(k+1)^3 - k^3] = 3\sum_{k=1}^{n} k^2 + 3\sum_{k=1}^{n} k + \sum_{k=1}^{n} 1$. Using (a) we can solve for $\sum k^2$.

d) $\displaystyle\sum_{k=1}^{n} \sin kx = \frac{\cos (x/2) - \cos (n+\frac{1}{2})x}{2 \sin (x/2)}$ g) $(n-1)2^{n+1} + 2$

12. The operations $\vee$ and $\wedge$ are commutative and associative. We also have the distributive laws: $(x \vee y) \wedge z = (x \wedge z) \vee (y \wedge z)$ and $(x \wedge y) \vee z = (x \vee z) \wedge (y \vee z)$. The latter two identities follow from the analogous set identities: $(A \cup B) \cap C = (A \cap C) \cup (B \cap C)$ and $(A \cap B) \cup C = (A \cup C) \cap (B \cup C)$, since $(-\infty, x \vee y] = (-\infty, x] \cup (-\infty, y]$ and $(-\infty, x \wedge y] = (-\infty, x] \cap (-\infty, y]$.

17. $0 \le \sum_{k=1}^{n} (a_k x + b_k)^2 = Ax^2 + 2Bx + C = Q(x)$, where $A = \sum_{k=1}^{n} a_k^2$, $B = \sum_{k=1}^{n} a_k b_k$, $C = \sum_{k=1}^{n} b_k^2$. The Cauchy-Schwarz inequality clearly holds for the case in which $a_1 = a_2 = \cdots = a_n = 0$. If the a_k's are not all zero, then $A > 0$. Since $Q(x) \ge 0$ holds for all $x \in \mathbf{R}$, it follows that $AC - B^2 \ge 0$, which is equivalent to the desired inequality. Equality implies that

$$0 = \frac{AC - B^2}{A} = Q(-\lambda) = \sum_{k=1}^{n} (-a_k \lambda + b_k)^2,$$

where $\lambda = B/A$. This in turn implies that

$$-a_k \lambda + b_k = 0 \quad \text{or} \quad b_k = \lambda a_k \quad (k = 1, \ldots, n).$$

18. b) Sum over the identity

$$a_j^2 b_k^2 + a_k^2 b_j^2 - (a_j b_k - a_k b_j)^2 = 2a_j b_j a_k b_k,$$

and observe that

$$\left(\sum_{j=1}^{m} A_j\right)\left(\sum_{k=1}^{n} B_k\right) = \sum_{j,k=1}^{m,n} A_j B_k.$$

21. Since

$$\frac{1 + \sqrt{5}}{2} > \frac{1+2}{2} > 1 \quad \text{and} \quad \left(\frac{1 + \sqrt{5}}{2}\right)^2 = \frac{3 + \sqrt{5}}{2} > \frac{3+2}{2} > 2,$$

we see that the inequality is satisfied when $n = 1, 2$. Assuming it to be true for $n - 1$ and n, we have

$$a_{n+1} = a_{n-1} + a_n < \left(\frac{1 + \sqrt{5}}{2}\right)^{n-1} + \left(\frac{1 + \sqrt{5}}{2}\right)^{n}$$

$$= \left(\frac{1 + \sqrt{5}}{2}\right)^{n-1}\left(\frac{3 + \sqrt{5}}{2}\right) = \left(\frac{1 + \sqrt{5}}{2}\right)^{n+1}.$$

By Exercise 20, the inequality must hold for all $n \in \mathbf{Z}^+$.

28. c) The equivalent dual problem is to show that among all triangles with a given perimeter an equilateral triangle has the greatest area. Since

$$(s - a) + (s - b) + (s - c) = s,$$

$(s - a)(s - b)(s - c)$ and hence A are greatest when $s - a = s - b = s - c$ or, equivalently, $a = b = c$.

31. f) Expand $(1 + 1)^n$ and $(1 - 1)^n$.

32. By the binomial theorem,

$$(k + 1)^{r+1} - k^{r+1} = \binom{r + 1}{1} k^r + \binom{r + 1}{2} k^{r-1} + \cdots + 1.$$

Summing over both sides, we get

$$(n + 1)^{r+1} - 1 = \binom{r + 1}{1} S_r(n) + \binom{r + 1}{2} S_{r-1}(n) + \cdots + n.$$

Solve for $S_r(n)$ and use the second form of the principle of induction.

Chapter 2

Section 2–2

1. Suppose that $(x, y) = (x', y')$. We must show that $x = x'$ and $y = y'$. We consider two cases.

   Case I. $x \neq y$. Then $\{x\} \neq \{x, y\}$, so that the set $(x, y) = \{\{x\}, \{x, y\}\}$ has two elements. Hence the set $\{\{x'\}, \{x', y'\}\} = (x', y') = (x, y)$ has two elements. This implies that $\{x'\} \neq \{x', y'\}$. Consequently, $x' \neq y'$. Since $\{x\}$ is an element of (x, y), it is also an element of (x', y'). Hence either $\{x\} = \{x'\}$ or $\{x\} = \{x', y'\}$. The latter is impossible, since $\{x\}$ contains one element while $\{x', y'\}$ contains two. Therefore, $\{x\} = \{x'\}$, which implies that $x = x'$. One can next argue that $\{x, y\} = \{x', y'\}$ and that $y = y'$.

   Case II. $x = y$. Easier.

2. mn

3. b) $(x_1, \ldots, x_n, x_{n+1}) = ((x_1, \ldots, x_n), x_{n+1})$

4. c)   e)   h)

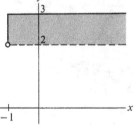

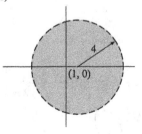

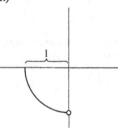

5. a) Circle with center $(-2, 3)$ and radius 5   b) $\{(-2, 3)\}$ c) $\varnothing$

Section 2–3

1. a) $(x_1, -y_1)$ c) $(-y_1, -x_1)$

3. $(\frac{9}{5}, -\frac{8}{5})$ 4. $4x + 5y = 0$

5. The equation of the circle is

$$\left(x - \frac{k^2}{1 - k^2}\right)^2 + y^2 = \left(\frac{k}{1 - k^2}\right)^2.$$

Section 2–4

1. e)

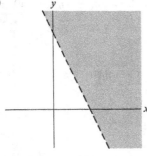

2. a) $2x - 3y - 7 = 0$
   b) $3x + 2y - 4 = 0$

3. The line clearly passes through P_1. Since the center of the circle is $(-A, -B)$, the slope of the radius through P_1 is $(y_1 + B)/(x_1 + A)$. The latter is seen to be the negative reciprocal of the alleged tangent line. (As usual, the argument breaks down if one of the lines is vertical.)

6. By the formula for the area of a triangle given in this section, such a triangle must have rational area. If s is the side length of such a triangle, then s^2 is an integer, so the area, $(s^2\sqrt{3})/2$, must also be irrational.

4. Let the equations of the circles be

$$x^2 + y^2 + 2Ax + 2By + C = 0$$

and

$$x^2 + y^2 + 2A'x + 2B'y + C' = 0.$$

Observe that

$$2(A - A')x + 2(B - B')y + C - C' = 0$$

is the straight line which passes through the points of intersection of the two circles. Write down the equation of the line which passes through the centers of the circles. Show that the two lines are perpendicular.

Chapter 3

Section 3–1

1. $f(6) = 3$, $f(7) = 7$, $f(16) = 2$, $f(75) = 5$, $f(13) = 13$, $f(52) = 13$. The range of f is the set of all prime numbers.

2. b) $\mathcal{D}_f = [-3, 3]$, $\mathcal{R}_f = [1, 4]$. f is even.

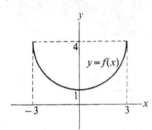

e) $\mathcal{D}_f = \mathbf{R}$, $\mathcal{R}_f = (-\infty, 0]$

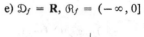
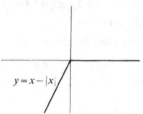

f) $\mathcal{D}_f = \mathbf{R}$, $\mathcal{R}_f = [0, +\infty)$

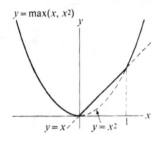

j) $\mathcal{D}_f = [0, +\infty)$, $\mathcal{R}_f = [0, +\infty)$

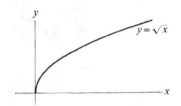

l) $\mathcal{D}_f = \mathbf{R}$, $\mathcal{R}_f = [0, 1]$. f is even and periodic with period π.

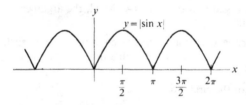

q) $\mathcal{D}_f = (-\infty, -1) \cup (-1, +1) \cup (1, +\infty)$, $\mathcal{R}_f = (-\infty, 0) \cup (0, +\infty)$. f is even.

u) $\mathcal{D}_f = \mathcal{R}_f = \mathbf{R}$. f is odd.

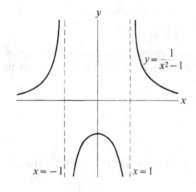

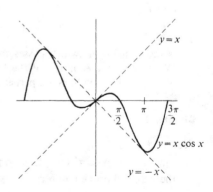

Section 3–2

2. a) g is an extension of f.   c) f is an extension of g.

3. a) $\frac{7}{2}$, $\pi/3$, $\pi/2$, $\frac{1}{4}$, $2/\sqrt{3}$

   b) $(-\infty, 0) \cup (0, +\infty)$, all real numbers except integral multiples of π, $(-\infty, 0) \cup (0, +\infty)$, $(-\infty, 0) \cup (0, +\infty)$, all numbers except integral multiples of π.

6. a) $3 + x^4$, even part; $-2x - 5x^7$, odd part of $3 - 2x + x^4 - 5x^7$. The even and odd parts of $\sin(x + \pi/3)$ are $\frac{1}{2}\sqrt{3}\cos x$ and $\frac{1}{2}\sin x$.

8. The zero element is the function which is identically zero. The unit element is the function which is identically one. Axiom M3 is satisfied iff S is a singleton set. For if S contains at least two points p_1 and p_2, then the function defined on S by

$$f(p) = \begin{cases} 0 & \text{if } p \neq p_1, \\ 1 & \text{if } p = p_1 \end{cases}$$

   is not the zero element and yet it has no multiplicative inverse.

9. $ad + b = bc + d$

10. Partial table. $g \circ h$ appears in the row labeled g and the column headed h.

	f_0	f_1	f_2	f_3	f_4	f_5
f_0						
f_1						
f_2		f_4				f_3
f_3			f_4			
f_4					f_3	
f_5			f_4			

   G1 is satisfied since the composition of functions is always an associative operation. The element f_0 satisfies G2. You will find that four elements are their own inverses. The equation may be solved as follows. Multiply first on the left by f_3^{-1} to get $f \circ f_4 = f_3^{-1} \circ f_5$. (Why?) Next multiply on the right by f_4^{-1}. The result is $f = f_2$.

Section 3–3

1. a) $\frac{1}{3}$ b) 0 c) 1 d) 0 e) 0 f) $1/(1 - x)$
   g) $\frac{1}{3}$ h) 1 i) $1/(1 - x)$ j) 0 k) $\frac{1}{4}$ l) $\max(a, a^{-1})$

2. c) Set $\frac{1}{2}\Delta_n = (\sqrt{n+1} - \sqrt{n})\sqrt{n+3} - \frac{1}{2}$. Show that

$$0 < \Delta_n = \frac{2\sqrt{1 + 3/n} - 1 - \sqrt{1 + 1/n}}{1 + \sqrt{1 + 1/n}} < \left(\sqrt{1 + \frac{3}{n}} - 1\right) + \left(\sqrt{1 + \frac{3}{n}} - \sqrt{1 + \frac{1}{n}}\right)$$

$$= \frac{3/n}{\sqrt{1 + 3/n} + 1} + \frac{2/n}{\sqrt{1 + 3/n} + \sqrt{1 + 1/n}} < \frac{5}{2n} < \frac{3}{n}.$$

   Given ϵ, let N be any positive integer greater than $3/\epsilon$.

3. b) $\lim_{n \to \infty} n^k r^n = 0$ whenever $|r| < 1$ and $k \in \mathbf{Z}^+$.

7. nth term $= a_n = 2^{1-2^{-n}}$. We have $2/\sqrt[n]{2} < a_n < 2$ $(n \in \mathbf{Z}^+)$. Apply Exercise 4 and pinch.

10. It converges to $(\sqrt{13} - 3)/2$.

12. If such a limit L existed, it would satisfy the equation $L^2 + 1 = 0$.

15. First establish the inequality

$$1 - \frac{a}{2} < \frac{1}{\sqrt{1+a}} < 1 - \frac{a}{2} + \frac{3a^2}{8} \qquad (a > 0).$$

Section 3-4

1. (a), (c), and (e) are unbounded.

2. The simplest proof uses Theorem 8. Suppose L and L^* are limits of a sequence $\{\alpha_n\}$. Set $\beta_n = \alpha_n$. Then $\lim \alpha = L$ and $\lim \beta = L^*$. Since $\alpha_n \leq \beta_n$, $L \leq L^*$. Since $\alpha_n \geq \beta_n$, $L \geq L^*$. Therefore, $L = L^*$.

3. Suppose A is a limit point of α. We may choose $n_1 \in \mathbf{Z}^+$ such that $|\alpha_{n_1} - A| < 1$. Since $|\alpha_n - A| < \frac{1}{2}$ holds for infinitely many positive integers n_1, we can choose an integer $n_2 > n_1$ such that $|\alpha_{n_2} - A| < \frac{1}{2}$. We next choose an integer $n_3 > n_2$ such that $|\alpha_{n_3} - A| < \frac{1}{3}$. Continuing, we get a sequence of positive integers $n_1 < n_2 < n_3 < \cdots$ such that $|\alpha_{n_k} - A| < 1/k$ for each $k \in \mathbf{Z}^+$. It follows that A is a limit of the subsequence $\alpha_{n_1}, \alpha_{n_2}, \ldots$

4. a) None   c) $-1, 1$ e) All nonnegative real numbers. A rigorous proof requires the use of the Archimedean ordering property.

7. We may define the integers by using the second form of the principle of induction. We let n_1 be the least positive integers such that $|a_{1,n_1} - A_1| < 1$. Having chosen

$$n_1 < n_2 < \cdots < n_k,$$

we define n_{k+1} to be the smallest positive integer such that

$$n_{k+1} > n_k \quad \text{and} \quad |a_{k+1,n_{1+k}} - A_{k+1}| < 1/(k + 1).$$

9. If x is rational, the sequence has only a finite number of distinct terms each of which occurs infinitely often. A rigorous proof of the assertion for irrational x requires machinery of Chapter 5. You should be able, nevertheless, to convince yourself on intuitive grounds that the statement is true.

Section 3-5

1. a) $\delta = \epsilon/3$ b) $\delta = \epsilon$ c) $\delta = \min (1, \epsilon/7)$ d) $\delta = \min (1, 3\epsilon)$
 e) $\delta = \frac{1}{2}$
 f) $\delta = \min (x_0 - [x_0], [x_0] + 1 - x_0) =$ distance from x_0 to the nearest integer
 g) $\delta = \epsilon^2$ h) $\delta = \epsilon$ i) $\delta = \epsilon/3$ j) $\delta = \min (1, 2\epsilon, \epsilon^2) = \min (1, \epsilon^2)$
 k) If $x_0 \neq 0$, $\delta = \min (|x_0|, |x_0|^{2/3}\epsilon)$. If $x_0 = 0$, $\delta = \epsilon^3$.

2. All except (b), (e), and (f)

3. d) The function oscillates infinitely near zero.

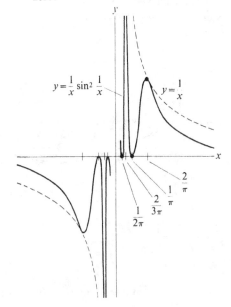

$$y = \frac{1}{x} \sin^2 \frac{1}{x}$$

$$y = \frac{1}{x}$$

e) Assume that a limit exists and deduce from the linking limit lemma that it must be equal to both 1 and 3.

g) Everything is amiss. The function fails to be defined throughout a deleted neighborhood of zero. Even if the function were defined at the points $2/\pi, -2/\pi, 2/3\pi, -2/3\pi, \ldots$, it could continue to oscillate infinitely.

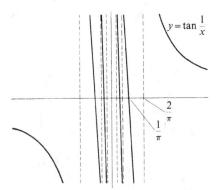

$$y = \tan \frac{1}{x}$$

5. a) 9   b) 0   c) $\frac{1}{2}$ d) -1 e) m/n f) 0   g) $\frac{1}{2}$
   h) $1/(4a\sqrt{a-b})$ i) 5   j) $\frac{2}{3}$ k) $\frac{1}{2}$ l) 0   m) $\frac{3}{2}$ n) 1   o) $\frac{1}{2}$
   p) The limit does not exist since the function is not defined for negative values of x. However, if x is replaced by $|x|$, the limit exists and is equal to zero. (Show that the function is in absolute value less than $\sqrt{|x|}$.)
   q) $-\sin a$ r) 0   s) $1/(4\sqrt{2})$ t) $\frac{1}{2}$

8. a) (I) $\Leftrightarrow$ (C), (VI) $\Leftrightarrow$ (F)
   c) Function (1) satisfies (D) only. Function (2) satisfies all conditions except (A). Function (4) satisfies (C), (D), and (E) only.

9. a) $\varnothing$, $[-1, 1]$, **R**
   b) Suppose L is a limit point of f at a. Set $\epsilon = \delta = 1/n$, where $n \in \mathbf{Z}^+$. Deduce that there exists a sequence $x_1, x_2, \ldots$ such that

$$0 < |x_n - a| < 1/n \quad \text{and} \quad |f(x_n) - L| < 1/n.$$

11. If $|\theta| \leq \pi/2$, argue that

$$1 - \cos \theta = \frac{\sin^2 \theta}{1 + \cos \theta} \leq \sin^2 \theta \leq \theta^2.$$

Show also by induction that $(1 - h)^n \geq 1 - nh$ if $0 < h < 1$. Deduce that for n sufficiently large,

$$1 \geq \left(\cos \frac{x}{n} \right)^n \geq 1 - \frac{x^2}{n}.$$

Pinch.

12. Show that $\Delta/\Delta' = (1 + \cos \theta)^2$.

Section 3-6

1. c) If r is the ratio of two relatively prime positive integers, and if the denominator is odd, then x^r is continuous on the entire line. The same is true if $r = 0$.

4. a) ± 2 are points of discontinuity.   c) None

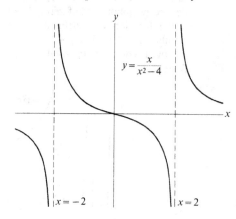

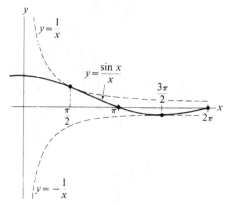

g) $f(x) = \begin{cases} -1 & \text{if } |x| < 1, \\ +1 & \text{if } |x| > 1, \\ 0 & \text{if } x = 0. \end{cases}$

The points of discontinuity are ± 1.

h) The points of discontinuity are $0, 1, -1$, $\frac{1}{2}, -\frac{1}{2}, \frac{1}{3}, \ldots$

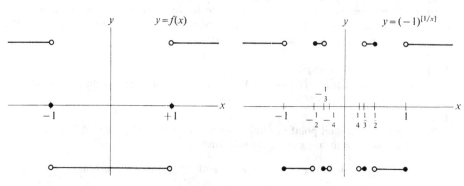

5. b) If the function is defined to have the value $\frac{2}{3}$ at 1, the function is everywhere continuous.

c) The function cannot be defined at zero so as to be continuous there.

6. $A = -1, B = 1$

7. b) *Note:* $f \vee g = \frac{1}{2}[f + g + |f - g|] = \frac{1}{2}[f + g + A \circ (f - g)]$, where $A(x) = |x|$. Use the closure properties of continuous functions.

10. The functions in (a), (c), and (e) are upper-semicontinuous at all points. (g) is upper-semicontinuous at all points except the integers.

11. b) Let $\epsilon > 0$ be given. Since f is upper-semicontinuous at x_0, we can choose a $\delta_f > 0$ such that

$$|x - x_0| < \delta_f \Rightarrow f(x) < f(x_0) + \epsilon.$$

Similarly, we can choose a $\delta_g > 0$ such that

$$|x - x_0| < \delta_g \Rightarrow g(x) < g(x_0) + \epsilon.$$

Set $\delta = \min(\delta_f, \delta_g)$. Then

$$|x - x_0| < \delta \Rightarrow \begin{cases} f(x) < f(x_0) + \epsilon \le (f \vee g)(x_0) + \epsilon, \\ g(x) < g(x_0) + \epsilon \le (f \vee g)(x_0) + \epsilon \end{cases}$$
$$\Rightarrow (f \vee g)(x) \le (f \vee g)(x_0) + \epsilon.$$

Thus $f \vee g$ is upper-semicontinuous at x_0.

Chapter 4

Section 4–1

1. a) 56 ft/sec, 48 ft/sec, 16 ft/sec
   b) 48 ft/sec, 32 ft/sec, -32 ft/sec
   c) $64t - 16t^2 = 64 - 16(t - 2)^2$. The maximum height is therefore 64 ft; it is attained after two seconds.
   d) $v(t) = 64 - 32t$. The ball has zero velocity when the maximum height is reached. The ball strikes the ground with the velocity -64 ft/sec. The sign of the velocity determines whether the ball is going up or down.

2. a) $3x^2 + 2$ c) $-1/x^2$ e) $1/(2\sqrt{x + 5})$

3. a) 0   b) 1   c) 0   d) 0   e) 1

4. a) $1/(3\sqrt[3]{x^2})$
   b) The domain of f' is the set of all nonintegers; f' is zero at all points of its domain.
   d) $f'(x) = \begin{cases} 1 & \text{if } x < 0, \\ 2 & \text{if } 0 < x < 3, \\ -1 & \text{if } x > 3. \end{cases}$

7. c) Yes. A proof may be given as follows.
   Let $\epsilon > 0$ be given. We set $v = v(t_0) + \epsilon$. Then the displacement of the second particle at time t will be

$$x = f(t_0) + [v(t_0) + \epsilon](t - t_0).$$

By (a), there exists a $\delta_1 > 0$ such that

$$f(t_0) + [v(t_0) + \epsilon](t - t_0) < f(t)$$

if $t_0 - \delta_1 < t < t_0$, and

$$f(t_0) + [v(t_0) + \epsilon](t - t_0) > f(t)$$

if $t_0 < t < t_0 + \delta_1$. Note that both displayed inequalities imply that

$$\frac{f(t) - f(t_0)}{t - t_0} - v(t_0) < \epsilon. \tag{*}$$

(In the first case $t - t_0$ is negative, so that inequalities are reversed upon division by $t - t_0$.) Thus (*) holds whenever $0 < |t - t_0| < \delta_1$.

By setting $v - v(t_0) - \epsilon$ and applying (b), we can show similarly that there exists a $\delta_2 > 0$ such that

$$-\epsilon < \frac{f(t) - f(t_0)}{t - t_0} - v(t_0) \tag{**}$$

whenever $0 < |t - t_0| < \delta_2$.

We now set $\delta = \min(\delta_1, \delta_2)$. If $0 < |t - t_0| < \delta$, then both (*) and (**) hold, so that

$$\left| \frac{f(t) - f(t_0)}{t - t_0} - v(t_0) \right| < \epsilon.$$

This proves that

$$v(t_0) = \lim_{t \to t_0} \frac{f(t) - f(t_0)}{t - t_0} = \lim_{t \to t_0} v(t, t_0)$$

and shows incidentally that the number $v(t_0)$ is unique.

Section 4–2

1. b) $x - 4y + 4 = 0$
2. a) $(0, 0)$ and $(\frac{3}{2}, \frac{27}{8})$
4. The point of tangency is $(3, -3)$. The origin is also a point of intersection.
6. c) The set is the line $y = -\frac{1}{2}$.
7. $a = 5, b = -12$, and $c = -3$

Section 4–3

1. Suppose that $x_0 \neq 0$ and that n is a negative integer. First observe that

$$\frac{d}{dx} (x^n) \bigg|_{x=x_0} = \lim_{x \to x_0} -\frac{x^{-n} - x_0^{-n}}{x - x_0} \cdot \frac{1}{x^{-n} x_0^{-n}}.$$

Since $-n$ is a positive integer, argue that the limit of the first factor is

$$\frac{d}{dx} (x^{-n}) \bigg|_{x=x_0} = (-n)x_0^{-n-1}.$$

6. c)
$$\frac{d}{dx} (\sin x^2) \bigg|_{x=x_0} = \lim_{h \to 0} \frac{\sin (x_0 + h)^2 - \sin x_0^2}{(x_0 + h)^2 - x_0^2} \cdot \frac{(x_0 + h)^2 - x_0^2}{h}.$$

Show that the limit of the second factor is $2x_0$. Use the substitution rule to show that the limit of the first factor is

$$\lim_{y \to x_0^2} \frac{\sin y - \sin x_0^2}{y - x_0^2} = \frac{d}{dy} (\sin y) \bigg|_{y=x_0^2} = \cos x_0^2.$$

Thus,

$$\frac{d}{dx} (\sin x^2) \bigg|_{x=x_0} = 2x_0 \cos x_0^2.$$

e)
$$\frac{d}{dx}(x \sin x)\Big|_{x=x_0} = \lim_{h \to 0} \frac{(x_0 + h)\sin(x_0 + h) - x_0 \sin x_0}{h}$$

$$= \lim_{h \to 0} \left[\sin(x_0 + h) + x_0 \frac{\sin(x_0 + h) - \sin x_0}{h} \right]$$

$$= \sin x_0 + x_0 \frac{d}{dx}(\sin x)\Big|_{x=x_0}$$

$$= \sin x_0 + x_0 \cos x_0.$$

Section 4–4

1. a) $4x^3 - 6x + 5$ c) $1/(2\sqrt{x})$
   e) $2/(3\sqrt[3]{x^2}) + \sqrt{2} \sec^2 x$ g) $(\sin x)/(2\sqrt{x}) + \sqrt{x}\cos x - 8/(5\sqrt[5]{x^3})$
   i) $2 \sin x \cos x$ k) $4 \sin^3 x \cos x$
   m) $5x^4 \sec x \tan x + x^5 \sec x \tan^2 x + x^5 \sec^3 x$
   p) Use the identity $\cos 2x = 2\cos^2 x - 1$ and the product rule.
   r) $\dfrac{(x + 2)\cos x - \sin x}{(x + 2)^2}$ t) $\dfrac{x(\cos x - \sin x) - \sin x - \cos x - 1}{(x - \cos x)^2}$
   v) $\dfrac{(1 - \sqrt[4]{x})[1/(2\sqrt{x}) + 1/(3\sqrt[3]{x^2})] + (1 + \sqrt{x} + \sqrt[3]{x})/(4\sqrt[4]{x^3})}{(1 - \sqrt[4]{x})^2}$
   x) Use (o), (p), and the quotient rule.

3. a) $2 \sin x \cos x$, $3 \sin^2 x \cos x$, $10 \tan^9 x \sec^2 x$

5. Let $f(x) = |x|$, and let $g(x) = -|x|$. Then the domain of $f' + g'$ is $(-\infty, 0) \cup (0, +\infty)$, whereas the domain of $(f + g)'$ is the entire real line.

7. If $x \neq 0$, we have

$$f'(x) = \frac{x \cos x - \sin x}{x^2},$$

by the rule for differentiating quotients. Use the inequality

$$0 < h - \sin h < \tan h - \sin h = (\tan h)(1 - \cos h),$$

where $0 < h < \pi/2$, to show that

$$f'(0) = \lim_{h \to 0} \frac{\sin h - h}{h^2} = 0.$$

Then argue that

$$0 = \sin x - \tan x \cos x < \sin x - x \cos x < (\sin x)(1 - \cos x),$$

if $0 < x < \pi/2$, and deduce from this that $\lim_{x \to 0} f'(x) = 0 = f'(0)$.

Section 4–5

1. a) $-2 \sin 2x$ c) $8(2 + 3x^2)(1 + 2x + x^3)^7$
   e) $\sqrt{1 - x^2} - x^2/\sqrt{1 - x^2}$
   g) $5 \tan(x/3 + x^2) + 5x(\frac{1}{3} + 2x) \sec^2(x/3 + x^2)$
   i) $-3 \cos 3x \sin(\sin 3x)$
   k) $(1/6\sqrt{x}) \sec^2(1 + \sqrt{x})[1 + \tan(1 + \sqrt{x})]^{-2/3}$

m) $2(7 + \sin x \sec^2 (\cos x)) \sin (7x - \tan (\cos x)) \cos (7x - \tan (\cos x))$

o) $-\sin (x\sqrt{1 + \tan^5 (3x \sin x)})$
$\times \{\sqrt{1 + \tan^5 (3x \sin x)} + \tfrac{5}{2}x \tan^4 (3x \sin x) \sec^2 (3x \sin x)$
$\times (3 \sin x + 3x \cos x)[1 + \tan^5 (3x \sin x)]^{1/2}\}$

4. a) All but the cosine function are infinitesimals.

$\sin x$	belongs to $\mathfrak{D}$ but not to $\mathfrak{o}$,
$|x|$	belongs to $\mathfrak{D}$ but not to $\mathfrak{o}$,
$(\sin x)/(1 + x^2)$	belongs to $\mathfrak{D}$ but not to $\mathfrak{o}$,
f	belongs to both $\mathfrak{D}$ and $\mathfrak{o}$,
$\sqrt[3]{x}$	belongs to neither $\mathfrak{D}$ nor $\mathfrak{o}$,
$\cos x - 1$	belongs to both $\mathfrak{D}$ and $\mathfrak{o}$,
g	belongs to both $\mathfrak{D}$ and $\mathfrak{o}$.

6. $\frac{7}{25}$ ft/sec

7. If we let x and θ be as indicated in the figure, then

$$\frac{dx}{dt} = -\frac{rx \sin \theta}{x - r \cos \theta}\frac{d\theta}{dt}.$$

The answer to the second part is $r\omega$.

8. 2.5 ft/sec

9. $HF/(H - F)$ ft/sec

10. $1/18\pi$ ft/sec, $\sqrt{2000/\pi}$ ft at least

11. $(-3, -\frac{16}{3})$

12. Let h and θ be as indicated in the figure. Then we have $vt = h \tan \theta$. Differentiating with respect to t, we get

$$v = h \sec^2 \theta \frac{d\theta}{dt},$$

so that the required angular velocity is

$$\frac{d\theta}{dt} = \frac{v}{h} \cos^2 \theta = \frac{vh}{h^2 + v^2 t^2}.$$

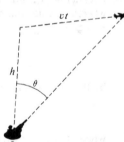

13. $8/9\pi$ ft/min

14. $\dfrac{dx}{dt} = \dfrac{1}{1 + \cos (s/r)}\dfrac{ds}{dt} = \dfrac{1}{2}\sec^2 \dfrac{s}{2r}\dfrac{ds}{dt}$

Section 4–6

1. a) $\left(\dfrac{40x}{10x^2 + 1} - \dfrac{9}{3x - 4} - \dfrac{10x}{x^2 - 1}\right)\dfrac{(10x^2 + 1)^2}{(3x - 4)^3(x^2 - 1)^5}$

b) $\dfrac{1}{3}\left(\dfrac{2x}{x^2 + 1} - \dfrac{1}{x} - \dfrac{2}{x + 2}\right)\sqrt[3]{\dfrac{x^2 + 1}{x(x + 2)^2}}$

c) $\left(\dfrac{1}{2x} + \dfrac{x}{1 - x^2} - \tan x\right)\dfrac{\sqrt{x}\cos x}{\sqrt{1 - x^2}}$

d) $\left(\dfrac{3}{4}\text{ctn } x + \dfrac{\sin x}{4(1 - \cos x)} - \dfrac{3}{x} - \dfrac{3}{2(3x + 5)}\right)\dfrac{(\sin x)^{3/4}(1 - \cos x)^{1/4}}{x^3\sqrt{3x + 5}}$

2. a) $-24 \sin 2x$ c) $-2 \sin x - x \cos x$ e) $(1 - x)^{-8}$

9. On the one hand,

$$D^n(x^{2n}) = n!\binom{2n}{n}x^{2n-n} = \frac{(2n)!}{n!}x^n.$$

On the other hand,

$$D^n(x^{2n}) = D^n(x^n \cdot x^n) = \sum_{k=0}^{n}\binom{n}{k}D^{n-k}(x^n)D^k(x^n)$$

$$= \sum_{k=0}^{n}\binom{n}{k}(n-k)!\binom{n}{n-k}x^k \cdot k!\binom{n}{k}x^{n-k} = n!x^n\sum_{k=0}^{n}\binom{n}{k}^2.$$

11. Show first that the identity holds when $n = 1, 2$. Assume true for $n - 1$ and n. Show that

$$(-1)^{n+1}D^{n+1}\left[x^n f\left(\frac{1}{x}\right)\right] = (-1)^{n+1}D^n\left[nx^{n-1}f\left(\frac{1}{x}\right) - x^{n-2}f'\left(\frac{1}{x}\right)\right]$$

$$= -\frac{n}{x^{n+1}}f^{(n)}\left(\frac{1}{x}\right) + (-1)^n D\left[\frac{1}{(-1)^{n-1}x^n}f^{(n)}\left(\frac{1}{x}\right)\right]$$

$$= \cdots$$

14. b) $\displaystyle\sum_{k=2}^{n}k(k-1)\binom{n}{k} = n(n-1)2^{n-2}$, $\displaystyle\sum_{k=1}^{n}k^2\binom{n}{k} = (n+1)n2^{n-2}$

16. a) Verification that $L + M$ is linear:

$$(L + M)(f + g) = L(f + g) + M(f + g) = L(f) + L(g) + M(f) + M(g)$$
$$= (L + M)(f) + (L + M)(g),$$
$$(L + M)(cf) = L(cf) + M(cf) = cL(f) + cM(f) = c(L + M)(f).$$

Linearity of L is essential to the first distributive law.

b) Let $f \in \mathcal{P}$. Then f may be written in the form

$$f(x) = \sum_{k=0}^{n}a_k x^k.$$

From the linearity of L and M we get

$$L(f) = \sum_{k=0}^{n}a_k L(x^k) = \sum_{k=0}^{n}a_k M(x^k) = M(f).$$

c) Prove the first assertion by induction on n. Prove Newton's formula by adapting the derivation in Example 1 of the text. The polynomial P may be obtained by first constructing a *difference table*.

k	$P(k)$	$\Delta P(k)$	$\Delta^2 P(k)$	$\Delta^3 P(k)$
0	2	0	6	6
1	2	6	12	
2	8	18		
3	26			

Hence by Newton's formula,

$$P(x) = 2 + \frac{0}{1!}x^{(1)} + \frac{6}{2!}x^{(2)} + \frac{6}{3!}x^{(3)}$$
$$= 2 + 3x(x - 1) + x(x - 1)(x - 2)$$
$$= x^3 - x + 2.$$

e) $[E^{-1}(f)](x) = f(x - 1)$. $(I - \Delta)^{-1} = I + \Delta + \Delta^2 + \cdots$ $(I + D)^{-1} = I - D + D^2 - D^3 + \cdots$

f) Prove first by induction that

$$\Delta^k(xf(x)) = x\,\Delta^k f(x) + k\,\Delta^{k-1}f(x).$$

Set $L = \Delta - \Delta^2/2 + \Delta^3/3 - \cdots$ Show next that

$$L(xf(x)) = xL(f(x)) + (I - \Delta + \Delta^2 - \cdots)f(x + 1) = xL(f(x)) + f(x).$$

Prove by induction that $L(x^n) = nx^{n-1} = D(x^n)$. Deduce that $L = D$.

Section 4–7

1. $(y - x)^2 = x^2 + 1$. $f'(2) = 1 + 2/\sqrt{5}$. $g'(-1) = 1 + 1/\sqrt{2}$
3. $dy/dx = [(\cos x - \sin x)y - \cos 2x]/(2y - \sin x - \cos x)$.
   When $x = \pi/4$, $y = 1/\sqrt{2}$ and the denominator above is zero.
4. a) $-\sqrt{(y/x)}$ c) $(\sin y)/(2 \sin 2y - x \cos y - \sin y)$
   e) $(2x - 2y - 3x^2 - 2xy)/(x^2 + 2x - 2y)$
5. a) $\sqrt{5x} - 2y = 1$ c) $y = 0$ and $(\pi - 2)(y - \pi) = 2x - 2\pi$
6. $\frac{4}{5}$, $-\frac{162}{125}$
8. Show first that

$$\sqrt{x^2 - 1} = (y^{1/m} - y^{-1/m})/2 \quad \text{and} \quad y^{1/m} = x + \sqrt{x^2 - 1}.$$

Differentiate to get

$$\sqrt{x^2 - 1}\,\frac{dy}{dx} = my.$$

Differentiate again and multiply by $\sqrt{x^2 - 1}$ to get the desired equation with $n = 0$. Then use Leibnitz's rule.

Chapter 5

Section 5–1

1. Yes. No.
2. Assuming the least upper bound axiom, we shall deduce the greatest lower bound axiom. Let A be a nonempty subset of **R** which is bounded below. We let S be the set of all lower bounds of the set A. Then S is nonempty (why?) and bounded above (by any member of A). Hence S has a least upper bound L. The number L is the greatest lower bound of the set A. (Why?)

4. Suppose S is a dense subset of **R**. Let $L \in \mathbf{R}$ be given. Since S is dense, there exists a number $x_1 \in S$ such that $L < x_1 < L + 1$. For the same reason there exists an $x_2 \in S$ such that $L < x_2 < \min (x_1, L + \frac{1}{2})$. Having selected x_1 and x_2, we choose $x_3 \in S$ such that $L < x_3 < \min (x_2, L + \frac{1}{3})$. Continuing in this way, we see that we can extract from S a sequence $x_1 > x_2 > x_3 > \cdots$ such that for each $n \in \mathbf{Z}^+$ we have $L < x_n < L + 1/n$. By the usual pinching argument, we conclude that $x_n \to L$. The convergence is clearly proper.

(Note that we have proved a stronger result than the one stated. A simpler procedure is to choose $x_n \in S$ such that $L < x_n < L + 1/n$.)

5. Modify the above discussion of Exercise 4.

6. a) Let M and N be upper bounds of A and B, respectively. Then $M + N$ is an upper bound of the set $A + B$. [If $z \in (A + B)$, then we may write $z = x + y$, where $x \in A$ and $x \in B$. Since $x \leq M$ and $y \leq N$, we have $z \leq M + N$.] Thus $A + B$ is bounded above.

Since lub A and lub B are in particular upper bounds of A and B, respectively, it follows by our previous remarks that lub A + lub B is an upper bound of $A + B$, so that

$$\text{lub } (A + B) \leq \text{lub } A + \text{lub } B.$$

Let $\epsilon > 0$ be given. We may then choose $x \in A$ and $y \in B$ such that $x > \text{lub } A - \epsilon/2$ and $y > \text{lub } B - \epsilon/2$. Then $x + y$ is a member of $A + B$ and

$$x + y > \text{lub } A + \text{lub } B - \epsilon.$$

Hence

$$\text{lub } A + \text{lub } B - \epsilon < x + y \leq \text{lub } (A + B).$$

Since $\epsilon > 0$ is arbitrary, we must have

$$\text{lub } A + \text{lub } B \leq \text{lub } (A + B).$$

This inequality, together with the inequality from the previous paragraph, shows that

$$\text{lub } (A + B) = \text{lub } A + \text{lub } B.$$

8. c) If $m < n$, then there exists a $k \in (\mathbf{Z}^+)_F$ such that $n = m + k$. Then $\varphi(n) = \varphi(m) + \varphi(k)$. Since $\varphi(k) \in (\mathbf{Z}^+)_K$, the element $\varphi(k)$ is positive. Hence $\varphi(m) < \varphi(n)$.

Section 5-3

1. Let a_n be the nth term of the sequence. Prove first by induction that $0 < a_n < 2$. Show next that

$$a_{n+1}^2 - a_n^2 = \tfrac{9}{4} - (a_n - \tfrac{1}{2})^2 > 0.$$

Apply the monotone convergence property to show the existence of the limit. (Compare with Exercise 8, Section 3-3.)

2. c) It suffices to show that $a_n b_n \to 1$. Now $a_n b_n$ is the sum of all numbers of the form

$$\frac{1}{i!} \frac{(-1)^j}{j!}$$

as i and j range independently through the integers $0, 1, 2, \ldots, n$. We group together terms for which $i + j$ has a common value. Thus

$$a_n b_n = \sum_{k=0}^{2n} \left({\sum_{i+j=k}}' \frac{1}{i!} \frac{(-1)^j}{j!} \right),$$

where the term in the parentheses indicates the sum of all numbers of the form

$$\frac{1}{i!} \frac{(-1)^j}{j!},$$

where (i, j) ranges over the set

$$\{(i, j) \in \mathbf{Z}^2 \mid 0 \le i \le n, 0 \le j \le n, \text{ and } i + j = k\}.$$

If $0 < k \le n$, show that

$${\sum_{i+j=k}}' \frac{1}{i!} \frac{(-1)^j}{j!} = \frac{1}{k!} \sum_{j=0}^{k} \binom{k}{j} (-1)^j = 0.$$

If $n < k \le 2n$, show that

$$\left| {\sum_{i+j=k}}' \frac{1}{i!} \frac{(-1)^j}{j!} \right| \le {\sum_{i+j=k}}' \frac{1}{i!} \frac{1}{j!} \le \sum_{j=0}^{k} \frac{1}{k!} \binom{k}{j} = \frac{1}{k!} 2^k.$$

Deduce that

$$|a_n b_n - 1| \le \frac{1}{(n+1)!} 2^{n+1} (2^n - 1),$$

and prove that the latter term tends to zero.

3. Show that $\{a_n\}$ is a strictly increasing sequence bounded above by b_1. Hence the sequence has a limit. Show similarly that $\{b_n\}$ has a limit. Show that the limits are the same by taking limits of both sides of the equation $b_{n+1} = (a_n + b_n)/2$.

5. Let $\epsilon > 0$ be given. There exists an $N \in \mathbf{Z}^+$ such that

$$n > N \Longrightarrow |x_n - L| < \epsilon.$$

(Why?) Set $M = \max \{\varphi^{-1}(n) \mid n \in \mathbf{Z}^+ \text{ and } n \le N\}$. (Why is this a legitimate definition?) Show that

$$m > M \Longrightarrow \varphi(m) > N \Longrightarrow |x_{\varphi(m)} - L| < \epsilon.$$

6. Show by induction that $|x_{n+1} - x_n| \le r^n |x_1 - x_0|$. If $m > n$, show that

$$|x_m - x_n| = \left| \sum_{k=n}^{m-1} (x_{k+1} - x_k) \right|$$

$$\le |x_1 - x_0|(r^n + r^{n+1} + \cdots + r^{m-1})$$

$$< |x_1 - x_0| r^n / (1 - r).$$

Deduce that $\{x_n\}$ is Cauchy.

9. Let $\{a_n\}$ be a sequence of real numbers. Let A be the set of all positive integers n such that the inequality $a_n \le a_k$ holds for infinitely many $k \in \mathbf{Z}^+$. Let B be the set of all positive integers n such that $a_n \ge a_k$ holds for infinitely many $k \in \mathbf{Z}^+$. Argue that $A \cup B = \mathbf{Z}^+$ and that one of the sets A or B has an infinite number of elements. If A is infinite, argue that $\{a_n\}$ contains a nondecreasing subsequence.

Chapter 6

Section 6–1

2. No. Yes, a strict relative minimum at zero.

4. No. Yes. No.

Section 6–2

3. Differentiability throughout the interior of the interval is essential, as evidenced by the function $|x|$ or $x^{2/3}$ on the interval $[-1, +1]$. Continuity at the endpoints is essential, as evidenced by the function $x - [x]$ on the interval $[2, 3]$.

4. Two

6. Set $L = \lim_{x \to x_0} f'(x)$. Let $\epsilon > 0$ be given. Choose $\delta > 0$ such that

$$0 < |x - x_0| < \delta \Rightarrow |f'(x) - L| < \epsilon.$$

If x belongs to the deleted neighborhood $V = \{x \mid 0 < |x - x_0| < \delta\}$, then by the mean-value theorem there exists a number ξ between x_0 and x, and hence also in V, such that

$$f(x) - f(x_0) = f'(\xi)(x - x_0).$$

Thus

$$\left| \frac{f(x) - f(x_0)}{x - x_0} - L \right| = |f'(\xi) - L| < \epsilon.$$

This proves that $f'(x_0) = L = \lim_{x \to x_0} f'(x)$.

Section 6–3

1. c) The function is strictly increasing throughout the intervals $[-1, 0]$ and $[1, +\infty)$. It is strictly decreasing throughout the intervals $(-\infty, -1]$ and $[0, 1]$.

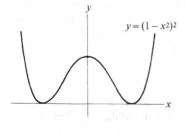

Ex. 1(c)

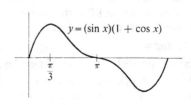

Ex. 1(e)

e) The function is increasing on intervals of the form $[-\pi/3 + 2n\pi, \pi/3 + 2n\pi]$, where n is an integer. The graph for $0 \le x \le 2\pi$ is shown in the figure.

g) The function is increasing on the intervals $(-\infty, -1]$ and $[1, +\infty)$. It is decreasing on the intervals $[-1, 0)$ and $(0, 1]$.

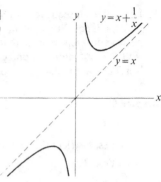

3. A precise answer to this exercise requires a careful notion of what is meant by a solution of the differential equation. We adopt the following definition. By a solution of the given equation we mean a function f such that the domain of f is an open interval I and for each x in I we have $xf'(x) + f(x) = 0$. If I is any open interval, then the function which is identically zero on I is a solution. If f is any other solution, then the domain of f is a subinterval of either $(-\infty, 0)$ or $(0, +\infty)$, and there exists a constant C such that $f(x) = C/x$ for all x in the interval.

4. e) If $0 \le x \le \pi/4$, then

$$0 \le S_3(x) - \sin x \le \frac{x^7}{7!} \le \frac{(\pi/4)^7}{7!} < 0.0005.$$

g) $-\frac{1}{6}, -\frac{1}{2}, -\frac{1}{3}$

6. An equivalent inequality is

$$\frac{\sin^2 x}{x^2 \cos x} > 1.$$

Show this by proving that the expression on the left is strictly increasing for $0 < x < \pi/2$.

Section 6–4

2. a) Strict relative maximum at 0, strict relative minimum at $\frac{3}{4}$
   b) Strict relative maximum at $a/2$
   c) Strict relative minimum at 2
   d) Strict relative maximum at all points of the form $\pi/3 + 2n\pi$, where n is an integer; strict relative minima at all points of the form $-\pi/3 + 2n\pi$, where n is an integer
   e) Same answer as (d) with "maxima" and "minima" interchanged
   f) Strict relative minimum at 27

3. -16 and 4

4. The side parallel to the river should be 500 ft.

6. The wire should not be cut at all for maximum area, but should be bent to form a circle. For minimum area, the side length of the square should be the length of the wire divided by $4 + \pi$.

8. There is exactly one real root if $D = q^2 + 4p^3 > 0$. There are three distinct real roots if $D < 0$. There are two distinct real roots if $D = 0$.

9. $x = (a_1 + a_2 + \cdots + a_n)/n$

13. The radius should be one-half the height.

15. For maximum volume, the height of the cylinder should be one-third the height of the cone. The situation regarding total lateral area (including the circular end pieces) is somewhat more delicate. Let r and h be the radius and height of the cone, respectively. If $2r < h$, then the lateral area is maximum when the radius of the cylinder is

$$hr/2(h - r).$$

If $2r \geq h$, then the lateral area is bounded above by $4\pi r^2$ and approaches this value as the height of the cylinder tends to zero.

17. $5\sqrt{5}$ ft

Section 6-5

1. d) Inflection points: $0, \pm\sqrt{3}$. The graph bends downward throughout the intervals $(-\infty, -\sqrt{3})$ and $(0, \sqrt{3})$; it bends upward throughout the intervals $(\sqrt{3}, 0)$ and $(\sqrt{3}, +\infty)$.

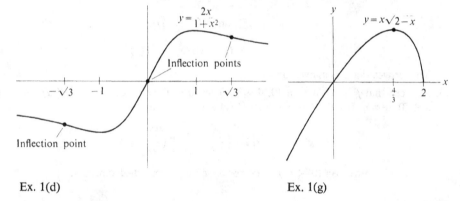

Ex. 1(d)   Ex. 1(g)

g) The graph bends downward at all points.

Section 6-6

2. a) Its second derivative is also of odd degree and hence must assume both positive and negative values. (See the proof of Proposition 6 in Section 5-2.)

   b) The polynomial $a_0x^4 + a_1x^3 + a_2x^2 + a_3x + a_4$ $(a_0 \neq 0)$ is convex iff $a_0 > 0$ and $8a_0a_2 \geq 3a_1^2$.

6. Suppose false. What would then be true of the tangent line to the graph at the origin?

7. d) Suppose $x, y \in I$ and $0 \leq \lambda \leq 1$. Since g is convex on I,

$$g((1 - \lambda)x + \lambda y) \leq (1 - \lambda)g(x) + \lambda g(y).$$

Since f is increasing,

$$(f \circ g)((1 - \lambda)x + \lambda y) \leq f((1 - \lambda)g(x) + \lambda g(y)).$$

Because f is convex, the right-hand side of the preceding inequality is less than or equal to $(1 - \lambda)f(g(x)) + \lambda f(g(y))$. Hence $f \circ g$ is convex on I.

   e) Not necessarily. The polynomials x^2 and x are both convex on the entire line, yet x^3 is not.

11. Prove by induction on n that for all x and y in I,

$$f((1 - \lambda)x + \lambda y) \le (1 - \lambda)f(x) + \lambda f(y),$$

where $\lambda = m2^{-n}$ and m is an odd integer less than 2^n. Such λ's form a dense subset of the interval $[0, 1]$. Hence, given any t in $[0, 1]$, we can choose a sequence of such numbers, say $\lambda_1, \lambda_2, \ldots$, such that $\lim \lambda_n = t$. (See Exercises 3 and 4 of Section 5–1.) Then for all x and y in I we have

$$\lim_{n \to \infty} [(1 - \lambda_n)x + \lambda_n y] = (1 - t)x + ty$$

and

$$f((1 - \lambda_n)x + \lambda_n y) \le (1 - \lambda_n)f(x) + \lambda_n f(y),$$

$n = 1, 2, \ldots$ Since f is continuous on I, it follows that

$$\begin{aligned}
f((1 - t)x + ty) &= \lim_{n \to \infty} f((1 - \lambda_n)x + \lambda_n y) \\
&\le \lim_{n \to \infty} [(1 - \lambda_n)f(x) + \lambda_n f(y)] \\
&= (1 - t)f(x) + tf(y).
\end{aligned}$$

This proves that f is convex on I.

13. Suppose that f is convex on $(0, +\infty)$. Then $0 \le f''(x) = 2\varphi'(x) + x\varphi''(x)$ for each $x \in (0, +\infty)$. If $y \in (0, +\infty)$, then $(1/y) \in (0, +\infty)$, and so

$$g''(y) = \frac{1}{y^3}\left[2\varphi'\left(\frac{1}{y}\right) + \frac{1}{y}\varphi''\left(\frac{1}{y}\right)\right] \ge 0.$$

Hence g is convex on $(0, +\infty)$. The converse may be handled similarly.

Section 6–7

1. a) $\frac{3}{4}-$, proper convergence   c) 0, improper convergence
   e) $+\infty$, proper convergence   g) $-\infty$, proper convergence
   i) -1, improper convergence   k) $-\infty$, proper convergence

2. a) $(\forall M)(\exists \delta > 0)(\forall x)(a < x < a + \delta \Rightarrow f(x) < M)$
   c) $(\forall M)(\exists N)(\forall x)(x < N \Rightarrow f(x) > M)$
   f) $(\exists M)(\exists \delta > 0)(\forall x)(a \le x < a + \delta \Rightarrow |f(x)| \le M)$

3. a) $(-11, -10), (-10\frac{1}{2}, -10), (-10\frac{1}{3}, -10), \ldots$ 10.

5. d) $f(x) = x^2 + \sin x,\ g(x) = x^2$

7. a) $f(x) = (1/x) \sin x^2$

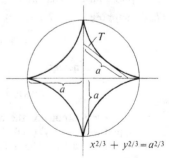

$x^{2/3} + y^{2/3} = a^{2/3}$

11. c) The x-coordinate of the inflection point is the root of the equation $x^3 - 6x + 6 = 0$. Its value to two decimal places is -2.85.

12.

e)

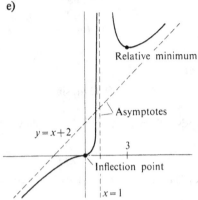

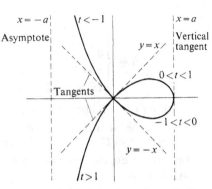

Strophoid: $(x^2+y^2)x - a(x^2-y^2) = 0$

$$x = a\frac{1-t^2}{1+t^2}, \qquad y = a\frac{t-t^3}{1+t^2}$$

16.

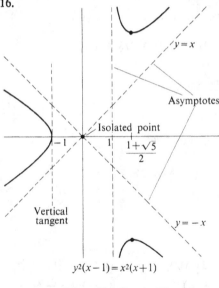

$y^2(x-1) = x^2(x+1)$

17.

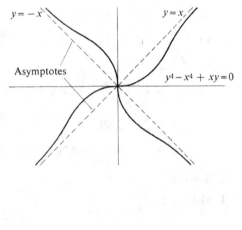

$y^4 - x^4 + xy = 0$

20. a) Set $\varphi(x) = f(x)f(-x)$. Then

$$\varphi'(x) = f'(x)f(-x) - f(x)f'(-x) = f(x)f(-x) - f(x)f(-x) = 0.$$

Hence φ is a constant function. Since $\varphi(0) = [f(0)]^2 = 1$, the constant must be one. If f were negative at some point x, then by the intermediate-value theorem there would exist a number X between 0 and x such that $f(X) = 0$. This is impossible, since it would imply that $0 = f(X)f(-X) = 1$. Since $f''(x) = f'(x) = f(x) > 0$ for all x, the function f is strictly increasing and convex on the entire real line.

b) Use Propositions 8 and 10 to prove the first part.

Chapter 7

Section 7–1

3. a) Let

$$f(x) = \begin{cases} -x & \text{if } -1 \le x \le 0, \\ 0 & \text{if } 0 < x \le 1, \\ x - 1 & \text{if } 1 < x \le 2, \\ 2x - 3 & \text{if } 2 < x \le 3. \end{cases}$$

c) $3x^{2/3}/2$
d) None
f) $x|x|/2$
g) None

Then f is an antiderivative of $[x]$ on $[-1, 3]$.

4. b) $\frac{1}{7}x^7 + \frac{1}{2}x^4 + x + C$ d) $\frac{2}{3}x^{3/2} - 2x^{1/2} + C$ f) $(1/a) \sin ax + C$
   i) $\tan t - t + C$ k) $(\sin^4 x)/4 + C$

5. Let $A(t)$ denote the area swept out in t seconds. Then $A'(t) = t^2/2$. It follows that $A(t) = t^3/6$.

Section 7–2

1. a) $20x^3\,dx$ c) $(\sin t + t \cos t)\,dt$ e) $2uv\,du + u^2\,dv + \sin u\,dv + v \cos u\,du$
   g) $uv\,dw + uw\,dv + vw\,du$

2. c) $\sin y\,dx + x \cos y\,dy + 2y \sin x\,dy + y^2 \cos x\,dx = 0$. Thus,

$$dy/dx = -(\sin y + y^2 \cos x)/(x \cos y + 2y \sin x).$$

3. b) $(\sin 3x^2)/6 + C$ d) $(1 + 3x^4)^{3/2}/8 + C$
   g) $\sin \theta - (\sin^3 \theta)/3 + C$ i) $\cos \sqrt{1 - x^2} + C$
   k) $\displaystyle\int \frac{\sin^3 x}{\cos^5 x}\,dx = \int \tan^3 x \sec^2 x\,dx = \frac{\tan^4 x}{4} + C$

4. a) $\frac{1}{2}\theta + \frac{1}{4}\sin 2\theta + C$
   c) $\sin^4 \theta = \left(\dfrac{1 - \cos 2\theta}{2}\right)^2 = \frac{1}{4} - \frac{1}{2}\cos 2\theta + \frac{1}{4}\cos^2 2\theta$

$$= \frac{1}{4} - \frac{1}{2}\cos 2\theta + \frac{1}{4} \cdot \frac{1 + \cos 4\theta}{2} = \frac{3}{8} - \frac{1}{2}\cos 2\theta + \frac{1}{8}\cos 4\theta$$

$$\int \sin^4 \theta\,d\theta = \frac{3}{8}\theta - \frac{1}{4}\sin 2\theta + \frac{1}{32}\sin 4\theta + C$$

Section 7–3

1. a) 9   b) $\frac{3}{2}$ c) $-\frac{1}{2}$
   d) 2   e) $(2\sqrt{2} - 5\sqrt{5})/3$ f) $1/(n + 1)$

3. a) $\sum_{k=1}^{8} \sqrt{k} - 20$ b) 1   c) 4   d) $-3(\sqrt{3} - \sqrt{2})$
   e) $(2\sqrt{2} + 5\sqrt{5})/3 - \frac{2}{3}$ f) $-3\sqrt{3}/8$ g) $4 - \sqrt[4]{8} - \sqrt{2}\sqrt[4]{3}$

   h) $\displaystyle\int_{-1}^{1/2} x^2\sqrt{1 - x^2}\,dx = \int_{-\pi/2}^{\pi/6} \sin^2 \theta \cos^2 \theta\,d\theta = \frac{1}{4}\int_{-\pi/2}^{\pi/6} \sin^2 2\theta\,d\theta$

$$= \frac{1}{8}\int_{-\pi/2}^{\pi/6} (1 - \cos 4\theta)\,d\theta = \frac{1}{8}(\theta - \frac{1}{4}\sin 4\theta)\Big|_{-\pi/2}^{\pi/6}$$

$$= \pi/12 - \sqrt{3}/64$$

   i) $x/2$ if $-\pi/2 < x < \pi/2$ j) $(\sqrt{2} + \sqrt{3})/8$
   k) $3 - 3\pi/4$ l) $\pi/6$ m) $\pi/8$

4. The formula is valid if f is Newton-integrable on $[a, b]$. The proof is as follows.
Let F be an antiderivative of f on the interval. Set $g(x) = f(x - c)$ and $G(x) = F(x - c)$.
Then G is an antiderivative of g on the interval $[a + c, b + c]$. [Note that $G'(x) = F'(x - c) = f(x - c) = g(x)$.] Thus

$$\int_{a+c}^{b+c} f(t - c)\, dt = \int_{a+c}^{b+c} g(t)\, dt = G(b + c) - G(a + c)$$
$$= F(b) - F(a) = \int_a^b f(t)\, dt.$$

6. Let G be an antiderivative of g on $[0, a]$. If $-a \leq x < 0$, we define $G(x) = G(-x)$.
One can then show that G, so extended, is an antiderivative of g on $[-a, a]$. [For instance, if $-a \leq x < 0$, then $G'(x) = -G'(-x) = -g(-x) = g(x)$ with at most a finite number of exceptions.] Hence

$$\int_{-a}^a g(t)\, dt = G(a) - G(-a) = 0.$$

9. $667 < \sum_{k=1}^{100} \sqrt{k} < 676$

Section 7–4

1. $2A/\omega$ 2. $\frac{4}{3}$ 3. $\frac{1}{6}$ 4. $\pi/2 - \frac{1}{3}$

5. $(4\sqrt{2} - 1)a^2/2$. The problem is somewhat
ambiguously worded. The answer given is
made on the assumption that the region in
question is that shown in the figure.

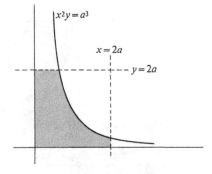

7. $\frac{16}{3}$

9. For the purposes of illustration suppose that the graph of f is as shown in Fig. (a).
Then the graphs of f^+ and f^- are given in Figs. (b) and (c). The integral $\int_a^b |f|$ represents
the area of the entire shaded region shown in Fig. (d). The integral $\int_a^b f$ is equal to the
sum of the areas of regions I and III minus the area of region II, that is, the "area above
the x-axis minus the area below the x-axis."

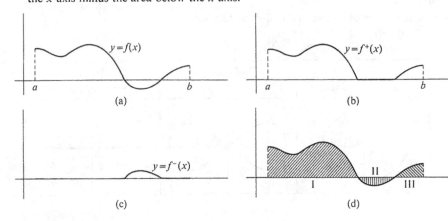

13. $4a^2/3$

17. The curve is shown in the figure. The required area
is $\frac{8}{15}$.

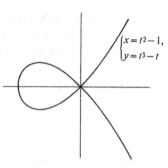

$$\begin{cases} x = t^2 - 1, \\ y = t^3 - t \end{cases}$$

Section 7–5

3. a) $[\sqrt{2}; 3\pi/4 + 2n\pi] = [-\sqrt{2}; -\pi/4 + 2n\pi]$, n an integer
   d) $[2; -\pi/6 + 2n\pi] = [-2; 5\pi/6 + 2n\pi]$, n an integer

4. a)   c)

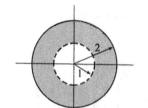

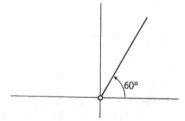

e)   g)

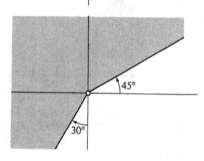

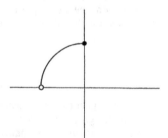

5. b) $(\cos \alpha)x + (\sin \alpha)y = \rho$

6. a) Use the law of cosines and half-angle formulas.
   b), c), and d)

$b = 2a$	$b = a$	$2b = a$
$PP_1 \cdot PP_2 = 4a^2$	$PP_1 \cdot PP_2 = a^2$	$PP_1 \cdot PP_2 = \frac{1}{4}a^2$

7. b)

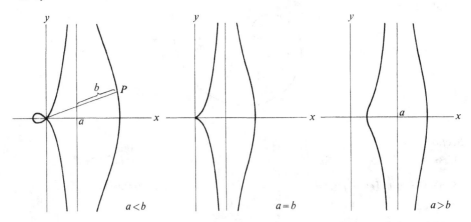

8. b)

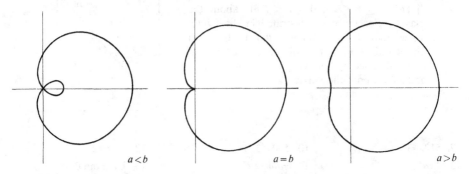

9. a) $16\pi^2$
   e) The area is $\pi a^2/2$.

   d) $3\pi a^2/2$
   h) The area is $\pi(2a^2 + b^2)/2$.

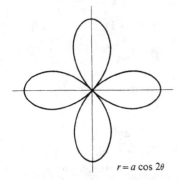

$r = a \cos 2\theta$

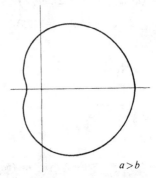

$a > b$

10. b) $3(\pi + 1)/2$, $(\pi - 3)/2$

11. a) $4\pi^3 a^2/3$
    b) $32\pi^3 a^2/3$
    c) $4n^3\pi^3 a^2/3$

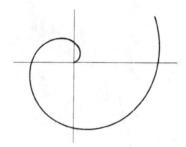

Section 7–6

1. $3\pi/10$
2. $8\pi^2 a^3$
3. $3\pi/10$
4. $32\pi a^2/105$

5. a) $V = \pi \int_a^b [g(t)]^2 f'(t)\, dt$

   b) $V = 2\pi \int_a^b f(t) g(t) f'(t)\, dt$

   c) $5\pi^2 a^3$

6. b) $\frac{2}{3}\pi \int_\alpha^\beta [f(\theta)]^3 \cos\theta\, d\theta$. Let $V(\theta_0)$ be the volume of the solid obtained by revolving the sector $\{[r;\theta] \mid \alpha \le \theta \le \theta_0,\ 0 \le r \le f\,\theta)\}$ about the x-axis. For the case illustrated in the figure, argue that $V(\theta_0 + \Delta\theta) - V(\theta_0)$ is bounded above and below by the volumes of the "conical hats" obtained by revolving the triangles $A'OB'$ and AOB about the x-axis.

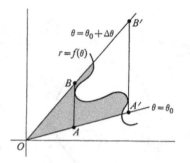

7. $512\pi/3$
10. $500\sqrt{3}/3$
12. a) $\pi/2$

13. $\frac{4}{3}\pi abc$
14. $\pi abh/2$
15. $\frac{2}{3}a^2 b \tan\theta$

Section 7–7

1. $P_{2\pi/3} = (a(2\pi/3 - \sqrt{3}/2), 3a/2)$
2. a) $6a$
   b) $4\dfrac{a^2 + ab + b^2}{a + b}$

3. a) $2\pi^2 a$
   c) $(8\sqrt{2197} - 125)/27$
6. a) $8(10\sqrt{10} - 1)/27$
   b) $\frac{55}{6}$

7. $8a$
12. $4\pi^2 ab$

13. $56\pi a^2/3$
14. $64\pi a^2/3$

15. $2\pi a^2(2 - \sqrt{2})$
16. $32\pi a^2/5$

Section 7–8

1. $(\frac{1}{5}, \frac{1}{5})$ and $\frac{1}{5}$

2. a) $(4a/3\pi, 4b/3\pi)$
   b) $\left(\dfrac{-4ab}{3\pi(4a - b)},\ \dfrac{-4b^2}{3\pi(4a - b)} \right)$

3. $\pi/2 + \frac{4}{5}$
4. $(5a/8, 0)$
5. $(\pi a, 5a/6)$

6. $(256a/315\pi, 256a/315\pi)$
9. $(0, 3\pi a/16)$
12. $(5a/6, 0)$

13. $(\sqrt{2}\,\pi a/8, 0)$

Section 7–9

1. $T = \dfrac{A}{5ka}\sqrt{\dfrac{2H}{g}}$

3. About 1.63×10^{11} kg

5. $\dfrac{g^2 M^3}{6m^2}$ ergs

6. About 1600 kg

7. b) $\gamma \omega^2 r^2 / 2$

8. About 1092 coulombs

9. $\dfrac{GmM}{s(s+l)}$

10. a) $\dfrac{GmMs}{\sqrt{(r^2+s^2)^3}}$, where s is the distance from the particle to the center of the ring

   b) If, say, $0 < s_1 < s_2$, then the work required to move the particle from a distance s_1 to a distance s_2 from the center of the ring is

   $$GmM\left(\frac{1}{\sqrt{r^2+s_1^2}} - \frac{1}{\sqrt{r^2+s_2^2}}\right).$$

   c) $\dfrac{2GmM}{r^2}\left(1 - \dfrac{s}{\sqrt{r^2+s^2}}\right)$, where m is the mass of the particle, M is the mass of the

   disk, r is the radius of the disk, and s is the distance from the particle to the center of the disk

11. The work done is $\pi/2$ along γ_1, $5\pi/2$ along γ_2, 8π along γ_3, $-7\pi/2$ along γ_4.

Chapter 8

Section 8–1

1. b) $x_0 = -8, x_1 = -5, x_2 = -2, x_3 = 0, x_4 = 1, x_5 = 3, x_6 = 4, x_7 = 5, x_8 = 6,$
   $x_9 = 7$

   $$\int_{-8}^{7} (f+g) = 27\tfrac{1}{2} \quad \text{and} \quad \int_{-8}^{7} fg = -11\tfrac{1}{2}$$

3. $f(x) = 0$ for all x [a, b] with at most finitely many exceptions.

6. b) We have

   $$\bar{I}(f+g) \le \bar{I}(f) + \bar{I}(g) = \int_a^b f + \bar{I}(g).$$

   To establish the reverse inequality, let $\epsilon > 0$ be given. Choose a step function h such that $g \le h$ on [a, b] and $\int_a^b h < \bar{I}(g) + \epsilon$. Then $f + h$ is Riemann-integrable and $f + h \ge f + g$ on [a, b]. Thus we have

   $$\int_a^b f + \bar{I}(g) + \epsilon \ge \int_a^b f + \int_a^b h = \int_a^b (f+h) = \bar{I}(f+h) \ge \bar{I}(f+g).$$

   Since $\epsilon > 0$ is arbitrary, $\int_a^b f + \bar{I}(g) \ge \bar{I}(f+g)$.

7. b) Let M be a bound of $|f|$ on the interval $[a, b]$. Let $\epsilon > 0$ be given. Choose c and d such that $a < c < d < b$, $c - a < \epsilon/3M$, and $b - d < \epsilon/3M$. Since f is Riemann-integrable on $[c, d]$, we can choose step functions g and h on $[c, d]$ such that $g \leq f \leq h$ on $[c, d]$ and $\int_c^d (h - g) < \epsilon/3$. Extend g to $[a, b]$ by defining it to have the value $-M$ on the intervals $[a, c)$ and $(d, b]$. Extend h similarly. Infer that $g \leq f \leq h$ on $[a, b]$ and that $\int_a^b (h - g) < \epsilon$. Deduce that f is Riemann-integrable on $[a, b]$.

Section 8–2

1. a) $I_{\pi,\sigma}(f) = 8.941$, $\|\pi\| = 0.6$, $I_{\pi,\sigma}(\varphi_J) = 0.5$

   b) $I_n(f) = 2 + 4(1 + 1/n) + \frac{4}{3}(1 + 1/n)(2 + 1/n)$, $\|\pi_n\| = 2/n$

   c) $I_n(f) = \dfrac{26}{1 + 3^{1/n} + 3^{2/n}}$, $\|\pi_n\| = 3(1 - 3^{-1/n})$

   d) $\|\pi_n\| = 1$ for each positive integer n. The functional I is linear, nonnegative, and monotone. It fails to be an integral since it does not behave properly on step functions. For instance,

$$I(\varphi_{[1,3/2]}) = 1 \neq \tfrac{1}{2}.$$

Section 8–3

3. It suffices to suppose that f is positive on the interval. (Why?) Let $\eta > 0$ be a constant such that $f(x) \geq \eta$ for all $x \in [a, b]$. Given $\epsilon > 0$, choose step functions g and h such that $\eta \leq g \leq f \leq h$ on $[a, b]$ and $\int_a^b (h - g) < \eta^2\epsilon$. Then $1/h \leq 1/f \leq 1/g \leq 1/\eta$ on $[a, b]$ and

$$\int_a^b \left(\frac{1}{g} - \frac{1}{h}\right) \leq \eta^{-2}\int_a^b (h - g) < \epsilon.$$

Deduce that $1/f$ is Riemann-integrable on $[a, b]$.

4. Let $\epsilon > 0$ be given. Choose step functions f^* and g^* such that $f \leq f^*$ and $g \leq g^*$ on $[a, b]$,

$$\int_a^b f^* < \bar{I}(f) + \epsilon/2 \quad \text{and} \quad \int_a^b g^* < \bar{I}(g) + \epsilon/2.$$

Then $f^* \vee g^*$ and $f^* \wedge g^*$ are step functions, $f \vee g \leq f^* \vee g^*$, and $f \wedge g \leq f^* \wedge g^*$ on $[a, b]$, and thus

$$\bar{I}(f \vee g) + \bar{I}(f \wedge g) \leq \bar{I}(f^* \vee g^*) + \bar{I}(f^* \wedge g^*)$$

$$= \int_a^b f^* \vee g^* + \int_a^b f^* \wedge g^* = \int_a^b (f^* \vee g^* + f^* \wedge g^*)$$

$$= \int_a^b (f^* + g^*) = \int_a^b f^* + \int_a^b g^* < \bar{I}(f) + \bar{I}(g) + \epsilon.$$

Since ϵ is arbitrary, we have

$$\bar{I}(f \vee g) + \bar{I}(f \wedge g) \leq \bar{I}(f) + \bar{I}(g).$$

8. a) Let $\pi = (x_0, x_1, \ldots, x_n)$. By the mean-value theorem for integrals,

$$\int_{x_{k-1}}^{x_k} f(x)\,dx = f(\xi_k)(x_k - x_{k-1})$$

for some ξ_k between x_{k-1} and x_k $(k = 1, 2, \ldots, n)$. Set $\sigma = (\xi_1, \xi_2, \ldots, \xi_n)$. Then

$$\int_a^b f(x)\, dx = \sum_{k=1}^n \int_{x_{k-1}}^{x_k} f(x)\, dx = \sum_{k=1}^n f(\xi_k)(x_k - x_{k-1})$$

$$= I_{\pi,\sigma}(f).$$

b) Let $\pi = (x_0, x_1, \ldots, x_n)$ and $\sigma = (\xi_1, \ldots, \xi_n)$. Choose $\sigma' = (\xi_1', \ldots, \xi_n')$ such that $\int_a^b f = I_{\pi,\sigma'}(f)$. Then

$$I_{\pi,\sigma}(f) - \int_a^b f = \sum_{k=1}^n [f(\xi_k) - f(\xi_k')](x_k - x_{k-1}).$$

By the mean-value theorem, we can write

$$f(\xi_k) - f(\xi_k') = f'(X_k)(\xi_k - \xi_k')$$

for some X_k between ξ_k and ξ_k'. Thus

$$|f(\xi_k) - f(\xi_k')| \le M(x_k - x_{k-1}) \le M\|\pi\| \qquad (k = 1, \ldots, n).$$

Hence

$$\left| I_{\pi,\sigma}(f) - \int_a^b f \right| \le \sum_{k=1}^n |f(\xi_k) - f(\xi_k')|(x_k - x_{k-1})$$

$$\le (b - a)M\|\pi\|.$$

9. c) $\pi/(2\sqrt{2})$

Section 8–4

2. a) $-\sqrt{25 - x^2}$ b), c), and d), $1/x$
   e) $-\sin x \sin (\cos^2 x) - 3x^2 \sin x^6$

3. $\dfrac{dy}{dx} = \dfrac{2 \sin xy + 2xy \cos xy + 4x^{-1} \sin (x^2 \sin y)}{y^{-1} \sin \sqrt{y} - 2x^2 \cos xy - 2 \operatorname{ctn} y \sin (x^2 \sin y)}$

5. b) $F_\delta'(x) = \dfrac{1}{2\delta} [f(x + \delta) - f(x - \delta)]$

   c) Since f is uniformly continuous on $[a - 1, b + 1]$, we can choose $0 < \delta < 1$ such that if x and y belong to $[a - 1, b + 1]$ and $|x - y| < \delta$, then $|f(x) - f(y)| < \epsilon$. Show that if $a \le x \le b$, then

$$|F_\delta(x) - f(x)| = \left| \frac{1}{2\delta} \int_{-\delta}^{\delta} [f(x + t) - f(x)]\, dt \right|$$

$$\le \frac{1}{2\delta} \int_{-\delta}^{\delta} |f(x + t) - f(x)|\, dt$$

$$< \frac{1}{2\delta} (\epsilon \cdot 2\delta) = \epsilon.$$

Section 8-5

3. a) Since the null set is disjoint from itself, $\mu(\varnothing) = \mu(\varnothing \cup \varnothing) = 2\mu(\varnothing)$. Thus $\mu(\varnothing) = 0$.

b) $\mu(A \cup B \cup C) = \mu(A \cup B) + \mu(C) - \mu((A \cup B) \cap C)$
$= \mu(A) + \mu(B) - \mu(A \cap B) + \mu(C) - \mu((A \cap C) \cup (B \cap C))$
$= \cdots$

9. A set $B \subset \mathbf{R}^2$ is Jordan-measurable iff for each $\epsilon > 0$ there exist blocks A and C such that $A \subset B \subset C$ and $m(C\backslash A) = m(C) - m(A) < \epsilon$.

Chapter 9

Section 9-1

1. a) Not invertible
   b) $f^{-1}(x) = \sqrt[3]{(x+2)/3}$
   c) Not invertible
   d) $f^{-1}(x) = 1 - x = f(x)$
   e) $f^{-1}(x) = 1/x = f(x)$
   f) $f^{-1}(x) = x - n$ if $2n \le x < 2n + 1$ and n is an integer
   g) $f^{-1}(x) = (\text{sgn } x)\sqrt{|x|}$
   h) $f^{-1}(x) = 1/(1 - x^2)$ if $x \ge 0$

3. $f^{-1}(x) = (-dx + b)/(cx - a)$

4. Assume that f is strictly increasing and convex on an interval I. Then $J = \mathfrak{R}_f$ is also an interval (by the intermediate-value theorem). Let y_1 and y_2 be elements of J, and let $0 \le \lambda \le 1$. Then $y = (1 - \lambda)y_1 + \lambda y_2$ also belongs to J. Set $x_1 = f^{-1}(y_1)$ and $x_2 = f^{-1}(y_2)$. Since f is convex, we have

$$f((1 - \lambda)x_1 + \lambda x_2) \le (1 - \lambda)f(x_1) + \lambda f(x_2).$$

Since f^{-1} is strictly increasing, we get upon application of f^{-1} to each side of the displayed inequality

$$(1 - \lambda)f^{-1}(y_1) + \lambda f^{-1}(y_2) = (1 - \lambda)x_1 + \lambda x_2 \le f^{-1}((1 - \lambda)f(x_1) + \lambda f(x_2))$$
$$= f^{-1}((1 - \lambda)y_1 + \lambda y_2).$$

Hence f^{-1} is concave on J.

7. If $g = f^{-1}$, then

$$g''(y) = -\frac{f''(g(y))}{[f'(g(y))]^3}.$$

9. The function f^{-1} is differentiable everywhere except at odd integral multiples of π.

10. Equality holds iff $b = f(a)$.

11. a) Suppose $(f, g) \sim (f^*, g^*)$, say $f = f^* \circ \varphi$ and $g = g^* \circ \varphi$. Then $f^* = f \circ \varphi^{-1}$ and $g^* = g \circ \varphi^{-1}$. Since φ^{-1} is strictly increasing, $(f^*, g^*) \sim (f, g)$.

b) $\varphi(\theta) = \tan(\theta/2)$ does the trick.

c) Use the closing remarks of Section 7-7.

Section 9-2

1. a) $\pi/3$
   d) $-\pi/6$
   f) $5/\sqrt{26}$

2. a) $\dfrac{1}{2\sqrt{x - x^2}}$
   b) $-\dfrac{1}{\sqrt{1 - x^2}}$
   c) $\text{Sin}^{-1} x$

e) $1/(x^2 + 2x + 2)$ g) $(\sqrt{\operatorname{ctn} x} + \sqrt{\tan x})/\sqrt{2}$

j) $\dfrac{1}{3(x^{2/3} + x^{4/3})}$

4. a) $\pi/2$ c) $\pi/12$ e) $-(\operatorname{Cos}^{-1} x)^2/2 + C$

g) $2\operatorname{Sin}^{-1}\sqrt{x} + C$

7.

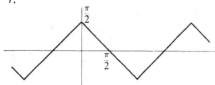

11. b) $T_3(x) = 4x^3 - 3x$
$T_4(x) = 8x^4 - 8x^2 + 1$

12. The constant is zero.

Section 9–3

1. a) $2x^{-1}\ln x$ c) $\ln x$ e) $\sec x$

g) $2\dfrac{e^x - e^{-x}}{(e^x + e^{-x})^2}$ i) $e^x/\sqrt{1 + e^{2x}}$ k) $(\pi + x\ln \pi)x^{\pi-1}\pi^x$

m) $2x(\ln 2)2^{x^2}$ o) $\frac{1}{2}(2 + \ln x)x^{\sqrt{x}-1/2}$ q) $[x^{-1} + (\ln x)(1 + \ln x)]x^{x^x+x}$

s) $[\cos x \operatorname{ctn} x - \sin x \ln (\sin x)](\sin x)^{\cos x} + [\cos x \ln (\cos x) - \sin x \tan x](\cos x)^{\sin x}$

u) $[\ln \ln x + (\ln x)^{-1} - 2x^{-1}\ln x]\dfrac{(\ln x)^x}{x^{\ln x}}$

5. a) $-\dfrac{1}{\ln 2}2^{-x} + C$ c) $\ln (\ln x) + C$ e) $\operatorname{Tan}^{-1}(\sin x) + C$

f) $\ln (\sin x) + C$ g) $\exp (\operatorname{Sin}^{-1} x) + C$

8. $y^{(2n+1)}(0) = 0,$ $y^{(2n)}(0) = [(n - 1)!]^2 2^{2n-1}$

10. $u_n(x) = x^n + \dfrac{n(n - 1)}{2}x^{n-2} + \dfrac{n(n - 1)(n - 2)(n - 3)}{2 \cdot 4}x^{n-4} + \cdots$

Section 9–4

3. 1771, 0.013   4. 24 hrs and 48 hrs

6. $f'(x) = \ln (1 + r/x) - \dfrac{(r/x)}{1 + (r/x)} > 0$

8. a) $\dfrac{d}{dx}\left[e^{-x}\left(1 + \dfrac{x}{n}\right)^n\right] = e^{-x}\left(1 + \dfrac{x}{n}\right)^{n-1}\left(\dfrac{x}{n}\right) \le e^{-x}\left(1 + \dfrac{x}{n}\right)^n\left(\dfrac{x}{n}\right) \le e^{-x}e^x\dfrac{x}{n} = \dfrac{x}{n}$

b) $A = (1 + 0.02)^{14} = (1 + 0.28/14)^{14} \approx e^{0.28} = 1.323 \dots$ Actually,

$$e^{0.28}\left(1 - \dfrac{(0.28)^2}{28}\right) < A < e^{0.28}.$$

Thus the error is less than

$$\dfrac{(0.28)^2}{28}e^{0.28} = 0.0037 \dots < 0.004.$$

12. Let p_0 be the pressure at the surface of the earth. Argue that the pressure at a height h is given by

$$p = p_0 - \int_0^h \sigma(x)\, dx.$$

Differentiate and apply Boyle's law to get

$$p' = -\frac{1}{a} p.$$

Deduce finally that $p = p_0 e^{-h/a}$.

13. $P_0 V_0 \ln (V/V_0)$

Section 9–5

2. $\sinh 2x = 2 \sinh x \cosh x,$

$\cosh 2x = \cosh^2 x + \sinh^2 x = 2 \cosh^2 x - 1 = 1 + 2 \sinh^2 x,$

$\cosh^2 x = \dfrac{1 + \cosh 2x}{2}, \qquad \sinh^2 x = \dfrac{-1 + \cosh 2x}{2}.$

If $u = \tanh (x/2)$, then

$$\cosh x = \frac{1 + u^2}{1 - u^2}, \qquad \sinh x = \frac{2u}{1 - u^2}.$$

6. a) $-\cos x\, \mathrm{csch}^2 (\sin x)$ c) $3 \tanh^2 x\, \mathrm{sech}^2 x$ e) $-2\, \mathrm{ctnh}\, x\, \mathrm{csch}^2 x$

Section 9–6

1. a) 0   b) $+\infty$ c) $\frac{1}{2}$ d) 0   e) $+\infty$ f) $1/e$
   g) 1   h) $\sqrt{e}$ i) e j) $\frac{3}{2}$ k) 1   l) e^2

2. The function is continuous, but not differentiable at zero. We have $f'_-(0) = 1$ and $f'_+(0) = 0$.

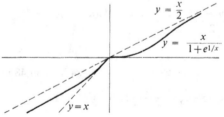

3. a)

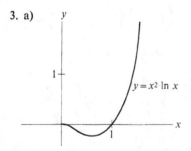

b)

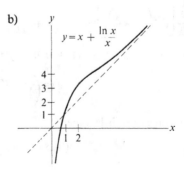

f)

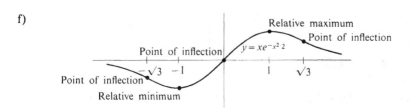

4. x^{2^x} grows more rapidly than x^{x^2}, and 2^{x^x} grows more rapidly than x^{2^x}.

5. x^x grows more rapidly than $a^x x^a$.

6. a) $\frac{1}{2}$ b) $\frac{1}{2}$ c) -2 d) 2   e) 2   f) 1   g) a/b if a and b are not both zero

Section 9–7

5. a) Use the inequality between weighted arithmetic and geometric means.

b) We have

$$\frac{\Pi_{k=1}^{r} a_{kj}}{\|a_1\|_{p_1}\|a_2\|_{p_2}\cdots\|a_r\|_{p_r}} = \prod_{k=1}^{r} \frac{a_{kj}}{\|a_k\|_{p_k}} \le \sum_{k=1}^{r} \frac{1}{p_k}\left(\frac{a_{kj}}{\|a_k\|_{p_k}}\right)^{p_k} \qquad (j = 1, \ldots, n).$$

Summing over the inequalities, we get

$$\frac{\sum_{j=1}^{n}\left(\Pi_{k=1}^{r} a_{kj}\right)}{\Pi_{k=1}^{r}\|a_k\|_{p_k}} \le \sum_{k=1}^{r} \frac{1}{p_k} = 1.$$

10. c) $(S/n)^S$

Section 9–9

2. $\cos 3\theta = \cos^3 \theta - 3\cos\theta\sin^2\theta$, $\cos 4\theta = \cos^4\theta - 6\cos^2\theta\sin^2\theta + \sin^4\theta$

4. The three cube roots of one are 1, $(-1 + \sqrt{3}i)/2$, and $(-1 - \sqrt{3}i)/2$.

7. b) Write the polynomial $P(z) = z^n - r^n$ in factored form. Take absolute values and interpret geometrically for the case in which z is a positive real number.

10. $\dfrac{d^n}{dx^n}(e^{az}\cos bx) = \dfrac{d^n}{dx^n}[\operatorname{Re} e^{(a+bi)x}] = \operatorname{Re}\left[\dfrac{d^n}{dx^n}e^{(a+bi)x}\right] = \operatorname{Re}(a + bi)^n e^{(a+bi)x}$

$$= \operatorname{Re}(re^{i\theta})^n e^{(a+bi)x} = \operatorname{Re} r^n e^{az} e^{i(b+n\theta)} = r^n e^{az}\cos(b + n\theta)$$

Miscellaneous Exercises

5. d)

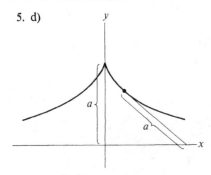

9. $(Gm^2/l^2)\ln\frac{4}{3}$

10. $H(\ln a - \ln c)/(\ln a - \ln b)$

Chapter 10

Section 10-1

1. $\frac{1}{10}(2x + 3)^{5/2} - \frac{1}{2}(2x + 3)^{3/2} + C$ 3. $e^{x^2}/2 + C$

5. $(\mathrm{Tan}^{-1} x)^2/2 + C$ 7. $-\sqrt{1 - x^2} + \mathrm{Sin}^{-1} x + C$

9. $-\cos\theta + \frac{2}{3}\cos^3\theta - \frac{1}{5}\cos^5\theta + C$ 11. $\frac{1}{7}\cos^7\theta - \frac{1}{5}\cos^5\theta = C$

13. $\ln|\sin\theta| + C$ 15. $\frac{1}{2}\ln\left|\dfrac{1 - \cos\theta}{1 + \cos\theta}\right| + C$

16. $-\mathrm{ctn}\,\theta - \csc\theta + C$ 19. $\ln(x - 2 + \sqrt{x^2 - 4x + 13}) + C$

21. $\frac{1}{2}\ln(x^2 - x + 1) + \dfrac{1}{\sqrt{3}}\mathrm{Tan}^{-1}\dfrac{2x - 1}{\sqrt{3}} + C$

23. $\mathrm{Sin}^{-1}(x - 1) + C$ 26. $\ln|\ln x| - \ln|\ln x + 1| + C$

29. $-\frac{1}{16}\cos 8x - \frac{1}{4}\cos 2x + C,\quad -\frac{1}{6}\sin 3x + \frac{1}{2}\sin x + C,\quad \frac{1}{4}\sin 2x + (\cos 2)/2x + C$

31. b) $\ln(e^x + e^{-x}) + C = \ln\cosh x + C'$

Section 10-2

1. $(x^2 - 2x + 2)e^x + C$ 3. $\frac{2}{3}x^{3/2}(\ln x - \frac{2}{3}) + C$

5. $x\,\mathrm{Sin}^{-1} x + \sqrt{1 - x^2} + C$ 7. $(x^2/2)\,\mathrm{Tan}^{-1} x - \frac{1}{2}x + \frac{1}{2}\mathrm{Tan}^{-1} x + C$

9. $-x^3\cos x + 3x^2\sin x + 6x\cos x - 6\sin x + C$

11. $x\tan x + \ln|\cos x| - x^2/2 + C$ 13. $-x\csc x + \frac{1}{2}\ln\left|\dfrac{1 - \sin x}{1 + \sin x}\right| + C$

15. $x\ln(1 + x^2) - 2x + 2\,\mathrm{Tan}^{-1} x + C$ 17. $\dfrac{x}{\ln 3}3^x - \dfrac{1}{(\ln 3)^2}3^x + C$

19. $2\sqrt{x + 1}\,\mathrm{Sin}^{-1} x + 4\sqrt{1 - x} + C$ 21. $\left(\dfrac{1}{\ln a}x^2 - \dfrac{2x}{(\ln a)^2} + \dfrac{2}{(\ln a)^3}\right)a^x + C$

23. $x(\mathrm{Sin}^{-1} x)^2 + 2\sqrt{1 - x^2}\,\mathrm{Sin}^{-1} x - 2x + C$

25. $\frac{1}{2}[(x^2 - 1)\sin x - (x - 1)^2\cos x]e^x + C$

27. $I = (x/2)(\cos\ln x + \sin\ln x),\quad J = (x/2)(\sin\ln x - \cos\ln x)$

30. e) $\frac{1}{2}\tan x\sec x + \frac{1}{2}\ln(\sec x + \tan x) + C$

33. a) Use the fundamental theorem of calculus to show that the derivatives of the two sides are equal.

Section 10-3

1. a) $\frac{1}{2}x^2 - x + 2\ln|1 + x| + C$ c) $\frac{1}{2}\ln(x^2 - 1) + C$

e) $\frac{1}{2}x^2 + 2x + \frac{1}{4}\ln|x + 1| + \frac{27}{4}\ln|x - 3| + C$

g) $\frac{1}{2}\ln\left|\dfrac{x^2 + 2x + 2}{2x - 3}\right| + \mathrm{Tan}^{-1}(x + 1) + C$

i) $\frac{1}{6}\ln\dfrac{(x + 1)^2}{x^2 - x - 1} + \dfrac{1}{\sqrt{3}}\mathrm{Tan}^{-1}\dfrac{2x - 1}{\sqrt{3}} + C$

k) $x + 4\ln|x - 2| - \dfrac{1}{x - 2} - \dfrac{1}{2}\mathrm{Tan}^{-1}\left(\dfrac{x}{2}\right) + C$

m) $\frac{1}{3} \text{Tan}^{-1} \frac{x}{3} - \frac{1}{2(x^2 + 9)} - \frac{1}{4(x^2 + 9)^2} + C$

o) $-\frac{1}{x^2} - \frac{3}{2} \text{Tan}^{-1} x - \frac{x}{2(x^2 + 1)} + C$

2. a) We use synthetic division.

$$
\begin{array}{r}
2 + 0 + 1 \ \underline{|2} \\
+ 4 + 8 \\
\hline
2 + 4 + 9 \ \underline{|2} \\
+ 4 \\
\hline
2 + 8
\end{array}
$$

$$\frac{2x^2 + 1}{(x - 2)^3} = \frac{2}{x - 2} + \frac{8}{(x - 2)^2} + \frac{9}{(x - 2)^3}.$$

Thus

$$\int \frac{2x^2 + 1}{(x - 2)^3} \, dx = 2 \ln |x - 2| - \frac{8}{x - 2} + \frac{9}{2(x - 2)^2} + C.$$

3. c) Clear the alleged equation of fractions. Show that the two resulting polynomials agree at the points $a_1, \ldots, a_n$. Examine degrees and apply (b). Reverse your steps.

Section 10–4

1. $\frac{2}{\sqrt{a^2 - b^2}} \text{Tan}^{-1} \frac{a \tan (x/2) + b}{\sqrt{a^2 - b^2}} + C$ if $|b| < |a|,$

$\frac{1}{\sqrt{b^2 - a^2}} \ln \left| \frac{a \tan (x/2) + b - \sqrt{b^2 - a^2}}{a \tan (x/2) + b + \sqrt{b^2 - a^2}} \right| + C$ if $|b| > |a|,$

$\pm \frac{1}{a} \tan \left(\frac{x}{2} \pm \frac{\pi}{4} \right)$ if $b = \mp a$

3. $\frac{1}{\sqrt{2}} \text{Tan}^{-1} \left(\frac{\tan x}{\sqrt{2}} \right) + C$

5. a) $\frac{1}{5}[x + 2 \ln |\cos x + 2 \sin x|] + C$

c) $2\sqrt{\tan x} + C$ e) $\frac{2}{5} \tan^{5/2} x + C$

g) $\ln |2 + \cos x| + \frac{4}{\sqrt{3}} \text{Tan}^{-1} \left(\frac{\tan (x/2)}{\sqrt{3}} \right) + C$

i) $\frac{1}{\sqrt{2}} \text{Tan}^{-1} (\sqrt{2} \tan x) + C$ k) $-\frac{1}{2} \left[\text{ctn} \, x + \frac{1}{\sqrt{2}} \text{Tan}^{-1} \left(\frac{x}{\sqrt{2}} \right) \right] + C$

m) $2 \left(\sin \frac{x}{2} - \cos \frac{x}{2} \right) + C$ if $\sin \frac{x}{2} + \cos \frac{x}{2} \geq 0,$ and

$-2 \left(\sin \frac{x}{2} - \cos \frac{x}{2} \right) + C$ otherwise

o) $\frac{1}{2} \ln (\sqrt{2} \tan x + \sqrt{1 + 2 \tan^2 x}) + C$

7. a) $\tanh x + C$ c) $\ln |\tanh (x/2)| + C$

e) $-\frac{1}{6} \text{ctnh}^3 \frac{x}{2} + \frac{1}{2} \text{ctnh} \frac{x}{2} + C$ g) $xe^x - e^x + \frac{1}{4}e^{2x} - \frac{1}{2}x + C$

9. a) $\dfrac{x}{\sqrt{a^2 - x^2}} - \operatorname{Sin}^{-1}\dfrac{x}{a} + C$

c) $-\dfrac{1}{\sqrt{2}}\ln\left|\dfrac{\sqrt{2 + x - x^2} + \sqrt{2}}{x}\right| + \dfrac{1}{2\sqrt{2}} + C$

e) $\frac{1}{2}[u\sqrt{a^2 - u^2} + a^2\operatorname{Sin}^{-1}(u/a)] + C$

g) $\dfrac{1 - \sqrt{x^2 + 2x + 2}}{x + 1} + \ln(x + 1 + \sqrt{x^2 + 2x + 2}) + C$

i) $\frac{1}{2}(3 - x)\sqrt{1 - 2x - x^2} + 2\operatorname{Sin}^{-1}\dfrac{x + 1}{\sqrt{2}} + C$

12. a) $-\dfrac{2(2a - 3x)\sqrt{(a + x)^3}}{15} + C$ b) $\ln\left|\dfrac{\sqrt{x} - 1}{\sqrt{x} + 1}\right| + C$

13. a) $3\sqrt[3]{x} + 6\sqrt[6]{x} + 6\ln|\sqrt[6]{x} - 1| + C$

c) $6[\frac{1}{9}(x + 1)^{3/2} - \frac{1}{8}(x + 1)^{4/3} - \frac{1}{7}(x + 1)^{7/6} - \frac{1}{6}(x + 1)$
$\qquad\qquad\qquad\qquad + \frac{1}{5}(x + 1)^{5/6} - \frac{1}{4}(x + 1)^{2/3}] + C$

14. a) $\ln\left|\dfrac{\sqrt{1 + x} - \sqrt{1 - x}}{\sqrt{1 + x} + \sqrt{1 - x}}\right| + 2\operatorname{Tan}^{-1}\sqrt{\dfrac{1 - x}{1 + x}} + C$

b) $\ln\dfrac{|u^2 - 1|}{\sqrt{u^4 + u^2 + 1}} + \sqrt{3}\operatorname{Tan}^{-1}\dfrac{1 + 2u^2}{\sqrt{3}} + C$, where $u = \sqrt[3]{\dfrac{1 - x}{1 + x}}$

Section 10–5

4. e

5. $\dfrac{2^{2n}}{\sqrt{\pi n}}$

Miscellaneous Exercises

1. $2e^{\sqrt{x}} + C$

3. $x - 2\sqrt{x} + 2\ln|1 + \sqrt{x}| + C$

5. $\frac{1}{8}\sin 2x - \frac{1}{4}x\cos 2x + C$

7. $\ln\dfrac{e^x}{e^x + 1} + C$

9. $\dfrac{1}{625}e^{5x}(125x^3 - 75x^2 + 30x - 6) + C$

11. $\ln|\sin x + \cos x| + C$

13. $\dfrac{1}{16}\left(\dfrac{\sqrt{x^2 + 4}}{x} - \dfrac{1}{3}\dfrac{(x^2 + 4)^{3/2}}{x^3}\right) + C$

15. $\ln|\ln\sin x| + C$

17. $x\operatorname{Sinh}^{-1}x - \sqrt{1 + x^2} + C$

19. $-\frac{1}{2}[\ln(1 + 1/x)]^2 + C$

21. $-6\ln(1 + 1/\sqrt[6]{x}) + C$

23. $-2e^{-\sqrt{x}}(x^{3/2} + 3x + 6x^{1/2} + 6) + C$

25. $x\ln(a^2 + x^2) - 2x + 2a\operatorname{Tan}^{-1}(x/a) + C$

27. $e^x\tan(x/2) + C$

28. $\pi a^3[(\sqrt{2}/8)\ln(2\sqrt{2} + 3) - \frac{1}{6}]$

29. $\frac{1}{2}(\pi/\sqrt{3} + 1 - \frac{2}{3}\ln 2)$

30. $\pi\sqrt{2}$

31. $\dfrac{|x|}{2p}\sqrt{x^2 + p^2} + \dfrac{p}{2}\ln\dfrac{|x| + \sqrt{x^2 + p^2}}{p}$

32. $1 + \frac{1}{2}\ln\frac{3}{2}$

33. a) Let $y = \pi/2 - x$ and show that

$$\varphi(x) = \int_{\pi/2}^{\pi/2-x} \ln \sin t \, dt.$$

Use the identity $\sin t = 2 \sin (t/2) \cos (t/2)$, etc.

35. $2\pi[\sqrt{2} + \ln (1 + \sqrt{2})]$

Chapter 11

Section 11-1

2. By the proof of the generalized mean-value theorem, we may write

$$(\Delta f)g'(\xi) = (\Delta g)f'(\xi),$$

from which the desired inequality immediately follows.

4. a) 0   c) 2   e) 1   g) 0
   i) $-\infty, +\infty$ k) $\frac{1}{3}$ m) $\frac{1}{2}$

5. a) $f'''(x)$ b) $f^{(4)}(x)$ 6. $\begin{vmatrix} f(x) & g(x) & \varphi(x) \\ f'(x) & g'(x) & \varphi'(x) \\ f''(x) & g''(x) & \varphi''(x) \end{vmatrix}$

You will probably find it convenient to use a rule for differentiating determinants, according to which the derivative of a determinant can be obtained by differentiating each row (or column) one at a time and adding the resulting determinants. Thus, the derivative of

$$\begin{vmatrix} u_1 & u_2 \\ v_1 & v_2 \end{vmatrix} \quad \text{is} \quad \begin{vmatrix} u_1' & u_2' \\ v_1 & v_2 \end{vmatrix} + \begin{vmatrix} u_1 & u_1 \\ v_1' & v_2' \end{vmatrix}.$$

7. a) 1   c) $+\infty$ e) $+\infty, -\infty$
   g) Does not exist   h) 0

9. a) 0   c) 0   e) 0
   g) $4a^2/\pi$ i) $(a + b + c)/3$ k) 1
   m) 1   o) 1
   q) 0 if $a > 0$, 1 if $a = 0$, and $+\infty$ if $a < 0$ s) 0   u) e^2

Section 11-2

2. The assertion is false. Consider the behavior of the function

$$f(x) = \begin{cases} x^{7/3} \sin (1/x^2) & \text{if } x \neq 0, \\ 0 & \text{if } x = 0 \end{cases}$$

at zero.

5. c) Note first that by setting $\xi = (1 - \theta)a + \theta x$ (where $0 < \theta < 1$), the Cauchy remainder takes the form

$$R_n(x) = \frac{f^{(n+1)}(a + \theta(x - a))}{n!} (1 - \theta)^n (x - a)^{n+1}.$$

In this particular instance we have

$$R_n(x) = (-1)^n x^{n+1} \frac{1}{1 + \theta x} \left(\frac{1 - \theta}{1 + \theta x} \right)^n.$$

Observe that

$$0 < \frac{1}{1 + \theta x} < \frac{1}{1 + x} \quad \text{and} \quad 0 < \frac{1 - \theta}{1 + \theta x} < 1.$$

Infer that $|R_n(x)| \leq |x|^{n+1}/(1 + x)$.

8. Show that if $|x| \leq 1$, then

$$R_n(x) \leq \frac{|x|^{n+1}}{(n + 1)(1 + x^2)^{(n+1)/2}} \leq \frac{1}{n + 1}.$$

9. a) Since the series converges for $|x| < 1$, we can write

$$r_n(x) = \frac{1 \cdot 3 \cdot 5 \cdots (2n + 1)}{2 \cdot 4 \cdot 6 \cdots 2n} x^{2n} s_n(x),$$

where

$$s_n(x) = \frac{1}{2n + 2} x^2 + \frac{2n + 3}{(2n + 2)(2n + 4)} x^4 + \frac{(2n + 3)(2n + 5)}{(2n + 2)(2n + 4)(2n + 6)} x^6 + \cdots$$

It can be shown that the coefficients are less than

$$\frac{1}{2}, \quad \frac{1 \cdot 3}{2 \cdot 4}, \quad \frac{1 \cdot 3 \cdot 5}{2 \cdot 4 \cdot 6}, \quad \cdots,$$

respectively. Thus

$$s_n(x) < 1 + \tfrac{1}{2} x^2 + \frac{1 \cdot 3}{2 \cdot 4} x^4 + \cdots = \frac{1}{\sqrt{1 - x^2}}.$$

(To show that

$$c_n = \frac{(2n + 3)(2n + 5)}{(2n + 2)(2n + 4)(2n + 6)} < \frac{1 \cdot 3 \cdot 5}{2 \cdot 4 \cdot 6}$$

for all positive integers n, observe that equality holds when $n = 0$ and that

$$\frac{c_{n+1}}{c_n} = \frac{2n + 7}{2n + 8} \cdot \frac{2n + 2}{2n + 3} < 1,$$

so that the inequality follows by induction.)

16. By formal multiplication,

$$\cos x \cosh x = \sum_{k=0}^{\infty} \frac{(-1)^k x^{2k}}{(2k)!} \cdot \sum_{k=0}^{\infty} \frac{x^{2k}}{(2k)!} = \sum_{k=0}^{\infty} c_k x^{2k},$$

where

$$c_k = \sum_{i+j=k} \frac{(-1)^i}{(2i)!(2j)!} = \frac{1}{(2k)!} \sum_{j=0}^{k} (-1)^j \binom{2k}{2j}$$

$$= \frac{1}{(2k)!} \operatorname{Re} (1 + i)^{2k} = \frac{1}{(2k)!} \operatorname{Re} (2i)^k.$$

In the last two expressions $i = \sqrt{-1}$. For $k = 0, 1, 2, \ldots$, the values of $(2i)^k$ are 1, $2i, -4, -8i, 16, 32i, \ldots$, while the values of Re $(2i)^k$ are 1, 0, -4, 0, 16, 0, $-32, \ldots$ Thus

$$\cos x \cosh x = 1 - \frac{4x^4}{4!} + \frac{16x^8}{8!} - \frac{32x^{12}}{12!} + \cdots$$

$$= \sum_{k=1}^{\infty} (-1)^{k+1} \frac{(4x^4)^{k-1}}{(4k-4)!}.$$

By Exercise 1, the latter series is the Maclaurin series for $\cos x \cosh x$. The fact that the series actually converges to $\cos x \cosh x$ can be shown by the method of Example 2.

18. We may write

$$f(a+h) = T_n(a+h) + \frac{f^{(n+1)}(a+\theta h)}{(n+1)!} h^{n+1}$$

$$= T_n(a+h) + \frac{f^{(n+1)}(a)}{(n+1)!} h^{n+1} + \frac{f^{(n+2)}(a+\theta' h)}{(n+2)!} h^{n+2}.$$

Thus

$$f^{(n+1)}(a+\theta h) - f^{(n+1)}(a) = \frac{h}{n+2} f^{(n+2)}(a+\theta' h).$$

Apply the mean-value theorem to the left-hand side, etc.

20. $P(x) = -2 - 6(x-2) + 6(x-2)^2 + (x-2)^3$. $P(2.003) = -2.0179$ (to four places)

21. The error is less than 0.00625.

25. Write

$$x^x = e^{x \ln x} = 1 + x \ln x + \frac{(x \ln x)^2}{2!} + \frac{(x \ln x)^3}{3!} + \cdots + \frac{(x \ln x)^n}{n!} + r_n(x).$$

Show that if $0 \le x \le 1$, then $-e^{-1} \le x \ln x \le 0$ and

$$|r_n(x)| \le \frac{e^{-(n+1)}}{(n+1)!}.$$

Apply $\int_0^1$ to each side of the equation above and let $n \to \infty$.

27. a) $-\frac{2}{3}$ c) $\frac{1}{2}$ e) ∞

29. a) None   b) Relative minimum   c) Relative maximum   d) None

Section 11–3

1. $p(x) = x^2 - 5x + 2$

3. Let $q(x)$ be the polynomial given by the Lagrange formula. By the theorem we know that the error $p(x) - q(x)$ can be written with a factor of the form $p^{(r+1)}(\xi)$. However, $p^{(r+1)}$ is identically zero.

5. a) $h = 0.02$ b) $h = 0.07$

8. a) $P(x) = f(x_0) + [f(x_1) - f(x_0)] \dfrac{x - x_0}{x_1 - x_0}$

$$+ [f(x_1) - f(x_0) - (x_1 - x_0) f'(x_0)] \frac{(x - x_0)(x - x_1)}{(x_1 - x_0)^2}$$

Section 11–4

1. a) $n \geq 129$ b) $n \geq 4$

6. b) $n(n + 1) \displaystyle\int_{-1}^{1} P_m(x)P_n(x) = \int_{-1}^{1} P_m(x) \dfrac{d}{dx}[(x^2 - 1)P_n'(x)]\, dx$

$$= (x^2 - 1)P_m(x)P_n'(x)\Big|_{-1}^{+1} - \int_{-1}^{1} (x^2 - 1)P_m'(x)P_n'(x)\, dx$$

$$= -\int_{-1}^{1} (x^2 - 1)P_m'(x)P_n'(x)\, dx = \cdots$$

$$= m(m + 1)\int_{-1}^{1} P_m(x)P_n(x)\, dx$$

Thus $\int_{-1}^{1} P_m P_n = 0$ if $m \neq n$.

c) Verify the first assertion by induction on n. Would the same result hold if $P_k(x)$ were any polynomial of degree k $(k = 0, 1, 2, \ldots)$?

Section 11–5

3. a) $x_{n+1} = x_n(2 - Ax_n)$

d) $x_0 = 0.3$, $x_1 = 0.355$, $x_2 = 0.367429$, $x_3 = 0.36787889$, $x_4 = 0.36787994$

4. b) Since $|\varphi'(x)| \leq 0.2$, the proof of Proposition 3 shows that the error after n iterations e_n must satisfy the inequality $e_n \leq (0.2)^{n+1}$. To obtain four-place accuracy, we shall require that $(0.2)^{n+1} \leq 5 \times 10^{-5}$. The smallest integer satisfying this inequality is 6. Thus six iterations will suffice.

Chapter 12

Section 12–1

1. Axis: $x = \dfrac{-B}{2A}$. Directrix: $y = \dfrac{4AC - B^2}{4A^2} - \dfrac{1}{4A}$

4. Maximum height: $(v_0 \sin \varphi)^2/(2g)$. Range: $(v_0^2 \sin 2\varphi)/g$

6. Eccentricity: $\sqrt{21}/5$. Foci: $(\pm\sqrt{21}, 0)$. Directrices: $x = \pm 25/\sqrt{21}$

11. Let P and Q be any two points on the ellipse. Argue that $|PQ|$ is less than or equal to both $\lambda = |FP| + |FQ|$ and $\lambda' = |F'P| + |F'Q|$, where F and F' are the foci of the ellipse. Deduce that $|PQ| \leq (\lambda + \lambda')/2$ and use the string property.

14. Neither set is convex, although the second set is the union of two convex sets, namely,

$$S_- = \{(x, y) \mid x \leq -a, (x/a)^2 - (y/b)^2$$
$$S_+ = \{(x, y) \mid x \geq a, (x/a)^2 - (y/b)^2 >$$

These sets are shown in the figure. The set S_+ consists of all points P such that

$$|F'P| - |FP| > 2a.$$

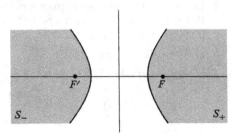

19. a) Let F and F' have the coordinates $(c, 0)$ and $(-c, 0)$, respectively. Then E and H have the equations

$$(x/a)^2 + (y/b)^2 = 1 \quad \text{and} \quad (x/A)^2 - (y/B)^2 = 1,$$

where $c^2 = a^2 - b^2 = A^2 + B^2$. Show that E and H have a unique point of intersection (ξ, η) in the first quadrant and that $A < \xi < a$. (Apply the intermediate-value theorem to the function

$$\varphi(x) = \frac{b}{a}\sqrt{a^2 - x^2} - \frac{B}{A}\sqrt{x^2 - A^2}$$

and observe that φ is strictly decreasing.) By subtracting the equations

$$(\xi/a)^2 + (\eta/b)^2 = 1 \quad \text{and} \quad (\xi/A)^2 - (\eta/B)^2 = 1,$$

show that

$$\frac{\xi^2}{a^2 A^2} - \frac{\eta^2}{b^2 B^2} = 0.$$

Write down the equations for the tangent lines to E and H at (ξ, η) and deduce that they are perpendicular.

b) If coordinates are introduced as in (a), then the statement is true provided that P does not lie on one of the coordinate axes.

23. Let F be a focus of the conic. For each point P on the conic, let P' be the unique point on the conic such that F lies on the chord PP'. You are to show that $|FP|^{-1} + |FP'|^{-1}$ is independent of P. Use polar coordinates and Exercise 21.

Section 12-2

1. Every vector in $\mathbf{R}^2$ can be written uniquely in the form $c_1\alpha + c_2\beta$. Specifically, $(x, y) = c_1\alpha + c_2\beta$, where $c_1 = (y - 2x)/7$ and $c_2 = (3x + 2y)/7$.

4. The vector of unit length having the same direction as α.

5. If the vertices are α, β, γ, then the center of gravity is $\frac{1}{3}(\alpha + \beta + \gamma)$.

7. When one vector is a *positive* scalar multiple of the other, that is, when the vectors have the same direction.

9. $T_A(\xi) = (7, 2)$. $\eta = (3, -2)$

10. $\begin{pmatrix} \frac{5}{7} & -\frac{2}{7} \\ \frac{1}{7} & \frac{1}{7} \end{pmatrix}$

12. $AB = \begin{pmatrix} 14 & 3 \\ 27 & -10 \end{pmatrix}$. $BA = \begin{pmatrix} -1 & 24 \\ 9 & 5 \end{pmatrix}$. $T_A T_B(\alpha) = (-8, -47)$

13. $\det A = 17$, $\det B = -13$, $\det AB = -221$

15. a) Let S and S' be the unique linear mappings which carry ϵ_1, ϵ_2 onto α, β, and α', β', respectively. Then $S' \circ S^{-1}$ is the required mapping.

16. $A^{-1} = \begin{pmatrix} \frac{7}{17} & -\frac{3}{17} \\ \frac{1}{17} & \frac{2}{17} \end{pmatrix}$. $B^{-1} = \begin{pmatrix} \frac{1}{13} & \frac{3}{13} \\ \frac{4}{13} & -\frac{1}{13} \end{pmatrix}$. $(AB)^{-1} = B^{-1}A^{-1} = \begin{pmatrix} \frac{10}{221} & \frac{3}{221} \\ \frac{27}{221} & -\frac{14}{221} \end{pmatrix}$

19. The mapping $A \to f_A$ is not one-to-one, since, for instance, every diagonal matrix of the form

$$\begin{pmatrix} a & 0 \\ 0 & a \end{pmatrix}$$

(where $a \neq 0$) is mapped onto the identity function.

Section 12–3

1. $T: (x, y) \to (3x + y + 2, x - 2y - 1)$. Let S be the linear transformation whose matrix is

$$\begin{pmatrix} 3 & 1 \\ 1 & 2 \end{pmatrix}.$$

Then $T = T_{(2,-1)} \circ S = S \circ T_{(1,-1)}$.

4. a) Matrix if $\varphi = \pi/4$: $\begin{pmatrix} 0 & 1 \\ 1 & 0 \end{pmatrix}$ b) Answer to question: yes

5. The fact that (a) $\Rightarrow$ (b) requires a straightforward induction argument. To prove that (b) $\Rightarrow$ (c), let $\alpha = T(0)$ and set $T^* = T_{-\alpha} \circ T$. Verify that T^* also preserves centroids and that it fixes the origin. If $\xi, \eta \in \mathbf{R}^2$ and $t \in \mathbf{R}$, argue that

$$T^*(t\xi) = T^*(1 - t)0 + t\xi) = (1 - t)T^*(0) + tT^*(\xi) = tT^*(\xi)$$

and

$$T^*(\xi + \eta) = 2T^*(\tfrac{1}{2}(\xi + \eta)) = T^*(\xi) + T^*(\eta).$$

Hence T^* is linear. Thus, $T = T_\alpha \circ T^*$ is the product of a linear transformation and a translation.

7. c) The inside of the triangle having $\zeta_0, \zeta_1, \zeta_2$ as vertices. The center of gravity of the triangle.
   d) Use (b) and the second part of (c).

10. (a) $\Rightarrow$ (b), since $\|T(\xi)\| = \|T(\xi) - T(0)\| = \|\xi - 0\| = \|\xi\|$.
    (b) $\Rightarrow$ (c), since $\langle \alpha, \beta \rangle = \tfrac{1}{4}[\|\alpha + \beta\|^2 - \|\alpha - \beta\|^2]$.
    (c) $\Rightarrow$ (d), since $\langle T(\epsilon_1), T(\epsilon_2) \rangle = \langle \epsilon_1, \epsilon_2 \rangle = 0$, $\|T(\epsilon_1)\|^2 = \langle T(\epsilon_1), T(\epsilon_1) \rangle = \langle \epsilon_1, \epsilon_1 \rangle = 1$, and similarly $\|T(\epsilon_2)\| = 1$.
    (d) $\Rightarrow$ (b). If $\alpha = a_1\epsilon_1 + a_2\epsilon_2$, then by Exercise 9(b), we have

$$\|T(\alpha)\|^2 = \|a_1 T(\epsilon_1) + a_2 T(\epsilon_2)\|^2 = \|a_1 T(\epsilon_1)\|^2 + \|a_2 T(\epsilon_2)\|^2 = a_1^2 + a_2^2 = \|\alpha\|^2.$$

Section 12–4

1. a) Ellipse; center is $(-3, 2)$; length of the semimajor axis is 10; length of the semiminor axis is $2\sqrt{5}$; eccentricity is $2\sqrt{5}/5$; foci are $(-3 \pm 4\sqrt{5}, 2)$; directrices are $x = -3 \pm 5\sqrt{5}$.
   c) The union of the intersecting lines $2x + 3y - 1 = 0$ and $x - 2y + 3 = 0$
   e) Hyperbola; asymptotes are $x = 0$ and $y = x$; eccentricity is $\sqrt{4 - 2\sqrt{2}}$; foci are

$$(\pm(2 - \sqrt{2})(1 + \sqrt{2})^{1/2}, \pm(2 - \sqrt{2})(1 + \sqrt{2})^{3/2});$$

directrices are $x + (1 + \sqrt{2})y = \pm(\tfrac{7}{2} + 2\sqrt{2})\sqrt{1 + \sqrt{2}}$.
   h) $\varnothing$

Section 12–5

7. Differentiate the identity $AA^{-1} = I$ using Exercise 5, and then multiply on the left by A^{-1}.

8. d) Let $\mathbf{a} = \int_a^b \mathbf{f}(t)\, dt$. If $\mathbf{a} = 0$, then the inequality clearly holds. If $\mathbf{a} \neq 0$, then we have

$$\left\| \int_a^b \mathbf{f}(t)\, dt \right\|^2 = \mathbf{a} \cdot \int_a^b \mathbf{f}(t)\, dt = \int_a^b \mathbf{a} \cdot \mathbf{f}(t)\, dt$$

$$\leq \int_a^b \|\mathbf{a}\|\, \|\mathbf{f}(t)\|\, dt = \|\mathbf{a}\| \int_a^b \|\mathbf{f}(t)\|\, dt,$$

and the desired inequality follows upon division by

$$\left\| \int_a^b \mathbf{f}(t)\, dt \right\| = \|\mathbf{a}\|.$$

e) Suppose γ is a piecewise smooth path joining two points P and Q in $\mathbf{R}^2$. Let γ be parametrized by the vector-valued function $\mathbf{f}(t)$ for $a \leq t \leq b$. Thus, $\mathbf{f}(a) = P$ and $\mathbf{f}(b) = Q$. Then

$$l(\gamma) = \int_a^b \|\mathbf{f}'(t)\|\, dt \geq \left\| \int_a^b \mathbf{f}'(t)\, dt \right\|$$

$$= \|\mathbf{f}(b) - \mathbf{f}(a)\| = |PQ|.$$

Section 12–6

1. The path has a singular point when $t = 0$. $\mathbf{V} = 2t\mathbf{i} + 3t^2\mathbf{j}$. $\mathbf{A} = 2\mathbf{i} + 6t\mathbf{j}$. Speed $= \|\mathbf{V}\| = |t|\sqrt{4 + 9t^2}$.

$$\mathbf{T} = \frac{2\,\mathrm{sgn}\,t}{\sqrt{4 + 9t^2}}\mathbf{i} + \frac{3|t|}{\sqrt{4 + 9t^2}}\mathbf{j},$$

$$\mathbf{N} = -\frac{3|t|}{\sqrt{4 + 9t^2}}\mathbf{i} + \frac{2\,\mathrm{sgn}\,t}{\sqrt{4 + 9t^2}}\mathbf{j}$$

$$\text{Tangential acceleration} = \mathbf{A} \cdot \mathbf{T} = \frac{(\mathrm{sgn}\,t)(4 + 18t^2)}{\sqrt{4 + 9t^2}},$$

$$\text{Normal acceleration} = \mathbf{A} \cdot \mathbf{N} = \frac{6|t|}{\sqrt{4 + 9t^2}},$$

$$\text{Curvature} = \frac{6}{|t|(4 + 9t^2)^{3/2}}.$$

6. b) Since only derivatives of s and y are involved, the expression for k is unaffected by replacing x by $x + a$ and y by $y + b$, where a and b are constants. Hence the formula is translation invariant. To prove rotation invariance, note that

$$k(t) = \frac{\mathbf{f}'(t) \times \mathbf{f}''(t)}{[\mathbf{f}'(t) \cdot \mathbf{f}'(t)]^{3/2}}$$

and use the rotation invariance of the dot and cross products. (See Exercise 6, Section 12–5, for instance.) Deduce finally that the expression is invariant under proper isometries.

7. It reverses the sign of the curvature.

11. d) $\rho^2 - 8as + s^2 = 0$.

Chapter 13

Section 13–1

2. Let S_j be the jth partial sum of the series $\sum_{k=1}^{\infty} a_k$, and let $T_{n,j}$ be the jth partial sum of the series $\sum_{k=n+1}^{\infty} a_k$. Observe that $T_{n,j} = S_{n+j} - S_n$. By letting $j \to \infty$, deduce that the tail is given by

$$T_n = \sum_{k=n+1}^{\infty} a_k = S - S_n,$$

where S is the sum of the series $\sum_{k=1}^{\infty} a_k$. Thus, $\lim_{n \to \infty} T_n = S - S = 0$.

Section 13–2

2. a) Converges, by comparison with the series $\sum k^{-2}$
   c) Converges, since the nth term is asymptotic to n^{-2}
   e) Converges iff $p > 1$. (The integral test is the simplest to apply in this case.)
   g) Converges, by the integral test, since

$$\lim_{R \to +\infty} \int_0^R xe^{-x^2}\, dx = 1$$

   and xe^{-x^2} is decreasing for $x > \sqrt{2}/2$
   i) Converges, by the comparison test, since the ratio between consecutive terms tends to $\frac{1}{4}$
   k) Diverges, since the nth term converges to one
   m) Converges. Let $a_n = (\ln n)^{-\ln n}$ and $b_n = n^{-2}$. Show that $a_n/b_n \to 0$. Hence the series converges by Exercise 1.
   o) Converges, since the limit superior of the nth root of the nth term is $(\sqrt{2} + 1)/3 < 1$
   q) Converges, since the nth term is asymptotic to n^{-2}
   s) Diverges, since the nth term converges to $\frac{1}{2}$
   u) Converges, by the root test
   w) Converges iff $p < \frac{1}{2}$. (Note that the nth term is asymptotic to $n^{p-3/2}/2$.)
   y) Diverges, since the nth term is asymptotic to n^{-1}

4. Suppose $r < 1$. Choose R such that $r < R < 1$. Then for sufficiently large n, we have $\sqrt[n]{a_n} < R$ or $a_n < R^n$. Thus the series $\sum a_k$ converges by comparison with the geometric series $\sum R^k$.
   Suppose $r > 1$. Then for infinitely many n, we have $\sqrt[n]{a_n} > 1$ or $a_n > 1$. Thus $\lim a_n \neq 0$. It follows that $\sum a_k$ diverges.

5. f) $\frac{3}{4}$
   i) Observe that

$$\mathrm{Tan}^{-1}\frac{1}{k^2 + k + 1} = \mathrm{Tan}^{-1}\frac{1}{k} - \mathrm{Tan}^{-1}\frac{1}{k+1}.$$

   j) The nth term is $b_n - b_{n+1}$, where

$$b_1 = \frac{1}{y-x}, \quad b_2 = \frac{x}{(y-x)y}, \quad b_3 = \frac{x(x+1)}{(y-x)y(y+1)}, \quad \cdots$$

7. d) Let α be the sequence $1, 0, 1, 0, \ldots$, and let β be the sequence $0, 1, 0, 1, \ldots$ Then $1 = \overline{\lim}\,(\alpha + \beta) < \overline{\lim}\,\alpha + \overline{\lim}\,\beta = 2$.

8. a) Let $A = \overline{\lim} \, (a_{n+1}/a_n)$. It suffices to show that for any given $\epsilon > 0$ we have

$$\overline{\lim} \, \sqrt[n]{a_n} < A + \epsilon.$$

Choose ϵ' such that $0 < \epsilon' < \epsilon$. Then for sufficiently large n,

$$a_{n+1}/a_n < A + \epsilon'.$$

By deleting, if necessary, a finite number of terms from the sequence, we may suppose that the latter inequality holds for every positive integer n. It then follows by induction that

$$a_n < K(A + \epsilon')^n,$$

where $K = a_1/(A + \epsilon')$. Thus, for each positive integer n.

$$\sqrt[n]{a_n} < K^{1/n}(A + \epsilon'),$$

from which it follows that

$$\overline{\lim} \, \sqrt[n]{a_n} \leq \overline{\lim} \, K^{1/n}(A + \epsilon') = A + \epsilon' < A + \epsilon.$$

12. a) Without loss of generality one may assume that the inequality $b_{n+1}/b_n \leq a_{n+1}/a_n$ holds for every positive integer n. (Why?) Show by induction that $b_n \leq Ca_n$, where $C = b_1/a_1$. Use the comparison test.

13. Suppose $a > 1$. Choose s such that $1 < s < a$, and set $b_n = n^{-s}$. Argue via Taylor's theorem that

$$\frac{b_{n+1}}{b_n} = 1 - \frac{s}{n} + \mathfrak{O}\left(\frac{1}{n^2}\right).$$

Argue that $b_{n+1}/b_n \geq a_{n+1}/a_n$ for n sufficiently large.

Section 13–3

2. Argue that for n sufficiently large the nth terms are in absolute value less than $|a_n|$ and $2|a_n|$, respectively.

12. a) Converges absolutely if $p > 1$, converges conditionally if $p \leq 1$ (by the Leibnitz alternating series test)
   c) Converges absolutely if x is an integral multiple of 2π, converges conditionally otherwise (by the Dirichlet test)
   e) Same as (c) (by Abel's test)
   g) Same as (c)
   i) The series is meaningless if β is a nonpositive integer. Otherwise the series converges absolutely if $|\alpha| < |\beta|$ and diverges if $|\alpha| > |\beta|$ (by the ratio test). If $\alpha = \beta$, the series diverges. If $\alpha = -\beta$, then the series converges conditionally, by the alternating series test.
   k) Converges conditionally (by Leibnitz's test)

14. Let $A = \sum a_k$. Clearly, A belongs to S. If, for some n,

$$a_n > \sum_{k=n+1}^{\infty} a_k,$$

argue that S contains no numbers between $A - a_n$ and A. Thus S cannot be an interval.

Suppose that for each positive integer n we have

$$a_n \le \sum_{k=n+1}^{\infty} a_k.$$

Use the technique of the proof of the rearrangement theorem to show that if $0 < s < A$, then a subseries may be found whose sum is s.

The geometric series satisfies the condition iff $\frac{1}{2} \le r < 1$. If $r = \frac{1}{2}$, note that this exercise proves, in effect, that every number has a binary expansion.

15. d) Use Minkowski's inequality and take limits.

Section 13–4

1. f) Suppose that f is Riemann-integrable on $[a, b]$. Then

$$\int_a^b f = \lim_{\|\pi\|\to 0} I_{\pi,\sigma}(f).$$

Thus, to a given $\epsilon > 0$, there corresponds a $\delta > 0$ such that

$$\left| I_{\pi,\sigma}(f) - \int_a^b f \right| < \epsilon$$

whenever $\|\pi\| < \delta$. Let π_0 be any partition of $[a, b]$ such that $\|\pi_0\| < \delta$, and let σ_0 be any selector for π_0. If $(\pi_0, \sigma_0) \prec (\pi, \sigma)$, then $\|\pi\| < \delta$, so that

$$\left| I_{\pi,\sigma}(f) - \int_a^b f \right| < \epsilon.$$

This proves that $\int_a^b f = \lim_{(\pi,\sigma)\in D} I_{\pi,\sigma}(f)$.

Suppose next that $\lim_{(\pi,\sigma)\in D} I_{\pi,\sigma}(f) = J$, where J is a real number. We must show that f is Riemann-integrable and that $\int_a^b f = J$. Let $\epsilon > 0$ be given. Then there exists a partition π of $[a, b]$ such that

$$|I_{\pi,\sigma}(f) - J| < \epsilon/3$$

for every selector σ for π. (Why?) Let h_π be the smallest step function such that $f \le h_\pi$ on $[a, b]$ and π fits h_π. [See Exercise 2(a), Section 8–2.] Argue that $\int_a^b h_\pi \le J + \epsilon/3$. Argue similarly that there exists a step function g_π such that $g_\pi \le f$ on $[a, b]$ and $\int_a^b g_\pi \ge J - \epsilon/3$. Thus $\int_a^b (h_\pi - g_\pi) \le 2\epsilon/3 < \epsilon$. Deduce that f is Riemann-integrable on $[a, b]$. Prove finally that $\int_a^b f = J$.

2. d) Let $D = R$, and let "$x \prec y$" mean "$x < y$." Then $\lim_{x\in D} f(x) = \lim_{x\to+\infty} f(x)$. If $f(x) = e^{-x}$, observe that $\lim_{x\in D} f(x) = 0$ and yet the net $\{f(x) \mid x \in D\}$ is not bounded.

4. Suppose that $\{x_\alpha \mid \alpha \in D\}$ is a net of real numbers and that $\lim_{\alpha\in D} x_\alpha = L$. Suppose that $\{\alpha_\beta \mid \beta \in D'\}$ is a net such that

a) for each $\beta \in D'$, $\alpha_\beta \in D$;

b) for each $\alpha \in D$, there exists a $\beta_0 \in D'$ such that

$$\beta_0 \prec \beta \Rightarrow \alpha \prec \alpha_\beta$$

for all $\beta \in D'$.

We then say that $\{x_{\alpha_\beta} \mid \beta \in D'\}$ is a *subnet* of $\{x_\alpha \mid \alpha \in D\}$. The analog of the substitution rule says that any such subnet also has L as a limit.

9. c) Let $S_{m,n} = \sum_{j,k=1}^{m,n} a_{jk}$, and let S be the sum of the double series. Let $\epsilon > 0$ be given. Choose N such that

$$m, n \geq N \Rightarrow S - \epsilon < S_{m,n} < S + \epsilon.$$

Letting $n \to \infty$, conclude that

$$S - \epsilon \leq \sum_{j=1}^{m} \left(\sum_{k=1}^{\infty} a_{jk} \right) \leq S + \epsilon,$$

provided $m \geq N$. Hence

$$\sum_{j=1}^{\infty} \left(\sum_{k=1}^{\infty} a_{jk} \right) = \lim_{m \to \infty} \sum_{j=1}^{m} \left(\sum_{k=1}^{\infty} a_{jk} \right) = S.$$

(Why?)

15. a) Prove the identity by induction on n. Assume true for n. We then have

$$\frac{n!}{x(x+1)(x+n)} = \binom{n}{0}\frac{1}{x} - \binom{n}{1}\frac{1}{x+1} + \binom{n}{2}\frac{1}{x+2} + \cdots$$

and (replacing x by $x + 1$)

$$\frac{-n!}{(x+1)(x+2)(x+n+1)} = -\binom{n}{0}\frac{1}{x+1} + \binom{n}{1}\frac{1}{x+2} + \cdots$$

Add the two equations and conclude that the identity holds for $n + 1$.

b) The identity holds provided x is not one of the integers $0, -1, -2, \ldots$

17. Observe that

$$\frac{x}{(1-x)^2} = \sum_{k=1}^{\infty} kx^k.$$

Hence

$$\frac{x^j}{(1-x^j)^2} = \sum_{k=1}^{\infty} kx^{jk} = \sum_{k=1}^{\infty} c_{jk}x^k,$$

where $c_{jk} = 0$ if k is not a multiple of j and $c_{jk} = m$ if $k = jm$. It follows that

$$\sum_{j=1}^{\infty} c_{jk} = \sigma(k)$$

($k = 1, 2, \ldots$). Thus, assuming the absolute convergence of the series involved, we have

$$\sum_{j=1}^{\infty} \frac{x^j}{(1-x^j)^2} = \sum_{j=1}^{\infty} \left(\sum_{k=1}^{\infty} c_{jk}x^k \right)$$

$$= \sum_{k=1}^{\infty} \left(\sum_{j=1}^{\infty} c_{jk} \right) x^k = \sum_{k=1}^{\infty} \sigma(k)x^k.$$

To justify the equality of the iterated series, observe that it suffices to show that the series $\sum \sigma(k)x^k$ converges absolutely for $|x| < 1$. Do this by comparing the series with the series

$$\sum \frac{k(k+1)}{2} x^k$$

and applying the ratio test to the latter.

Section 13-5

1. $\int_0^1 f_n(x)\, dx = n^2/[(n+1)(n+2)]$. Thus,

$$\lim_{n\to\infty} \int_0^1 f_n(x)\, dx = 1, \qquad \text{whereas} \qquad \int_0^1 \left[\lim_{n\to\infty} f_n(x)\right] dx = 0.$$

2. The sequence converges pointwise to the function

$$f(x) = \begin{cases} 0 & \text{if } |x| < 1, \\ \tfrac{1}{2} & \text{if } |x| = 1, \\ 1 & \text{if } |x| > 1. \end{cases}$$

Since the function f is discontinuous at ± 1, the convergence cannot be uniform throughout any neighborhood of these points.

4. Using Taylor's theorem, show that $-\tfrac{1}{2} < h < \tfrac{1}{2} \Rightarrow h - 2h^2 \le \ln(1+h) \le h$. Let $[a, b]$ be any closed bounded interval. Choose a positive integer N such that $-\tfrac{1}{2} < a/N < b/N < \tfrac{1}{2}$. If $n \ge N$ and $a \le x \le b$, show first that

$$x - 2x^2/n \le n \ln(1 + x/n) \le x.$$

By taking exponentials, deduce that

$$0 \le e^x - (1 + x/n)^n \le e^x(1 - e^{-2x^2/n}) < e^b(1 - e^{-k/n}),$$

where $k = 2 \max (a^2, b^2)$. The latter expression tends to zero as $n \to \infty$, and it is also independent of x. This proves that $(1 + x/n)^n$ converges to e^x uniformly on $[a, b]$.

Convergence is not uniform, however, over the entire line. In fact, *no sequence* of polynomials can converge uniformly to e^x over the entire line. Otherwise, we could certainly find a polynomial P such that $|P(x) - e^x| < 1$ for every real number x. Such a polynomial would have to be nonconstant and, consequently, we would then have $\lim_{x\to-\infty} |P(x)| = +\infty$. Since, however, $\lim_{x\to-\infty} e^x = 0$, it must follow that

$$\lim_{x\to-\infty} |P(x) - e^x| = +\infty,$$

which violently contradicts the assertion that $|P(x) - e^x| < 1$ for every real number x.

8. c) Let $f_n(x) = x + 1/n$. Then f_n converges uniformly to the identity function. However, f_n^2 does not converge uniformly over the entire line.

10. a) Consider closure under taking uniform limits. Suppose that $\{f_n\}$ is a sequence of regulated functions and that $f_n \to f$ uniformly on $[a, b]$. We must show that f is regulated, that is, that it can be uniformly approximated by step functions. Let $\epsilon > 0$ be given. Choose n so large that $\|f - f_n\| < \epsilon/2$. Since f_n is regulated, we can choose a step function g such that $\|f_n - g\| < \epsilon/2$. Deduce that $\|f - g\| < \epsilon$.

13. Formally,

$$\frac{1}{2} - \frac{1\cdot 3}{2\cdot 4} + \frac{1\cdot 3\cdot 5}{2\cdot 4\cdot 6} - \cdots = \frac{2}{\pi} \int_0^{\pi/2} (\cos^2 x - \cos^4 x + \cos^6 x - \cdots)\, dx$$

$$= \frac{2}{\pi} \int_0^{\pi/2} \frac{\cos^2 x}{1 + \cos^2 x}\, dx = \frac{2 - \sqrt{2}}{2}.$$

The difference between the nth partial sum and the alleged sum is in absolute value

$$\frac{2}{\pi} \int_0^{\pi/2} \frac{\cos^{2n+2} x}{1 + \cos^2 x} \, dx < \frac{2}{\pi} \int_0^{\pi/2} \cos^{2n+2} x \, dx = \frac{1 \cdot 3 \cdot 5 \cdots (2n+1)}{2 \cdot 4 \cdot 6 \cdots (2n+2)}.$$

Show that the latter term is asymptotic to $1/\sqrt{\pi n}$. (Use the corollary to Wallis' product, page 476.)

14. The Weierstrass M-test applies in all instances.

16. If $a > 0$, then

$$x \geq a \Rightarrow |ke^{-kx}| \leq ke^{-ak}.$$

Since the series $\sum ke^{-ak}$ converges via the integral test, it follows from the M-test that the given series converges uniformly on the interval $[a, +\infty)$. Since each term is continuous, the function f is continuous on $[a, +\infty)$. Since $a > 0$ was selected arbitrarily, we see that f is defined and continuous on the interval $(0, +\infty)$. $\int_{\ln 2}^{\ln 3} f(x)\, dx = \frac{1}{2}$

23. False. Consider $\sum x^n/n$ on the interval $[0, 1)$.

Section 13–6

1. a) $\{3\}$ b) $[-1, +1]$
    c) $[-2 - e^{-1}, -2 + e^{-1})$ (Use Stirling's formula.)
    d) $\{1\}$ e) $(-1, +1)$ f) $(-1 - e^{-1}, -1 + e^{-1})$ g) $(-1, +1)$

3. a) Prove first that if the series converges absolutely at the point s_0 and if $s_0 < s$, then the series also converges absolutely at the point s. Let γ_a be the greatest lower bound of the set of points at which the series converges absolutely (where $+\infty$ are permissible values for γ_a).
    b) Show that under the conditions stated the terms of the series are unbounded.

10. $y = \displaystyle\sum_{k=0}^{\infty} \frac{x^k}{(k!)^2}$

11. $\ln (x + \sqrt{1 + x^2}) = x - \dfrac{1}{2} \dfrac{x^3}{3} + \dfrac{1 \cdot 3}{2 \cdot 4} \dfrac{x^5}{5} - \dfrac{1 \cdot 3 \cdot 5}{2 \cdot 4 \cdot 6} \dfrac{x^7}{7} + \cdots$

Section 13–8

1. a) and b) $(0, +\infty)$
    c) Any open interval not containing zero
    d) Every open interval contained in the domain of the function
    e) Same as (c)   f) and (g) $(-1, +1)$

8. Let $s_0 > 0$. Choose S such that $1 < S < s_0$. If $s > 1$, we have

$$\zeta(s) = \sum_{k=1}^{\infty} e^{(-\ln k)s_0} \left(\sum_{j=0}^{\infty} \frac{[(-\ln k)(s - s_0)]^j}{j!} \right).$$

If $|s - s_0| \leq s_0 - S$, and if we replace the terms of the above iterated series by their absolute values, show that the resulting iterated series converges and has sum less than or equal to $\zeta(2s_0 - S)$. Apply the Fubini theorem and conclude that $\zeta(s)$ has a power series expansion about s_0 which is valid for $|s - s_0| \leq s_0 - S$.

10. e) The domain of f is $[e^{-(1/e)}, e^{1/e}]$; the range of f is $[e^{-1}, e]$. The function f^{-1} is a restriction of $x^{1/x}$. Both f and f^{-1} are analytic throughout the interior of their domains.

Before attempting this problem the reader might review Proposition 3, Section 11–5 and its proof.

Begin by graphing the function $(\ln x)/x$; let h be its restriction to the interval $(0, e]$. Observe that $f(x)$, if it exists, satisfies the equation $[\ln f(x)]/f(x) = \ln x$; deduce that $x \le e^{1/e}$ and that f must be a restriction of $F = h^{-1} \circ \ln$. If $0 < x \le e^{1/e}$, set $\varphi_x(y) = x^y$. Observe that $F(x)$ is the unique fixed point of φ_x and that $F(x) = f(x) = \lim_{n \to \infty} \varphi_x^n(1)$, whenever the limit exists. (Here φ^n denotes the nth iterate of the function φ.) Show that the limit exists if $1 < x < e^{1/e}$. [If $1 < y < F(x)$, show that $0 < \varphi_x'(y) < 1$ and then use the mean-value theorem to prove by induction that $\{\varphi_x^n(1)\}$ is a strictly increasing sequence bounded above by $F(x)$.] Show similarly that the limit exists if $e^{-(1/e)} \le x \le 1$. If $0 < x < e^{-(1/e)}$, show that $|\varphi_x'(F(x))| > 1$. Deduce that $\{\varphi_x^n(1)\}$ cannot converge to $F(x)$.

Section 13–9

7. a) Use the theorem concerning best least-squares approximations.
   c) Let $\epsilon > 0$ be given. Choose f_n such that $\|f - f_n\|_2 < \epsilon/2$. Choose a function φ of the form $\varphi = \sum_{k=1}^m a_k \varphi_k$ such that $\|f_n - \varphi\|_2 < \epsilon/2$. (Why can φ be so chosen?) Argue that $\|f - \varphi\|_2 < \epsilon$ and apply (a).

9. a) The expansion is valid if $x = 0$ or $x = 1$ by inspection. It is not valid for negative values of x. (The left-hand side is even, while the right-hand side is not.)
   c) The sum is $\frac{1}{3}(\frac{1}{30} - x^2 + 2x^3 - x^4)\pi^4$ $(0 \le x \le 1)$.

12. d) The radius of convergence of the series for $\pi x \operatorname{ctn} \pi x$ is

$$\lim_{k \to \infty} \sqrt[2k]{\frac{(2k)!}{(2\pi)^{2k}|B_{2k}|}} = \lim_{k \to \infty} [2\zeta(2k)]^{-1/2k} = 1,$$

since $2 < 2\zeta(2p) < 4$.

15. Use Exercise 14 with $p = 2$ and apply the Weierstrass M-test.

16. The theorem on pointwise convergence does not apply since the function is unbounded.

23. Let

$$\sigma_n = \frac{A_0}{2} + \sum_{k=1}^n (A_k \cos kx + B_k \sin kx).$$

By the proof of the theorem concerning best least-squares approximations,

$$\|f - \sigma_n\|_2^2 - \|f - s_n\|_2^2 = \frac{1}{2}(A_0 - a_0)^2 + \sum_{k=1}^n [(A_k - a_k)^2 + (B_k - b_k)^2].$$

Show that $A_k = (k + 1)a_k/(n + 1)$ and $B_k = (k + 1)b_k/(n + 1)$.

24. a) $\pi/3$ b) $\pi/(2\sqrt{3})$ g) $\pi^3/32$

Section 13–10

8. d) Make the substitution $u = t^2$ and use the fact that $\Gamma(\frac{1}{2}) = \sqrt{\pi}$.

IMPORTANT FORMULAS

I. Some trigonometric identities

Rational relationships:

$$\tan x = \frac{\sin x}{\cos x}, \qquad \text{ctn } x = \frac{1}{\tan x} = \frac{\cos x}{\sin x},$$

$$\sec x = \frac{1}{\cos x}, \qquad \csc x = \frac{1}{\sin x}.$$

Pythagorean identities:

$$\sin^2 x + \cos^2 x = 1, \qquad \tan^2 x + 1 = \sec^2 x, \qquad \text{ctn}^2 x + 1 = \csc^2 x.$$

Complementary and supplementary relationships:

$$\sin x = \cos (\pi/2 - x) = \sin (\pi - x),$$
$$\cos x = \sin (\pi/2 - x) = -\cos (\pi - x),$$
$$\tan x = \text{ctn } (\pi/2 - x) = -\tan (\pi - x),$$
$$\text{ctn } x = \tan (\pi/2 - x) = -\text{ctn } (\pi - x).$$

Identities involving two angles:

$$\sin (x \pm y) = \sin x \cos y \pm \cos x \sin y,$$
$$\cos (x \pm y) = \cos x \cos y \mp \sin x \sin y,$$
$$\tan (x \pm y) = \frac{\tan x \pm \tan y}{1 \mp \tan x \tan y},$$
$$\sin x \pm \sin y = 2 \sin \tfrac{1}{2}(x \pm y) \cos \tfrac{1}{2}(x \mp y),$$
$$\cos x + \cos y = 2 \cos \tfrac{1}{2}(x + y) \cos \tfrac{1}{2}(x - y),$$
$$\cos x - \cos y = -2 \sin \tfrac{1}{2}(x + y) \sin \tfrac{1}{2}(x - y),$$
$$\sin x \cos y = \tfrac{1}{2}[\sin (x + y) + \sin (x - y)],$$
$$\cos x \cos y = \tfrac{1}{2}[\cos (x + y) + \cos (x - y)],$$
$$\sin x \sin y = \tfrac{1}{2}[\cos (x - y) - \cos (x + y)].$$

Multiple angle formulas:

$$\sin 2x = 2 \sin x \cos x,$$
$$\cos 2x = \cos^2 x - \sin^2 x = 2 \cos^2 x - 1 = 1 - 2 \sin^2 x,$$
$$\sin 3x = 3 \sin x - 4 \sin^3 x,$$
$$\cos 3x = 4 \cos^3 x - 3 \cos x,$$
$$\sin 4x = 8 \cos^3 x \sin x - 4 \cos x \sin x,$$
$$\cos 4x = 8 \cos^4 x - 8 \cos^2 x + 1,$$
$$\cos nx = T_n(\cos x), \qquad \text{where } T_n \text{ is the nth-order Chebyshev}$$
$$\text{polynomial (pages 383 through 384),}$$
$$\cos^2 x = \frac{1 + \cos 2x}{2},$$
$$\sin^2 x = \frac{1 - \cos 2x}{2}.$$

Half-angle formulas:

If $t = \tan(\theta/2)$, then

$$\cos \theta = \frac{1 - t^2}{1 + t^2}, \qquad \sin \theta = \frac{2t}{1 + t^2},$$
$$d\theta = \frac{2 \, dt}{1 + t^2}, \qquad t = \frac{1 - \cos \theta}{\sin \theta} = \frac{\sin \theta}{1 + \cos \theta},$$
$$\tan \tfrac{1}{2}(x + y) = \frac{\sin x + \sin y}{\cos x - \cos y},$$
$$\csc x = \operatorname{ctn}(x/2) - \operatorname{ctn} x.$$

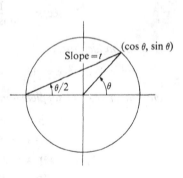

Sum formulas:

$$\tfrac{1}{2} + \cos x + \cos 2x + \cdots + \cos nx = \frac{\sin (n + \tfrac{1}{2})x}{2 \sin (x/2)},$$
$$\sin x + \sin 2x + \cdots + \sin nx = \frac{\cos (x/2) - \cos (n + \tfrac{1}{2})x}{2 \sin (x/2)}.$$

Triangle relationships:

$$c^2 = a^2 + b^2 - 2ab \cos \gamma \qquad \text{(law of cosines),}$$
$$\frac{\sin \alpha}{a} = \frac{\sin \beta}{b} = \frac{\sin \gamma}{c} \qquad \text{(law of sines).}$$

II. Differentiation formulas

General formulas (loosely stated):

$$\frac{d}{dx}(u+v) = \frac{du}{dx} + \frac{dv}{dx},$$

$$\frac{d}{dx}(uv) = u\frac{dv}{dx} + v\frac{du}{dx},$$

$$\frac{d}{dx}\left(\frac{u}{v}\right) = \frac{v(du/dx) - u(dv/dx)}{v^2},$$

$$\frac{dz}{dx} = \frac{dz}{dy}\frac{dy}{dx}, \qquad \frac{dx}{dy} = \frac{1}{dy/dx},$$

$$\frac{d^n}{dx^n}(uv) = \sum_{k=0}^{n}\binom{n}{k}\frac{d^{n-k}u}{dx^{n-k}}\frac{d^k v}{dx^k} \qquad \text{(Leibnitz's formula)}.$$

Special formulas:

$$\frac{d}{dx}(x^n) = nx^{n-1},$$

$$\frac{d}{dx}(\sin x) = \cos x, \qquad\qquad \frac{d}{dx}(\sinh x) = \cosh x,$$

$$\frac{d}{dx}(\cos x) = -\sin x, \qquad\qquad \frac{d}{dx}(\cosh x) = \sinh x,$$

$$\frac{d}{dx}(\tan x) = \sec^2 x, \qquad\qquad \frac{d}{dx}(\tanh x) = \operatorname{sech}^2 x,$$

$$\frac{d}{dx}(\sec x) = \sec x \tan x, \qquad\qquad \frac{d}{dx}(\operatorname{sech} x) = -\operatorname{sech} x \tanh x,$$

$$\frac{d}{dx}(\csc x) = -\csc x \operatorname{ctn} x, \qquad\qquad \frac{d}{dx}(\operatorname{csch} x) = -\operatorname{csch} x \operatorname{ctnh} x,$$

$$\frac{d}{dx}(\operatorname{Sin}^{-1} x) = \frac{1}{\sqrt{1-x^2}}, \qquad \frac{d}{dx}(\sinh^{-1} x) = \frac{1}{\sqrt{1+x^2}},$$

$$\frac{d}{dx}(\operatorname{Cos}^{-1} x) = -\frac{1}{\sqrt{1-x^2}}, \qquad \frac{d}{dx}(\operatorname{Cosh}^{-1} x) = \frac{1}{\sqrt{x^2-1}},$$

$$\frac{d}{dx}(\operatorname{Tan}^{-1} x) = \frac{1}{1+x^2}, \qquad \frac{d}{dx}(\tanh^{-1} x) = \frac{1}{1-x^2},$$

$$\frac{d}{dx}(e^x) = e^x, \qquad\qquad \frac{d}{dx}(a^x) = (\ln a)a^x,$$

$$\frac{d}{dx}(\ln x) = \frac{1}{x}, \qquad\qquad \frac{d}{dx}(\log_a x) = \frac{1}{x}\log_a e.$$

III. Expressions for the remainder in Taylor's theorem

$$f(x) = \sum_{k=0}^{n} \frac{f^{(k)}(a)}{k!} (x - a)^k + R_n(x),$$

where

$$R_n(x) = \frac{1}{n!} \int_a^x f^{(n+1)}(t)(x - t)^n \, dt \qquad \text{(integral remainder)},$$

$$R_n(x) = \frac{f^{(n+1)}(\xi)}{(n + 1)!} (x - a)^{n+1} \qquad \text{(Lagrange remainder)}$$

for some ξ between a and x, and

$$R_n(x) = f^{(n+1)}(\xi) \frac{(x - \xi)^n}{n!} (x - a) \qquad \text{(Cauchy remainder)}$$

for some ξ between a and x. (See pages 493, 494, and 504.)

Table 1

NATURAL TRIGONOMETRIC FUNCTIONS

Angle Degree	Angle Radian	Sine	Co-sine	Tan-gent	Angle Degree	Angle Radian	Sine	Co-sine	Tan-gent
0°	0.000	0.000	1.000	0.000					
1°	0.017	0.017	1.000	0.017	46°	0.803	0.719	0.695	1.036
2°	0.035	0.035	0.999	0.035	47°	0.820	0.731	0.682	1.072
3°	0.052	0.052	0.999	0.052	48°	0.838	0.743	0.669	1.111
4°	0.070	0.070	0.998	0.070	49°	0.855	0.755	0.656	1.150
5°	0.087	0.087	0.996	0.087	50°	0.873	0.766	0.643	1.192
6°	0.105	0.105	0.995	0.105	51°	0.890	0.777	0.629	1.235
7°	0.122	0.122	0.993	0.123	52°	0.908	0.788	0.616	1.280
8°	0.140	0.139	0.990	0.141	53°	0.925	0.799	0.602	1.327
9°	0.157	0.156	0.988	0.158	54°	0.942	0.809	0.588	1.376
10°	0.175	0.174	0.985	0.176	55°	0.960	0.819	0.574	1.428
11°	0.192	0.191	0.982	0.194	56°	0.977	0.829	0.559	1.483
12°	0.209	0.208	0.978	0.213	57°	0.995	0.839	0.545	1.540
13°	0.227	0.225	0.974	0.231	58°	1.012	0.848	0.530	1.600
14°	0.244	0.242	0.970	0.249	59°	1.030	0.857	0.515	1.664
15°	0.262	0.259	0.966	0.268	60°	1.047	0.866	0.500	1.732
16°	0.279	0.276	0.961	0.287	61°	1.065	0.875	0.485	1.804
17°	0.297	0.292	0.956	0.306	62°	1.082	0.883	0.469	1.881
18°	0.314	0.309	0.951	0.325	63°	1.100	0.891	0.454	1.963
19°	0.332	0.326	0.946	0.344	64°	1.117	0.899	0.438	2.050
20°	0.349	0.342	0.940	0.364	65°	1.134	0.906	0.423	2.145
21°	0.367	0.358	0.934	0.384	66°	1.152	0.914	0.407	2.246
22°	0.384	0.375	0.927	0.404	67°	1.169	0.921	0.391	2.356
23°	0.401	0.391	0.921	0.424	68°	1.187	0.927	0.375	2.475
24°	0.419	0.407	0.914	0.445	69°	1.204	0.934	0.358	2.605
25°	0.436	0.423	0.906	0.466	70°	1.222	0.940	0.342	2.748
26°	0.454	0.438	0.899	0.488	71°	1.239	0.946	0.326	2.904
27°	0.471	0.454	0.891	0.510	72°	1.257	0.951	0.309	3.078
28°	0.489	0.469	0.883	0.532	73°	1.274	0.956	0.292	3.271
29°	0.506	0.485	0.875	0.554	74°	1.292	0.961	0.276	3.487
30°	0.524	0.500	0.866	0.577	75°	1.309	0.966	0.259	3.732
31°	0.541	0.515	0.857	0.601	76°	1.326	0.970	0.242	4.011
32°	0.559	0.530	0.848	0.625	77°	1.344	0.974	0.225	4.332
33°	0.576	0.545	0.839	0.649	78°	1.361	0.978	0.208	4.705
34°	0.593	0.559	0.829	0.675	79°	1.379	0.982	0.191	5.145
35°	0.611	0.574	0.819	0.700	80°	1.396	0.985	0.174	5.671
36°	0.628	0.588	0.809	0.727	81°	1.414	0.988	0.156	6.314
37°	0.646	0.602	0.799	0.754	82°	1.431	0.990	0.139	7.115
38°	0.663	0.616	0.788	0.781	83°	1.449	0.993	0.122	8.144
39°	0.681	0.629	0.777	0.810	84°	1.466	0.995	0.105	9.514
40°	0.698	0.643	0.766	0.839	85°	1.484	0.996	0.087	11.43
41°	0.716	0.656	0.755	0.869	86°	1.501	0.998	0.070	14.30
42°	0.733	0.669	0.743	0.900	87°	1.518	0.999	0.052	19.08
43°	0.750	0.682	0.731	0.933	88°	1.536	0.999	0.035	28.64
44°	0.768	0.695	0.719	0.966	89°	1.553	1.000	0.017	57.29
45°	0.785	0.707	0.707	1.000	90°	1.571	1.000	0.000	

Table 2

EXPONENTIAL FUNCTIONS

x	e^x	e^{-x}	x	e^x	e^{-x}
0.00	1.0000	1.0000	2.5	12.182	0.0821
0.05	1.0513	0.9512	2.6	13.464	0.0743
0.10	1.1052	0.9048	2.7	14.880	0.0672
0.15	1.1618	0.8607	2.8	16.445	0.0608
0.20	1.2214	0.8187	2.9	18.174	0.0550
0.25	1.2840	0.7788	3.0	20.086	0.0498
0.30	1.3499	0.7408	3.1	22.198	0.0450
0.35	1.4191	0.7047	3.2	24.533	0.0408
0.40	1.4918	0.6703	3.3	27.113	0.0369
0.45	1.5683	0.6376	3.4	29.964	0.0334
0.50	1.6487	0.6065	3.5	33.115	0.0302
0.55	1.7333	0.5769	3.6	36.598	0.0273
0.60	1.8221	0.5488	3.7	40.447	0.0247
0.65	1.9155	0.5220	3.8	44.701	0.0224
0.70	2.0138	0.4966	3.9	49.402	0.0202
0.75	2.1170	0.4724	4.0	54.598	0.0183
0.80	2.2255	0.4493	4.1	60.340	0.0166
0.85	2.3396	0.4274	4.2	66.686	0.0150
0.90	2.4596	0.4066	4.3	73.700	0.0136
0.95	2.5857	0.3867	4.4	81.451	0.0123
1.0	2.7183	0.3679	4.5	90.017	0.0111
1.1	3.0042	0.3329	4.6	99.484	0.0101
1.2	3.3201	0.3012	4.7	109.95	0.0091
1.3	3.6693	0.2725	4.8	121.51	0.0082
1.4	4.0552	0.2466	4.9	134.29	0.0074
1.5	4.4817	0.2231	5	148.41	0.0067
1.6	4.9530	0.2019	6	403.43	0.0025
1.7	5.4739	0.1827	7	1096.6	0.0009
1.8	6.0496	0.1653	8	2981.0	0.0003
1.9	6.6859	0.1496	9	8103.1	0.0001
2.0	7.3891	0.1353	10	22026	0.00005
2.1	8.1662	0.1225			
2.2	9.0250	0.1108			
2.3	9.9742	0.1003			
2.4	11.023	0.0907			

Table 3

NATURAL LOGARITHMS OF NUMBERS

n	$\log_e n$	n	$\log_e n$	n	$\log_e n$
0.0	*	4.5	1.5041	9.0	2.1972
0.1	7.6974	4.6	1.5261	9.1	2.2083
0.2	8.3906	4.7	1.5476	9.2	2.2192
0.3	8.7960	4.8	1.5686	9.3	2.2300
0.4	9.0837	4.9	1.5892	9.4	2.2407
0.5	9.3069	5.0	1.6094	9.5	2.2513
0.6	9.4892	5.1	1.6292	9.6	2.2618
0.7	9.6433	5.2	1.6487	9.7	2.2721
0.8	9.7769	5.3	1.6677	9.8	2.2824
0.9	9.8946	5.4	1.6864	9.9	2.2925
1.0	0.0000	5.5	1.7047	10	2.3026
1.1	0.0953	5.6	1.7228	11	2.3979
1.2	0.1823	5.7	1.7405	12	2.4849
1.3	0.2624	5.8	1.7579	13	2.5649
1.4	0.3365	5.9	1.7750	14	2.6391
1.5	0.4055	6.0	1.7918	15	2.7081
1.6	0.4700	6.1	1.8083	16	2.7726
1.7	0.5306	6.2	1.8245	17	2.8332
1.8	0.5878	6.3	1.8405	18	2.8904
1.9	0.6419	6.4	1.8563	19	2.9444
2.0	0.6931	6.5	1.8718	20	2.9957
2.1	0.7419	6.6	1.8871	25	3.2189
2.2	0.7885	6.7	1.9021	30	3.4012
2.3	0.8329	6.8	1.9169	35	3.5553
2.4	0.8755	6.9	1.9315	40	3.6889
2.5	0.9163	7.0	1.9459	45	3.8067
2.6	0.9555	7.1	1.9601	50	3.9120
2.7	0.9933	7.2	1.9741	55	4.0073
2.8	1.0296	7.3	1.9879	60	4.0943
2.9	1.0647	7.4	2.0015	65	4.1744
3.0	1.0986	7.5	2.0149	70	4.2485
3.1	1.1314	7.6	2.0281	75	4.3175
3.2	1.1632	7.7	2.0412	80	4.3820
3.3	1.1939	7.8	2.0541	85	4.4427
3.4	1.2238	7.9	2.0669	90	4.4998
3.5	1.2528	8.0	2.0794	95	4.5539
3.6	1.2809	8.1	2.0919	100	4.6052
3.7	1.3083	8.2	2.1041		
3.8	1.3350	8.3	2.1163		
3.9	1.3610	8.4	2.1282		
4.0	1.3863	8.5	2.1401		
4.1	1.4110	8.6	2.1518		
4.2	1.4351	8.7	2.1633		
4.3	1.4586	8.8	2.1748		
4.4	1.4816	8.9	2.1861		

* Deduct 10 from each of the first nine values in column two.

INDEXES

INDEX

INDEX OF NOTATION

Logic

$P \wedge Q$	P and Q, 30
$P \vee Q$	P and/or Q, 31
$\sim P$	not P, 30
$P \Rightarrow Q$	P implies Q, 32
$P \Leftrightarrow Q$	P if and only if Q, 32
$(\forall x)$	for each x, 27
$(\exists x)$	there exists an x such that, 27

Sets

$A \cup B$	union of A and B, 2
$A \cap B$	intersection of A and B, 2
$A \backslash B$	relative complement of B in A, 2
A', A^c	complement of A, 3
$p \in A$	p is a member of A, 2
$p \notin A$	p is not a member of A, 2
$A \subset B$	A is a subset of B, 2
$\mathcal{S}(S)$, 2^S	power set of S, 4
$A \times B$	Cartesian product of A and B, 57
(x, y)	ordered pair, 57
$(x_1, x_2, \ldots, x_n)$	ordered n-tuple, 63
$\{x \mid \ldots\}$	set of all x such that $\ldots$, 2
$\{a, b, c, \ldots\}$	set consisting of $a, b, c, \ldots$, 1
$\displaystyle\bigcup_{\lambda \in \Lambda} E_\lambda$	union of an indexed family of sets, 37
$\displaystyle\bigcap_{\lambda \in \Lambda} E_\lambda$	intersection of an indexed family of sets, 37

Intervals, 20f

$[a, b]$	$\{x \in \mathbf{R} \mid a \leq x \leq b\}$; closed interval
(a, b)	$\{x \in \mathbf{R} \mid a < x < b\}$; open interval
$[a, b), (a, b]$	$\{x \in \mathbf{R} \mid a \leq x < b\}$, $\{x \in \mathbf{R} \mid a < x \leq b\}$; half-open (or half-closed) interval
$[a, +\infty), (-\infty, a], \ldots$	$\{x \mid x \geq a\}$, $\{x \mid x \leq a\}, \ldots$; unbounded interval

Special sets

$\varnothing$	null set, 4
$\mathbf{Z}^+$	set of positive integers, 6
$\mathbf{Z}$	set of integers, 6
$\mathbf{Z}_m$	integers modulo m, 10f, 15
$\mathbf{Q}$	set of rational numbers, 6
$\mathbf{R}$	set of real numbers, 8
$\mathbf{R}^+$	set of positive real numbers, 16
$\mathbf{R}^2$	$\mathbf{R} \times \mathbf{R}$, 58
$\mathscr{B}$	class of all bounded sequences, 104
$\mathscr{N}$	class of all null sequences, 104
$\mathscr{B}_{x_0}$	class of all functions bounded near x_0, 123
$\mathscr{N}_{x_0}$	set of all functions f such that $\lim_{x \to x_0} f(x) = 0$, 123
$\mathscr{F}_{x_0}$	set of all real-valued functions defined near x_0, 123
$\overline{\mathbf{R}}$	extended real line, 228
$\mathscr{B}([a, b])$, $\mathscr{B}$	set of all bounded real-valued functions on the interval $[a,b]$, 311
$\mathscr{R}([a, b])$, $\mathscr{R}$	set of all Riemann-integrable functions on the interval $[a, b]$, 315
$\mathscr{J}(\mathbf{R})$	family of Jordan-measurable subsets of $\mathbf{R}$, 331, 340
$\mathscr{J}$	family of Jordan-measurable subsets (usually of $\mathbf{R}^2$), 346
$\mathscr{B}_0$	family of all blocks, 343

Special functions

exp	exponential function, 387
ln	natural logarithm, 385
Ln	Ln $x = $ ln $|x|$, endpapers
Sin, Tan, . . .	restrictions of sin, tan, . . . , 374ff
Sin^{-1}, Tan^{-1}, . . .	inverse trigonometric functions, 374ff
T_n	nth Chebyshev polynomial, 383
P_n	nth Legendre polynomial, 164, 521f
B_n	nth Bernoulli polynomial, 672f
J_n	nth Bessel function, 611, 657ff
Γ	gamma function, 699ff
ζ	Riemann zeta function, 612

Constants

e	base of the natural logarithms, 389 (The letter "e" is also occasionally used to denote the eccentricity of a conic section.)
π	ratio of circumference to diameter, 437 (The letter "π" is also used to denote a partition of an interval.)
γ	Euler's constant, 603, 700 (The letter "γ" is also used for sequences (Chapter 3), for paths (Chapters 7 and 12), vectors (early part of Chapter 12), and angles.)

Miscellaneous

$\sqrt{x}$	*positive* square root of x, 21
$\displaystyle\sum_{k=1}^{n} a_k$	sum of $a_1, a_2, \ldots, a_n$, 43
$\displaystyle\prod_{k=1}^{n} a_k$	product of $a_1, a_2, \ldots, a_n$, 43
$x \vee y = \max(x, y)$	the larger of the numbers x and y, 50
$f \vee g$	pointwise supremum of the functions f and g, 128
$x \wedge y = \min(x, y)$	the smaller of the numbers x and y, 50
$f \wedge g$	pointwise infemum of the functions f and g, 128
$n!$	n factorial, the product of the first n integers, 54
$\dbinom{n}{k}$	the binomial coefficient $\dfrac{n(n-1)\cdots(n-k+1)}{1\cdot 2\cdot 3\cdots k}$, 54
$\gcd(a, b)$	greatest common divisor of a and b, 56
$\lvert P_1 P_2\rvert,\ d(P_1, P_2)$	distance between two points P_1 and P_2, 58, 274
$f\colon A \to B,\ A \xrightarrow{f} B$	f is a function mapping A into B, 79
$\mathfrak{D}_f$	domain of the function f, 80
$\mathfrak{R}_f$	range of the function f, 80
$f \mid S$	restriction of the function f to the set S, 87
$f \circ g$	composition product of the functions f and g, 87
$[x]$	greatest integer in x, 111
$\displaystyle\lim_{n\to\infty} a_n = L,\ a_n \to L$	L is the limit of the sequence $\{a_n\}$, 94, 229
$\displaystyle\lim_{x\to a} f(x) = L,$ "$f(x) \to L$ as $x \to a$"	the limit of $f(x)$ as x approaches a is L, 110, 228
$+\infty,\ -\infty,\ a+,\ a-$	for interpretations as used in limit statements, see page 228
$f'(x_0)$	derivative of f at x_0, 131f
$\Delta y/\Delta x$	difference quotient, 132
dy/dx	derivative of y with respect to x, 133
$\dfrac{dy}{dx}\bigg\rvert_{x=x_0}$	derivative at x_0, 132
f'	derivative of f, 133
$\lambda(AB),\ \lambda(A, B)$	slope of the line through the points A and B, 136, 221
$D,\ \dfrac{d}{dx},\ \dfrac{d}{dt},\ \cdots$	derivative operator, 156
$\Delta,\ \Delta_h$	difference operator, 157
$\mathcal{L}$	logarithmic derivative operator, 156
$x^{(n)}$	$x(x-1)(x-2)\cdots(x-n+1)$, if x is a real number, 156
$f^{(n)}$	nth derivative of the function f, 158
$D^n,\ \left(\dfrac{d}{dx}\right)^n,\ \dfrac{d^n}{dx^n},\ \cdots$	nth iterate of the derivative operator, 158
$f',f'',f''',f^{\mathrm{iv}},\ldots$	consecutive derivatives of f, 158

lub S	least upper bound of the set S, 172
glb S	greatest lower bound of the set S, 174
$K^+(f)$	set of points above the graph of f, 220
$K^-(f)$	set of points beneath the graph of f, 226
$\int f(x)\,dx$	antiderivative of f, 247
du	differential of u, 253
$f(x)\,dx$	first-order differential, 253
$\int_a^b f(x)\,dx,\ \int_a^b f$	Newton or Riemann integral of f over $[a, b]$, 255, 315
$F(x)\big|_{x=a}^{x=b}$	the difference $F(b) - F(a)$, 256
φ_S	characteristic function of the set S, 330
$A(S),\ m(S)$	area or Jordan content of a set S, 263, 345
f^+ and f^-	the positive and negative parts of the function f, 270
$[r;\theta]$	the point with polar coordinates r and θ, 271
$V(S)$	volume of the set S, 281
B_x	cross section of the set B, 281, 346
$\widehat{P_1P_2}$	path length from P_1 to P_2, 287
$l(\gamma)$	length of the path γ, 288
$M_{x=x_0}(S)$	moment of the plane lamina S about the line $x = x_0$, 295
$(\bar{x}(S), \bar{y}(S)),\ (\bar{x}, \bar{y})$	center of gravity of the plane lamina S, 299
$\bar{I}(f),\ \overline{\int_a^b} f$	upper Riemann integral of f on $[a, b]$, 313, 330
$\underline{I}(f),\ \underline{\int_a^b} f$	lower Riemann integral of f on $[a, b]$, 313
$\|\pi\|$	norm of the partition π, 320
$I_{\pi,\sigma}(f)$	Riemann sum, 321
$I_{\Pi,\Sigma}$	definite integral obtained as a limit of Riemann sums, 322
$l^*(A)$ and $l_*(A)$	outer and inner one-dimensional Jordan contents of A, 331, 340
$l(A)$	one-dimensional Jordan content (length) of A, 331, 340
$m^*(A)$ and $m_*(A)$	outer and inner two-dimensional Jordan contents of A, 344
$m(A)$	two-dimensional Jordan content (area) of A, 345
$\bar{A}$	closure of the set A, 353
A^0	interior of the set A, 353
∂A	boundary of the set A, 354
$\Omega(f, I)$	oscillation of f on I, 357
$\omega_f(x)$	oscillation of f at x, 357
$S(\epsilon)$	the set $\{x \mid \omega_f(x) \geq \epsilon\}$, 358
$I_{\pi,\sigma,\sigma'}(f, g)$	a Riemann-like sum, 363
f^{-1}	inverse of the function f, 365
$\sum_{k=1}^{\infty} a_k$	infinite series, 379
$\|\mathbf{a}\|$	norm of the vector $\mathbf{a}$, 542
$\|f\|$	uniform norm of the function f, 384
$\|f\|_2$	L_2-norm of the function f, 677

INDEX

IMPORTANT INTEGRALS (Continued)

62. $\int \dfrac{dx}{a+bx-cx^2} = \dfrac{1}{\sqrt{b^2+4ac}} \text{Ln} \dfrac{\sqrt{b^2+4ac}-b+2cx}{\sqrt{b^2+4ac}+b-2cx} + C.$

63. $\int \dfrac{dx}{\sqrt{a+bx-cx^2}} = \dfrac{1}{\sqrt{c}} \text{Sin}^{-1} \dfrac{2cx-b}{\sqrt{b^2+4ac}} + C.$

64. $\int \dfrac{dx}{\sqrt{a+bx+cx^2}} = \dfrac{1}{\sqrt{c}} \text{Ln}(2cx+b+2\sqrt{c}\sqrt{a+bx+cx^2}) + C.$

65. $\int \sqrt{a+bx+cx^2}\,dx = \dfrac{2cx+b}{4c}\sqrt{a+bx+cx^2}$
$\qquad - \dfrac{b^2-4ac}{8c^{3/2}}\text{Ln}(2cx+b+2\sqrt{c}\sqrt{a+bx+cx^2}) + C.$

66. $\int \sqrt{a+bx-cx^2}\,dx = \dfrac{2cx-b}{4c}\sqrt{a+bx-cx^2}$
$\qquad + \dfrac{b^2+4ac}{8c^{3/2}}\text{Sin}^{-1}\dfrac{2cx-b}{\sqrt{b^2+4ac}} + C.$

67. $\int \dfrac{x\,dx}{\sqrt{a+bx-cx^2}} = \dfrac{-\sqrt{a+bx-cx^2}}{c} + \dfrac{b}{2c^{3/2}}\text{Sin}^{-1}\dfrac{2cx-b}{\sqrt{b^2+4ac}} + C.$

68. $\int \dfrac{x\,dx}{\sqrt{a+bx+cx^2}} = \dfrac{\sqrt{a+bx+cx^2}}{c}$
$\qquad - \dfrac{b}{2c^{3/2}}\text{Ln}(2cx+b+2\sqrt{c}\sqrt{a+bx+cx^2}) + C.$

69. $\int \sin^2 ax\,dx = \dfrac{1}{2a}(ax - \sin ax \cos ax) + C.$

70. $\int \cos^2 ax\,dx = \dfrac{1}{2a}(ax + \sin ax \cos ax) + C.$

71. $\int \sin^n x\,dx = -\dfrac{\sin^{n-1}x \cos x}{n} + \dfrac{n-1}{n}\int \sin^{n-2}x\,dx + C.$

72. $\int \cos^n x\,dx = \dfrac{\cos^{n-1}x \sin x}{n} + \dfrac{n-1}{n}\int \cos^{n-2}x\,dx + C.$

73. $\int \tan^n x\,dx = \dfrac{\tan^{n-1}x}{n-1} - \int \tan^{n-2}x\,dx + C.$

74. $\int \text{ctn}^n x\,dx = -\dfrac{\text{ctn}^{n-1}x}{n-1} - \int \text{ctn}^{n-2}x\,dx + C.$

75. $\int \sec^2 x\,dx = \tan x + C.$

91. $\int x \,\text{Tan}^{-1}x\,dx = \tfrac{1}{2}[(x^2+1)\text{Tan}^{-1}x - x] + C.$

92. $\int x^n e^{ax}\,dx = \dfrac{x^n e^{ax}}{a} - \dfrac{n}{a}\int x^{n-1}e^{ax}\,dx + C.$

93. $\int x^n \text{Ln}\,x\,dx = x^{n+1}\left[\dfrac{\text{Ln}\,x}{n+1} - \dfrac{1}{(n+1)^2}\right] + C.$

94. $\int x^n \text{Ln}^m x\,dx = \dfrac{x^{n+1}\text{Ln}^m x}{n+1} - \dfrac{m}{n+1}\int x^n \text{Ln}^{m-1}x\,dx + C.$

95. $\int e^{ax}\sin nx\,dx = \dfrac{e^{ax}(a\sin nx - n\cos nx)}{a^2+n^2} + C.$

96. $\int e^{ax}\cos nx\,dx = \dfrac{e^{ax}(n\sin nx + a\cos nx)}{a^2+n^2} + C.$

97. $\int \dfrac{dx}{a+be^{nx}} = \dfrac{1}{an}[nx - \text{Ln}(a+be^{nx})] + C.$

98. $\int \dfrac{dx}{ae^{nx}+be^{-nx}} = \dfrac{1}{n\sqrt{ab}}\text{Tan}^{-1}\left(e^{nx}\sqrt{\dfrac{a}{b}}\right) + C.$

99. $\int \dfrac{xe^x\,dx}{(1+x)^2} = \dfrac{e^x}{1+x} + C.$

100. $\int \dfrac{dx}{a+b\cos x} = \dfrac{2}{\sqrt{a^2-b^2}}\text{Tan}^{-1}\dfrac{\sqrt{a^2-b^2}\tan\frac{x}{2}}{a+b} + C,$ if $a^2 - b^2 > 0,$

$\qquad = \dfrac{1}{\sqrt{b^2-a^2}}\text{Ln}\dfrac{a+b+\sqrt{b^2-a^2}\tan\frac{x}{2}}{a+b-\sqrt{b^2-a^2}\tan\frac{x}{2}} + C,$ if $b^2 - a^2 > 0.$

101. $\int \dfrac{dx}{a+b\sin x} = \dfrac{2}{\sqrt{a^2-b^2}}\text{Tan}^{-1}\left(\dfrac{a\tan\frac{x}{2}+b}{\sqrt{a^2-b^2}}\right) + C,$ if $a^2 > b^2,$

$\int \dfrac{dx}{a+b\sin x} = \dfrac{1}{\sqrt{b^2-a^2}}\text{Ln}\dfrac{a\tan\frac{x}{2}+b-\sqrt{b^2-a^2}}{a\tan\frac{x}{2}+b+\sqrt{b^2-a^2}} + C,$ if $a^2 < b^2.$

102. $\int \dfrac{dx}{a^2\cos^2 x + b^2\sin^2 x} = \dfrac{1}{ab}\text{Tan}^{-1}\left(\dfrac{b\tan x}{a}\right) + C.$